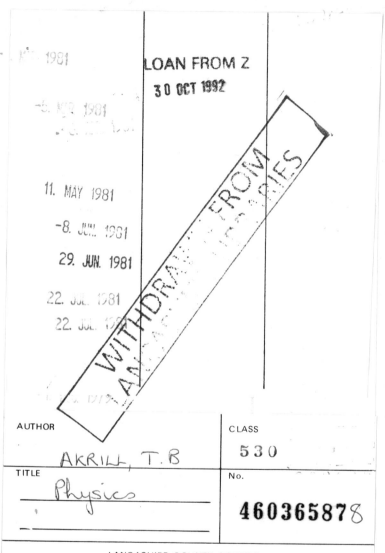

AUTHOR	CLASS
AKRILL, T.B	530
TITLE	No.
Physics	460365878

Physics

Physics

T. B. Akrill
G. A. G. Bennet
C. J. Millar

Edward Arnold

First published 1979
by Edward Arnold (Publishers) Ltd.
41 Bedford Square, London WCIB 3DQ

British Library Cataloguing in Publication Data

Akrill, Timothy Bryan
  Physics.
  1. Physics
  I. Title   II. Bennet, George Arthur Grey
  III. Millar, Christopher John
  530   QC21.2

  ISBN 0-7131-0297-7

Set in Hong Kong by Filmset Limited and
Printed in Great Britain by Fletcher & Sons Ltd, Norwich

46036587

A companion book of questions is also available
under the title *Practice in Physics*

Preface

A new physics

The task of the textbook author is to re-think the familiar material of his subject, so that it may come alive for the next generation of students and their teachers. He should anticipate developments so as to put into print what his colleagues are beginning to look for. He should help to put aside the lumber that accumulates in syllabuses (and is sometimes enshrined in our apparatus); he should steer the language of science away from the sort of technical jargon that hinders clear thinking. And so he should bring clarity to areas of confusion, and reveal afresh the inner coherence of the subject.

This book is the product of our combined creative efforts along these lines in the past three years. We offer it to our fellow teachers and their students as the outcome of our collective experience.

Physics for understanding

The student we have in mind is one who has recently tackled a science examination at 16+, and is embarking on a more advanced course. We have tried to use a style of language and treatment suitable for this level, particularly in those parts of the subject that might be tackled first. This is meant to be a book for the student to have by him throughout the course, and to use both at home and at college or school.

The book is based on the requirements of present teaching syllabuses, but we do not try to be encyclopaedic in this respect, and so risk overloading the text with a mass of fringe material. At the same time we aim to be forward-looking, and are confident that what we cover will have a central place in the next generation of physics courses. Above all we have given ourselves room to explain and illustrate the *basic concepts* of the subject, such as force, internal energy and p.d., which are often not well understood by a student entering an advanced physics course. About one-third of the book is occupied with such fundamental material.

We are not afraid to sacrifice rigour of treatment where this might get in the way of the student grasping a new idea. Sometimes we establish a theoretical point only by experimental test. Sometimes we rely simply on numerical illustration. We avoid almost all algebraic manipulation. The aim is *understanding*. But we claim never to be merely superficial; and at some points, in order to encourage clear thinking, we give a more thorough treatment than is usually adopted at this level.

The following are a few examples which illustrate our approach.

(a) In our treatment of forces we are careful always to make clear what body acts on what other body with what type of force; we use phrases like 'the gravitational pull of A on B'. Our force diagrams are *free-body diagrams*, in which the body or part-body whose motion or equilibrium we are considering is shown detached from its actual surroundings.

(b) We develop fully the concepts of quanta and energy levels. But we do not attempt superficial explanations of, for instance, the band theory of solid conduction, for which a proper quantum-mechanical treatment would really be necessary. We present conduction by positive holes as an experimental fact without trying to produce a theoretical basis for it.

(c) We give complete descriptive explanations of the properties of materials in terms of their molecular structure. But we regard the detailed measurement of these properties as mostly rather unimportant.

(d) In discussing energy conversions we are very precise. Our emphasis here is on the *internal energy* of a body and the changes in this quantity as a result of heating or working. We avoid using the word 'heat' as a noun; and we consider that this brings a long-overdue clarity to this area of the subject.

(e) We explain electric and magnetic fields in terms of E and B only. Gauss's theorem for electric fields and the continuity of magnetic flux we give without any detailed mathematics as an exercise for the imagination, as we believe that they are particularly helpful in providing a mental framework within which these fields may be visualised.

Flexibility

We have felt the need to develop a unified approach to the whole subject, as a glance at the contents page will show. The teacher will be able to detect here the traditional blocks of material into which physics has been divided (properties of matter, electricity, light, etc), but we have tried to avoid allowing the subject to fall into such isolated sections, preferring to show how the same ideas recur in many different places, and how the parts of the subject interact with each other.

Yet within this coherent approach we have arranged the book to be as flexible as possible, leaving the individual teacher free to plan the order in which different parts of the subject are taken. It is possible using this text to start a course at a number of different topics – current electricity (chapter 7), heating matter (chapter 9), or optics (chapter 25) – as well as simply beginning in chapter 1 with elementary mechanics. At all these points the treatment is suitable for the student who has achieved only a fair standard in mathematics and physical science at 16+.

There is a companion book of questions called *Practice in Physics*. We have planned this to provide, for each section of the book, exercises to help the student in his understanding of the subject matter of that section.

Practical work

We expect any physics course to include regular laboratory work. We assume that all physics students at this level will be gaining working experience of certain basic equipment (such as the c.r.o.). But the text does not attempt to be a laboratory manual, and so we do not give many of the practical details needed when handling the actual apparatus. Rather we write for the student who knows the apparatus, but has probably not got it in front of him at the time he is using this book.

We have been careful to provide examples of clear yet simple line drawings, since proficiency for the student in this respect is an objective worth reaching in itself, as well as being a necessary ingredient of good examination answers. We have used photographs where these impart immediacy and where they convey information that a student will not readily acquire otherwise (e.g. in some mechanics demonstrations). In electrical diagrams we have followed British Standard 3939 (Graphical Symbols); we personally have found no difficulty in making the change to 'rectangular' resistances.

Mathematics

We have been careful to adjust the mathematical level of the text to that which is now expected of the average student at 16+. We do not rely on any great expertise in handling algebra. In explaining a new type of calculation we generally prefer using numbers with their units, rather than proceeding at once to the deduction of a formula.

We do not use any calculus. But we have allowed ourselves the convenience of using the symbol Δ to express 'an increase in . . .'. Also we adopt the notation dv/dt, dI/dt, etc, to express the rate of change of v, I, etc. Appendix B on useful mathematics explains this point as well as several other related matters about gradients and rates of change.

In writing down expressions to be calculated we *always include the units with the numbers*, so that the reader should never be in doubt about units. In some ways this is cumbersome; but we consider it good practice for students to do the same. We assume that students will use electronic calculators, and therefore do not include intermediate steps in calculations.

The use of SI is now universal in scientific literature. The nature of this system is sufficiently explained for the purposes of this book in appendix A, which also includes the definitions of the SI base units. We have followed throughout the recommendations of the Symbols Committee of the Royal Society, both in the choice of symbols and in the presentation of graphs and tables; and we believe that students should learn to be equally precise in these matters.

January 1979

T. B. Akrill
G. A. G. Bennet
C. J. Millar

Contents

Acknowledgements

The publishers would like to acknowledge the following for providing photographs:

T B Akrill: 22.9a and b
Anderson & Neddermeyer (1936) *Phys. Rev.* **50**, 261: 19.17
P M S Blackett (1923) *Proc. Roy. Soc. A* **103**, 78: 19.16
P M S Blackett (1925) *Proc. Roy. Soc. A* **107**, 360: 19.13
P M S Blackett & D S Lees (1932) *Proc. Roy. Soc. A* **136**, 325: 19.14, 19.18
Professor Fowler taken from *Mechanics Vibrations and Waves* Akrill & Millar (John Murray): 13.17b, 3.18
J K Boggild taken from *An Atlas of Typical Expansion Chamber Photographs* Genter, Maier-Leibnitz & Bothe (Pergamon Press 1954): 19.5
Sir Lawrence Bragg & J F Nye (1947) *Proc. Roy. Soc. A* **190**: 4.12, 4.16, 5.13
Bromhead (Bristol) Ltd: 18.2a
J Cobby & C Mijovic, North London Collegiate School: 13.9b
College Physics 4th ed Sears, Zemansky & Young (Addison-Wesley 1960): 27.14
C B Daish: 25.14, 28.22
Ealing Beck Ltd: 13.9a, 28.17
Professor H E Edgerton: 3.11, 3.14
Elementary Classical Physics 2 Weidner & Sells (Allyn & Bacon Inc.) © Bausch and Lomb: 27.8
Professor A P French 24.5 and also from *Mechanics Vibrations and Waves* Akrill & Millar (John Murray): 22.11
G R Graham, Cambridgeshire College of Arts and Technology, *Physics Education* 7 no 6, July 1972: 27.4c, 27.12, 27.16
Griffin & George Ltd: 22.1b
Philip Harris, Shenstone: 12.7a
The Jeremiah Horrocks and Wilfred Hall Observatories: 12.13b
Jodrell Bank: 28.13
G G Scott *Journal of Applied Physics* **28**, Dec. 1957: 17.11
Professor A Keller: 4.22
Kodansha Ltd: 1.2b, 1.10a and b, 22.2
Leybold Heraeus Ltd: 18.3
Sir James Menter: 5.15
C J Millar: 4.11, 4.15, 4.20
Modern Physics D E Caro, J A McDonell & B M Spicer (Edward Arnold): 19.19
Mullard Ltd: 21.17

National Physical Laboratory (crown copyright): 7.4
Nuffield Advanced Science (1971) *Physics Students' Book, Unit 1, Materials and Structure* (Longman): 27.24b
Nuffield Advanced Science (1971) *Physics Teachers' Guide, Unit 1, Materials and Structure* (Longman): Professor E H Andrews 4.23; Natural Rubber Producers Ltd: 4.23c; Pilkington Brothers Ltd: 4.23b, 4.25
Physics Students' Book and Teachers' Guide, Unit 10, Waves, Particles and Atoms (Longman): 28.20
Physics, Concepts and Models Wenham, Dorling, Snell & Taylor (Addison-Wesley): L Meitner, provided by E N de C Andrade 19.15; B Taylor 3.8
Physics for Students of Science and Engineering (Halliday & Resnick, John Wiley & Sons Inc.): 28.19
Physics is Fun: Book 3 J T Jardine (Heinemann Educational Books Ltd): 2.4
Courtesy of Polaroid (UK) Ltd: 28.16
PSSC Physics, 4th ed © (D C Heath & Co 1976): 23.6a and b, *1st ed* 3.16a
The Project Physics Course (Holt Rinehart and Winston 1970): 12.13a, 18.10, 25.1a, 25.7b, 25.13
Quantum Physics E H Wichmann © Dr W Hines & Professor W Knight, Berkeley. Used with permission of McGraw-Hill Book Company: 4.23a
Radioactivity J L Lewis & E J Wenham (Longman Physics Topics): 19.6
Ripple Tank Studies of Wave Motion W Llowarch (Oxford University Press 1961): 25.15
Scientific American September 1967, courtesy of G C Smith: 5.11
The Science Museum (crown copyright): 19.6 (top), 19.10
Seeing Beyond the Visible ed A Hewish (Hodder & Stoughton Educational): 4.17
Special Relativity A P French (Thomas Nelson and Sons Ltd) reproduced from the film 'The Ultimate Speed': 3.17a
Three Phases of Matter Walton © (McGraw-Hill Book company (UK) Ltd): 4.24
Unilever: 6.30
R H Wallace *Understanding and Measuring Vibrations* (Wykeham 1970): 22.12
K W C Watson, Clifton College: 7.11, 14.14, 15.13a, 17.3b, 18.6, 18.11b, 20.9, 20.8, 21.6, 21.7
Waves D C Chaundy (Longman): 27.2a
C T R Wilson (1923) *Proc. Roy. Soc. A* **104** Plate 5: 19.7

The publishers and authors would also like to thank the Parkway Group for their care and patience in preparing the illustrations.

1 Describing motion

1.1 Measuring speed

To find the speed at which an object is moving you need to know how long it takes to move a known distance. Its *average speed* is then defined by the equation

$$\text{average speed} = \frac{\text{distance moved}}{\text{time taken}}$$

If we measure distances s from a fixed starting line then the increase in distance over which the speed is found is written as Δs (appendix B1). Similarly the time taken is written as Δt, so that

$$v_{av} = \frac{\Delta s}{\Delta t}$$

The unit of speed is the metre per second (m s^{-1}). Speeds are often quoted in other units, e.g. miles per hour, but for all calculations in physics we must first convert to the basic SI unit. It is useful to know that 50 m.p.h. $\approx$ 80 km h^{-1}, and is nearly 25 m s^{-1}.

Measuring speed gives us a chance to think about the errors involved in measuring length and time as well as the experimental techniques available. In what follows we assume that when a number of measured physical quantities are multiplied or divided, the *uncertainty* (expressed as a percentage error) in the final result is dominated and decided by the uncertainty in the *least precise* of the experimental measurements.

Measuring speed over a fixed distance

Figure 1.1 shows two types of experimental arrangement. In (a) a moving object (here an air gun pellet) is timed over a fixed distance AB which is about 2 m. This distance can be measured with a ruler; an uncertainty of ± 10 mm would be only $\pm 0.5 \%$. The time interval can be measured to the nearest millisecond using the 1000 Hz oscillator of the scaler timer. As, however, in this case Δt is only of the order of 10 ms there is an uncertainty of about $\pm 10 \%$ in Δt, and thus in v. Mechanical make-and-break circuits of this sort are commonly used to measure time intervals of up to a few seconds. The larger Δt the more precisely it can be measured and the better the calculated value for the speed.

Photoelectric switching is another common way of measuring time intervals and hence speed. In figure 1.1b a card attached to a moving object, here an air track glider (page 10), is arranged to interrupt a light beam. While the beam is broken the electric clock runs. The length of the card may be 100 mm and can be measured with a ruler; an uncertainty of ± 1 mm would represent $\pm 1 \%$. The clock hand completes one sweep in 1.00 s so that the time interval can be measured to better than a centisecond. For Δt of the order of 0.30 s there is uncertainty of about $\pm 3 \%$, so that the uncertainty in v will be dominated by that in Δt, and is about $\pm 3 \%$.

The scaler timer and the electric clock used in these experiments must themselves be accurate and this may

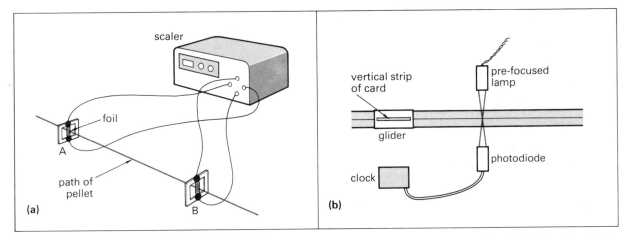

Figure 1.1 (a) A millisecond scaler-timer used to measure how long the pellet takes to travel from A, where it breaks a thin strip of foil, to B, where it breaks a second strip of foil. (b) A centisecond electric clock used to measure for how long the air track glider breaks the light beam from the lamp to the photodiode.

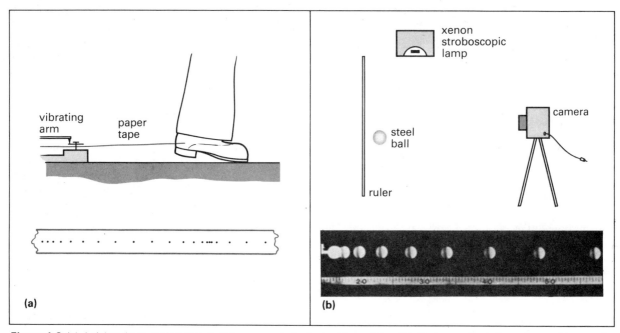

Figure 1.2 (a) A ticker-timer used to analyse the movement of a foot while walking. (b) Stroboscopic photography used to record the motion of a falling ball. Below each diagram is shown the result of a typical experiment.

depend on the frequency of the mains being exactly 50 Hz. One other 'clock' sometimes used for very short time intervals Δt is a cathode ray oscilloscope with a calibrated time base (page 340). Estimates of Δt of only a few tens of microseconds are possible using this instrument. For longer time intervals you can use a hand-operated stop watch or stop clock, the requirement being that Δt should be long compared with your reaction time which is about 0.2 s.

Measuring speed during a fixed time interval

Figure 1.2 shows two types of experimental arrangement. In (a) the moving object is the foot of a person as he walks. The foot has attached to it a length of tape which passes under a carbon-paper disc. The disc is struck at regular intervals by a blunt metal pin which is forced to oscillate by a mains-powered electromagnet. The resulting dots on the paper tape are thus produced every 0.020 s. To find the average speed of the foot between any two adjacent dots we need to measure the distance Δs between the dots. If the maximum value of Δs in this experiment is about 50 mm, then the uncertainty in this may be ± 1 mm or $\pm 2\%$. Taken together with any doubt about the 0.020 s we are unlikely to be able to measure the maximum speed of the foot during a step as precisely as this, and anyway the issue is confused by the fact that the foot moves up and down a little. When a *ticker tape* technique of this sort is used to measure the average speed of an object which is moving with a uniform speed then the uncertainties can be reduced by analysing the tape over longer Δt, e.g. measuring Δs over a ten-space length of tape.

In figure 1.2b a freely falling steel ball is *photographed stroboscopically*. The lamp flashes intermittently and a single long exposure is taken. The lamp, a small xenon discharge tube, is 'on' for about 10 μs per flash and can operate at frequencies between 1 and 250 s^{-1}. Providing the falling ball moves close to a metre rule or other calibrated scale, measurements from the photograph give the distance moved between flashes. Assuming the stroboscope to be correctly calibrated the precision with which v can be found depends on the uncertainty in Δs, which will vary with the size and quality of the print. As with the tape we can not expect an uncertainty of better than a few percent. The xenon lamp can be replaced by a bright filament lamp and a disc stroboscope with slits placed immediately in front of the camera aperture. The disc should be run by a synchronous motor so that its speed is accurately known. The time of each exposure is now longer and so the photograph is less sharp.

► If a twelve-slit motor-driven stroboscope disc is rotating at (5.0 ± 0.1) rev s^{-1} and two adjacent exposures of a moving ball are measured to be (75 ± 5) mm apart, calculate the average speed of the ball and the uncertainty in this value.

The time between exposures is

$$\Delta t = \left(\frac{1}{12} \times \frac{1}{5.0} \right) s = \frac{1}{60} s$$

$$\Rightarrow \quad v_{av} = \frac{\Delta s}{\Delta t} = \frac{75 \text{ mm}}{1/60 \text{ s}}$$

$$= 4500 \text{ mm s}^{-1} \text{ or } 4.5 \text{ m s}^{-1}$$

The percentage error in Δt is the same as that in the rotational speed

i.e. uncertainty in $\Delta t = \dfrac{0.1}{5.0} \times 100\% = 2\%$

The uncertainty in $\Delta s = \dfrac{5}{75} \times 100\% \approx 7\%$

The uncertainty in this experiment is thus dominated by that in Δs. As 7% of 4.5 is about 0.3 the calculated value of v_{av} should be quoted as

(4.5 ± 0.3) m s^{-1} ◄

There are other indirect ways of establishing the speed of an object. In a police speed trap the electromagnetic Doppler effect is used (page 368), and for charged particles the speed can be calculated from an analysis of observed collisions. We can measure how fast continents are drifting apart (a few mm per year!) or the speed of cosmic ray particles (up to 99.99% of the speed of light).

1.2 Velocity, a vector quantity

It matters whether you walk *towards* or *away from* a cliff edge at a given speed. We say that the speed of the motion is the same in each case but that the *velocity* is different. To define velocity we must include the direction of the motion: velocity is a *vector* quantity. All vector quantities obey a special rule for addition and subtraction. Force and magnetic field strength are other examples of vector quantities. Physical quantities which obey the normal rules of arithmetic are called *scalars*. Time and energy are examples of scalar quantities.

Adding velocities

Figure 1.3 illustrates how two velocities are added. You can find the size and direction of the resultant velocity, 3.6 m s^{-1}

at an angle of 34° below the horizontal, either (i) by scale drawing, in which the velocities to be added are represented by lines proportional to their lengths; or (ii) by calculation, which is easy here, since the velocities to be added are perpendicular to one another.

$$v^2 = (3.0 \text{ m s}^{-1})^2 + (2.0 \text{ m s}^{-1})^2$$

hence $v = 3.6$ m s^{-1}

and $\tan \theta = \dfrac{2.0}{3.0} \quad \Rightarrow \quad \theta = 34°$

This vector $v = 3.6$ m s^{-1} at an angle of 34° below the horizontal is known as the *resultant* of the two original vectors. Notice that the vectors in this example have been 'slid' so that they can be added head-to-tail. What we are saying is that the resulting motion is as though the body first moved horizontally and then vertically: we would get the same resultant velocity had we added them in the other order.

Subtracting velocities is an extension of the addition process. The velocity being subtracted is first reversed in direction (-20 m s^{-1} due north $\equiv +20$ m s^{-1} due south) and then added as above. Subtraction is illustrated in the example which follows. Once the problem is properly understood all we need is a sketched triangle showing the relative directions of the two vectors to be added and the resultant vector.

► A car is initially travelling at a speed of 20 m s^{-1} along a road in a direction 20° S of E. A few minutes later it is still travelling at the same speed but is now moving in a direction 20° N of E. What is the change of velocity of the car?

The *change* of velocity means the final velocity minus the initial velocity. Figure 1.4 shows the problem and the

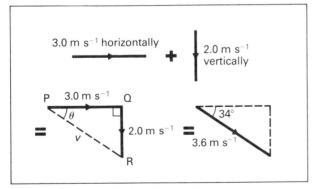

Figure 1.3 Vector addition: if *PQ* and *QR* represent the two velocities to be added in size and direction, then the resultant velocity is represented in size and direction by *PR*.

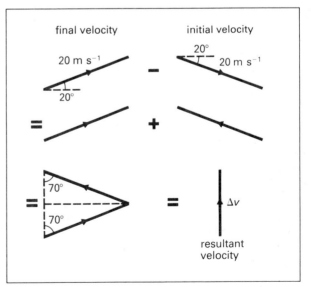

Figure 1.4 Vector subtraction: $\Delta v = 2(20 \text{ m s}^{-1})\cos 70°$.

solution. Clearly the change of velocity is due north and is of size Δv, where

$$\Delta v = 2(20 \text{ m s}^{-1}) \cos 70°$$
$$= 14 \text{ m s}^{-1}$$

◀

Resolving velocities

Just as two velocities can be added to yield a single result-ant velocity, so it is sometimes convenient to break one velocity up into two parts. Where these are perpendicular the process is called *resolving* and the two velocities are called the two *resolved parts* or perpendicular components of the original velocity.

If we want to resolve v in figure 1.5 along the x- and y-axes, then the resolved parts v_x and v_y are given by

$$v_x = v \cos \theta$$
$$v_y = v \sin \theta$$

or, if v_x and v_y are known, then

$$v^2 = v_x{}^2 + v_y{}^2$$

$$\text{and} \quad \tan \theta = \frac{v_y}{v_x}$$

The process of resolving is useful with all vector quantities and is often the key to seeing how a problem about velocity or force can be solved. Thus a pull of 800 newtons (800 N) acting at 30° above the horizontal can be resolved into

(i) a horizontal pull of (800 N) cos 30° = 690 N
(ii) a vertical pull of (800 N) sin 30°   = 400 N

We must stress that the methods of calculating *all* vectors are the same as those used here for velocity vectors. When we meet any new quantity we shall state whether it is a vector or a scalar.

Relative velocity

All velocities are relative. Suppose you are on a boat which is moving at 4.0 m s^{-1}. You do not walk along the boat *at*, (say) 0.8 m s^{-1}, but at a velocity of 0.8 m s^{-1} relative to the boat and at a velocity of 4.8 m s^{-1} or 3.2 m s^{-1} relative to the water on which the boat is floating, and at a velocity of ... the Earth, and ... the Sun, ... galaxy ... !! All velocities are relative.

Calculating relative velocities is straightforward in situa-tions similar to that described in the example above. But when, for example, two cars A and B are approaching a crossroads on a collision course the calculation is more difficult. Their relative speed of approach is found as follows. Imagine you are in car A, then the velocity of car B relative to you is the *vector* difference $v_B - v_A$. Equally the velocity of car A relative to car B is $v_A - v_B$. The problem now becomes one of vector subtraction. For all problems in-volving collisions the nature of the impact depends critically on the relative velocity of car and car, car and lamp-post etc. Thus two cars travelling in adjacent lanes on a motorway at

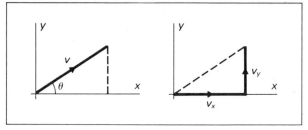

Figure 1.5 Resolving a velocity; $v_x = v \cos \theta$ and $v_y = v \sin \theta$.

25 m s^{-1} and 26 m s^{-1} respectively will come to little harm if they touch (relative velocity 1 m s^{-1}), but there is a barrier between the two opposite fast lanes to ensure that two cars moving with these speeds in opposite directions do *not* touch (relative velocity 51 m s^{-1}).

Displacement

The average velocity v of a particle is defined by an equation just like that for average speed:

$$v_{av} = \frac{\Delta s}{\Delta t}$$

where Δs is the *displacement* of the particle in the time interval Δt. Displacement, a vector quantity, measures the separation of two points A and B in size and direction. A person might undergo a displacement of 3.3 m vertically upwards in going to bed. If the journey took him 30 s then his average *velocity* would be 3.3 m/30 s = 0.11 m s^{-1} vertically upwards. But in going upstairs he may have walked a distance of 18 m, so that his average *speed* is 18 m/30 s = 0.60 m s^{-1}. In travelling round and round a circular track a runner can cover a great distance but whenever he is at the starting point his displacement is zero.

A man once bet that he would drive a car, starting at the Brooklands road-racing circuit (about 35 km from London) and finishing at Piccadilly Circus and that he would achieve an average speed (not velocity) of 60 m.p.h. You can probably now guess how he won his bet.

The *instantaneous velocity* of a body is the value obtained when the time interval Δt is made very small; we then write

$$v = \frac{ds}{dt} \qquad \text{(appendix B3)}$$

i.e. velocity is the *rate of change* of displacement. For motion in a straight line it is equal to the gradient of the displace-ment-time graph, as explained in appendix B3.

1.3 Acceleration

A body accelerates whenever its velocity changes, whether the change is a change in its size or its direction. The *average acceleration* a_{av} of an object is defined by the equation

$$a_{av} = \frac{\Delta v}{\Delta t} = \frac{v_2 - v_1}{\Delta t}$$

where Δv is the change of velocity of the object in the time interval Δt. The unit of acceleration is the m s^{-2}. Acceleration is a vector quantity and its direction is that of Δv, the change of velocity. Referring to the example on page 3, if the car's change of velocity occurred in 4 minutes, then the average acceleration would be

$$a_{av} = \frac{14 \text{ m s}^{-1}}{240 \text{ s}} = 0.058 \text{ m s}^{-2} \text{ due north}$$

In this example the speed of the car was unchanged: the acceleration was wholly the result of changes in its direction of motion. Sometimes the whole of the acceleration results from a change in speed: consider a sprinter in a 60 m dash who accelerates from rest to a speed of 8.4 m s^{-1} in 2.4 s. He moves in a straight line and his acceleration is

$$a_{av} = \frac{(8.4 - 0) \text{ m s}^{-1}}{2.4 \text{ s}}$$

$$= 3.5 \text{ m s}^{-2} \text{ along the track}$$

To measure a_{av} a record of the vector velocity at various times is needed. The techniques shown in figure 1.2 can provide this information in the laboratory.

The instantaneous acceleration of a body is the value obtained when the time interval Δt is made very small; we then write

$$a = \frac{dv}{dt} \qquad \text{(appendix B3)}$$

i.e. acceleration is the rate of change of velocity. For linear motion it is equal to the gradient of the velocity-time graph.

▶ A piece of ticker tape like that shown in figure 1.2a is analysed to give the following information:

dot	0	5	10	15	20	25	30	35	40	45
distance moved/mm	0	7	80	254	488	802	1133	1347	1380	1388

Estimate the maximum acceleration of the foot.

Each 5-space length of tape represents a time interval of $5 \times 0.02 \text{ s} = 0.1 \text{ s}$ so that the average speeds v_{av} over each 5-space length of tape are:

v_{av}/mm s^{-1}	70	730	1740	2340	3140	3310	2140	330	80

The *greatest* acceleration occurs when the speed changes most rapidly: here this is between speeds of 2140 mm s^{-1} and 330 mm s^{-1}, giving an average acceleration there of

$$a = \frac{(0.33 - 2.14) \text{ m s}^{-1}}{0.1 \text{ s}}$$

$$= -17.1 \text{ m s}^{-2}$$

The maximum acceleration of the foot thus occurs as it slows down prior to touching the ground and in this example is about 17 m s^{-2} in the opposite direction to the movement of the foot.   ◀

Motion in a straight line: graphs

In the above example a set of numerical data enables us to visualise the motion of a walking foot to some extent. But it is much easier to visualise the motion if the information is presented in graphical form. Three types of graph are useful and it is important to be able to transform from one to another.

Displacement-time graphs

The gradient of an *s-t* graph at any point is by definition the velocity of the body at that point. Where the gradient is constant then so is the velocity; if the gradient is zero (a horizontal line on the graph), the velocity is zero, i.e. the body is stationary. The graph can cross the time-axis; if it does so from above the time-axis, this means that the motion is backwards towards the place from which the displacement is measured.

Velocity-time graphs

The gradient of a *v-t* graph at any point is by definition the acceleration of the body at that point. Where the gradient is zero there is zero acceleration, and where negative the body is decelerating (or perhaps speeding up in the *negative direction*), etc.

The *area* between the graph-line and the time-axis measures the *change of displacement* of the body. Consider figure 1.6: during the time interval $t = 0$ to $t = 1.5$ s the graph shows that the body's velocity was constant at 20 m s^{-1}. The displacement during this time interval is therefore $(20 \text{ m s}^{-1})(1.5 \text{ s}) = 30$ m. The large shaded area

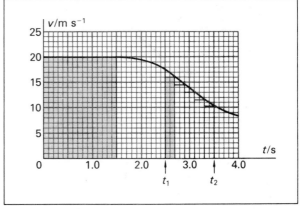

Figure 1.6 A *v-t* graph: an area between the graph-line and the time-axis represents a change of displacement.

has an area of 12 large squares. As each large square represents $(5 \text{ m s}^{-1})(0.5 \text{ s}) = 2.5 \text{ m}$, we see that the displacement can be found from the area, i.e. $12 \times 2.5 \text{ m} = 30 \text{ m}$. When the velocity changes *continuously* the area under the graph-line is perhaps the only way of calculating the displacement. For example between t_1 and t_2 in figure 1.6 the area under the graph-line is approximately equal to the sum of the areas of the rectangles ($\Delta s = v \Delta t$) and so represents the displacement. By making the rectangles narrower and narrower their total area more and more closely approximates to the area under the graph-line. The area from t_1 to t_2 is 4 large squares plus about 37 small squares, i.e. 137 small squares. One small square represents 0.1 m, so that the displacement between 2.5 s and 3.5 s is 13.7 m.

When the velocity remains positive, the area under the graph-line is equal to the distance travelled, as it is in figure 1.6. If part of the graph-line dips below the time-axis (as in figure 1.7a) the area is negative and so is the corresponding displacement.

Figure 1.7 shows two sets of graphs for motion in a straight line. In (a), looking at the bottom graph first:
(i) the *acceleration* is *constant* and *negative*; it is g for a freely falling body. ($g = -10 \text{ m s}^{-2}$ as here we are taking 'upwards' to be positive and it does not depend on whether the body is moving up or down).

(ii) the *velocity* gets steadily less but is instantaneously zero when the stone is at the top of its flight at $t = t_1$.
(iii) the area under the v-t graph-line from $t = 0$ to $t = t_2$ is zero (area below time-axis is negative); at that moment the stone once again passes the thrower and has zero *displacement*, even though it has perhaps travelled a considerable distance up and down.

In (b), looking at the top graph first:
(iv) the *displacement* increases throughout the journey it increases steadily between $t = t_3$ and $t = t_4$.
(v) the *velocity* is constant between these times.
(vi) the area under the v-t graph-line is equal to the train's net *displacement* (here equal to the distance covered).

Motion in a straight line: equations

Figure 1.8 shows a v-t graph for a body which is moving in a straight line *with constant acceleration*. Using the notation from the figure we can deduce that during the time interval t:

$$a = \frac{v - v_0}{t} \qquad \text{(the gradient)} \qquad (1)$$

and $$s = \left(\frac{v + v_0}{2}\right)t \qquad \text{(the shaded area)} \qquad (2)$$

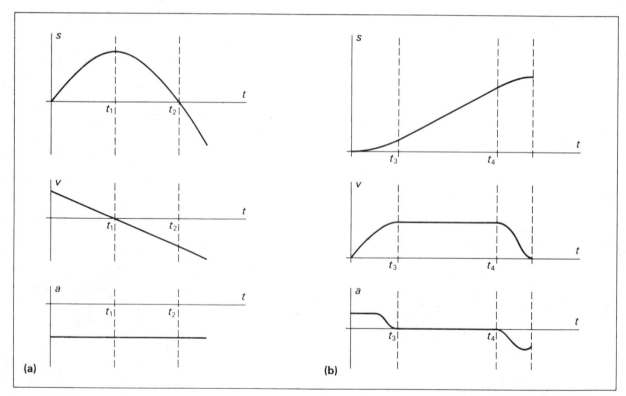

Figure 1.7 Linear motion graphs for (a) a stone thrown vertically upward by a child standing on a wall, and (b) an underground train moving from one station to the next.

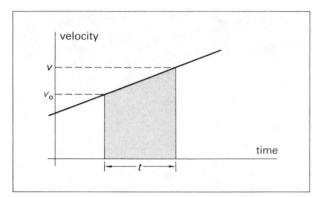

Figure 1.8 A v-t graph for motion with constant acceleration a. Here $a = (v - v_0)/t$ and the change of displacement during the time interval t is $s = \frac{1}{2}(v_0 + v)t$.

Rearranging (1) gives

$$v = v_0 + at \tag{3}$$

and multiplying (1) by (2) gives

$$as = \left(\frac{v - v_0}{t}\right)\left(\frac{v + v_0}{2}\right)t = \frac{v^2 - v_0^2}{2}$$

i.e. $\quad v^2 = v_0^2 + 2as \tag{4}$

Substituting from (3) into (2) gives

$$s = \left(\frac{v_0 + v_0 + at}{2}\right)t$$

i.e. $\quad s = v_0 t + \frac{1}{2}at^2 \tag{5}$

These equations, which all stem from (1) and (2), enable a range of problems to be solved. They are vector equations: one direction along the line must be chosen as positive for all of s, v_0, v and a. For an object accelerating from rest equation (5) becomes $s = \frac{1}{2}at^2$.

▶ A train slows down at a constant rate from 50 m s^{-1} to 10 m s^{-1} in 80 s. Find its acceleration and the distance travelled in this time. We shall take the positive direction to be in the direction of the train's motion.

$$a = \frac{v - v_0}{t} = \frac{(10 - 50) \text{ m s}^{-1}}{80 \text{ s}}$$

$$= -\frac{40}{80} \text{ m s}^{-2} = -0.50 \text{ m s}^{-2}$$

$$s = \frac{(10 + 50) \text{ m s}^{-1}}{2} \times 80 \text{ s}$$

$$= 30 \times 80 \text{ m} = 2400 \text{ m} \qquad ◀$$

1.4 Free fall

Any object thrown upwards or downwards or simply released close to the Earth's surface will, in the absence of air resistance, move with a constant downward acceleration of about 10 m s^{-2}, which is called g. It is useful and important to know g precisely at a given place. Stroboscopic photographs or ticker tapes for bodies moving freely in a vertical line can provide the information from which we can measure g. Two more precise methods designed to achieve an uncertainty of only about $\pm 1\%$ are outlined below. Whether the object moves up *or* down we say it is in a state of *free fall*.

A direct determination of g

Figure 1.9 shows an arrangement for measuring the time taken for a steel ball to fall a measured distance from rest. For a fall of about 1 m, $t \approx 500$ ms and can be read from the clock to ± 2 ms. Several values of t should be taken for the same h, which can itself be measured with a ruler to at most ± 4 mm. As the initial velocity of the ball is zero, we have

$$h = \frac{1}{2}gt^2$$

from which g can be found. The uncertainties in the experiment include the action of the two-way switch and the electromagnet-release and trap-door-opening action. By plotting a graph of $\sqrt{h}$ against t for different values of h and measuring the gradient of this graph a value of g, free from these *systematic errors*, can be achieved. The gradient is $\sqrt{g/2}$, and should give g with an uncertainty of less than $\pm 1\%$, i.e. 9.8 m s^{-2} rather than 9.7 m s^{-2} or 9.9 m s^{-2}. A recent determination at the National Physical Laboratory

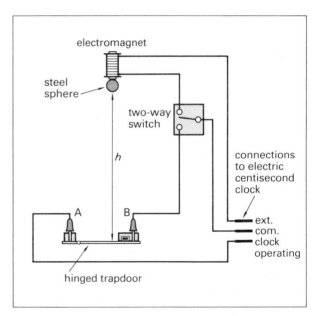

Figure 1.9 An arrangement for measuring the free-fall acceleration g of a steel sphere. AB are the terminals of an impact switch, the hinge is on side A and on side B the trap door is held closed by a tiny magnet.

 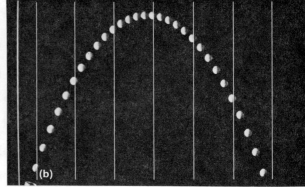

Figure 1.10 Stroboscopic photographs of freely falling objects. The horizontal lines in (a) and the vertical lines in (b) demonstrate that the vertical and horizontal motions are independent. In (a) the balls were rolled off separately and so the horizontal lines don't quite cut the balls in the same place.

using a freely-falling ball gave a value of g, the free fall acceleration in a vacuum, of $(9.812\,603 \pm 0.000\,002)$ m s^{-2}.

A simple-pendulum determination of g

On page 303 it is shown that the period of oscillation T of a simple pendulum of length l is given by $T = 2\pi\sqrt{l/g}$. A graph of T^2 against l for a range of values of l will have a gradient of $4\pi^2/g$, from which g can be found. The theory assumes a rigid support, no air resistance and *small* amplitude swings and all these can be achieved with a small metal sphere supported by a cotton thread which swings through about 5°. Clearly several measurements of nT (where $nT \approx 50$ s) should be taken for each value of l. The number of oscillations n should be counted from the moment when the pendulum passes through the vertical position; and l must be properly defined at the support, perhaps by holding the thread between two flat pieces of wood. The length of the pendulum should be measured to the centre of the swinging sphere. Again a careful experiment will give g to better than $\pm 1\%$.

Projectiles

When an object falls freely its vertical motion is *independent* of its horizontal motion. This is illustrated and supported by the evidence of figure 1.10a, in which three ballbearings are rolled off the end of a runway with different horizontal speeds. The stroboscope clearly demonstrates that each ball arrives at a given vertical level after the same time interval. By considering the motion of a ball which is imagined to drop simultaneously with zero horizontal velocity we can therefore use the linear motion equations of the previous section, with constant acceleration g, to make calculations about the vertical motion of the projected ball. In figure 1.10b the further point is made that the horizontal motion is motion at constant *speed*. These ideas enable us to solve many problems about the motion of projectiles such as basketballs or bombs, or problems

about electrons moving in uniform electric fields (page 236). Having calculated the horizontal or vertical displacements or velocities (s_x and s_y or v_x and v_y) we can then, if need be, find the *resultant* displacement or velocity of the projectile at any given time.

The path of a projectile is a *parabola*: this looks to be the case in figure 1.10b and we can show that, with the origin at the apex of the parabola,

$$s_x = v_0 t \quad \text{and} \quad s_y = \tfrac{1}{2}gt^2$$

i.e. $\quad s_y = \dfrac{g}{2v_0{}^2}s_x{}^2$

where v_0 is the horizontal velocity of the projectile. This equation is that of a parabola.

In real situations a projectile's motion is influenced by *air resistance*. The upper curve in figure 1.11 shows the vertical and horizontal displacements of a ball projected without air resistance at just under 45° above the horizontal at a speed of 30 m s^{-1}. It would have a *range* of almost 100 m. The actual path for a ball projected in the same way in air is shown by the lower curve. In this example the ball in the lower curve has a mass of about 0.015 kg and a diameter of about 120 mm; it could be a tennis ball. Not only is the

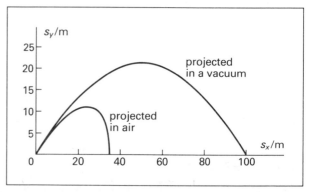

Figure 1.11 The effect of air resistance on a tennis ball.

range and maximum height much reduced by the air resistance but the projectile's path is asymmetric – the ball comes down more steeply than it rises. The upper curve is the same for *all* masses and sizes of projectile with this initial velocity, but the lower curve depends on both factors. Obviously a cricket ball would follow a path which was closer to the upper curve and a ball of cotton wool would have a range of only a few metres. For very long range missiles (e.g. ICBMs), g is not constant in either size or direction. Both of these factors make the projectile path non-parabolic, even if now the effect of air resistance is small.

▶ An electron moving at a speed of 2.0×10^7 m s^{-1} enters a region in which it experiences a constant acceleration of 1.0×10^{15} m s^{-2} perpendicular to its initial direction of motion. In what direction will it be moving after 5.0×10^{-9} s? (These numbers are of the right order for an electron passing between the Y-plates of a laboratory c.r.o.)

We can treat the motion in two parts: let us call the initial direction of motion the x-direction.

In 5×10^{-9} s the electron accelerates in the y-direction from zero to a velocity of

$$v_y = (v_0 + at) = (1.0 \times 10^{15} \text{ m s}^{-2})(5 \times 10^{-9} \text{ s})$$
$$= 5 \times 10^6 \text{ m s}^{-1}$$

It meanwhile continues in the x-direction at a velocity of 2.0×10^7 m s^{-1}. It is thus moving at an angle θ to its initial direction of motion where

$$\tan \theta = \frac{5.0 \times 10^6}{2.0 \times 10^7} \quad \Rightarrow \quad \theta = 14° \qquad \blacktriangleleft$$

2 Momentum and force

2.1 Mass and momentum

When we look at colliding trucks on a railway siding we see that it is not only their velocities which decide how they behave. Their behaviour also depends on the quantity of matter they contain. Two identical empty trucks contain the same quantity of matter (the same number of each type of atom and molecule which go to make them) and we say that they have the same *mass* or *inertia*. A truck full of coal has a greater quantity of matter than an empty truck and thus a greater mass. In a collision the full truck undergoes a smaller change of velocity than the empty one, i.e. the mass or inertia m of a body (in this case the truck) is a measure of its reluctance to change velocity in a collision.

We define the relative masses m_1/m_2 of two objects which act like particles by the relation

$$\frac{m_1}{m_2} = -\frac{\Delta v_2}{\Delta v_1} \tag{1}$$

where Δv_1 represents the change of velocity of body 1 as a result of an interaction (collision) with body 2, which itself suffers a change of velocity Δv_2; m_1/m_2 thus measures the ratio of the reluctances of the bodies to changes in their velocities – the ratio, that is, of their inertias. A pair of these books has twice the mass of a single one. If you press a spring between a pile of two books and one book and then let go, the single book moves away initially with twice the speed of the two books.

To achieve a scale of masses it is only necessary to select one body and agree to call its mass 1 unit. In SI the unit is the *kilogram* (kg). Any other mass m can then in principle be measured by observing interactions, e.g. collisions and explosions – see below. In equation (1) Δv_1 and Δv_2 are vectors; this is why there is a minus sign, for if one body accelerates in the positive direction, the other has negative acceleration.

Rearranging (1) gives, for a straight-line or head-on collision

$$m_1 \Delta v_1 + m_2 \Delta v_2 = 0$$
$$\Rightarrow \quad m_1(v_1 - u_1) + m_2(v_2 - u_2) = 0$$
$$\text{or} \quad m_1 u_1 + m_2 u_2 = m_1 v_1 + m_2 v_2$$

where u_1, u_2 are the initial velocities and v_1, v_2 are the final velocities of m_1, m_2 along the line. This is sometimes expressed by saying that when two or more bodies interact

$$\text{the sum of all the products } (mv) = \text{constant} \tag{2}$$

The product mv of mass and velocity of a body is called the *linear momentum* or simply the momentum of the body and equation (2) is called the *principle of conservation of linear momentum*. The unit of momentum is the kg m s^{-1} which can also be written as N s (page 16). Because velocity is a vector quantity so is momentum, its direction being that of the velocity.

In practice masses can be compared by much simpler methods than by analysing collisions for we find that the weight of a body at a given place is proportional to its mass. We can therefore compare masses by comparing weights using ordinary balances. A very sensitive test that mass is proportional to weight comes from the identical behaviour of different objects in an orbiting satellite.

Experiments with air-track gliders

The linear air track allows us to study collisions in one dimension in a situation where there are no other horizontal influences on the interacting objects. The gliders sit astride a track which consists of a tube into which air is pumped. The air continuously emerges through rows of tiny holes and supports the glider on a cushion of air. The gliders are not therefore in contact with the track and move freely along it. Their speeds can be measured using the techniques outlined in figure 1.1b (page 1). Suppose two photocell switches and electric clocks are available.

(a) We can set glider 1 in motion towards glider 2, which is stationary and has no card attached (figure 2.1a) and arrange for them to stick together. Plasticine will be adequate for this. We can show that the momentum before the collision (all in glider 1) is equal to the momentum after the collision (in the coupled gliders).

(b) We can arrange for both the gliders to be stationary and then have them fly apart, each recoiling from the other (figure 2.1b). A light spring attached to one glider can provide the necessary energy, release being achieved by burning through the cotton connecting thread. We can show that the gliders move apart with equal and opposite momenta so that the total momentum is still zero after the explosion (as it was before the string was burned).

In both situations the necessary speeds can be measured to about $\pm 5\%$ and the conservation principle demonstrated. Of course the gliders and photocells could be replaced by dynamics trolleys and ticker tape. Trolleys often have built-in springs to facilitate experiments of type (b) and pins and corks can be used to ensure that two trolleys stick together as in (a). These experiments are both examples of things which happen in practice. Thus (a) is similar to what happens when a car or lorry collides with a stationary vehicle, when a man jumps into a small boat or when a bullet hits a cowboy (in the better gangster or cowboy films a wire is attached to the victim's waistcoat and given a sudden jerk as the shot is fired). (b) is similar

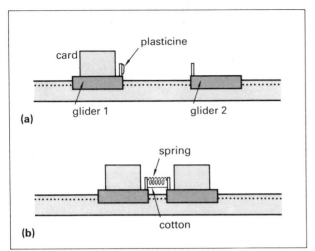

Figure 2.1 Apparatus for experiments involving collisions and explosions on an air track.

to what happens in all sorts of *recoil* situations, for example, when two skaters push each other apart or when a gun fires a shell.

In solving problems using the principle of conservation of momentum you must be sure that in the direction in which momentum conservation is to be applied either there are no external forces acting or the time for the interaction is very small. It is useful to sketch the situation both before and after the interaction, adding all the relevant data to the diagram.

▶ A pair of skaters, a man of mass 65 kg and a woman of mass 50 kg, are together moving in a straight line across the ice at a speed of 6.0 m s^{-1}. They push each other apart along their line of motion so that after they are separate the man is moving in the same direction at 4.0 m s^{-1}. What is the woman skater's new velocity?

Assuming that the ice exerts no horizontal force on either skater, we can say that (figure 2.2)

momentum before = momentum after
$$(65 + 50) \times 6.0 \text{ kg m s}^{-1} = (65 \times 4.0 \text{ kg m s}^{-1})$$
$$+ (50 \text{ kg} \times v)$$
$$\therefore \quad 50 v = (115 \times 6.0 - 65 \times 4.0) \text{ m s}^{-1}$$

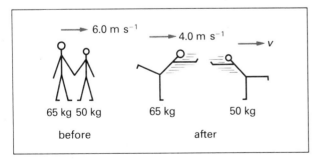

Figure 2.2 Two moving skaters push each other apart as they cross the ice; the positive direction is taken to be to the right.

$$= 430 \text{ m s}^{-1}$$
which gives $\quad v = 8.6 \text{ m s}^{-1}$ to the right

We could also use the equation (1) which defines mass to solve this problem:

$$\frac{m_1}{m_2} = -\frac{\Delta v_2}{\Delta v_1}$$

i.e. $\quad \dfrac{65 \text{ kg}}{50 \text{ kg}} = -\dfrac{\Delta v_{\text{woman}}}{(4.0 - 6.0) \text{ m s}^{-1}}$

$\therefore \quad \Delta v_{\text{woman}} = \dfrac{65 \times 2.0 \text{ m s}^{-1}}{50}$

$$= 2.6 \text{ m s}^{-1}$$

and the woman's final velocity is $(6.0 + 2.6) \text{ m s}^{-1} = 8.6 \text{ m s}^{-1}$. ◀

▶ A ball of mass 0.80 kg is kicked at a speed of 12 m s^{-1} to the right and hits a wooden box of mass 12 kg which is at rest on a frictionless floor. The box moves off with an initial velocity of 1.4 m s^{-1} to the right after the collision. What is (i) the new velocity of the ball and (ii) the change of momentum of (a) the ball, (b) the box, as a result of the collision?

(i) Referring to figure 2.3, and applying the principle of conservation of momentum, we get

$$12 \times 0.80 \text{ kg m s}^{-1} = (0.80 \text{ kg} \times v) + (12 \times 1.4 \text{ kg m s}^{-1})$$
$$\therefore \quad 0.8 v = (9.6 - 16.8) \text{ m s}^{-1} = -7.2 \text{ m s}^{-1}$$
which gives $\quad v = -9.0 \text{ m s}^{-1}$

The minus sign tells us that the ball has rebounded from the box and is now travelling to the left.

(ii) The change of momentum of the box is

$$\Delta(m_2 v_2) = m_2 \Delta v_2 = 12 \text{ kg} (1.4 - 0) \text{ m s}^{-1}$$
$$= 16.8 \text{ kg m s}^{-1} \text{ to the right}$$

while that of the ball is

$$\Delta(m_1 v_1) = m_1 \Delta v_1 = 0.8 \text{ kg} (-9 - 11) \text{ m s}^{-1}$$
$$= -16.8 \text{ kg m s}^{-1} \text{ to the right}$$
$$\text{i.e. } 16.8 \text{ kg m s}^{-1} \text{ to the left}$$

On page 31 there is an energy analysis of this and the previous example. ◀

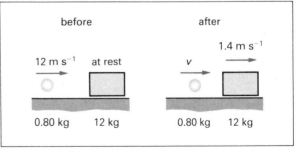

Figure 2.3 A collision between a ball and a box. The positive direction is taken to be to the right; if the ball is moving to the left after the collision then v must to be negative.

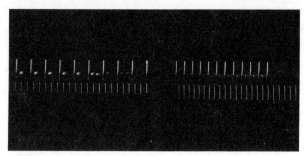

Figure 2.4 Analysing an elastic collison between two air track gliders using stroboscopic photography.

To analyse more complicated interactions than those described in figure 2.1 the technique of stroboscopic photography can be used. Figure 2.4 shows the result of a collision between two air track gliders (or it could be any two bodies) which approach one another along a line and rebound. Each glider has a straw attached to it: as they approach, the tops of the straws are seen and at the moment of collision a shutter is raised which hides the top but reveals the bottom of the straws. The photograph can be analysed using arbitrary units of mass and speed to illustrate that momentum is conserved, due account being taken of the signs of the velocities. The analysis of two-dimensional interactions is considered on page 32.

2.2 Forces

We all know what forces are. They are *pushes* and *pulls*. The SI unit of force is the *newton* (N), named after Sir Isaac Newton (1642–1727). This unit is defined on page 14. For the moment, however, it is enough to know, for instance, that the pull of the Earth on a mass of 60 kg is about 600 N and that typical pushes and pulls made by the hand are between 0.1 N (pushing the key of an electronic calculator) and 100 N (pulling open a massive door).

All the apparently different types of force can be traced to only three basic interactions: gravitational, electromagnetic and nuclear. Of these the *nuclear* force exists only between the parts of the nucleus and becomes negligible when these particles are more than about 10^{-15} m apart. Of the other two, *gravitational* forces are much weaker than *electromagnetic* forces. To quantify this consider two protons: the electric force which each exerts on the other is about 10^{36} times greater than the gravitational forces which they exert on each other. Thus although all bodies do attract each other gravitationally (chapter 12), these forces are only noticeable if one of the bodies is very massive e.g. like the Earth or the Moon.

Everyday *contact forces* are electromagnetic in origin (we refer to electromagnetic rather than electric as the forces between moving charged particles are called magnetic forces). When we shake hands or stamp on the ground it is the electromagnetic interaction (push or pull) between groups of atoms which we feel.

Representing forces

It is *always* possible to describe a force by the phrase

the push (or pull) of A on B

where A is a body exerting a force on body B. E.g. we talk of 'the push of the pen on the paper' or 'the pull of the lamp on its cable.' The consistent use of this sort of phrase prevents statements like 'the force of friction' or 'the pull of gravity' and makes us identify B (the body on which the force acts) and A (the body which exerts the force). To identify B is often the key to starting a problem and to thinking clearly about it. For this reason, if we are drawing a sketch to show the forces acting on body B, it is best not to include any other bodies in the diagram. Such a diagram is called a *free-body diagram*. Figure 2.5a shows an example of a simple situation involving forces. Suppose we wish to consider the forces acting on the box; a free-body diagram of the box is drawn (b) showing the push P of the ground, the pull W of the Earth, and the pull T of the rope – these all being pushes or pulls acting *on the box*.

If possible the forces in a free-body diagram should be drawn so that their lengths are proportional to the size of the forces. There are other bodies in figure 2.5 in which we may be interested. You should try to draw free-body diagrams for the pulley block and for the man.

Newton's first law

In section 2.1 a body is said to change its momentum only

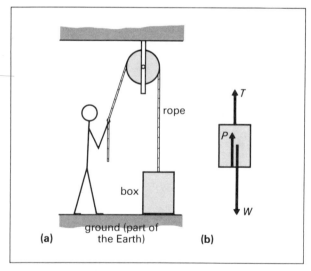

Figure 2.5 (a) A man tries to lift a box using a single fixed pulley. (b) is a free-body diagram of the box: T is the pull of the rope on the box, P is the push of the ground on the box and W is the pull of the Earth on the box.

as a result of an interaction with another body, i.e. when a second body exerts a push or pull on it. This is known as *Newton's first law of motion* and is usually stated as follows:

> *a force is required to change the velocity of a body, that is, to alter its speed or the direction of its motion.*

It is sometimes called the law of inertia and was almost fully formulated by Galileo Galilei (1564–1642). A body is said to be *in equilibrium* if it is either at rest (relative to the Earth's surface) or moving with a constant speed along a straight line. The box in figure 2.5 is in equilibrium and so we can deduce that $T + P = W$. If a car is moving along a straight road at 25 m s^{-1} there may be a wind resistance force (the push of the air on the car) of 2000 N. As the car is in equilibrium there must be a forward driving force (the push of the road on the wheels of the car) which is also 2000 N.

Adding forces

Forces obey the rules of vector calculations, that is they can be added and subtracted using the triangle method described in section 1.2, and they can be resolved. We shall for the moment be concerned with forces acting on bodies without producing rotation or tending to produce rotation. The bodies, e.g. a car or a box, can be treated as if they were single particles.

To show experimentally that forces obey the rules of vector addition consider the arrangement shown in figure 2.6. In (a) two forces, F_1 and F_2, pulling on the ring are measured by spring balances P and Q and their directions

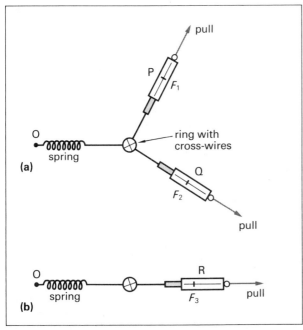

Figure 2.6 Investigating how forces add.

noted. The position of the centre of the cross-wires is marked and P and Q are now replaced by a single spring balance R. This is used to measure the size of the single force F_3 which will return the centre of the cross-wires to the marked position. Hence the force which produces an extension of the spring identical to that produced by the vector sum $F_1 + F_2$ is found. F_1 and F_2 can now be added by scale drawing and their resultant compared in size and direction with the measured F_3. A single experiment may not be conclusive but a number of trials should verify that force obeys the rules for vector addition. It may be convenient to replace the spring balances by pulleys over which hang bodies of known weight.

Some common forces

Figure 2.5 showed three forces acting on a box. Forces of each type are commonly met in problems in mechanics and they are referred to in a general context as tension T, weight W, and perpendicular contact force (or reaction) P.

Tension

Any flexible rope (or chain or string) being used to pull a body is said to be in a state of tension. The molecules of the rope are very slightly further apart then they would be if it was slack. The tension in the rope at a point X is the size of the force which one part of the rope cut at the point X exerts on the other part. For chains or thick ropes such as that shown in figure 2.5 the tension varies from place to place, but this is often ignored and the tension is taken to be constant. For light threads and strings it is a reasonable assumption. Where the rope ends, the pull of the rope on the object to which it is attached is equal in size to the tension in the rope at the point of connection, e.g. T in figure 2.5. When a rope passes round a free-running pulley or a smooth post the tension is usually taken to be the same on both sides of the pulley or post. This again is only approximately true in practice. Pulleys can thus change the direction of a force without altering its size and it is for this reason that they are useful. In figure 2.5, if the pull of the man on the rope is 70 N downwards, then the tension in the rope is 70 N and the pull of the rope on the box is 70 N upwards.

Weight

The Earth exerts an attractive force, a gravitational pull, on every particle. This pull is directed towards the centre of the Earth. For an extended body the resultant of all the tiny forces is a *single* pull called its weight. Whichever way the body is turned its weight is found always to act through a particular point; this is called the *centre of gravity* of the body (page 169). The weight of a body varies slightly from place to place on the Earth's surface but the variations are always less than 0.5%, so we shall usually treat the weight of a body on the Earth's surface as effectively constant.

Perpendicular contact force

Any two bodies pressed together by some external agency (the gravitational pull between the Earth and the box in figure 2.5) repel one another. The repulsion is the result of the compression of the surface layers of material, i.e. the surface molecules are very slightly closer together than when they were not pressed together. The contact force is the resultant of all the tiny pushes and can be represented by a single force which is perpendicular to the surfaces in contact.

If a body is being pushed against a surface in such a way as to make it slide over the surface the contact force will no longer be perpendicular to the surface. It is however convenient to talk of the two resolved parts of the contact force in this case: the *perpendicular* push of the surface on the body and the *tangential* (frictional) push of the surface on the body. The perpendicular push we will call a *perpendicular contact force*. See section 2.4 for friction forces.

Equilibrium

If a body is in equilibrium then it is necessary that the vector sum of the forces acting on it should be zero. The sum of the resolved parts of the forces acting on it in any convenient direction must therefore also be zero. In order to *show* that a particle is in equilibrium it is necessary to show that the sum of the resolved parts of the forces acting on it is zero in any three directions (which must not lie in one plane). For a set of coplanar forces (usually the case in the examples we shall consider) we need only resolve in two directions.

▶ A kite of weight 2.0 N is held stationary by the push of the wind and the pull of the string on it. The tension in the string, which makes an angle of 35° to the vertical at the place where it meets the kite, is 9.0 N. What is the size and direction of the push of the air on the kite?

Figure 2.7 shows a sketch of the kite and a free-body diagram of it represented as a particle (as it does not tend to rotate). The pull of the Earth on the kite and the pull of the string on the kite are drawn and the unknown push of the air on the kite is shown as P at an angle θ above the horizontal. We can see that P must be upwards and to the right to achieve equilibrium as the sum of the resolved parts of the forces must be zero both vertically and horizontally. Resolving horizontally

$$P \cos \theta - (9.0 \text{ N}) \sin 35° = 0$$
so that $\qquad P \cos \theta = 5.17 \text{ N} \qquad$ (3)

Resolving vertically

$$P \sin \theta - (9.0 \text{ N}) \cos 35° - 2.0 \text{ N} = 0$$
so that $\qquad P \sin \theta = (7.37 + 2.0) \text{ N} = 9.37 \text{ N}$

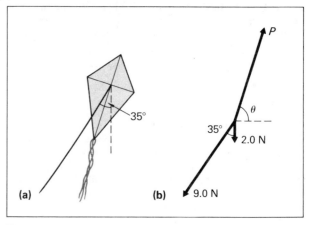

Figure 2.7 A kite and the forces on it. (b) is a free-body diagram of the kite.

Thus $\quad \dfrac{P \sin \theta}{P \cos \theta} = \dfrac{9.37}{5.17} = 1.81$

i.e. $\qquad \tan \theta = 1.81 \quad \Rightarrow \quad \theta = 61°$

Substituting in (3)

$$P \cos 61° = 5.17 \text{ N}$$
gives $\qquad P = 10.7 \text{ N}$

Thus the push of the air on the kite is 10.7 N at an angle of 61° above the horizontal. Notice that if a triangle were drawn to scale to add the 2.0 N and the 9.0 N forces, P must equal their resultant in size but be *opposite* to it in direction to maintain equilibrium. ◀

2.3 Newton's second law

Sir Isaac Newton *defined* force in his laws of motion. The first law is given in the previous section. *Newton's second law* tells us more than what a force does, it tells us how to *measure* forces, and is usually stated as follows:

> the rate of change of linear momentum of a body is equal to the resultant force acting on it and takes place in the direction of this force.

In symbols $\quad \dfrac{\mathrm{d}(mv)}{\mathrm{d}t} = F_{\text{res}} \qquad$ (4)   (appendix B3)

where v and F_{res} are parallel vectors. The unit of force is therefore the $\text{kg m s}^{-1} \div \text{s}$ or kg m s^{-2}. This, being something of a mouthful, is called a *newton* (N).

$$1 \text{ N} \equiv 1 \text{ kg m s}^{-2}$$

In many practical situations we are interested in the motion of a body of constant mass m (perhaps an egg or a car). In these cases rate of change of momentum can be written as mass times acceleration, for

$$\frac{\mathrm{d}(mv)}{\mathrm{d}t} = m\frac{\mathrm{d}v}{\mathrm{d}t} = ma \qquad \text{(constant } m\text{)}$$

i.e. $\qquad ma = F_{\text{res}} \qquad\qquad\qquad$ (5)

The gravitational field

The most commonly met force-acceleration situation is that for freely falling bodies near the Earth's surface. The vertical acceleration is $g \approx 10$ m s^{-2} and at one place this is constant for all bodies. Photographs like that shown in figure 1.10a but using balls of different mass confirm this experimentally. Rearranging equation (5) we get, for any free-falling body of weight W and mass m ·

$$g = \frac{F_{\text{res}}}{m} = \frac{W}{m}$$

In this equation g has units N kg^{-1}; you will see that it is a constant which describes the strength of the *gravitational field* or *g-field* at the chosen place. From the definition of the newton we see that 10 N kg$^{-1} \equiv 10$ m s^{-2}, so that the weight of a body of mass 60 kg is about 600 N, of a body of mass 0.5 kg about 5 N and so on.

A careful measurement of the free-fall acceleration g thus helps us to set up a scale of forces using a set of known masses. For if a body of mass 1.00 kg is being supported in equilibrium by a piece of string then the upward pull of the string on the body must equal the downward pull of the Earth on the body. (We are ignoring the rotation of the Earth in saying that the body is in equilibrium.) This downward force is the same as when the body is in free fall, i.e. mg. The tension in the string is thus 1.00 kg $\times$ 9.81 m s^{-2} = 9.81 kg m s^{-2} or 9.81 N. This and similarly derived forces could, for example, be used to calibrate a spring balance or a lever balance. The fact that $m \propto W$ also enables us to *compare* masses by comparing weights with a beam balance.

Testing Newton's second law

We can (i) measure the acceleration of a body of fixed mass acted on by different forces or (ii) measure the acceleration produced by the same force on bodies of different mass. Trolleys, ticker-timers and elastic are needed (figure 2.8). To compensate for frictional forces, the experiments are best performed on an adjustable slope arranged so that a trolley travels down it at a constant speed after being given an initial push. The trolley is then in equilibrium so that from Newton's first law, $W + N + X = 0$ (vector addition) i.e. a scale drawing of W, N and X would form a closed triangle. When this is achieved the slope is said to be *friction-compensated*.
(i) To produce various forces accelerating a trolley we can use a set of identical elastic threads. When stretched to a given length, e.g. the length of the trolley, they can give forces (the pull of the thread on the trolley) of P, $2P$, $3P$ etc. The acceleration a can be found from the tape (page 5) and

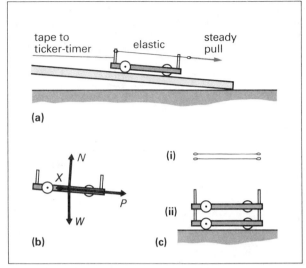

Figure 2.8 (a) The experimental set-up for $ma = F$ experiments. (b) shows a free-body diagram of one trolley being pulled down a friction-compensated slope and (c) illustrates how (i) the force and (ii) the mass can each be doubled.

a graph drawn of a against P. If this is a straight line through the origin we know that $a \propto P$ for a fixed mass.
(ii) By stacking identical trolleys on top of one another we can use a single elastic thread giving a constant force P to accelerate successive bodies of mass m, $2m$, $3m$ etc. This time a graph of a against $1/m$ can be drawn. If it is a straight line through the origin we know that $a \propto 1/m$ for fixed P. Such experiments are very imprecise and we can expect to measure the acceleration to perhaps $\pm 10\%$.

A more careful measurement of the effect of a constant force on a fixed mass is possible. An air-supported puck is pulled across a flat surface and photographed stroboscopically. The constant force can be provided by attaching the puck to one end of a thread which, after passing over a pulley, is connected to a hanging lump. Although the tension in the thread is not in this case quite equal in size to the weight of the lump, it is constant.

In none of these experiments is the law verified; they show only that it is a consistent statement of what happens in such cases.

Applying Newton's second law

In solving problems using $ma = F_{\text{res}}$ it is best to start with a as it is the direction of the acceleration which is usually known. The forces can then be resolved in this chosen direction and the resolved parts added to give the resultant force producing the acceleration, i.e.

$ma =$ sum of forces resolved parallel to a

In applying the law you should:
(i) choose a body (as a first step this act of choosing cannot

be emphasised too strongly).
(ii) draw a free-body diagram of the chosen body, marking all the external forces on it. Ensure that each is a genuine force on the body B by expressing it as the pull (or push) of something on B.
(iii) mark the acceleration alongside the body.
(iv) in numerical problems ensure that all forces are expressed in newtons.
(v) apply Newton's second law in the form given above.

► A man stands in a lift which is moving upwards and slowing down at 0.80 m s^{-2}. If his mass is 70 kg, what is the push of the floor on his feet? Take $g = 10$ m s^{-2}.

Figure 2.9 shows the situation and a free-body diagram of the man; for it is a force *on* the man which we have to find. Using $ma = F_{res}$ we have

$$(70 \text{ kg})(0.80 \text{ m s}^{-2}) = 700 \text{ N} - P$$

$$\therefore \quad P = (700 - 56) \text{ N} = 644 \text{ N} \quad \blacktriangleleft$$

► Two blocks of mass 5 kg and 20 kg rest in contact on a frictionless horizontal surface. A horizontal force of 100 N is exerted on the 5 kg block so as to accelerate them both. Find (i) the acceleration and (ii) the perpendicular contact push of the 5 kg block on the 20 kg block.
(i) To find the acceleration we choose the pair of blocks as a *single body*: figure 2.10a shows the external pushes and pulls acting on it.
Applying Newton's second law horizontally then gives

$$(25 \text{ kg}) a = 100 \text{ N}$$

$$\therefore \quad a = \frac{100 \text{ N}}{25 \text{ kg}} = 4.0 \text{ m s}^{-2}$$

(ii) Figure 2.10b shows a free-body diagram of the 20 kg block alone (picking a body is the vital first move) for we are now interested in the forces acting *between* the blocks. Using Newton's second law horizontally then gives

$$(20 \text{ kg})(4.0 \text{ m s}^{-2}) = P$$

$$\therefore \quad P = 80 \text{ kg m s}^{-2} = 80 \text{ N}$$

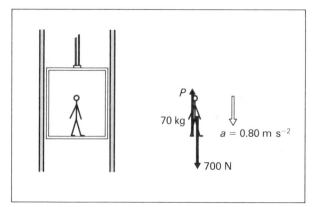

Figure 2.9 The forces on a man in a lift. As the lift accelerates downwards *P* will be less than his weight 700 N.

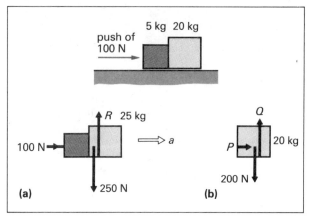

Figure 2.10 The choice of a body for which to draw a free-body diagram is critical: (a) and (b) show two choices used to analyse the initial situation.

and so the push of the 5 kg block on the 20 kg block is 80 N.   ◄

Impulse

Returning to the statement of Newton's second law (equation (4) page 14), this can be rewritten as (F for F_{res})

$$\Delta(mv) = F\Delta t$$

i.e. the change of momentum of a body is equal to the product of the resultant external force F and the time Δt for which it acts. The product $F\Delta t$ is called the *impulse* of the force.
Thus

impulse = change of momentum

can be regarded as an alternative statement of Newton's second law. It is useful in problems involving steady forces acting for known lengths of time, for instance in calculating the effect of firing a rocket. Suppose a rocket motor is switched on for 3.0 s and during that time produces a steady thrust of 12 000 N, then the impulse is 36 000 N s. (The N s, *not* N/s, is identical to our earlier unit of momentum, the kg m s^{-1}. It should be clear why this is so.) If the rocket has a mass of 2000 kg (assumed constant), then the change of velocity is given by

$$\Delta v = \frac{36\,000 \text{ N s}}{2000 \text{ kg}} = 18 \text{ m s}^{-1}$$

This might be a course correction thrust on an Apollo moon shot; obviously the direction of F and of the change of velocity would be critical. Where F and Δt are not known at every instant, (e.g. when a racket hits a tennis ball), then the impulse of the push of the racket on the ball can still be found by calculating the change of momentum of the ball. If an electric circuit can be arranged to give a measure of the time interval Δt during which the racket is in contact with the ball, then the average value of F can be found. Sudden large forces which last for small time intervals are sometimes themselves referred to as *impulsive* forces.

► An oxygen molecule has a mass of 5.3×10^{-26} kg. If it is moving at a speed of 450 m s^{-1} directly towards a wall and rebounds with the same speed, what is the impulse which the wall gives to the molecule?

Initial momentum of molecule
$$= (5.3 \times 10^{-26}\ \text{kg})(450\ \text{m s}^{-1})$$
$$= 2.4 \times 10^{-23}\ \text{kg m s}^{-1} \text{ towards the wall}$$

Final momentum of molecule $= -2.4 \times 10^{-23}$ kg m s^{-1} (i.e. away from the wall).

Calling towards the wall the positive direction, the change of momentum of the molecule is

$$\Delta(mv) = (-2.4 \times 10^{-23} - 2.4 \times 10^{-23})\,\text{kg m s}^{-1}$$
$$= -4.8 \times 10^{-23}\ \text{kg m s}^{-1}$$

The impulse of the wall on the molecule is thus

$$F\Delta t = -4.8 \times 10^{-23}\ \text{kg m s}^{-1}$$
$$\text{or} \quad -4.8 \times 10^{-23}\ \text{N s}$$

i.e. 4.8×10^{-23} N s away from the wall. ◄

This impulse-change of momentum way of using Newton's second law is often the best method of giving a physical explanation in answer to questions such as 'Why are eggs packed in crushable egg-boxes?' or 'How should you try and catch a cricket ball?' In each case we are dealing with an object, a dropped egg or a moving ball, which has to have its momentum reduced to zero. $\Delta(mv)$ is predetermined for us and so, therefore, is the impulse $F\Delta t$ that is needed. Clearly the larger Δt the smaller F, the force which is used to slow down the egg or the ball. Thus the egg is put in a box which will crush slightly on hitting the ground; the cricketer allows his hands to 'give' with the ball as he catches it. In both cases Δt is many times greater than the value it might have had were the egg not boxed or were the ball to strike a rigid hand.

Packaging eggs to prevent breakage during handling is illustrative of a much more significant problem: the packaging of human beings in motor cars. If the car is to crash, let us say head-on into a large concrete block, then there are two main factors which affect the safety of the driver. Firstly the front of the car will crush, acting like the egg box (but nevertheless producing very high decelerations ≈ 400 m s^{-2} for impact velocities below 20 m s^{-1}). Secondly, the passenger is not snugly fitted into this box with moulded supporting materials all around him. He is sitting some 0.5 m away from a solid wind-screen, dash-board and steering wheel.

The second factor is illustrated by graphs A and B of figure 2.11 which show the variation of the force F acting on the driver against time t (note the scales). For curve A the driver is wearing a seat belt and F is the pull of the belt on his chest. For curve B he is not wearing a seat belt and F is the push of the windscreen, etc. on his chest. The impulse must be the same in each case as $F\Delta t = \Delta(mv) = 60$

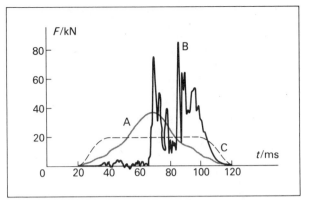

Figure 2.11 F-t curves for a driver in a car which hits a massive concrete block. Curve B is for the case when the driver is not wearing a seat belt and C represents the ideal (the smallest maximum F) with a seat belt.

kg $\times$ 20 m s^{-1} = 1200 N s, regardless of how he is brought to rest. The area between graph-line A and the time-axis must therefore be the same as the area between graph-line B and the time-axis.

In the design of *seat belts* the objective is to tie the passenger to the car body and to hold him with a restraining force something like the dashed graph-line C in figure 2.11. To achieve this the crumpling characteristics of the front of the car can be altered, as large a space as possible between the seated driver and the windscreen provided, and drivers encouraged to wear their seat belts as tightly as is comfortable.

2.4 Animal and vehicle propulsion

No animal or vehicle can pull *itself* along: an *external* force is always needed. If you doubt this perhaps you could try to lift yourself up to the ceiling (by your shoelaces perhaps). No, you cannot have a rope for then it would be the pull of the rope which was lifting you; nor a stepladder, for then it would be the push of the steps, and so on. The external forces result from the animal or vehicle itself exerting a force on the ground, water or air with which it is in contact. Let us first of all concentrate on the case where the movement is over a solid surface.

Newton's third law

Newton was the first to appreciate that whenever two bodies interact, *two* forces are involved. One force acts on each body and there is a simple relationship between them at every instant of the interaction, which can be expressed as follows:

if body A exerts a force F on body B, then body B exerts a force −F on body A, that is, a force which has the same size but is opposite in direction.

This statement is called *Newton's third law* and implies that a single force is an impossibility. Forces always occur in pairs, one force acting on each of the two interacting bodies. They will never therefore *both* appear on the same free-body diagram.

The law holds *at every instant* of the interaction. Consider the situation of figure 2.12 which shows a boy landing on the ground (part of the Earth) having jumped off a wall. Two free-body diagrams are necessary as we are interested in the forces on the boy and the forces on the ground.

In this example, by Newton's third law

$$P = -P' \quad \text{and} \quad W = -W'$$

and the graphs illustrate that this is true at every instant during the interaction. Some other examples:

(i) If the Earth pulls you down with a force of 600 N, then you pull the Earth up with a force of 600 N. If the Earth's surface pushes your feet up with a force of 650 N, then your feet are pushing the Earth down with a force of 650 N. Notice that the resultant force on *you* is 50 N upward; you must be accelerating upward. Perhaps you are in the process of jumping.

(ii) When you step on an egg the push of your foot on the egg is equal in size but opposite in direction to the push of the egg on your foot. The fact that the egg breaks (and not your foot) is irrelevant. We should need to know about the other forces acting, and the properties of the materials of eggshells and of feet to explain why it is the egg that breaks.

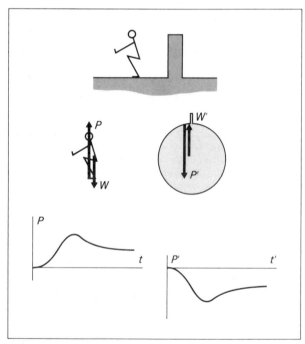

Figure 2.12 Newton's third law is true at every instant, e.g. throughout the jump and landing of the boy.

Newton's third law and the principle of conservation of momentum are equivalent statements of the same physical result. The principle of conservation of momentum is equivalent to the (vector) relationship

$$m_1 \Delta v_1 + m_2 \Delta v_2 = 0$$

for an interaction between two objects of constant mass m_1 and m_2. Suppose that the interaction takes a time interval Δt, then using the impulse-momentum relation, we can write

$$F_1 \Delta t + F_2 \Delta t = 0$$
$$\text{or} \quad F_1 = -F_2$$

which is Newton's third law, F_1 being the push or pull of body 2 on body 1 and F_2 the push or pull of body 1 on body 2. This is a vector relation so we see that F_1 and F_2 have the same size but are opposite in direction. The above argument holds over any small time interval Δt during the interaction. By reversing the steps of the argument, the principle of conservation of momentum follows from Newton's third law.

Frictional forces

Frictional forces occur in a wide variety of situations but always act so as to oppose or prevent the onset of relative motion between the two solid surfaces which are in contact. Though this seems to imply that friction will always be a nuisance, quite the reverse is true as our ability to move about on the Earth's surface is almost entirely the result of frictional forces. Let us consider the forces acting on a motor car which is driving at a steady speed along a straight road.

Figure 2.13 shows a representative free-body diagram for the car and a separate diagram of the ground on which only the contact forces are shown. The car shown has a rear-wheel drive. The *driving force* is F, the push of the ground on the rear wheels of the car. F is a third-law force and is equal in size to F', the push of the rear wheels of the car on the ground. F' in turn results from the car engine attempting to rotate the rear wheels at a greater speed than they would if they rolled forward. F and F' are thus a pair of oppositely directed frictional forces which prevent the wheel slipping on the road surface. The frictional force which opposed the motion of the wheel becomes the car's external driving force. F is called a *static* friction force.

The size of F (and F') depends upon the nature of the road and wheel surfaces. A lack of tread on the tyres and ice or an oily film on the road will reduce the maximum possible value of F before slipping occurs. It also depends on the size of the perpendicular contact push M of the ground on the rear wheels (M is equal in size to the third-law force M', the perpendicular contact push of the rear wheels of the car on the ground). The larger M is, the larger the size of F before slipping occurs. In an ordinary car $M + N = W$, the pull of the Earth on the car. If it were a

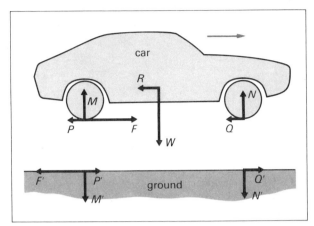

Figure 2.13 A free-body diagram of a rear-wheel drive car and an incomplete free-body diagram of the ground (which is part of the Earth).

racing car it might have an inverted 'wing' or aerofoil placed over the rear wheels, which makes $M + N > W$, and so achieves a greater frictional force F.

In figure 2.13 two other frictional forces are shown, P and Q together with their third-law counterparts P' and Q'. These are forces of *rolling* friction. They are also described as the frictional push of the ground on the wheels but must be distinguished from F as they cannot be utilised as driving forces. Rolling friction results from the deformation of the wheel and the ground and has the effect of making the wheels always seem to be rolling slightly uphill. The other force opposing motion is an air resistance force R, the push of the air on the car. Forces like this are discussed on page 68.

If the car is to accelerate then the engine must attempt to rotate the rear wheels faster. F' and F, the forward push of the ground on the car, thus increase. To brake the car the discs or drums attempt to prevent the wheels rotating. Because of the inertia of the car the direction of F is reversed so that the braking force is the (external) backward push of the ground on the car.

The above discussion is based entirely on a car. But the operation of all wheel-driven vehicles can be understood similarly. Electric or diesel trains, motorcycles and bicycles (an example follows at right), powered lawn mowers and track-laying tanks all use the same principle, the main difference being the nature of the energy source. Moreover animals, both man and four-legged beasts, all achieve a forward motion by pushing the ground backwards and relying on Newton's third law of motion. In all such propulsion momentum is conserved if we take into account not only the momentum of the vehicle or animal but also the momentum of the ground which is part of the Earth. As the mass of the Earth is so huge we do not notice it recoiling when we accelerate forward. But it does recoil, an effect which can be demonstrated in principle by standing on a large trolley and attempting to move forward along it or off it.

The nature of friction

So far we have concentrated on friction as a force which opposes the onset of relative motion. Friction leads to wear, and this leads us to look more closely at the nature of surface contact at a microscopic level.

When two surfaces are pushed together (perhaps because the upper is pulled down by the Earth and the lower is pushed up by the ground) the tiny irregularities (called asperities) of which the surface consists, make a *real* area of contact A' much less than the *apparent* area of contact A. Figure 2.14a illustrates this point. The materials deform (plastically if they are metals) until the local pressure p' at the asperities has a value which the microscopic structure can support. $p' = N/A' = $ constant, where N is the size of the perpendicular contact force between the surfaces. Note that A' is independent of A.

Relative motion between the two materials in a direction perpendicular to N, i.e. sliding, now leads to a bumpy ride at the microscopic level. The asperities have to ride up and over each other, and in metals cold welding and the shearing of these welds is continually taking place (figure 2.14b). We can find tiny pieces of one metal in the surface of the other after sliding. The tangential frictional push F of one surface on another is likely to be proportional to the number of the interacting asperities and therefore to their total area A'. We can thus predict that when sliding is taking place

(i) F is independent of A, the apparent area of contact,
(ii) $F \propto N$, as $F \propto A'$ and $A' \propto N$ as argued above.

Both these deductions can be tested experimentally by dragging a block over a surface with a spring balance, and are found to be approximately true.

To reduce F for a given value of N the surfaces are lubricated. The liquid layer now tends to prevent the solid asperities from touching and the force required to shear the liquid is smaller than the dry sliding frictional force.

▶ A bicycle and its rider have a combined mass of 90 kg. Assuming that the maximum frictional force between the rear wheel and the ground is 80% of the perpendicular contact force at that wheel, estimate the maximum possible acceleration of the bicycle.

Let us suppose that the size of the perpendicular contact push of the ground on the wheel is $N = M = 450$ N at each wheel, taking g to be 10 m s^{-2} (figure 2.15). The maximum frictional push of the ground on the wheel is thus

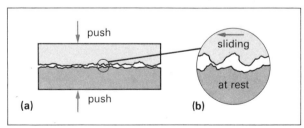

Figure 2.14 The nature of sliding friction. In (b) a small part of the surface is shown much magnified.

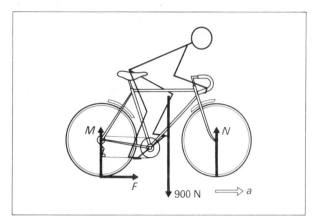

Figure 2.15 The external forces acting on an accelerating bicycle and rider.

$$F = 0.8 \times 450 \text{ N} = 360 \text{ N}$$

and the acceleration (assuming rolling friction and air resistance to be negligible) is given by Newton's second law

$$(90 \text{ kg}) a = 360 \text{ N} \quad \Rightarrow \quad a = 4 \text{ m s}^{-2}$$

But this is not the maximum possible acceleration. To increase a, the rider may try to lift the front wheel clear of the ground so that M becomes 900 N (which leads to $a = 8 \text{ m s}^{-2}$), or he may even, by 'bouncing' his bicycle, make M greater than 900 N for a fraction of a second. It is anyway reasonable to assume that his maximum acceleration is much greater than the 4 m s^{-2} calculated above. ◄━━

2.5 Jets and rockets

Using Newton's second law as in equation (4) (page 14) we can analyse systems which do not have constant mass, e.g. a firework rocket.

Consider first a stationary hovering helicopter of mass 800 kg, see figure 2.16. By Newton's *first* law we see that $P + W = 0$, i.e. P and W are each 8000 N in size, taking g as 10 m s^{-2}. P is the upward push of the air on the helicopter. By Newton's *third* law the push of the helicopter on the air is $P' = 8000$ N downwards. Thus to hover the helicopter projects air downwards at such a rate as to achieve, by Newton's *second* law, a rate of change of momentum of the air of 8000 N, i.e.

$$F = \frac{\mathrm{d}(mv)_{\text{air}}}{\mathrm{d}t} = 8000 \text{ N}$$

if the air is moving at constant speed v

$$v\left(\frac{\mathrm{d}m}{\mathrm{d}t}\right)_{\text{air}} = 8000 \text{ N}$$

Suppose the air is projected at a speed of 25 m s^{-1}; then we have

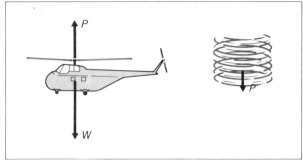

Figure 2.16 A helicopter and the air it projects downwards. The push P' of the helicopter on the air is, by Newton's third law, equal in size to the push P of the air on the helicopter.

$$\left(25\frac{\text{m}}{\text{s}}\right)\left(\frac{\mathrm{d}m}{\mathrm{d}t}\right)_{\text{air}} = 8000 \text{ N}$$

$$\Rightarrow \quad \frac{\mathrm{d}m}{\mathrm{d}t} = \frac{8000 \text{ kg m s}^{-2}}{25 \text{ m s}^{-1}} = 320 \text{ kg s}^{-1}$$

As air has a density of about 1 kg m^{-3}, the helicopter blades are gathering and projecting about 320 m^3 of air each second. If the area swept by the blades is A then we have

$$A \times 25 \text{ m s}^{-1} = 320 \text{ m}^3 \text{ s}^{-1}$$

which gives $A = 12.8$ m^2. The length of the blades is given by $\pi r^2 = 12.8$ m^2 which gives $r = 2.0$ m.

Note that in this argument each of Newton's laws is used. The second law is applied to the air and not to the helicopter. In a similar way we can analyse the action of jet engines and ships' propellers where in each case a fluid medium is projected backwards at a steady speed. In these examples the mass of the vehicle itself remains unchanged.

Jet engines

In principle a jet engine is very simple:
(i) air is taken in at the front; this air is compressed;
(ii) fuel (usually paraffin vapour) is mixed with the compressed air and ignited. In the resulting explosion the air expands and,
(iii) the exhaust gases rush out of the back, some of their momentum being used to turn a turbine which drives the compressor (in a turbo-jet) or the compressor and a propeller (in a turbo-prop). See figure 2.17.

The turbo-jet is the basic propulsion unit for modern jet aeroplanes. The jet engine uses the oxygen in its air intake as well as the fuel it carries to cause the explosive expansion which ensures that the exhaust gases rush out at a much higher speed than the air enters. If the difference is 300 m s^{-1} and the mass of gas passing through the system is 400 kg s^{-1} (ignoring the paraffin's mass which is relatively small) the resulting thrust = (300 m s^{-1})(400 kg s^{-1}) = 120 000 kg m s^{-2} or 120 kN.

► A horizontal jet of water emerges from a nozzle of cross sectional area A at a speed v. It strikes a vertical wall

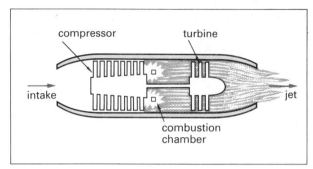

Figure 2.17 A schematic diagram of a turbo-jet engine.

and does not rebound. If the density of water is ρ, what is the push of the water on the wall?

The push *of the wall on the water*, by Newton's second law, is given by F, the rate of change of momentum of the water.

$$F = \frac{\mathrm{d}(mv)}{\mathrm{d}t} = v\frac{\mathrm{d}m}{\mathrm{d}t} = v\rho\frac{\mathrm{d}V}{\mathrm{d}t}$$

where $\mathrm{d}V/\mathrm{d}t$ is the volume of water hitting the wall per unit time. But $\Delta V = vA\Delta t$

$$\therefore \quad \frac{\mathrm{d}V}{\mathrm{d}t} = vA$$

Hence $F = v^2\rho A$

The push of the water on the wall $F' = -F$ is thus, by Newton's third law, also of size $v^2\rho A$. ◀

Rocket propulsion

A jet engine collects air but a rocket carries all its propellent materials with it. Figure 2.18a is a schematic diagram of a liquid-fuel rocket motor. The free-body diagram (b) shows the rocket accelerating upwards near the Earth's surface. The mass of the rocket is m and will decrease as the exhaust gases, ejected at v_0 relative to the rocket, escape at a rate $\mathrm{d}m/\mathrm{d}t$. The push F' of the rocket on the gases is given by

$$F' = \frac{\mathrm{d}(mv)}{\mathrm{d}t} = v_0\frac{\mathrm{d}m}{\mathrm{d}t}$$

and by Newton's third law, the *thrust* (the push of the exhaust gases on the rocket) $F = -F'$ and is thus of size $v_0\mathrm{d}m/\mathrm{d}t$. Applying Newton's second law in the form $ma = F_{\mathrm{res}}$ to the rocket over a time interval during which m can be assumed to be constant

$$ma = F - mg$$

$$m\frac{\mathrm{d}v}{\mathrm{d}t} = v_0\frac{\mathrm{d}m}{\mathrm{d}t} - mg$$

This equation tells us that to accelerate the rocket upwards at all from the Earth's surface ($\mathrm{d}v/\mathrm{d}t > 0$), we must have

$$v_0\frac{\mathrm{d}m}{\mathrm{d}t} > mg$$

The chemistry of the combustion process limits v_0 to about 5×10^3 m s^{-1}. The rocket engineer can achieve astonishingly high values of $\mathrm{d}m/\mathrm{d}t$. A Saturn V rocket system will eject more than 10^4 kg s^{-1} at take-off thus giving an initial thrust of more than $(5 \times 10^3$ m s$^{-1})(10^4$ kg s$^{-1}) = 5 \times 10^7$ kg m s^{-2} or 50 MN. Such a thrust is greater than the pull of the earth on a mass of 5×10^6 kg or 5000 tonnes. For a rocket of initial mass 4000 tonnes the initial upward acceleration would be 2.5 m s^{-2}.

As the rocket rises so m decreases and thus the acceleration increases. Unfortunately even if all the mass of the rocket were fuel then it would be completely burnt after 5×10^6 kg/10^4 kg s$^{-1} = 500$ s. And there is no point in designing a rocket system with (i) no payload and (ii) no spare fuel (perhaps for injection into Earth orbit or simply as a safety factor).

To improve the ratio of payload to take-off mass, multi-stage rocket systems are used. Each stage is a complete rocket motor with fuel, oxidant and reaction chambers. The dumping of used fuel containers and reaction chambers more than makes up for the need to carry the extra motor systems.

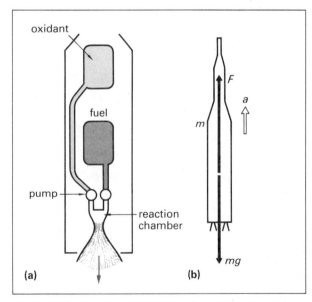

Figure 2.18 (a) A schematic diagram of a rocket motor, and (b) a free-body diagram of a two stage rocket vehicle.

3 Energy and its conservation

3.1 Work and power

Forces can alter the motion of a body and we have found how to calculate the effect that a resultant force can have on the body's momentum ($F\Delta t = m\Delta v$). It is also useful to consider the product $F\Delta s$, where Δs is the displacement of the force, because this helps us to understand changes in what we call the *energy* of the body. For instance when F and Δs are in the same direction the body speeds up; we say it gains *kinetic energy*: but when F and Δs are perpendicular the speed and therefore the kinetic energy of the body are unchanged. To find out what energy 'is' we need first to define *work*. Consider a body which undergoes a displacement Δs as shown in figure 3.1 while a constant force F acts on it. The work W_{AB} done by the force is defined by the equation

$$W_{AB} = F\Delta s \cos \theta$$

where θ is the angle between F and Δs. We can think of $F\Delta s \cos \theta$ either as the force times the resolved part of the displacement in the direction of F or as the displacement times the resolved part of the force in the direction of Δs. W_{AB} is a scalar quantity. The unit of work is the N m which is called a *joule* (J) after J.P. Joule (1818–1889), who first classified the relation between heating a body and doing work on it. Figure 3.2 shows four particular, yet common situations in which we can discuss this definition and relate the work done by a force to the concept of energy.
(a) The pull of the string on the lump is *always* perpendicular to the direction of motion ($\cos \theta = 0$). The work done by T is thus zero, or, as we sometimes say, T does no work on the lump. T does, of course, change the momentum of the lump as it is the force which alters the direction of motion. If the lump swings so as to have moved a vertical distance of 0.8 m, then if $W = 2.5$ N the work done by W is simply

$$(2.5 \text{ N})(0.8 \text{ m}) = 2 \text{ J}$$

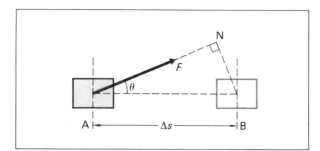

Figure 3.1 The definition of work.

In this case where the lump follows a curved path, but where the force considered is constant, it is easier to find the resolved part of the displacement than to try to resolve the force during each stage of the motion. The work done by W speeds up the lump (increasing its kinetic energy).
(b) The work done by the perpendicular contact force N on the sledge is zero as N is perpendicular to the displacement. The work done by W can be easily calculated: e.g. if $W = 90$ N and it is a 45° slope, then in sliding 10 m the work done is

$$(90 \text{ N}) (10 \text{ m}) \cos 45° = 640 \text{ J}$$

It does not matter whether you think of the force or the displacement as being resolved in this case. This 640 J will, by itself, speed up the sledge – but there is a retarding force F. The work done by this frictional push of the slope on the sledge as the sledge slides is *negative*, for the angle between F and the displacement is 180°, and cos 180° = −1. Thus if in this case $F = 64$ N, the work done by F on the sledge is

$$(64 \text{ N}) (10 \text{ m}) \cos 180° = -640 \text{ J}$$

This will, by itself, slow the sledge down; and we will find that the temperature of the sliding surfaces rises. In our example the sum of the work done by the forces happens to be zero (640 J − 640 J) so that the sledge moves at a steady speed down the slope.
(c) The work done by a force is zero if there is no motion of the body, i.e. $\Delta s = 0$. The pull of the man on the barbell does not lift it and so P *does no work*, nor do W or N. The man becomes tired because the many skeletal muscles in his arms, back and legs are continually relaxing and contracting in order to maintain the pull P.
(d) The work done by M, N and W on the car is zero because it moves horizontally. The work done by the frictional push of the ground on the car is also zero; the force F does not move in the direction of the displacement as the wheel does not slide on the road, so that $\Delta s = 0$ for F. The forces which do the work are forces *internal* to the car, pushes on pistons and axles. It is from these forces that the energy of the car comes and which lead to the wheels pushing the road backwards and hence (page 18) leads to the *external* propelling force F. If the car is moving at any speed there will be a further resistive force, the push of the air on the car. This force will do negative work on the car and the effect will be to warm up the air.

Work done by a variable force

With a variable force we can calculate the value of $F\Delta s \cos \theta$

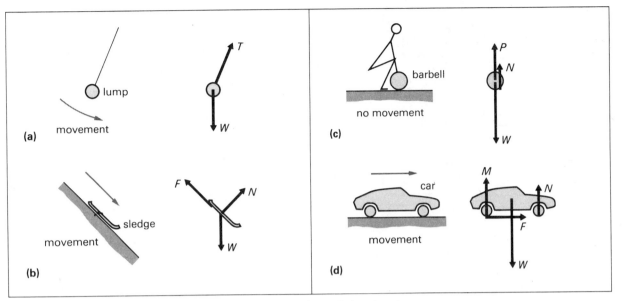

Figure 3.2 Some situations and free-body diagrams to illustrate the work done by a force.

for a *small* displacement and then sum the values over the path we wish to consider.

As in section 1.3 where a displacement for a journey with a varying velocity was summed graphically, so the area between the graph-line and the *s*-axis of a graph of *F* against *s* represents the work done. Consider the case of a spring for which the extending force *F* is proportional to the extension *x*,

$$F \propto x \quad \text{or} \quad F = kx$$

where *k* is a constant called the *stiffness* of the spring. Let us consider the work done on the spring by the extending force. Figure 3.3 shows how this force varies with *x* and we could extend the line beyond the origin if we wanted to consider compressions as well as extensions. For an extension from 0 to x_1, we have

$$W_{\text{OA}} = \text{heavily shaded area}$$

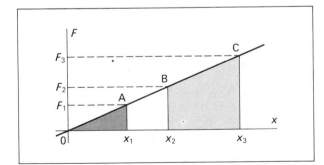

Figure 3.3 Work done by a force which varies linearly with the distance moved.

$$= \tfrac{1}{2}F_1 x_1 \ (\text{or } \tfrac{1}{2}kx_1{}^2)$$
$$= (\text{average force})(\text{displacement OA})$$
Similarly $\quad W_{\text{BC}} = \text{lightly shaded area}$
$$= \tfrac{1}{2}(F_2 + F_3)(x_3 - x_2)$$
$$= (\text{average force})(\text{displacement BC})$$

In both cases the work done is given by the area under the graph of force against extension. This is generally true when both the force and the displacement are along a line e.g. when a rubber band (for which *F* is not proportional to *x*) is stretched or a foam cushion compressed. We say that the work done increases the energy stored in the spring, rubber band or foam cushion.

▶ A box is being dragged along the floor by a rope which makes an angle of 30° to the horizontal. If the weight of the box is 400 N, the tension in the rope 160 N and the sliding frictional force 105 N, how much work is done on the box in dragging it 4.0 m?

We can calculate the work done by each force on the box:
(i) pull of the Earth: (400 N)(4.0 m) cos 90° = 0
(ii) frictional push of the ground: (105 N)(4.0 m) cos 180° = −420 J
(iii) pull of the rope: (160 N)(4.0 m) cos 30° = 554 J
(iv) normal contact push of the ground (the size of this force is not given): *N* (4.0 m) cos 90° = 0.
The 554 J would speed up the box, while the −420 J would slow it down and the surface of the box and ground become warm. The total work done by all the forces is (554 − 420) J = 134 J and this determines by how much the box speeds up. It is equal to the work done on the box by the resultant force; a result which is sometimes useful when we are interested only in changes of speed of the box. ◀

▶ The following table gives values for *F*, the pull of the Earth on a satellite of constant mass 5000 kg, as a

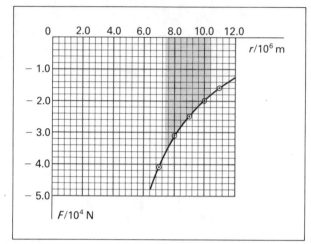

Figure 3.4 A graph showing the pull F of the Earth on a satellite of mass 5000 kg at different distances r from the Earth's centre. F is negative as it is towards the centre of the Earth while r is measured away from the centre.

function of its distance r from the centre of the Earth. The radius of the Earth is 6.4×10^6 m.

$F/10^4$ N	4.1	3.1	2.5	2.0	1.6
$r/10^6$ m	7.0	8.0	9.0	10.0	11.0

What is the work done by the pull of the Earth on the satellite as it moves from 4000 km above the Earth's surface to 1000 km above its surface ?

Figure 3.4 shows the graph of F against r with the shaded area representing the work done on the satellite as it moves between $(4000 + 6400)$ km $= 10.4 \times 10^6$ m and $(1000 + 6400)$ km $= 7.4 \times 10^6$ m from the Earth's centre.

The shaded area = 98 small squares

Each small square represents $(0.2 \times 10^4 \text{ N})(0.4 \times 10^6 \text{ m})$
$$= 8.0 \times 10^8 \text{ J}$$

Therefore work done $= 98(8.0 \times 10^8 \text{ J})$
$$= 7.8 \times 10^{10} \text{ J}$$

This work done is *independent* of the path of the satellite; only its initial and final distances from the Earth's centre are important.

An approximate answer could have been obtained by saying that the 'average' force is roughly

$$\tfrac{1}{2}(3.6 + 1.8) \times 10^4 \text{ N} = 2.7 \times 10^4 \text{ N}$$

and so the work done $= (2.7 \times 10^4 \text{ N})(3 \times 10^6 \text{ m})$
$$= 8.1 \times 10^{10} \text{ J}$$

This is less than 4% bigger than the work calculated from the graph, so that to this accuracy the approximation is a reasonable one.   ◀

Power

How quickly a force does work is often more significant than the total amount of work done. The rate at which an agent does work is called the power of the agent.

$$P = \frac{dW}{dt} \qquad \text{(appendix B3)}$$

P is a scalar quantity with unit J s^{-1} which is called a *watt* (W) after the British engineer James Watt (1736–1819) who first developed the steam engine. kW and MW are common multiples and the kW h $(= 3.6 \times 10^6$ J) is often used in commercial contexts as a convenient unit of energy (appendix A4). When work W is done by a force F, $W = F\Delta s$, and P can be expressed as

$$P = F\frac{\Delta s}{\Delta t}$$

or we can write $P = F\dfrac{ds}{dt} = Fv$

where F is the resolved part of the force in the direction of the velocity v.

▶ A sack of mass 45 kg is lifted 6.0 m in 10 s at a steady speed. What is the average power needed?
Either: work done $= (450 \text{ N})(6.0 \text{ m}) = 2700$ J

$\Rightarrow$ power $= \dfrac{dW}{dt} = \dfrac{2700 \text{ J}}{10 \text{ s}} = 270$ W

Or   speed $= \dfrac{6.0 \text{ m}}{10 \text{ s}} = 0.60$ m s^{-1}

$\Rightarrow$ power $= Fv = (450 \text{ N})(0.60 \text{ m s}^{-1})$
$$= 270 \text{ W} \qquad ◀$$

3.2 Kinetic energy

The effect of doing a fixed amount of work W_{AB} on bodies of different mass can be investigated by 'firing' an air track glider from a simple elastic catapult (figure 3.5). The elastic is pulled back to A and the speed v which the glider acquires measured using a photocell and electric clock or scaler placed beyond B. The mass m of the glider is altered and the experiment repeated. W_{AB} is the same each time because the distance AB is the same each time. An analysis of the results, e.g. by plotting graphs, reveals that $v^2 \propto 1/m$, or $mv^2 = $ constant, which can be proved theoretically as follows:

Suppose that a constant force F accelerates a body of mass m from a speed v_0 to a speed v. In so doing the body is displaced a distance s.

The work done on the body $= Fs = mas$ (as $F = ma$). From the equations on page 7 we have

$$v^2 = v_0{}^2 + 2as$$

so that $as = \tfrac{1}{2}v^2 - \tfrac{1}{2}v_0{}^2$

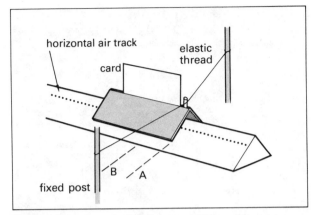

Figure 3.5 An air track glider is catapulted from rest at A to a speed v at B.

$$mas = \tfrac{1}{2}mv^2 - \tfrac{1}{2}mv_0{}^2$$

$$\text{and} \quad W_{AB} = \tfrac{1}{2}mv^2 - \tfrac{1}{2}mv_0{}^2 = \Delta(\tfrac{1}{2}mv^2) \quad (1)$$

This equation is called the *work-energy equation*. In the experiment described above the work done was constant and v_0 was zero, so that we should have found that $\tfrac{1}{2}mv^2$ was constant. An amount of work Fs has accelerated the body to a speed v; we say that the body gains a store of energy, *kinetic energy*, of amount $\tfrac{1}{2}mv^2$. A moving body of mass m can do work as it slows down. In coming to rest from a speed v it can do $\tfrac{1}{2}mv^2$ of work on some other object and in so doing loses its store of kinetic energy.

We define the kinetic energy E_k of the moving body by the equation

$$E_k = \tfrac{1}{2}mv^2$$

Kinetic energy is a scalar quantity and is measured in joules (J). The work-energy equation, equation (1), can now be read as

work done on a body = change in its kinetic energy

and figure 3.6 illustrates the content of this statement.

The work-energy equation was derived using a constant force F, but it holds in all cases, v_0 and v being the initial and final speeds. Thus in the example on page 24, the change of kinetic energy of the satellite is 7.8×10^{10} J. If its initial speed had been 9000 m s^{-1} we could find its final speed (its mass was 5000 kg).

$$7.8 \times 10^{10}\text{ J} = \tfrac{1}{2}(5000\text{ kg})v^2 - \tfrac{1}{2}(5000\text{ kg})(9000\text{ m s}^{-1})^2$$

i.e. $(2500\text{ kg})v^2 = (7.8 + 20.3) \times 10^{10}$ J

$$\Rightarrow \quad v = \sqrt{\left(\frac{28.1 \times 10^{10}}{2500}\right)}\text{ m s}^{-1}$$

$$= 10\,600\text{ m s}^{-1}$$

Whenever we can see that *a body has the ability to do work* we say that *the body possesses energy*. The sort of test we need is to ask the question, 'Could I, in principle, by any

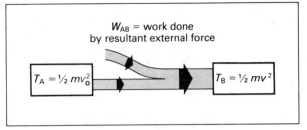

Figure 3.6 An energy flow diagram for the work-energy equation.

series of devices (engines, pulleys etc.) arrange for the body to push a piston or pull a rope?' A moving truck could readily be made to do either and we say that it has energy of motion – kinetic energy. Water in a stream could do either, perhaps by letting it first turn a paddle wheel. In the rest of this chapter we will be largely concerned with identifying and defining other forms of energy.

▶ Figure 3.7 shows a rough slope up which a body of mass 2.0 kg is moving with a speed of 12 m s^{-1}. The frictional push of the slope on the block is 2.0 N. Find the speed of the block when it has travelled 10 m along the slope and has thus risen 4.0 m vertically.

Taking $g = 10$ m s^{-2}, the pull of the Earth on the block is 20 N. The work done on the block
(i) by the pull of the Earth is $-(20\text{ N})(4.0\text{ m}) = -80$ J
(ii) by the perpendicular contact push R of the slope is zero
(iii) by the frictional push of the slope is $-(2.0\text{ N})(10\text{ m}) = -20$ J

Using $W = \tfrac{1}{2}mv^2 - \tfrac{1}{2}mv_0{}^2$

We have $\quad -100\text{ J} = \tfrac{1}{2}(2.0\text{ kg})v^2 - \tfrac{1}{2}(2.0\text{ kg})(12\text{ m s}^{-1})^2$

$$\Rightarrow \quad 44\text{ J} = \tfrac{1}{2}(2.0\text{ kg})v^2$$

so that $\quad v = 6.6$ m s^{-1} ◀

Internal work

There are occasions when the kinetic energy of a body changes but no external force can be found which moves through any measurable distance, i.e. which is doing work

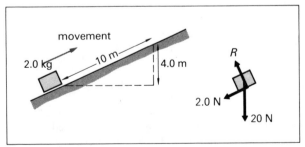

Figure 3.7 A body slides up a slope: both the pull of the Earth and the frictional push of the slope on it do negative work and thus slow it down.

on the body, for instance a boy who jumps into the air or a car which goes faster. In neither case do the propulsive forces (the push of the ground) move and hence they do no work. We have to look inside the bodies concerned, the boy and the car, to find internal forces which are moving and doing work – the push of the gases on a piston in the car, for example. It is the work done by these forces that alters the kinetic energy of the boy or the car. To include these cases the work-energy equation should be restated

$$W_{AB}(\text{external forces}) + W(\text{internal forces}) = \Delta(\tfrac{1}{2}mv^2)$$

where $\Delta(\tfrac{1}{2}mv^2)$ is the change of kinetic energy of the body.

3.3 Potential energy

Gravitational potential energy

When a body of mass m moves in the Earth's gravitational field (page 162) the pull mg of the Earth does work on it. For a vertical displacement Δh between two levels A and B, the work done by mg is

$$W_{AB} = mg\Delta h$$

If g is constant we can write

$$W_{AB} = \text{change in } (mgh) \qquad (2)$$

The change in (mgh) is called the change in the body's *gravitational potential energy* (usually abbreviated g.p.e.). Thus a book of mass 0.3 kg which drops 0.8 m from table to floor loses 2.4 J of g.p.e. (taking $g = 10 \text{ m s}^{-2}$ or 10 N kg^{-1}). It will gain 2.4 J of g.p.e. when it is lifted or thrown back onto the table. As it fell down the book gained kinetic energy (k.e.) and we can relate changes of g.p.e. and k.e. using the work-energy equation.

In order to be consistent about signs and to avoid having to use words like gains and losses consider equation (2) again. If the body moves down the work done by mg is positive ($\theta = 0°$ in $F\Delta s \cos\theta$) and there is a loss of g.p.e. Similarly if the body moves up the work done by mg is negative ($\theta = 180°$) and there is a gain of g.p.e. Equation (2) should therefore be written

$$W_{AB} = -\Delta(mgh)$$

From the work-energy equation we know that $W_{AB} = \Delta(\tfrac{1}{2}mv^2)$, the change in the body's kinetic energy, so we have

$$-\Delta(mgh) = \Delta(\tfrac{1}{2}mv^2)$$

or $\Delta(mgh) + \Delta(\tfrac{1}{2}mv^2) = 0$ (3)

Equation (3) is a *conservation of energy* rule. If the k.e. decreases then the g.p.e. must increase by an equal amount etc. The rule holds when a body is falling freely in the Earth's gravitational field (see figure 1.10 page 8 for example). It can also be applied when other forces are acting, provided these forces do no work.

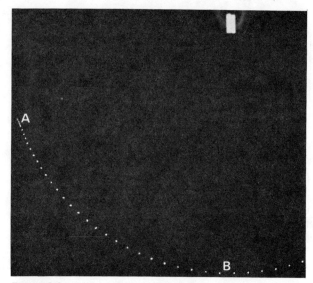

Figure 3.8 A stroboscopic photograph of a simple pendulum. Scale: 1 mm on the photograph represents 5 mm.

► Figure 3.8 is a stroboscopic photograph of a swinging pendulum. The flash rate was 100 Hz and the pendulum 0.325 m long (the photo is printed to one fifth scale). If the mass of the pendulum bob was 0.10 kg, show that $\Delta(mgh) + \Delta(\tfrac{1}{2}mv^2) = 0$ for the pendulum bob. Consider the portion of the path A to B . Let us take $g = 9.8 \text{ m s}^{-2}$.

(i) g.p.e.: $\quad \Delta h = -5 \times 41 \text{ mm}$
$\qquad\qquad\qquad = -0.205 \text{ m}$
$\Rightarrow \qquad \Delta(mgh) = (0.10 \text{ kg})(9.8 \text{ m s}^{-2})(-0.205 \text{ m})$
$\qquad\qquad\qquad = -0.201 \text{ J}$

No other forces do work on the pendulum bob, since the tension in the string is always perpendicular to the motion of the bob.

(ii) k.e.: the speed at A is zero,

$$\text{at B} \quad v = \frac{5 \times 12.0 \text{ mm}}{3 \times 0.01 \text{ s}} = 2.00 \text{ m s}^{-1}$$

$$\therefore \quad \Delta(\tfrac{1}{2}mv^2) = \tfrac{1}{2}(0.10 \text{ kg})(2.00 \text{ m s}^{-1})^2$$
$$= 0.200 \text{ J}$$

So we see that 0.201 J of gravitational potential energy are lost while 0.200 J of kinetic energy are gained which, to better than 1%, confirms our expectations. The speed at B was averaged over 0.03 s and, of course, there is some air resistance. Note that m, the mass of the pendulum bob, was not really required as it appears in both expressions for energy change. ◄

Elastic potential energy

When a body moves under a resultant force which is the push or pull of a spring on it, the spring does work on it.

The body might be a frictionless trolley oscillating on a horizontal table. If the spring obeys Hooke's law ($F \propto x$) then the work done in stretching the spring can be written as $\Delta(\frac{1}{2}kx^2)$ (page 23). Thus the work done by the spring on the body when the spring alters its extension is

$$W_{AB} = \text{change in } (\tfrac{1}{2}kx^2)$$

The change in $(\frac{1}{2}kx^2)$ is called the change in the body's *elastic potential energy* (usually abbreviated e.p.e.). Again to be consistent with signs we write

$$W_{AB} = -(\tfrac{1}{2}kx^2)$$

for if the spring does work on the body decreasing its extension from $x = 0.2$ m to $x = 0.1$ m, for instance, the body *gains* kinetic energy.

From the work-energy equation we know that $W_{AB} = \Delta(\frac{1}{2}mv^2)$, so we can relate changes of e.p.e. and k.e.

$$-\Delta(\tfrac{1}{2}kx^2) = \Delta(\tfrac{1}{2}mv^2)$$
$$\text{or} \quad \Delta(\tfrac{1}{2}kx^2) + \Delta(\tfrac{1}{2}mv^2) = 0 \qquad (4)$$

Equation (4) is another *conservation of energy* rule. If the e.p.e. of a spring decreases (e.g. when a stone is fired from a catapult) then its k.e. must increase by an equal amount.

Mechanical energy

Equations (3) and (4) can be combined to produce a single conservation principle for mechanical energy.

$$\Delta(\tfrac{1}{2}mv^2) + \Delta(mgh) + \Delta(\tfrac{1}{2}kx^2) = 0$$

This equation says that the total change of energy of a body under the action of gravitational and elastic forces is zero. Written in terms of energies rather than changes in energy

$$\begin{array}{ccc} \text{k.e.} & +\ \text{g.p.e.} & +\ \text{e.p.e.} & = \text{constant} \\ (\tfrac{1}{2}mv^2) & (mgh) & (\tfrac{1}{2}kx^2) \end{array}$$

i.e. *the total mechanical energy of a body*, on which the only work done is by the pull of the Earth and the push or pull of springs, *is constant*. The mathematical expressions we have given apply when
(i) g is constant (for g.p.e.)
(ii) k is constant (for e.p.e.)
but the general statement still hold when g and k are not constant, although then the values of g.p.e. and e.p.e. are more difficult to calculate. A body might move a long way from the Earth and a perfectly elastic band might not obey Hooke's law when it is stretched, but we can still use the principle of conservation of mechanical energy.
▶ A very bouncy ball is dropped onto a fixed flat surface. It bounces and is caught. Describe the energy changes of the ball.

The best descriptions will involve some sort of sketch showing the energy changes in diagrammatic or graphical form. Figure 3.9 gives three possible answers. In (a) energy transfers are shown as completed steps while (b) shows the relative proportions of the three forms of energy as the bouncing proceeds. In (b) the horizontal axis roughly indicates the time scale. In both it is assumed that

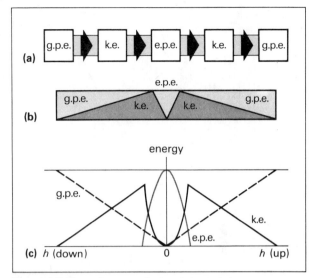

Figure 3.9 Three diagrammatic descriptions of the energy changes as a very bouncy ball falls, bounces and rises.

mechanical energy is conserved, though in practice this is an approximation as the ball is likely to warm up slightly as it bounces. Figure 3.9c is an attempt to represent the energy changes graphically as a function of the height h of the ball above the surface. The precise shape of the curved parts of the graph is not predictable but if the sum of the mechanical energies is assumed constant, the k.e. + g.p.e. + e.p.e. should add to give the horizontal line at the top of the graph. ◀

Potential energy – distance graphs

If we call the gravitational potential energy of a body E_g, then

$$E_g = mgh$$
$$\text{and} \quad \Delta E_g = \Delta(mgh) = mg\Delta h \qquad \text{(page 26)}$$
$$\text{Thus} \quad mg = \frac{\Delta E_g}{\Delta h}$$

i.e. the pull mg of the Earth on the body is equal to the gradient of a graph of g.p.e. against h. This statement is the converse of that used in the example on page 24, where the area under a graph of gravitational force against distance from the centre of the Earth was taken to be equal to the work done on the body, i.e. to its change of g.p.e. Both ideas are generally valid even when the force is not constant: thus where we have a potential energy E_p, the associated force F is given in size by

$$F = \frac{\Delta E_p}{\Delta x}$$

where distances x are measured parallel to F. Figure 3.9c illustrates the result for gravitational potential energy, as

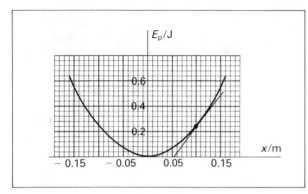

Figure 3.10 A graph of e.p.e. against displacement for an oscillating air track glider. The tangent is drawn at $x = 0.10$ m.

the gradient of the g.p.e. against h graph is constant, $F = mg$ being constant for the falling ball.

▶ An air track glider of mass 0.22 kg is attached to two springs the other ends of which are fixed. The springs are parallel to the air track. Figure 3.10 shows a graph of the elastic potential energy of the glider as it oscillates between two points 320 mm apart. Calculate (i) the speed of the glider when it is 100 mm from its equilibrium position, (ii) the restoring force (the pull of the springs) on the glider when it is 100 mm from its equilibrium position, and (iii) the acceleration of the glider as it passes through this position.

(i) When $x = 0.10$ m, $E_p = 0.24$ J. The *total* mechanical energy of the glider is equal to 0.64 J, the e.p.e. at the extremes ($x = 0.16$ mm) of the oscillation. Thus at $x = 0.10$ m the k.e. of the glider $= (0.64 - 0.24)$ J $= 0.40$ J. If the speed of the glider at this moment is v, then

$$\tfrac{1}{2}mv^2 = 0.40 \text{ J}$$

Therefore $v^2 = \dfrac{2(0.40 \text{ J})}{0.22 \text{ kg}} = 3.6 \text{ J kg}^{-1}$

and $v = 1.9 \text{ m s}^{-1}$

(ii) A tangent drawn at $x = 0.10$ m (figure 3.10) has a gradient equal to the restoring force F at this point, i.e.

$$F = \dfrac{(0.48 - 0) \text{ J}}{(0.150 - 0.055) \text{ m}} = \dfrac{0.48 \text{ J}}{0.095 \text{ m}}$$

$\Rightarrow$ $F = 5.1$ N

(iii) The accleration of the trolley at this position is thus, using Newton's second law,

$$\dfrac{5.1 \text{ N}}{0.22 \text{ kg}} = 23 \text{ m s}^{-2} \qquad \blacktriangleleft$$

3.4 The principle of conservation of energy

A man on a bicycle free-wheeling along a level road has kinetic energy. If he brakes suddenly and comes to rest mechanical energy is *not* conserved but the brake blocks and wheel rims become hot. Whenever solid surfaces rub against one another, as here, they warm up and mechanical energy is lost. We say that the brake blocks and wheel rims are gaining *internal energy* (page 102), or random thermal energy as it is sometimes called. The system, the man on the bicycle in this case, does not lose any energy if we take mechanical *and* internal energy into account.

Consider another example: a pole vaulter, figure 3.11 runs down the runway (he then has k.e.), plants his pole and swings into it bending the pole (he and the pole then have k.e. + e.p.e.). The pole straightens as he rises (he then has g.p.e. + k.e.), he just clears the bar (he then has mainly g.p.e.) and falls towards the pit (he then has k.e.). It thus seems possible to see the vault as an example of the conservation of mechanical energy. After landing the vaulter and the pit gain some internal energy. But where does his initial kinetic energy as he runs down the runway come from? And does he not use his muscles during the vault itself? In asking these questions we have taken it for granted that we do not expect energy to disappear. On the contrary we expect that

the total energy of a closed system is constant.

It took scientists a long time to develop the concept of energy and ways of measuring the different forms it takes. For the pole vault, the closed system is the vaulter, his pole

Figure 3.11 A multiflash photograph of a pole vaulter.

and the ground (which is part of the Earth). His initial k.e. came from chemical changes in his body which enabled his muscles to do internal work. We say that he transforms *chemical energy* to kinetic energy. At all stages of the vault his body transforms chemical energy to internal energy so that he becomes warmer as a result of the effort involved in the vault.

Forms of energy

The forms of energy which are met in this book are:
(i)   Mechanical energy: kinetic energy
                            gravitational potential energy
                            elastic potential energy
                            sound energy
(ii) Energy in matter: chemical energy
                            nuclear energy *
                            internal energy
(iii) Electromagnetic energy: electrical energy
                            magnetic energy
                            radiant energy

*Note: This is often wrongly called atomic energy

When mechanical energy is transformed to internal energy the process can be seen in terms of the nature of the contact between two surfaces. Before the surfaces come into contact all the molecules of the moving body are travelling in the same direction. As the surfaces slide on one another this energy becomes *random* energy associated with individual molecules or groups of molecules. The forward motion of the body is reduced and the surfaces become slightly warmer. It was established nearly 150 years ago that a given rise in temperature in a particular body could always be achieved by converting a given quantity of mechanical energy whether the transformation or conversion was achieved by rubbing, by stirring or in any other way. Scientists thus learned to follow changes of internal energy quantitatively using the techniques of *calorimetry* outlined in chapter 9. Most of the other non-mechanical forms of energy listed are measured by first converting the energy to internal energy. Though it is possible to convert all forms of energy to internal energy *completely*, we are limited in the efficiency with which the reverse can be achieved, e.g. in power stations or petrol engines – this limitation, or irreversibility, is considered in chapter 11.

	energy from	energy to	efficiency/% (approx.)
night storage heater	electrical	internal	100
dry cell or battery	chemical	electrical	90
small electric motor	electrical	mechanical	65
steam turbine	internal	mechanical	45
liquid fuel rocket	chemical	kinetic	45
internal combustion engine	chemical	mechanical	25
fluorescent lamp	electrical	radiant	25
solar cell	radiant	electrical	10
thermocouple	internal	electrical	8

Devices which transform energy from one form to another are called *energy transducers*. The table below gives a variety of them and many are described in detail elsewhere in the book. An efficiency of 80% means that 80% of the energy input has been usefully converted to the desired form, the other 20% being converted mainly to internal energy and, sometimes, some sound energy. When it becomes internal energy it is usually of little further use.

Chemical energy and fuels

Chemical energy is stored in coal, oils and natural gas. These form a group of primary energy sources called the *fossil fuels*. Many attempts have been made to estimate the available store of fossil fuels in the Earth's surface. Table 3.1 gives some approximate statistics. The first two columns show how fossil fuel reserves are believed to be distributed between the solid, liquid, and gaseous states. Such estimates involve a lot of guesswork but the pattern is clear. Gas and oil are relatively easy to recover, at least on the continental shelves, as they flow to the surface and do not have to be dug out manually. The last two columns show that we are now relying more heavily on liquid and gaseous fuels in 1972 than 40 years earlier. This is partly because they are easy to obtain, but also because they are much more convenient to use.

	readily recoverable supplies %	recoverable at twice present cost %	fuel used worldwide in 1932 %	fuel used worldwide in 1972 %
coal and lignite	88	77	72	32
petroleum liquids	7	13	14	44
natural gas	5	10	6	21
other sources	—	—	8	3

Table 3.1 Fossil fuel reserves (first two columns) and world fuel uses. The figures in each column are percentages for that column. The other sources were mainly wood in 1932 and hydroelectric or nuclear in 1972.

The most significant things to be learned from statistics such as these, however, is that at present rates we are going to run out of oil and gas in the not very distant future. To appreciate this we need to talk about quantities of energy and not rely simply on relative amounts, the percentages of table 3.1. The numbers given in table 3.1 are percentages of total energies as follows:

recoverable supplies	3000×10^{20} J
supplies recoverable at no more than twice the present cost of extraction	300×10^{20} J
fuel used worldwide in 1932	0.7×10^{20} J
fuel used worldwide in 1972	2.1×10^{20} J

Thus, for example, the total energy value of all recoverable supplies of natural gas is 5% of 3000×10^{20} J or 150×10^{20} J, and is only 10% of 300×10^{20} J or 30×10^{20} J if we limit ourselves to extracting only those supplies which are readily accessible. The implications of these numbers are immense. Consider petroleum liquids:

Readily recoverable reserves

$$= \frac{13}{100}(300 \times 10^{20} \text{ J}) \approx 40 \times 10^{20} \text{ J}$$

World use of oil during 1972

$$= \frac{44}{100}(2.1 \times 10^{20} \text{ J}) \approx 0.9 \times 10^{20} \text{ J}$$

Therefore at the 1972 rate of use, the readily available reserves of oil will last for a further

$$\frac{4}{0.09} \approx 45 \text{ years!}$$

This is not as ridiculous an estimate as it may seem for although we may ultimately extract five times as much petroleum liquid as in the calculation (including such sources as tar-sands and shales), the world consumption of oil was rising in 1972 at about 7% per year. Taking everything into account it is estimated that by the year 2030 Man will have extracted and used 90% of the petroleum liquids available to him.

▶ Describe the energy changes which occur when a man sprints uphill from rest.

A diagrammatic description is given, see figure 3.12. The principle of conservation of energy is implied in the thickness of the bars which are proportional to the energy transformed. ◀

Nuclear energy; fission and fusion

All the energy sources discussed or mentioned so far have originated in sunlight. We use mainly radiant energy which has been transformed to chemical energy and stored over long periods of time in fossil fuels. Some other possible sources are
(i)   direct solar energy
(ii) wind energy
(iii) wave energy
(iv) hydroelectricity
(v)   tidal energy
(vi) geothermal energy, and so on.
These are often called *alternative energy sources*. They avoid the atmospheric pollution produced by the burning of fossil fuels and the problems of waste disposal posed by the use of nuclear fuels. (ii), (iii) and (iv) derive from sunlight stored over short periods of time, (v) is transformed from gravitational energy and (vi) is transformed from nuclear energy in the Earth's interior. If the best use were made of all these alternative energy sources then it is conceivable that a large proportion of the world's supply of energy could

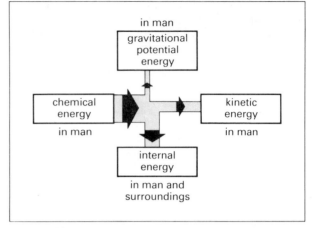

Figure 3.12 An energy-flow diagram for a man sprinting uphill from rest.

derive from them, particularly if cheap and efficient methods of transforming solar energy directly into electrical energy are developed.

The fuels of the early part of the twenty-first century are, however, likely to be uranium and plutonium – that is, *nuclear energy* will become our primary energy source. Nuclear energy is transformed when there are changes in the nuclear composition of matter, just as chemical energy is transformed when matter undergoes chemical changes, such as oxidation. Nuclear energy may be released in two ways, (i) by the breakdown of the nuclei of heavy elements, called *fission*, and (ii) by the building up of heavier nuclei from light ones, called *fusion* (page 266).

The fission of uranium-235 is used in the earliest type of nuclear power station. The kinetic energy of the fission products is converted to internal energy and then used to generate steam and so drive turbines. A few kilograms of uranium are sufficient to produce about 10^{13} J of energy which is equivalent to a power output of 100 MW for a day. Mined uranium is, however, a limited resource and can only delay a genuine energy crisis by tens rather than hundreds of years. A second type of nuclear power station, the fast breeder reactor, uses a mixture of uranium-235 and plutonium-239 as its primary fuel; but from a blanket of uranium-238, used as a containing material, it produces more of the fissionable plutonium-239. The reactor can 'breed' more fuel, in this case plutonium-239, than it uses itself. This does not contradict the principle of conservation of energy, for what happens could be likened to rolling a massive stone to the edge of a high cliff. Although the action of rolling the stone takes up some energy, the potential energy then released by the stone, if it were to fall off the cliff, could be very many times greater.

The energy transformed in the fusion of light nuclei has so far not been achieved in a controlled way but it is hoped that the nuclear power stations of the mid twenty-first century will contain fusion reactors. Man's only success in tapping this store of energy has been in the hydrogen bomb.

If a deuterium-deuterium ($^2_1$H $-$ $^2_1$H) fusion reactor could become our basic power station unit, it is estimated that the energy released by using only 1 % of the deuterium in the world's oceans would amount to about 500 000 times the energy of the world's initial supply of fossil fuels!

On page 34 the principle of conservation of mass – energy is stated as $\Delta E = c^2 \Delta m$. A mass Δm is equivalent to energy ΔE, c^2 ($\approx 9 \times 10^{16}$ m^2 s^{-2} or 9×10^{16} J kg^{-1}) being the exchange rate. In chemical reactions Δm is a very, very small fraction of m, the mass of coal etc. which burns; but in nuclear reactions Δm is a much larger fraction (10^{-4} to 10^{-5}) of the mass of the fissionable material. All energy conversions ΔE could be expressed as mass conversions Δm or vice-versa, but we are used to thinking of nuclear energy as mass conversion and chemical energy as energy conversion. The Sun, which is responsible for almost all our available energy sources on Earth, is a vast fusion bomb and we think in terms of the *mass* converted – about four million tonnes per second. When the Sun's radiant energy is converted to chemical energy in a plant, however, we would give an *energy* conversion rate (in milliwatts).

In any series of energy transformations there is a tendency, as the series proceeds, for more and more of the initial energy to be converted to internal energy. This internal energy is then diffused by conduction and radiation so that a larger and larger number of atoms or molecules share the initial energy. At every stage energy is conserved but the diffusion or spreading of the energy leads to a situation where it becomes more and more difficult to convert the internal energy back into, for instance, mechanical energy. The limited way in which we could make use of the warmth of the oceans is discussed on page 150.

3.5 Collisions

The principle of conservation of linear momentum in section 2.1 allows us to analyse interactions such as collisions or explosions. Energy also is conserved in an isolated system: consider the examples on page 11.

(a) *The skaters*: their kinetic energy before they separate

$$= \tfrac{1}{2}(65 \text{ kg} + 50 \text{ kg})(6.0 \text{ m s}^{-1})^2 = 2070 \text{ J}$$

and their kinetic energy after separation

$$= \tfrac{1}{2}(65 \text{ kg})(4.0 \text{ m s}^{-1})^2 + \tfrac{1}{2}(50 \text{ kg})(8.6 \text{ m s}^{-1})^2$$
$$= 520 \text{ J} + 1850 \text{ J} = 2370 \text{ J}$$

The mechanical energy (kinetic) after separation is 300 J *more* than the mechanical energy before separation. The principle of conservation of energy tells us that 300 J of chemical energy is converted to mechanical energy as they separate. They push themselves apart. Figure 3.13a is an energy flow diagram for this explosive situation.

(b) *The ball and box*: proceeding as above, k.e. before collision

$$= \tfrac{1}{2}(0.80 \text{ kg})(12 \text{ m s}^{-1})^2 = 58 \text{ J}$$

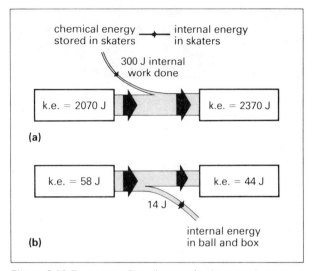

Figure 3.13 Two energy flow diagrams for the example on page 11.

and k.e. after collision

$$= \tfrac{1}{2}(0.80 \text{ kg})(9.0 \text{ m s}^{-1})^2 + \tfrac{1}{2}(12 \text{ kg})(1.4 \text{ m s}^{-1})^2$$
$$= 32 \text{ J} + 12 \text{ J} = 44 \text{ J}$$

The mechanical energy after the collision is 14 J less than the mechanical energy before the collision. The principle of conservation of energy tells us that 14 J of mechanical energy is converted to internal energy as the ball (and to a lesser extent the box) deform and recover during the interaction (figure 3.13b). When some mechanical energy is converted to internal energy we say that the interaction is *inelastic*. The maximum conversion to internal energy occurs when the relative velocity of two colliding objects is zero after the interaction, i.e. when they stick together. Such a wholly inelastic collision is used in figure 2.1a, page 11, to illustrate momentum conservation and is a very common type of interaction. If the initial velocities and masses of the two bodies involved in such a collision are known, then the principle of conservation of momentum alone is enough to predict their combined velocity after the interaction.

Elastic collisions

An interaction in which mechanical energy is conserved is called a *perfectly elastic* collision or, more simply, an *elastic* collision. The elastic forces do not in such cases convert mechanical energy to internal energy during the process of deformation and recovery; and the total kinetic energy of the system before the collision is equal to the total kinetic energy of the system after the collision. The gliders photographed in figure 2.4 bounce in what is virtually a perfectly elastic way. For such collisions in one dimension the principles of (i) conservation of linear momentum and (ii) conservation of mechanical energy are sufficient to

Figure 3.14 A stroboscopic photograph of a golf ball being hit by a wooden club.

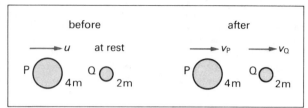

Figure 3.15 A perfectly elastic head-on collision.

predict the velocities of the bodies after the interaction if their masses and their velocities before the interaction are known. (i) will give a relation between v_1 and v_2, the final velocities of the two colliding bodies, and (ii) will give a relation between v_1^2 and v_2^2. These two relations enable us to find v_1 and v_2 in the next example.

The case where a body of mass m_1 moving with velocity u which directly strikes a stationary body of mass m_2 is of special interest. If the collision is perfectly elastic we can show that the velocities of the bodies after the collision, v_1 and v_2, depend upon the ratio of their masses. In particular

	m_1/m_2	v_1	v_2
(a)	$\gg 1$	$\approx u$	$\approx 2u$
(b)	1	0	u
(c)	$\ll 1$	$\approx -u$	≈ 0

(a) The maximum velocity v_2 of the body which is struck is only twice that of the striking body. Thus in kicking a football or hitting a golf ball the ball cannot have a speed greater than twice that of the foot or club-head respectively. The stroboscopic photograph of figure 3.14 illustrates the golfing case, the ball moving away at a speed about 1.5 times that of the club-head before impact.

(b) When the masses are equal the moving body stops and *all* its momentum and kinetic energy are transferred to the body which was struck. Two air track gliders of equal mass and fitted with repelling magnetic buffers illustrate this case as do two snooker balls, (providing the moving ball is skidding and not rolling). The energy transfer in this case is the maximum possible.

(c) If the struck body is very massive, (e.g. the Earth against which is bouncing a very elastic ball), the recoil of the massive body is undetectable and the velocity of the moving body is reversed. The moving body loses very little of its initial kinetic energy as a result of the collision. If an electron hits an atom of helium, the ratio $m_1/m_2 \approx 10^{-4}$, so that an electron could bump elastically into hundreds of helium atoms without losing much kinetic energy. This is of great significance in interpreting ionisation and excitation experiments with gases. A single inelastic interaction could however transfer most of the electron's kinetic energy to the atom.

◄ A ball P of mass $4m$ makes a head-on collision with a ball Q of mass $2m$. If the collision is perfectly elastic what is the ratio of the speeds of the balls after the collision?

Figure 3.15: by the principle of conservation of momentum

$$4mu = 4mv_P + 2mv_Q$$

By the principle of conservation of mechanical energy

$$\tfrac{1}{2}(4m)u^2 = \tfrac{1}{2}(4m)v_P^2 + \tfrac{1}{2}(2m)v_Q^2$$

Eliminating m and substituting for v_Q from the first equation in the second gives

$$2u^2 = 2v_P^2 + (2u - 2v_P)^2$$

Thus $0 = 6v_P^2 - 8uv_P + 2u^2$

$\Rightarrow$ $0 = (3v_P - u)(v_P - u)$

which gives either $v_P = u$ (i.e. there is no collision, for if $v_P = u$ then $v_Q = 0$), or

$$v_P = \frac{u}{3} \text{ and } v_Q = \frac{4u}{3}$$

so that $v_Q/v_P = 4$. Notice that after the collision their relative velocity $v_P - v_Q = u/3 - 4u/3 = -u$, which is the same size as their relative velocity $u_P - u_Q = u - 0 = u$ before it. Elastic collisions always have the effect of reversing the direction of the *relative velocity* without altering its size. ◄

Collisions in two dimensions

We will consider only collisions in which one of the bodies is initially at rest. Figure 3.16a shows a stroboscopic photograph of an off-centre collision between two balls of equal mass; (b) shows how we apply the principle of conservation of linear momentum to it. Suppose that the mass of each ball in this experiment is 0.20 kg, the flash rate 20 s^{-1} and the photograph is reproduced 1/10th actual size. (You should appreciate that the numbers used will not affect the mechanics of the ensuing discussion.) The sizes of the momenta are, to two significant figures

$$m_1 u_1 = (0.20 \text{ kg})(2.47 \text{ m s}^{-1}) = 4.9 \text{ N s}$$
$$m_1 v_1 = (0.20 \text{ kg})(1.14 \text{ m s}^{-1}) = 2.3 \text{ N s}$$
$$m_2 v_2 = (0.20 \text{ kg})(2.15 \text{ m s}^{-1}) = 4.3 \text{ N s}$$

(We obtain the speeds by measuring values of Δs and Δt from the photograph.) As $\theta = 90°$ on the photograph, the triangle formed by the momentum vectors will have sides which are related by

$$(m_1 v_1)^2 + (m_2 v_2)^2 = (m_1 u_1)^2 \qquad \text{(Pythagoras)}$$

Evaluating the left hand side gives $(2.3^3 + 4.3^2)$ N^2 s^2 = 23.8 N^2 s^2, and the square root of this is 4.9 N s, the same

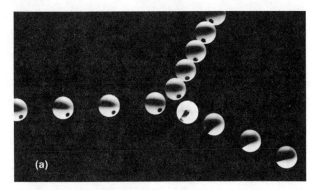

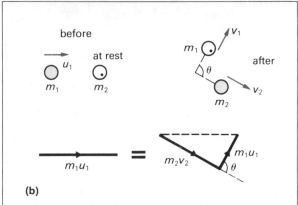

Figure 3.16 (a) A two dimensional collision between a moving puck and a stationary puck of the same mass. (b) Adding the momentum vectors for the collision.

as the measured value of $m_1 u_1$ above. We see that (vector) momentum *is* conserved.

Whenever $m_1 = m_2$, θ is always 90° in elastic two-dimensional collisions whatever the other angles involved. In figure 19.13 (page 260) $\theta \approx 90°$; (the plane of the collision is not quite at right angles to the axis of the camera) so we can conclude that the mass of the moving α-particle is the same as that of the struck helium nucleus. Such cloud chamber evidence is a key factor in identifying unknown particles. When $m_1 > m_2$, then $\theta < 90°$, and when $m_1 < m_2$, then $\theta > 90°$, which enables us to deduce something about the ratio of the masses of striking particles and struck nuclei in the other photographs of figure 19.14 and 19.16.

3.6 High-energy physics

When particles are accelerated to high speeds their behaviour is not that predicted by the rules of mechanics developed so far in this book. Figure 3.17a records a pulse of electrons at the beginning and end of an 8.4 m flight path in a vacuum. The time base of the oscilloscope is set at 3.0×10^{-8} s cm^{-1}, so that the electrons took about 3.3×10^{-8} s to travel 8.4 m, which gives a speed of just

under 2.6×10^8 m s^{-1}. The mass of an electron is 9.1×10^{-31} kg so that, using our rules, the kinetic energy of each electron is

$$\tfrac{1}{2}mv^2 = \tfrac{1}{2}(9.1 \times 10^{-31} \text{ kg})(2.6 \times 10^8 \text{ m s}^{-1})^2$$
$$\approx 3 \times 10^{-14} \text{ J}$$

But the experimenter (the photograph is taken from a film made by W. Bertozzi at M.I.T.) transferred 8×10^{-14} J to each electron in accelerating it from rest before measuring its speed as above. The energy was certainly all transferred to the electron and yet 3×10^{-14} J $\neq 8 \times 10^{-14}$ J by any stretch of the imagination. Our rule for calculating kinetic energy *does not work* at very high speeds.

Figure 3.17b shows a collision between an electron moving in from the left and a stationary electron. The tracks are in a thin photographic emulsion and the electrons are almost in the plane of the photograph (taken by P.H. Fowler of Bristol University). The angle between the tracks after the collision can be measured: it is about 20°. However, the rule about elastic collisions between particles of equal mass was that the angle after collision is always 90°, if the particles obey the principles of conservation of momentum and energy. We have no reason to doubt that the collision in figure 3.17b is not an elastic one and yet 20° is much less than 90°. Again our rules are found wanting at very high speeds.

Let us accept these two experimental results and ask what may have 'gone wrong'. In the first experiment, giving kinetic energy to the electron does not produce the expected increase in speed; in the second the incident electron behaves like a particle of mass greater than that of an electron at rest. In both cases a high-energy electron seems to gain in mass (for this would also 'explain' the rise in kinetic energy in (a)). Higher energy does give higher speed, up to a point, but it also gives higher inertia. The Bertozzi experiment goes

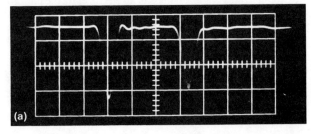

(a)

(b)

Figure 3.17 Two pieces of evidence which cast doubts on the usefulness of the rules of Newtonian mechanics. Measurements from both suggest electrons which are moving at very high speeds have a larger inertia than stationary ones.

on to show that, no matter how much energy you give to an electron you cannot accelerate it beyond a speed of 3×10^8 m s^{-1} (the speed of light).

Special relativity

The rules of high speed or *relativistic* mechanics were developed by Albert Einstein (1879–1955) in the early years of the twentieth century. In them the kinetic energy E_k of a particle of *rest mass* m_0 is given by

$$E_k = (m - m_0)c^2 \qquad (5)$$

where its mass m varies with its speed v relative to the observer in the following way

$$m = \frac{m_0}{(1 - v^2/c^2)^{\frac{1}{2}}} \qquad (6)$$

c being the speed of light and m_0 the value of m at $v = 0$. At low speeds $v \ll c$ and we can write $m \approx m_0(1 + v^2/2c^2)$ So that $E_k \approx \frac{1}{2}m_0 v^2$, as in Newtonian mechanics. We need only use (5) and (6) when $v > 10^8$ m s^{-1}. The highest speed at which man has so far travelled is $\approx 10^4$ m s^{-1}, so that we can see that the ordinary Newtonian rules continue to be adequate for ordinary purposes.

Changes in kinetic energy ΔE_k, can from (5) be seen to be given by $\Delta E_k = c^2 \Delta m$, and this is a special case of the more general *principle of conservation of mass-energy*

$$\Delta E = c^2 \Delta m \qquad (7)$$

Suppose, for example, a nucleus of rest mass 2.0×10^{-25} kg emits a γ-ray photon (page 261) of energy 2.7×10^{-14} J. This photon energy is equivalent to a mass of

$$\frac{2.7 \times 10^{-14} \text{ J}}{c^2} = \frac{2.7 \times 10^{-14} \text{ J}}{(3.0 \times 10^8 \text{ m s}^{-1})^2} = 3.0 \times 10^{-31} \text{ kg}$$

or about a third of the rest mass of an electron. As the photon is moving at 3.0×10^8 m s^{-1} and thus has a momentum mc ($= 9.0 \times 10^{-23}$ N s) the nucleus must recoil. The principle of conservation of momentum predicts a speed v for the nucleus given by

where $(2.0 \times 10^{-25} \text{ kg})v = 9.0 \times 10^{-23}$ N s

so $v = 450$ m s^{-1}, which is a low speed and thus the

recoil kinetic energy has negligible mass equivalent. We can say that mass is conserved if we use the mass of the photon or we can say that energy is conserved if we use the rest energy of the nucleus

$$c^2 m_0 = (3.0 \times 10^8 \text{ m s}^{-1})^2 (2.0 \times 10^{-25} \text{ kg})$$
$$= 1.8 \times 10^{-8} \text{ J}$$

We usually do neither but refer to mass-energy as being conserved, remembering the equivalence implied by equation (7). A further example of mass-energy accounting is given on page 262 where α-decay is considered.

The quantities of energy handled in everyday life are far too small to involve detectable changes of mass. Thus 1 kg of water needs 4.2×10^5 J of energy to raise its temperature from 0°C to 100°C. Equation (7) predicts that its mass will increase accordingly by Δm, where

$$\Delta m = \frac{4.2 \times 10^5 \text{ J}}{(3.0 \times 10^8 \text{ m s}^{-1})^2} = 5 \times 10^{-12} \text{ kg}!$$

Two further consequences of the equivalence of mass and energy are that:
(i) if energy has inertia then *light exerts a pressure*. There is direct evidence for this from observations of comet tails, which point away from the Sun (though there are other forces from a sort of solar 'wind' of particles which affect the tail), and
(ii) matter, that is particles with measurable rest masses, *can be produced from energy*. Figure 3.18 shows a high-energy cosmic ray particle which has hit a nucleus in the Earth's atmosphere. The shower of particles produced has a total rest mass which is many times the rest mass of the incident particle.

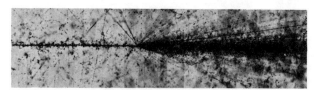

Figure 3.18 Matter from energy; a high energy cosmic ray particle collides with a stationary nucleus.

4 Structure of matter

4.1 Atoms and molecules

The first list of elements was drawn up by Lavoisier, and it was therefore he who was the first to make the study of chemistry systematic. Even so, his list included such items as light and caloric, and a number of other substances which he himself recognised might later be discovered to be capable of being split further. Dalton made the study quantitative: studying the masses of elements which combined with each other, he first stated the laws of definite proportions and multiple proportions, which he saw could be explained if the material of an element consisted of *atoms* – indivisible identical particles. He thus arrived at the relative atomic masses of about 20 elements. Several of these are quite different from the values we accept to-day, for two reasons: (a) the unreliability of some of the measurements he was working with, and (b) his sensible but sometimes false assumptions about the ways in which the atoms combined. For example, he assumed that one atom of hydrogen and one atom of oxygen combined to form one atom of water. His ideas were not readily accepted, and indeed there was no direct evidence for the existence of these atoms.

Avogadro was on surer ground with his experiments on the combining *volumes* of gases: making the assumption that equal volumes of gases contained, under the same conditions, the same number of what he called *molecules*, he was able to deduce the relative atomic masses of elements which were readily available in gaseous form. These values, which did not depend on assumptions about the ways in which atoms combined, are generally closer to the values accepted today. A synthesis of Dalton's ideas of atoms, and their relative masses, and Avogadro's ideas of molecules and their volumes, began to make sense to many scientists in the 1830's and 1840's. Doubts did linger, even into the twentieth century: the last eminent scientist to state publicly that he was not convinced that there were atoms was Ostwald in 1904.

The atom

The structure of the atom, and even the question of whether it had a structure at all, was a problem with which no progress was made until the very end of the nineteenth century, when the first experiments which identified the electron were performed. We shall leave until later in this book a detailed consideration of the work of Thomson, Rutherford, Moseley and others, who probed the atom and produced the information about it which we now have. At this stage we shall merely summarise some simple facts about the atom. You may already know these, but in any case the justification for them will appear later.

(i) an atom has a diameter of about 10^{-10} m. Inside it, at its centre, is a *nucleus*, whose diameter is only about 1/10 000 of that of the atom. If the full stop at the end of the last sentence represents the nucleus, the edge of the atom is about 5 page-widths away. This space is empty, except for the electrons.

(ii) the nucleus contains two kinds of particle, *protons* and *neutrons*. These nuclear particles are called *nucleons*. They have very nearly the same mass, and each is about 2000 times more massive than an electron, so that the nucleus, despite its small size, contains nearly all the mass (always at least 99.9 %) of the atom. The material of the nucleus is unimaginably dense (about 10^{16} kg m^{-3}).

(iii) Protons and electrons pull on each other with *electrical* forces. Protons have positive electric charge and electrons have negative electric charge. Particles with similar charges repel each other, and particles with different charges attract each other, so the protons in the nucleus are pushing each other apart. The reason most nuclei do not disintegrate is because all nucleons attract each other with another and different kind of force which we call a *nuclear* force, so an equilibrium is established. *Radioactivity* occurs when the number of protons and neutrons are such that a permanent equilibrium is not possible.

(iv) The protons in the nucleus attract the electrons outside the nucleus, but the electrons do not fall into the nucleus because they are moving around it in *orbitals*. The size of the electric charges on a proton and an electron are the same, so that when there are as many electrons in the orbitals as there are protons in the nucleus the total charge is zero, and the atom is *neutral*. Normally atoms are neutral, and the forces which they then exert on each other are small (but we shall discuss this later). From time to time an atom may gain or lose an electron, and will therefore have a net electric charge. In this state it is called an *ion*. Hydrogen and the metals tend to lose electrons and form positive ions. The non-metals tend to gain electrons, and form negative ions.

Many research workers have performed an experiment in which they have placed a drop of a fatty acid (an oil) on a water surface. The oil spreads to produce a roughly circular patch of oil on the surface. Drops of the same size always produce a patch of the same size and it is therefore assumed that the oil has spread until it can spread no further, and that the patch is then just one molecule thick.

In a particular experiment using palmitic acid ($C_{16}H_{32}O_2$) 1.00 cm³ of the oil was added to 499 cm³ of benzene. (The point of doing this is to produce a dilute solution which is more bulky and therefore easier to measure. The benzene dissolves in the water, so there is no effect on the size of the patch.) One drop of the diluted oil was placed on the water surface, and the diameter of the circle was found to be 222 mm. 100 such drops were later found to have a volume of 4.72 cm³. Estimate the thickness of the layer.

One drop had a volume of $\frac{4.72}{100} \times 10^{-6}$ m³

and the volume of palmitic acid in such a drop was

$$\frac{1}{500} \times \frac{4.72}{100} \times 10^{-6} \text{ m}^3 = 9.44 \times 10^{-11} \text{ m}^3$$

The area of the patch was $\pi(0.111 \text{ m})^2 = 3.87 \times 10^{-2} \text{ m}^2$

so the thickness of the layer $\left(= \frac{\text{volume}}{\text{area}}\right)$

$$= \frac{9.44 \times 10^{-11} \text{ m}^3}{3.87 \times 10^{-2} \text{ m}^2} = 2.44 \times 10^{-9} \text{ m}$$

If we assume that each molecule of palmitic acid consists of a long chain of 16 carbon atoms, and that these molecules are standing on end on the water surface, the diameter of a carbon atom is about 1/16 of the thickness of the layer, i.e. about 1.6×10^{-10} m.

Electrons

The precise position of an electron is essentially unknowable. All we can state is the *probability* of finding it at a given position. Electron position in an atom or molecule is best represented by diagrams of varying density; a region of high density shows where the electrons are most likely to be. The bands of high density round a nucleus are called electron orbitals. All that can be stated with certainty is the number of electrons in each orbital. Figure 4.1 is such a deliberately vague representation of an aluminium atom, with electrons occurring in groups of 2, 8 and 3. These numbers, at least, are exact! Remember that the diagram, like others to come, is a two-dimensional representation of a three-dimensional situation. The regions in which the electrons are found are called shells or *orbitals*. Table 4.1 shows the orbitals in which the electrons are found for the first 13 elements. It can be seen that the first orbital is 'full' when it contains two electrons, and the second orbital is 'full' when it contains 8 electrons: there are similar limits to the capacity of higher orbitals.

Molecules

Atoms of elements join with atoms of other elements to form *compound* substances. For example, atoms of the elements carbon and oxygen combine to form the compound substances carbon monoxide (CO) and carbon dioxide (CO_2). A *molecule* of carbon monoxide is the smallest possible amount of that substance: The atoms of some elements combine with other identical atoms to form molecules and do not exist as single atoms, e.g. the hydrogen molecule H_2. The term *molecule* is therefore used as an all-embracing description of the smallest particle of any substance, element or compound.

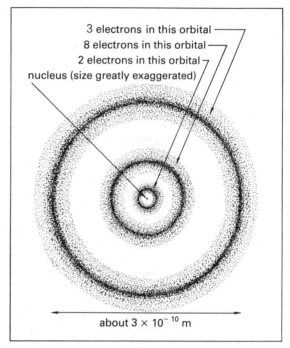

Figure 4.1 An atom of aluminium.

Atomic number and atomic masses

The *chemical behaviour* of an atom (i.e. which atoms it joins to form molecules) is decided by the number of electrons which the neutral atom has, and particularly by the number in the outermost orbital. Referring to table 4.1 we should therefore expect lithium and sodium to behave similarly (each has one electron in its outermost orbital), but not identically (each has a different number of electrons). The number of electrons in the neutral atom (= the number of protons in the nucleus) is called its *atomic number Z*.

We have already stated that the mass of an atom is almost equal to that of the nucleons it contains. The modern practice is to measure atomic masses relative to that of the commonest sort of *carbon* atom (whose nucleus contains 12 nucleons). This is arbitrarily given a relative atomic mass of 12 precisely. Thus if the mass of one of these atoms is m_c, one-twelfth of this is called the *unified atomic mass constant* m_u: that is $m_c = 12m_u$.

Other atoms will not have masses which are whole numbers of m_u: e.g. the mass of the commonest atoms of hydrogen and oxygen are $1.0078m_u$ and $15.995m_u$ respectively. If m_a is the mass of any atom, the number obtained by dividing m_a by m_u is called the *relative atomic mass A_r* of the atom, which is therefore defined by the equation

$$A_r = \frac{m_a}{m_u}$$

The atomic mass m_c of carbon-12 is found to be 19.92×10^{-27} kg, so

element	atomic number Z	electrons in shells K L M	relative atomic mass A_r	percentage abundance	proton number Z	neutron number N	mass number A
H hydrogen	1	1	1.0078	99.985	1	0	1
(D deuterium)			2.0141	0.015	1	1	2
He helium	2	2	3.0161	0.000 13	2	1	3
			4.0026	99.999 87	2	2	4
Li lithium	3	2 1	6.015	7.42	3	3	6
			7.016	92.58	3	4	7
Be beryllium	4	2 2	9.012	100.00	4	5	9
B boron	5	2 3	10.013	19.6	5	5	10
			11.009	80.4	5	6	11
C carbon	6	2 4	12	98.892	6	6	12
			13.003	1.108	6	7	13
N nitrogen	7	2 5	14.003	99.635	7	7	14
			15.000	0.365	7	8	15
O oxygen	8	2 6	15.995	99.759	8	8	16
			16.999	0.037	8	9	17
			17.999	0.204	8	10	18
F fluorine	9	2 7	18.998	100.00	9	10	19
Ne neon	10	2 8	19.992	90.92	10	10	20
			20.994	0.257	10	11	21
			21.991	8.82	10	12	22
Na sodium	11	2 8 1	22.990	100.00	11	12	23
Mg magnesium	12	2 8 2	23.985	78.60	12	12	24
			24.986	10.11	12	13	25
			25.983	11.29	12	14	26
Al aluminium	13	2 8 3	26.982	100.00	13	14	27

Table 4.1 The first 13 elements in the periodic table, with their relative atomic masses and percentage abundances of the atoms with stable nuclei. Other varieties of these atoms exist, which have unstable nuclei, e.g. the carbon atom which has 6 protons and 8 neutrons in its nucleus.

$$m_u \left(= \tfrac{1}{12} m_c\right) = 1.66 \times 10^{-27} \text{ kg}$$

We have already mentioned that the mass m_H of the commonest type of hydrogen atom is $1.0078 m_u$, so the mass of one of these hydrogen atoms is very nearly equal to m_u. As the commonest hydrogen atom has only one electron, the mass of its nucleus (a single proton) m_p also is very nearly equal to m_u. So, incidentally, is the mass of a neutron, m_n and so to three significant figures we can write

$$m_H = m_p = m_n = m_u = 1.66 \times 10^{-27} \text{ kg}$$

The *relative molecular mass* M_r of a molecule is defined in a similar way to A_r:

$$M_r = \frac{m_m}{m_u}$$

where m_m is the mass of the molecule.

For all species of nucleus the relative atomic mass is always close to a whole number; this is the number of nucleons in the atom, and is referred to as its *nucleon* or *mass number A*. (N.B. *Nucleon number* has no units and is necessarily a whole number, but *atomic mass* is measured in kg.) Thus the *nucleon number A* is the sum of the *proton number Z* (which is the same as the atomic number) and the *neutron number N*.

That is $A = Z + N$

Isotopes

The simplest possible atom is that of *hydrogen* consisting of a single proton round which orbits a single electron. The nucleus of *deuterium* contains one proton and one neutron and since one electron only is required to balance this electrically, it has the same chemical properties as ordinary hydrogen. It is sometimes called *heavy hydrogen*. It occurs as about 1 part in 4000 of naturally occuring hydrogen.

The atoms of a given chemical element must all contain the same number of protons. With the light elements in the table 4.1 the number of neutrons is approximately equal to the number of protons; but in most cases there are several possible alternatives. Groups of atoms having the same atomic number but differing atomic masses are known as *isotopes*. Samples of a given element derived from natural sources consist of mixtures of the possible isotopes of the element; the proportions are found to vary little from one sample to another. The measurement of relative atomic mass by chemical methods therefore gives a figure that is an average for the naturally occurring mixture of isotopes. This figure is not necessarily close to a whole number. For instance table 4.1 shows that natural *magnesium* is a mixture of isotopes of mass numbers 24, 25 and 26, the lightest providing nearly 80% of the total. The mixture is in such proportions that the relative atomic mass is 24.32. It is this quantity that must be used in chemical calculations involving magnesium.

The mole

It is often convenient to measure the amount of a substance in terms of the number of individual particles it contains. The unit of *amount of substance n* reckoned in this way is called the *mole* (abbreviated to *mol*).

The mole is the amount of substance of a system which contains as many elementary particles as there are atoms in 0.012 kg of carbon-12.

We can at once say that the *molar mass M* of carbon atoms (i.e. the mass-per-mole of carbon atoms) is 0.012 kg mol^{-1} or 12 g mol^{-1}. Note that the units of molar mass are *not* those of mass. Approximately, the mass of a hydrogen atom is 1/12 that of an atom of carbon-12, so the molar mass of hydrogen atoms is 0.0010 kg mol^{-1} or 1.0 g mol^{-1}. The mass of a hydrogen molecule is twice that of a hydrogen atom, so the molar mass of hydrogen molecules is 0.0020 kg mol^{-1} or 2.0 g mol^{-1}. We have obtained these molar masses M from the equation

$$M = A_r \text{ (or } M_r) \times 0.001 \text{ kg mol}^{-1}$$
$$= A_r \text{ (or } M_r) \times 1.0 \text{ g mol}^{-1}$$

Since we can measure the mass of a carbon-12 atom (we gave it earlier as 19.92×10^{-27} kg) we can find how many atoms there are in 0.012 kg (= 12 g) of carbon-12, i.e. in one mole. If we divide the molar mass of carbon-12 atoms by the mass of a single atom, then

$$\frac{0.012 \text{ kg mol}^{-1}}{19.92 \times 10^{-27} \text{ kg}} = 6.02 \times 10^{23} \text{ mol}^{-1}$$

This quantity is called the *Avogadro constant L*. So one mole of bricks is 6.02×10^{23} bricks. Of course we use the idea of a mole only when we deal with particles of atomic or molecular size, since only then do we have the large numbers which justify using the mole. Here are some examples of its use:

(i)   the mass of 2 moles of hydrogen atoms
      = 2.0 mol $\times$ 6.02 $\times$ 10^{23} mol^{-1} $\times$ 1.0078 $\times$ 1.66 $\times$ 10^{-27} kg
      = 2.01 $\times$ 10^{-3} kg
(ii) the mass of 0.10 mol of molecules of sodium fluoride (NaF, see table 4.1 for values of A_r)
      = 0.10 mol $\times$ 6.02 $\times$ 10^{23} mol $\times$ (18.998 + 22.990) $\times$ 1.66 $\times$ 10^{-27} kg
      = 4.20 $\times$ 10^{-3} kg
(iii) the electric charge of one mole of electrons
      = 1.0 mol $\times$ 6.02 $\times$ 10^{23} mol^{-1} $\times$ 1.60 $\times$ 10^{-19} C
      = 9.63 $\times$ 10^{4} C

Note that we *must* say what particles we are considering: we cannot say, e.g., one mole of hydrogen. One mole of hydrogen molecules has twice the mass of one mole of hydrogen atoms.

▶ If the pressure of a gas depends on the number of molecules present, find the mass of neon (a monatomic gas) which would be needed to produce the same pressure as 4.0 g of hydrogen in the same space at the same temperature. (Molar masses of hydrogen molecules and neon atoms are 2.0 g mol^{-1} and 10 g mol^{-1} respectively.)

To obtain the same pressure we need the same number of molecules, *i.e. the same number of moles*. We have 4.0 g of hydrogen, i.e. 2.0 moles of hydrogen molecules, so we also need 2.0 moles of neon atoms. Since its molar mass is 10 g mol^{-1}, we need 20 g of neon to produce the same pressure as 4.0 g of hydrogen in the same space at the same temperature. ◀

In this example we obtained the number of moles, the amount of substance n, by dividing the mass m of the substance by the molar mass M. This is true generally,

i.e. $$n = \frac{m}{M}$$

▶ Assuming that a copper atom can be thought of as a cubical block, and that solid copper consists of such blocks stacked together in regular rows and layers, calculate the width d of such an atom. (For copper, $A_r = 63.5$, density = 8.93×10^3 kg m^{-3}; $L = 6.02 \times 10^{23}$ mol^{-1}.)

One mole of copper has a mass of 63.5 g, so 6.02×10^{23} atoms have a mass of 63.5×10^{-3} kg

$$1 \text{ atom has a mass of } \frac{63.5 \times 10^{-3} \text{ kg}}{6.02 \times 10^{23}}$$

so its volume $d^3 = \dfrac{63.5 \times 10^{-3} \text{ kg}}{6.02 \times 10^{23} \times 8.93 \times 10^3 \text{ kg m}^{-3}}$

i.e. $d = 2.3 \times 10^{-10}$ m

So one of these 'cubical block' atoms has a width of about 2×10^{-10} m. We should need to know how the atoms are packed together in solid copper before we could improve on this estimate. This sort of calculation is used in reverse to calculate the Avogadro constant. We need to know the arrangement of the atoms, and the spacing of the layers: this can be found using X-ray diffraction (see page 362). ◀

4.2 Atomic forces

We begin by considering the forces which hold atoms together.

Ionic bonding

The single electron in the third orbital of the sodium atom is only loosely bound to the nucleus and is easily removed. The atom becomes an ion, Na$^+$. The second orbital of the

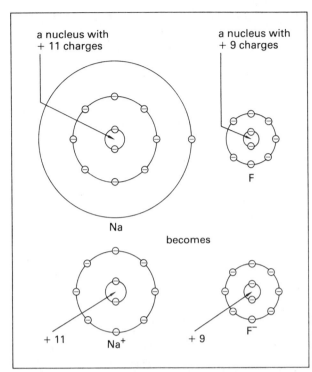

Figure 4.2 An atom of sodium donates an electron to a fluorine atom. They are now ions, and attract each other. You should note the different sizes of the atoms and ions.

fluorine atom contains 7 electrons, and needs one more to fill that orbital and make it stable. If it acquires an electron (possibly, though not necessarily, from the sodium atom we have been considering) it will become a fluorine ion F^-. A sodium ion and a fluorine ion will attract each other electrically to produce a molecule of sodium fluoride, NaF, as shown in figure 4.2. This *ionic bonding* is the common mechanism for those atoms which can easily lose or gain one or two electrons. These ions can attract others of opposite sign and form crystals of sodium fluoride: in this the individual molecules lose their identity. In sodium fluoride, for example, each Na^+ ion is surrounded by 6 F^- ions, and each F^- ion is surrounded by 6 Na^+ ions, and it would be pointless to try to label particular pairs of ions as molecules. Some molecules have their atoms arranged in such a way that they exert similar electrical forces on each other. The water molecule is an important example. These are called *polar* molecules, and are said to be *polarised*.

Covalent bonding

Two hydrogen atoms combine to form a hydrogen molecule, but for reasons of symmetry, if nothing else, the bonding cannot be ionic. The two atoms share their electrons so that each has access to both. Figure 4.3a illustrates this, and figure 4.3b illustrates the formation of a molecule of methane (CH_4) by this *covalent bonding*. In a sense the

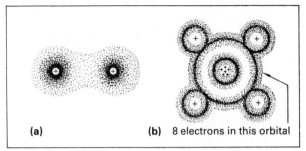

Figure 4.3 (a) The hydrogen molecule: each hydrogen atom has a share in both electrons. (b) The methane molecule: the carbon atom has four electrons which it shares with the hydrogen atoms, and each hydrogen atom has one electron which it shares with the carbon atom. This is a two-dimensional representation of a molecule which through symmetry would have the hydrogen atoms at the corners of a tetrahedron.

electrons act like a kind of glue between the other, positive, parts of the atom, pulling them together.

Metallic bonding

The atoms of metals (which are the elements with one or more loosely bound electrons in their outermost orbitals) are bound together by the sharing of their electrons indiscriminately. This bonding is similar to covalent bonding, but the electrons do not bond any two particular ions; a far greater number of sites is available. The electrons are free to move about in the space between the ions. Again the positive ions are held together, as in the covalent bond, by the negatively charged electrons moving between them.

Van der Waals forces

Two neon atoms, which have full outer orbitals, cannot be bound together by any of the previous methods. However, the electrons are in different places at different times, so that the atoms become instantaneously but temporarily polarised: see figure 4.4, where there are instantaneously more electrons on the left-hand side than on the right-hand side. Figure 4.4b shows the electric field around it. This polarised atom can then polarise other neighbouring atoms, and these atoms attract each other. Other atoms which have

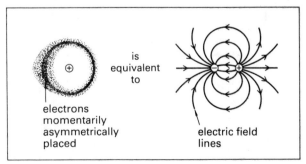

Figure 4.4 The origin of Van der Waals forces.

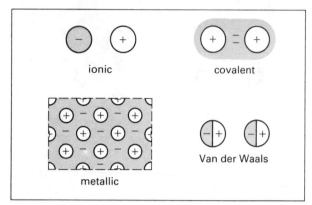

Figure 4.5 A simple summary of the different types of bond.

full orbitals are those of helium and argon, for example, but obviously other kinds of atom in already existing molecules can attract each other in this way (see page 47 where the alignment of polymeric chains is discussed). Molecules, too, can push or pull on each other with these forces. For example, we might have a hydrogen molecule (itself a covalent joining of two hydrogen atoms) attracted by such forces to another hydrogen molecule. These *Van der Waals forces* are weak, and the ionic, covalent and metallic bonds are always stronger. But nevertheless these forces are important because they affect the physical behaviour of gases, especially when the temperature is low or the pressure high (see page 139).

We summarise these four types of bond in figure 4.5. The gravitational force of attraction between atoms, incidentally, is *very* small and entirely negligible. The energy needed to separate atoms held by Van der Waals forces (the weakest electrical forces) is typically 10^{30} times greater than the

energy which would be needed to separate them if the atoms were attracted only by gravitational forces.

What keeps atoms apart?

If there are forces of attraction between atoms and molecules, there must also be forces of repulsion, or else matter would collapse into the very dense nuclear matter which we have already mentioned. We can see that where there is an ionic bond the electrostatic force of attraction will pull the ions together until the electron clouds repel each other, but beyond this there is little we can say at this level, except that the mechanism which provided the covalent and metallic bonding and the Van der Waals forces is also responsible for the repulsion when the separation of the atoms is reduced. The important thing for you to realise is that there *must* be forces of repulsion if there are forces of attraction, otherwise all matter would collapse.

Molecules

The cause of *intermolecular* forces is the same as the cause of interatomic forces and from now on we shall refer to molecules rather than atoms. We shall use the term molecule to refer to, e.g. (i) a sodium fluoride 'molecule' in a solid, even though each Na^+ ion is surrounded by six identical F^- ions, and it is impossible to associate a Na^+ ion with a particular F^- ion, (ii) the single atoms of helium in the gas phase and (iii) the ions of copper in a solid copper block. In doing this we are ignoring the fact that some of these molecules may have nothing like the spherical symmetry of the atoms of which they consist. The larger and less symmetrical they are, the further our simple statements will be

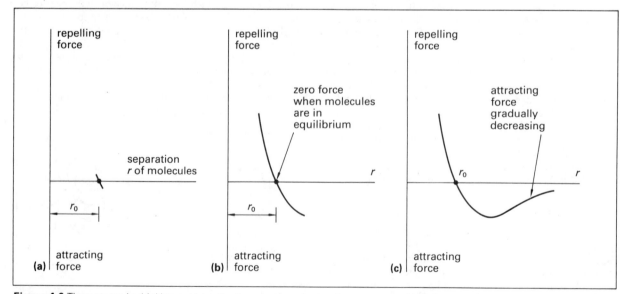

Figure 4.6 Three stages in thinking out how the force between molecules must vary with their separation.

from reality, but at this stage we *must* make simplifications to be able to get anywhere at all.

How the forces vary with molecular separation

So far we have given reasons for thinking that molecules both push and pull on each other. How much further can we go? We know that

(i) molecules close together in a solid seem to be in equilibrium (on the average). If we consider one particular molecule, the repelling forces on it from neighbouring molecules must be equal to the attracting forces.

(ii) it is hard to push molecules in a solid closer together. Therefore as they come closer together the repelling forces must increase. It is also hard to pull the molecules of a solid further apart. So as they move further apart the attracting forces also must increase.

(iii) we can break a copper rod by pulling on its ends, but we cannot break it by pushing its ends inwards. The repelling forces must, therefore, continue to increase as the molecules come closer together, but the attracting forces must reach a limit.

If we had to draw a graph which showed how the resultant force on a molecule varied with its distance from a neighbour (when that distance is small) it would look something like that shown in figure 4.6a. (b) and (c) show what happens when the molecules are not so close together. The greater difficulty of compressing, rather than stretching, is shown by the steepness of the graph for $r < r_0$. The separation of the atoms when they are in equilibrium has been labelled r_0: this distance is *defined* to be the diameter of the molecule.

How does the graph continue? Presumably when $r < r_0$ the graph continues to rise, steeply, but when $r > r_0$, what happens? We know that there is a limit to the size of the attracting forces, since we can pull molecules apart, i.e. we can pull them with forces larger than those with which they can pull on each other. We also know gases consist of molecules which are moving about without apparently exerting any force on each other except when they are very close. Thus the complete picture must be something like that shown in figure 4.6c.

We have already seen that the bonding of molecules may be of at least four different types, so that one law of force cannot possibly describe all four, but the following example considers a law of force which successfully describes the behaviour of one kind of molecule.

► Consider two neon atoms near each other. There will be Van der Waals forces F_a which pull them together, but the electron clouds will push them apart with a repelling force F_r. Both forces decrease with distance apart of the atoms.

Suppose $F_a \propto \dfrac{1}{r^7}$ and $F_r \propto \dfrac{1}{r^{13}}$

We know that $F_a = F_r$ at $r = r_0$ (the atoms are in equilibrium): call this force F. Tabulate values of F_a and F_r at values of r on both sides of r_0, at intervals of $0.1r_0$, giving

r	F_r	F_a	$F_r + F_a$
$0.80r_0$	18.2 F	-4.8 F	13.4 F
$0.90r_0$	3.9 F	-2.1 F	1.8 F
$1.00r_0$	1.0 F	-1.0 F	0
$1.05r_0$	0.53F	$-0.71F$	$-0.18F$
$1.10r_0$	0.29F	$-0.51F$	$-0.22F$
$1.20r_0$	0.09F	$-0.28F$	$-0.19F$
$1.30r_0$	0.03F	$-0.16F$	$-0.13F$

Table 4.2 The variation of $F_r + F_a$ with separation r of the atoms.

the values of F_a a negative sign, and find the resultant force $F_r + F_a$. Plot the results on a graph.

$F_r = F$ when $r = r_0$. To find F_r when $r = 0.90r_0$ we divide F by $(0.90)^{13}$ (F must increase as r decreases) and get $3.9F$. Similarly when $r = 0.80r_0$ we get F_r by dividing F by $(0.80)^{13}$ and get $18.2F$. We can see that F_r is increasing rapidly. To obtain values of F_a we divide F by $(0.90)^7$, $(0.80)^7$, etc. The results are tabulated (with extra values calculated for $r = 1.05r_0$) in table 4.2, and the graph of $F_r + F_a$ against r is plotted in figure 4.7a. We see that this *model* of the repelling and attracting forces does meet our needs. In figure 4.7b we have drawn the force-separation graph for a 'hard sphere' collision such as we might expect between two infinitely hard billiard balls. There is *no* force for $r > r_0$, and then a very large repelling force for $r = r_0$. This is *not* how atoms behave. ◄

Potential energy

Suppose one of the molecules is fixed in position at the origin of the graph, and that the other moves towards it from an infinite distance. Let us say that the potential energy of this molecule is zero at an infinite distance (it is an arbitrary choice, but agrees with the similar choice we make in the gravitational field). As it approaches the fixed molecule, it *loses* potential energy, and continues to lose it as long as the resultant force on it is attractive. Therefore the potential energy will vary with separation as shown by the section AB of the graph in figure 4.8b. However, when

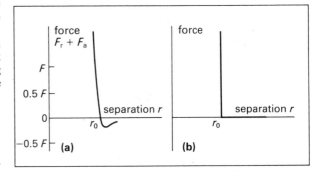

Figure 4.7 (a) Shows what our model predicts (b) shows the force-separation graph for 'hard spheres'.

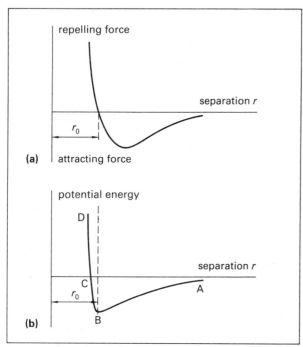

Figure 4.8 (a) The variation of intermolecular force with the separation of the molecules, and (b) the variation of the potential energy of a molecule with separation of two molecules. The minimum potential energy occurs when the molecules are a distance r_0 apart, i.e. when the molecules exert no resultant force on each other.

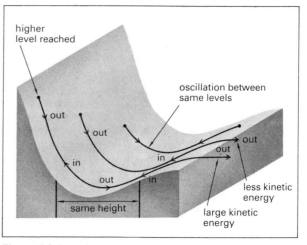

Figure 4.9 A gravitational analogy for the intermolecular potential energy. The two nearer particles represent gas molecules: the other one represents a molecule in a solid.

the separation of the molecules has become less than one diameter r_0 the resultant force becomes repulsive and the potential energy begins to increase, as shown by section BC of the graph. Because the repulsive force becomes very large when the molecule has moved a very short distance further, the potential energy soon becomes zero and even positive, as shown by section CD of the graph. The exact form of the curve can be calculated from a knowledge of the exact variation of the resultant force with separation.

What the potential energy curve predicts

We shall not consider the potential energy curve itself immediately, but shall first look at a situation which you will realise will serve as an analogy. Let us consider the frictionless surface whose shape is shown in figure 4.9. If a particle enters from the right with some kinetic energy it will slide down the slope and up the other (steep) side. It will then stop, slide down again through the trough and out of the picture to the right. All such particles which enter will emerge again with unchanged kinetic energy, although the less the kinetic energy with which they enter, the smaller the height to which they will rise on the steeper slope.

A particle which is *placed* (initially stationary) at a point on a sloping part of the surface will oscillate between that point and the corresponding point at the same height on

the other slope. If such an oscillating particle can somehow gain extra kinetic energy, its oscillations will have a larger amplitude; if it is given enough kinetic energy, it will slide out of the trough altogether, to the right. If the particle loses all its kinetic energy, it will settle down in the bottom of the trough.

This situation is a *potential well* in a gravitational field, and by now you will probably have realised that it is a good *analogy* for the force fields which molecules create around themselves. The shape of the surface is similar to the shape of the potential energy curve, and the behaviour of molecules is similar to the behaviour of particles on the surface. So we can expect the following:

(i) molecules with little kinetic energy will be oscillating with small amplitudes about a mean position one diameter from a nearest neighbour.

(ii) molecules which are given more kinetic energy will absorb it and merely make oscillations of greater amplitude.

(iii) molecules which are given even more kinetic energy will escape from their neighbours and will remain separate.

(iv) molecules which come from a distance and collide with another molecule will be repelled without losing kinetic energy.

In (i) and (ii), of course, the substance of which the molecules form part is a *solid*, and in (iii) and (iv) it is a *gas*. There is no information which we can get from the potential energy curve about the *liquid* phase. Before we consider these three phases in more detail, let us look at a two-dimensional representation of the difference between the state of the molecules in the three phases (figure 4.10): this was drawn by James Joule as long ago as 1847. It is possible to make three-dimensional diagrams to illustrate the same idea, but it is perhaps dangerous to try too hard, since we must remember that molecules are not spherical bodies with definite edges, and in any case these models of the three phases cannot illustrate the incessant movement of the molecules.

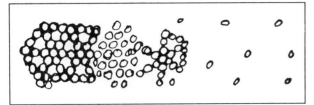

Figure 4.10 Joule's two-dimensional representation of the difference between the three phases. Note the partial order which he has correctly attributed to the liquid phase.

4.3 The solid phase

Many of the predictions of the potential energy curve are borne out by our observations on solids. The molecules do have fixed positions: we know this because in the solid phase substances have shape. The molecules *are* close together: we know this because it requires such large forces to move them even closer. For example, it would need a force of the order of 10^5 N to shorten a copper bar of about 100 mm² cross-sectional area (about the area of a finger-nail) by 1%. The molecules *are* vibrating: we know this, because the molecules must have energy of some sort (solids can become hotter or colder) and as the molecules do not move about bodily, they must be oscillating – there is a continual interchange between kinetic and potential energy. The *arrangement* of the molecules in a solid substance is such an important topic that it will be dealt with in a separate paragraph.

Crystals

The molecules in a solid are often arranged in a regular manner: much more so than is sometimes realised. Such solids are called crystals, and we describe their appearance as *crystalline*. There may be a million or more molecules occurring in regular rows and the crystal is said to have *long-range order*. Figure 4.11 is a photograph showing how a crystalline substance, mica, can be easily split. The smoothness of the face is not the result of careful polishing, nor is the sharpness of the edges the result of careful sawing: the surfaces and edges were produced by *cleaving* a larger (and probably rather irregular) crystal. Figure 4.11 also illustrates the cleaving of a crystal: the edge of a razor blade is pressed against the crystal where the cleavage is expected to form, and no great force is then needed to separate one part of the crystal from another. These newly formed surfaces occur naturally, and it is these *naturally-occurring plane faces* which are the distinguishing feature of crystals. Crystals seem to be rare, and large ones will be found only in geological museums (where they are worth seeing) but small ones are very common. A close look at some grains of salt, or sugar, shows that these substances, like all salts, and metals, and some non-metals, are crystalline. Indeed there are only a few types of substance (the glasses and the polymers) which are not crystalline. The smallness of most crystals usually arises because in a solidifying liquid the

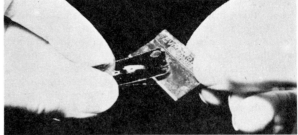

Figure 4.11 Cleaving a piece of crystalline mica. The razor blade is slicing between layers of atoms which were laid down by nature. The cleaved sheets can be cleaved again (and again) if the razor blade has a fine enough edge.

growth of particular crystals is hindered by the nearness of other growing crystals, or the edge of the vessel. But crystals of side about 10 mm can be grown by anyone with a little care and patience.

Another characteristic of crystalline solids is their *sharp melting-points*. Indeed, the melting-points of particular metals are used as fixed points in defining the International Practical Temperature Scale, and are quoted to five significant figures.

Molecular patterns in a crystal

Crystallography, the study of the different arrangements of atoms in crystals, is a science of its own, and here we shall only give some indication of how different crystalline shapes arise, using models to help us. One such model is the *bubble raft*. A detergent solution is placed in a shallow dish, and a piece of glass tubing with its end drawn out into a jet is used to blow bubbles in the surface of the liquid.

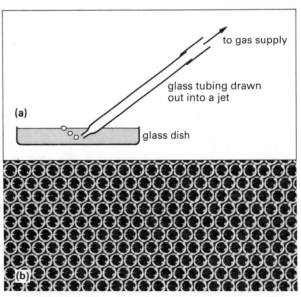

Figure 4.12 (a) Apparatus for producing a bubble raft; (b) is a photograph of the result.

To get a steady stream of bubbles all of the same size, the jet should be held at a constant depth below the surface of the liquid and connected to a constant-pressure source (e.g. a bench gas supply). Figure 4.12 shows the apparatus and the result (which is best displayed by placing the dish on an overhead projector). The bubbles congregate on the surface in a pattern in which they fit together as closely as possible. The bubbles make a good model of molecular behaviour since in their equilibrium positions they, like the molecules, are under the influence of two opposing forces – the (attracting) surface tension forces and the (repelling) air pressure forces.

Figure 4.13a shows a very simple (and therefore probable) way for some bubbles to come together to form a group with straight boundaries and we can see that the angles between the boundary edges are either 60° or 120°. Obviously the bubbles will serve as a good model only for single atoms, and we can imagine that the angles would not be as simple in a crystal which is formed from molecules each containing several atoms. Even the bubbles can come together to produce boundaries which make angles other than 60° or 120°, as can be seen in figure 4.13b. The problems of the different shapes which can be formed using the same building blocks are not easy to unravel and this is not the place to do so. Looking at the bubble raft we can at least more easily believe that molecules will fit themselves together in a particular way to give smooth surfaces and characteristic angles between those surfaces.

A three-dimensional model can be made from polystyrene spheres: again a model helps us to see the different arrangements when molecules come together. For example, if we start to build a model by piling up spheres on a base of spheres arranged in a triangular pattern (see figure 4.14) it is not at all obvious, until we have actually built the model, that there exist many different planes of atoms within a crystal. The photographs in figure 4.15 show the same model from two different angles, and we can see a 'hexagonal' layer and a 'square' layer of spheres. We can see that the square layer is not as closely-packed as the hexagonal layer. Because in the model as a whole the spheres are uniformly densely packed, the fact that the square *layer* is not closely-packed means that the square layers (the one we can see and the other parallel layers beneath it) are closer together than the hexagonal layers. This spacing of the layers is of great importance in X-ray crystallography, and the use of a model like this is a great help to anyone who is studying the way in which X-rays are reflected.

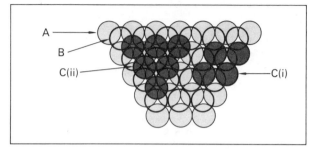

Figure 4.14 Layer A is laid down first, and layer B follows. The next layer C can be put in either of two positions: (i) directly above those in layer A, or (ii) in the other hollows.

Polycrystalline materials

We have already mentioned that when a liquid freezes the crystals which form are likely to be small because crystal growth has started simultaneously in many different places.

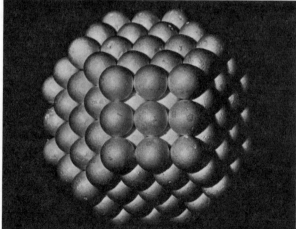

Figure 4.15 Photographs from different angles of the same 'crystal'. The existence of different types of layer can be clearly seen.

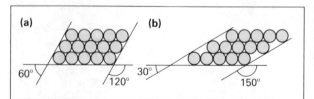

Figure 4.13 Ways in which bubbles can come together in a bubble raft, showing different angles which are possible between the boundaries.

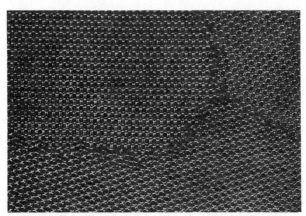

Figure 4.16 A photograph of a bubble raft model of grain boundaries in a polycrystalline material.

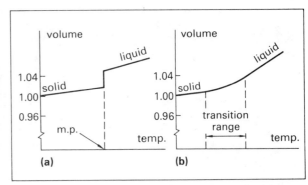

Figure 4.18 Graphs of volume against temperature for (a) a crystalline solid (b) a glass.

A metal or salt in the solid phase will therefore consist of many regions in which the orderliness is perfect, but there is no relation between the orderliness of one region and the orderliness of the next. These regions are called *grains*. The bubble raft model illustrates this very well, and in figure 4.16 we can see the grain boundaries clearly. Occasionally the grains in metals are large, and the grain boundaries can easily be seen: for example, on the surface of a galvanised iron bucket or a worn brass door handle or, as in the photograph in figure 4.17, on the surface of an aluminium sheet. The existence and size of the grains has an important bearing on the mechanical properties of materials which we shall be discussing next.

Glasses

There are some substances which, unlike crystals, pass from

Figure 4.17 Crystals in a specially-treated sheet of aluminium. The boundaries between the grains can be clearly seen. Within the boundaries the atoms are regularly arranged.

the liquid to the solid state and vice-versa almost imperceptibly: these *glasses*

(i)   do not have a sharp melting-point

(ii) do not undergo a sudden change of density (see figure 4.18)

(iii) do not release energy (latent heat: see page 111) when they melt.

These observations suggest that there is little change in the structure of a glass when it changes from the liquid to the solid phase: in the solid phase it retains the lack of orderliness which characterises all liquids, and the solid substance therefore does not have the characteristic shape of a particular crystalline solid. It simply sets into whatever shape is determined by the forces which are acting on it while it solidifies. Glasses behave in this way because they are very viscous when liquid, and become increasingly so as they cool: the molecules of these viscous liquids are not able to find the right sites which would enable them to crystallise into long-range orderliness. Figure 4.19 represents molecules moving freely to find a site which will enable them to take part in the repetitive nature of a crystalline structure. Molecules of a glass are brought to a standstill wherever they are, although there is more chance of them finding an order-creating site if the liquid is cooled slowly.

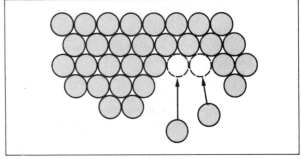

Figure 4.19 Molecules moving freely to fill vacant sites to maintain the regularity in a crystalline solid. The molecules of a glass do *not* have the chance to do this.

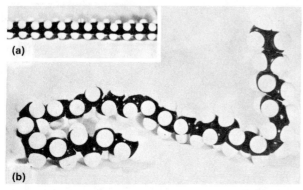

Figure 4.20 (a) A photograph of a model of part of a polythene molecule. The white spheres represent the hydrogen atoms which are attached to the chain of carbon atoms; (b) shows how the part of the molecule might be twisted. But remember that there might be 20 000 to 30 000 carbon atoms in the chain.

Polymers

Very many organic materials, and an increasing number of man-made materials, are formed from molecules which consist of very long chains of atoms. One of the simplest of these *polymers* (from the Greek words for many-parts) is poly(ethene) or *polythene*. As its name suggests, it is formed by linking together many molecules of ethene (C_2H_4) to form a long chain of perhaps 20 000 carbon

atoms to each of which are attached two hydrogen atoms. Its length might therefore be of the order of 10^{-6} m, and its relative molecular mass about 300 000. Figure 4.20a shows a photograph of a model of a part of this molecule: however, the molecule is not rigid, and in practice the chain will be twisted (figure 4.20b), doubled back on itself, and in something of a tangle. Indeed it is the tangling of the chains which gives a polymeric material much of its strength.

Polymers include many natural materials such as hair, wool, silk, cotton and cellulose (the main ingredient of wood), but since about 1920, when polymer science began to expand, many new polymeric materials have been created, some of which serve as alternatives to natural polymers and others provide us with totally new materials. The properties of these materials differ so widely that it is impossible to make any deductions about polymers from looking at their properties as a class of solid: what we shall do is to explain their properties in terms of their molecular structure. Table 4.3 lists some of the commoner man-made polymeric materials.

Thermoplastics

Typically these materials exist in a glassy state at low temperatures, soften and become flexible and rubbery when they have passed the *glass-transition temperature* T_g and at higher temperatures soften and melt. In the glassy state the polymeric chains are trapped in fixed positions at random (as are the much smaller molecules in an inorganic glass).

date of discovery	common name	present -day uses
	thermoplastics	
1868	cellulose nitrate (CN)	table-tennis balls, cycle mudguards
1927	cellulose acetate (CA)	photographic film, recording tape, spectacle frames
1936	polymethyl methacrylate, perspex (PMMA)	car rear lamp lenses, perspex sheets, windscreens
1938	polystyrene (PS)	model kits, yogurt cartons
1938	polyamides, nylon	hinges, curtain rail fittings, clothing (in fibre form)
1939	polyvinyl chloride (PVC)	gramophone records, guttering and down pipes
	polyester film	magnetic recording tape
1939	polyurethane foam (PU)	upholstery filling
1942	low-density polythene (LDPE)	waste disposal bags, milk crates, detergent squeeze bottles
	high-density polythene (HDPE)	domestic bleach bottles
1943	PTFE	non-stick saucepans, bearings
1948	acrylonitrile-butadiene-styrene (ABS)	food mixer casings, radio knobs, oil trays, shoe heels
1953	expanded polystyrene	ceiling tiles, packing, insulation
1957	polypropylene (PP)	chairs, car engine fans, toys, pudding basins
	thermosets	
1909	phenol formaldehyde (PF), bakelite	dark-coloured electric plug tops
1929	urea formaldehyde (UF)	white electric plug tops
1939	melamine formaldehyde (MF)	tableware, kettle handles, laminates, paints
1942	polyester resin	(with glass fibre) car and boat bodies, repair kits

Table 4.3 Dates of discovery of, and some uses of, common man-made polymeric materials.

As the temperature rises some parts of the chains (being held only by relatively weak Van der Waals forces) become able to rotate relative to other parts, and at still higher temperatures softening and melting occur as the molecules gain enough energy to be able to move independently of each other. There is no sharp melting point; it softens and continues to do so as the temperature rises, until it is liquid. There will, however, be a discontinuity (in, e.g. a graph of volume against temperature) at what is called the melting-point T_m if the material has crystallised: see figure 4.21.

Crystallisation occurs when chains align themselves alongside each other: clearly it is most likely to happen when the chains have a simple repeating pattern, and when they are not branched or with bulky groups of atoms attached to them. Crystallisation in polymers is usually only partial: figure 4.22 is an electron microscope photo-graph of some polythene, in which *crystallites* (regions of crystallisation) can be seen. Incidentally the alignment of the chains is at right-angles to the plane of the photograph, and because the chains are far longer than the depth of the crystals, a particular chain must double back on itself many times to help form the crystal. The aligned chains are held together only weakly, by Van der Waals forces (and hence they allow the material to flex), but nevertheless the strength and rigidity of the material is increased. Table 4.4 shows how some of the physical properties of polythene vary with the percentage of crystallisation. It is interesting to note that the more orderly packing of the chains results in an increase in density.

Rubbers (or *elastomers*) are not really a separate class of polymer: they are thermoplastic polymers whose chains are very weakly bound together, and *any* thermoplastic polymer passes through a rubbery range when it is heated, as we have already seen. When we think of a material as being rubbery, it merely means that we usually meet it at a temperature which is above its glass-transition tempera-ture T_g: e.g. the T_g for natural rubber is $-75°C$. This is

properties (at 293 K)	percentage crystallisation			
	65	75	85	95
density/kg m^{-3}	0.91	0.93	0.95	0.97
tensile strength/10^6 Pa	13	18	22	31
softening point/K	363	388	393	398
relative rigidity	1	3	4	5

Table 4.4 The variation of some physical properties of polythene with percentage crystallisation.

lower than normal air temperatures so we think of it as a rubber. Perspex on the other hand, has $T_g = 100°C$, so that we normally find it glassy – but it will be rubbery if we heat it above $100°C$.

Thermosets

Some other polymers are hard and rigid: when heated they do not soften, but eventually decompose. To create these materials, chemicals are mixed to produce the polymers, and then the polymer is either heated strongly, or a catalyst is added, so that chemical (covalent) bonds or cross-links are formed between the polymeric chains (compare this with the weak Van der Waals forces in thermoplastics). This chemical action is not reversible: the material cannot be softened by re-heating. The amount of energy needed to break a covalent bond cannot be got by heating. A good example of this is the *vulcanisation* of rubber. Natural

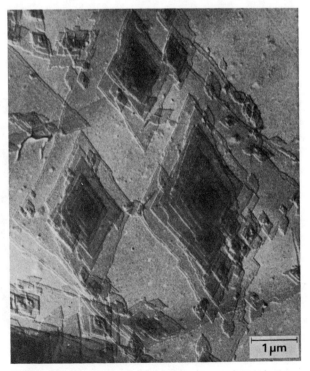

Figure 4.22 Single crystals of polythene as seen in an electron microscope (magnification 18 000 ×).

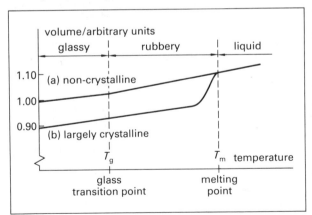

Figure 4.21 The variation of volume with temperature for a typical thermoplastic polymer which (a) is non-crystalline and (b) has largely crystallized. The melting point T_m has no significance for (a): the glass transition point T_g has little significance for (b).

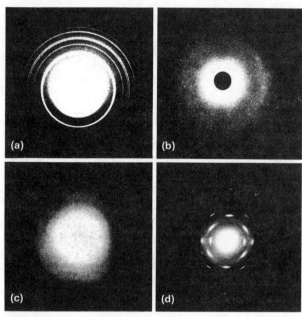

Figure 4.23 Diffraction photographs for (a) a crystalline material (tin) (b) a glass (c) unstretched rubber and (d) stretched rubber. We can clearly see the orderliness introduced into rubber by stretching it.

rubber is a flexible thermoplastic material, and a disadvantage of it is that it does not readily return to its original state when a deforming force is removed. The addition of sulphur atoms links the polymer chains together and produces a stiffer, more resilient material. Further addition of sulphur produces the very hard material known as ebonite.

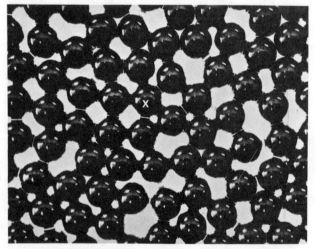

Figure 4.24 An instantaneous snapshot of a two-dimensional liquid model. The ball marked X, although vibrating, will not move far from that position until it acquires an abnormal amount of kinetic energy.

X-ray diffraction photographs

In chapter 27 we explain how X-rays can be used to produce diffraction patterns which give us some idea of the degree of order in a solid material. The four photographs in figure 4.23 show us (a) the orderliness of the arrangement of the atoms in a crystal, (b) and (c) the lack of order in a glass and a rubber, and (d) the orderliness which can be introduced into a rubber by stretching it. Further evidence for the orderliness of stretched thermoplastic materials is provided by a very simple experiment; if we take a piece of rubber (e.g. from a balloon) or a piece of polythene (e.g. from a plastic bag) and stretch it, and push a pin through it while it is stretched, it splits parallel to the direction of stretching. This suggests that the polymeric chains have been aligned parallel to the direction of stretching.

4.4 The liquid and gaseous phases

We observe that a substance in the liquid phase is usually only slightly less dense than it is in the solid phase (although water is an exception). A typical metal, for example, is only 3 per cent less dense as a liquid. A liquid is not rigid but is hard to compress. This all suggests that the molecules are nearly as close together in a liquid as they are in the solid phase, but that they do not just vibrate about a fixed position. They must be free to move about (or at least groups of molecules must be free to move about) but presumably they still vibrate, as they are still close enough to their neighbours for this to have to happen.

There will not be the long-range order which exists in crystals, where there might be rows of about a thousand regularly arranged molecules (in each direction), but there may well be *short-range order* extending over a few near neighbours. The study of liquids has not advanced nearly as far as that of solids and gases. Much of the work now being done consists of making models of liquids in order to investigate how far from a particular molecule other molecules can be expected to be. One such two-dimensional model consists of ball-bearings coated with a very viscous oil: a photograph (figure 4.24) taken while they are being made to vibrate shows some orderliness. Most of the ball-bearings have five near neighbours, which compares with six near neighbours which a ball-bearing would have in a close packed two-dimensional arrangement. Figure 4.25 shows a diffraction photograph for water: the broad ring shows that there is some orderliness. We can say that liquids are temporarily 'polycrystalline' in *small* regions, but the molecules making up the crystalline regions are continually changing places.

We can think of a molecule in a liquid as being like a man in a dense football crowd. Individual members of the crowd are never still. They are always moving backwards and forwards, but never moving very far, unless they make a determined effort to change neighbours or change places.

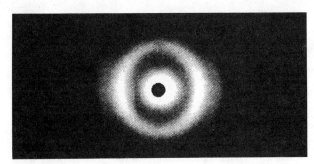

Figure 4.25 An X-ray diffraction photograph for water.

The gaseous phase

A substance in the gaseous phase is much less dense than it is in either the liquid or the solid phases – perhaps by a factor of 10^3. It is also highly compressible, and fills the whole of any closed space in which it is put. We can also observe

(i) *Brownian motion*, first noticed in 1827 by Robert Brown (although his first observations were made on small particles suspended in a *liquid*). This can be seen through a microscope when ash particles are blown into an enclosed space. The ash particles have a jerky, random motion (see figure 4.26): it is as if they were being pushed about by something we cannot see, which we believe to be air molecules. The molecules are too small to see, even under a microscope, so presumably they must be moving fast (and randomly) to be able to produce their effect.

(ii) the *diffusion* of one gas into another. If liquid bromine is released at the bottom of a vertical glass tube containing air, it evaporates. The brown gas is seen to make its way slowly up the tube. We know that the bromine molecules must in fact be moving at high speeds because when released into a similar evacuated tube, they reach the top in an immeasurably short time. So when diffusing they must be taking a very irregular path, colliding with many air molecules, since their upward progress is so slow. We can deduce from measurements of the time taken etc. that the average distance travelled by the bromine molecules between collisions (the *mean free path*) is of the order of 10^{-7} m, and that they make of the order of 10^9 collisions each second. These observations suggest that in the gaseous phase molecules are relatively far apart (but not as much as 10^{-7} m, the mean free path, because the molecules will not collide with their nearest neighbours: the molecules may be about 10^{-9} m apart, i.e. about 10 times the diameter of an atom or a molecule). They are moving rapidly, independently and randomly.

Because the molecules of a gas do move independently of each other, except when they make collisions, a gas is a very simple form for a substance to have. As a result, much quantitative work has been done to explain how the pressure, volume and temperature of a gas are related – so much so that this will be dealt with in chapter 10 later in this book.

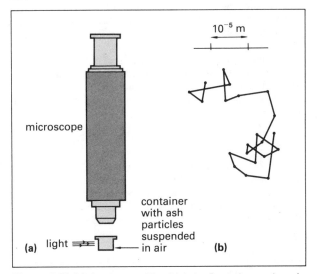

Figure 4.26 (a) Apparatus with which the Brownian motion of ash particles in air can be observed. (b) A typical series of positions of one ash particle, at time intervals of one second. The particle does *not* move straight along the lines: between the positions shown it will travel a random path similar to that shown in the diagram.

5 Performance of materials

5.1 Materials in tension

We have so far considered the *structure* of materials: we now examine their performance. A solid can be formed into many different shapes, and have forces exerted on it in many different ways: we shall consider materials which have been shaped into rods, wires or strips, and which are pulled lengthways. The specimen is then in a state of *tension*, and the forces are called *tensile forces*. Figure 5.1 shows a specimen of negligible weight hung vertically with a load of weight W at its lower end. The free-body diagrams show that the tension throughout the specimen is W, i.e. at any point on the specimen such as A, the part below A exerts a pull W on the part above A, and the part above A exerts a pull W on the part below A. Even if the cross-sectional area varied, the tension would be W throughout the specimen.

Stress

An important quantity for a specimen will be its ultimate tensile force, the force which will break it; this is a measure of its *strength*. But if we want to compare the strengths of two materials we need to consider the force per unit area. We define the *tensile stress* σ in a specimen by the equation

$$\sigma = \frac{F}{A}$$

where F is the tensile force and A is the cross-sectional area. Suppose, for example, that a mass of 2.0 kg is hung on the end of a vertical copper wire of diameter 0.50 mm. Since the mass is in equilibrium, the tension in the wire is numerically equal to the weight of the mass (≈ 2.0 kg $\times$ 10 N kg^{-1}), so that the stress σ is given by

$$\sigma = \frac{20 \text{ N}}{\pi (0.25 \times 10^{-3} \text{ m})^2}$$

$$= 1.0 \times 10^8 \text{ N m}^{-2}$$

We see that the unit of stress is the N m^{-2}, which is given the name *pascal* (Pa) (after the French scientist Blaise Pascal, who in 1650 first investigated the transmission of pressure in fluids). *The ultimate tensile stress* (i.e. the stress at which it breaks) of copper is between 2×10^8 Pa and 4×10^8 Pa (depending on the method of production) so that we can safely hang the mass of 2.0 kg on that wire.

When materials are stretched their cross-sectional area decreases, sometimes considerably. It is not usually easy to measure the new cross-sectional area, so that the *nominal stress* or engineering stress is used: it is defined by the equation

$$\text{nominal stress} = \frac{\text{tensile force}}{\text{original cross-sectional area}}$$

Where the wire thins the actual stress will be greater than the nominal stress.

Strain

We have considered the strength of a material: we need to know also how much it stretches when a stress is applied to it. The extension will depend on the material, but it will also depend on the length of the specimen, so that to make fair comparisons we measure not the extension, but the *fractional* extension, or *tensile strain* ε which occurs. This is defined by the equation

$$\varepsilon = \frac{\Delta l}{l}$$

where Δl is the extension which occurs from an original length l. Because it is a *fractional* extension, tensile strain has no unit. Now we can define the tensile *stiffness* of a material by the equation

$$E = \frac{\text{tensile stress}}{\text{tensile strain}}$$

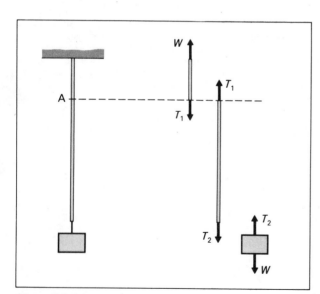

Figure 5.1 A load of weight W is hung from a wire. At any point in the wire (e.g. A) the tension T_1 is equal to W.

E is called the *Young modulus* after Thomas Young (1773–1829) and its unit is the same as that of tensile stress, the pascal.

▶ An electric power cable has a length of 100 m, and consists of an aluminium wire of diameter 5.0 mm surrounded by six steel wires of diameter 2.0 mm. It is hung between two pylons. The sag of the cable is such that the tension in it has an average value of 800 N. What is the extension of the cable and the average stress in the wires? (The Young modulus for the steel and for aluminium is 2.0×10^{11} Pa and 7.0×10^{10} Pa respectively.)

We use the equation

$$E = \frac{\text{tensile stress}}{\text{tensile strain}} = \frac{F/A}{\Delta l/l} = \frac{Fl}{A\Delta l}$$

for the different wires separately. We know that for all seven wires the extension Δl is the same. We have

$$E_s = \frac{F_s l}{A_s \Delta l} \text{ and } E_a = \frac{F_a l}{A_a \Delta l}$$

where the suffixes s and a refer to the steel and the aluminium respectively. We also know that $F_s + F_a = 800$ N, since with the two wires running parallel the total extending force is the sum of the forces in the individual wires.

So $\qquad \dfrac{E_s A_s \Delta l}{l} + \dfrac{E_a A_a \Delta l}{l} = 800$ N

or $\qquad \dfrac{\Delta l}{l}(E_s A_s + E_a A_a) = 800$ N

Hence $\qquad \Delta l = \dfrac{(800 \text{ N})l}{E_s A_s + E_a A_a}$

$E_s A_s = (2.0 \times 10^{11} \text{ N m}^{-2})\{6 \times \pi(1.0 \times 10^{-3} \text{ m})^2\}$
$\qquad = 3.77 \times 10^6$ N

$E_a A_a = (7.0 \times 10^{10} \text{ N m}^{-2})\{\pi(2.5 \times 10^{-3} \text{ m})^2\}$
$\qquad = 1.37 \times 10^6$ N

Therefore $\qquad \Delta l = \dfrac{(800 \text{ N})(100 \text{ m})}{(3.77 + 1.37)10^6 \text{ N}}$

$\qquad\qquad = 1.56 \times 10^{-2}$ m $= 16$ mm

The stress $= E \times$ strain.

The strain $= \dfrac{1.56 \times 10^{-2} \text{ m}}{100 \text{ m}} = 1.56 \times 10^{-4}$

Therefore, in the steel,

stress $= (2.0 \times 10^{11} \text{ Pa})(1.56 \times 10^{-4})$
$\qquad = 3.1 \times 10^7$ Pa

and in the aluminium,

stress $= (7.0 \times 10^{10} \text{ Pa})(1.56 \times 10^{-4})$
$\qquad = 1.1 \times 10^7$ Pa

The ultimate tensile stress of the steel is about 2×10^8 Pa, and that of the aluminium is about 5×10^7 Pa, so these calculated stresses are reasonably safe, though winds may produce additional forces which would increase the tensions and therefore the stresses. ◀

Measurement of the Young modulus

We hang up a wire from a convenient beam in the ceiling and load it at the other end. The wire needs to be as long as possible so that the extension will be as large as possible. The extension is in any case so small that we need a sensitive measuring device: a vernier scale is just sensitive enough, and it is convenient to support it on a second wire hung parallel to the first wire from the same support, as in figure 5.2. At the same time this solves the problem of what to do about any sagging of the supporting beam (which would exaggerate the extension readings): if the wire supporting the vernier is supported from the same point the vernier sags by the same amount.

Masses are gently placed on a carrier at the lower end of the wire under test, and the extension read from the vernier scale. Table 5.1 shows a typical initial set of readings in column 1. There is a sudden jump from 4.7 mm to 6.8 mm: probably a kink in the wire (likely to occur if the wire has been unwound from a reel) has there been straightened out. The same wire could be used again, and then the readings in column 2 might be obtained. These increase uniformly and are satisfactory for our purpose, and with them a force-extension graph may be plotted, as in figure 5.3. From the average slope of this graph we can calculate a value of $F/\Delta l$. We shall be using the expression

$$E = \frac{\text{tensile stress } \sigma}{\text{tensile strain } \varepsilon} \text{ where } \sigma = \frac{F}{A} \text{ and } \varepsilon = \frac{\Delta l}{l}$$

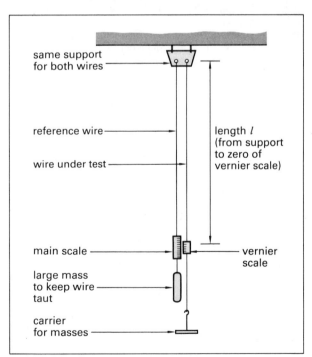

Figure 5.2 The measurement of the Young modulus for a wire.

load m/kg	tensile force F/N	vernier readings/mm (1)	vernier readings/mm (2)	extension/mm in column (2)
0	0	3.5	3.5	0
2	20	4.7	4.7	1.2
4	39	6.8	5.8	2.3
6	59	8.1	7.1	3.6
8	78	9.3	8.3	4.8
10	98	10.5	9.5	6.0

Table 5.1 The result of loading a wire with masses on two different occasions. In (1) a kink in the wire causes a sudden jump in the extension. In (2) the readings are satisfactory.

so that $\quad E = \dfrac{Fl}{A\Delta l} = \left(\dfrac{F}{\Delta l}\right)\left(\dfrac{l}{A}\right)$

We therefore need to measure the original length l (which can be measured sufficiently sensitively with a metre rule) and the cross-sectional area A. For this a micrometer screw gauge must be used to measure the diameter in several different places to take account of any possible thinning, or any ovality, of the wire. Hence we can calculate E for the material of the wire.

Distinction between stiffness and strength

It is clear from the definition that a material is regarded as stiff if a particular stress produces a small strain. The stiffness of a material, and its strength, are different, and independent, properties. Figure 5.4 shows stress-strain graphs for two materials A and B (the material breaks at the final point on each graph): A is twice as stiff as B, but B is twice as strong as A.

◄—— An aluminium wire of length 1.0 m has a cross-sectional area of 0.10 mm², and a chromium wire of length 2.0 m has a cross-sectional area of 0.050 mm². The Young

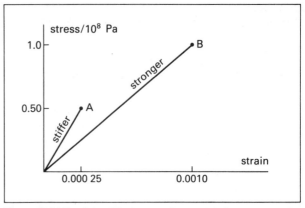

Figure 5.4 Wire A is stiffer because its graph is steeper. Wire B is stronger because it does not break until more stress is applied.

modulus of chromium is four times larger than that of aluminium. When the same mass is hung from each wire, which has the larger extension?

The chromium wire is *twice* as long and has *half* the cross-sectional area. These factors would, together, make the extension of the chromium wire *four* times larger. But the Young modulus of chromium is four times greater, so that chromium is four times stiffer and therefore the extensions of the two wires are the same.   ◄——

The tensile stress, and the stiffness, of a material are also important in a beam or a cantilever. Figure 5.5 shows that these devices have regions which are in tension and compression, and a designer of any large structure must take this into account.

Stress-strain graphs

Let us now consider the actual performance of five different specimens, chosen to represent most of the commoner possibilities: a glass rod, a copper wire, an iron *whisker*, (a crystalline growth of length about 20 mm, and diameter about 10^{-6} m), a length of rubber, and a strip of polythene. Figure 5.6 shows graphs of stress against strain for the five specimens: there are two graphs for the copper wire since

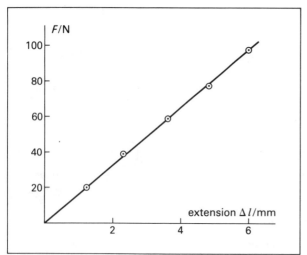

Figure 5.3 Plotting the results of a Young modulus measurement. The slope of the graph $(F/\Delta l) = (98\text{ N})/(6.0\text{ mm})$ $= 1.6 \times 10^4$ N m⁻¹.

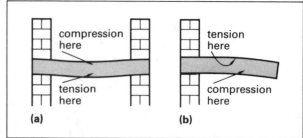

Figure 5.5 The existence of tension and compression in (a) a beam (b) a cantilever. The amount of sag is greatly exaggerated.

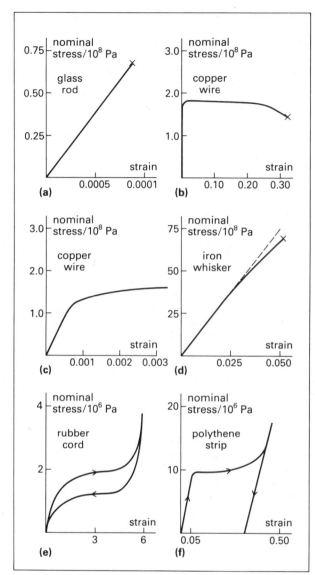

Figure 5.6 Stress-strain graphs for (a) a glass rod (b) and (c) a copper wire (d) an iron whisker (e) a rubber cord and (f) a polythene strip. Note the very different scales. The cross X marks the point where the specimen breaks. These are *typical* results only, and the stress is the *nominal* stress.

the one with the larger strain scale cannot show enough detail where the strain is small. Before considering the graphs individually, there are some points you should note:
(i) these are *typical* results: the behaviour of a specimen depends on its purity, its previous treatment, and many other factors.
(ii) the scales on the axes cover very different ranges.
(iii) we cannot usually tell from the graph whether the specimen is behaving *elastically*, that is, whether the

specimen will return to its original shape when the stress is removed. Elasticity is obviously a very important property for a material to have: a bridge must not only be strong (i.e. it must be able to support a large load) and stiff (i.e. it must be able to support a large load without much deformation) but it must also be elastic (i.e. it must return to its original position when the load is removed). A material which does not do so is said to be behaving *plastically*; both copper and polythene (of the materials selected) behave plastically if the stress is high enough.
(iv) the slope of the graph gives the stiffness, or Young modulus, of the material. We can see that different materials have very different, and not always constant, stiffnesses. Table 5.2 gives the values E of the Young modulus (which can be deduced from the graphs) and also values of the ultimate tensile stress σ_u. Where the stiffness is constant, i.e. the strain is proportional to the stress, the material is said to be obeying *Hooke's law*, named after Robert Hooke (1635–1703) though he stated the law in the form *extension is proportional to force*. (He performed his experiments on helical springs, for which, if the forces are not too large, his statement of the law is also true.) If a material has a constant stiffness calculations can be made to predict the strain which a particular stress will cause.

(a) the glass rod

We can see, from figure 5.6 and table 5.2 that glass is about half as strong as our representative metal (copper in wire form) and about half as stiff: it also behaves elastically. It would therefore seem to be a useful structural material, but the abrupt end to the graph shows that it is brittle, and is therefore not *tough*. Toughness is yet another property of materials, and is distinct from strength and stiffness. We shall discuss this further in the next example.

(b) and (c) the copper wire

The graph for the copper wire is more complicated: see also figure 5.7, which shows the graph in more detail. For small stresses the copper wire obeys Hooke's law, i.e. it has a constant stiffness: in figure 5.7 A marks the *limit of proportionality*. Up to and slightly beyond this point the wire behaves elastically, until the *elastic limit* B is reached. The stiffness then decreases considerably. In fact on a graph of *nominal* stress the stiffness *appears* to be negative, but the actual stress *is* increasing, though not by very much.

specimen	$E/10^6$ Pa	$\sigma_u/10^6$ Pa
glass (rod)	75 000	75
copper (wire)	130 000 where constant	200
iron (whisker)	150 000	7500
rubber (cord)	≈ 0.7 average value	4
polythene (strip)	200 where constant	20

Table 5.2 The values of the Young modulus E and the ultimate tensile stress σ_u which can be deduced from the graphs in figure 5.6.

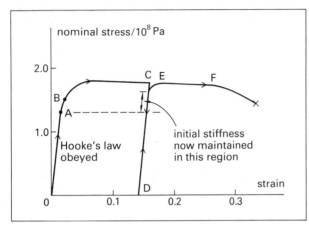

Figure 5.7 An enlargement of part of the graph of figure 5.6b. In the plastic region the curve slopes downwards only because it is the nominal stress which is being plotted: the *actual* stress is increasing, and the material breaks when the actual stress is highest. Note what happens when the wire is unloaded at C and then reloaded: the limit of proportionality, and the elastic limit, are both higher. The material is said to have been *work-hardened*.

The stiffness decreases considerably. If the stress is removed at a point such as C the wire does *not* return to its original length, but the graph follows a line parallel to AO, and retains a permanent strain OD. If a stress is then applied to the wire again, the graph retraces the line from D towards C until it curves round to rejoin the curve which would have passed direct from C to E if the original loading had been maintained. It can be seen that when reloaded the wire obeys Hooke's law again up to larger stresses than before, and its stiffness remains at its original high value up to larger stresses. This effect, which is known as *work-hardening*, obviously increases the usefulness of a material. The change in direction of the curve at F is the result of a constriction or *neck* occurring at one point in the wire. At this particularly narrow part the actual stress is very high, and here the wire will eventually break. The plastic behaviour of the copper makes it a tough material: copper objects are not designed so that plastic flow is likely to occur, but the ability to behave plastically is useful in an emergency. A copper saucepan, if dropped, or a steel motor car body, if it collides with a wall, will only be dented: a glass bowl will break. We say that copper is *ductile*: it can be drawn, rolled and hammered into shape. It is typical of many metals; and it is particularly the toughness of metals (not their strength or stiffness, which other materials also possess) which makes them so useful.

(d) the iron whisker

The graph for the iron whisker is unusual in that the ultimate tensile stress, and consequent strain, are very high for a metallic substance. We expect this to be related to the fact that it has grown in nearly perfect conditions, and perhaps

does not possess the flaws of bulk material. The graph for an iron wire would follow the same line, but would end when the stress had reached about 1.5×10^8 Pa and the corresponding strain was 0.001, which are values 50 times smaller than the final values for the whisker. The whisker has the same stiffness as an iron wire, but is much stronger. It behaves elastically, despite the curving of the graph near its end: the curving here is very small, and does not indicate that plastic deformation is occurring.

(e) the rubber cord

Rubber is unusual in that it can undergo *very* large strains and still behave elastically, although as the graph shows, the stress-strain curve for unloading is not the same as the graph for loading. This effect is called *hysteresis*. Nevertheless, the rubber returns to its original length when the stress is finally removed. Since the strain is so large, and the volume of the rubber is constant, the cross-sectional area decreases considerably, and plotting the nominal stress makes rubber appear less stiff than it actually is. With a strain of 6, the cord is 7 times longer than it was, its cross-sectional area is 1/7 of its original value, and the actual stresses are 7 times greater than the nominal stresses. We notice too that the stiffness is not constant: it increases considerably towards the end of the graph.

(f) the polythene strip

The strip could be cut from the low density polythene which is used to make plastic bags. We see that its stiffness is constant at first: in this region it also behaves elastically. Then there is a sudden increase in strain for virtually no increase in stress: what happens is illustrated in figure 5.8. The part which has narrowed becomes no narrower, but it does become longer, and all the extension occurs in this section. When the whole strip has narrowed the material becomes about as stiff as it was originally. If at this stage the stress is removed, the strip is left with a large permanent strain.

5.2 A molecular view

We shall now see whether we can explain the observed performance of different materials, using our model of molecular behaviour. It will be justified as a good model by the extent to which it can provide *explanations*. Here is a summary of the behaviour of solid materials, together with the explanation which the model provides.
(i) They obey Hooke's law (i.e. have a constant stiffness) up to a point, when failure of some kind usually occurs. The material breaks, or deforms considerably. Whiskers of a material do not have a constant stiffness, and are stronger. What happens on the large scale, to a material, happens at the molecular level too: when we increase the length of a copper wire, we are increasing the distance

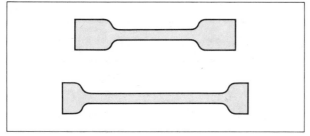

Figure 5.8 The behaviour of polythene when it is stressed. The narrow part becomes longer without becoming any narrower.

between the molecules, *and the strain is the same*, whether we measure it by the fraction $\Delta l / l_0$ or the fraction $\Delta r / r_0$ where l_0 is the original length of the wire and r_0 the original separation of the molecules. At first sight the graph of figure 4.7a suggests that the strain is *not* proportional to the stress – until we remember that we expect Hooke's law to be obeyed for very small strains only – up to about 0.001 for most materials. A closer look at the graph shows that the graph *is* very nearly straight for strains of this order: it is so nearly straight that at a strain of 0.001 the difference between the value of F on the curve, and the value of F on a tangent to the curve through the point $(r_0, 0)$ is only 1%. So our model explains why Hooke's law will be obeyed, as nearly as we shall be able to tell, for strains of up to 0.001. Incidentally, the fact that the force-separation graph is a straight line through the point $(r_0, 0)$ into the compression region explains why materials have a constant stiffness in compression, and the same stiffness as they have in tension.

In figure 5.9a part of the force-separation graph *for molecules* has been isolated, and in (b) it has been inverted and enlarged, and its axes re-labelled, so that it is a stress-strain graph *for the material*. We see that the curving of the force-separation graph for the molecules predicts a decrease in the slope of the stress-strain graph for the

material, and therefore a decrease in the stiffness of the material, which should be just noticeable when a strain of about 0.001 has been reached (as we have already said). From then on the stiffness should steadily decrease further until at a strain of about 0.1 (10%) the material should break, since at about this strain the maximum attractive force between the molecules has been reached, and if in pulling the material we exceed this limit, the molecules separate completely. This behaviour did occur, at least to some extent, with the iron whisker whose stiffness decreased steadily up to a strain of about 0.05, when it broke. It did not occur with the glass rod and the copper wire: presumably the brittle fracture of the glass and the ductile yielding of the copper occurred through some fault in the material. We shall deal with these *failure mechanisms* later.

(ii) The rubber cord was an exception in that its stiffness increased considerably as the strain increased. Rubber is a polymer and we know that the long-chain molecules are not usually arranged in an orderly way. They double back on themselves, and are tangled. Figure 5.10 gives some idea of how three such molecules might be arranged. Stretching a rubber cord merely straightens out the long chains, and it needs little force to do this (in the absence of cross-linking, which makes some other polymers stiff): also, a considerable extension can be produced. However, when the chains have been straightened, the length of the rubber can be increased only by increasing the separation of the molecules, and it is as hard to do this with rubber as it is with any other substance. We could expect the rubber cord then to be flexible up to *much* larger strains than other materials, but to be stiff eventually.

(iii) *Materials are elastic for small stresses.* When external forces are used to keep some material stretched, its molecules are kept in equilibrium at a greater separation than usual. The forces of attraction between the molecules are then greater than usual. When the external forces are removed, the molecular forces pull the molecules back to their normal positions again. This elasticity can occur even when the material does not have a constant stiffness (e.g. with the iron whisker and the rubber cord). Some specimens (e.g. the glass rod, the iron whisker and the rubber cord) are elastic up to the point where they break: we shall discuss the *plasticity* of the copper wire when we discuss its failure mechanism.

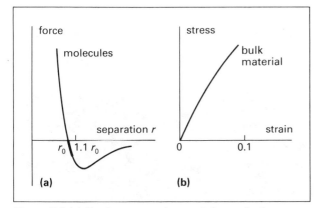

Figure 5.9 (a) The force-separation graph for a pair of molecules. In (b) the part of that graph from $r = r_0$ to $r = 1.1 r_0$ has been inverted and enlarged, and the axes re-labelled stress and strain. We can see that we can expect a stiffness which is initially constant, but which gradually decreases.

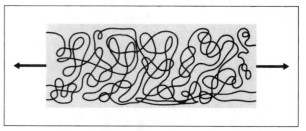

Figure 5.10 A possible arrangement for three long-chain rubber molecules. Very little force need be exerted on the ends of the specimen to produce a large extension.

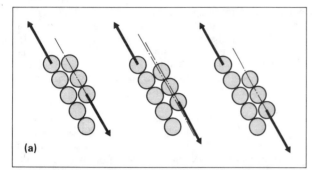

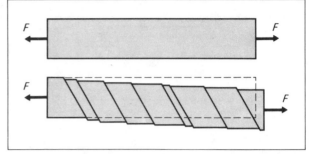

Figure 5.12 Under the action of two equal and opposite forces *F*, layers of molecules slip over each other, thus producing an elongation, and a series of parallel steps or ridges on the surface of the material.

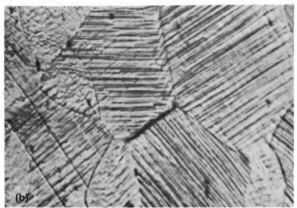

Figure 5.11 (a) The force needed to slide one layer of molecules over another may not be very large. (b) A photograph (magnified 50 times) of the surface of a piece of aluminium in which layers of molecules have slid over each other. Within each grain we can see the parallel rows of steps formed as a result.

5.3 Failure mechanisms

There is nothing in our model which explains why, with a relatively small stress, the glass rod breaks and the copper wire flows plastically: these specimens ought to behave like the iron whisker.

Plastic flow

Part of the explanation is that in a crystalline or poly-crystalline material it is easier to slide whole planes of molecules over each other than it is to pull molecules directly apart. Figure 5.11 shows what would happen if two layers of molecules slid over each other. The upper layer has slid over the lower layer and has settled down again in a similar position.

Crystalline materials do have layers of molecules which slide over each other in this way: figure 5.11b is a photo-graph of the surface of a piece of aluminium in which this has happened. It shows the rows of parallel steps formed within the grains of the aluminium after the layers have slipped over each other. Figure 5.12 explains why the slipping results in considerable elongation. The idea that layers of molecules slide over each other goes some way

towards explaining why crystalline materials are not as strong as we could expect them to be if they could be broken only by the molecules being pulled directly apart.

Dislocations

But in practice however we find that crystalline materials flow under stresses which are perhaps 10 or 100 times smaller than the stresses which we can calculate should be needed: something must be helping the layers of molecules to slide. In 1934 G.I. Taylor put forward the idea of *dislocations*. There are different types of dislocation: the simplest to describe and explain is shown in the bubble raft photograph of figure 5.13 where one row of molecules ends abruptly in the middle of one grain of a crystal. Where this happens it is possible for plastic deformation to occur relatively easily, since only one molecule need move at a time, and therefore plastic flow will occur *when the applied forces are relatively small*. Figure 5.14 explains how a dislocation moves when forces are applied. Figure 5.15 shows a photo-graph of a dislocation taken with an electron microscope. It was possible to do this only because of the large spacing of the molecules in this crystal. An analogy of the dislocation explanation may be seen in the pulling of a large carpet across a floor. It is hard to pull the whole carpet across at once. A technique which requires smaller forces is to make a ruck at one edge, and to move that ruck across the floor a little at a time.

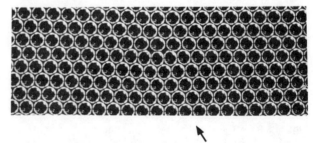

Figure 5.13 A photograph of a bubble raft model of a dislocation. You will find the dislocation most easily if you raise the page to eye level, and look along the rows of bubbles.

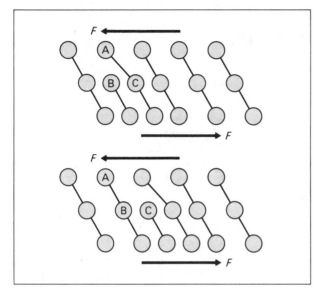

Figure 5.14 How a dislocation moves. Under the action of the pair of forces *F*, molecule A is pushed to the left, and a bond is formed between it and B instead of between it and C. The dislocation has now moved one molecular diameter to the right, and this will continue to happen until it reaches the edge of the grain, or another dislocation.

Dislocations in the arrangement of the molecules within the grains of a polycrystalline material are inevitable. What we described on page 43 is the ideal situation: in practice the arrangement, even within a grain, will not be perfect. Molecules will not all come together in the right place at the right time. Since the dislocations are inevitable, we make the best of the situation: it is found that if we increase the number of dislocations (which we can do by stressing the material) the materials becomes stronger. If there are more dislocations, the dislocations do not enable the layers of molecules to slip very far before their progress is impeded by other layers sliding in other directions within the same grain. We say that the dislocations become entangled. This is what happens when a material like copper is work-hardened. The number of dislocations is increased and it therefore becomes a stiffer and stronger material. Another way in which the movement of dislocations can be increased is the introduction of atoms of another element. These will generally be of a different size, and this interrup-

tion in the regularity of the layers prevents the dislocations from moving far. Work-hardening, and the deliberate introduction of foreign atoms, are two examples of ways in which we now *design* materials to do a particular job. Engineers used to have to make do with whatever materials occurred naturally, or whose composition and structure occurred by chance.

Fracture

The failure of the glass rod was quite different from the yielding of the copper wire. A.A. Griffith (1893–1963) suggested in 1934 that the cause of all brittle failures is a *crack* within, or on the surface of a material. The crack may be very narrow: indeed it is more dangerous if it is. The theory of this is too advanced for this book, but anyone who has watched a glass-cutter at work will know that he runs his cutter very lightly over the glass to score it, and the glass then breaks easily when tapped or flexed. A deeper, blunter, scratch makes the glass harder to break. The danger of a crack can be seen from figure 5.16. The shaded area has a greater stress in it than the area to the left of it because there is more force exerted on the molecules at the tip of the crack. For this reason a crack is sometimes called a *stress-multiplier*. The molecular bonds at the tip of the crack are more likely to break than any others, and thus the crack spreads through the material, the local stress becoming even greater as the tip of the crack moves on. Newly-drawn fibres of brittle materials are strong because cracks are initially absent, but handling them, and chemical action, will cause cracks to form before long and even these fibres will become susceptible to brittle fracture.

It is clearly not possible for glassy, non-crystalline, materials to yield, since this flow mechanism occurs only when there is a long-range order of planes of molecules which can slip over each other (and more easily still when

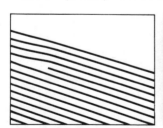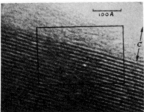

Figure 5.15 An electron microscope photograph of a dislocation in a crystal of platinum phthalocyanine.

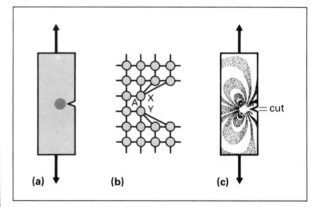

Figure 5.16 In (a) the stress is greatest in the shaded area. (b) shows why the force in bond A will be large: the force in this bond has to balance the forces exerted on molecules X and Y from above and below. In (c) photoealstic analysis shows, by the closeness of the spacing of the interference fringes, the high stress at the tip of a cut in the material.

there are dislocations). But why do crystalline materials seem not to be susceptible to failure through cracks? It might be thought that if a crack existed on the surface of a copper wire it, like glass, would also break in a brittle manner. When a crack does occur in a ductile material, the material yields at the tip of the crack, and thus the stress is relieved by being shared among many more molecular bonds, and the crack will probably not extend further. This argument is summed up in table 5.3.

There is no rigid dividing line between materials which are ductile and those which are brittle. Crystalline materials possess different amounts of ductility, and any of them may behave in a brittle manner under different conditions (e.g. at a lower temperature).

	glasses	crystalline materials
brittle fracture?	yes	no, because flow at the tip of the crack relieves the stress concentration
ductile yielding?	no, because there are no planes of molecules which could slip over each other	yes

Table 5.3 A summary of why different failure mechanisms predominate in different types of material.

6 Fluid behaviour

6.1 Pressure

A block resting on a horizontal surface (figure 6.1) exerts a vertical downward force on the surface, and the surface exerts a vertical upward force on the block. These forces (which are equal as stated in Newton's third law) are called the *perpendicular contact forces*. If the block is placed on another of its faces, of different area, these forces will stay the same size as before, but the area of contact will be different and the force on each unit area of surface will change. If the area of contact is smaller, there will be a greater risk of damage to the surface (this is why we use our heels to make a dent in firm ground, and why pins have sharp points). We define the *pressure p* by the equation

$$p = \frac{\text{perpendicular contact force}}{\text{area of contact}}$$

For example, consider a man of mass 80 kg standing on the ground. Suppose the area of contact of his feet with the ground is 160 cm² ($= 160 \times 10^{-4}$ m²). Since the man is in equilibrium, the perpendicular contact force is the same size as his weight, so the pressure p is given by

$$p = \frac{(80 \text{ kg}) (9.8 \text{ N kg}^{-1})}{160 \times 10^{-4} \text{ m}^2}$$

$$= 4.9 \times 10^4 \text{ N m}^{-2}$$

Note that the unit of pressure is the N m^{-2}. It is the same as the unit of stress, since that also is a force per unit area (page 50). Again we call this unit the *pascal* (Pa).

Pressures are exerted not only by solids but also by liquids and gases. We shall be considering the cause of gaseous pressure in detail later. Here it will be enough to say that it is caused by the bombardment of a surface by the molecules of the gas. A model can be used to demonstrate this. It consists of a vertical transparent cylinder (figure 6.2) which contains a cardboard disc whose mass is about one

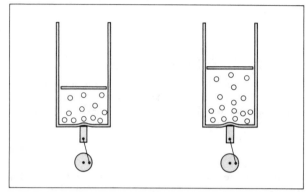

Figure 6.2 A model of the way in which gas molecules behave. In (b) the ball-bearings are moving faster, and are producing the same pressure in a larger volume.

gram. A motor drives a crank which pushes a piston against the flexible bottom of the cylinder. This agitates the ball bearings which are in the cylinder, and they move fast. They hit the underside of the disc. Each time a ball bearing hits this disc it exerts a force on it (the ball has its momentum changed, so there must be a force on the ball and a force on the disc). The size of the force depends on the speed of the balls (the balls have different, unpredictable speeds) and the impacts do not occur at regular intervals, so the upward force on the disc is not constant, but even with only 50 balls in the cylinder the impacts keep the disc steady to within a few millimetres of its average position. If there were more, smaller, balls the force would be even more constant.

Figure 6.3 shows the graph of how the force exerted on the disc might vary with time for the two situations shown in figure 6.2. In (b) the impulses are larger but occur less

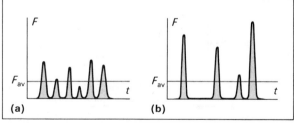

Figure 6.3 Graphs showing how the forces exerted by the ball-bearings in figure 6. might vary with time. The area under each spike measures the impulse of the force; the total area in each graph is the same, and is equal to the area beneath the grey line.

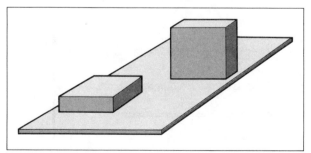

Figure 6.1 The same block exerts the same force on the table, but different pressures.

often: the average force exerted by the balls is the same in (a) and (b), as it must be to balance the weight of the disc.

This is a model of the way in which gas molecules exert a force, and the way in which the gas, as a whole, exerts a pressure. In the volume of the model there might be 10^{23} gas molecules, so we can understand why gases exert pressures in which we can detect no fluctuation even though the pressure is the result of impacts of varying size which occur randomly.

Atmospheric pressure

We live at the bottom of a sea of air: the molecules are kept near the surface of the Earth by the gravitational forces pulling on them (very few are ever moving fast enough to escape). The pressure decreases with height above the surface because at greater heights the weight of the air above is less. There is less to support: figure 6.4 illustrates this.

The pressure does not decrease *linearly* with height for two reasons. Air is compressible, and so the density decreases with height, and, secondly, the temperature varies. The variation of temperature (in no regular way) makes it impossible to state laws about the variation of pressure with height. In any case, for *small* differences in height the change of pressure is small compared with atmosphere pressure, e.g. in moving from the floor to the ceiling of an average room the decrease in pressure would be about 40 Pa, which is about 0.04% of atmospheric pressure.

A *barometer* is used to measure atmospheric pressure. An aneroid barometer consists of a sealed cylinder containing gas at low pressure. It is constructed in such a way that its length changes when the atmospheric pressure changes, and a system of levers and cogwheels makes a pointer move over a scale graduated in units of pressure. This barometer is convenient to use, but initially it has to be calibrated. A *mercury* barometer is less convenient but does not need to be calibrated. We shall discuss this in more detail later.

Atmospheric pressure varies not only with height above the surface of the Earth but also with time at a particular point. These changes are linked with meteorological changes, and are the basis of modern weather forecasting. *Standard atmospheric pressure* is defined to be $1.013\,25 \times 10^5$ Pa, and from day to day the variation is very rarely more than $\pm 5\%$ of this value.

Pressure at a point in a fluid

Consider a small surface of area A in a fluid. The pressure in the fluid at that point will exert a force F on the surface, and the pressure p at a point in a fluid is defined by the equation

$$p = \frac{F}{A}$$

The area A does not need to be horizontal. It could be vertical, or inclined at any angle to the horizontal. In all cases the force F is the same size. We say that pressure has no direction. It exerts a force at right angles to *any* surface on which it acts.

Pressure change with depth

How does the pressure change as we move deeper in a liquid? In figure 6.5 we outline a cylinder of cross-sectional area A and height Δh in liquid of density ρ. The weight W of this cylinder is given by

$$W = (\text{volume})(\text{density})(g\text{-field strength})$$

That is $W = \Delta h A \rho g$

Suppose the pressures at the top and bottom of the cylinder are p_1 and p_2 respectively. The cylinder will be in equilibrium. Considering the free-body diagram for the cylinder we have

$$p_1 A + W = p_2 A$$
$$(p_2 - p_1) A = \Delta h A \rho g$$
$$\Delta p = \rho g \Delta h$$

where Δp is the difference in pressure between the two

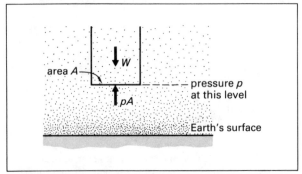

Figure 6.4 The pressure p in the atmosphere decreases with height, because at greater heights there is a smaller weight W to support.

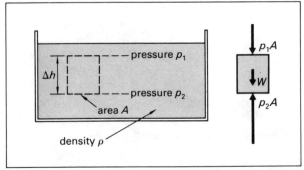

Figure 6.5 The pressure difference Δp is equal to $\rho g \Delta h$.

levels. So the increase of pressure = ρg × the increase in depth. The area of the cylinder, or of the vessel, is irrelevant: this makes the result very useful. In using it we assume that the density of the liquid is uniform (i.e. that its temperature is uniform, and the pressure does not become so great that the liquid becomes compressed).

We have already noted that with gases, when Δh is small, Δp is negligible. When Δh is large, their density is not usually constant, so that this result cannot be used with gases.

▶ A vessel has the air removed from it and is sealed by a valve which can withstand a pressure of 650 kPa. How deep in sea water of density 1.02×10^3 kg m^{-3} can it be lowered if no water is to enter?

Suppose the pressure at the surface is standard atmospheric pressure = 101 kPa. The vessel can be lowered to a depth Δh which increases the pressure by 549 kPa. Using $\Delta p = \rho g \Delta h$, we have

$$5.49 \times 10^5 \text{ Pa} = (1.02 \times 10^3 \text{ kg m}^{-3})(9.8 \text{ N kg}^{-1})\Delta h$$
$$\Rightarrow \qquad \Delta h = 55 \text{ m}$$

A descent of 55 m means an increase in pressure of about $5\frac{1}{2}$ 'atmospheres'. It is useful to remember that a descent of about 10 m in water produces an increase in pressure equal to atmospheric pressure. ◀

The pressure at all points at the same horizontal level in a liquid or gas (in equilibrium) is the same. Consider figure 6.6 which shows a horizontal tube connecting two points X and Y: if the pressure at X were greater than the pressure at Y there would be a resultant force in the direction XY on the liquid in the tube, and this is clearly impossible if the liquid is in equilibrium. The fact that the pressures at X and Y are equal means that they must be equally far below the surface, which means that the surface of a liquid *in equilibrium* must be horizontal.

We can use the result $\Delta p = \rho g \Delta h$ to calculate Δp between two points in the same liquid which are not in the same vertical line.

▶ One arm of a U-tube containing oil of density 800 kg m^{-3} is connected to a gas supply, and the other is left open to the atmosphere. The difference in level of the oil in the two sides of the tube is 0.20 m. By how much does the pressure of the gas exceed atmospheric pressure?

Consider figure 6.7a. The pressure at X is atmospheric. Y is a vertical distance of 0.20 m below X and so the additional pressure there is given by

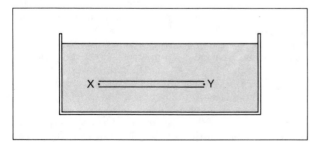

Figure 6.6 The pressures at X and Y must be the same, or else there would be liquid accelerating along the tube.

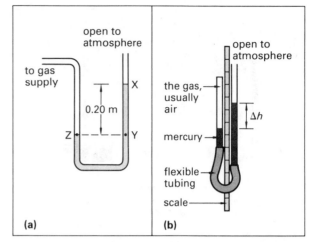

Figure 6.7 Two manometers: (a) is being used to measure the pressure of a gas supply, and (b) is being used in an experiment to verify Boyle's law.

$$\Delta p = (800 \text{ kg m}^{-3})(9.8 \text{ N kg}^{-1})(0.20 \text{ m})$$
$$= 1.6 \text{ kPa}$$

(i.e. about 1.6% of atmospheric pressure). The pressure at Z must be equal to the pressure at Y, since Z and Y are at the same horizontal level, and it is possible to pass through the liquid from Y to Z. The pressure at Z must be equal to the pressure in the gas (assumed to be everywhere in the tube the same, since the gas density is relatively small, and the vertical distances are small).

Note that (i) the cross-sectional areas of the tubes are not important. There would be the same difference in level even if the arms had had different cross-sectional areas, and (ii) a reason for using oil rather than water (density 1000 kg m^{-3}) might have been that its smaller density gives a larger difference in levels, which makes the device more sensitive. The difference in levels, using water, would have been 0.16 m. ◀

Manometers and barometers

The device described in the last example is called a *manometer* and conveniently measures the *difference* in pressure between two regions. For example, one simple piece of apparatus for verifying Boyle's law (see page 125) is shown in figure 6.7b. The volume of the gas is measured directly in its graduated tube, and its pressure is calculated by adding Δp ($= \rho g \Delta h$, where ρ is the density of mercury, and Δh the difference in levels) to the measured value of atmospheric pressure.

When we want to use this kind of device to measure atmospheric pressure we need to have zero pressure (and therefore no gas) in one of the arms of a manometer. This kind of *barometer* always uses mercury as the liquid because its high density makes the differences in level as

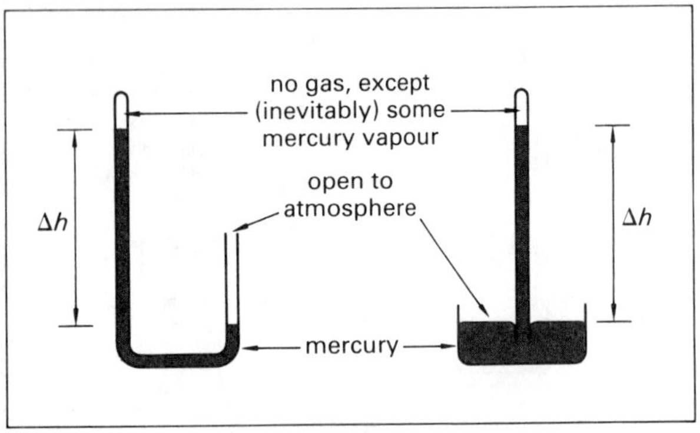

Figure 6.8 Two forms of a simple mercury barometer. The atmospheric pressure $= \rho g \Delta h$, where ρ is the density of the mercury.

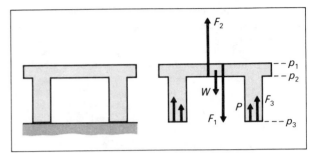

Figure 6.9 A situation diagram and a free-body diagram for a table resting on horizontal ground. F_1, F_2 and F_3 (the vertical pushes of the air on the table) are much larger than P or W.

small as they can be. Even so the difference in level (typically about 0.75 m) is large enough to make a mercury barometer cumbersome, which is why the aneroid barometer is often used. Two versions of a mercury barometer are shown in figure 6.8. You should realise that they are theoretically identical. The second one is more convenient, as it needs just a straight tube and a bowl to hold the mercury. Inevitably the space above the mercury in the closed end is not a true vacuum, because the mercury evaporates. However, at room temperature the pressure exerted by this vapour is about 0.2 Pa, so that the error is very small (it is equivalent to about 1/1000 mm of mercury!). The space is described as a *Torricellian vacuum* after Torricelli, an Italian who invented the barometer in 1643. The *Fortin* barometer is a mercury barometer with additional refinements: a vernier device to make the barometer more sensitive, and a means of ensuring that the level of the mercury in the bowl is constant.

Safety precautions

A disadvantage of using mercury is that it is a cumulative poison, which can be absorbed by the human body both by inhaling the vapour, and also if the liquid comes into contact with the skin. It should therefore *never* be touched or handled. A tray should always be used with mercury apparatus, so that any liquid spilt can be contained in a small area, and afterwards collected. Apparatus containing the liquid should be fitted with rubber bungs, and kept in a separate well-ventilated cupboard.

Upthrusts

We have already noted that pressure has no direction: it exerts its force at right angles to *any* surface. What effect does this have on an ordinary object like a table, which is surrounded by air at atmospheric pressure? Figure 6.9 shows a table resting on horizontal ground, and a free-body diagram for the table (although the horizontal pressure forces have been omitted, since these balance).

There are the two forces which we would normally consider: the pull W of the Earth, and the push P of the ground. There are also three pressure forces: F_1 on the top of the table (as a result of the pressure p_1 at that level), F_2 on the underside of the table, and F_3 on the underside of the legs (this force *does* exist because there is certainly air between the legs and the floor). Are these three forces of negligible size? Is that why we usually ignore them? No! If the table has an area of about 2 m², W and P will each be about 500 N, whereas F_1, for example, will be about 200 000 N. F_1, F_2 and F_3 are usually ignored because F_1 is very nearly equal to $F_2 + F_3$. $F_2 + F_3$ is *slightly* greater than F_1 because F_2 and F_3 are caused by the pressures p_2 and p_3, which are slightly greater than p_1 (and which act on the same total area as p_1 does). The difference between $F_2 + F_3$ and F_1 would (in this case) be less than 1 N, which can certainly be ignored compared with W and P.

But the air *does* produce a resultant upward push on the table, however small, and there must always be such an *upthrust* when any body is immersed in a gas or a liquid. We can see at once that the upthrust is likely to be far larger in a liquid, since the pressure differences are larger. We shall now show that we can deduce the size of the upthrust *in any situation* very simply.

Archimedes's principle

Let us imagine some empty space in a fluid, into which we could put different bodies (of the right size and shape to fill the space), as shown in figure 6.10. For simplicity, suppose the empty space has a cylindrical shape, and let the pressures at the top and bottom be p_1 and p_2. The forces which will act on *whatever* is put into the space will be $p_1 A$ (downwards) and $p_2 A$ (upwards), as shown in the free-body diagram. The fluid therefore produces a resultant push of $p_2 A - p_1 A$ upwards: this is the force on anything put into this space (of the right size and shape). Normally there is simply more of the same fluid in the space, and it is in equilibrium, so that the upward push $p_2 A - p_1 A$ must equal the weight of the fluid which is normally there. This proof can easily be extended to bodies of any size and shape, and so we have Archimedes's principle:

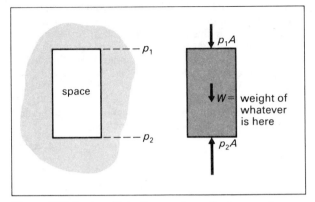

Figure 6.10 The free-body diagram shows the forces acting on *anything* put into the space.

The upthrust on a body immersed in a fluid is equal and opposite to the weight of the fluid displaced by the body.

It is rare for the upthrust to keep a totally immersed body in equilibrium: usually the weight of the body is greater than the upthrust, so that if these are the only two forces acting on the body it accelerates downwards (e.g. a stone in air or water). If the upthrust is greater than the weight, the body accelerates upwards until (i) if it is in a liquid it reaches the surface, where it floats in equilibrium with enough of itself immersed for the upthrust to be equal to its weight, (ii) if it is in the atmosphere it reaches a level where the density of the surrounding air has fallen enough for the upthrust to equal its weight, (iii) some other force (e.g. the downward push of the top of a closed vessel) is exerted. Figure 6.11 illustrates these three situations.

▶ A *hydrometer* is an instrument for measuring the density of liquids. A simple form of it is shown in figure 6.12. The stem has a constant cross-sectional area, and it is loaded (e.g. with lead) at its bottom so that it floats upright. Suppose that the hydrometer has 120 mm of stem immersed when it is floating in a liquid of density 920 kg m^{-3} and that in another liquid it floats with 11 mm less of its stem immersed. What is the density of the other liquid?

The hydrometer has a constant weight: when it floats with less of its stem immersed it must be floating in a denser liquid so that the upthrust will still be equal to its weight. The original upthrust (= weight of liquid displaced, by Archimedes's Principle)

= volume immersed × density × g-field strength
= $(0.120 \text{ m})A(920 \text{ kg m}^{-3})g$

In the new liquid the upthrust

= $(0.109 \text{ m})A\rho g$

where ρ is the density of the new liquid. These two upthrusts must be equal (because they are equal to the weight of the hydrometer) so

$(0.120 \text{ m})(920 \text{ kg m}^{-3}) = (0.109 \text{ m})\rho$
$\Rightarrow \qquad\qquad \rho = 1.01 \times 10^3 \text{ kg m}^{-3}$ ◀

Markings are made on the stem of a hydrometer (as

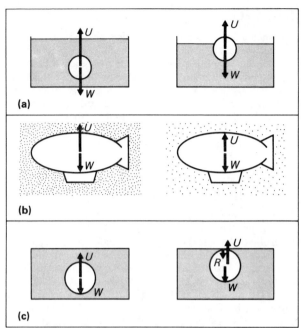

Figure 6.11 Each left-hand diagram shows the initial situation, where the upthrust U of the fluid is greater than the pull W of the Earth on the object. In (a) the object finally floats, with less weight of liquid displaced, so U is less. In (b) the airship rises until the density of the surrounding air is small enough for the weight of the air displaced to equal W. In (c) the object accelerates upwards until it can be pushed down by the top of the vessel: then $U = R + W$.

shown in figure 6.12c) so that the user can measure the density of a liquid simply be observing which mark is level with the surface of the liquid in which it is floating. A hydrometer can be made more sensitive by using a thinner stem. Then two particular markings will be further apart. To cover a large range of densities a sensitive hydrometer would have to be very long, so in practice different hydrometers are used to cover small ranges of density.

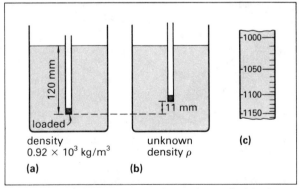

Figure 6.12 A simple hydrometer floating in two liquids of different density; (c) shows typical markings on a hydrometer. The unit would be kg m^{-3}. The markings are not evenly spaced.

6.2 Flow

We shall now consider fluids which are moving. In this section we shall assume that the flow is *streamline* or *laminar*, i.e. that layers of fluid slide over each other, with the velocity constant at a particular point in the fluid. We shall also assume first that the fluids are not *viscous*, i.e. that when the layers of fluid slide over each other they do not impede each other. Of course all fluids are viscous: this is why the coffee in a cup does not spin round endlessly when we stop stirring.

Bernoulli's equation

Figure 6.13 shows a pipe which becomes narrower: all the pipe is on the same horizontal level. The area changes from A_1 to A_2. The speed v_2 of the fluid at the far end must therefore be greater than the speed v_1 of the fluid at the near end, in order that the rate of flow of mass should be the same at both ends of the pipe. So the fluid is accelerating. Since the pipe is all at the same horizontal level the only force which is available to accelerate the fluid is that produced by the pressure in the fluid. We can therefore deduce that the pressure at the near end of the pipe is greater than the pressure at the far end.

Suppose the densities of the fluid at the near and far ends of the pipe are ρ_1 and ρ_2 respectively. In practice we shall assume that $\rho_1 = \rho_2$, even for gases: this will be a good approximation unless the pressures p_1 and p_2 are very different.

In a time Δt the fluid moves a distance $v_1 \Delta t$ at the near end of the pipe, and a distance $v_2 \Delta t$ at the far end. The masses of fluid which move into and out of the pipe are equal, as we have noted above, and are given by $\rho A_1 v_1 \Delta t$ and $\rho A_2 v_2 \Delta t$.

So $A_1 v_1 = A_2 v_2$

At the near end of the pipe the fluid outside the pipe is pushing the fluid into the pipe with a force $F_1 = p_1 A_1$, and the work done by this force in the time Δt is (force × displacement) $p_1 A_1 v_1 \Delta t$. At the other end of the pipe the fluid outside the pipe exerts a similar force, but in the opposite direction to the flow, so the work done by this force is negative: it is $-p_2 A_2 v_2 \Delta t$. The total work done by these forces is equal to the change in kinetic energy of the mass of fluid which passes through the pipe in the time Δt, so using the work-energy theorem we have

$$p_1 A_1 v_1 \Delta t - p_2 A_2 v_2 \Delta t = \tfrac{1}{2}(\rho A_2 v_2 \Delta t)v_2^2 - \tfrac{1}{2}(\rho A_1 v_1 \Delta t)v_1^2$$

But $$A_1 v_1 = A_2 v_2$$

so we can write $p_1 - p_2 = \tfrac{1}{2}\rho v_2^2 - \tfrac{1}{2}\rho v_1^2$

or $$p + \tfrac{1}{2}\rho v^2 = \text{constant}$$

This analysis can be extended to take account of situations where there is a difference in height between the ends of the pipe: then the equation becomes

$$p + \tfrac{1}{2}\rho v^2 + \rho g h = \text{constant}$$

where h is the height of the pipe above some fixed level, and this equation is known as *Bernoulli's equation*, after Daniel Bernoulli (1700–1782) who first derived it.

Practical applications

The fact that the pressure in fast-moving fluid is lower than the pressure in the same fluid moving slowly, and the fact that the pressure in a fluid moving at any speed is less than the pressure in the same fluid when it is stationary is of enormous importance. Figure 6.14 shows seven tubes connected to the upper and lower surfaces of an *aerofoil*. It is designed so that air flows faster over the upper surface than over the lower surface, so, as we should expect, the four tubes (A, B, C, D) connected to the upper surface show

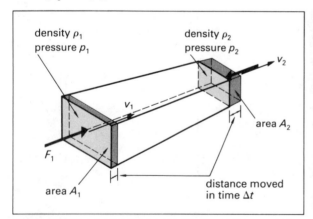

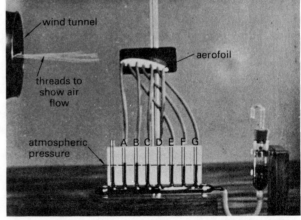

Figure 6.13 The shaded mass enters the pipe at the near end in a time Δt: in the same time the same mass (also shaded) leaves the pipe at the far end. For this to be possible the fluid must move faster at the far end.

Figure 6.14 A wind tunnel directs air over the upper and lower surfaces of an aerofoil. Tubes A, B, C, D are connected to the upper surface; tubes E, F, G to the lower surface. The main effect is a reduction in pressure at the upper surface.

that the pressure there is less than atmospheric. The three (E,F,G) connected to the lower surface show that the pressure there is either atmospheric, or slightly more than atmospheric: the net effect is an upward force on the aerofoil. Aerofoils are used for the wings of aircraft: then the aircraft moves through stationary air, but it is the *relative* motion which is important, and the pressure difference produces the upward force which balances the weight of the aircraft.

Other practical applications

(i) The *carburettor* and bunsen burner. The air in a carburettor is made to move through a constriction so that it moves faster and the atmospheric pressure can push petrol vapour into the faster-moving, lower-pressure air. In the *bunsen burner* air is similarly pushed into a faster-moving gas stream.

(ii) The *filter pump*. This is similar to the carburettor, except that the moving fluid is water, and it is the gas from some apparatus which is pushed into the moving water stream.

(iii) *Spinning tennis balls* move in a curved path through air. Their rough surface drags round the air near it, and produces a difference in speed of the air on opposite sides of the ball. *Golf balls* rise up in the air far more than they would if the club did not make them spin. The dimples on the ball drag the air, and this makes the ball swerve upwards.

(iv) The *Venturi meter*, the working of which is explained in the next example.

◄ Oil of density 850 kg m⁻³ is made to flow along a horizontal pipe whose cross-sectional area changes from 50 mm² to 25 mm². Manometers (using the same oil) are attached to the two sections of the pipe, and the difference in levels of the oil in them is 40 mm. Find the rate of flow of mass of the oil in the pipe.

Suppose the speeds of the oil are v_1 and v_2, respectively, in the parts of the pipe which have cross-sectional areas of 50 mm² (A_1) and 25 mm² (A_2). Since $A_1 = 2A_2$, $v_2 = 2v_1$. Since the pipe is horizontal we have

$$p_1 + \tfrac{1}{2}\rho v_1{}^2 = p_2 + \tfrac{1}{2}\rho v_2{}^2$$

or $\quad p_1 - p_2 = \tfrac{1}{2}\rho(v_2{}^2 - v_1{}^2)$

From the manometer levels we have

$$p_1 - p_2 = \rho g \Delta h$$
$$= (850 \text{ kg m}^{-3})(9.8 \text{ N kg}^{-1})(40 \times 10^{-3} \text{ m})$$

so that

$$850 \times 9.8 \times 40 \times 10^{-3} \text{ Pa} = \tfrac{1}{2}(850 \text{ kg m}^{-3})(4v_1{}^2 - v_1{}^2)$$

or $\quad v_1{}^2 = \dfrac{2 \times 9.8 \times 40 \times 10^{-3}}{3} \text{ m}^2 \text{ s}^{-2}$

$$v_1 = 0.51 \text{ m s}^{-1}$$

The rate of flow of mass is given by

$$\rho A_1 v_1 = (850 \text{ kg m}^{-3})(50 \times 10^{-6} \text{ m}^2)(0.51 \text{ m s}^{-1})$$
$$= 2.2 \times 10^{-2} \text{ kg s}^{-1}$$

This *Venturi meter* is widely used for measuring the rate of flow of volume or mass in pipes. ◄

6.3 Viscosity

We now consider the fact that fluids are viscous, and to fix our minds on something concrete let us think about water flowing steadily in a rectangular channel. The water will not flow at all if there is not a difference of pressure between the ends of the channel, so in figure 6.15 the surface of the water is shown to be sloping (though the diagram exaggerates the slope which would be necessary to maintain a reasonable flow of a liquid like water which is not very viscous). The fact that the pressure difference is not accelerating the water, but maintaining its average velocity constant, tells us that there must be resisting forces. These resisting forces arise because the upper layers of water are moving faster than the lower layers: in fact the layer next to the bottom of the channel is at rest. The enlargement in figure 6.15 shows the forces on two layers of water. There we can see that the resultant of the forces on one layer caused by adjacent layers is in the opposite direction to the flow: if it were not for the pressure difference between the two ends of the channel, the layers would slow down and stop.

They would stop because these *viscous forces* which the layers of water exert on each other do negative work and convert the kinetic energy of the water into internal energy. Even when the flow is steady these forces are converting mechanical energy into internal energy: in figure 6.15, where there is no gain of kinetic energy by the water between the time it enters and the time it leaves the channel, the gravitational potential energy which the water initially had has been converted into internal energy.

Viscous forces and frictional forces. We can think of viscous forces as being due to 'frictional' forces between neighbouring layers of fluid, but there is an important

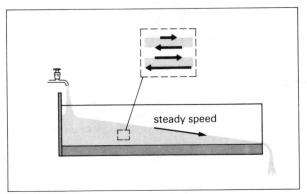

Figure 6.15 Water flowing in a rectangular channel at a steady speed. The difference in pressure between the ends of the channel provides the force which balances the viscous forces and enables the water to have a constant velocity. The enlargement shows the forces on two adjacent layers of water.

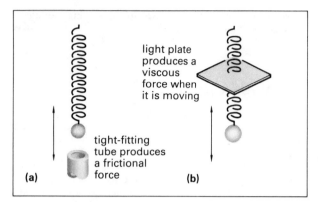

Figure 6.16 Two methods of reducing the number of oscillations of a bob on a spring. Using a frictional force, in (a), there is a risk that the bob may come to rest in the wrong position.

fluid	$\eta/\text{N s m}^{-2}$ 293 K	373 K
hydrogen	8.8×10^{-6}	10.4×10^{-6}
air	1.8×10^{-5}	2.2×10^{-5}
mercury	1.6×10^{-3}	1.2×10^{-3}
ethanol	1.2×10^{-3}	
water	1.0×10^{-3}	0.28×10^{-3}
olive oil	8.4×10^{-2}	
glycerol	1.5	

Table 6.1 Values of viscosity η at two different temperatures for some representative fluids.

difference between viscous and frictional forces. Viscous forces cannot exist when there is no flow; frictional forces can exist when there is relative motion between the two surfaces *and* when there is not. Figure 6.16 shows two methods of reducing the number of oscillations made by a body before it comes to rest in its equilibrium position.

In (a) the bob on the end of the spring moves in a tight-fitting tube: the *frictional* force will certainly slow down the motion, but we cannot be sure that there is not a frictional force still acting when the bob has finally stopped, so that the bob may not come to rest in its correct position. In (b) a light horizontal plate has been fitted to the spring, so that it has to push through the air as the bob oscillates. This will produce a *viscous* force, which will not exist when the bob has stopped. There is therefore no risk of the bob stopping in the wrong position. This use of viscous forces is called *damping*.

Values of viscosity

In this book we need not concern ourselves with the *definition* of viscosity, as it involves some difficult concepts. We shall, however, be using several relationships in which the *dynamic viscosity* η appears. In these relationships you should notice the effect which the value of η has on, e.g. the viscous force, or the rate of flow of a fluid. The value of η, for a particular fluid, varies with temperature, and table 6.1 gives values for some representative liquids and gases. Notice that the viscosity of the liquids decreases with temperature, but the viscosity of the gases increases with temperature. This is generally true.

Flow in pipes

The rectangular channel in figure 6.15 was a conveniently simple situation, but most fluids flow in pipes. It is sometimes important that the flow should be *streamline* and not *turbulent*. If the flow is turbulent there is a greater rate

of conversion of mechanical energy to internal energy. The condition for flow to be streamline in a pipe of circular cross-section is that the average velocity of the liquid defined by

$$v_{\text{av}} = \frac{\text{rate of flow of volume}}{\text{cross-sectional area}}$$

must not exceed a certain *critical velocity* v_c which is given (for tubes of circular cross-section) by

$$v_c = (Re)\frac{\eta}{\rho a}$$

where (Re) is a constant known as *Reynolds's number*. The expression is usually written as

$$(Re) = \frac{\rho a v_c}{\eta}$$

where (Re) must not be more than about 1000 for tubes of circular cross-section. Greater values of (Re) would make the flow turbulent. (Similar expressions, with different critical values of (Re), exist for other situations.) Reynolds's number will be low if with a particular fluid we use a tube of small diameter and a small difference of pressure across its ends (since this gives a small value of v).

► What is the maximum rate of flow of volume of water in a circular tube of radius 5.0 mm, if the flow is to be streamline, and the temperature is such that the density and viscosity of water are 1000 kg m^{-3} and 1.5×10^{-3} N s m^{-2} respectively?

We have $(Re) = \dfrac{\rho a v_c}{\eta}$

If (Re) must not exceed 1000,

$$\frac{(1.0 \times 10^3 \text{ kg m}^{-3})(5.0 \times 10^{-3} \text{ m} \times v_c)}{1.5 \times 10^{-3} \text{ N s m}^{-2}} < 1000$$

or $v_c < 0.30$ m s^{-1}

The rate of flow of volume

$$= \pi(5.0 \times 10^{-3} \text{ m})^2 \times (0.30 \text{ m s}^{-1})$$
$$= 2.4 \times 10^{-5} \text{ m}^3 \text{ s}^{-1}$$ ◄

Rate of flow of volume

Having ensured that the flow is streamline it is next im-

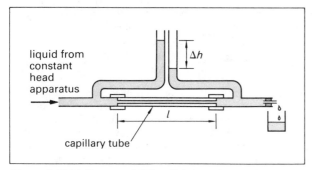

Figure 6.17 Making use of Poiseuille's equation to measure the viscosity of a liquid of low viscosity.

portant to know what the rate of flow of volume V/t will be. *Poiseuille's equation* states that for a circular tube of internal radius a and length l, with a pressure difference of Δp between its ends, the rate of flow of volume V/t is given by

$$\frac{V}{t} = \frac{\pi a^4 \Delta p}{8\eta l}$$

This equation enables us to measure viscosities of the order of 10^{-3} N s m^{-2} if we set up a suitable practical situation, as shown in figure 6.17. The constant head device ensures that the liquid flows into the tube at a constant rate, and provides a pressure difference Δp across the length l of the tube, which is measured with a manometer: $\Delta p = \rho g \Delta h$, where ρ is the density of the liquid and Δh the difference in level in the two limbs of the manometer. We can easily measure V/t by measuring the volume of liquid collected in a known time. The radius a of the tube must be measured very carefully, since it is raised to the fourth power, and this is not easy, as the tube must be very narrow (a *capillary* tube) if the rate of flow is to be small enough for it to be streamline. The best that can be done is to measure the average value of a^2, by finding the mass of a mercury thread (see page 73) in a capillary tube. If the thread has a length L, the mass m of the thread is given by $m = \pi a^2 L \rho_{Hg}$, where ρ_{Hg} is the density of mercury. m and L are measured, and ρ_{Hg} is known. Hence a^2 can be calculated. This is the most precise way of measuring the average radius of a capillary tube. Then all the quantities in Poiseuille's equation are known, except η, and this can be calculated.

Viscous force on falling spheres

When a sphere falls *slowly* through a fluid it pulls cylindrical layers of the fluid with it. These exert an equal and opposite force on the sphere; *Stokes's equation* states that the resisting force F is given by

$$F = 6\pi a \eta v$$

where a is the radius of the sphere, η the viscosity of the fluid and v the speed of the sphere. We emphasise that the

speed of the sphere must be low (see the next example): also, if the fluid is contained in a tube, the tube must be of large radius (20 times greater than the radius of the sphere if the error is not to exceed 10 %), and the flow of fluid past the sphere must not be turbulent (the expression for Reynolds's number in this situation is the same as before, i.e. $\rho a v_c / \eta$, but now its value must be less than 1). If a sphere is released from rest in a fluid it will initially accelerate, but the viscous force will increase as the speed increases, until the forces on the sphere are balanced and the sphere reaches its (constant) *terminal speed.*

A small ball-bearing can be dropped in a tall wide cylinder of the liquid. The forces acting on the ball-bearing are:
(i) upwards: the Archimedean upthrust U of the liquid
(ii) upwards: the viscous force F of the liquid
(iii) downwards: the pull W of the Earth
When the ball-bearing reaches its terminal speed (which will happen within a few mm) it is in equilibrium, and so

$$F + U = W$$

i.e. $6\pi a \eta v + \frac{4}{3}\pi a^3 \rho_2 g = \frac{4}{3}\pi a^3 \rho_1 g$

where ρ_1 and ρ_2 are the densities of steel and the liquid respectively, so

$$6\pi a \eta v = \frac{4}{3}\pi a^3 (\rho_1 - \rho_2) g$$

This equation can be used to calculate η for viscosities ≈ 1 N s m^{-2}. The apparatus is shown in figure 6.18.

Horizontal lines are ruled at equal vertical intervals, and the ball-bearing is timed over each of the two intervals. These times should be the same if it has reached its terminal speed. If the ball-bearing is *dropped* into the liquid it may collect an air bubble. It should be released from below the surface of the liquid to prevent this happening. The experiment will be repeated with ball-bearings of the same diameter. The diameter of the ball-bearings is guaranteed by the manufacturer, so it need not be measured, although obviously it must be known. Their total mass may be used to find the density of the steel. If the density of the liquid is then measured all the quantities except η in the equation will be known, and so η can be calculated.

► A steel ball-bearing of radius 2.0 mm is released from rest in a large tank of glycerol (whose density and viscosity at this temperature are 1.3×10^3 kg m^{-3} and 1.50 N s m^{-2}

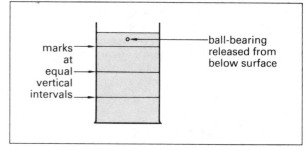

Figure 6.18 Making use of Stokes's equation to measure the viscosity of a liquid of high viscosity.

respectively). Find the terminal speed of the ball-bearing and check that the flow of glycerol past the ball-bearing is streamline. The density of steel is 7.8×10^3 kg m^{-3}.

We can apply the equation given in the previous paragraph. Dividing both sides by πa, and rearranging, we have

$$v = \frac{2a^2 g (\rho_1 - \rho_2)}{9\eta}$$

$$= \frac{2(2.0 \times 10^{-3} \text{ m})^2 (9.8 \text{ N kg}^{-1})(7.8 - 1.3) \times 10^3 \text{ kg m}^{-3}}{9 \times 1.5 \text{ N s m}^{-2}}$$

$$= 3.8 \times 10^{-2} \text{ m s}^{-1}$$

For this value of v Reynolds's number (Re) is given by

$$(Re) = \frac{\rho a v}{\eta}$$

$$= \frac{(1.3 \times 10^3 \text{ kg m}^{-3})(2.0 \times 10^{-3} \text{ m})(3.8 \times 10^{-2} \text{ m s}^{-1})}{1.5 \text{ N s m}^{-2}}$$

$$= 6.3 \times 10^{-2}$$

which is less than 1, so that the flow is streamline.   ◀

Another resisting force

A body falling through a fluid will also experience a resisting force for another reason. Relative to the body, the fluid is moving upwards and hitting its underside. There is therefore a force on the fluid, and, by Newton's third law, a force on the body. This is very similar to the jet of water striking a surface: this situation was discussed on page 21 and the equation derived there ($F = \rho A v^2$) can be used here, though we should not assume that all the fluid stops in its relative motion when it meets the body. So the force is $k\rho A v^2$ where k is a number less than 1.

In our example of the falling sphere we can perhaps assume that one-tenth of the glycerol stops in its relative motion when it meets the sphere. This resisting force F will then be given by

$$F = 0.1 \rho A v^2$$

$$= 0.1(1.3 \times 10^3 \text{ kg m}^{-3})\{\pi(2.0 \times 10^{-3} \text{ m})^2\} \times (3.8 \times 10^{-2} \text{ m s}^{-1})^2$$

$$= 2.4 \times 10^{-6} \text{ N}$$

Which is small compared with the weight of the ball-bearing ($= 2.6 \times 10^{-3}$ N). But even doubling the radius of the ball-bearing, which would mean that its terminal speed would be greater, would increase the force caused by the impact of the glycerol by so much that it could no longer be ignored.

6.4 Dimensions

All physical quantities have *dimensions* which can be expressed in terms of the dimensions of a few fundamental physical quantities (which include mass, length and time). For example, the dimensions of volume are the dimensions of length cubed.

We write this   [volume] = L^3
Similarly   [speed] = LT^{-1}

You will notice that to obtain the dimensions of a quantity you need only replace the unit (expressed in terms of the fundamental units kg, m, s) by the symbols M, L, T. For example, the unit of force is the newton. What are the equivalent fundamental units? We use the equation $F = ma$: the left-hand side has the newton as its unit, and the right-hand side has kg $\times$ m s^{-2} as its unit. So the newton is equivalent to kg m s^{-2}, and [force] = MLT^{-2}. Alternatively we could say that [mass] = M, [acceleration] = LT^{-2}, so that

$$[\text{force}] = [\text{mass}][\text{acceleration}]$$
$$= \text{MLT}^{-2}$$

Checking equations

This idea helps us to check the correctness of equations. If you think, perhaps, that the centripetal acceleration a is given by v^2/r (with the usual notation) you can write down the dimensions of each side of the equation.

$$[a] = \text{LT}^{-2} \quad \text{and} \quad [v^2/r] = (\text{LT}^{-1})^2 \div \text{L} = \text{LT}^{-2}$$

Clearly you *might* be right, but you will see that this method cannot tell you about any dimensionless numbers which might appear (e.g. $[v^2/2r]$ also has dimensions LT^{-2}): you still have to know some physics!

Predicting the form of equations

We can use this idea to predict the form of equations which would be too difficult to derive formally. This is of course very similar to the checking technique which we have described above. The only difference is that we are now not sure which quantities might be relevant. In particular there are two equations which have appeared in this section for which we were not able to give a proof: Stokes's equation and Poiseuille's equation. They *can* be deduced, but the mathematics involved is beyond the scope of this book. For both we shall need to know [viscosity]. Its unit has been given as

$$\text{N s m}^{-2} = (\text{kg m s}^{-2})(\text{s m}^{-2})$$
$$= \text{kg m}^{-1} \text{ s}^{-1}$$

so $[\eta]$ = ML^{-1} T^{-1}

Stokes's equation. We guess that the resisting force might be proportional in some simple way to the viscosity η of the fluid, the speed v of the sphere, and something to do with the size of the sphere: let us say the radius a, for simplicity. We write $F = k(\eta)^x (v)^y (a)^z$, where k is a dimensionless constant and x, y, z are the powers to which η, v, a might be raised in the equation. Equating the dimensions we have

$$MLT^{-2} = (ML^{-1}T^{-1})^x(LT^{-1})^y(L)^z$$
$$= M^xL^{-x}T^{-x} L^yT^{-y} L^z$$
$$= M^xL^{-x+y+z} T^{-x-y}$$

and we make use of the fact that the two sides of this equation must have the same dimensions. We equate the powers of the separate dimensions:

$$M: 1 = x, \quad L: 1 = -x+y+z, \quad T: -2 = -x-y$$

The solution of these equation is: $x = 1$, $y = 1$, $z = 1$, so that

$$F = k\eta va$$

where k is a number which we cannot find by this method. This equation agrees with our earlier statement $F = 6\pi a\eta v$. You should notice that this method cannot deal with situations where a quantity depends on more than three others, since in mechanics there are only three dimensions, and therefore we can produce only three equations. If we had thought, not unreasonably perhaps, that F depended also on the density ρ of the fluid we should have had three equations with four unknown quantities, which we could not have solved.

Poiseuille's equation. We assume that the rate of flow of volume V/t is in some way simply proportional to the pressure difference Δp, the viscosity η, the length l of the tube and the radius a of the tube. But we now have four quantities. However, we can guess that perhaps it is the *pressure gradient* $\Delta p/l$ which matters, and not the separate values of Δp and l, and we can therefore work with the three quantities $\Delta p/l$, η and a.

We write $\quad \dfrac{V}{t} = k\left(\dfrac{\Delta p}{l}\right)^x (\eta)^y (a)^z$

We have $[V/t] = L^3T^{-1}$ and $[\Delta p] = [F] \div [A] = MLT^{-2} \div L^2$. So we have $[\Delta p] = ML^{-1}T^{-2}$ and $[\Delta p/l] = ML^{-2}T^{-2}$. Equating the dimensions of the two sides of the equation

$$L^3T^{-1} = (ML^{-2}T^{-2})^x(ML^{-1}T^{-1})^y(L)^z$$

Equating the powers of the separate dimensions we have

$$M: 0 = x+y, \quad L: 3 = -2x-y+z, \quad T: -1 = -2x-y$$

which gives $x = 1$, $y = -1$, $z = 4$, so that

$$\frac{V}{t} = k\frac{\Delta p}{l} \times \frac{a^4}{\eta}$$

which agrees with our earlier statement $\dfrac{V}{t} = \dfrac{8\pi a^4 \Delta p}{\eta l}$.

6.5 Surface energy

Imagine a volume of liquid (as shown in figure 6.18) which is split into two parts. There are now two new surfaces, and to create them molecules had to be pulled apart. The *free surface energy* γ of the liquid is defined by the equation

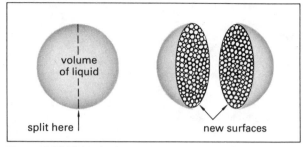

Figure 6.19 Molecules have to be separated in order to form new surfaces.

$$\gamma = \frac{\Delta W}{\Delta A}$$

where ΔW is the additional *surface potential energy* which the surface molecules now have, and ΔA is the newly-created surface area. The units of γ are J m^{-2} (or the equivalent N m^{-1}).

We include in ΔW any energy which flows in from the surroundings to keep the temperature constant. You should not confuse this surface potential energy with gravitational potential energy. It is not because they are *higher* that the molecules have more potential energy, but because they are now in the surface of the liquid. In figure 6.19 the molecules have not gained height; they have merely been separated.

We should expect the size of γ to depend on the temperature of the liquid. At a higher temperature the molecules have more kinetic energy of their own, and it is easier to pull them apart. The free surface energy does decrease with temperature. For water, for example, it decreases from 7.6×10^{-2} J m^{-2} at 273 K to 5.9×10^{-2} J m^{-2} at 373 K.

The formation of drops

If molecules in the surface of a liquid have extra potential energy we might expect that, other things being equal, liquids would form themselves into shapes such that their surface area was a minimum, so that the surface potential energy would be a minimum. We do indeed find that small amounts of liquid form spherical drops, a sphere being the shape which has the least surface area for a given volume. With larger amounts other forces (e.g. gravitational forces) distort the shape, so it seems that the molecular forces are important when the size is small.

▶ Calculate the surface potential energy (s.p.e.) at 293 K of a spherical drop of water of radius (i) 10 mm, (ii) 1.0 mm, if at that temperature the free surface energy of water is 7.3×10^{-2} J m^{-2}. Also, supposing that the drops are placed on a horizontal table, find their gravitational potential energy (g.p.e.) relative to the table surface before they begin to collapse. Density of water $= 1.0 \times 10^3$ kg m^{-3}. Comment on your result.

We can calculate the s.p.e. by finding the surface area A: then the s.p.e. $\Delta W = \gamma \Delta A$. We can find the g.p.e. from the equation $W = mgr$, where $m =$ volume $V \times$ density

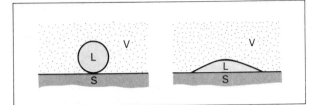

Figure 6.20 A liquid drop collapses. Changes occur in the surface potential energy at all three boundaries (L-V, S-V and S-L).

(and $V = 4\pi r^3/3$). The results of these calculations are tabulated below:

r/mm	ΔA/m²	V/m³	m/kg	s.p.e./J	g.p.e./J
1.0	1.3×10^{-5}	4.2×10^{-9}	4.2×10^{-6}	9.2×10^{-7}	4.1×10^{-8}
10	1.3×10^{-3}	4.2×10^{-6}	4.2×10^{-3}	9.2×10^{-5}	4.1×10^{-4}

We can see that the s.p.e. of the smaller drop is larger than its g.p.e., but the reverse is true of the larger drop. We would expect the smaller drop to remain nearly spherical, but the larger drop to collapse. It seems that the relative importance of molecular forces and gravitational forces depends on the size of the drop.

The spreading of liquids

If a liquid has least surface potential energy when its surface area is smallest, why does it ever spread (e.g. water does when spilt on a table)? The drop loses gravitational potential energy, which is available to be converted to the extra surface potential energy, but there is another reason. *All surfaces and interfaces have surface potential energy*, not only the boundary between a liquid and its vapour which is all we have so far considered. When a liquid spreads over a surface (see figure 6.20) the situation is therefore quite complicated, and here all we need say is that it *is* possible for a liquid-vapour surface to increase in area (and therefore have an increase in surface potential energy) *if* another surface (e.g. the solid-vapour surface) decreases in area (and therefore has a decrease in surface potential energy). The liquid spreads for as long as spreading reduces the total potential energy of the system.

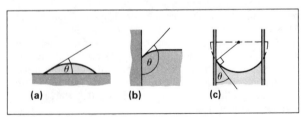

Figure 6.21 θ is the angle of contact. In (a) and (c) the liquid might be water, touching glass. In (b) it is mercury touching glass.

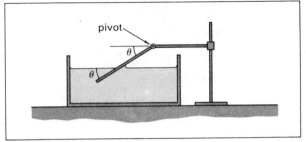

Figure 6.22 The principle of the tilting-plate method of measuring the angle of contact.

When a liquid surface meets a solid surface the *angle of contact* is defined as the angle between the solid surface and the tangent to the liquid surface, measured in the liquid. Some examples are shown in figure 6.21.

It can be measured by the tilting-plate method: figure 6.22 shows the principle. A plate of the chosen solid material is placed at an angle in the liquid, and rotated until the liquid surface is horizontal right up to the solid surface. The angle which the plate then makes with the horizontal is measured: this is the angle of contact of that liquid with that solid.

Surface tension

The fact that the molecules of a liquid cling together in drops suggests that the surface of a liquid is in a state of tension. That this is so can easily be demonstrated with the apparatus shown in figure 6.23. The U-shaped wire frame is placed in a horizontal plane, and the wire AA' is placed on top, and left free to slide along the frame. A thread T is attached to the wire and runs over a fixed wire to a light scale pan. A film of detergent solution is formed in the space enclosed by the wire and the frame, and it is found that the thread T must be pulled to maintain the equilibrium of the wire. If the wire is a few cm long, the mass needed in the scale pan is about 10 g. The size of the force does *not* depend on
(i) the thickness of the film; the solution can be allowed to evaporate, thinning the film, and there is no change in the force needed.

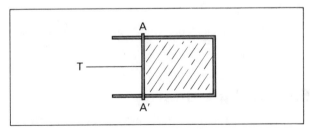

Figure 6.23 The wire AA' is laid across the U-shaped wire frame. A thread T runs over a fixed wire or pulley and is attached to a light scale pan on which masses can be placed.

(ii) the area of the film; the same load in the scale pan will keep the wire in equilibrium in different positions along the frame.

So the surface tension in a liquid is *not* like the tension in a rubber sheet, which would depend on both its thickness and the extension.

A sufficient explanation of surface tension would be to show that in the surface of a liquid the molecules are further apart than they normally are, in equilibrium. There are two arguments to show that this must be so.

Firstly, consider the two molecules A and B in figure 6.24a. They attract and repel each other, and in equilibrium are a certain distance apart. A third molecule C is added. A and C attract each other and repel each other (though the attracting force, at this distance, is larger than the repelling force) so the three molecules move closer together. If we imagine other molecules D, E, F etc. brought up to A, B, C we can imagine that when molecules are completely surrounded by others, as they are in the body of any liquid, they are closer together than when they are in the surface and *not* completely surrounded by others.

Secondly, realise that the molecules in the surface of a liquid are not always the same molecules – they are changing places with other molecules in the body of the liquid. Once the surface has been established, and is in equilibrium, the molecules in the surface must be moving into the body of the liquid at the same rate as molecules are moving out of the body of the liquid into the surface. It must be easier for a molecule in the surface (e.g. A in figure 6.24d) to move into the liquid than for a molecule in the body of the liquid (e.g. B) to move out, because there are no molecules behind A to pull it back. The rates of flow into and out of the surface can be equal only if there are fewer molecules per unit volume in the surface than in the body of the liquid.

There is experimental evidence to show that the surface effects in a liquid are confined to a layer just two or three molecules deep.

It is convenient to define a quantity called the *surface tension* of a liquid. Consider a line of length l drawn on the surface of a liquid (figure 6.25). The surface layer on one side of the line pulls with a force F on the surface layer on the other side of the line (and *vice-versa*, by Newton's third law). The surface tension of the liquid is defined by the equation

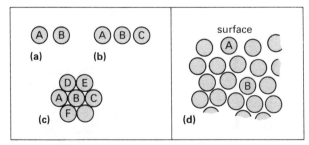

Figure 6.24 Explanations of why the molecules in the surface of a liquid are further apart than they are in the body of the liquid.

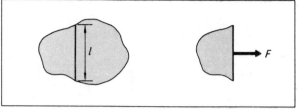

Figure 6.25 The surface tension of the liquid is defined by F/l, where F is the force the right-hand part exerts on the left-hand part (or vice-versa).

$$\text{surface tension} = \frac{F}{l}$$

It has the unit N m^{-1} (it is a force per unit length, not a force, despite its name). This unit is the same as the J m^{-2}, the unit of free surface energy, and we shall now show that these two quantities are equivalent.

Equivalence of surface tension and free surface energy

Referring back to figure 6.23, suppose that a force F moves the wire a distance Δx at a steady speed (so that the wire is kept in equilibrium). The force F is, by the definition of surface tension, equal to $2(\text{surface tension})l$, where l is the length of the film attached to the wire; the factor 2 is included because there is an upper and a lower surface to this film. This force does work $F\Delta x$ and increases the free surface energy of the film by an amount $2\gamma\Delta A$ (where $\Delta A = l\Delta x$, and the factor 2 is again included because there are two surfaces). So equating $F\Delta x$ and $2\gamma\Delta A$ we have

$$2(\text{surface tension})l\Delta x = 2\gamma l\Delta x$$

or $\text{surface tension} = \gamma$

i.e. the surface tension, or force per unit length, is equivalent to the free surface energy, or energy per unit area. The surface tension is sometimes a more convenient concept and we shall use it in the next paragraph.

The excess pressure in a drop of liquid

If the liquid in a small drop has a spherical shape because it is keeping its surface a minimum, we can assume that the pressure in it is greater than the pressure outside it. Suppose that the excess pressure is Δp in a drop of radius r, and consider the drop split in half, as shown in figure 6.26a. The *surface* of the other half of the drop pulls on this half all round its circumference.

$$\text{the force} = \text{surface tension} \times \text{length} = \gamma(2\pi r)$$

The equilibrium of this half of the drop is maintained by the resultant force produced by the excess pressure in the drop. The pressure pushes outwards on the hemispherical

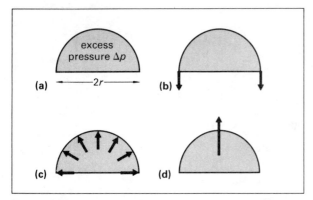

Figure 6.26 (a) Shows half the drop (b) shows the surface tension force acting round the circumference of the half-drop, (c) represents the forces pushing outwards on the inside surface of the drop, because of the excess pressure in the drop and (d) shows the resultant of these pressure forces.

area ($2\pi r^2$) of the inside of the drop, but the resultant force, which is, by symmetry, upwards in figure 6.26d, can be shown to be equal to the excess pressure Δp multiplied by the *projected* area, i.e. the area of cross-section of the hemisphere, πr^2. So equating these two forces we have

$$\gamma(2\pi r) = \Delta p(\pi r^2)$$

or $$\Delta p = \frac{2\gamma}{r}$$

This result holds for a whole drop, or any spherical liquid surface of radius r. The pressure is, of course, larger on the concave side of the surface.

   'Soap' bubbles have two surfaces, and the excess pressure in them is therefore $4\gamma/r$, since the surface tension force will be double.

▶ Find the pressure in an air bubble of radius 0.10 mm which is 0.15 m below the surface of some water (at 373 K) in a beaker. Take atmospheric pressure to be 1.013×10^5 Pa: at this temperature the free surface energy of water is 5.9×10^{-2} J m^{-2} and its density = 958 kg m^{-3}.

   The pressure at the level of the bubble is greater than that at the surface by an amount $\rho g\Delta h$, so the pressure there

$$= 1.013 \times 10^5 \text{ Pa} + (958 \text{ kg m}^{-3})(9.8 \text{ N kg}^{-1})(0.15 \text{ m})$$
$$= 1.013 \times 10^5 \text{ Pa} + 1408 \text{ Pa}$$
$$= 1.013 \times 10^5 \text{ Pa} + 0.014 \times 10^5 \text{ Pa}$$
$$= 102.7 \text{ kPa}$$

The excess pressure inside the bubble equals $2\gamma/r$

$$\frac{2\gamma}{r} = \frac{2 \times 5.9 \times 10^{-2} \text{ J m}^{-2}}{10^{-4} \text{ m}}$$

$$= 1.2 \text{ kPa}$$

Therefore the pressure in the bubble is 103.9 kPa. We shall see in Chapter 9 that the bubble cannot grow (i.e. the liquid cannot boil) unless the vapour pressure of the water is equal to this value, which it cannot be unless the temperature is

rather more than 373 K (about 0.7 K more). Smaller bubbles have a higher pressure in them: you can check that a bubble of radius 0.010 mm will have a pressure of 114.5 kPa, and a bubble of that size will not grow unless the water temperature is about 376.4 K. So *superheating* (the raising of the temperature above that at which it should boil) occurs if there are not enough large bubbles to allow boiling to occur.   ◀

Capillary rise

One important example of spreading occurs in narrow vertical tubes. Again the system adjusts so that the potential energy of the whole system is a minimum. We can identify the following changes in figure 6.27a:
(i)   an increase in the g.p.e. of the liquid
(ii) a small increase in the s.p.e. of the liquid-vapour boundary (because it is no longer entirely horizontal)
(iii) an increase in the s.p.e. of the solid-liquid boundary
(iv) a decrease in the s.p.e. of the solid-vapour boundary.
The height of rise can be calculated by considering these energy changes, but there is a simpler way.

   Let us first note that the inner radius r of the tube and the radius of curvature R of the spherical surface of the liquid in the tube will be different (unless the angle of contact is zero, and fortunately it *is* zero for many solid-liquid boundaries). Figure 6.27b shows that r and R are related by $r = R\cos\theta$. The pressure at A in figure 6.27a is atmospheric (since it is level with the open surface of the liquid outside the tube). But in moving upward a distance h from A (i) the pressure *decreases* by $\rho g h$, where ρ is the density of the liquid, and (ii) it *increases* by $2\gamma/R$ as it crosses the liquid boundary of radius R (excess pressure on inner side of single spherical surface $= 2\gamma/R$). However on that side of the surface the pressure is atmospheric, so the *decrease* of $\rho g h$, and the *increase* of $2\gamma/R$ must be the same size. So

$$\rho g h = \frac{2\gamma}{R} \text{ where } R = \frac{r}{\cos\theta}$$

so that $$h = \frac{2\gamma\cos\theta}{\rho g r} \tag{1}$$

Note (i) for a particular liquid h will be large when r is small,

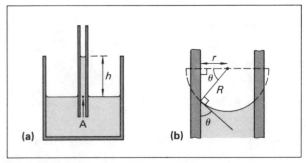

Figure 6.27 Capillary rise; (b) shows that $r = R\cos\theta$.

e.g. for water ($\theta \approx 0°$, $\gamma \approx 7 \times 10^{-2}$ J m^{-2}, $\rho = 10^3$ kg m^{-3}) in a capillary tube of radius 0.10 mm, $h = 140$ mm. For $r = 1.0$ mm, $h = 14$ mm. (ii) the derivation does not depend on the inner radius of the tube being uniform. The relevant radius r is therefore the radius at the meniscus. (iii) capillary rise occurs, for example, where oil rises in the wick of an oil lamp or water rises in brick walls: the small gaps between the particles of the brick act as capillary 'tubes'. Hence the need for damp-proof courses in buildings.

Measurement of γ

The capillary rise of liquids provides a basis for measuring γ. A clean, preferably newly-drawn, tube is used and rinsed with the liquid. (Any trace of detergent will seriously affect the result.) The tube is placed vertically with its lower end in the liquid. A glass scale is fixed to the tube, and used to measure the difference in level h between the bottom of the meniscus in the tube and the surface in the vessel containing the liquid. The values of ρ and g will be known, and θ can be measured as described above. Ideally we need to measure r where the meniscus was. The tube can be broken at this point, but the difficulties of measuring the radius (e.g. with a travelling microscope) make it preferable to use the mercury thread method (see page 67) to find the average radius of the tube for a few cm on either side of this point. (Manipulation of the thread in the tube will enable us to see if the tube is uniform, anyway: if it is not, the length of the thread will alter.) Knowing h, ρ, g, r and θ we can now use equation (1) above to calculate γ.

▶ A capillary tube of internal radius 0.30 mm is placed vertically in water, with 40 mm of the tube above the water surface. How high does the water rise in the tube? Angle of contact = 0°, $g = 9.8$ N kg^{-1}, density of water = 1000 kg m^{-3}, γ for water at this temperature = 7.3 $\times 10^{-2}$ N m^{-1}.

We use $h = \dfrac{2\gamma \cos \theta}{\rho g r}$

$\Rightarrow \qquad h = \dfrac{2(7.3 \times 10^{-2} \text{ N m}^{-1})(1)}{(1000 \text{ kg m}^{-3})(9.8 \text{ N kg}^{-1})(0.30 \times 10^{-3} \text{ m})}$

$\qquad = 5.0 \times 10^{-2}$ m = 50 mm

which is, of course, impossible if there is only 40 mm of the tube above the water surface, what does happen? The water rises to the top of the tube and stops, with the radius of the meniscus increased to such a value that equilibrium *is* possible. The pressure difference across the surface of the meniscus is then small enough to be equal to the pressure difference in a height of 40 mm of water. The situation will be like that shown in figure 6.27b, with the meniscus at the top of the tube instead. Let us find the value of θ using $h = (2\gamma \cos \theta/\rho g r$ again, but this time knowing that $h = 40$ mm. The calculation gives $\theta = 36°$ ◀

Soap films

Although films of water and other pure liquids cannot exist,

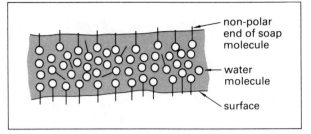

Figure 6.28 How soap (or detergent) molecules arrange themselves near a liquid surface.

films of soap solution (or detergent) *can* exist, as anyone who has blown a soap bubble knows. This is because the long soap molecules align themselves at right-angles to the surface. They have a polar group of atoms at one end which is attracted to the (polar) water molecules, as shown in figure 6.28. This reduces the surface tension (because the forces between the non-polar groups, which are in the surface, are smaller than those between the polar water molecules which were previously there). This makes a *stable* film. If the film *is* stretched slightly, a water molecule (more mobile than a soap molecule) appears at the surface and therefore the surface tension increases until a soap molecule has had time to move to fill the gap. Vertical films are possible: consider a soap solution film formed on the vertical wire frame in figure 6.29. The weight of the film means that the surface tension must increase steadily from the bottom to the top of the film. $S_1 = W + S_2$, i.e. S_1 is greater than S_2. This variation in surface tension would be impossible with a pure liquid like water: it *is* possible with a soap solution where the concentration of soap molecules adjusts itself to provide the varying force.

Detergency

Dirt is often helped to attach itself to crockery and clothes by fat and grease. The non-polar ends of soap (or detergent) molecules attach themselves to the grease and the material, and bring water molecules in between the grease and the material so that the dirt and grease is rolled up and washed

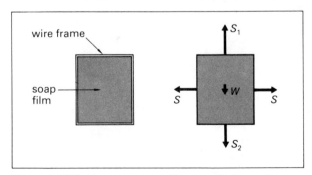

Figure 6.29 The forces on a soap film on a wire frame. The forces S are equal and opposite. $S_1 = W + S_2$, so S_1 is greater than S_2.

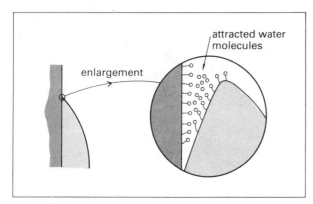

Figure 6.30 A detergent at work.

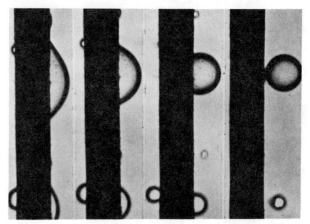

Figure 6.31 Spots of grease being rolled up and washed away by a detergent solution.

away: see figures 6.30 and 6.31. The free surface energy of a soap solution is less than that of water: hence it spreads more easily. Look back at figure 6.20: when a liquid's free surface energy is lowered it can increase its surface area (and hence its surface potential energy) if it covers up a solid and reduces the solid's surface potential energy. We want (i) to *increase spreading* in order to remove grease from dishes. The large-scale view of what is happening in figure 6.30 is that when the liquid has a small angle of contact it is able to penetrate between the solid and the grease, and (ii) to *reduce spreading* in order to waterproof materials. This we do by putting polish on a car's paintwork, or when we coat fabric with silicones for use in a waterproof coat.

7 Electric circuits

7.1 Electric current

There are many substances in which there are charged atomic particles that can wander freely through the material. In such a substance electric charge can be transferred from one point to another by a general drift of the charged particles within it. Such a movement of electric charge is called an *electric current*. Materials through which an electric current will pass are known as *conductors*. Those substances in which there are no charged particles that are free to move are known as *non-conductors* or *insulators*.

The electrical forces between charged particles (page 186) are so great that it is impossible to unbalance the charge in a piece of matter to any appreciable extent. Any tendency for the charge of one sign to pile up at some point generates forces of repulsion that soon bring the process to a halt. A continuous flow of electric charge can happen only in a closed path or *circuit*, as it is called. The circuit must also include some device such as an electric cell or dynamo, which maintains the circulation of electric charge. This it does by an internal transference of electrons from one of its terminals to the other; in a cell this is brought about by chemical processes. One terminal gains a surplus of electrons and so carries a *negative* charge. At the other terminal there is a corresponding lack of electrons and a net *positive* charge. When a conducting path is provided between the terminals the electrical forces tend to even out the distribution of charge by causing it to flow; the cell or dynamo then acts to maintain the initial distribution of charge by a fresh transference of electrons between the terminals *inside* the device. A continuous circulation is thus created (figure 7.1).

The direction of the current

The same electric current may be carried in various ways in different parts of the same circuit; at one point there may be a movement of positively charged particles in one direction; at another there may be an equal movement of negatively charged particles in the opposite direction – the net transference of charge is the same in both cases; or there may be simultaneous movement of positive and negative charges in opposite directions.

By general agreement the *positive* direction of electric current is taken to be that in which a positive charge would flow. An arrow drawn on a conductor indicates this positive direction; it implies that, if the current at that point is actually a flow of negatively charged particles (e.g. electrons), then the movement of the particles is opposite to that of the arrow. Figure 7.1b shows how this is done on a conventional circuit diagram with a single cell. Note that in the symbol for a cell the *longer stroke* is taken to be its positive terminal.

When there is an electric current in a circuit, two physical effects are observed:

(i) *A heating effect.* The movements of the charged particles in a conductor are hindered by their interactions with stationary particles of matter in their paths. The energy of the moving charged particles is converted in collisions to internal energy, increasing the random oscillations of the particles of the conductor; in other words the temperature of the conductor rises.

(ii) *A magnetic effect.* This may be demonstrated by making an electric current pass through a conductor near a small pivoted magnet (a compass). In most positions close to the circuit the compass will be deflected slightly from its usual north-pointing direction. Likewise a wire carrying a current is acted on by a force in the neighbourhood of a magnet (page 205). These effects are the result of another type of force that acts between the fundamental particles of matter; it occurs between *charged particles in motion*, and acts in addition to the electrical forces already considered (page 35). Such a force may be demonstrated by passing a large current through two parallel flexible conductors (figure 7.2). If the currents are in the same direction (a), the magnetic forces act so as to pull the conductors together. When the currents are in opposite directions (b) the conductors are pushed apart. Such effects are considered in detail in chapter 16.

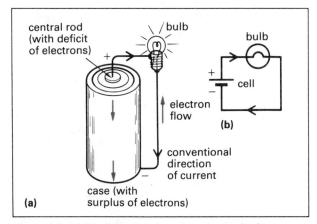

central rod (with deficit of electrons)

bulb

bulb

+

cell

–

electron flow

(b)

conventional direction of current

–

case (with surplus of electrons)

(a)

Figure 7.1 (a) The flow of electrons in a simple circuit. Conventionally, the direction of the current is taken to be that in which positively charged particles move, opposite to the direction of the electron flow. (b) The same circuit using conventional symbols.

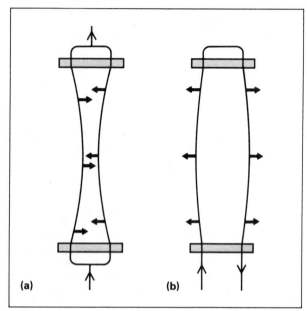

Figure 7.2 (a) Currents in the same direction: the wires attract each other; (b) Currents in opposite directions: the wires repel each other.

Meters

A common form of instrument for measuring an electric current consists of a small coil pivoted between the poles of a magnet; the current to be measured passes through this coil. The reading of the instrument depends on the twisting effect of the magnetic forces on the coil. (Details of the construction of such *moving-coil meters* are considered in chapter 16, page 207).

Readings taken with moving-coil meters in branching circuits show that the current measured by these instruments is constant at all cross-sections of a circuit; this includes cross-sections where the circuit divides into several paths. Consider, for instance, the simple branching circuit of figure 7.3. First, we notice that, whatever the settings of the two current controllers (or *rheostats*), the readings of the meters M and N are always identical. The current I_3 from the positive terminal of the cell divides into two parts, I_1 through meter K with its rheostat, and I_2 through the other meter L. These currents recombine to form the current I_3 through meter N. The currents may be varied in any manner we wish with the two rheostats; but for all settings of the rheostats we find that

$$I_1 + I_2 = I_3$$

In other words the current through the cross-section X-X is the same as that through the cross-section Y-Y (or through any other cross-section of the circuit).

Other observations lead us to believe that the current I recorded by a moving-coil meter is a measure of the *rate of flow of electric charge* Q through it. For instance, we find that an electric current in a solution of any salt is accom-

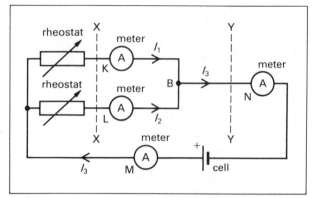

Figure 7.3 The current is the same at all cross-sections of a branching circuit; in this case $I_1 + I_2 = I_3$.

panied by a transport of actual matter through it (page 79). In this case we know that the charge is carried through the solution by the charged particles (ions) in it; and we find that the rate of transport of matter is directly proportional to the current. So we can write (see appendix B3)

$$I = \frac{dQ}{dt} \tag{1}$$

If we have a *steady* current I, then the total charge Q that passes any cross-section of the circuit in time t is given by

$$I = \frac{Q}{t}$$

or $Q = It$ \hfill (2)

At any junction-point in a circuit (such as B in figure 7.3) the rate at which charge enters the junction ($I_1 + I_2$) is equal to the rate at which charge leaves it (I_3). This result is sometimes known as *Kirchhoff's first law of electric circuits*. It can also be called the law of conservation of electric charge; electric charge cannot either disappear or come into being at any point in a circuit.

The ampere

The unit of electric current is called the *ampere* (A) after André Ampère the French scientist who in 1825 established the laws governing the forces that act between electric currents. The ampere is one of the *base units* of the Système International (SI). It is defined in terms of the magnetic force that acts between current-carrying conductors (see appendix A2). The definition envisages an idealised geometrical arrangement of two infinite, straight, parallel wires one metre apart. Currents in the same direction in these would cause them to attract one another (as in figure 7.2a). If the currents are both 1 A (exactly), then the force per metre length acting on one wire caused by the current in the infinite length of the other is defined to be 2×10^{-7} N.

This is an imaginary measurement which we do not expect

7.2 Currents in solids

The most important class of conductors is the *metals*. In any solid the atoms themselves must be fixed at permanent sites in the material. But some of the electrons of each atom are partly shared with neighbouring atoms, providing the chemical bonds that bind the material together. Only a small number of the outermost electrons of each atom are involved in these chemical bonds. These are known as the *valence electrons*. In most non-metals the valence electrons remain fixed within particular groups of atoms, and no general movement of electrons through the material is possible. But with metals the bonds between atoms are such that some of the valence electrons are free to move from one atom to the next right through the crystal lattice. Typically an average of about one electron per atom is free, and the movement of these constitutes the electric current in the metal.

As well as conduction of a current through a solid by movement of free electrons, it is also possible for charge to be carried through it by means of unfilled vacancies in the electron structure of the material. These gaps in the electron structure are called *holes*. It is possible for a *bound* electron of a neighbouring atom to move into such a hole; the electron still remains a bound one (in its new site), but the hole is transferred to the atom just vacated. In this way a hole (which behaves as a centre of positive charge) may be transferred from one atom to the next, and so on right through the material. Some conduction in solids is thus by the movement of the positively charged holes, which behave very much as though they were positively charged electrons. These two different types of conduction are referred to as *n-type* (when the conduction is by *negative* electron movement) and *p-type* (when the conduction is by transfer of *positive* holes). In many metals (e.g. copper and silver) the conduction is *n*-type. But there are some whose conduction is mostly *p*-type (e.g. zinc).

Semiconductors

In a metal about one valence electron per atom is free to move through the crystal lattice to conduct electric currents. In an insulator there are virtually no free electrons. Intermediate between these lies the class of materials known as *semiconductors*, such as silicon and germanium. In these only a very small proportion of the valence electrons are free to move through the crystal lattice. In a pure (or *intrinsic*) semiconductor only about 1 atom in 10^{10} contributes a charge carrier (electron or hole) to the crystal. Semiconductors gain their useful properties from the controlled addition during manufacture of minute traces of impurity. An impurity level of only 1 part in 10^7 can increase the conductivity a thousand times; and by choice of the type of impurity the conduction can be made either *n*-type or *p*-type. Semiconductor devices, such as transistors (page 289) and integrated circuits (page 293), are made by fashioning crystals of silicon with intricate microscopic patterns of *n*-type and *p*-type regions within them.

Figure 7.4 The National Physical Laboratory current balance. In this picture the outer, fixed pair of coils has been lowered on the left-hand side so that one of the inner suspended coils may be seen. The masses are lifted onto or off the small scale pans above the coils by means of rods controlled from outside the case.

to perform in practice. But the laws discovered by Ampère enable us to use the definition to calculate the forces acting between other, more practical, arrangements of conductors.

The instrument used in the standardising laboratories for measuring a current in amperes is the *current balance*. This makes use of a system of cylindrical coils. The beam of the balance, supported on its knife-edges at the centre, carries at each end a cylindrical coil (figure 7.4). These coils are joined in series, and the connection to them is by means of wires arranged to exert negligible forces on the beam. A set of fixed single-layer cylindrical coils surrounds the suspended coils, concentric with them. All the coils, suspended and fixed, are joined in series with one another in such a way that the forces operating between them will combine to turn the beam in one direction. The beam is kept in equilibrium when the current is switched on by adding weights to the small scale pans. In this way the forces which the coils exert on one another are measured in terms of a known mass, and so the current in the coils is found. The precision attainable at present is about 5 parts in 10^6.

Equation (2) above shows that the unit of *electric charge* is the A s. But it is convenient to have a special name for the unit, and it is called the *coulomb* (C) after Charles de Coulomb who first investigated the forces between electric charges in 1785.

Thus $1 \text{ C} = 1 \text{ A s}$
and $1 \text{ A} = 1 \text{ C s}^{-1}$

Measurements show that the charge on one electron is -1.60×10^{-19} C (page 193). Or we may say that the charge on 6.24×10^{18} electrons is one coulomb.

Magnetic materials

We have seen how magnetic forces arise between conductors carrying electric currents. It is not perhaps obvious how the properties of *magnets* can derive from the interactions of electric currents. But a little thought will show that on an atomic scale there are never-ending circulations of electric charge taking place in *all* materials. Electrons in rapid motion round a nucleus constitute an electric current. Indeed it is now understood that electron, proton, and neutron must all be regarded as being permanently in a state of *spin*, which causes a magnetic effect even when they are not otherwise in motion. We therefore expect all matter to show magnetic effects; and such is the case, though with most materials the effects are very weak (page 232). It so happens that in iron and a few other substances the interactions between atoms magnify the effects considerably. This aspect of magnetism is explained more fully in chapter 17 (page 232).

The speed of the current carriers

Consider a length of conductor of cross-sectional area A containing n free electrons per unit volume (figure 7.5). The number of electrons in the conductor remains constant; when a current passes in it, electrons enter it at one end and the same number leave at the other. The motions of individual electrons are doubtless somewhat irregular, but suppose their average speed is v. Then the number of electrons passing a point such as Y in time Δt is the number contained in the part of the conductor between X and Y of length $v\Delta t$.

As the volume between X and Y $= Av\Delta t$
the number of electrons contained $= nAv\Delta t$

and so the total charge ΔQ passing the point Y in time Δt is given by

$$\Delta Q = nAve\Delta t$$

where $e =$ the electronic charge. The electric current I in the conductor is the rate of passage of electric charge;

so $\quad I = \dfrac{\Delta Q}{\Delta t}$

$\quad\quad = nAve$

$\Rightarrow \quad v = \dfrac{I}{nAe}$ \hfill (3)

A good conducting material is one for which n is large and v therefore small; at low speeds there is less opposition to the movement of the electrons. Copper is one of the best conductors, having about 10^{29} free electrons per cubic metre.

In a simple circuit of copper conductors joined in series the current I is the same at all cross-sections. However, the speed of the electrons varies from one part of the circuit to another, being inversely proportional to the cross-sectional

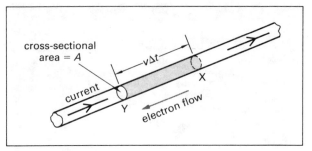

Figure 7.5 Calculating the speed of the charge carriers in a conductor.

area A. In a wire of relatively large cross-section v is small and there is very little of the kinetic energy of the electrons transformed to internal energy of the wire. But if the circuit includes a length of fine wire such as the filament of a light bulb, v is there much greater; and the rate of conversion of kinetic energy of the electrons to internal energy of the wire is larger; the wire thus becomes hot. But the values of the speed v are always quite small, as the following example makes clear.

▶ A copper fuse wire of cross-sectional area 2.0×10^{-7} m² (i.e. 0.2 mm² or about 0.5 mm diameter) can carry a maximum current of 5.0 A without melting. Estimate the average speed of the electrons in it when the current reaches this level.

Using equation (3) above and taking $n = 1.0 \times 10^{29}$ m⁻³ for copper, we have

$$v = \frac{5.0\ \text{A}}{(1.0 \times 10^{29}\ \text{m}^{-3})(2.0 \times 10^{-7}\ \text{m}^2)(1.6 \times 10^{-19}\ \text{C})}$$

$$= 1.6 \times 10^{-3}\ \text{m s}^{-1} = 1.6\ \text{mm s}^{-1} \quad \blacktriangleleft$$

Although the electrons move so slowly, the electrical forces that set them in motion are propagated round the circuit with great rapidity – in fact with very nearly the speed of light (3×10^8 m s⁻¹). If a current of 1 A passed steadily in a wire like the above in a transatlantic cable (3000 km long), it would take about 300 years for a given electron to traverse the ocean. Even in the circuit of a small electric torch it is doubtful whether a given electron could pass right round the circuit before the battery is run down. And yet a small displacement of the electrons at the London end of a transatlantic cable sends a wave of electrical forces along it, and a similar displacement of the electrons in New York occurs within 0.01 s!

7.3 Currents in liquids

Many chemical compounds when they go into solution break into oppositely charged ions, which then move independently in the solution, and are therefore available to conduct an electric current through it. Such solutions are called *electrolytes*. Solutions of many inorganic compounds (common salt, sulphuric acid, etc.) are examples of this type of liquid. There are also many substances (sugar, for instance) that dissolve without splitting up into ions. Non-ionic solutions do not conduct electric currents, and are called *non-electrolytes*.

In an electrolyte conduction takes place through the simultaneous movement of positive and negative ions in opposite directions. Suitable conducting plates, called *electrodes*, must of course be provided for the current to enter and leave the solution. The electrode towards which the positive ions travel is called the *cathode*; and the other, towards which the negative ions travel, is called the *anode* (figure 7.6a). At the surfaces of the electrodes the mechanism of conduction changes from *electronic* to *ionic*. This can happen in one of two ways: either some of the ions are discharged, electrons being transferred between ions and electrode; or else fresh ions are formed from the material of the electrode, and these pass into solution. Chemical changes are therefore observed at the surfaces of the electrodes as long as the current is passing. Either substances in solution are liberated (as deposits on the electrodes or as bubbles of gas), or else the surfaces of the electrodes are eaten away. This chemical process is called *electrolysis*.

Figure 7.6b shows a simple example of electrolysis, namely that of copper sulphate solution with copper electrodes. The copper sulphate solution consists of the hydrated positive ion $Cu(H_2O)_4^{2+}$ and the negative ion SO_4^{2-}. Also the water itself provides small quantities of the ions $(H_3O)^+$ and OH^-. The positive ions travel towards the cathode; at the surface of the cathode the copper ions are discharged, and a layer of metallic copper is deposited. At the surface of the anode fresh copper ions are formed from the copper of the anode. The concentration of the copper ions in the electrolyte thus remains constant as the

electrolysis proceeds; and the net result is just the transfer of copper from the anode to the cathode. The chief practical importance of this process lies in its ability to purify the copper; the impurities in the copper of the anode are not deposited on the cathode. There is a large demand for electrolytically purified copper, since this degree of purity is needed for electric cables and connectors.

Specific charge

The passage of electric charge through an electrolyte is associated with a transport of actual matter. Each ion carries a definite charge and has a definite mass, so that the mass liberated or deposited at the surface of an electrode is proportional to the total charge passed. This is called *Faraday's first law of electrolysis*. The charge that passes per unit mass deposited is therefore a constant for a particular substance, and is a property of the individual ion concerned; it is called its *specific charge*. Thus, if an ion of mass m_i carries a charge Q_i, we have for this substance

$$\text{specific charge} = \frac{Q_i}{m_i}$$

A simple electrolytic experiment enables us to determine the specific charge of the ions concerned. We have to pass a steady current I through the electrolyte for time t; and the mass m deposited is found by weighing the electrode before and after the electrolysis. Then

$$\text{specific charge} = \frac{\text{total charge passed}}{\text{total mass deposited}}$$

$$= \frac{It}{m} \qquad (4)$$

It is interesting to note that such an experiment enables us to measure an *atomic* property with ordinary *large-scale* laboratory equipment (ammeter, clock and balance).

Notice also that, if we pass the same total charge through a number of different types of electrolytic tank, the mass m liberated or deposited is inversely proportional to the specific charge of the ions involved. This is called *Faraday's second law of electrolysis*.

▶ In the electrolysis of copper sulphate solution with copper electrodes a steady current of 1.20 A is passed for 45 minutes. The mass of the cathode is then found to have increased by 1.06 g. Calculate the specific charge of the copper ion (i) from these figures, (ii) from the known values of the electronic charge e and the unified atomic mass constant m_u (page 36), given that the relative atomic mass of copper is 63.6.

(i) Using equation (4) above

$$\text{specific charge} = \frac{(1.20 \text{ A})(45 \times 60 \text{ s})}{1.06 \times 10^{-3} \text{ kg}}$$

$$= 3.06 \times 10^6 \text{ C kg}^{-1}$$

(ii) We take the values $e = 1.60 \times 10^{-19} \text{ C}$,

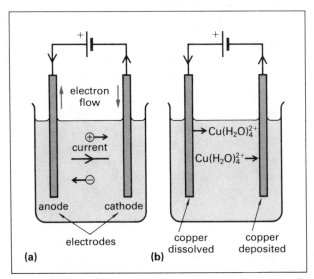

 labels: electron flow; current; ⊕→; ←⊖; anode; cathode; electrodes; →$Cu(H_2O)_4^{2+}$; $Cu(H_2O)_4^{2+}$→; copper dissolved; copper deposited; **(a)**; **(b)**

Figure 7.6 (a) Electrolysis: the diagram shows the meanings of the terms used. (b) In the electrolysis of copper sulphate solution with copper electrodes the net effect is the transport of copper from anode to cathode.

$m_u = 1.66 \times 10^{-27}$ kg. The copper ion carries a double electronic charge, and so

$$Q_i = 2 \times 1.60 \times 10^{-19} \text{ C}$$

The difference between the mass m_i of the ion and the mass m_a of the neutral atom is negligible; so that we can write

$$m_i = m_a = A_r m_u \qquad \text{(page 36)}$$
$$= 63.6 \times 1.66 \times 10^{-27} \text{ kg}$$
$$\Rightarrow \quad \text{specific charge} = \frac{2 \times 1.60 \times 10^{-19} \text{ C}}{63.6 \times 1.66 \times 10^{-27} \text{ kg}}$$
$$= 3.03 \times 10^6 \text{ C kg}^{-1}$$

This agrees with our measured value in (i) to within 1%, which is well within the limits of uncertainty of such an experiment. ◄

Once we know the charge carried by particular kinds of ion we can use experiments of this kind to obtain values of relative atomic masses. For instance, from an electrolytic experiment in which hydrogen is liberated we may derive the specific charge of the hydrogen ion (or proton). This gives

$$\frac{e}{m_p} = 9.58 \times 10^7 \text{ C kg}^{-1}$$

Since we have $e = 1.60 \times 10^{-19}$ C, this gives the mass m_p of the proton as

$$m_p = \frac{1.60 \times 10^{-19} \text{ C}}{9.58 \times 10^7 \text{ C kg}^{-1}}$$
$$= 1.67 \times 10^{-27} \text{ kg}$$

Similar measurements can be made of the masses of many different kinds of ions. These agree well with measurements made with a mass spectrometer (page 248).

7.4 Electrical energy

The circulation of electric charge in a circuit leads to the continual conversion of electrical potential energy to internal energy in all the conductors of the circuit. It may also result in the production of other forms of energy (e.g. mechanical energy in an electric motor). All this energy must be supplied by the cell, dynamo, etc, that maintains the movement of charge. Such devices for generating electric current are therefore primarily means of converting energy to electrical form from some other form. Thus an electric cell is a means of producing electrical potential energy from chemical potential energy; a dynamo produces it from kinetic energy, a microphone from the vibrational energy of a sound wave, a photocell from light energy, and so on. In the first instance, such a device produces electric *potential* energy; this it does through its ability to set up an uneven distribution of electric charge, by transferring electrons internally from one of its terminals to the other (figure 7.1). In this way electrical forces are brought into

play which act on the charged particles in the terminals, and this charge thus has electrical potential energy. When a conducting path is provided between the terminals, the free charges in the conductors are set in motion; the potential energy is changed to kinetic energy, and then into random internal energy of vibration of the atoms in the conductors.

This is described by saying that the cell or dynamo produces an electromotive force, usually abbreviated to *e.m.f.* The current is maintained in the circuit by the ability of the cell or dynamo to continue to separate charge. If the charge separated is Q and the electrical potential energy produced is W, then we define the e.m.f. E by

$$E = \frac{W}{Q} \quad \text{Joules.} \quad \frac{}{It}$$

The unit of e.m.f. is therefore the J C^{-1}; but it is convenient to have a special name for this, and we call the unit the *volt* (V) after Alessandro Volta (1745–1827), who was the first to investigate the properties of electric cells.

Thus $1 \text{ V} = 1 \text{ J C}^{-1}$

The e.m.f. of a cell depends on its chemical composition. For a Leclanché dry cell (as used in torches) the e.m.f. is about 1.5 V. This is not affected by the physical size of the cell. However, the charge separation in a cell is the result of chemical changes in it; and the total charge that may be separated by it is therefore a function of the quantity of chemicals it contains. A large torch cell stores proportionately more energy than a small one and lasts longer for a given current.

► A small electric cell in a torch can maintain a current of 0.20 A for about 2 hours. Estimate the total energy converted in the torch in this time.

The charge Q that passes in the circuit in 2 hours is given by

$$Q = It$$
$$= (0.20 \text{ A})(2 \times 3600 \text{ s})$$
$$= 1440 \text{ C}$$

The total energy W converted is given by

$$W = EQ$$
$$= (1.5 \text{ J C}^{-1})(1440 \text{ C})$$
$$= 2.16 \times 10^3 \text{ J} = 2.16 \text{ kJ}$$

This is a very expensive form of energy, about 10 000 times more expensive than electrical energy supplied by the mains, and vastly more expensive than the energy in a lump of coal. ◄

Potential difference

An electric cell makes charges of opposite sign separate between its terminals. This causes electrical forces to act on any charged particles nearby. We describe this by saying that there is an *electric field* in the space between the

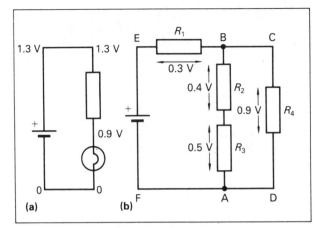

Figure 7.7 (a) The potentials at various points in a simple circuit. (b) The potential differences across the components making up a circuit; the sum of the p.d.'s round any closed loop of the circuit (such as ABCDA or ABEFA) is always zero.

terminals. If there is a complete conducting path between them, the electric field acts inside the conductors forming the circuit, setting the free charges in them in motion. (The field also exists in the air and other insulators outside the conductors; but here it is not effective in causing a current, since there are no free charges to be set in motion.) Inside the conductors the field does work on the free charged particles, first accelerating them, and so causing electrical potential energy to be converted into kinetic energy and then into internal energy (and perhaps other forms of energy as well).

To describe such an electric field we attach to each point of the circuit a figure, called its *potential*, so that the direction of the current (in the conventional sense) is from points at higher potential to points at lower potential (figure 7.7a). Although only *differences* of potential are of any significance in electric circuits, it is sometimes convenient to have a conventional *zero of potential* with reference to which we calculate the potential of other points. Many pieces of electrical equipment are operated with some part connected to the Earth. It is then usual to regard the Earth as being at zero potential. But the choice is quite arbitrary, and we are always free to fix the zero (if we need one at all) to suit our convenience (page 188). In figure 7.7a we have chosen the negative terminal of the cell as the arbitrary zero of potential, and the potentials of other points in the circuit are marked beside them with reference to this.

The *potential difference* (p.d.) between two points in a circuit is an indication of the strength of the electric field acting to drive electric charge from one point to the other. The greater the potential difference, the greater the energy converted per unit charge passing through the part of the circuit between the points. If the charge that passes is Q, and the work done by the forces of the field W, then the p.d. V between the points is defined by

$$V = \frac{W}{Q}$$

The unit of electrical potential difference is therefore the same as the unit of e.m.f., namely the J C^{-1} or volt (V).

In figure 7.7a a cell of e.m.f. 1.5 V maintains a current in a small bulb and resistor joined in series. The p.d. between the terminals of the bulb is 0.9 V; between the ends of the resistor the p.d. is (1.3 V − 0.9 V) = 0.4 V. (We assume that there is negligible p.d. between the two ends of any length of wire connecting the components.) Notice that the p.d. between the terminals of the cell is *less than its e.m.f.* This is always the case when a cell is maintaining a current in a circuit; some of the electrical energy made available to the whole circuit by the cell is converted to internal energy in the cell itself as the charge passes through it. In the present case, for every coulomb of electric charge that flows in the circuit 1.5 J of energy is converted in the cell from chemical potential energy to electrical energy. However, 0.2 J of this is converted to internal energy as the charge passes through the cell itself (whose temperature therefore rises). Only (1.5 J − 0.2 J) = 1.3 J of this energy is given to the rest of the circuit; 0.4 J is converted to internal energy in the resistor, and 0.9 J in the bulb. The use of p.d.s and e.m.f.s in this way enables us to see how the energy balance in an electric circuit works out (in accordance with the principle of conservation of energy, page 28). Thus in the above case, for each coulomb of charge that passes,

internal energy in cell	0.2 J
internal energy in resistor	0.4 J
internal energy in bulb	0.9 J
electrical energy converted from chemical energy in cell	1.5 J

You will notice that the conservation of energy requires the p.d. across components *in series* to be the *sum* of the p.d.s across them individually; e.g. for the bulb and resistor above 0.9 V + 0.4 V = 1.3 V across the pair of components. But across components *in parallel* there is *the same p.d.* (e.g. there is the same p.d. of 1.3 V across the cell and across the resistor-bulb combination in figure 7.7a).

Figure 7.7b shows a more elaborate network of resistors and the p.d.s across each of them. The resistors R_2 and R_3 are in series, and the p.d. between B and A is therefore 0.5 V + 0.4 V = 0.9 V (B being positive with respect to A). The resistor R_4 is in parallel with $R_2 + R_3$, and the p.d. across R_4 is therefore 0.9 V also (C being positive with respect to D). Resistor R_1 is in series with the combination of R_2, R_3 and R_4, and the p.d. across this (E being positive with respect to B) is therefore added to 0.9 V to obtain the p.d. between E and A, namely 1.2 V. This is also the p.d. across the cell.

Notice what happens if we add up the p.d.s (taking careful account of signs) all the way round *a closed loop* in the circuit, such as the loop ABCDA:

p.d. between B and A	$\begin{cases} 0.5 \text{ V} \\ +0.4 \text{ V} \end{cases}$
p.d. between D and C	-0.9 V
sum	0

Similarly for the loop ABEFA that includes the cell:

p.d. between B and A	0.9 V
p.d. between E and B	+0.3 V
p.d. between F and E	−1.2 V
sum	0

The sum of the p.d.s round any closed loop must always be zero. This result is sometimes known as *Kirchhoff's second law of electric circuits*; it is the statement in electrical terms of the principle of conservation of energy.

Electric field strength

If we have a current in a conductor of constant cross-section, the potential decreases uniformly along its length from its positive end to its negative end. The electrical force acting on a charged particle in the conductor is proportional to the *potential gradient* along it (page 189). This quantity is the electric field strength in the conductor. Thus, if the p.d. between the ends of the conductor is V and its length l, we have

$$\text{potential gradient} = \frac{V}{l}$$

For instance, if the resistor in figure 7.7a is a straight wire of length 2 m, then the electric field strength in it is

$$\frac{0.4 \text{ V}}{2 \text{ m}} = 0.2 \text{ V m}^{-1}$$

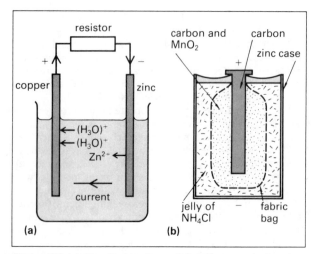

Figure 7.8 (a) The simple cell, consisting of copper (+) and zinc (−) electrodes in dilute sulphuric acid. (b) A section through the sealed type of Leclanché cell used in torches.

The significance of this for electric circuits is that in most conductors the average speed v of the electrons is directly proportional to the potential gradient along them.

$$v = u_e \text{ (potential gradient)}$$

where u_e is a constant for the electrons in a given substance (at given temperature) known as the *electrical mobility*. On page 78 we worked out the average speed v of the electrons in a copper fuse wire; this came to 1.6 mm s^{-1}. The electric field strength required to move the electrons at this speed at room temperature is about 0.4 V m^{-1}, giving

$$u_e = \frac{1.6 \times 10^{-3} \text{ m s}^{-1}}{0.4 \text{ V m}^{-1}}$$

$$\approx 4 \times 10^{-3} \text{ m}^2 \text{ s}^{-1} \text{ V}^{-1}$$

But u_e decreases quite rapidly as the temperature rises.

7.5 Electric cells

If any two different metals are dipped into a tank of any electrolyte, and an electrical connection is made between them, it is found that an e.m.f. acts so as to produce a current in the circuit so formed. The resulting electrolysis is accompanied by the usual chemical changes, and the energy of the source of e.m.f. is derived from these. Such an arrangement is called an *electric cell*. For instance, we may make a *simple cell* out of electrodes of copper and zinc in an electrolyte of dilute sulphuric acid (figure 7.8a). The e.m.f. is about 1 V. When the circuit is completed through a resistor, and a current passes, hydrogen is evolved at the surface of the copper electrode and zinc goes into solution from the other (as Zn^{2+} ions).

Table 7.1 shows the chemical composition and e.m.f. of a number of common types of cell. The Leclanché sealed cell (shown in section in figure 7.8b) is commonly used in

type	electrodes +	−	electrolyte	e.m.f./V
simple cell	copper	zinc	sulphuric acid	1.0
Leclanché sealed cell	carbon	zinc	ammonium chloride	1.5
mercury cell	steel	zinc	potassium hydroxide	1.4
Weston standard cell	mercury	cadmium-mercury	cadmium(II) sulphate	1.018 6 (at 293 K)
lead-acid storage cell	lead(IV) oxide	lead	sulphuric acid	2.0
nife storage cell	nickel oxide	iron	potassium hydroxide	1.2
nicad storage cell	nickel oxide	cadmium	potassium hydroxide	1.2

Table 7.1 The composition of different types of cell.

torches, calculators, etc. Its e.m.f. is about 1.5 V, but this falls to some extent as the materials are used up. The mercury cell is about twice as expensive, but lasts about twice as long as a Leclanché cell of the same size; it also maintains a nearly constant e.m.f. throughout its life. It is widely used in hearing aids and other small electronic devices. The Weston cadmium cell has been specially designed as the working standard of e.m.f., against which other laboratory instruments may be tested. Its e.m.f. is 1.018 6 V at a temperature of 293 K (20°C), and varies little with temperature. Carefully used its e.m.f. will remain constant to within 1 part in 10^6 for several years. But it preserves this constancy only as long as the current taken from it is kept less than about 10 μA at all times.

Storage cells

In these types of cells the action is reversible. A current driven through it in opposition to its e.m.f. re-forms the original materials of the electrodes, and the cell can therefore be used again and again. It is in fact a device for the conversion of electrical energy to chemical potential energy. The commonest type is the lead-acid storage cell, whose e.m.f. is about 2.0 V; a 12 volt car battery has 6 of these in series. It is arranged so that the dynamo driven by the engine produces a current through it in opposition to its e.m.f., converting electrical energy to chemical potential energy in it. This energy is then available to work the lights and other electrical systems of the vehicle. A current of several hundred amperes may be taken from a lead-acid cell for a short time; this may be necessary to drive the starter motor when the engine is cold.

The nickel-iron (nife) storage cell has similar properties; its e.m.f. is rather less (1.2 V), and this falls considerably during use. Its virtue lies in being about half as massive as a lead-acid cell for a given quantity of energy stored. The nickel-cadmium (nicad) cell is much more expensive, but it maintains its e.m.f. nearly constant in use. It is used to provide 'rechargeable' batteries for electric shavers and calculators.

Batteries

Electric batteries consist of several electric cells joined together. There are two ways of doing this:
(i) *Cells in series.* In this case the same charge Q must pass through each cell. The total energy given to the circuit is the sum of the energies provided by each cell. In figure 7.9a the total energy produced $= E_1Q + E_2Q + E_3Q + E_4Q$

$$\text{the total e.m.f.} = \frac{\text{total energy}}{Q}$$
$$= E_1 + E_2 + E_3 + E_4$$

Most electric cells have e.m.f.s between 1 V and 2 V. By joining a number of them in series in this way batteries of

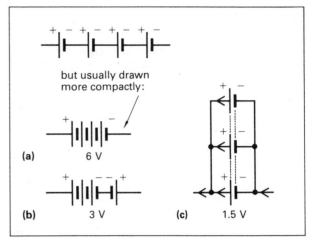

Figure 7.9 (a) A 6 V battery of four cells in series. (b) One cell is incorrectly inserted. (c) Cells in parallel; this gives a greater current than one cell, but the same e.m.f.

large e.m.f. may be made. For instance a small radio or calculator may need a 6 volt supply. For this purpose we need a battery of 4 cells (each of e.m.f. 1.5 V) joined in series. This is manufactured as a single unit of e.m.f. 6.0 V.

However, many calculators are designed to have a battery of 4 *separate* cells in series. It is then important to insert each of these the right way round. If one of the cells is reversed so that its e.m.f. acts in opposition to the others, the resultant e.m.f. of the combination (figure 7.9b) is 3.0 V. The effect of driving current through a cell in opposition to its e.m.f. is to reverse the energy conversion that usually takes place in it; i.e. chemical energy is produced from the electrical energy made available by the other cells. If it is a storage cell, the original material is re-formed in it, and the cell is 'recharged' (page 95); but the effect is more complicated in other kinds of cell.
(ii) *Cells in parallel.* Suppose we have 3 cells of equal e.m.f. E joined in parallel as in figure 7.9c. In this case part of the total current passes through each cell. The resulting battery is the same as though we had *one* cell with three times as much material in the electrodes and electrolyte. The resultant e.m.f. is the same as that of one of the cells. The advantage of this arrangement is that it can provide a larger current than a single cell.

Thermocouples

If we form a circuit of two different metals A and B (figure 7.10a), a small e.m.f. E is found to act in it when the junctions of the metals are at different temperatures. A pair of junctions of this kind is known as a *thermocouple*. This e.m.f. can be regarded as the difference of two e.m.f.s E_1 and E_2 acting in the junctions; (there is also a small additional e.m.f. in the individual metals A or B where there is a temperature gradient along them, but we shall ignore this). Thus

$$E = E_1 - E_2$$

A thermocouple is a form of heat engine (page 146), converting internal energy to electrical energy. When a charge Q flows in the circuit, at the hot junction internal energy $E_1 Q$ is transformed to electrical energy; but at the cold junction (where the current flows in opposition to the junction e.m.f. E_2) electrical energy $E_2 Q$ is transformed to internal energy. The rest of the electrical energy, $(E_1 - E_2)Q = EQ$, is available in the electric circuit; in principle it could be used to do useful work (e.g. by turning an electric motor). In practice the resultant e.m.f. E is far too small to provide useful amounts of energy unless many thermocouples are joined in series. For one thermocouple the e.m.f. is not more than a few mV for a temperature difference ΔT between the junctions of 100 K. Figure 7.10b shows the variation of e.m.f. with temperature difference for a thermocouple in which A = constantan, and B = copper, the cold junction being kept in melting ice. Thermocouples provide convenient types of thermometer for some purposes (page 124).

7.6 Electrical power

The rate at which energy is converted from one form to another in some mechanical or electrical device is called its *power* (page 24). Thus, a device of power P operating for time t converts a quantity of energy W given by

$$W = Pt$$

The unit of power is the J s^{-1}, called the *watt* (W) after James Watt (page 24). So

$$1\,\text{W} = 1\,\text{J s}^{-1}$$

When there is an electric current in a circuit component, the p.d. V across it is the energy converted within it per unit charge that passes. The current I is the electric charge passing per unit time. Therefore the product VI is the energy converted per unit time, i.e. the power P of the device; and so

$$P = VI \tag{5}$$

By the same reasoning the power P_{tot} converted to electrical energy by a source of e.m.f. E is given by

$$P_{tot} = EI$$

This is the total power converted to electrical form by the source of e.m.f.; part of it is lost as internal energy within the source of e.m.f. (e.g. a battery), the rest is delivered to the circuit external to the source of e.m.f. For instance, in the circuit of figure 7.7a above, suppose the current is 0.5 A; then

$$P_{tot} = (1.5\,\text{V})(0.5\,\text{A})$$
$$= 0.75\,\text{W}$$

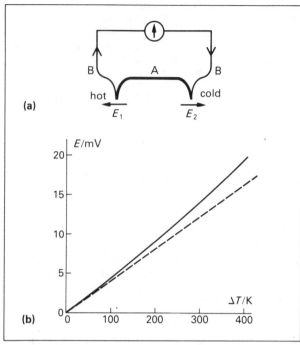

Figure 7.10 (a) A thermocouple, consisting of two dissimilar metals; when the junctions are at different temperatures a small e.m.f. is produced. (b) The variation of e.m.f. E with temperature difference ΔT for a copper-constantan thermocouple with the cold junction at 0°C.

The power P converted in the bulb to internal energy (and light) is given by

$$P = (0.9\,\text{V})(0.5\,\text{A})$$
$$= 0.45\,\text{W}$$

Lamp filaments

When there is an electric current in a wire, internal energy is produced in it and its temperature rises until the rate of loss of energy from its surface is equal to the rate at which it is supplied. The temperature then remains steady. For the filament of an electric lamp this final temperature is well above 2300 K. To obtain maximum luminous efficiency the temperature of a lamp filament should be as high as possible. Tungsten is therefore used, since it has a melting point of about 3700 K. However, even tungsten cannot be used in a vacuum above 2400 K if serious evaporation of the filament is to be avoided. Most lamps are nowadays filled with an inert gas. This reduces evaporation and enables temperatures of 2600 K or more to be reached. Also the tungsten that does evaporate is now carried away by convection and deposited on the top of the bulb instead of producing a uniform blackening of the glass, as in the vacuum type.

However, the loss of energy from the filament is so

increased by the presence of the gas that a straight wire would need to be excessively fine and fragile to reach the required temperature. To get over this difficulty a coiled filament is used. The tight coiling reduces the exposed surface area, and so enables the same temperature to be reached with a more robust wire. This also makes for a more compact light source, a matter of some importance in projector and headlamp bulbs.

An even higher filament temperature, and therefore greater luminous efficiency, is obtained in the *quartz-iodine lamp*, in which an ingenious chemical process recycles the evaporated tungsten. These lamps are filled with iodine vapour, and the bulb itself is made of fused quartz. The tungsten evaporated from the filament diffuses to the walls of the bulb, and in contact with the quartz forms the compound, tungstate(VI) iodide, which is a vapour. This compound diffuses back to the high temperature filament, where it is decomposed depositing the tungsten once more on the filament. The life of the filament is thus considerably prolonged despite its high temperature.

Fuses

The wires used for joining up an electric circuit must be thick enough so that the temperature they reach is normally very little above that of the surroundings. In mains installations a fuse is always included at some point in each circuit. The function of this is to protect the circuit against the effects of excessive currents such as can arise from accidental short circuits. It usually consists of a short length of relatively fine wire contained in a porcelain or glass tube; its diameter is selected so that it quickly reaches its melting point and breaks the circuit if the current exceeds the maximum safe value.

In a modern installation a separate fuse is provided inside each plug used to connect any appliance. Other larger fuses are provided centrally in the wiring of the

power and lighting circuits where they leave the main distribution box of the system.

Alternating current

We have dealt so far only with cases in which the currents are steady and in one direction: this is called *direct current* (d.c.). For large-scale power distribution there are, however, many advantages in using *alternating current* (a.c.). In this the direction of the e.m.f. acting in the circuit reverses many times a second, and so therefore does the current.

The variation of p.d. with time (the waveform) may follow almost any repetitive pattern; but the most important for practical purposes is the *sinusoidal* waveform in figure 7.11. This is the waveform to which the public electricity supply approximates closely. The sinusoidal waveform is so called because the variation of p.d. V with time t is according to the sine function:

$$V = V_0 \sin 2\pi f t$$

where V_0 and f are constants. Since the maximum and minimum values of $\sin \theta$ are $+1$ and -1, V_0 in this equation is the maximum value of the p.d. in either direction. It is usually known as the *peak value*.

A sine function goes through a complete cycle of variation each time the angle increases by 2π radians; for the angle $2\pi f t$ this happens f times per second. The constant f is thus the number of oscillations per second, and is called the *frequency* of the oscillation.

Similarly a sinusoidal alternating current I of the same frequency f can be expressed by the equation

$$I = I_0 \sin 2\pi f t$$

where I_0 is the peak value of the current.

The unit of frequency is called the *hertz* (Hz) after Heinrich Hertz (1857–1894) who first studied the properties of oscillatory electric discharges. Thus

$$1 \text{ Hz} = 1 \text{ s}^{-1}$$

The frequency of the public electricity supply in Great Britain (and in most of Europe) is 50 Hz. (In the U.S.A. a frequency of 60 Hz is used.)

The time taken for one complete oscillation of an alternating supply is known as the *period T* of the oscillation. For the public supply in Great Britain we therefore have

$$T = \frac{1}{f} = \frac{1}{50 \text{ Hz}} = 0.020 \text{ s}$$

The time interval T is marked in on the oscilloscope trace of figure 7.11 (which is in effect a graph of p.d. against time).

For many applications it does not matter whether alternating or direct current is used. Electric heating and lighting, for instance, are unaffected by the direction of the current; electrical energy is converted to internal energy in a conductor whichever way the current passes. The power converted in a resistor fluctuates 100 times per second with the 50 Hz mains; but this is too rapid to allow much

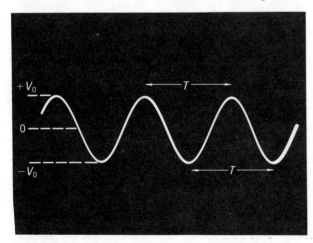

Figure 7.11 An oscilloscope trace of a sinusoidal waveform; the peak value of the alternating p.d. is V_0, and the period of oscillation is T.

oscillation of the temperature of an electric fire element or the filament of a bulb. We can detect the flickering of an electric light (e.g. with a hand-held stroboscope), but it does not normally trouble us; it can even be useful (page 156).

For some purposes direct current is essential, e.g. for electrolytic processes and for electronic equipment (radio, television, etc). But in such cases simple and efficient devices have been developed for *rectifying* the alternating current to give direct current at any desired p.d. (page 287). The generation and distribution of alternating current is so much cheaper and more efficient than direct current that its use for large-scale power purposes is now universal.

8 Electrical resistance

8.1 Ohm's law

An electric current I is produced in a conductor by applying a potential difference V across it. The relationship between the two quantities for metal conductors was first investigated by Georg Ohm is 1826.

The current in a particular metallic conductor depends on a number of factors besides the p.d. across it; the most important of these is the temperature. But such things as the elastic strain to which it is subjected, the illumination of its surface, or indeed almost any of its physical conditions, may also affect the current. *Ohm's Law* states that if all these were kept constant, the current I in a given metallic conductor is directly proportional to the p.d. V across it.

In electrolytic conduction the connection between p.d. and current is complicated by e.m.f.s that arise at the electrodes, accompanying the interconversion of electrical and chemical energy (page 80). The same situation arises in other cases where the conductor is itself the site of an e.m.f. (e.g. the armature coil of a motor or the junctions of a thermocouple). For these reasons Ohm's law applies strictly only to steady currents in metallic conductors which are not themselves the site of any e.m.f.

The quantity V/I is a constant for a given metallic conductor under steady physical conditions. It is known as its *resistance R*. Thus

$$R = \frac{V}{I}$$

Even when we are dealing with a circuit component for which Ohm's law does not apply it is convenient to define V/I as the resistance of the component; in such a case the resistance R varies with the current.

The unit of resistance is therefore the V A^{-1}; but it is

convenient to have a separate name for this much-used unit, and it is called the *ohm* (symbol Ω, the Greek letter 'omega').

Thus $\quad 1\,\Omega = 1\,\text{V A}^{-1}$

It equally follows from Ohm's law that the reciprocal quantity I/V is a constant for a given metallic conductor; this is known as its *conductance G*. The unit of conductance is the Ω^{-1}, which is also called the *siemens* (S).

Thus $\qquad 1\,\text{S} = 1\,\Omega^{-1} = 1\,\text{A V}^{-1}$

And we have $\quad G = \dfrac{1}{R} = \dfrac{I}{V}$

The relationship between current and applied p.d. for a coil of wire (or any other suitable component) may be investigated with the simple arrangement shown in figure 8.1a. (Note that the symbol for a resistor is now a long rectangle. An older symbol, still widely used, is the zig-zag line shown alongside in figure 8.1b.) The applied p.d. may be varied in equal steps (provided the cells are identical) by selecting different tapping points on the battery, and in each case the current through the resistor is noted. If Ohm's law applies, a graph of current against p.d. will be a straight line through the origin (appendix B2). The same graph is obtained whichever way round the resistor is joined in the circuit. If the e.m.f. E of each cell is known (and the current taken is small), then with n cells in use the p.d. applied is nE; and the gradient of the graph line is equal to $1/R$. The battery and ammeter must be of types that introduce negligible resistance into the circuit.

▶ What p.d. must be applied to a coil of wire of resistance $500\,\Omega$ to produce a current in it of 3.0 mA?

The p.d. V is given by

$$\begin{aligned} V &= IR \\ &= (3.0 \times 10^{-3}\,\text{A})(500\,\text{V A}^{-1}) \\ &= 1.5\,\text{V} \end{aligned}$$

This is about the e.m.f. of a single dry cell. With such a small current as this the difference between the p.d. across the cell and its e.m.f. is negligible. ◀

Fixed resistors

A number of different methods of construction are used commercially; the values obtainable cheaply cover the whole range from about $0.5\,\Omega$ to $20\,\text{M}\Omega$. A selection of types is shown in section in figure 8.2. The best quality are the *wire-wound* type (a). Commercial resistors generally use *nichrome* wire (the same material as is used in electric fire

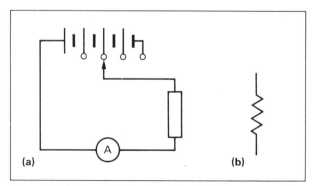

(a)   (b)

Figure 8.1 (a) Investigating the relationship between current and p.d. for a resistor. (b) The older symbol for a resistor.

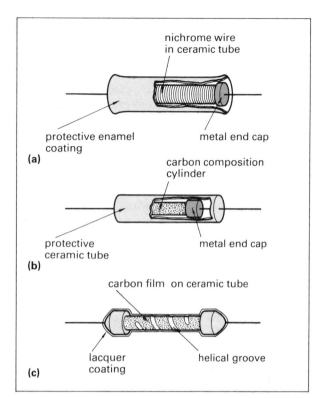

Figure 8.2 Various types of resistor in section: (a) wire-wound, (b) carbon composition, (c) carbon film.

elements). The wire is enamelled in this case to insulate the turns from one another. If the resistor is to be used with alternating currents it is necessary for it to produce very little magnetic field, otherwise it causes other effects in the circuit besides its intended function of resisting the current (page 227). With large resistances, containing long lengths of wire, the wire is wound in sections, these being wound alternately in opposite directions so that their magnetic fields cancel out. This is called a *non-inductive* winding.

The cheapest form of resistor is the *carbon composition* type (b), consisting of a short rod of fire clay mixed with powdered carbon. But the resistances of these are apt to vary with time, and they are used only for low-precision purposes. A resistance with better stability is the *carbon film* type (c). This consists of a fine film of carbon deposited on a ceramic rod. A helical groove is then cut in the surface to bring the resistance of the unit to the required value. Another type of resistor (not shown) with high stability is the *metal oxide* type, consisting of a film of a tin oxide deposited on a ceramic rod. All commercial resistors have their surfaces covered by a thick coating of lacquer or epoxy resin to protect the delicate wires or films against mechanical damage or chemical corrosion.

Variable resistors

The simple ones used in radio circuits consist of a strip of

carbon composition in the form of a part of a circle, to which a sliding connection is made by means of a rotating arm; fixed connections are made to the two ends of the strip. A similar wire-wound design (figure 8.3a) has the wire wound on thin card, which is then bent into an arc of a circle a few centimetres in diameter; the rotating arm makes contact with the turns along the edge of the card. For dealing with larger currents a straight ceramic tube about 200 mm long is wound with the resistance wire (figure 8.3b), and the sliding contact is carried on a metal bar parallel with the tube. You should examine the types to be found in a laboratory.

A variable resistor can be used in circuits in two ways. (a) *As a rheostat* – for controlling the current in a low-resistance device, such as a light bulb. Only one of the end terminals is used in this case (figure 8.4a); often the circuit symbol for this is simplified as shown in the alternative diagram alongside.

➤ A rheostat (variable between 0 and 10 Ω) is used as in figure 8.4a to control the brightness of a bulb; the cell is a lead-acid type of e.m.f. 2.0 V (and negligible resistance), and the ammeter is also of negligible resistance. At the minimum setting of the rheostat the bulb is at full brightness and the current is 0.25 A. With maximum rheostat resistance

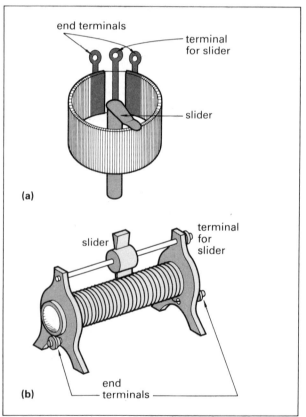

Figure 8.3 Wire-wound variable resistors: (a) a small circular radio type; (b) a large heavy-duty type on a straight tube.

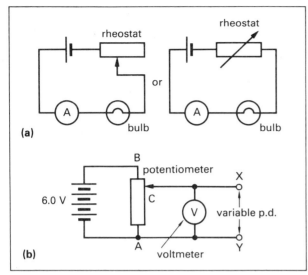

Figure 8.4 Two ways of using a variable resistor: (a) as a rheostat for controlling a current over a limited range, (b) as a potentiometer to provide a p.d. variable from zero upwards.

the bulb scarcely glows red and the current is 0.15 A. What is the resistance of the bulb at these two settings?

At maximum brightness the only significant resistance in the circuit is that of the bulb, and the full p.d. of 2.0 V is applied across it. Its resistance R_1 is therefore given by

$$R_1 = \frac{2.0\ \text{V}}{0.25\ \text{A}} = 8.0\ \Omega$$

At minimum brightness we have

$$\text{p.d. across rheostat} = (0.15\ \text{A})(10\ \Omega)$$
$$= 1.5\ \text{V}$$

and so, p.d. across bulb = 2.0 V − 1.5 V = 0.5 V

and the bulb resistance R_2 is now given by

$$R_2 = \frac{0.5\ \text{V}}{0.15\ \text{A}} = 3.3\ \Omega$$

The use of the variable resistor as a rheostat in this example allows the current to be controlled only in the range 0.15 A to 0.25 A; but this is sufficient to cover a wide range of lamp brightness. ◀

(b) *As a potentiometer* – for controlling the p.d. across some device. A battery may be joined to the two end terminals of the variable resistor (figure 8.4b), so that the battery p.d. is applied across the whole length of the resistor. Any required fraction of the total p.d. can then be tapped off between the sliding contact and one end of the resistor. This arrangement enables the p.d. to be varied continuously right down to zero.

Thus with the sliding contact C at the top end of the potentiometer (at point B) the full p.d. of 6.0 V is obtained at the terminals X and Y. With the slider at the other end (at point A) the p.d. is zero. If the potentiometer is a *linear*

one, the resistance between A and C is directly proportional to the distance of the slider from A. Half-way along its track it divides the potentiometer into two halves of equal resistance; provided the current taken by the voltmeter, etc, through the contact C is negligible, the p.d. between B and C is equal to that between C and A, namely 3.0 V. Three-quarters of the way from A to B the p.d. between C and A is 3/4 of the total, namely 4.5 V; and so on. This simple reasoning applies only as long as the current taken through the contact C is small compared with the current in the potentiometer, as the following example shows.

▶ In figure 8.4b the resistance of the linear potentiometer (between A and B) is 1000 Ω. The voltmeter is found to read 3.0 V with the slider 3/5 of the way from A to B. (i) What current is taken by the voltmeter? (ii) What is the voltmeter resistance?

(i) The voltmeter reading shows that the p.d. between C and A is 3.0 V. Since the total p.d. across the potentiometer is 6.0 V, the p.d. between B and C is also 3.0 V. Now

$$\text{resistance of CA} = \frac{3}{5} \times 1000\ \Omega = 600\ \Omega$$

$$\text{resistance of BC} = \frac{2}{5} \times 1000\ \Omega = 400\ \Omega$$

Therefore   current in CA $= \dfrac{3.0\ \text{V}}{600\ \Omega} = 5.0 \times 10^{-3}\ \text{A}$

$$= 5.0\ \text{mA}$$

and   current in BC $= \dfrac{3.0\ \text{V}}{400\ \Omega} = 7.5\ \text{mA}$

The current arriving at the point C must be equal to the total current leaving C (Kirchhoff's first law, page 76). Therefore the current I taken by the voltmeter through the slider C is given by

$$I = 7.5\ \text{mA} - 5.0\ \text{mA} = 2.5\ \text{mA}$$

(ii) The resistance R of the voltmeter is given by

$$R = \frac{3.0\ \text{V}}{2.5 \times 10^{-3}\ \text{A}}$$

$$= 1200\ \Omega$$

This is about the same size as the potentiometer resistance and in this case the current takes by the voltmeter significantly affects the p.d. between C and A. ◀

Characteristics

A potentiometer (and voltmeter) can be used to provide a continuously variable p.d. to apply across the resistor and ammeter in the experiment of figure 8.1a. With this improvement it is instructive to repeat the experiment not only for coils of wire at steady temperatures (figure 8.5a) for which Ohm's law applies, but also for a number of components for which it does not apply.

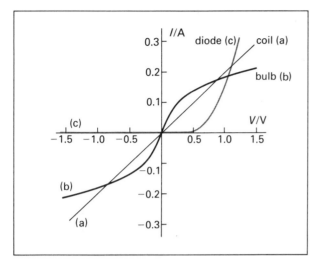

Figure 8.5 The characteristics of various circuit components: (a) a coil of wire, (b) a bulb, (c) a semiconductor diode (silicon).

A fine wire, e.g. an electric bulb filament. In this case the temperature changes considerably with the current, and the resistance does not remain constant (figure 8.5b).

A thermistor. This is a temperature-dependent resistor (made of semiconducting substances), whose resistivity decreases rapidly with rising temperature. Touching the thermistor or blowing on it drastically alters the current in it. Also the energy converted in the thermistor by the current itself raises the thermistor's temperature.

A semiconductor diode (page 285). This has unidirectional properties; it conducts current in one direction only; in the reverse direction its resistance is very high. Even in the conducting direction the current is very far from being proportional to the p.d. (figure 8.4c).

There is no single constant that describes the properties of these components. Rather we must display them by plotting a graph of current I against p.d. V and, if the properties are unsymmetrical (as in the semiconducting diode), then the graph must extend also to negative values of V and I. Such curves are called the *characteristics* of the components concerned.

Energy conversion in resistors

The electrical power P converted in a resistor (or any other circuit component) is given by

$$P = VI \quad \text{(page 84)} \tag{1}$$

But the resistance R is given by

$$R = \frac{V}{I}$$

Substituting for V we have

$$P = I^2 R$$

Thus, the rate of production of internal energy in a resistor is proportional to the *square* of the current in it, a result known as *Joule's law.*

Substituting for I in equation (1) above, we have

$$P = \frac{V^2}{R}$$

So we have three alternative expressions for the rate of conversion of energy P:

$$P = VI = I^2 R = \frac{V^2}{R}$$

The last two of these formulae can apply only to the production of internal energy in resistors. But the first formula, not involving the resistance R, may be used to give the rate of production of energy of any form in a circuit component in which a current I passes and across which the potential difference is V.

For example, if a current of 0.50 A passes through the armature of a small motor across which the p.d. is 20 V, the total power supplied is 10 W; this includes internal energy and mechanical energy. If the armature resistance is 12 Ω, then the rate of production of *internal energy* is given by

$$I^2 R = (0.50 \text{ A})^2 (12 \text{ }\Omega) = 3.0 \text{ W}$$

and the rate of production of *mechanical energy* is given by

$$VI - I^2 R = 10 \text{ W} - 3.0 \text{ W} = 7.0 \text{ W}$$

8.2 Combinations of resistors

Many circuits contain networks of resistors and cells; and we must be able to calculate the combined effective resistance of various arrangements.

(a) *Resistors in series* (figure 8.6a). In this arrangement there is the same current I in each of the resistors R_1, R_2 and R_3. The combined series resistance R_{ser} is given by

$$R_{ser} = \frac{V}{I}$$

where V is the total p.d. across the combination. It is the sum of the separate p.d.s across the individual resistors (page 81), and so

$$V = IR_1 + IR_2 + IR_3$$
$$= I(R_1 + R_2 + R_3)$$
$$\Rightarrow \quad R_{ser} = R_1 + R_2 + R_3$$

(b) *Resistors in parallel* (figure 8.6b). In this arrangement the same p.d. V exists across all the resistors, R_1, R_2 and R_3 (page 81). The combined parallel resistance R_{par} is given by

$$R_{par} = \frac{V}{I}$$

where I is the total current through the combination. This

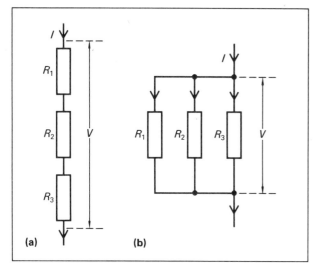

Figure 8.6 (a) Resistors in series: there is the same current in all of them, but the p.d.s across them add up. (b) Resistors in parallel: there is the same p.d. across all of them, but the currents in them add up.

is the sum of the separate currents in the individual resistors, and so

$$I = \frac{V}{R_1} + \frac{V}{R_2} + \frac{V}{R_3} = V\left(\frac{1}{R_1} + \frac{1}{R_2} + \frac{1}{R_3}\right)$$

$$\Rightarrow \quad \frac{I}{V} = \frac{1}{R_{par}} = \frac{1}{R_1} + \frac{1}{R_2} + \frac{1}{R_3}$$

This result may be expressed by saying that the combined *conductance* of conductors in parallel is the sum of their separate *conductances*, since conductance is the reciprocal of resistance. Thus

$$G_{par} = G_1 + G_2 + G_3$$

In two particular cases the result may be simplified, making for greater speed in calculation.

(i) For *n equal* resistances R in parallel we have

$$R_{par} = \frac{R}{n}$$

(ii) For *two* resistances only in parallel, we have

$$R_{par} = \frac{R_1 R_2}{R_1 + R_2} \quad \left(= \frac{product}{sum}\right)$$

▶ Work out the combined resistance of each of the arrangements of standard resistance coils shown in figure 8.7.

(a) $\dfrac{1}{R_{par}} = \dfrac{1}{3\,\Omega} + \dfrac{1}{4\,\Omega} + \dfrac{1}{6\,\Omega} = \dfrac{9}{12}\,\Omega^{-1}$

$\Rightarrow \quad R_{par} = \dfrac{12}{9}\,\Omega = 1.33\,\Omega$

Notice that with resistances in parallel the combined

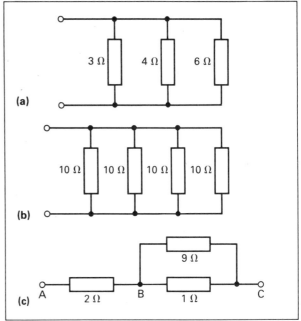

Figure 8.7 Combinations of resistors.

resistance is less than the *smallest* resistance involved.
(b) In this case the resistances are equal; so we have

$$R_{par} = \frac{10\,\Omega}{4} = 2.5\,\Omega$$

(c) Between B and C we have two resistances in parallel, and so

$$R_{par} \left(= \frac{product}{sum}\right) = \frac{9\,\Omega \times 1\,\Omega}{9\,\Omega + 1\,\Omega}$$

$$= 0.9\,\Omega$$

This resistance is in series with $2\,\Omega$; and so the combined resistance $R_{AC} = 2\,\Omega + 0.9\,\Omega = 2.9\,\Omega$. ◀

Measuring instruments

Ammeters

The basis of construction of an ammeter is an ordinary moving-coil meter (page 207). A typical meter of this kind may be designed by the manufacturer to give full-scale deflection (f.s.d.) for a current of 1 mA, and its resistance may be 150 Ω (the coil has many turns of very fine wire). This is then adapted to the range of current required in the ammeter by joining a low resistance, called a *shunt*, in parallel with it. Most of the current through the combination passes through the shunt, and just a small fixed fraction of it passes through the meter (figure 8.8a).

For instance, suppose we want to adapt the meter just

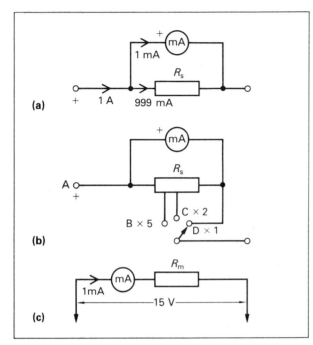

Figure 8.8 (a) The use of a shunt to enable a 1 mA meter to read currents up to 1 A. (b) A multi-range ammeter shunt. (c) The use of a series multiplier to enable the same meter to be used to read p.d.s up to 15 V

described to make an ammeter reading up to 1A. This means that the shunt resistance R_s must be chosen so that when 1 mA passes through the meter, 999 mA passes through the shunt; the total current through the combination is then 1000 mA ($= 1$ A). There is the same p.d. across shunt and meter, so that the currents in them are inversely proportional to their resistances. So we can write

$$\frac{999 \text{ mA}}{1 \text{ mA}} = \frac{150 \,\Omega}{R_s}$$

$$\Rightarrow \quad R_s = \frac{150 \,\Omega}{999} = 0.15 \,\Omega$$

Since the resistances of meter and shunt are in the ratio 999 : 1, the total current is always 1000 times the current in the meter; for instance, if the meter registers 0.7 mA, the total current is 0.7 A, etc.

If the same instrument is to be used on a number of different ranges, it is possible to provide it with a 'universal' shunt of the type shown in figure 8.8b. The resistance R_s is first adjusted to provide the lowest required range of the ammeter; connection for this range is made through the switch contact D. Further tappings are then made at points along R_s, as shown, to give the higher ranges. The ammeter is always joined in a circuit so that the current enters at the positive terminal A; the other terminal is then selected according to the range required.

An ammeter should have as low a resistance as possible, so that the potential difference across it is always small.

The p.d. across a good modern instrument at full-scale deflection is much less than 0.1 V, corresponding to a shunt resistance for a 1 A range of less than 0.1 Ω.

Voltmeters

The basis of construction of a voltmeter is again an ordinary moving-coil meter. Any such meter can be used, as it stands, as a voltmeter; its reading is proportional to the current in it, which in turn is proportional to the potential difference across it. For instance, the meter used above, giving full-scale deflection at 1 mA and of resistance 150 Ω, can be used as a voltmeter, as it stands, since the p.d. across it at full-scale deflection is

$$(1.0 \times 10^{-3} \text{ A})(150 \,\Omega) = 0.15 \text{ V}$$

and the scale could be marked with divisions labelled 0 to 0.15 V. But if we need to adapt it to read a higher range of p.d. than this we have to connect a high resistance in series with it, called a *multiplier* (figure 8.8c).

For instance, suppose we want to make this meter into a voltmeter reading up to 15 V. The total resistance of meter and multiplier in series must then be such that a current of 1 mA passes through them when the p.d. across the combination is 15 V; and so

$$\text{total resistance} = \frac{15 \text{ V}}{1.0 \times 10^{-3} \text{ A}}$$

$$= 15\,000 \,\Omega$$

The resistance R_m of the multiplier is therefore given by

$$R_m = 15\,000 \,\Omega - 150 \,\Omega = 14.85 \text{ k}\Omega$$

Resistance measurement

The direct measurement of a resistance R consists in finding the value of the steady current I in it and the corresponding p.d. V across it. Then

$$R = \frac{V}{I}$$

We may use an ammeter and voltmeter to do this with the circuit of figure 8.9. The resistance actually obtained from the meter readings in this experiment is that of the resistor and voltmeter in parallel. Knowing the voltmeter resistance we can then work out the required resistance R. For instance, suppose the resistance given by the meter readings is 140 Ω, and the voltmeter resistance is known to be 10 kΩ. Then

$$\frac{1}{140 \,\Omega} = \frac{1}{R} + \frac{1}{10\,000 \,\Omega}$$

$$\text{giving} \quad R = \frac{(10\,000 \,\Omega)(140 \,\Omega)}{10\,000 \,\Omega - 140 \,\Omega}$$

$$= 142 \,\Omega$$

There is thus a correction of 2 Ω to the measured resistance.

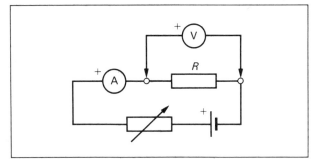

Figure 8.9 Measuring a resistance with ammeter and voltmeter; allowance may have to be made for the resistance of the voltmeter.

Clearly, if the voltmeter resistance is very large compared with R, the correction is unimportant. For instance, with a voltmeter resistance of 100 kΩ you can check that the correction in the above case would be negligible (at least to 3 significant figures, which is the maximum precision obtainable with ordinary dial-reading instruments).

A good moving-coil voltmeter should have as high a resistance as possible, so that the current in it is a negligible fraction of that in the circuit component across which it is connected. A good modern instrument for a 2 V range has a resistance of about 20 kΩ, and takes a current of only 50 μA at full-scale deflection; such an instrment is often described as having a resistance of 10 kΩ V^{-1}.

When even this current would be too great in a particular situation, we make use of a different kind of instrument called an *electrometer* (page 182). This is a meter joined to an electronic amplifying circuit. Its deflection is proportional to the input p.d., but the input resistance is usually more than 10^{12} Ω. The current taken from almost any component across which it is connected is quite negligible. For some purposes we can use a cathode-ray oscilloscope (c.r.o.) as a voltmeter; the deflection of the spot of light is proportional to the p.d. between the deflector plates, and again the input resistance is very high ($\approx 10^7$ Ω).

The Wheatstone bridge.

For the quick and accurate *comparison* of resistances the Wheatstone network or *bridge* is widely used (figure 8.10). The four resistances R_1, R_2, R_3, R_4 are joined in a quadrilateral ABCD. A source of e.m.f. is connected across one diagonal AC, and a sensitive meter and tapping key across the other BD. One or more of the resistances is adjusted until no deflection of the meter can be detected when the tapping key is pressed. The bridge is then said to be *balanced*. When this condition holds, it may be shown that the relation connecting the four resistances is

$$\frac{R_1}{R_2} = \frac{R_3}{R_4}$$

Thus, if R_3 is a known resistance, and the ratio of R_1 to R_2 is known, R_4 may be found. The proof is as follows.

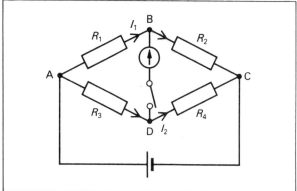

Figure 8.10 The Wheatstone bridge network for measuring resistances; when the bridge is balanced: $R_1/R_2 = R_3/R_4$.

When the bridge is balanced, there is no current in the meter; therefore there is the same current in R_1 and R_2; let this be I_1. Likewise there is the same current in R_3 and R_4; let this be I_2. Also, with no current in the meter, B and D must be at the same potential, so that the p.d.s across R_1 and R_3 are the same.

Therefore $I_1 R_1 = I_2 R_3$

Similarly, the p.d.s across R_2 and R_4 are the same;

and so $I_1 R_2 = I_2 R_4$

Dividing, $\dfrac{R_1}{R_2} = \dfrac{R_3}{R_4}$

A Wheatstone network may be made from standard laboratory components. Standard resistance boxes are made in a variety of patterns, employing either plugs or switches to bring into use the sets of coils in the box so as to give any desired value of resistance. Figure 8.11 shows the arrangement used in a *decade resistance box*, whose total resistance can be switched in steps of 1 Ω from 0 upwards. The resistances are arranged in banks of units, tens, hundreds, etc, of ohms, which are selected by means of switches, as shown.

Suppose we need to set up a bridge to measure a resistance

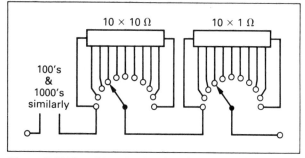

Figure 8.11 The wiring of a decade resistance box that gives values from zero upwards in steps of 1 Ω.

of about 50 Ω. This unknown resistance is connected in the network as R_4, and a standard decade resistance box is used as R_3. We could take two fixed standard resistors for R_1 and R_2, choosing values of 100 Ω and 10 Ω respectively. If R_4 is about 50 Ω, the bridge should then be balanced for some value of R_3 in the neighbourhood of 500 Ω; suppose we find the actual value to be 512 Ω at balance. Then we know that the unknown resistance R_4 is 51.2 Ω.

In a practical measurement it is usual to include in series with the meter a protective resistor to prevent damaging the instrument by excessive currents while preliminary attempts are being made to find the balance condition. The protective resistor is removed from the circuit when the balance condition is nearly reached, so that the maximum sensitivity of the meter is available for the final adjustment.

8.3 Battery resistance

When a current is taken from a cell (or any other source of e.m.f.), it is found that the p.d. across it falls. This is just what we expect from considerations of energy, as we have already seen in the previous chapter (page 81). The way in which the p.d. across a cell varies with the current through it may be investigated with the circuit shown in figure 8.12a. If too large a current is taken from the cell, chemical changes occur round the electrodes and the e.m.f. changes, an effect known as *polarisation*. We need to make sure that the currents taken are small enough to avoid this; in other words the voltmeter should be watched before each obser-

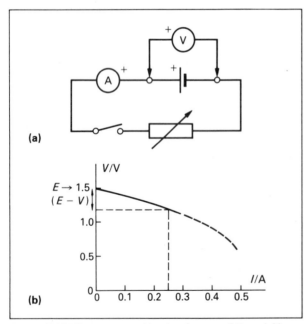

(a)

(b)

Figure 8.12 When a current I is taken from a cell the p.d. V across it falls on account of its internal resistance.

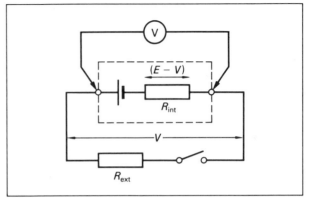

Figure 8.13 The internal resistance of a cell can be represented as a resistance in series with it. When the switch is closed in this circuit, the drop in p.d. enables us to calculate R_{int} in terms of R_{ext}.

vation to see that its reading is not slowly falling. The graph of figure 8.12b shows how the p.d. V varies with the current I (for small enough currents). For zero current the p.d. is equal to the e.m.f. E; the voltmeter should take negligible current, and therefore reads the e.m.f. when the switch is open. The straight line graph shows that the *drop in p.d.* $(E - V)$ is proportional to the current.

Thus $(E - V) = IR_{int}$

where R_{int} is a constant for the cell. A source of e.m.f. thus behaves as though it included an *internal resistance* R_{int} joined in series with it (figure 8.13). It is as though the observed drop in p.d. $(E - V)$ occurs across this. The internal resistance R_{int} is of course an integral part of the source of e.m.f. and can never be separated from it. But for the purpose of circuit analysis we may represent it separately as in figure 8.13, provided we remember that the p.d. V across the cell is actually that across cell and internal resistance combined. We therefore write

$$R_{int} = \frac{(E - V)}{I} \tag{1}$$

The p.d. V across the cell is also the p.d. across the the *external resistance* R_{ext} of the circuit outside the cell. We can therefore write

$$R_{ext} = \frac{V}{I} \tag{2}$$

Substituting for V/I in equation (1) and re-arranging, we have

$$(R_{ext} + R_{int}) = \frac{E}{I}$$

Now $(R_{ext} + R_{int})$ is the *total resistance* of the circuit. If we represent this by R_{tot}, we have

$$R_{tot} = \frac{E}{I} \tag{3}$$

Equations (1), (2) and (3) above are of essentially the same form. They show the application of the concept of resistance to the analysis of the electrical behaviour of different parts of a circuit: (1) of the source of e.m.f., (2) of the external circuit joined to the source, and (3) of the complete circuit of which the source of e.m.f. is part. There is no need for you to trouble to remember these formulae as separate things. They are all examples of the concept of resistance:

$$\text{resistance} = \frac{\text{p.d.}}{\text{current}}$$

You need only to be clear in your mind which resistance you are considering, and therefore which p.d. and current go with it.

▶ A dry cell of e.m.f. 1.5 V and internal resistance 2.0 Ω is connected to a torch bulb whose resistance at its operating temperature is 5.5 Ω. What is the p.d. across the cell? At what rate is energy converted (i) from chemical energy to electrical energy in the cell, (ii) to internal energy in the cell, (iii) to internal energy in the bulb?

The total resistance R_{tot} of the circuit is given by

$$R_{tot} = 5.5\,\Omega + 2.0\,\Omega = 7.5\,\Omega$$

From equation (3),

$$\text{current } I = \frac{E}{R_{tot}} = \frac{1.5\text{ V}}{7.5\,\Omega}$$
$$= 0.20\text{ A}$$

Now considering the internal resistance of the cell only, from equation (1),

$$(E - V) = IR_{int} = (0.20\text{ A})(2.0\,\Omega)$$
$$= 0.40\text{ V}$$

Hence   p.d. V across the cell $= 1.5\text{ V} - 0.40\text{ V} = 1.1\text{ V}$

Alternatively, we could have reasoned from equation (2)

$$\text{p.d. } V\text{ across the bulb} = IR_{ext} = (0.20\text{ A})(5.5\,\Omega)$$
$$= 1.1\text{ V}$$

which, of course, is the same as the p.d. across the cell. (i) The rate at which energy is converted in the cell P_{tot} (page 84) is given by

$$P_{tot} = EI = (1.5\text{ V})(0.20\text{ A})$$
$$= 0.30\text{ W}$$

This is the rate at which energy is converted from chemical to electrical form in the cell. Some of this energy is converted forthwith to internal energy because of the internal resistance R_{int} of the cell.

(ii)   The power P_{int} converted in this way is given by

$$P_{int} = I^2 R_{int} = (0.20\text{ A})^2(2.0\,\Omega)$$
$$= 0.080\text{ W}$$

(iii)   The power P_{ext} supplied to the bulb is given by

$$P_{ext} = VI = (1.1\text{ V})(0.20\text{ A})$$
$$= 0.22\text{ W}$$

And we observe in passing that $P_{tot} = P_{int} + P_{ext}$ ◀

Measuring internal resistance

Figure 8.13 shows a simple method of doing this. We use a voltmeter to measure the p.d. V across the cell when a small *known* resistance R_{ext} is joined across it as in the figure. With the switch open the high-resistance voltmeter indicates the e.m.f. E of the source. Then when the switch is closed the p.d. across the source falls, and the reading V is taken. We then have for the complete circuit

$$E = I(R_{ext} + R_{int})$$

and for the external circuit

$$V = IR_{ext}$$

Dividing these equations,

$$\frac{E}{V} = \frac{R_{ext} + R_{int}}{R_{ext}} = 1 + \frac{R_{int}}{R_{ext}}$$

$$\Rightarrow \quad R_{int} = R_{ext}\left(\frac{E}{V} - 1\right)$$

If there is any significant polarisation of the cell, this will be shown up by a steady fall in the voltmeter reading after the switch is closed. The reading required to give the internal resistance in this case is that obtained *immediately* after closing the switch.

To work out the combined internal resistance of a battery of several cells we treat them as a series or parallel combination of equal internal resistances. The rules for working out the combined e.m.f. of a battery have already been discussed (page 83).

Consider for instance a battery of four identical dry cells each of e.m.f. 1.5 V and internal resistance 0.6 Ω. Joined in series, their combined e.m.f. is

$$1.5\text{ V} \times 4 = 6.0\text{ V}$$

and their combined internal resistance is

$$0.6\,\Omega \times 4 = 2.4\,\Omega$$

Joined in parallel, the e.m.f. is the same as that of one cell, namely 1.5 V; but the internal resistance is much less, being that of four 0.6 Ω resistances in parallel, namely

$$0.6\,\Omega \div 4 = 0.15\,\Omega$$

Recharging storage cells

In certain types of cell the action is reversible, e.g. the lead-acid cell or the nife cell (page 83). If the current is driven through these in opposition to the e.m.f., the chemical action in the cell is reversed and the ingredients are returned to their original condition. This is called *recharging* the cell, though it should more correctly be described as 're-energising' it; electrical energy in the circuit is changed into chemical energy in the cell – subsequently to be made available again as electrical energy when the cell drives a current in the usual direction.

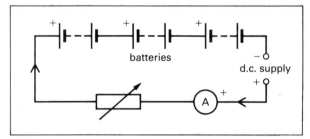

Figure 8.14 Charging three car batteries from a d.c. supply, the p.d. of the supply must be greater than the combined battery e.m.f.s.

Figure 8.14 shows the sort of circuit that must be used, for example to recharge a set of car batteries. The d.c. supply must have a p.d. greater than the e.m.f. of the whole battery if it is to drive current backwards through it. Because of the low internal resistance of such batteries a series resistance is included to limit the current to the required value.

Suppose the batteries in figure 8.14 have a total e.m.f. of 36 V, and that a 50 V d.c. supply is used. Then to obtain a charging current of 2.0 A we can calculate the resistance required as follows. (We can ignore the internal resistance of the car batteries, which is probably less than 0.01 Ω each.)

The resultant e.m.f. in the circuit

$$= 50 \text{ V} - 36 \text{ V} = 14 \text{ V}$$

Hence the series resistance R required is given by

$$R = \frac{14 \text{ V}}{2.0 \text{ A}} = 7.0 \ \Omega$$

The total rate of supply of energy

$$= \text{(p.d. of the supply)} \times \text{(current)}$$
$$= (50 \text{ V})(2.0 \text{ A}) = 100 \text{ W}$$

The rate at which energy is stored in the battery

$$= \text{(e.m.f. of battery)} \times \text{(current)}$$
$$= (36 \text{ V})(2.0 \text{ A}) = 72 \text{ W}$$

The rest of the power (28 W) is the rate at which electrical energy is being converted to internal energy in the resistor.

8.4 Resistivity

The resistance of a conductor at a given temperature depends on its length and cross-sectional area, and on the material of which it is composed. Simple experiments show that for a uniform conductor the resistance is directly proportional to its length l and inversely proportional to its cross-sectional area A. And this is what we expect on theoretical reasoning. The resistance R of such a conductor can therefore be written

$$R = \frac{\rho l}{A}$$

which is an equation that defines ρ, a property of the material of the conductor called its *resistivity*. We can see that the unit of resistivity is the Ω m.

▶ Calculate the resistance of a copper cable 2.5 km long consisting of 6 strands of wire each of diameter 0.60 mm, if the resistivity of copper is $1.7 \times 10^{-8} \ \Omega$ m.

The length l and radius r of the wire must first be expressed in m. Thus

$$l = 2.5 \times 10^3 \text{ m}$$
$$r = 3.0 \times 10^{-4} \text{ m}$$

Then the resistance R of one strand is given by

$$R = \frac{\rho l}{\pi r^2} = \frac{(1.7 \times 10^{-8} \ \Omega \text{ m})(2.5 \times 10^3 \text{ m})}{\pi (3.0 \times 10^{-4} \text{ m})^2}$$
$$= 150 \ \Omega$$

The resistance of the 6 strands in parallel is therefore 1/6 of this, namely 25 Ω. ◀

The reciprocal of the resistivity of a material is known as its *conductivity* σ. Thus

$$\sigma = \frac{1}{\rho}$$

and we have $\quad R = \frac{1}{\sigma} \frac{l}{A}$ (4)

We can account for the conductivity of a material in terms of its atomic properties in the following way.

The characteristic property of a conducting solid is that it contains free charge carriers, either electrons or positive holes (page 77); these are not fixed at any particular site in the crystal lattice, but are free to wander through it in any direction. When there is no electric field in the material, they carry a share of the internal energy of the substance, and move at random rapidly in all directions; but there is no net movement of electric charge. They form in fact a kind of electron 'gas' confined within the crystal structure. When a p.d. is applied between the two ends of a conductor, an electric field is produced in it that acts on the free charge carriers causing them to gain energy. Thus a slow drift of the electron gas through the material takes place which is superimposed on the rapid random movement and constitutes an electric current.

If the p.d. between the ends of the conductor is V, the electric field strength is the potential gradient V/l along it (page 82). If u_e is the electric mobility of the charge carriers (page 82), their average drift speed v is given by

$$v = u_e \frac{V}{l}$$

But we have already shown (page 78) that the average speed v of the charge carriers in a current I is given by

$$v = \frac{I}{nAe}$$

where n is the number of charge carriers per unit volume of the material, and e is the electronic charge. Hence

$$\frac{u_e V}{l} = \frac{I}{nAe}$$

Re-arranging $\frac{V}{I} = R = \frac{1}{u_e ne}\frac{l}{A}$

Comparing this with equation (4), we have

$$\frac{1}{\sigma}\frac{l}{A} = \frac{1}{u_e ne}\frac{l}{A}$$

$$\Rightarrow \qquad \sigma = u_e ne$$

The conductivity σ of a material thus depends on the mobility u_e and the charge carrier density n. Metals all have large values of n (about 10^{29} m^{-3}), and therefore have high conductivities (which means low resistivities). At 293 K (20°C) silver is the best conductor, closely followed by copper (table 8.1). The resistivity of a pure metal is considerably increased by even small traces of impurity; for this reason the copper used for electrical connecting wire is always electrolytically purified (page 79). Aluminium has about twice the resistivity of copper. But since its density is about a third as much, an aluminium conductor of given resistance is much lighter (and cheaper) than a copper one. The overhead power lines of the national electricity grid are made of aluminium; the cables are given a steel core for strength, and the aluminium strands are twisted round this. Other metals have resistivities rather greater, but all of the same order of magnitude. Alloys generally have greater resistivities than pure metals. But non-metallic substances have vastly greater resistivities, almost all of them more than 10^4 Ω m. And the materials used for the insulation of electric circuits have resistivities between 10^{13} Ω m and 10^{16} Ω m.

Intermediate between these extremes comes the important class of materials known as *semiconductors* (silicon, germanium, etc). In these substances only a very small proportion of the valence electrons is free to move through the crystal lattice; the resistivity at 293 K of silicon is 4×10^3 Ω m, and of germanium 0.7 Ω m. However, these figures are obtained only with the purest specimens; and the addition of minute traces of impurities may increase the number of free charge carriers in the crystal lattice considerably, and so decrease the resistivity. By contrast, impurities in a metal do not much affect the number of free electrons, but have a marked effect on their freedom of movement, and therefore increase the resistivity.

Temperature effects

When the temperature of a metal is raised, the thermal vibrations of the atoms increase. This increases the interaction of the electrons with the crystal lattice, and therefore reduces their mobility. The resistivity of the metal therefore increases with temperature.

An opposite effect occurs in an electrolyte. In this case raising the temperature decreases the viscosity of the liquid (page 66), allowing the ions to move more freely through it. A temperature rise therefore decreases its resistivity. Semiconductors show a very considerable decrease of resistivity when the temperature rises. In these materials greater thermal vibration of the atoms sets free more electrons (and holes) in the crystal lattice for electrical conduction, i.e. n increases. The *conductivity* is found to rise approximately *exponentially* with temperature (appendix B7). You may test this with a thermistor (which is a resistor made of semiconducting oxides); between 0°C and 100°C the resistance of a thermistor decreases by a factor of 20 or 30.

The resistance of a resistor made of any pure metal is found to increase nearly *uniformly* with temperature. If the resistance is R_0 at 0°C and R_θ at some other Celsius temperature θ, then the fractional change in resistance between 0°C and θ is $(R_\theta - R_0)/R_0$, and this quantity is approximately proportional to θ. We can therefore write

$$\frac{R_\theta - R_0}{R_0} = a\theta$$

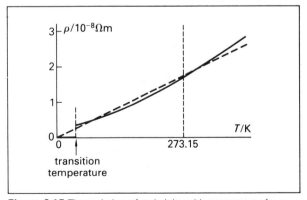

Figure 8.15 The variation of resistivity with temperature for a typical pure metal that becomes superconducting near the absolute zero; above the transition temperature the resistivity is approximately proportional to the kelvin temperature.

metal	resistivity/10^{-8} Ω m	temp.coeff./10^{-3} K^{-1}
pure metals:		
silver	1.63	4.0
copper	1.69	4.3
gold	2.42	4.0
aluminium	3.21	3.8
platinum	11.0	3.8
iron	14.0	6.2
lead	20.8	4.3
mercury	94.1	0.9
alloys:		
nichrome	130	0.17
eureka, constantan	49	0.02
manganin	44	0.02

Table 8.1 Resistivities of common metals (at 293 K).

This is an equation that defines a, a constant for the material of the resistor known as its *temperature coefficient of resistance*. We can re-arrange the above expression to give a formula for the resistance R_θ at any Celsius temperature θ:

$$R_\theta = R_0 \left(1 + a\theta\right)$$

Table 8.1 shows the electrical properties of a number of metals. While the resistivities of *pure* metals vary over a range of more than 50 to 1, their temperature coefficients of resistance show surprisingly little variation; they are all fairly close to 3.7×10^{-3} K^{-1}, which is the same as the coefficient of expansion of a gas ($1/273$ K^{-1}, page 119). In other words the resistance of a pure metal resistor is approximately proportional to its kelvin temperature T. If we plot a graph of resistivity against T we obtain a curve that lies close to a straight line through the origin (figure 8.15).

At temperatures near the absolute zero some metals are found to lose the property of resistance altogether and become perfect conductors. They are then said to be *superconducting*. The onset of superconductivity occurs at a sharply defined transition temperature (figure 8.15), which falls when the specimen is in a magnetic field. We can have an electric current in a superconductor without any energy being transformed to internal energy. Electric currents have been caused to circulate in rings of superconducting material (by momentarily changing the magnetic field inside them, page 222), and have continued for periods of several months without further assistance from any source of e.m.f. In this conduction there seems to be no interaction whatever between the crystal lattice of the material and the electrons that carry the current through it.

For alloys, temperature coefficients of resistance are generally less than for pure metals, as shown in table 8.1. The three alloys shown there are all used for constructing wire-wound resistors.

Semiconductors have large *negative* temperature coefficients of resistance, usually about -0.06 K^{-1}; but no single constant can be quoted for these because of the highly non-uniform variation of resistance with temperature that they show.

8.5 Potentiometer circuits

A variable resistor connected up as a potentiometer (page 89) forms the basis of a class of circuit that may be used for comparing potential differences. These circuits may also be adapted for current and resistance measurement. Under the best conditions they enable these measurements to be made with an uncertainty of only of 1 part in 10^6, and this is therefore the only method used for *very* precise d.c. measurements.

In its simplest form (figure 8.16a) a potentiometer circuit consists of a uniform resistance wire AB stretched over a scale; through this wire a steady current is passed from a d.c. supply (e.g. a 2.0 V lead-acid cell). The potential

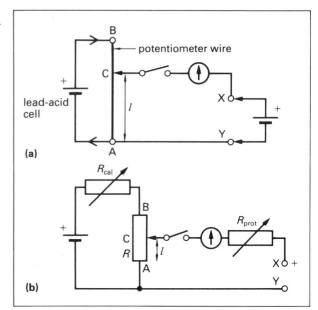

Figure 8.16 (a) The principle of a simple potentiometer circuit for comparing potential differences. (b) A potentiometer circuit based on a linear radio-type potentiometer; the circuit is calibrated with R_{cal} so that it reads p.d.'s up to 1.5 V directly.

difference V to be found is connected at X and Y. It might, for instance, be the p.d. across a small torch cell (about 1.5 V). Its negative terminal (at Y) is joined to the negative end A of the wire; and its positive terminal (at X) is joined through a sensitive meter and a tapping key switch to the contact maker C.

It is convenient in discussing such circuits to take the zero of potential at one end of the potentiometer wire AB; we take A (and Y) as our zero of potential. Then the potential of X is 1.5 V. The potential of the contact maker C will be somewhere between 0 (at A) and 2.0 V (at B). If C is near A, it will be at a lower potential than X, and there will be a current through the meter (when the key is pressed). Similarly, if C is close to B, it will be at a higher potential than X, and the current through the meter will be in the opposite direction. Three-quarters of the way from A to B the potential of C will be $+1.5$ V, the same as X; there is then no current through the meter, and its pointer will not even flicker when the tapping key is pressed. The potentiometer circuit is then said to be *balanced*. The right position for C to balance any given p.d. applied between X and Y is thus found very quickly by trial and error. The length of wire l between the points A and C is then read off from its scale. If the wire is uniform, we may write

$$V \propto l$$

To compare two p.d.s V_1 and V_2 we join them in turn between the points X and Y, and find the corresponding balance lengths l_1 and l_2. Then, provided the potential difference across the wire produced by the lead-acid cell has not changed between the two observations, we have

$$\frac{V_1}{V_2} = \frac{l_1}{l_2}$$

When the circuit is balanced, no current passes through the contact maker C, and the current in the two parts BC and CA of the potentiometer wire are the same; this is not true for an unbalanced position of C. Likewise no current passes (at balance) through X or Y. If we have a cell joined to X and Y, then the p.d. between X and Y (at balance) is precisely equal to its e.m.f. By using a standard cell (e.g. the Western cell, page 83) to provide one of the two p.d.s, we may find other p.d.s by direct comparison with the working standard of e.m.f.

We may also use a potentiometer circuit with the terminals X and Y joined to two points between which we want to know the p.d. in some other circuit (thus using it like a sort of voltmeter). Then at balance no current passes between C and X or between A and Y, and the potentiometer circuit has no effect whatever on the currents or p.d.s in the circuit under test.

The wire AB with its sliding contact C in figure 8.16a is in effect a variable resistor connected up as a potentiometer (page 89). We may if we wish use a simple radio-type potentiometer for this type of circuit. These potentiometers are not designed as precision devices, and even the so-called *linear* kind are often lacking in uniformity. Also in the wire-wound type the sliding contact arm can make contact only at a particular point on each turn of the wire round the card that holds it. The resistance therefore changes in a series of small steps in such a potentiometer, rather than continuously. However, if the wire is fine enough this effect is scarcely noticeable. A carefully selected high quality radio-type potentiometer may be used to build a useful and compact potentiometer circuit for measurement of p.d.s.

The practical arrangement is shown in figure 8.16b. A protective resistor R_{prot} is joined in series with the meter to protect it from excessive currents while preliminary attempts are being made to find the balance point; it does not affect the position of balance, but only the sensitivity of the meter. It can then be reduced to zero to give maximum meter sensitivity for the final adjustment. The variable resistor R_{cal} is used to adjust the current in the potentiometer wire so that the scale of the instrument reads p.d.s directly. For instance, the potentiometer knob can be equipped with a large dial reading from 0 to 150. We can then plan to adjust R_{cal} so that the total p.d. across AB is 1.5 V exactly; each division of the scale then represents 0.01 V. We can make this adjustment using a standard cell. The e.m.f. of the Weston cell is 1.019 V (to four significant figures); the potentiometer dial is therefore set to 101.9 divisions. The instrument is then adjusted with the rheostat R_{cal} until it is balanced at this setting of the potentiometer; and so it is calibrated to read 0.01 V per division. The calibration must of course be repeated each time the instrument is used, since the p.d. of the d.c. supply is liable to change. Other p.d.s may now be joined to the points X and Y instead of the standard cell; the potentiometer is quickly balanced, and the p.d. is read off from the dial.

Calibrating a voltmeter

The potentiometer circuit can be regarded as a sort of voltmeter. But even in its simplest forms it can be much more sensitive and precise than an ordinary moving-coil voltmeter. It may therefore be used to calibrate other kinds of voltmeter, and to measure the errors in their scales.

The range of p.d.s that may be measured with our simple circuit of figure 8.16b is from 0 to 1.5 V. If the voltmeter to be calibrated covers a larger range than this, we can use several lead-acid cells in series to provide the p.d. across AB. But the following example shows another way of adapting the potentiometer method for measuring relatively large p.d.s (in this case up to 25 V); it also shows the inadequancy of ordinary voltmeters for certain types of measurement.

► Figure 8.17 shows a *fixed* potential divider ACB, consisting of two fixed resistors BC and CA as shown, joined in series across a battery of e.m.f. 24.0 V. Calculate the p.d. between C and A as measured (i) with a potentiometer circuit, (ii) with a voltmeter of resistance 20.0 kΩ.

We assume the battery resistance is negligible compared with the total resistance joined across it, so that the full p.d. of 24.0 V is produced between B and A.
(i) When the potentiometer is balanced, it draws no current through its terminals X and Y (figure 8.16b); joined between A and C in figure 8.17 it takes no current from the resistance chain. The p.d.s across the resistors are therefore in the ratio of their resistances, so that

$$\frac{\text{p.d. between C and A}}{\text{p.d. between B and A}} = \frac{3.00 \text{ k}\Omega}{47.0 \text{ k}\Omega + 3.00 \text{ k}\Omega}$$

$$\Rightarrow \quad \text{p.d. between C and A} = \frac{3.00}{50.0} \times 24.0 \text{ V}$$

$$= 1.44 \text{ V}$$

and this is the p.d. as measured by the potentiometer.
(ii) When the voltmeter is joined to C and A, its resistance of 20.0 kΩ is in parallel with the resistance of 3.00 kΩ. Hence their combined resistance R_{par} is given by

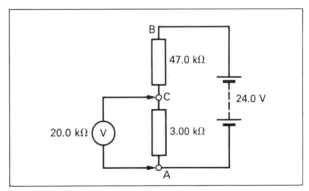

Figure 8.17 The use of a fixed potential divider. A potentiometer circuit can be used to measure the true p.d. between C and A, but a voltmeter alters the p.d. it is supposed to measure.

$$R_{par} = \frac{3.00 \text{ k}\Omega \times 20.0 \text{ k}\Omega}{23.0 \text{ k}\Omega} = 2.61 \text{ k}\Omega$$

This is now the resistance between C and A. The total resistance between B and A is now

$$47.0 \text{ k}\Omega + 2.61 \text{ k}\Omega = 49.6 \text{ k}\Omega$$

Hence   p.d. between C and A $= \dfrac{2.61}{49.6} \times 24.0 \text{ V}$

$$= 1.26 \text{ V}$$

and this is the p.d. measured by the voltmeter.

This example shows clearly the difficulty of using a voltmeter for measuring p.d.s in a circuit where the resistances of the components are comparable with that of the voltmeter. The voltmeter resistance would need to be more than 2 MΩ in this case before it would give the p.d. between C and A as 1.44 V. The potentiometer method does not suffer from this defect. An electrometer (whose input resistance is more than 10^{12} Ω) also avoids this kind of difficulty.

We could use the fixed potential divider of figure 8.17 to adapt our simple 1.5 V potentiometer to measure p.d.s up to 25 V (joined between B and A). The resistances have been chosen so that the p.d. between B and A is always 50/3 times that between C and A (which the potentiometer circuit can be used to measure). This principle can be adapted to handle any required range of high p.d.s.

Measuring small p.d.s

The sensitivity of a potentiometer may be increased almost indefinitely by joining extra resistors in series with the potentiometer; the extra resistance counts in effect as an extension of the potentiometer wire. Suppose we are wanting to use a potentiometer circuit to measure the e.m.f. of a thermocouple, which lies in the range 0 to 15 mV. If we contrive to make the p.d. across the potentiometer 15 mV, then the whole scale of the instrument will be available for the measurement we have in mind. This can be done with the basic circuit of figure 8.16b by using a very much larger resistance in place of R_{cal} to reduce the p.d. across the potentiometer to the required 15 mV. But then we have to find a way of calibrating the instrument against the standard cell.

Figure 8.18 shows how we may do this. A decade resistance box R' is joined in series with the potentiometer R, and its value is adjusted (after calculation) so that the p.d. across it is just less than the e.m.f. of the standard cell (1018.6 mV); for instance we could choose R' so that the p.d. across it is 1014 mV, while the p.d. across R is to be 15 mV. Then we have

$$\frac{R'}{R} = \frac{1014 \text{ mV}}{15 \text{ mV}}$$

Thus, if the total resistance of R is 100Ω, for example, then we want R' to be adjusted to the value

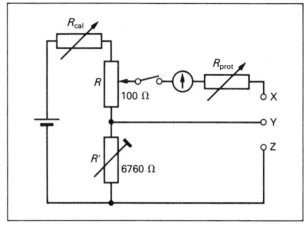

Figure 8.18 A potentiometer circuit adapted for measuring the e.m.f. of a thermocouple.

$$R' = \frac{1014}{15} \times 100 \text{ Ω} = 6760 \text{ Ω}$$

We then balance the standard cell across R' together with 4.6 mV of the p.d. across R (since 1014 mV + 4.6 mV = 1018.6 mV). We connect it between X and Z and adjust the dial of the potentiometer to 46 (on our scale from 0 to 150). We now adjust R_{cal} to balance the potentiometer. We are thus assured of our 15 mV across R, and can use the instrument with the thermocouple connected to the terminals X and Y. Each division of the dial then corresponds to 0.1 mV.

Current measurement

For this purpose the current to be found is arranged to pass through a known resistance R_s (figure 8.19). The p.d. V across this is measured with a potentiometer in the usual way by comparison with the e.m.f. of a standard cell. The current I is then given by

$$I = \frac{V}{R_s}$$

This can be compared with the ammeter reading to give the error at any chosen point of its scale. The standard resistance R_s should be chosen so that the p.d. across it is about 1 V (in the range of our basic potentiometer).

For precise work wire-wound resistors to be used with a potentiometer in this way are made with *four* terminals, as in figure 8.19. Two of these, known as the *current terminals*, are at the ends of the resistance wire, and are used to pass the current into and out of it. The other two, known as the *potential terminals*, are joined to two points P and Q on the resistance wire; and it is between these points that the resistor has its specified resistance. The potential terminals are joined to the points X and Y of the potentiometer circuit; the p.d. between these is equal to

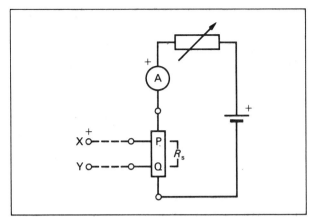

Figure 8.19 Calibrating an ammeter using a standard resistor in series with it; the resistance shown is of the four-terminal type.

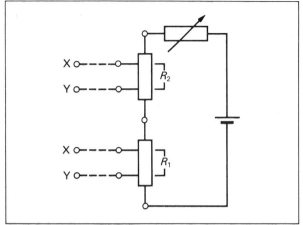

Figure 8.20 Comparing resistances with a potentiometer circuit; the p.d.s across the resistors are in the ratio of their resistances.

that between P and Q on the resistance wire, since at balance no current passes through the wires PX or QY and there is therefore no potential drop along them. An ordinary two-terminal resistor suffers from the uncertain resistance of the end connectors through which the current is passing. This uncertainty can amount to about 0.005 Ω even with heavy brass connectors, and can be much more than this if there is dirt on them.

Resistance measurement

The potentiometer circuit provides the most accurate method of comparing two resistances. The arrangement used is shown in figure 8.20. For precision work both resistors should be of the four-terminal type. A suitable current I is passed through the two resistors R_1 and R_2 connected in series. The potentiometer circuit terminals X and Y are joined in turn across each resistor, and the corresponding readings of p.d. are taken. Provided the currents have not changed between the two measurements, we then have

$$\frac{R_1}{R_2} = \frac{V_1}{V_2}$$

Thus if R_2 is known then R_1 can be measured. Only the ratio of the two p.d.s need be found, so that the potentiometer circuit does not need to be calibrated with a standard cell for this type of measurement.

We choose the current I so that the larger of the two p.d.s is just under 1.5 V. However, it may well be that the currents must be much smaller than this to avoid overheating the coils. In any case the measurements should be repeated with various values of current through the resistors. Only by doing this can we be sure that the currents are not causing changes in the resistances.

9 Heating solids and liquids

9.1 Temperature and internal energy

The temperature of a body is a measure of 'how hot it is'. The higher the temperature of the body, the more energy its molecules have. Instruments for measuring temperature are called *thermometers*. You will be familiar with the common mercury-in-glass type. Such an instrument is calibrated with the Celsius (°C) scale of temperature, defined by

> freezing point of water = 0°C
> boiling point of water = 100°C

(when the freezing and boiling occur under specified physical conditions).

The temperature of a body is raised by adding energy to it, and lowered by removing energy from it. In fact it is impossible to remove all the energy from the molecules. The lowest temperature, i.e. the temperature at which most energy has been removed, is called *absolute zero* (−273.15°C) and the remaining energy is called *zero-point energy*. Figure 9.1a shows a potential energy-separation graph, and a molecule in a body at absolute zero, making an oscillation of very small amplitude. Figure 9.1b shows a molecule in a body at a higher temperature: now when the molecule has minimum potential energy it has more kinetic energy. With more kinetic energy the molecule can move further from its mean position until it stops, and is brought back to its mean position, through which it passes, since there it has its maximum kinetic energy. It continues to oscillate, and the potential energy and kinetic energy of the molecule are continually interchanging.

More convenient, for scientific purposes, than the Celsius scale of temperature is the kelvin scale (K) in which absolute zero = 0 K. On this scale the degree marks on a thermometer are the same distance apart as on the Celsius scale i.e.

> 0°C = 273.15 K
> 100°C = 373.15 K

Internal energy, heat and work

Internal energy is the name we give to the total of the kinetic and potential energies of the molecules of a substance. It has the symbol U, and changes in it are therefore symbolised by ΔU. We can change the internal energy of a body in two ways. We can *heat* it; that is, we can put it near another hotter body, and energy will flow from one to the other. The symbol for this sort of energy transfer is ΔQ.

We cannot talk about 'the heat' Q in a body; we mean the internal energy of the body. 'Heating' is simply a word to tell us that internal energy transfers are occurring because there is a difference of temperature. You will have to accept that there has been much confusion about how the word

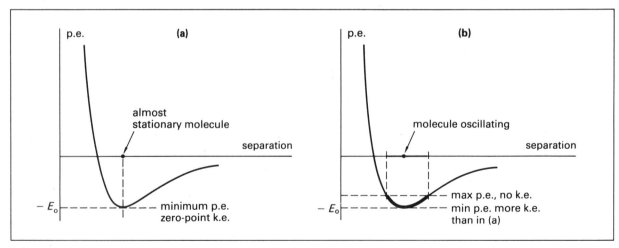

Figure 9.1 Potential energy-separation curves, (a) shows a molecule of a substance at absolute zero. In its mean position it has only zero-point k.e. in its mean position, which allows it to make oscillations of only very small amplitude. In (b) the molecule is from a warmer substance, and the molecule has more k.e. in its mean position, and oscillates over the much larger range shown.

'heat' should be used. It is safe to say that it should never be used as a noun, but only as a verb. We can say, e.g., that a gas flame is heating some water, but not that the gas flame is giving heat to the water. It is giving *energy* to the water, and the process is *heating* (not *working*). Unfortunately the way in which the word heat has been used in the past has led to the retention of the word (as a noun) in the quantities *specific heat capacity* and *specific latent heat*. These really mean specific internal energy capacity, and specific latent internal energy, but you can see that these phrases would be cumbersome. To avoid using 'heat' as a noun we shall refer to these quantities as s.h.c. and s.l.h. (rather as e.m.f. is now used to avoid saying the word 'force' in electro-motive force).

We could also raise the internal energy of a body by doing *work* and the symbol for energy transferred in this way is ΔW. For example, the brake shoes sliding on the brake drums of a car will raise the internal energy of both the shoes and the drums. The molecules of the brake shoes and brake drums were already vibrating before the brakes were applied. The velocities of the molecules in their vibrations were not entirely random, since superimposed on the random vibrations which they had when they were at rest was the forward velocity of the car. Afterwards the molecules vibrate with more kinetic energy, and now the velocities are entirely random. The kinetic energy of the car (an *ordered* form of energy) has become the random kinetic energy of the molecules (a *disordered* form of energy). Some examples of the two processes, heating and working, are given in table 9.1.

situation	process by which energy is transferred	type and site of energy	
		initially	finally
electric bar 'radiant' heater in a room	heating	internal energy of element	internal energy of room
man turning a blunt drill in a block of wood	working	chemical energy of man	internal energy of drill and wood
car stopping by being braked	working	k.e. of car	internal energy of brakes and surroundings
man walking out of warm room	heating	internal energy of man	internal energy of surroundings

Table 9.1 Some examples of heating and working. Notice that when there is *heating*, the initial and final forms are both internal energy, which is transferred because one body is hotter than the other. When there is *working*, the (disordered) internal energy appears at the expense of some ordered form of energy. Chapter 3 also contains some examples of energy transfer, and uses energy-flow diagrams.

First law of thermodynamics

Of course the two processes, heating and working, can occur simultaneously: for example, a refrigerator could be removing 20 J of internal energy from a block of wood while a man is turning a blunt drill in it and therefore doing 30 J of work. The net gain of internal energy would be 10 J. This is generalised in the *first law of thermodynamics*, which states

$$\Delta Q = \Delta U + \Delta W$$

where the sign convention is:

ΔQ = energy transferred *to* the body by heating
ΔU = internal energy *gain*
ΔW = energy transferred *by* the body by working

In our example, therefore, $\Delta Q = -20$ J, $\Delta W = -30$ J, so

$$\Delta U = (-20 \text{ J}) - (-30 \text{ J})$$
$$= +10 \text{ J}$$

the positive sign indicating a *gain* of internal energy, which tallies with the result we had previously.

9.2 Expansion

In figure 9.2 we see some very simple apparatus which can be used to show that a metal bar becomes longer when it is heated. The bar is fixed at the left-hand end, and at the right-hand end rests on a needle placed on a block of wood. The needle passes through one end of a pointer (such as a drinking straw). When the bar is heated (e.g. with a Bunsen burner) the right-hand end of the bar moves outwards and makes the needle roll along the block of wood: the movement of the needle is indicated by the rotation of the straw. The device of the needle and straw is necessary because the amount of expansion is small: an iron bar of length 0.5 m would increase in length by 0.5 mm if its temperature were raised by 100 K. The bar also increases in cross-sectional area, in proportion, but we choose to study the increase in *length* because the increase in radius of the bar would be far too small to be observable by any simple means. Figure 9.3 illustrates the two-dimensional expansion of a body:

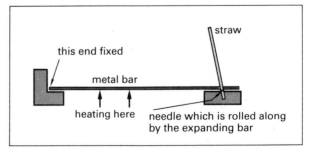

Figure 9.2 Some simple apparatus which shows that the length of a metal bar increases by a small amount when it is heated.

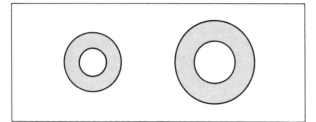

Figure 9.3 All the dimensions of a body increase (in the same proportion) when it expands. The hole becomes larger. (In this diagram the amount of expansion is *greatly* exaggerated.)

the diameter of the hole in the ring increases as much as the other linear dimensions. Quantitative experiments (with more precise apparatus!) show that the increase in length Δl is proportional to the original length l_0 and to the rise in temperature $\Delta \theta$ (provided $\Delta \theta$ is not too large). We can write

$$\Delta l \propto l_0 \Delta \theta$$

and if we include a factor a to allow for different materials expanding differently we have

$$\Delta l = a l_0 \Delta \theta$$

and this equation defines the *linear expansivity* a of a substance. It can be seen that the linear expansivity is the *fractional* increase in length per unit rise in temperature, so that the units of length do not appear in the unit of a: we could write the unit as $m\ m^{-1}\ K^{-1}$, but it is simpler just to write K^{-1}.

Values of expansivity

The values of a lie between $4 \times 10^{-6}\ K^{-1}$ and $70 \times 10^{-6}\ K^{-1}$ for all metals and figure 9.4 is a plot which shows that the higher the melting point of a metal, the lower its linear expansivity. The linear expansivity of a substance is not a constant: it increases with temperature, as is shown (for copper) in figure 9.5.

Molecular explanation

We can understand these facts by looking at the potential energy-separation curve for a pair of molecules. As the temperature of the substance rises, and the molecules have more energy, they make oscillations of greater amplitude, as in (b) rather than (a) of figure 9.6. But the potential energy curve is not symmetrical, so that as the amplitude increases, the centre of the oscillation moves further outwards from its original position, and so, on average, the *molecules are further apart*, and what is true for a pair of molecules will be true for the bulk material: the substance will *expand*. We can also see in figure 9.6c that the asymmetry of the curve means that the locus of the mid-points of lines like AA′, BB′ is not straight: the linear expansivity will therefore

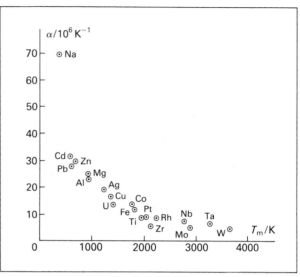

Figure 9.4 The correspondence between a at 298 K and T_m for some metals. We can see that the higher the melting point, the lower the linear expansivity.

rise as the temperature rises. Figure 9.6d shows the curve for a substance with a higher melting-point (it has a deeper trough in its curve, because the substance has to be given more energy to enable it to melt). The deepness of the trough makes the curve more symmetrical, and so the linear expansivity will be smaller.

We have now described, using our model of molecular behaviour
(i)   why substances expand
(ii)   why the linear expansivity increases with temperature
(iii) why susbtances with a high melting point have low linear expansivities.
► A rectangular block of copper which measures $100.00\ mm \times 80.00\ mm \times 60.00\ mm$ has its temperature raised from 300 K to 500 K. Deduce the average linear expansivity of copper over this temperature range (from figure 9.5) and find the new volume of the copper block.

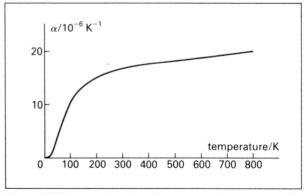

Figure 9.5 The variation of the linear expansivity of copper with temperature.

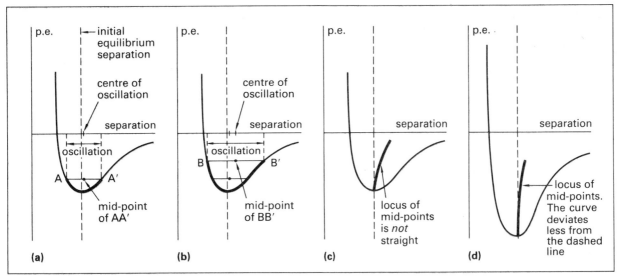

Figure 9.6 An explanation of the expansion of substances.

Hence find the increase in volume, and the cubic expansivity γ (defined by the equation $\Delta V = \gamma V_0 \Delta \theta$, where the symbols have the obvious meanings) of copper over this temperature range. Comment on your result.

The mean linear expansivity is

$$\approx \tfrac{1}{2}(16.7 + 18.3) \times 10^{-6} \text{ K}^{-1}$$
$$= 1.75 \times 10^{-5} \text{ K}^{-1}$$

Using $\Delta l = \alpha l_0 \Delta \theta$ we have for the 100.00 mm side

$$\Delta l = (1.75 \times 10^{-5} \text{ K}^{-1})(100.00 \text{ mm})(200 \text{ K})$$
$$= 0.35 \text{ mm}$$

similarly for the 80.00 mm side $\Delta l = 0.28$ mm

and for the 60.00 mm side $\Delta l = 0.21$ mm

so the new volume

$$V_1 = (100.35 \text{ mm})(80.28 \text{ mm})(60.21 \text{ mm})$$
$$= 485\,058 \text{ mm}^3$$

The initial volume

$$V_0 = 480\,000 \text{ mm}^3$$

so that the increase in volume

$$\Delta V = 5058 \text{ mm}^3$$

and the cubic expansivity

$$\gamma \left(= \frac{\Delta V}{V_0 \Delta \theta} \right) = \frac{5058 \text{ mm}^3}{(480\,000 \text{ mm}^3)(200 \text{ K})}$$

$$= 5.27 \times 10^{-5} \text{ K}^{-1}$$

This is very nearly equal to $3a$ ($3a = 5.25 \times 10^{-5}$ K^{-1}), and it is in general a good approximation to say that the cubic expansivity γ is three times the linear expansivity a. It is of course the cubic expansivity which must be used

when we deal with liquids and gases, as we shall do next.

The cubic expansivities (see the example for the definition) of *liquids* at 300 K are in the range 0.2×10^{-3} K^{-1} to 2.0×10^{-3} K^{-1} (i.e. typically 10 times greater than those of solids) and they increase with temperature, for the same reason as those of solids do. All *gases* have very nearly the same cubic expansivity: at 300 K it is 3.66×10^{-3} K^{-1} (i.e. higher than that of common liquids). The cubic expansivity of a gas is not, however, a concept which is much used, because the volume is so much affected by pressure changes also, as we shall see in chapter 10.

9.3 Heat capacity

Figure 9.7 shows a metal block into which fits an immersion heater and a thermometer. If the immersion heater is switched on, and the thermometer is read at intervals of time, we can obtain graphs like those shown in figure 9.8, with the same immersion heater placed in three different blocks. We see that the different blocks have different *capacities* for absorbing internal energy. The 1.2 kg aluminium block absorbs it well (i.e. the temperature does not rise much), the 0.60 kg aluminium block less well (actually there is twice the temperature rise), and the 1.2 kg copper block is the worst. The idea of *heat capacity* (h.c.) is a useful one, and we define the h.c. C of *a body* by the equation

$$C = \frac{\Delta Q}{\Delta \theta}$$

where $\Delta \theta$ is the temperature rise when an amount of energy ΔQ is transferred by heating (you will realise that we should really call C the internal energy capacity). For example,

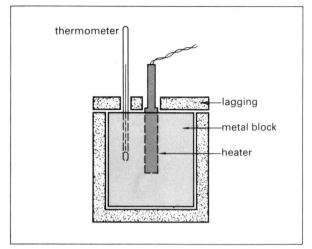

Figure 9.7 Some apparatus which can be used to find how the temperature of a metal block rises when a heater is placed in it. The lagging reduces the amount of heating of the surroundings by the block.

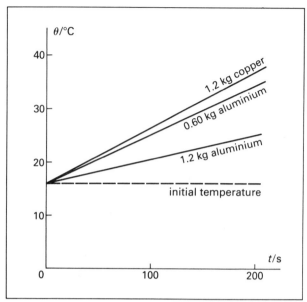

Figure 9.8 Graphs which show the temperature rise with time, when the same heater is placed in three different metal blocks.

if the power of the immersion heater was 48 W, in 200 s it would transfer 9600 J ($= 48 \text{ J s}^{-1} \times 200 \text{ s}$). From the graph we can see that for the 1.2 kg block of aluminium the temperature rise was about 9 K. So from the definition, the h.c. C of that block is given by

$$C = \frac{9600 \text{ J}}{9 \text{ K}} = 1.06 \times 10^3 \text{ J K}^{-1}$$

The h.c. of the 0.60 kg aluminium block would be only half as much, i.e. $5.3 \times 10^2 \text{ J K}^{-1}$, and the h.c. of the 1.2 kg copper block would be

$$\frac{9600 \text{ J}}{21 \text{ K}} = 4.6 \times 10^2 \text{ J K}^{-1}$$

Specific heat capacity

It would also be useful to define the *specific heat capacity* (s.h.c.) of a substance, e.g. the s.h.c. of aluminium. By this we mean the h.c. per unit mass, so the definition of s.h.c. c is

$$c = \frac{\Delta Q}{m \, \Delta\theta}$$

where ΔQ and $\Delta\theta$ are defined as before, and m is the mass of substance. From this definition, and using the data for the 1.2 kg block of aluminium in our experiment, we have, for aluminium,

$$c = \frac{9600 \text{ J}}{(9 \text{ K})(1.2 \text{ kg})}$$

$$= 8.9 \times 10^2 \text{ J K}^{-1} \text{ kg}^{-1}$$

If we had taken the data for the 0.60 kg block of aluminium we should have got the same result, since we are now

dealing with *aluminium* and not with a particular aluminium block. (You should check that this is true, bearing in mind the difficulty of reading the temperature rise from the graph.)

We can use the equations $\Delta Q = C\Delta\theta$ and $\Delta Q = cm\Delta\theta$ to find the amounts of energy which must be given to a body to produce particular temperature rises (or how much must be removed to produce a particular temperature fall). We cannot use either equation to calculate how much 'heat' there is 'in' a body (see page 102): we can use them only to calculate internal energy *changes*.

Energy losses to the surroundings

In the previous paragraphs we have assumed that the blocks kept all the energy given to them by the heaters. In practice the blocks will start to heat the surroundings as soon as their temperature rises above that of the surroundings, and so all the h.c.s and s.h.c.s obtained were overestimates of the actual values, since the temperature rises would have been greater if there had been no energy losses.

Some *lagging* would normally be placed round the blocks, as shown in figure 9.7, to reduce the loss of energy: a suitable material would be expanded polystyrene (see page 118).

▶ An electric kettle has a 2750 W element, and a h.c. of 530 J K^{-1}. If 1.7 kg of water is placed in it (s.h.c. of water = 4200 J K^{-1} kg^{-1}), how long will it take for the temperature of the water (and the kettle) to rise from 20°C to 100°C?

We shall begin by combining the equations $\Delta Q = C\Delta\theta$ and $\Delta Q = cm\Delta\theta$ (since the temperature rise is the same for both objects). The energy ΔQ to be supplied is given by $\Delta Q = (C + cm)\Delta\theta$

$$= \{530 \text{ J K}^{-1} + (4200 \text{ J K}^{-1} \text{ kg}^{-1})(1.7 \text{ kg})\} 80 \text{ K}$$
$$= 6.14 \times 10^5 \text{ J}$$

If it is being supplied at a rate of 2750 J s^{-1}, it will take a time t given by

$$t = \frac{6.14 \times 10^5 \text{ J}}{2750 \text{ J s}^{-1}}$$

$$= 223 \text{ s}$$

The kettle will be heating the surroundings, so this is an underestimate of the time needed. It is obviously sensible to make kettles of a material which has a low density and a low s.h.c. (since the kettle always has to be given energy too), and also sensible never to use more water than we need, especially as its high s.h.c. means that a lot of energy is needed to raise its temperature. ◄

Rates of supply of energy

If we consider that the supply of energy ΔQ, and the consequent rise in temperature $\Delta \theta$, occur in a time Δt, we can rewrite the equation $\Delta Q = cm\Delta\theta$ as

$$\frac{\Delta Q}{\Delta t} = cm \frac{\Delta \theta}{\Delta t}$$

where $\Delta Q/\Delta t$ and $\Delta \theta/\Delta t$ mean the average *rates* of increase of energy and rise of temperature, respectively. We write the instantaneous rates as

$$\frac{dQ}{dt} = cm \frac{d\theta}{dt} \qquad (1)(\text{appendix B3})$$

This equation expresses a useful idea: e.g. if we have a body of known heat capacity which has been heated and is being allowed to cool, and we plot a graph of its temperature against time, we can find the rate of fall of temperature at any point and can calculate the rate at which it is losing energy. Figure 9.9 shows such a *cooling curve*: suppose the h.c. of the body is 1400 J K^{-1}, and we want to know its rate of loss of energy at a temperature of 60°C. We draw the tangent to the curve at this temperature: this gives

$$\frac{d\theta}{dt} = \frac{-65 \text{ K}}{780 \text{ s}} = -8.33 \times 10^{-2} \text{ K s}^{-1}$$

Substituting in the above equation we have

$$\frac{dQ}{dt} = (1400 \text{ J K}^{-1})(-8.3 \times 10^{-2} \text{ K s}^{-1})$$

$$= -117 \text{ J s}^{-1}$$

(the negative sign merely indicates that the body is losing energy).

► The kettle used in the example on page 106 is again used to boil some water. It is switched off at the maximum temperature reached (99°C) and allowed to cool. It is found that the temperature falls 4 K in 5 minutes. What is the rate of loss of energy?

The total h.c. C of the kettle and its contents is given by

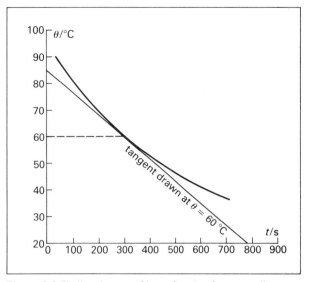

Figure 9.9 Finding the rate of loss of energy from a cooling curve. We find $d\theta/dt$ from the graph, by drawing the tangent at the point where we want to find the rate of loss of energy: then $dQ/dt = C(d\theta/dt)$.

$$C = 530 \text{ J K}^{-1} + (1.7 \text{ kg} \times 4200 \text{ J K}^{-1} \text{ kg}^{-1})$$
$$= 7670 \text{ J K}^{-1}$$

From equation (1) we have

$$\frac{dQ}{dt} = (7670 \text{ J K}^{-1})\left(\frac{4 \text{ K}}{300 \text{ s}}\right)$$

$$= 102 \text{ W} \qquad ◄$$

Most solid substances have s.h.c.s within the range 100 J K^{-1} kg^{-1} to 1000 J K^{-1} kg^{-1}, but most common liquids have s.h.c.s of about 200 J K^{-1} kg^{-1}, that of water (4200 J K^{-1} kg^{-1}) being unusually high. The s.h.c.s of gases vary greatly with the gas: for air it is about 1000 J K^{-1} kg^{-1}.

The high value of the s.h.c. of such a common substance as water has many consequences in everyday life. On a global scale, the oceans rise and fall in temperature much more slowly than does the land, so that the temperature of places on the Earth depends partly on whether they are surrounded by sea or by land. For example, taking the monthly averages of temperature at London (England) and Orenburg (USSR), which have similar altitudes and the same latitude, the differences between the highest and lowest monthly averages at London and Orenburg are 11 K and 40 K respectively. On a smaller scale, the high s.h.c. of water is responsible for the relative coolness in summer of coastal areas, and their relative mildness in winter. On a still smaller scale, the high s.h.c. of water means that a particular heater will take a long time to raise the temperature of water, but that water will not fall in temperature quickly, either. This has obvious significance for people taking baths, or drinking coffee.

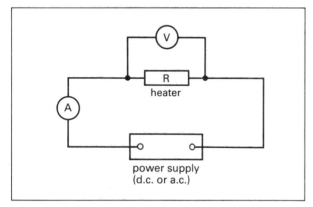

Figure 9.10 The electrical circuit needed to measure the power of a heater. The voltmeter V should have a resistance which is high compared with that of the heater, R.

Measuring s.h.c.

Electrical temperature-raising method

We have already described this method in principle on page 106 and in figure 9.7. It can be used whether the substance is in solid or liquid form although of course a container will be needed if the substance is liquid. Figure 9.10 shows the electrical circuit for this and any other experiment in which electrical heating is used. The ammeter A measures the current I in the heater, and the voltmeter V measures the p.d. V across the heater. If the experiment runs for a time t, the energy given to the substance is VIt. If the h.c. of the apparatus (the heater, thermometer, container, if any, and lagging) is C, and the mass and s.h.c. of the substance are m and c respectively, and the temperature rise is $\Delta\theta$, we can write

$$VIt = (C + cm)\Delta\theta$$

from which we can calculate c.

The h.c. C of the apparatus can be calculated, or measured in a separate experiment. Some allowance should be made for the energy which will have been given to the surroundings. A suitable technique for this is to measure the temperature of the substance while it is being heated and while it is cooling after the current has been stopped. A graph similar to that shown in figure 9.11a will be obtained: an idealised form of this is shown in figure 9.11b. The rate of loss of energy from the apparatus depends on how much hotter it is than the surroundings. During the heating its *average* excess temperature was $\frac{1}{2}(\theta_2 - \theta_1)$. During cooling its excess temperature was $(\theta_2 - \theta_1)$, approximately; so it lost energy twice as fast while it was cooling as it did, on average, while it was being heated. So if the heating time was t, it lost as much energy in a time $\frac{1}{2}t$, while cooling. If we find the fall in temperature in a time $\frac{1}{2}t$, this is the correction we should make to the final recorded temperature. This is easier to do in practice than it is to describe: see the next example.

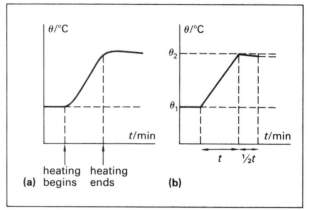

Figure 9.11 (a) Shows the curve which would be obtained in practice, and (b) shows the idealised curve into which it should be converted. The fall in temperature in the time $\frac{1}{2}t$ is the correction to be added to the maximum recorded temperature θ_2.

An alternative to making this correction is to start with the apparatus a few degrees below room temperature, and to heat it until its temperature is the same amount above room temperature. Then the gain of energy from the surroundings during the first half of the heating will exactly balance the loss of energy to the surroundings during the second half.

Mechanical temperature-raising method

If the substance is in solid form a cylindrical block of it must be obtained. If it is liquid, some of it is placed in a thin-walled hollow cylindrical metal container. Either the block or the cylindrical container is mounted as shown in figure 9.12, and thermally insulated from the rest of the apparatus. Rope or string is wrapped round it as shown, with a mass of at least 1 kg suspended from the lower end of the rope. The upper end of the rope is attached to a spring balance. A pulley on the shaft is used to rotate the block (or cylinder) against the frictional force which the rope exerts on the block. The free-body diagram in figure 9.12b shows that the forces on the rope are
(i) the pull S of the spring balance
(ii) the frictional push F of the block on the rope (by Newton's third law equal and opposite to the force F doing work on the block)
(iii) a pull W on the lower end of the rope (equal to the weight of the mass).

We can see that $F = W - S$

The work done by this frictional force when the block (of circumference l) makes n revolutions is $(W - S)nl$ and if we measure the mass m of the block, and its temperature rise $\Delta\theta$ we can use $\Delta Q = cm\Delta\theta$ to show that the s.h.c. of the substance is given by

$$c = \frac{(W - S)nl}{m\Delta\theta}$$

A correction should be made for the loss of energy to the surroundings during the experiment, so the load should be

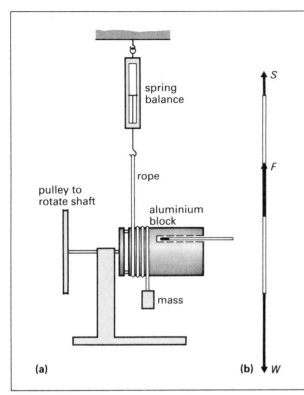

(a)   **(b)**

Figure 9.12 (a) Apparatus by means of which work can be done (by frictional forces) on an aluminium block. (b) A free-body diagram for the rope which is wrapped round the block: $F = W - S$.

removed when the n revolutions have been completed, and the rotation should be continued for half as long as the work was being done. The conditions under which the cylinder now loses energy will then be the same as during the experiment itself.

▶ A mass of 250 g of glycerol was placed in a cylindrical brass container of mass 140 g: the s.h.c. of brass is 370 J K^{-1} kg^{-1}. Its circumference was 0.24 m. A rope was wound round the cylinder as shown in figure 9.12 and loaded with a mass of 1.0 kg. The cylinder was rotated steadily and performed 2000 revolutions in 4 minutes. During the experiment the spring balance reading was 1.5 N. The following temperatures were recorded (room temperature was 17.5°C) during and after the experiment:

t/min	0	1	2	3	4	5	6
θ/°C	17.5	18.9	20.5	22.0	23.5	23.3	23.1

Calculate the s.h.c. of glycerol.

In one revolution the work done by the frictional force of the rope on the drum cylinder was (9.8 N − 1.5 N)(0.24 m) = 1.99 J, so that altogether 2000 × 1.99 J = 3980 J of work was done on the cylinder. There was a fall in temperature of 0.4°C in the 2 minutes after the load was removed, so that

is the correction to be made to the value of 23.5°C recorded after 4 minutes' working. The corrected temperature rise is (23.9 − 17.5) K = 6.4 K, so

$$3980 \, J = \{(370 \, J \, K^{-1} \, kg^{-1} \times 0.14 \, kg) + c(0.25 \, kg)\} \, (6.4 \, K)$$
$$\Rightarrow \quad c = 2.3 \times 10^3 \, J \, K^{-1} \, kg^{-1} \quad \blacktriangleleft$$

Continuous flow method

The liquid or gaseous substance flows past a heater continuously and the substance rises in temperature as it does so. The various parts of the apparatus, however, reach different *steady* temperatures (after an initial heating period) and so none of the energy is being absorbed by the apparatus: its h.c. is therefore irrelevant. Thus one uncertainty is removed, since no measurements of the mass of the apparatus, or the s.h.c. of its material, need be made.

One form of the apparatus is shown in figure 9.13. The current I in the heating element is measured, and also the p.d. V across its ends. The fluid is passed slowly and steadily through the apparatus, and the rate of heating and the rate of flow are adjusted until a temperature difference $\Delta\theta$ is achieved which is large enough to be measured precisely. The mass m of fluid which passes through the apparatus in a time t is measured, and we can say that

$$VIt = cm\Delta\theta$$

where c is the s.h.c. of the fluid. An evacuated jacket is placed around the apparatus to reduce the energy loss to the surroundings. In theory it is possible to repeat the experiment in such a way that the small remaining energy loss can be measured and eliminated but in practice with this simple apparatus it is usually hard to get the rate of flow of fluid steady (and hence the temperature rise constant) and therefore very difficult to get the same temperature rise on two separate occasions. It is therefore best not to try to eliminate the energy loss. Also the only fluid of which there can usually be a plentiful supply is water − but this method has been used successfully.

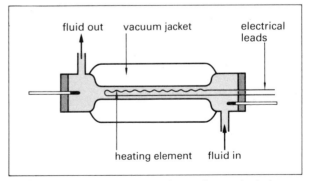

Figure 9.13 Apparatus for the continuous flow method of measuring s.h.c. Temperatures in the apparatus are steady and so the h.c. of the apparatus is irrelevant.

Variation of s.h.c.

The s.h.c.s of substances vary with temperature, and figure 9.14 shows the variation for copper. You may think that this variation is very like the variation of the linear expansivity of copper with temperature, and in fact it is. Figure 9.15 shows the two curves on the same axes, with one vertical axis scaled down so that at 400 K (chosen arbitrarily) the values coincide. It can be seen that, although not identical, the two graphs vary similarly throughout the temperature range. The similarity can be explained by reference to the potential energy-separation graph, but to do so would be beyond the scope of this book. We mention it only to reinforce the idea that many apparently separate properties of a substance can be explained by considering a single idea, the potential energy of the molecules.

The value of the s.h.c. also depends on whether it is measured at constant pressure or constant volume: these values are written as c_p and c_V. For solids and liquids it is always c_p which is measured and quoted, since it would be very hard to increase the pressure enough to keep a constant volume. In any case c_p and c_V have very similar values for a particular solid or liquid. However, we have much greater freedom with gases: we can measure their s.h.c.s under constant pressure or constant volume or under any other conditions which we choose to specify, and the values are significantly different – though c_p and c_V are the only values ever quoted.

c_p is always greater than c_V, for two reasons. When a substance expands at constant pressure
(i) it does *external* work in pushing back the atmosphere
(ii) work is done *internally* because the molecules are further apart and have more intermolecular potential energy.
The energy needed to do this work has to be supplied to the substance *in addition to* the energy needed to raise the temperature of the substance, so ΔQ in the equation $c = \Delta Q / m\Delta\theta$ will be larger when the volume is allowed to increase at constant pressure.

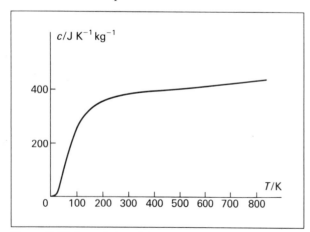

Figure 9.14 The variation of the s.h.c. of copper with temperature. Other solid substances produce similar curves, but flatten out at lower or higher temperatures.

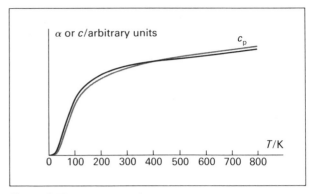

Figure 9.15 The variation of linear expansivity a and s.h.c. c_p with temperature T. The vertical scales have been adjusted so that the curves coincide at $T = 400$ K. The agreement is good.

The internal work is greater than the external work for solids and liquids, but the total amount is not in any case large compared with the energy needed to raise the temperature, so c_p is not much larger than c_V (2% or 3% for copper at room temperature). For gases the external work is far greater than the internal work, and is comparable with the energy needed to raise the temperature of the gas, so c_p is much larger than c_V (about 40% for air at room temperature).

► A cylinder of cross-sectional area 0.20 m² contains 0.16 kg of neon gas at a temperature of 300 K. If the specific heat capacity of neon at constant volume is 628 J K⁻¹ kg⁻¹, how much energy must be supplied to raise the temperature to 500 K? The operation is repeated with the piston enclosing the gas allowed to move freely: it is found that it moves outwards through a distance of 0.60 m. Under these conditions how much energy must be supplied to raise the temperature by the same amount, and what is the s.h.c. of neon at constant pressure? Take atmospheric pressure to be 100 kPa.

Using $\Delta Q = c_V m\Delta\theta$, for the first (constant volume) operation we have

$$\Delta Q = (628 \text{ J K}^{-1} \text{ kg}^{-1})(0.16 \text{ kg})(200 \text{ K})$$
$$= 2.0 \times 10^4 \text{ J}$$

When the piston is allowed to move, the gas does work to push back the atmosphere. The force F which it exerts is given by $F = pA$, where p is the pressure of the gas (= atmospheric pressure) and A is the cross-sectional area of the piston. So

$$F = (1.0 \times 10^5 \text{ Pa})(0.20 \text{ m}^2) = 2.0 \times 10^4 \text{ N}$$

The work ΔW done by the piston is given by $\Delta W = F\Delta x$ where Δx is the displacement of the piston. So

$$\Delta W = (2.0 \times 10^4 \text{ N})(0.60 \text{ m}) = 1.2 \times 10^4 \text{ J}$$

On this occasion the energy supplied to the gas must not only raise its temperature by 200 K (and we have just seen that this energy is 2.0×10^4 J – and it will always require just this much energy to raise the temperature of the gas by

200 K) but must also supply the energy to do the work needed to push back the atmosphere. So this time

$$\Delta Q = 2.0 \times 10^4 \text{ J} + 1.2 \times 10^4 \text{ J} = 3.2 \times 10^4 \text{ J}$$

Since under the conditions of this second experiment we have $\Delta Q = c_p m \Delta \theta$, we can write

$$c_p = \frac{\Delta Q}{m \Delta \theta}$$

$$= \frac{3.2 \times 10^4 \text{ J}}{(0.16 \text{ kg})(200 \text{ K})}$$

$$= 1000 \text{ J K}^{-1} \text{ kg}^{-1}$$

We have assumed that when the gas expands the only work done is the *external* work: negligible work is done in moving the molecules further apart.

Molar heat capacity

Table 9.2 shows, in the first two columns, the values of s.h.c. (at 273 K) and the molar mass of a few elements, and the third column gives the product of these two quantities, which is the heat capacity per mole of the substance, or *molar heat capacity* (m.h.c.). The interesting result is that the m.h.c. is nearly the same for all these elements, provided the temperature is high enough: we have seen in figure 9.14 that the s.h.c. decreases to zero as the temperature falls. Indeed we can see in table 9.2 that for silicon the s.h.c. does not reach a high enough value until about 1000 K. What interpretation can be placed on this product? We see that, provided the temperature is high enough, it takes just as much energy (about 25 J) to raise the temperature of 1 mole of aluminium by 1 K as it does to raise the temperature of 1 mole of copper by 1 K, despite their different *masses*. The same amount of energy is needed per molecule: it is the number of molecules in the body which matters. Since we know that there are about 6×10^{23} atoms in every mole of an element, we can see that it takes

$$\frac{25 \text{ J K}^{-1} \text{ mol}^{-1}}{6 \times 10^{23} \text{ atom mol}^{-1}}$$

$$\approx 4 \times 10^{-23} \text{ J K}^{-1} \text{ atom}^{-1}$$

element	c_p/J K^{-1}kg^{-1}	M_m/kg mol^{-1}	$c_p \times M_m$ = m.h.c/ J K^{-1} mol^{-1}
sodium	1184	0.0230	27.2
aluminium	877	0.0270	23.7
iron	437	0.0558	24.4
copper	380	0.0635	24.2
lead	126	0.2072	26.2
silicon (273 K)	669	0.0281	18.8
silicon (350 K)	762	0.0281	21.4
silicon (1000 K)	938	0.0281	26.3

Table 9.2 Values of s.h.c., molar mass, and their product, the m.h.c. Values of s.h.c. (and therefore those of m.h.c.) are given for 273 K. Additional values are given for silicon.

So that if we want to raise the temperature of a solid substance which contains 10^{26} atoms by 50 K, we know we shall need

$$(4 \times 10^{-23} \text{ J K}^{-1} \text{ atom}^{-1})(10^{26} \text{ atoms})(50 \text{ K})$$

$$= 2 \times 10^5 \text{ J}$$

without needing to know what the substance was. This is a remarkable unifying concept.

The agreement between the values of molar heat capacity is in fact better than appears at first sight. We have been using the constant pressure specific heat capacities, and these include the energy supplied to allow the substance to expand. There is also the point that the value of s.h.c. becomes constant at different temperatures (called the *Debye temperature*) for different substances. When allowance is made for these facts we find that the value of the molar heat capacity is very nearly 25.0 J K^{-1} mol^{-1} for all solid substances.

9.4 Latent heat

If we put a thermometer and a heater in a block of ice (well below 0°C), and switch the heater on, the temperature of the ice will rise, as we should expect. However, after a time the temperature will stop rising and will remain steady. If we look at the ice we see that it is melting. When the ice has completely melted (to water) the temperature will start rising again, until once again the temperature remains steady for a time. The water is then boiling and the temperature of the vessel will remain steady until all the water has evaporated. Figure 9.16 is a graph which shows this behaviour. All substances behave similarly, unless the pressure is low enough for the substance to change from solid to gas directly, i.e. *sublime*, without passing through the liquid phase.

The energy which we supply to change the phase of the substance (e.g. change ice to water) is called *latent heat L*,

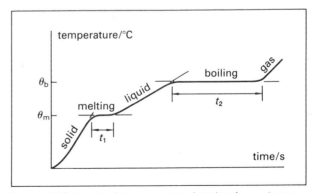

Figure 9.16 A graph of temperature against time for a substance which begins as a solid and ends as a gas. θ_m and θ_b are the melting and boiling points of the substance. The rate of heating is constant.

i.e. hidden heat, since it seems to disappear because the temperature does not rise while it is being supplied. We can define the *specific latent heat* (*s.l.h.*) *l* by the equation

$$l = \frac{\Delta Q}{m}$$

where ΔQ is the energy needed to change the phase of a mass m of the substance. The unit is the J kg^{-1}. We can refer to the s.l.h. of melting (or fusion), the s.l.h. of vaporisation, or the s.l.h. of sublimation, whichever is appropriate. For any normal substance the latent heat is large compared with the energy needed to change the temperature of the substance by, for instance, 100 K. E.g. for copper, the energy needed to raise the temperature of 10 g of copper by 100 K is (using $\Delta Q = cm\Delta\theta$) typically

$$(400 \text{ J K}^{-1}\text{ kg}^{-1})(10^{-2}\text{ kg})(100\text{ K}) = 400\text{ J}$$

but the energy needed to melt 10 g of copper is 2060 J, and the energy needed to evaporate 10 g of molten copper is 48 000 J.

When a gas condenses to form a liquid the latent heat is released again (so water vapour at 100°C is much more dangerous than water at 100°C) and when liquid freezes to form a solid the latent heat is again released. We can put this another way: energy needs to be removed from a gas in order to condense it, and from a liquid in order to freeze it. Water (e.g. a puddle on the ground) does not completely freeze as soon as the air or ground temperature falls below 0°C; it takes some time for the latent heat to be transferred from the water at 0°C to the colder surroundings, and some of the water will still be liquid (at 0°C) for some time after the surrounding temperature has fallen below 0°C.

▶ Consider again the kettle described in the example on page 106. If the water boils at 100°C, how much longer will it take for all the water to evaporate?

The s.l.h. of vaporisation of water (at 100°C) is 2.3×10^6 J kg^{-1}. We are not concerned now with the heat capacity of the kettle or the s.h.c. of the water, since no temperature changes occur. The energy ΔQ needed to evaporate the water ($= ml$) is given by

$$\Delta Q = (1.7\text{ kg})(2.3 \times 10^6\text{ J kg}^{-1})$$
$$= 3.9 \times 10^6\text{ J}$$

If it is being supplied at a rate of 2750 J s^{-1}, the time taken t is given by

$$t = \frac{3.9 \times 10^6\text{ J}}{2750\text{ J s}^{-1}}$$

$$= 1422\text{ s}$$

far longer than the water took to reach boiling point. The latent heat is a very significant quantity. ◀

Molecular explanation

When the phase changes, the energy which we supply has two separate jobs to do. To melt a solid, some of the molecules have to be pulled away from their neighbours (as we mentioned on page 110), so that the structure no longer has any rigidity. To pull any molecules further apart work must be done: we say that *molecular bonds are broken*. Of course the *bonds* are not real, like chains or ropes or even springs, but it is convenient to use this phrase to describe the electrical links holding the molecules together. In a liquid there are groups of molecules. In any one group the molecules are held together, but the groups are not fixed in position relative to other groups: hence the lack of shape of a liquid. And the population of a group does not always consist of the same individual molecules: there will always be molecules detaching themselves from one group and attaching themselves to another. But once a liquid has been formed from a solid the number of broken bonds is constant, on average. When a liquid evaporates, all of the remaining bonds must be broken. Since melting means the breaking of relatively few bonds we should expect the s.l.h. of melting to be less than the s.l.h. of vaporisation. The relative sizes of l_m and l_v will give us some idea of the proportion of the molecular bonds broken when the solid melts. Table 9.3 gives values for some very different substances. We can see that for these substances l_m is always much less than l_v.

Secondly there is nearly always expansion when a solid melts (ice is the only common exception) and always an expansion when a liquid evaporates. This expansion means that the atmosphere must be pushed back to make room, and energy is needed for this too. The example which follows, however, shows that this energy is, even for the liquid-gas change, a small fraction of the total latent heat. Latent heat must be supplied mainly to break molecular bonds.

▶ 1.0 g of water at 373 K (density = 958 kg m^{-3}) is contained by a piston in a cylinder whose cross-sectional area is 50 cm^2. The piston is free to move in the cylinder, and the pressure outside it is standard atmospheric pressure. Energy is supplied to the water so that it evaporates. How much of this energy is needed to push back the atmospheric air? Give your answer as a percentage of the total energy supplied. For water at 373 K $l_v = 2.3 \times 10^6$ J kg^{-1}; standard atmospheric pressure = 101 kPa; density of water vapour at this pressure and a temperature of 373 K = 0.0588 kg m^{-3}.

We first calculate the length of cylinder occupied by (i) the water, and (ii) the water vapour.

(i) volume of water

$$V = \frac{m}{\rho} = \frac{1.0 \times 10^{-3}\text{ kg}}{958\text{ kg m}^{-3}} = 1.04 \times 10^{-6}\text{ m}^3$$

	l_m/kJ kg^{-1}	l_v/kJ kg^{-1}
argon	30	163
carbon dioxide	189	932
copper	205	4790
mercury	11	296
sulphur	44	300
tungsten	192	4350
water	333	2260

Table 9.3 Some values of l_m and l_v. In all cases l_m is much less than l_v, so relatively few molecular bonds are broken when a substance melts.

length of cylinder

$$l = \frac{V}{A} = \frac{1.04 \times 10^{-6} \text{ m}^3}{50 \times 10^{-4} \text{ m}^2} = 2.09 \times 10^{-4} \text{ m}$$

(a negligible distance)

(ii) volume of water vapour

$$V = \frac{m}{\rho} = \frac{1.0 \times 10^{-3} \text{ kg}}{0.0588 \text{ kg m}^{-3}} = 1.70 \times 10^{-3} \text{ m}^3$$

length of cylinder

$$l = \frac{V}{A} = \frac{1.70 \times 10^{-3} \text{ m}^3}{50 \times 10^{-4} \text{ m}^2} = 0.34 \text{ m}$$

As the liquid evaporates it exerts a constant force on the piston, since the pressure in the cylinder remains equal to the pressure on the outside of the piston, which is atmospheric.

$$F = pA = (1.01 \times 10^5 \text{ Pa})(50 \times 10^{-4} \text{ m}^2)$$
$$= 5.05 \times 10^2 \text{ N}$$

The distance which the piston moves while the water evaporates

$$= 0.34 \text{ m} - 2.09 \times 10^{-4} \text{ m} = 0.34 \text{ m}$$

So the work done W is given by

$$W = (5.05 \times 10^2 \text{ N})(0.34 \text{ m}) = 172 \text{ J}$$

The total energy supplied $=$

$$l_v m = (2.3 \times 10^6 \text{ J kg}^{-1})(10^{-3} \text{ kg}) = 2300 \text{ J}$$

Of the total energy supplied, the percentage which is used to push back the atmospheric air is

$$\frac{172}{2300} \times 100\% = 7.5\%$$

In the solid-liquid change of phase the percentage is even smaller, since the volume change is much smaller. Thus most of the latent heat is used to separate the molecules from each other.

The energy supplied during melting or evaporation changes the relative positions of the molecules: their potential energy increases, but not their kinetic energy, since the temperature remains constant until the whole of the solid has melted, or the whole of the liquid has evaporated.

Bonding energy

Can we get any idea of the amount of energy which is needed to pull two molecules apart from each other? This would be the quantity E_0 in the graphs of figure 9.1, since if this were added to a molecule in (a) the potential energy would increase to zero and the molecule would be free from its neighbours. Suppose a molecule is surrounded by n others when its substance is in the liquid phase. To evaporate one mole of the liquid (containing L molecules) $\frac{1}{2}nL$ molecular bonds would have to be broken (the factor $1/2$ is included because otherwise each bond would be counted twice) and if E_0 is the energy needed to break one bond the

total energy needed is $\frac{1}{2}nLE_0$. One mole of water has a mass of 0.018 kg, and it takes 2.3×10^6 J to evaporate 1 kg of water (at 373 K). So we need

$$(2.3 \times 10^6 \text{ J kg}^{-1})(0.018 \text{ kg mol}^{-1}) = 4.14 \times 10^4 \text{ J mol}^{-1}$$

and we can write

$$4.14 \times 10^4 \text{ J mol}^{-1} = \frac{1}{2}nLE_0$$

Now $L = 6.02 \times 10^{23} \text{ mol}^{-1}$, and n might be about 10 for a liquid, so

$$E_0 = \frac{2 \times 4.14 \times 10^4 \text{ J mol}^{-1}}{10 \times 6.02 \times 10^{23} \text{ mol}^{-1}}$$

$$= 1.4 \times 10^{-20} \text{ J for water}$$

This can be only a rough estimate, and it will be different for other substances.

For example, the same calculation applied to liquid argon at its boiling point (87 K) gives $E_0 = 2.2 \times 10^{-21}$ J which is smaller, and therefore bears out our earlier statements that Van der Waals bonding (which is what holds argon atoms together in the solid and liquid phases) is relatively weak.

Since molecules do evaporate spontaneously at the liquid's boiling point we can assume that this value (2.2×10^{-21} J) is comparable with the kinetic energy of a typical atom in gaseous argon. At room temperature the k.e. will only be about three times larger, about 7×10^{-21} J. We can compare this with the energy needed to ionise an argon atom (see page 246) which is 2.5×10^{-18} J, i.e. about 300 times greater. The chance of ionisation occurring as a result of collisions between argon atoms at room temperature is therefore very small.

Evaporation

Not all the molecules of a liquid have the same kinetic energy. There will be some with more than, and some with less than, the average kinetic energy. At any temperature there will be some at the surface of the liquid which have enough kinetic energy to escape completely from the other molecules. This escape of molecules from the surface of the liquid is called *evaporation*. (Even when the substance is in the solid phase there will be some molecules with enough energy to escape completely: this escape is called *sublimation*.)

What affects the rate at which a liquid evaporates? The rate will increase if the temperature rises, i.e., if the average kinetic energy of the molecules increases. Then there will be more molecules which have enough energy to escape. Another way of increasing the rate of evaporation is to increase the surface area of the liquid. If the surface area is doubled, for example, there will be twice as many molecules near the surface of the liquid, and therefore twice as many molecules able to escape. Evaporation is a *surface* effect. Liquid in a shallow dish will evaporate faster than the same volume of liquid in a flask or bottle. A way of effectively increasing the surface area is to blow a stream of air through the liquid: the liquid evaporates into the air bubbles. A third way of increasing the rate of evaporation is to place the surface of the liquid in a draught of air (or other gas). Some of the escaping liquid molecules would normally return to the liquid, by collision with randomly-moving air molecules, for example. In a draught the air molecules moving horizontally in the air stream will knock some of these molecules away from the liquid surface.

Inevitably the more energetic molecules are those which escape; the average kinetic energy of those remaining will therefore be less. In other words the temperature of the liquid will fall. The lower the temperature, the lower the rate of evaporation, as we discussed in the previous paragraph. But there will still be some evaporation, and therefore a further fall in temperature. What eventually happens? Usually the evaporating liquid is in contact with surroundings at the original temperature of the liquid, so energy flows into the liquid as soon as its temperature falls below that of the surroundings. Eventually this happens as a rate which just makes good the loss of energy by evaporation. A steady temperature is produced and therefore a constant rate of evaporation.

You should realise that if only molecules with an average kinetic energy escaped there would be a loss of energy, but no fall in temperature, since the average kinetic energy would be unchanged. The mechanism by which evaporation does occur, however, means that there is not only a loss of energy but also a fall in temperature. The human body relies on this mechanism for the regulation of body temperature. When muscular activity raises body temperature sweat, glands produce water on the surface of the body. When the water evaporates the loss of energy lowers the temperature. It would, however, be dangerous to stand in a draught with one's skin wet because the increased rate of evaporation would produce too large a fall in temperature. Evaporation is an effective way of transferring energy, because of the large values of s.l.h. of liquids.

Measurement of s.l.h.

Melting

One method is to melt a mass m of the substance and place it in a container with a heater and a thermometer. With the heater disconnected the liquid is allowed to freeze and cool. If the melting point of the solid is θ_m a constant temperature bath kept at the temperature θ_m is now placed around the container. The immersion heater is switched on, and the current I passing through it, and the potential difference V between its terminals, noted. The temperature of the solid begins to rise, and is measured at regular intervals of time. At a temperature θ_m the temperature will remain steady while the solid is melting, and will begin to rise again only when all the substance is liquid. A graph like the first part of figure 9.16 will be obtained. We need to measure the time t corresponding to t_1 on that graph. An amount of electrical energy VIt has been converted into

internal energy, and by definition the s.l.h. of melting of the substance is VIt/m. There is no exchange of energy between the substance and its surroundings, since the immediate surroundings are at the same temperature θ_m as the substance during the time t. Nor need we measure the heat capacity of the container, heater, and thermometer, since during the time t their temperature was constant. The relative simplicity of this method is the result of the presence of the constant temperature bath. It will not be possible to provide this at all temperatures. The following example describes an alternative.

▶ A student wishes to find the s.l.h. of melting of lead, but he cannot provide a constant temperature bath at the melting point of lead (601 K). He melts 1.20 kg of lead in a container which can exchange energy with the surroundings, removes the immersion heater, and allows the lead to freeze, and then fall further in temperature. He measures the temperature of the lead at regular time intervals, and obtains the graph shown in figure 9.17. He estimates that the time $t = 350$ s, and that the values of $\Delta\theta$ and Δt, from which he can measure the rate of fall of temperature *as soon as* the liquid has completely frozen, are 18 K and 50 s respectively. He also knows that the s.h.c. of the solid lead at 601 K is 142 J K⁻¹ kg⁻¹, and that the heat capacity of the container is 50 J K⁻¹. What value does he obtain for the s.l.h. of melting of lead?

While the lead was freezing it *was* losing energy to the surroundings (its temperature did not fall while its latent heat was being released to make good the loss to the surroundings). The rate at which it was losing energy to the surroundings depended only on its temperature, so that the rate of loss of energy while freezing must be equal to its initial rate of loss of energy $d\theta/dt$ when it had all frozen, before its temperature had fallen appreciably. The equation $\Delta Q = C\Delta\theta$ can be modified to give

$$\frac{\Delta Q}{\Delta t} = C\frac{\Delta\theta}{\Delta t}$$

and we have $C = (1.20 \text{ kg} \times 142 \text{ J K}^{-1} \text{ kg}^{-1}) + 50 \text{ J K}^{-1}$
$= 220 \text{ J K}^{-1}$, $\Delta\theta = 18$ K, and $\Delta t = 50$ s, so

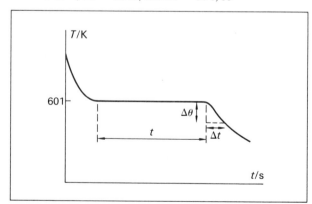

Figure 9.17 The cooling of molten lead: its freezing point is 601 K, and at this temperature its temperature will not fall further until all the lead has frozen.

$$\frac{\Delta\theta}{\Delta t} = (220 \text{ J K}^{-1})\left(\frac{18 \text{ K}}{50 \text{ s}}\right)$$

$$= 79.3 \text{ W}$$

This was the rate at which the latent heat of the molten lead was being released for 350 s, so that the latent heat of the lead is given by

$$L = (79.3 \text{ W})(350 \text{ s})$$
$$= 2.78 \times 10^4 \text{ J}$$

and the s.l.h. l of lead is given by

$$l = \frac{2.78 \times 10^4 \text{ J}}{1.20 \text{ kg}}$$

$$= 2.3 \times 10^4 \text{ J kg}^{-1}$$

Vaporisation

It is possible to use a continuous flow method, with the advantage that the h.c. of the apparatus need not be measured. One form of the experiment is shown in figure 9.18. A thermos flask is half-filled with the liquid whose s.l.h. of vaporisation l is required, and is fitted with a heater and an outlet tube. It is then inverted, and the outlet tube is connected to a condenser, as shown. The condenser must have an adequate supply of cold water so that all the emerging vapour is condensed. The heater is connected to the supply, and is allowed to heat the liquid until it boils and seems to be producing vapour at a steady rate. A beaker is then placed to collect a mass m of condensed vapour in time t. The experiment should then be run for a further time t to check that the mass collected is again m.

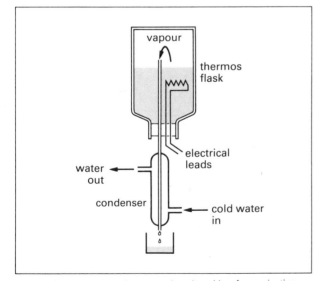

Figure 9.18 Apparatus for measuring the s.l.h. of vaporisation of a liquid.

If the current in the heater is I, and the p.d. across its terminals is V,

$$VIt = ml + Q$$

where Q is the loss of energy to the surroundings. This loss of energy must now be found from a second experiment in which different values of V and I are used, *for the same time*, so that a different mass m is collected. The *rate* of loss of energy from the thermos flask is always the same, since it is always at the same temperature, so in the same time the loss of energy will be the same. So we have

$$V_1 I_1 t = m_1 l + Q$$
$$V_2 I_2 t = m_2 l + Q$$

Subtracting, $(V_1 I_1 - V_2 I_2)t = (m_1 - m_2)l$

or $$l = \frac{(V_1 I_1 - V_2 I_2)t}{m_1 - m_2}$$

Neither mass nor the difference $(m_1 - m_2)$ should be so small that the possible uncertainty in l might be large.

9.5 Energy transfer

In the last section we saw that evaporation transfers energy from the evaporating liquid to the surroundings. This section is about other methods of energy transfer. It is mainly about *conduction*, although a second method, *convection*, is briefly mentioned first. A third method, *radiation*, is described in chapter 28. These three processes, which all occur because of temperature differences, can legitimately be described as *heating* processes.

Because a substance generally expands when its temperature rises, its density decreases. If the fluid in a container is heated at the bottom, the fluid at the bottom will rise bodily to the top, and the upper, cooler, fluid will fall to the bottom. This is why kettles always have their heaters at the bottom, and, conversely, why refrigerators always have the freezing compartment at the top. This bodily movement of a fluid, which transfers energy, is called *natural convection*. It is of enormous importance (the large-scale movements, as a result of convection, of the atmosphere are the cause of most meteorological changes) but the quantitative treatment of it is very complicated, and beyond the scope of this book.

Conduction is the transfer of energy through a substance *without* the bodily movement of the substance. It is this process which, for example, makes the metal handle of a kettle hot when only the water in it and the lower part of the kettle are being given energy. To study this quantitatively we need to consider, initially at least, as simple a situation as possible, and so we shall think about a bar of the material, of constant cross-sectional area, surrounded by material (*lagging*) which does not allow energy to pass in or out from the sides: see figure 9.19. If the energy is supplied to one end, e.g. by placing the end in contact with a

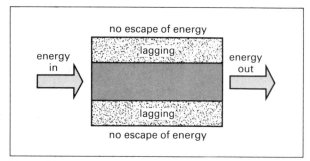

Figure 9.19 We deal only with the simplest situation, where the energy flow is parallel to the length of the bar, and none escapes through the sides. Then, when a steady state has been reached, the rates of input and output of energy are equal.

metal plate containing a heater, energy will flow into the bar because of the difference of temperature. Initially the energy will raise the temperature of the part of the bar nearest the input. When that is hot, energy will pass to the next part of the bar (because of the temperature difference) and so on until energy is emerging from the far end of the bar. If that end of the bar is in contact with the surrounding air, the energy will be conducted to the layers of air in contact with the end of the bar, and then convected away by movement of the air. Although the energy which first enters the bar raises its temperature, the temperatures of different parts of the bar eventually become steady. At any particular point the temperature then has a constant value, although these values decrease continuously from one end of the bar to the other. Then the rate at which energy enters the bar is equal to the rate at which it leaves. None of the energy is then being used to raise the temperature of the bar, so that its heat capacity is irrelevant.

Thermal conductivity

What decides the rate at which energy enters, passes through and leaves the bar (these rates are all identical, once a steady state has been reached)? Experiments show that the following changes would all *increase* the rate of flow:
(i) lowering the temperature of the far end of the bar, or in some other way making the temperature difference $\Delta\theta$ between the ends greater
(ii) increasing the cross-sectional area A of the bar
(iii) decreasing the length l of the bar.
We expect the rate of flow of energy dQ/dt (where Q is the energy which flows past any point) to depend on the temperature difference, the cross-sectional area and the length in this manner:

$$\frac{dQ}{dt} \propto \frac{A\Delta\theta}{l}$$

and this can be verified by experiment. The rate of flow of energy will also depend on the material used, and to allow for this, and at the same time to convert the proportionality into an equation, we write

$$\frac{dQ}{dt} = \frac{\lambda A\Delta\theta}{l}$$

where λ is the *thermal conductivity* of the material and is *defined* by this equation. Its unit can be deduced from the equation and is the $W\ m^{-1}\ K^{-1}$. Metals are the best thermal conductors: table 9.4 gives the values for some typical materials (at 273 K).

The thermal conductivities of *metals* generally decreases slightly with temperature: e.g., for copper at 373 K its value is 395 $W\ m^{-1}\ K^{-1}$. At very low temperatures the thermal conductivity may be very high indeed.

▶ A room has one wall which is an external wall. It measures 5.0 m × 3.0 m, and is 225 mm thick, and is made of brick. Find the rate of flow of energy through the wall if the temperature in the room is 20°C when the temperature outside is 3°C. Assume that each wall surface differs in temperature by 3 K from the air near it. Discuss ways in which the rate of flow of energy might be reduced.

Applying the thermal conductivity equation to the wall we have $\lambda \approx 1\ W\ m^{-1}\ K^{-1}$ (for brick, from table 9.4), $A = 15\ m^2$, $\Delta\theta = 17°C - 6°C = 11\ K$ and $l = 0.225$ m, so that

$$\frac{dQ}{dt} = \frac{(1\ W\ m^{-1}\ K^{-1})(15\ m^2)(11\ K)}{0.225\ m}$$

$$= 7.3 \times 10^2\ W$$

To keep the temperature in the room at 20°C internal energy must be converted (e.g. from chemical energy, or electrical energy) in the room at a rate of 7.3×10^2 W. You might like to note that once the temperature of a room is steady, the heater in it is no longer heating the room, but the air outside it!

To satisfy the Building Regulations in Great Britain (in 1979) it is necessary for the value of λ/l (the U-value) for the wall to be less than 1.0 $W\ m^{-2}\ K^{-1}$, whereas the brick wall in our example has a value of 4.4 $W\ m^{-2}\ K^{-1}$. But the inside of the wall is usually lined with plasterboard, and cavity walls (i.e. a sandwich of air of about 50 mm thickness between two brick walls of the same total thickness as the original one) are often used. There are no solid materials which are, in themselves *good* thermal insulators. Table 9.4

material	λ/W m^{-1} K^{-1}
copper	403
aluminium	236
iron	84
brick	~ 1
glass	~ 1
water	0.56
polythene	~ 0.4
air	0.24
plaster	0.13
expanded polystyrene	~ 0.01

Table 9.4 The thermal conductivities (at 273 K) of some common materials. The worst conductor is only about 1000 times worse than the best, but the worst metallic conductor is 10 times better than the best non-metallic conductor.

shows that the thermal conductivity of the poorest conductors is only about 10^3 less than that of the good conductors. But air is a worse conductor than most solids, so we can use materials which have air trapped in them (trapped so that convection cannot occur). Breeze blocks (made of clinker: $\lambda \approx 0.4\,\mathrm{W\,m^{-1}\,K^{-1}}$) or aerated concrete blocks ($\lambda \approx 0.2\,\mathrm{W\,m^{-1}\,K^{-1}}$) could be used for the inner wall, and the cavity could be filled with foamed plastic. Notice that we have not assumed that the surface of the wall is at the same temperature as the bulk of the air near it. It cannot be, since energy is being transferred from the air to the wall, or vice-versa.

Analogy between thermal and electrical conduction

There is a good analogy between the thermal conductivity equation and the electrical equation which can be formed by putting together the equations

$$I = \frac{V}{R} \quad \text{and} \quad R = \frac{\rho l}{A}\left(= \frac{l}{\sigma A}\right)$$

These two equations give $I = V\sigma A/l$, which can be written (where Q is the electric charge which flows in time t, and ΔV is the p.d. across the length l of the conductor)

$$\frac{\mathrm{d}Q}{\mathrm{d}t} = \frac{\sigma A \Delta V}{l}$$

which is of exactly the same form as

$$\frac{\mathrm{d}Q}{\mathrm{d}t} = \frac{\lambda A \Delta \theta}{l}$$

The quantities, too, have similar properties. Thus, just as electrical charge Q flows because of a potential difference ΔV, so energy Q flows because of a temperature difference $\Delta \theta$. The rate of flow is directly proportional to the cross-sectional area A in each case, and inversely proportional to the length l. σ and λ, the electrical and thermal conductivities, have the same function in each equation. It may sometimes help you to make use of this analogy: when you try to think what will happen in an electrical situation, it may help you if you can think what would happen in a thermal situation – or *vice-versa*.

Measurement of the thermal conductivity of solid materials

We use the equation

$$\frac{\mathrm{d}Q}{\mathrm{d}t} = \frac{\lambda A \Delta \theta}{l}$$

and whether the material is a good or a bad conductor we need to make sure that our experiment is performed in such a way that the necessary conditions are fulfilled. In particular we must make sure that
(i) while the temperatures at different points in the specimen will (and must) be different, the temperature at any parti-

cular point is constant. This condition can be met by waiting for a time long enough for this to happen, and it is tested by measuring the temperature at different points at regular time-intervals.
(ii) the flow of energy is parallel to the length of the conductor. This condition is met (at least approximately) by lagging the sides of the bar to prevent energy escaping to the surroundings, although if the length is very short and its cross-sectional area large (as it must be to get a reasonable rate of flow of energy through a poor conductor) the lagging is less necessary.

Good conductors

The usual apparatus found in school laboratories was originally designed by Searle. The material is in the form of a solid bar (nearly always of copper) (see figure 9.20); its length is about 300 mm and its diameter d about 100 mm. An electrical heater heats one end of the bar uniformly over its cross-section, and a measurement of the p.d. V across the heater terminals, and the current I in the heater, enables the rate of supply of energy VI to be calculated. At the other end thin copper tubes, soldered to the bar to make good thermal contact, are supplied with cold water (from a constant-head apparatus to ensure a steady flow). The rate of flow is measured by collecting the mass m which emerges in a time t. The temperatures θ_3 and θ_4 of the water before and after it has passed through the tubes are measured by mercury thermometers. This supply of cold water produces the necessary temperature difference between the ends of the bar; without this the energy would not flow. At two points a distance l apart (about 100 mm), fine holes are drilled radially into the bar, and thermocouples measure the temperature difference $\Delta \theta$ across the length l. If these holes are narrow they form a relatively

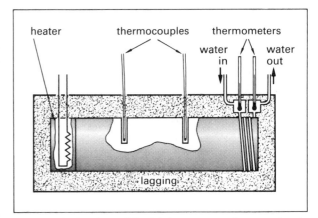

Figure 9.20 One form of Searle's bar apparatus for measuring λ for a good conductor. The holes for the thermocouples have been exaggerated so that the thermocouples can be shown: they should be as narrow as possible. The side of the lagging can usually be removed so that measurements can be made on the bar.

small proportion of the cross-sectional area of the bar and so do not greatly disturb the straight-line energy flow. The current in the heater is switched on, the water in the cooling tubes is started, and thermometer readings taken at regular time intervals until each is constant. If the s.h.c. of water is c, the rate of removal of energy from the cooler end is

$$\frac{mc(\theta_4 - \theta_3)}{t}$$

and this will equal the rate of input VI only if the lagging is perfect. In practice the average of these two readings is calculated: this will be dQ/dt in the thermal conductivity equation. In that equation we also know l, $\Delta\theta$ and A (this from a knowledge of the diameter d of the bar, measured with vernier callipers) and so λ can be calculated.

Poor conductors

Table 9.4 shows that some materials have thermal conductivities which may be 1000 times smaller than those of good conductors. So the length of the specimen must be small (perhaps less than 1 mm) and its cross-sectional area A as large as is convenient to ensure that energy will flow through the specimen at a reasonable rate without having to provide a very large temperature difference $\Delta\theta$. The specimen is therefore in the form of a thin disc (probably circular). The original form of this apparatus was designed by Lees. Two identical discs are made, and a heater sandwiched between them (figure 9.21), which ought to heat them uniformly over their cross-sectional area, and it is therefore contained in a thin-walled cylindrical box with brass top and bottom. Other brass plates are fitted on the outer faces of the specimens. A thermocouple junction is placed on the surface of each brass plate, and a third in the heater box, so that the

average temperature difference $\Delta\theta$ across the specimen can be measured. (The temperatures of the inner and outer faces of the brass are so nearly the same that it does not matter where the thermocouple junctions are put: the next example makes this clear.) Lagging surrounds the edges of the apparatus as shown in the figure but not over the outer brass plates: energy must escape from them to the surrounding air. The heater is switched on, and the thermocouples are read at regular intervals until they give constant readings. The p.d. V across the heater is measured, and the current I in it: the rate of supply of energy is then VI. From the symmetry of the apparatus we can say that half of this energy passes through each specimen to the outer brass plate to the surrounding air: thus dQ/dt in the thermal conductivity equation is $VI/2$. We can also measure l (using a micrometer screw gauge), A (using a ruler to measure the diameter of the discs) and $\Delta\theta$, so that we can calculate λ.

▶ In the apparatus shown in figure 9.21 the specimens are sheets of rubber 0.50 mm thick, and the brass plates are 5.0 mm thick. The lagging can be assumed to be perfect. The diameter of both is 100 mm. If the power of the heater is 50 W, and the temperature in the heater box is 105°C, find the temperatures of the inner and outer surfaces of the brass plates. The thermal conductivities of brass and rubber are 128 W m⁻¹ K⁻¹ and 0.15 W m⁻¹ K⁻¹ respectively.

Figure 9.22 shows the situation: suppose the temperatures required are θ_1 and θ_2 respectively. Using the thermal conductivity equation for the flow of energy through the rubber, we have

$$25\ \text{W} = \frac{(0.15\ \text{W m}^{-1}\ \text{K}^{-1})\{\pi(0.050\ \text{m})^2\}(105°\text{C} - \theta_1)}{0.50 \times 10^{-3}\ \text{m}}$$

which gives $\theta_1 = 94.4°\text{C}$

Similarly, for the brass, we have

$$25\ \text{W} = \frac{(128\ \text{W m}^{-1}\ \text{K}^{-1})\{\pi(0.050\ \text{m})^2\}(94.4°\text{C} - \theta_2)}{5.0 \times 10^{-3}\ \text{m}}$$

which gives $\theta_2 = 94.5°\text{C}$

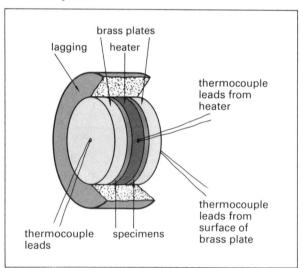

Figure 9.21 One form of Lees's disc apparatus for measuring λ for a poor conductor. The energy flows from the heater symmetrically through the two specimens and into the brass plates and hence into the surrounding air.

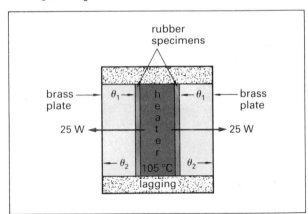

Figure 9.22 The heater has a power of 50 W. As it is assumed to be perfectly lagged at the edges, energy flows through each specimen at a rate of 25 W.

Clearly the temperature is very much the same throughout one of the brass plates, so fixing a thermocouple junction to the outside of a brass plate, rather than to the inside, introduces very little error. Notice that because of the low thermal conductivity of rubber there needs to be a much larger temperature difference across it than across the brass to pass energy at the same rate through both – despite the fact that the brass is 10 times thicker. ◀

The thermal conductivity of liquids and gases

The most effective way of passing energy through a liquid or a gas is by convection. Transfer purely by conduction can be observed only if the energy is supplied at the top of a fluid and we are careful to prevent convection, since gases and non-metallic liquids are relatively poor conductors. The thermal conductivity of water (at 273 K) is 0.56 W m^{-1} K^{-1}, and the thermal conductivity of many organic liquids (at 273 K) is about 0.2 W m^{-1} K^{-1}. The thermal conductivity of many gases (including oxygen and nitrogen, and therefore air) is also about 0.2 W m^{-1} K^{-1} at 273 K. The thermal conductivity of liquids and gases can be measured by a modified form of Lees's apparatus which must have the specimen horizontal: a very thin layer of the fluid is used, to minimise convection.

Mechanisms of thermal conduction

Metals and metallic alloys contain electrons some of which are free to move at random (at a speed of about 10^6 m s^{-1}) within the ionic lattice. If one end of a bar is heated, the electrons (and the ions) at that end gain energy. The electrons inevitably move away from that end, and their places are taken by slower electrons from the far end: this is simply a diffusion process, such as is occurring all the time, whether or not there are electrons of different speeds at the two ends.

Another mechanism allows *non-metals* to conduct. This can involve only the atoms, since in non-metals there are no free electrons. At the heated end the atoms gain vibrational energy, and this is passed on to the neighbouring atoms by elastic waves like sound waves: these do therefore have the speed of sound in the material (about 5000 m s^{-1}), but the frequency is much higher (typically 10^{13} Hz). The atoms can absorb and emit energy only in quanta of size hf (see page 242): by analogy with quanta of light (photons) these quanta are called *phonons*. The energy carried by the phonons does not reach the far end as fast as their high speed would suggest, since they are readily absorbed by the atoms and re-emitted (i.e. they are scattered) because they

metal	λ/10²W m⁻¹K⁻¹		σ/10⁷Ω⁻¹ m⁻¹		$\frac{\lambda}{\sigma}$/10⁻⁶WΩK⁻¹	
	273 K	573 K	273 K	573 K	273 K	573 K
silver	4.28	4.07	6.8	3.0	6.3	14
copper	4.03	3.81	6.5	2.8	6.3	14
gold	3.19	2.99	4.9	2.2	6.5	14
aluminium	2.36	2.33	4.0	1.7	5.9	14
magnesium	1.57	1.50	2.5	1.0	6.2	15
iron	0.84	0.56	1.1	0.32	7.4	18
lead	0.36	0.32	0.52	0.20	6.9	16
mercury	0.078	0.12	0.11	0.078	7.3	15

Table 9.5 Values of λ, σ and λ/σ for eight metals at 273 K and 573 K. Notice that except for mercury, which is a liquid at these temperatures, both λ and σ decrease as the temperature rises, though there is a much greater change in σ. The Wiedemann-Franz law states that the ratio λ/σ is a constant for pure metals at a particular temperature, and is proportional to the temperature. These data show that it is approximately true.

inevitably have similar frequencies to the atoms (unlike the phonons of audible elastic waves). So the phonons *diffuse* through the material, and the randomness of this process means that even fast particles can take a long time to cover a short distance. Electrons, in metals, have frequencies much higher than those of the atoms, and so are weakly scattered; their diffusion rate is much higher, and so metals are generally better thermal conductors than non-metals.

But although the best conductors are metals, there are many non-metals which conduct almost as well, particularly crystalline materials which have light atoms and strong bonds, so that the speed of their phonons is high. The phonons do not succeed in transferring much energy in metals because the free electrons scatter them.

It is not surprising that the good electrical conductors are also good thermal conductors, since it is the free electrons, present in metals, which provide both conduction mechanisms, and this correspondence is not only qualitatively true. The Wiedemann-Franz law states that

$$\frac{\lambda}{\sigma} = (\text{constant})\, T$$

for all metals. Table 9.5 gives some values of λ and σ, and λ/σ for some metals at 273 K and 573 K.

The worst electrical conductors, however, are about 10^{24} times worse than the best, since the only method of charge conduction is by free electrons. The worst thermal conductors are only about 10^3 times worse than the best, since there are other mechanisms available. In solids phonons can carry energy, and in liquids and gases molecular collisions can transfer energy. The molecular collisions, which are the only available mechanism for energy transfer in a fluid at rest, cannot be expected to be very effective.

10 Thermometers and the ideal gas

10.1 Measuring temperature

Our senses give us a qualitative idea of hot and cold, but we need to *measure* the degree of hotness, or *temperature*, quantitatively. Any instrument which does this is called a *thermometer*. The thermometer is placed in contact with the body whose temperature we wish to find and left there until the temperature reading is steady. The thermometer and the body are then in *thermal equilibrium*, and there will be no *net* exchange of energy between the thermometer and the body. (The body will still be giving energy to the thermometer, but the thermometer will be giving just as much energy to the body.) We must in practice be careful that the thermometer is not simultaneously receiving energy from some other body e.g. a thermometer placed in direct sunlight will receive energy from the Sun as well as from the surrounding air, and so will not give us the temperature of the air. To make it easy to use, a thermometer
(i) should have a small h.c. so that it does not absorb much energy from the body whose temperature is required.
(ii) should have a casing which is a good thermal conductor, or else thermal equilibrium will take a long time to establish.

Defining a temperature scale

Many different types of thermometer are possible: any property which varies with temperature can be used as the basis of a thermometer. Here are a few properties, all of which have been used: the length of a column of liquid in a tube, the resistance of a coil of wire, the e.m.f. of a thermocouple, the saturation vapour pressure of a liquid, the pressure or the volume of a fixed mass of gas. What is the best choice?

Let us start with a simple one: the length of a mercury thread in a glass tube, the basis of the common mercury-in-glass thermometer. To make use of this
(i) we need to choose two *fixed points* (easily reproducible temperatures) and give them numbers. These points are often the temperature at which ice and water co-exist in equilibrium, called 0° on the Celsius scale (the *ice-point*), and the temperature when water and its vapour co-exist in equilibrium, called 100° on the Celsius scale (the *steam-point*). In both cases the water must be pure, and the pressure standard atmospheric pressure.
(ii) we need to measure the length of the thread at these two fixed points. The scale of a Celsius thermometer is then divided into 100 equal parts between these fixed points. These markings on the thermometer stem are not in fact necessary. All we need to know is the length of the mercury

thread at the two fixed points and at the unknown temperature. Other kinds of thermometer which make use of a different property (e.g. the e.m.f. of a thermocouple) are not usually marked with a scale of temperatures, and we then have to calculate the temperature from the values of the property at the two fixed points and at the unknown temperature. The example shows how we should proceed with an unmarked mercury thermometer.

► The length of a mercury thread is measured when it is at 0°C, 100°C and some other unknown temperature θ. It is found to be 23 mm, 195 mm and 168 mm at these temperatures. What is θ?

The length of the thread increases from 23 mm to 195 mm (i.e. 172 mm) when the temperature rises by 100°C, and from 23 mm to 168 mm (i.e. 145 mm) when the temperature rises by θ. Therefore

$$\theta = \frac{145}{172} \times 100°C$$

$$= 84.3°C \qquad \blacktriangleleft$$

We can write this out more formally. If X is the property by which the temperature is being measured, e.g. the length of a mercury thread, the e.m.f. of a thermocouple, or the resistance of a coil of wire, the temperature θ is defined by this equation

$$\theta = \frac{X_\theta - X_0}{X_{100} - X_0} \times 100°C$$

You should check that this definition, with X the length of the mercury thread, could be used to solve the problem above. It is easier to use your common sense, rather than the definition, in a practical problem, but we state the definition for the sake of completeness. The graph of figure 10.1 gives a pictorial explanation of this definition. When X_0 and X_{100} have been plotted, a straight line drawn through these points defines the temperatures other than 0°C and 100°C. When X_θ is measured, θ can be read from the graph.

Choice of property for a thermometer scale

This definition can, of course, be used with any of the properties which earlier we suggested could be used to measure temperatures. Would they all give the same result? Not necessarily. If we used the resistance of a wire we might have found that the resistances at 0°C, 100°C and the same unknown temperature were 4.52 Ω, 6.17 Ω and 5.92 Ω. A similar calculation (as in the above example) now gives $\theta = 84.8°C$. Neither is wrong: *there is no reason why the two thermometers should agree* (except, of course, at 0°C and

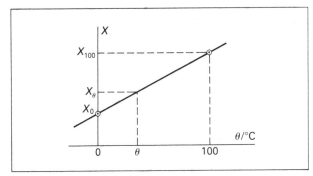

Figure 10.1 An illustration of the way in which temperatures are calculated. When X_θ has been measured, θ can be calculated.

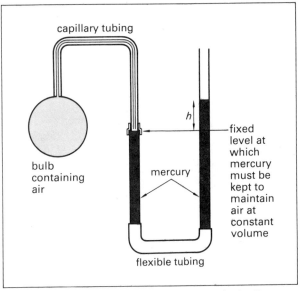

Figure 10.2 A simple form of gas thermometer. In the diagram the pressure of the air is atmospheric pressure plus the pressure of a column of mercury of height h.

100°C). In other words, the resistance of a coil of wire does not change uniformly with temperature as measured by a mercury thermometer, or, alternatively, the length of a mercury thread does not change uniformly with temperature as measured by a resistance thermometer. Neither can claim to be the correct thermometer. 84.3°C is the temperature on the mercury-in-glass thermometer scale, and 84.8°C is the temperature on the resistance thermometer scale. This is, of course, not a satisfactory situation, but it is not an easy problem to resolve. Is there any thermometer which is a less arbitrary choice than the others? One property, the increase with temperature of the pressure of a gas kept at constant volume, has long been known (since about 1800) not to depend (much) on the choice of gas used: e.g., between 0°C and 100°C the pressure of a fixed mass of gas (kept at constant volume) increases 1.37 times whether it is hydrogen or helium, oxygen or nitrogen. There are slight differences but to three significant figures the ratio is the same. So perhaps we should use this property, the pressure p, which means that our definition of the temperature θ changes to

$$\theta = \frac{p_\theta - p_0}{p_{100} - p_0} \times 100°C \qquad (1)$$

The constant-volume gas thermometer

In the simple form of constant-volume gas thermometer shown in figure 10.2, the gas is usually air and is contained in a glass bulb. The bulb is connected by capillary tubing to a mercury manometer, which traps the gas and measures its pressure. When the temperature of the bulb changes, the volume of the gas will tend to change, but it is kept at constant volume by raising or lowering the mercury reservoir. Its pressure is measured by adding to atmospheric pressure the pressure of a column of mercury of height h. The temperature θ of a substance is measured by placing the bulb in thermal equilibrium with the substance and measuring the pressure p. The bulb is then placed successively at the ice-point and the steam-point, and the corresponding gas pressures p_0 and p_{100} are measured. Equation (1) is then used to calculate θ. There are several obvious weaknesses in this simple version of the gas thermometer:

(i) there is a *dead space* between the bulb and the fixed mercury level. The gas in this dead space is at some temperature between that of the bulb and the surrounding air, whereas it should ideally be at the temperature of the bulb. The dead space should be made as small as possible: hence the use of capillary tubing.
(ii) the volume of the bulb and the capillary tubing will themselves increase with increasing temperature.
(iii) the density of the mercury in the manometer changes with temperature, and a correction must be made for this.
These weaknesses can be removed by careful design, or by making corrections, but all gas thermometers necessarily suffer from the disadvantage of having a bulb which is large and cumbersome and which has a high h.c. However, the gas thermometer is the *standard* thermometer, and it is this which is used in standards laboratories to measure the temperatures of the subsidiary fixed points such as the freezing point of silver. Such a measurement might take months, but when made it will be accepted as *the* temperature of the freezing point of silver.

There are other gas thermometers which are used in industry: these usually consist simply of a bulb (of porcelain, platinum or iridium – for temperatures of up to 1350 K, 1750 K and 2250 K respectively) connected to a Bourdon type of pressure gauge. These are not as accurate as those used in standards laboratories, but they are simpler to use, and have a very wide range.

Ideal-gas temperatures

A further improvement in the defining of temperature can be introduced by specifying that the pressure in the constant-

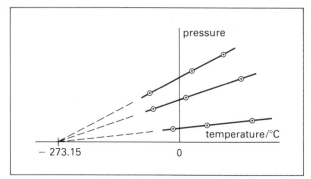

Figure 10.3 Measurements of pressure, using different amounts of different gases, suggest that the pressure would be zero at −273.15 °C (if the gas had not previously liquefied or solidified).

volume gas thermometer should be low: the lower the pressure, the more closely do thermometers using different gases agree. So the procedure is this: with the pressure p_0 of the gas at, (say), 60 kPa a measurement is made of the unknown temperature θ. Suppose the calculated value is 16.63°C. Further measurements are made with $p_0 = 40$ kPa and 20 kPa (by removing some of the gas in the bulb) and the calculated values might now be 16.60°C and 16.57°C. We notice that the lower the value of p_0, the smaller the calculated value of θ, and that as p_0 approaches zero, θ approaches 16.54°C. If we had used a different gas in the bulb (e.g. hydrogen instead of oxygen) we might have obtained successive readings of 16.51°C, 16.52°C and 16.53°C *but these also tend to the same value* (16.54°C) *as the pressure p_0 tends to zero*. We assume that as the pressure becomes lower the gases are losing their individual characteristics and are behaving increasingly like an *ideal gas* (for a definition see page 125) and so a temperature measured in this way is known as an *ideal-gas scale temperature*. To remind ourselves that we need to follow this procedure of lowering the pressure by stages we re-write equation (1) as

$$\theta = \lim_{p_0 \to 0} \frac{p_\theta - p_0}{p_{100} - p_0} \times 100°C \qquad (2)$$

You will, however, notice that any one measurement, using a gas thermometer, gives a good approximation to the ideal-gas scale temperature, and the more so as the pressure is reduced.

The present-day definition of temperature

In 1954 the position was reviewed. Measurements of the pressures of real gases had long ago suggested that there is a temperature at which, if the gas had not already liquefied or solidified, the pressure exerted by it would have been zero. This temperature is 273.15 Celsius degrees below 0°C and is the same, whichever gas is used, provided the initial pressure of the gas is very low. This is illustrated in figure 10.3. Since it is impossible to imagine a gas exerting a negative pressure, this temperature of −273.15°C must be the lowest temperature attainable, and could be thought of

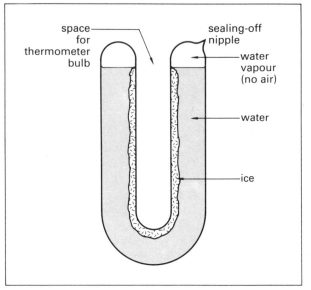

Figure 10.4 A triple-point cell. The only temperature at which the ice, the water and the water vapour can co-exist in equilibrium is 273.16 K.

as the *absolute* zero compared with the arbitrary zero of the Celsius scale, below which it is obviously possible to get. So the new lower fixed point was defined to be the temperature at which an ideal gas would exert no pressure. (There is much other evidence that this zero is absolute.) At the same time it was agreed that the old fixed points, the ice-point and the steam-point, could not be sufficiently easily reproduced precisely, and a new fixed point was introduced, the *triple-point of water*. This is the temperature (0.01°C) at which water can exist in equilibrium with ice and its own vapour, and it is fairly easy to construct and maintain a triple-point cell (figure 10.4) into which a thermometer can be placed.

What equation would now be used to define temperature on this ideal-gas scale? We use a new symbol T for temperature measured on this scale. Figure 10.1 gave a pictorial description of our first definition of temperature; now, similarly, we plot the points (0, 0) and (T_{tr}, p_{tr}) on the graph in figure 10.5 and draw a straight line to join them. What number shall we give to T_{tr}? We could choose any simple number like 100 or 1000, but if we want to keep the degrees on this scale the same size as Celsius degrees we must make the number 273.16, since there are 273.16 Celsius degrees between absolute zero and T_{tr}. Now if we measure a pressure p_T at some unknown temperature T we see that on this ideal-gas scale

$$T = \frac{p_T}{p_{tr}} \times 273.16 \text{ K}$$

where we use a new unit, the kelvin (K) for temperatures on this scale. Using a real-gas thermometer we must write

$$T = \lim_{p_{tr} \to 0} \frac{p_T}{p_{tr}} \times 273.16 \text{ K} \qquad (3)$$

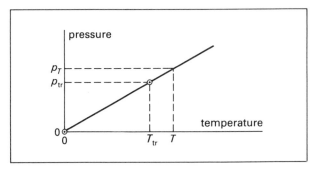

Figure 10.5 An illustration of the way in which temperatures are calculated, using the present-day definition of temperature. When p_T has been measured, T can be calculated.

since the behaviour of the real gases approaches that of an ideal gas only at sufficiently low pressures. So definition (3) is the current definition of ideal-gas scale temperature, and in scientific use the word *temperature* now means 'temperature as defined by the ideal-gas scale' and this is how we shall use it from now on. The symbol is T and the unit is K. We shall still sometimes use θ and °C when we want to refer to a Celsius temperature measured with some arbitrary property like the length of a mercury thread.

▶ A constant-volume hydrogen thermometer is to be used to measure the normal melting point (n.m.p.) of zinc. Three measurements are made of the pressure of the gas when the thermometer bulb is in contact with a mixture of solid and liquid zinc in equilibrium, for three different

initial pressures (at the triple-point) of the gas. The results are

initial pressure/kPa at triple-point	pressure/kPa at the n.m.p. of zinc
90.000	228.160
60.000	152.124
30.000	76.071

What is the n.m.p. of zinc on the ideal-gas scale?

Using the equation $\quad T = \dfrac{p_T}{p_{tr}} \times 273.16$ K

for the first measurement, we have

$$T = \frac{228.160}{90.000} \times 273.16 \text{ K}$$

$$= 692.49 \text{ K}$$

The second and third measurements give 692.57 K and 692.65 K, and we see that the values are increasing steadily, as p_{tr} is reduced, and we can deduce that the measured temperature tends to 692.73 K as p_{tr} tends to zero. This, then, is the n.m.p. of zinc on the ideal-gas scale of temperature. Figure 10.6 illustrates this process of extrapolation to find the ideal-gas scale temperature, and shows the results which would have been obtained using other gases in the thermometer. ◀

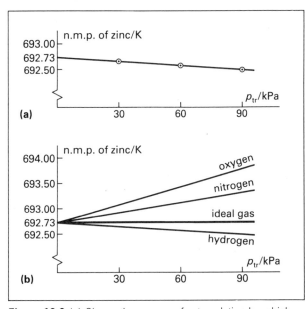

Figure 10.6 (a) Shows the process of extrapolation by which a temperature (on this occasion the n.m.p. of zinc) is measured on the ideal-gas scale using a real gas (hydrogen) thermometer. (b) Shows that the same value (692.73 K) would have been obtained for the n.m.p. of zinc, using other gases in the thermometer.

Some other types of thermometer

(a) The liquid-in-glass thermometer

The straight-tube type, probably the commonest of all thermometers because it has the advantage of being portable and direct-reading, is shown in figure 10.7a. It can be made sensitive by using a tube of small diameter, and a bulb which holds a large amount of liquid. The mass of liquid should not be too large, however, or else the time taken for the thermometer to come to thermal equilibrium will be increased. This time can be reduced by making the bulb walls thin. Mercury is liquid in the range 234 K to 632 K (under atmospheric pressure) so that it has a convenient (though small) range.

This type of thermometer can be calibrated by marking the position of the end of the mercury thread when the bulb is at the ice-point and at the steam-point. The length between these marks is then divided into, say, 100 equal divisions, and the scale is probably continued for a little way on the other side of the 0 and 100 marks. The glass tube is usually evacuated, to avoid the high pressures which occur if there is air in the tube to be compressed when the mercury expands. Sometimes, however, a gas such as nitrogen is deliberately introduced in order to raise the pressure and prevent the mercury boiling. Under these conditions (using specially strengthened tubes) the range of the thermometer may be raised to about 1000 K.

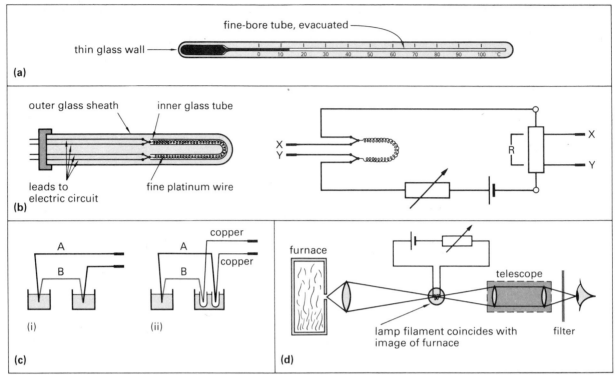

Figure 10.7 Thermometers: (a) shows a mercury-in-glass thermometer (b) shows a platinum resistance thermometer, with the circuit. The leads X and Y measure, in succession, the p.d.s across the platinum wire and across the standard resistor R, using a potentiometer. (c) A thermocouple thermometer: (i) is the simplest arrangement, using two metals A and B. But the two wires of A will have to make a junction with copper somewhere in the circuit, and if these junctions are at different temperatures, an additional e.m.f. will be produced. To avoid this, the arrangement shown in (ii) is sometimes used. (d) A disappearing-filament type of optical pyrometer.

(b) resistance thermometers

The electrical resistance of a metal (particularly platinum because it can be obtained to a high degree of purity, reproducibly) or of a semi-conductor (such as carbon or germanium) may be used to measure temperature. On grounds of sensitivity platinum is preferred at temperatures above about 80 K: below this temperature the semi-conductors are more sensitive (though less accurate than platinum).

A platinum resistance thermometer is shown in figure 10.7b, together with the electrical circuit needed. A very small current is passed through the platinum wire, and through a standard resistor R. Leads X and Y then enable the p.d.s. across these resistances to be measured with a potentiometer (see page 98). The current has to be small to minimise the conversion of energy in the platinum, so the p.d. to be measured is small, but the potentiometric method is sensitive enough for the measurement to be accurate. The actual resistance material is protected by a sheath, which might be 30 or 40 mm long for platinum, and only 10 mm long for the semi-conductors.

(c) thermocouple thermometers

If wires of two metals (or alloys) are joined as shown in figure 10.7c(i) and the junctions are at different temperatures θ_1 and θ_2, an e.m.f. is produced which, as its size depends on the temperature difference between the junctions, may be used to measure temperature. In practice the arrangement shown in figure 10.7c(ii) is often used, the introduction of extra junctions between the copper wires and wires A and B being unimportant, since these junctions are at the same temperature. The e.m.f. is measured using a potentiometer adapted to measure small e.m.f.s, as described on page 100. Many different pairs of metals (or alloys) may be used: copper-constantan, iron-constantan and chromel-alumel are common. Any one thermocouple has a wide range (typically 1000 K) but the advantage of all thermocouple thermometers is the low heat capacity of their junctions: they can respond quickly to a change in temperature.

(d) pyrometers

The thermometers so far described must all be placed in

contact with the substance whose temperature is to be measured: none of them, therefore, can be used to measure temperatures of much more than 2000 K. In this region *pyrometers* are used to analyse the visible radiation coming from the substance. The colour and intensity of this radiation is, for a particular source, characteristic of its temperature (see page 376). The commonest type is the *disappearing-filament pyrometer*, figure 10.7d. A wire (the filament) has a current passed through it until it glows. It is then placed (together with a narrow band filter) between the observer and the object (e.g. a furnace) whose temperature is required, and the current adjusted until the filament cannot be seen against the background of the object. The current in the filament is then measured and used to determine the temperature. Other pyrometers measure temperature by measuring the total radiation emitted by the source.

Comparability of thermometer readings

How are measurements of ideal-gas scale temperature made with thermometers other than gas thermometers? Let us take a platinum resistance thermometer as an example. Suppose we measure the resistance of the platinum wire at the triple-point of water (273.16 K by definition) and find that this is 7.56 Ω. We then use it to measure the temperature of, for example, melting tin, and find that the resistance then is 14.29 Ω. Suppose we modify the equation

$$T = \frac{p_T}{p_{tr}} \times 273.16 \text{ K}$$

to read $T = \frac{R_T}{R_{tr}} \times 273.16 \text{ K}$

where R_{tr} and R_T are the resistances of the platinum wire at the triple-point and the unknown temperature T respectively. We then have

$$T = \frac{14.29 \ \Omega}{7.56 \ \Omega} \times 273.16 \text{ K}$$

$$= 516.30 \text{ K}$$

But the n.m.p. of tin is a subsidiary fixed point and is known to be 505.00 K. What has gone wrong? We have been wrong in writing $T = (R_T/R_{tr}) \times 273.16$ K, as this is equivalent to assuming that the resistance of platinum is zero at absolute zero, and is proportional to the temperature T – and there is no reason why these things should be true. As we saw earlier (page 120) there is no reason why two thermometers, using different properties, should agree. The value of 516.30 K is the temperature on the scale of the resistance thermometer but we have given reasons for preferring the ideal-gas scale temperature.

To measure temperature, using anything other than a gas thermometer with readings extrapolated to zero pressure, a rather more complicated technique is used which forms the basis of the International Practical Temperature Scale (IPTS) which we shall not consider here.

10.2 Boyle's law and the ideal gas

Robert Boyle, experimenting on what he called the 'spring of air' in 1660, discovered that the pressure p and the volume V of a fixed mass of gas were inversely proportional to each other, or

$$p \propto \frac{1}{V} \quad \text{or} \quad V \propto \frac{1}{p} \quad \text{or} \quad pV = \text{constant} \qquad (1)$$

provided the temperature of the gas was kept constant.

We can use the apparatus shown in figure 10.8a to test this relationship (Boyle's law) in the laboratory, though unfortunately it does not allow us to use a very wide range of pressures (perhaps from 50 kPa to 150 kPa). The air is trapped in the left-hand tube. The length l of this air column enables us to calculate the volume V of the gas if we know the cross-sectional area of the tube. The pressure p of the air is atmospheric pressure plus the pressure of a mercury column of height h. If we plot p against V we obtain a hyperbola. It is more useful to try to obtain a straight line; to obtain this we can plot p against $1/V$ (appendix B2). Figure 10.8b, c shows graphs which might be obtained for 1.0×10^{-3} mol of air at room temperature. We should obtain exactly the same result for any other gas if we used the same amount of substance, (i.e. 1.0×10^{-3} mol). This implies that the pressure of the gas is proportional to the number of molecules present, since a particular amount of substance (i.e. a definite number of moles: see page 38) always contains the same number of molecules.

What would be the result if we used twice the amount of substance (i.e. 2.0×10^{-3} mol of air)? Clearly for the pressure to be the same the volume would need to be twice as great, and figure 10.8b, c also shows the result of an experiment with twice the amount of air. We conclude that we can write Boyle's law in the form

$$pV \propto n \qquad\qquad (T \text{ constant) (2)}$$

where n is the amount of substance.

You should realise that Boyle's law is an idealisation of the way in which gases behave. It holds well for air at room temperature, and for oxygen and nitrogen separately, and for other gases like hydrogen or helium. But at lower temperatures, or higher pressures, the law does not hold even for these gases. Boyle's law is like Hooke's law or Ohm's law; a statement which is useful as a working rule over a limited range of conditions.

The ideal gas

We define an *ideal gas* as one for which Boyle's law is exactly true, for all temperatures and pressures.

How does the product pV for an ideal gas depend on its temperature? We must be very careful here. By 'temperature' we mean the ideal-gas scale temperature, so *by definition* p is proportional to T at constant volume. We cannot 'discover' by experiment whether p is proportional to T: the *definition* of temperature enables us to write

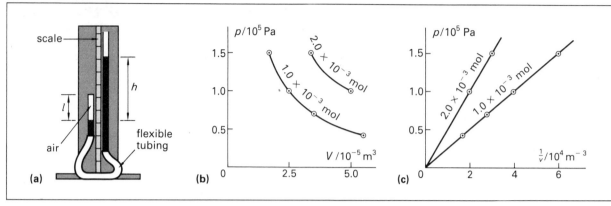

Figure 10.8 (a) A simple apparatus for testing the truth of Boyle's law. The two tubes are raised and lowered to produce different pressures and volumes. (b) and (c) show the sort of graphs which might be produced for two different amounts of gas. The temperature of the gas should be kept constant throughout.

$$pV \propto nT \qquad (3)$$

Of course, if we made ourselves a mercury-in-glass thermometer, using the ice-point and the steam-point, we could perform a genuine experiment to see whether the pressure p of a gas varied linearly with the Celsius mercury-in-glass scale temperature θ. We should find, within the limits of experimental error, that it did; it was indeed this sort of result which, a hundred years ago, suggested the use of the pressure of a gas as a thermometric property. The statement that $p \propto \theta$ was known as Joly's law, and the corresponding statement that $V \propto \theta$ was known as Charles's Law, but since the introduction of ideal-gas temperature scales they have no significance. If we did perform such an experiment it would not be so much a test of whether p was proportional to T as a test of whether the mercury-in-glass scale temperatures corresponded to ideal-gas scale temperatures. Figure 10.9 sets out the relationships between the definitions and the deductions.

Gas constants

The constant in equation (3) is something which can be measured experimentally – but obviously we cannot use an ideal gas for our experiment! We have to use a real gas and measure the values of p, V, n and T at successively decreasing pressures. Figure 10.10 shows that although at a particular pressure the value of pV/nT is (very slightly) different for different real gases, the value of pV/nT to which all tend, as p approaches zero, is the same. This value is labelled R and called the *molar gas constant*:

$$\frac{pV}{nT} = R$$

Measurements give $R = 8.3143$ J K^{-1} mol^{-1}.

These measurements suggest that an ideal gas is one whose behaviour is the behaviour to which a real gas approaches as the pressure approaches zero. But even at atmospheric pressure many real gases behave almost indistinguishably from the ideal gas: see figure 10.10 which shows three gases

for which the value of pV/nT differs from R by less than 0.1% when the pressure is atmospheric.

We can rewrite $pV = nRT$ in two useful ways: where m is the mass of gas, M the molar mass, and N the number of molecules

$$pV = \frac{m}{M}RT \quad \text{and} \quad pV = \frac{N}{L}RT$$

The second of these can be rewritten $pV = NkT$

where $k = \dfrac{R}{L} = \dfrac{8.31 \text{ J K}^{-1} \text{ mol}^{-1}}{6.02 \times 10^{23} \text{ molecules mol}^{-1}}$

$$= 1.38 \times 10^{-23} \text{ J K}^{-1} \text{ molecule}^{-1}$$

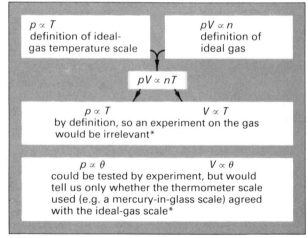

Figure 10.9 The logic of some of the relationships between p, V and T. In each proportionality it is assumed that the other variables are being kept constant.
(*if the gas being tested is at the same very low pressure as the gas in the thermometer bulb of a gas thermometer – using the same gas – whose readings are being extrapolated to zero pressure).

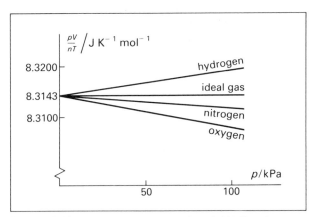

Figure 10.10 Whatever the gas, the value of pV/nT tends to 8.3143 J K^{-1} mol^{-1} as the pressure is reduced to zero. These lines give the data for these gases at 273 K: at other temperatures the lines would be different, but they would still tend to the same value, 8.3143 J K^{-1} mol^{-1}.

k is called Boltzmann's constant. The unit (molecule^{-1}) is not usually included (i.e. $k = 1.38 \times 10^{-23}$ J K^{-1}) but we have added it here to point out that if R is the molar gas constant, k is the *molecular gas constant* (i.e. the gas constant for one molecule). Since R and L are both experimentally-determined quantities, so is k.

Explanations

We should now be asking ourselves *why* real gases obey Boyle's law (very nearly). We know (page 49) that a gas consists of molecules in rapid, random motion: what is it about the behaviour of molecules which makes $pV = $ constant, even though the behaviour of individual molecules is entirely chaotic and unpredictable? We have already explained why gases exert a pressure (page 59) and it is fairly easy to get a quantitative idea of what is happening. If the volume of the container is reduced, the molecules collide with the walls of the container more often, so that the force on the walls is greater. It is reasonable to suppose that if the volume of the space were halved, the collisions would occur twice as often, so that the pressure would be doubled: we should expect the product pV to be constant. Indeed, the model described on page 59 can be used to predict this.

But we can do better than this: we know that the pressure of a gas must be caused by the molecules, which have mass and which are moving, hitting the walls of the container. Let us see if we can find out how the pressure depends on these properties of the molecules. Now we know that a molecule can be a very complex particle, which exerts attracting and repelling forces on other molecules, so we shall begin by making things easy for ourselves by assuming that they are a bit simpler than this. We know that they are small compared with most containers, and we know

that the attracting force is negligible unless the molecules are very close to each other. So let us first assume that the molecules behave like tiny, smooth, elastic spheres. Then in their collisions with the walls there will be no change in their kinetic energy. They have *translational* k.e. (i.e., k.e. as a result of their movement from place to place). Although they may exchange energy with each other if and when they collide, the total of this translational k.e. will be constant. This is the *internal energy U* of this simplified (ideal) gas. The procedure of making a model of a gas, like this, is called the *kinetic theory of gases*. For the first people to think along these lines (in the mid-nineteenth century) it was very much a step in the dark, since the ideas of molecular motion were then relatively new.

The kinetic theory of gases

Let us consider a single molecule of mass m moving with speed v in the direction shown in the rectangular box of sides a, b and c in figure 10.11. Its velocity can be resolved into three parts parallel to the axes x, y and z (and the sides of the box): v_x, v_y and v_z. We shall look at what happens in the x-direction. In that direction the molecule has momentum $+mv_x$; after a collision with face X it has momentum $-mv_x$. It continues to move with velocity $+v_x$ or $-v_x$, whatever the sizes of v_y or v_z. Since it has a constant speed in the x-direction it collides with face X at *regular* time intervals Δt given by

$$\Delta t = \frac{2a}{v_x}$$

The average force F_{av} on the molecule because it hits face X is equal to its rate of change of momentum there. Since the change of momentum is $-2mv_x$ and the change happens at time intervals Δt,

$$F_{av} = \frac{\Delta(mv)}{\Delta t}$$

$$= \frac{-2mv_x}{2a/v_x}$$

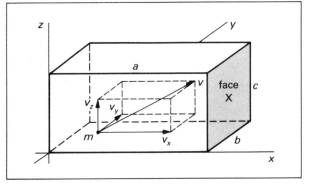

Figure 10.11 A molecule of mass m has a speed v in the direction shown. Its velocity can be resolved into three parts v_x, v_y, v_z. $v_x{}^2 + v_y{}^2 + v_z{}^2 = v^2$, since v is represented by the diagonal of the dashed rectangular solid inside the box.

$$= -\frac{mv_x^2}{a}$$

By Newton's third law the average force *on the wall* is $+mv_x^2/a$. But there are other molecules too, with different x-velocities. As we do not know what these velocities are we can say only that the total average force F on the face is given by

$$F = \text{the sum of } N \text{ terms like } \frac{mv_x^2}{a}$$

if there are N molecules in the box. The terms will all have different sizes, but the size of m and a is the same for each term, so we can write

$$F = \frac{m}{a} \times \text{the sum of } N \text{ terms like } v_x^2$$

We cannot find the sum of these terms, because we do not know what their sizes are, but we can write the sum as

$$= N \times \text{the average (or mean) of the different terms in } v_x^2$$

$$= N\overline{v_x^2}$$

So $\quad F = \dfrac{Nm\overline{v_x^2}}{a}$

Referring to figure 10.11 again we see that for the molecule shown there

$$v^2 = v_x^2 + v_y^2 + v_z^2$$

Using suffixes 1, 2, 3 ... N for the N different molecules in the box,

$$v_1^2 = v_{x1}^2 + v_{y1}^2 + v_{z1}^2$$
$$v_2^2 = v_{x2}^2 + v_{y2}^2 + v_{z2}^2 \text{ etc.}$$
$$v_N^2 = v_{xN}^2 + v_{yN}^2 + v_{zN}^2$$

Adding up these equations we get the sum of all the v^2 terms, the sum of all the v_x^2 terms, the sum of all the v_y^2 terms, and the sum of all the v_z^2 terms, and each of their sums is equal to N times the average value. So

$$N\overline{v^2} = N\overline{v_x^2} + N\overline{v_y^2} + N\overline{v_z^2}$$
or $\quad \overline{v^2} = \overline{v_x^2} + \overline{v_y^2} + \overline{v_z^2}$

The motion is random, with no preferred direction, so that the averages of the squares of the speeds in the different directions must be the same:

$$\overline{v_x^2} = \overline{v_y^2} = \overline{v_z^2} \text{ and } \overline{v_x^2} \text{ (or } \overline{v_y^2} \text{ or } \overline{v_z^2}) = \tfrac{1}{3}\overline{v^2}$$

Therefore $\quad F = \dfrac{Nm}{a} \times \tfrac{1}{3}\overline{v^2}$

The area of the face X, on which this force is exerted, is bc, so that the pressure p at face X is

$$p = \tfrac{1}{3}\frac{Nm}{abc}\overline{v^2}$$

and since the volume V of the box is abc, this can be written as

$$p = \tfrac{1}{3}\frac{Nm}{V}\overline{v^2}$$

Nm is the total mass of the molecules, so that the density ρ of the gas is given by $\rho = Nm/V$, so that we have

$$pV = \tfrac{1}{3}Nm\overline{v^2} \quad \text{or} \quad p = \tfrac{1}{3}\rho\overline{v^2}$$

We made some very sweeping (and rather crude) assumptions about what the molecules of a gas are like. Mostly they were approximations: e.g. we said that one molecule does not attract another. That is true provided they are at least two or three molecular diameters apart. We said that the molecules were tiny: real molecules in a gas at atmospheric pressure occupy about 0.01 % of the volume of the container, so again our approximation is reasonable. We assumed that collisions were elastic: we had to do this, because we were about to calculate what would happen to a particular molecule, and we had to know how it would behave – but now we could say that all we need to assume is that on the average the collisions with the walls and each other are elastic. This is reasonable because we find that in a thermally isolated container the pressure of a gas does not decrease with time (as it would if the molecules were continually losing their translational k.e.). Incidentally their collisions with each other do not affect our argument, since in any collision the momentum which one molecule loses is gained by the other, so momentum is being transferred across the box at the same rate, whether or not there are collisions between molecules.

It seems our simple model is quite sensible, but the real test is how well its result agrees with experiment. We shall, later, compare its predictions with the experimental evidence. In this next example we calculate the root mean square speed (we shall explain this term in the example) of air molecules so that we can get some immediate results from the theory.

► Estimate the *root mean square speed (r.m.s.)* of the air molecules in the atmosphere at sea level.

We know that the atmospheric pressure is usually about 100 kPa: rough measurements of the density of atmospheric air (e.g. by finding the mass of a known volume of it) give a value of about 1.2 kg m^{-3}.

We have $\quad p = \tfrac{1}{3}\rho\,\overline{v^2}$

so $\quad \overline{v^2} = \dfrac{3p}{\rho}$

With our data $\quad \overline{v^2} = \dfrac{3 \times 1.0 \times 10^5 \text{ N m}^{-2}}{1.2 \text{ kg m}^{-3}}$

$$= 2.5 \times 10^5 \text{ m}^2 \text{ s}^{-2}$$

This is the *mean square speed*: i.e., the average of the squares of the different speeds of the molecules. The root mean square (r.m.s.) speed is the square root of this, and is given by

$$v_{rms} = \sqrt{\overline{v^2}} = 5 \times 10^2 \text{ m s}^{-2}$$

We have made some progress: two *macroscopic* measurements (i.e. measurements of the bulk properties of materials, in this case pressure and density) have enabled us to calculate a *sub-microscopic* quantity (a property of the molecules which we could not have investigated even with a microscope) namely the r.m.s. speed of a molecule. Air is, of course, a mixture of different kinds of molecule. What we have calculated is a typical speed: we shall see later what decides the speeds of different molecules. You may think that this calculated speed is very high and therefore unlikely. But sound travels in air because molecules of air move, and the speed of sound in air is about 340 m s^{-1}, which is comparable with the calculated 500 m s^{-1} for the speeds of the molecules.

It is tempting to think that $v_{rms} = \bar{v}$, i.e. that we can take the square root of $\overline{v^2}$, and hence find the average speed $\bar{v}$. But the proof given above shows that we cannot do this: in it we had to take the average or mean of the different *squared speeds*. It is easy to check that $v_{rms} \neq \bar{v}$ by considering just 3 molecules. If $v_1 = 300$ m s^{-1}, $v_2 = 500$ m s^{-1}. $v_3 = 700$ m s^{-1},

$$\bar{v} = 500 \text{ m s}^{-1}$$

but

$$v_{rms} = \sqrt{\frac{300^2 + 500^2 + 700^2}{3}} \text{ m s}^{-1}$$

$$= 404 \text{ m s}^{-1}$$

a totally different value.

10.3 Consequences of kinetic theory

Temperature and the speeds of molecules

If we put together the equations $pV = \frac{1}{3}Nm\overline{v^2}$ and $pV = NkT$ we get

$$\tfrac{1}{3}Nm\overline{v^2} = NkT$$

Dividing both sides by N, and multiplying both sides by 3/2, we get

$$\tfrac{1}{2}m\overline{v^2} = \tfrac{3}{2}kT$$

i.e. the mean translational k.e. of each of the molecules of an ideal gas is $\frac{3}{2}kT$, and here we have a *molecular interpretation of temperature*.

The equation

$$\text{mean translational k.e. of an ideal-gas molecule} = \tfrac{3}{2}kT$$

tells us that the temperature of an ideal gas is proportional to the mean translational k.e. of its molecules: e.g. if the temperature of an ideal gas increases from 300 K to 600 K we should expect the mean k.e. of one of its molecules to

double too. Figure 10.10 showed that real gases behave very nearly like an ideal gas, so that we can apply this result to real gases too, provided that the pressure is not too high or the temperature too low.

► Find the mean translational k.e. of a carbon dioxide molecule at a temperature of 290 K, and hence its r.m.s. speed at this temperature. For carbon dioxide, $M_r = 44$; and $m_u = 1.66 \times 10^{-27}$ kg (page 37).

$$\begin{aligned}
\tfrac{1}{2}m\overline{v^2} &= \tfrac{3}{2}kT \\
&= \tfrac{3}{2}(1.38 \times 10^{-23} \text{ J K}^{-1})(290 \text{ K}) \\
&= 6.0 \times 10^{-21} \text{ J}
\end{aligned}$$

This will be the mean translational k.e. of a molecule of *any gas* at 290 K; the expression $\frac{3}{2}kT$ depends only on temperature. We could say at once, for example, that the mean translational k.e. of a hydrogen molecule at this temperature is also 6.0×10^{-21} J. Since for carbon dioxide, $m = 44 \times 1.66 \times 10^{-27}$ kg,

$$\overline{v^2} = \frac{2 \times 6.0 \times 10^{-21} \text{ J}}{44 \times 1.66 \times 10^{-27} \text{ kg}}$$

$$= 1.64 \times 10^5 \text{ m}^2 \text{ s}^{-2}$$

$$v_{rms} = 4.1 \times 10^2 \text{ m s}^{-1}$$

This is not the r.m.s. speed of any molecule at 290 K; since the k.e. is the same, the smaller the mass, the higher the speed, e.g., a hydrogen molecule ($M_r = 2$) would have a r.m.s. speed given by

$$\overline{v^2} = \frac{2 \times 6.0 \times 10^{-21} \text{ J}}{2 \times 1.66 \times 10^{-27} \text{ kg}}$$

$$v_{rms} = 1.9 \times 10^3 \text{ m s}^{-1}$$

Dalton's law of partial pressures

We can extend the ideas of this last example. In a mixture of gases the molecules of different kinds will have the same mean translational k.e. (since they are all part of a mixture at a particular temperature). The translational k.e. is shared equally among the molecules which are present, no matter what kind they are. For example, in a room which contains molecules of oxygen ($M_r = 32$), nitrogen ($M_r = 28$) and water vapour ($M_r = 18$), all will have the same mean translational k.e. but their mean speeds will be different. This can be extended to larger molecules, specks of dust and ash – for example, the ash particles which are sometimes used to demonstrate Brownian motion (page 49).

► Estimate the mass of an ash particle used in a Brownian motion experiment; take $k = 1.38 \times 10^{-23}$ J K^{-1}.

Observation shows that an ash particle might appear to move about 1 mm in 1 second. But this 1 mm is from the beginning to end of a random zig-zag path, so its mean speed might be about 10 mm s^{-1} i.e. it might have a mean speed of about 10^{-2} m s^{-1}. In this rough estimate we need not distinguish between $\bar{v}$ and v_{rms}, so let us say $v_{rms} = 10^{-2}$ m s^{-1}. We have

$\frac{1}{2}m\overline{v^2} = \frac{3}{2}kT$ for any molecule or small particle

If room temperature is about 300 K, we can write

$$m = \frac{3kT}{\overline{v^2}}$$

$$= \frac{3(1.38 \times 10^{-23} \text{ J K}^{-1})(300 \text{ K})}{10^{-4} \text{ m}^2 \text{ s}^{-2}}$$

$$\approx 10^{-16} \text{ kg}$$

This is a reasonable result. It is less than the mass of a typical oil droplet ($m \approx 10^{-14}$ kg) in a Millikan experiment (page 191). There the Brownian motion of the droplet can just be observed, but it is not as noticeable as with the ash particles.

How can we predict the pressure caused by a mixture of gases? Imagine molecules of two kinds present together in a box. Each kind of molecule occupies the whole volume V of the box, and each has the same mean translational k.e., so that the molecules make a contribution to the pressure which is proportional to their numbers. If we have numbers of molecules N_A and N_B of two different gases A and B, each produces a pressure p_A and p_B given by

$$p_A V = N_A kT \quad \text{and} \quad p_B V = N_B kT$$

The pressure p in a gas which contains $N_A + N_B$ molecules is given by

$$pV = (N_A + N_B)kT$$

so we see that

$$p = p_A + p_B$$

i.e. the pressure produced by a mixture of gases at a particular temperature is the sum of the pressures which each gas would produce if it occupied the whole volume by itself at that temperature. This is known as *Dalton's law of partial pressures*. It does not, of course, apply if the gases react chemically, since the number of molecules may then change. It is well supported by experiment.

Graham's law of diffusion

At a particular temperature any molecule has the same mean translational k.e. In a process like diffusion, therefore, the most massive molecules will diffuse less readily, because their speeds will be less. If we have a mixture of gases A and B enclosed in a container with a porous wall, and the rates of diffusion of the two gases are r_A and r_B, we should expect

$$\frac{r_A}{r_B} = \frac{\bar{v}_A}{\bar{v}_B}$$

$$= \frac{v_{A(rms)}}{v_{B(rms)}}$$

and since $\frac{1}{2}m_A\overline{v_A^2} = \frac{1}{2}m_B\overline{v_B^2}$

(see

$$\frac{r_A}{r_B} = \sqrt{\frac{m_B}{m_A}}$$

This, too, is well supported by experiment, and is known as *Graham's law of diffusion*.

The different rates of diffusion of gases consisting of molecules of different mass is used to separate gases. Obviously it is most effective when the gases have very different masses, but it is even used to separate uranium hexafluoride (UF_6) molecules, formed from the isotope $^{238}_{92}U$, from those molecules formed from the isotope $^{235}_{92}U$ (see page 262). The relative atomic mass of fluorine is 19, so the relative molecular masses of the two molecules are 352 and 349 respectively. If a mixture of the two molecules is allowed to diffuse through a porous wall, the less massive molecules will diffuse 1.004 times faster than the more massive molecules. Only 0.7% of naturally occurring uranium is $^{235}_{92}U$ but this is the isotope which will allow chain reactions to occur and make an atomic bomb feasible: separation by diffusion, repeated many times, is able to produce a mixture which contains the 40% of $^{235}_{92}U$ which is necessary for a chain reaction.

The Maxwellian distribution

Before we go further, let us stop and think what we are now doing. We have put together the ideal-gas definition $pV = nRT$ ($= NkT$) and the kinetic theory result $pV = \frac{1}{3}Nm\overline{v^2}$. The result of this, and any further results, can be tested by experiment (of which Dalton's law and Graham's law suggest examples). One obvious experiment is to measure the speeds of the molecules of a gas at a particular temperature. At any particular temperature, of course, all the different molecules have different speeds. In 1859 Clerk Maxwell derived an expression for the distribution of speeds, which is illustrated in figure 10.12 for 1 000 000 oxygen molecules at temperatures of 300 K and 600 K. The meaning of the graph can be understood by looking at one point on it, e.g. the speed 750 m s^{-1}. The graph shows that, at 300 K, 500 oxygen molecules (out of the total population of 1 000 000) have speeds between 749.5 m s^{-1} and 750.5 m s^{-1}; at 600 K there are 1200 molecules within this speed range. The way in which the speeds are distributed among the molecules is called the *Maxwellian distribution*: for this sort of distribution, incidentally, $v_{rms} = 1.09\bar{v}$. This result depends on the ideas of the kinetic theory, and it can be verified by direct measurement of the speeds of molecules. One of the more recent methods has used the apparatus shown diagrammatically in figure 10.13.

The substance is heated in the oven, and atoms emerge from the slit, moving almost horizontally, at different speeds. The only force on the atoms is their weight, so they move in parabolic paths. Some of them pass through the second slit and arrive at the detector, which is a heated tungsten wire which can be moved up and down. The wire ionises the atoms if they strike it, and these ions are collected by a plate at a negative potential. This current I is a measure of the number of atoms arriving at the detector. The ion

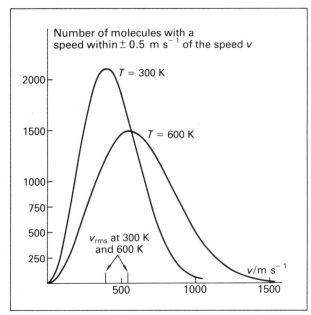

Figure 10.12 The Maxwellian distribution for 1 000 000 oxygen molecules at two temperatures. At any particular whole-number speed the height of the curve represents the number of molecules have a speed within ± 0.5 m s^{-1} of that speed (e.g. between 344.5 m s^{-1} and 345.5 m s^{-1} at 345 m s^{-1}). The sum of the number given by these heights will be 1 000 000.

current is measured when the detector is at different depths d below the line through the slits, and a graph of I against d is plotted. Figure 10.14 shows a typical set of points, together with the curve predicted by Maxwell for this experiment. (This curve is not the same shape as that in figure 10.12, because the y-axis now represents distance fallen d, which will be small when v is large.) The agreement between the experimental results and the kinetic theory predictions is obviously excellent.

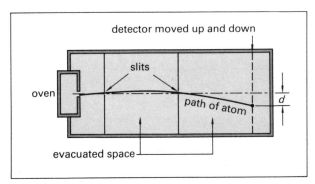

Figure 10.13 A sketch of apparatus which can be used to measure the speeds of atoms. The length of the apparatus is about 2.5 m. The faster the atom, the smaller the depth d which it falls.

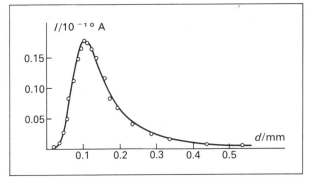

Figure 10.14 Results of an experiment using the apparatus shown in figure 10.13. The points marked o are found by experiment: the smooth curve is the prediction.

10.4 The internal energy of a gas

The internal energy U of a substance is defined as the sum of the kinetic and potential energies of its molecules. We do not include any k.e. (or p.e.) which the gas as a whole might have, through being in a container which is moving, or because of the position of the gas container. In the case of the ideal gas of our simple kinetic theory U is easy to calculate. It is simply the number of molecules multiplied by their mean translational k.e. In our assumptions we made it impossible for our molecules to have p.e., or any other forms of k.e. So

$$U = N(\tfrac{3}{2}kT)$$
$$= \tfrac{3}{2}nRT$$

If we heat the gas while its volume remains constant, the energy transferred by heating ΔQ is equal to the gain in internal energy of the gas ΔU since no work is done if there is no volume change. Therefore

$$\Delta Q = \Delta U$$
$$= \tfrac{3}{2}nR\Delta T$$

Molar heat capacity

Earlier (page 110) we defined the s.h.c. c_V for a substance under constant-volume conditions, and we now define the molar heat capacity (m.h.c.), under constant-volume conditions, C_V in an analogous way:

$$C_V = \frac{\Delta Q}{n\Delta T}$$

where ΔT is the temperature change when an amount of substance n has energy ΔQ supplied by heating. Compare $c_V = \Delta Q/m\Delta\theta$: we are now using ΔT instead of $\Delta\theta$; there is no significance in this. It is merely that when we deal with gases the convenient temperature is T. We have replaced m by n, but otherwise the defining equations are the same. Using this definition we see that the m.h.c. C_V of an ideal gas is given by

$$C_V = \frac{\frac{3}{2}nR\Delta T}{n\Delta T}$$

$$= \frac{3}{2}R$$

$$= 12.47 \text{ J K}^{-1} \text{ mol}^{-1}$$

What are the values of m.h.c.s of real gases at room temperature? Table 10.1 lists those of some common gases at 300 K. We see that the prediction of our theory has been entirely fulfilled for argon and helium, but not for the other gases. But we note that argon and helium are monatomic gases and our model was of simple spherical atoms. The other gases are diatomic and perhaps we should not expect the model to work well for them. We shall be discussing diatomic gases further, but we notice now that their m.h.c.s are nearly the same. The fact that the m.h.c.s are the same for these groups of gases is the reason why we often use the *molar* h.c. for gases: it should have the same value, no matter which monatomic, or which diatomic, gas it is. The *specific* h.c.s of helium and argon would *not* be the same.

▶ Calculate the s.h.c.s of helium ($A_r = 4$) and argon ($A_r = 40$); $C_V = 12.47 \text{ J K}^{-1} \text{ mol}^{-1}$.
For helium, $A_r = 4 \Rightarrow M_m = 4 \text{ g mol}^{-1} = 4 \times 10^{-3}$ kg mol^{-1}.

Since $C_V = 12.47 \text{ J K}^{-1} \text{ mol}^{-1}$

$$c_V = \frac{12.47 \text{ J K}^{-1} \text{ mol}^{-1}}{4 \times 10^{-3} \text{ kg mol}^{-1}}$$

$$= 3.1 \times 10^3 \text{ J K}^{-1} \text{ kg}^{-1}$$

We need not repeat the whole calculation to find c_V for argon, since its relative atomic mass is 10 times that for helium. This means that c_V will be ten times smaller, i.e., for argon $c_V = 3.1 \times 10^2 \text{ J K}^{-1} \text{ kg}^{-1}$. The specific heat capacities are very different. In general

$$C_V = c_V M_m$$ ◀

The average value of C_V for the three *diatomic gases* in table 10.1 is 20.84 J K^{-1} mol^{-1}. The fact that it is higher than the value for the monatomic gases means that it takes more energy to raise the temperature of a diatomic gas. What can we deduce from this? One possibility is that the diatomic gas molecules have just as much *translational* k.e. as the monatomic molecules but also have some *rotational* k.e. (which they can have because of their dumb-bell shape). Support for this comes from the fact that if we lower the temperature of all these gases, the m.h.c. of monatomic gases does not change, but the m.h.c. of diatomic gases falls to 12.47 J K^{-1} mol^{-1}. It is as if at low temperatures all gases have only translational k.e. but above a certain temperature the diatomic gases can have rotational k.e. as well. Figure 10.15 shows that the temperature at which this begins to happen for hydrogen is about 100 K.

We can go on to examine why the average value of the m.h.c. for diatomic gases is about 20.84 J K^{-1} mol^{-1}: this

gas	C_V/J K^{-1} mol^{-1}
argon (Ar)	12.47
helium (He)	12.47
hydrogen (H$_2$)	20.53
nitrogen (N$_2$)	20.87
oxygen (O$_2$)	21.12

Table 10.1 Values of C_V for some common gases at 300 K. The two monatomic gases have the same value of C_V, and the three diatomic gases have nearly the same value of C_V.

is approximately $5R/2$ ($5R/2 = 20.79 \text{ J K}^{-1} \text{ mol}^{-1}$) so perhaps we could write U_H (for hydrogen) as $U_H = N(3kT/2 + kT)$, where the two terms account for translation and rotation respectively, and accept that C_V for hydrogen is $5R/2$. A plausible theory is provided by an extension of the idea of equipartition of energy, which we met in a simple form on page 129, and which we now discuss further.

Degrees of freedom

The monatomic molecule in our box had three different directions in which it could move, or three *degrees of freedom*. Suppose that its mean *translational* k.e., $3kT/2$, is shared *equally* between these three degrees of freedom (just as, earlier, the total translational k.e. of a collection of molecules was shared *equally* between the different kinds). A diatomic molecule could have k.e. because of *rotation* about *two* of its axes (the x- and z-axes shown in figure 10.16) so it has two further degrees of freedom. It is reasonable to think that it could have little k.e. through rotation about the y-axis, since the masses of the atoms lie on this line. If each of these two degrees of freedom has, on average, as much energy as is associated with each of the three translational degrees of freedom (more equipartition), the average energy of this molecule is $5kT/2$. If the molecule can vibrate, along the y-axis in figure 10.16, it can have both k.e. and p.e. associated with this vibration, and if each of these further two degrees of freedom has, on average,

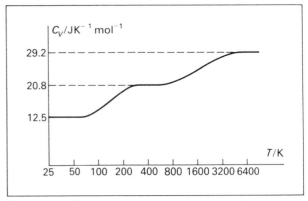

Figure 10.15 The variation of C_V with temperature for hydrogen: 12.5 J K^{-1} mol^{-1} = $3R/2$, 20.8 J K^{-1} mol^{-1} = $5R/2$, 29.2 J K^{-1} mol^{-1} = $7R/2$. Note the logarithmic spacing of the marks on the temperature axis.

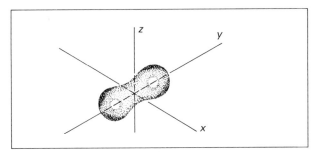

Figure 10.16 The three axes of a hydrogen molecule. If we remember that nearly all the mass of an atom lies in the nucleus, we can see that nearly all the molecule's mass lies on the *y*-axis, as there cannot be much k.e. of rotation about that axis.

an energy $kT/2$, the energy of the molecule is now $7kT/2$. These values of the average energy of a diatomic molecule, $3kT/2$, $5kT/2$, and $7kT/2$ in different circumstances are in fact matched by values for C_V of $3R/2$, $5R/2$ and $7R/2$ for hydrogen at successively higher temperatures. Also, a molecule in a solid could be thought of as being able to vibrate in each of three directions, and so having 6 degrees of freedom, and therefore an average energy of $3kT$, and a value for C_V of $3R$, i.e. 24.9 J K^{-1} mol^{-1}. This agrees well with the value most metals have at room temperature, as we saw on page 111.

It would be tempting to think that the principle of equipartition of energy 'explains' the values of C_V which we obtain for gases and solids, but it gives no hint as to why the diatomic molecules rotate, but do not vibrate, at room temperature, or why C_V for solids decreases as the temperature approaches absolute zero. This cannot be explained without the help of the quantum theory, which is beyond the scope of this book, but you should note that although the transition in figure 10.15 is not sudden for the bulk gas, there is a sudden transition for individual molecules. An individual diatomic molecule either has an energy of $3kT/2$ *or* an energy of $5kT/2$ *or* an energy of $7kT/2$. It cannot have intermediate values, such as $3.19kT$. It can absorb rotational and vibrational energy only in lumps or *quanta* (page 242).

Internal energy changes

The internal energy U of an ideal monatomic gas depends only on its temperature T, as shown by the equation $U = 3nRT/2$. It follows that $\Delta U = 3nR\Delta T/2$, and we have seen that $C_V = 3R/2$, so

$$\Delta U = nC_V\Delta T$$

provides us with a way of calculating the change of internal energy when an ideal monatomic gas has its temperature changed. This statement is true *even for processes where the volume is not constant*, so it is very useful. The next example will make this clear. In this case we shall be thinking about gas maintained at constant pressure while its tem-

perature changes; it is enclosed in a cylinder fitted with a frictionless piston, as shown in figure 10.17. Outside the cylinder the pressure is the atmospheric pressure. If no other forces act on the frictionless piston, the pressure in the cylinder must always be the same as the pressure outside it when it is in equilibrium.

Heating and working

The internal energy U of a gas can be changed in two quite different ways, as we saw on page 102: there can be *heating* or *working*. Figure 10.17 will act as a reminder that either the gas flame (supplying energy ΔQ because of the temperature difference between the flame and the gas in the cylinder) or the piston (supplying energy ΔW because the individual gas molecules are made to go faster if the piston moves inwards and hits them) can produce a change ΔU in the internal energy of the gas, and so raise its temperature. You must be careful to appreciate this distinction between heating and working. In the system shown, work can also be done by the gas when the piston moves outwards. In the equation $\Delta Q = \Delta U + \Delta W$ the value of ΔW would then be positive.

You will find that we shall use the word *reversible* to describe all the processes which follow, e.g. 'the gas is heated reversibly', or 'provided the process is reversible'. This is necessary. We shall explain what it means later, on page 137.

We shall need to know how to calculate the work done on, or by, the gas when the piston moves. The simplest situation is where energy is supplied reversibly to the gas, with the piston free to move. Then provided the changes happen slowly enough for us to be able to say that the piston is always in equilibrium, the pressure will be the same (atmospheric) on both sides of the piston, and this will be a constant-pressure process. If (figure 10.18a) this pressure is p_1, the area of the piston is A, and the displacement of the piston is Δx, the work done ΔW is given by

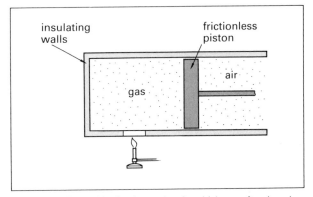

Figure 10.17 The idealised container in which we often imagine a gas to be. The walls are perfect insulators, except where energy is allowed to enter from the flame. If the piston moves inwards the piston does work *on* the gas (ΔW is negative); if the piston moves outwards the work is done *by* the gas on the piston (ΔW is positive).

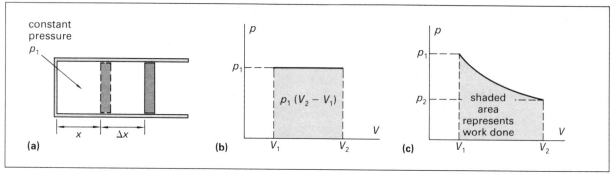

Figure 10.18 (a) Illustrates a constant-pressure expansion, and (b) shows the p-V graph for this process. The work done is represented by the area under the graph: since p is constant, it is easily calculated to be $p_1(V_2 - V_1)$. In (c) there is no simple relationship between pressure and volume, and so the work done has to be estimated from the area beneath the graph.

$$\Delta W = \text{force} \times \Delta x$$
$$= p_1 A \times \Delta x$$
$$= p_1 \Delta V$$
$$= p_1(V_2 - V_1)$$

where V_1 and V_2 are the initial and final volumes. Figure 10.18b shows how this work ΔW can be represented by the area under the graph of p against V. This is a simple calculation, because the pressure is constant. If the pressure had initially been greater than atmospheric pressure (perhaps because the piston had been held stationary while energy was supplied to the gas) and the piston then moved outward, it would have been more difficult to calculate the work done by the gas, but figure 10.18c shows a possible curve for the variation of p with V. The work done is again represented by the area beneath the graph.

▶ 0.050 mol of an ideal monatomic gas is enclosed in a cylinder fitted with a frictionless piston which is free to move. It is heated reversibly, and the gas temperature rises from 290 K to 350 K. (i) How much more internal energy does the gas now have, (ii) how much work has been done by the gas and (iii) how much energy has been supplied by heating? $R = 8.31$ J K^{-1} mol^{-1}, and atmospheric pressure $= 100$ kPa.

For an ideal monatomic gas, $C_V = 3R/2 = 12.47$ J K^{-1} mol^{-1}. We know that $\Delta U = nC_V\Delta T$ even though the volume is not constant, so

(i) $\Delta U = (0.050 \text{ mol})(12.47 \text{ J K}^{-1} \text{ mol}^{-1})(60 \text{ K})$
$$= 37.4 \text{ J}$$

(ii) The pressure of the gas has remained constant at 100 kPa (if the heating was done slowly, so that the piston was always in equilibrium). To calculate ΔW, we need to calculate ΔV. We can find the original and final volumes; rewriting $pV = nRT$ we have $V = nRT/p$, so

$$V_1 = \frac{(0.50 \text{ mol})(8.31 \text{ J K}^{-1} \text{ mol}^{-1})(290 \text{ K})}{10^5 \text{ N m}^{-2}}$$

$$= 1.20 \times 10^{-3} \text{ m}^3$$

and $$V_2 = \frac{(0.50 \text{ mol})(8.31 \text{ J K}^{-1} \text{ mol}^{-1})(350 \text{ K})}{10^5 \text{ N m}^{-2}}$$

$$= 1.45 \times 10^{-3} \text{ m}^3$$

so $\Delta V = 0.25 \times 10^{-3} \text{ m}^3$

and $\Delta W (= p\Delta V) = (10^5 \text{ N m}^{-2})(0.25 \times 10^{-3} \text{ m}^3)$
$$= 24.9 \text{ J}$$

(iii) We know that $\Delta Q = \Delta U + \Delta W$
so ΔQ, the energy supplied by heating $= 37.4$ J $+ 24.9$ J $= 62.3$ J Notice that the gas was heated, but not all of the energy supplied was used to increase the internal energy (and therefore the temperature) of the gas. Some of it was used to do work in pushing back the atmospheric air so that the gas in the cylinder could stay at its original pressure.

If the piston had been fixed, how much energy would need to have been supplied by heating if the temperature rise was to be the same? Not so much, since now no work would have been done by the gas. We should have had to heat it only enough to supply the gas with enough internal energy to raise its temperature from 290 K to 350 K. Then $\Delta Q = \Delta U = 37.4$ J. It would have been a constant-volume process, to which not only $\Delta U = nC_V\Delta T$, but also $\Delta Q = nC_V\Delta T$ applies. ◀

C_p and C_V

In that example we calculated ΔQ indirectly, using $\Delta Q = \Delta U + \Delta W$. We could have taken a short cut if we had the value of the *constant-pressure* m.h.c., C_p, which is defined in a similar way to C_V. There is a simple relationship between C_p and C_V, which we shall now derive.

Consider a *constant-pressure* process, like the one in the last example. If we *define* C_p by the equation

$$C_p = \frac{\Delta Q}{n\Delta T}$$

where ΔQ is the energy supplied by heating to raise an amount of substance n by a temperature ΔT, we can say that

$$\Delta Q = nC_p\Delta T$$

We *always* have $\Delta U = nC_V \Delta T$

and $\Delta W = p\Delta V$

But $pV = nRT$

and with p, n and R constant we can write

$p\Delta V = nR\Delta T$

(see figure 10.19 for a graphical illustration of this).

Using $\Delta Q = \Delta U + \Delta W$

we can write $nC_p\Delta T = nC_V\Delta T + nR\Delta T$

i.e. $C_p = C_V + R$

This is very useful, since it implies that if $C_V = 3R/2$, $C_p = 5R/2$. In the last example we could (since that was a constant-pressure process) have obtained the answer to (iii) by using

$$\begin{aligned} Q &= nC_p\Delta T \\ &= (0.050 \text{ mol})(\tfrac{5}{2} \times 8.31 \text{ J K}^{-1} \text{ mol}^{-1})(60 \text{ K}) \\ &= 62.3 \text{ J} \end{aligned}$$

C_V and C_p are the two *principal* m.h.c.s of a gas. Obviously a particular temperature change ΔT could require any amount of heating, depending on the circumstances. The next example makes that clear. It is convenient to refer to the ratio of the two principal m.h.c.s, and label the ratio γ

$$\gamma = \frac{C_p}{C_V}$$

For a monatomic ideal gas γ is 5/3.

Earlier (page 132) we saw good agreement between the predicted value of C_V for an ideal gas, and C_V for monatomic real gases (and for diatomic real gases, too, if we extended our theory to include the possibility of the molecules absorbing energy by rotating and vibrating). We now see that the values of C_p should be greater than the values of C_V by R, so that C_p should be $5R/2$ ($= 20.79 \text{ J K}^{-1} \text{ mol}^{-1}$) for monatomic gases at all temperatures, and $5R/2$, $7R/2$ $9R/2$ for diatomic gases, the value rising as the temperature of the gas rises. Table 10.2 gives measured values of C_p for some inert monatomic gases at 273 K, some elements at temperatures at which they are monatomic gases, and some diatomic gases at 298 K. We can deduce from

substance	T/K	C_p/J K^{-1} mol^{-1}
ideal monatomic gas		**20.79**
helium (He)	273	20.8
argon (Ar)	273	20.79
neon (Ne)	273	20.8
mercury (Hg)	1500	20.8
sodium (Na)	1500	20.8
zinc (Zn)	1500	20.8
ideal diatomic gas		**29.10**
carbon monoxide (CO)	298	29.14
hydrochloric acid (HCl)	298	29.14
hydrogen (H$_2$)	298	28.84
nitrogen (N$_2$)	298	29.12
oxygen (O$_2$)	298	29.37

Table 10.2 Values of C_V for some gases, showing the good agreement between predictions (the values given for ideal gases) and experiment.

the values of C_p for these diatomic gases that their molecules are not rotating or vibrating at this temperature. The agreement between experimental and predicted values is excellent.

10.5 Isothermal and adiabatic processes

Isothermal processes

We have been thinking about two of the processes through which a gas can pass: heating at constant volume, and heating at constant pressure. There are two more processes which we shall particularly consider. The first of these is an *isothermal* process, i.e. one in which the temperature of the gas does not change. For this sort of process $pV =$ constant so that if we draw the p-V graph for the process it will be identical with one of the Boyle's law p-V graphs, i.e. a hyperbola. Such a line is called an isothermal.

It should at once be clear that in an isothermal process there can be no change in the internal energy of the gas ($\Delta T = 0 \Rightarrow \Delta U = 0$). So if we let some gas expand, and do work, it must be heated by the surroundings (or by us) at an equal rate, so that there is no change in its internal energy. If $\Delta U = 0$, $\Delta Q = \Delta W$.

▶ Suppose we have 0.050 mol of an ideal monatomic gas enclosed in a cylinder fitted with a frictionless piston which is initially fixed in such a way that the pressure of the gas is 200 kPa. Its temperature is equal to that of its surroundings, and is 290 K. The cylinder is in good thermal contact with its surroundings, so that the gas is heated by the surroundings so that its temperature remains constant. When the piston is released slowly enough for to process to be reversible, how much energy do the surroundings provide by heating? Atmospheric pressure = 100 kPa, $R = 8.31 \text{ J K}^{-1} \text{ mol}^{-1}$.

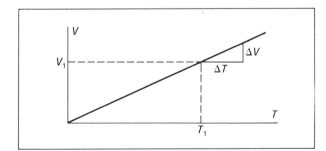

Figure 10.19 The ratio $\Delta V/\Delta T$ is the same as the ratio of any particular volume and temperature, e.g. V_1/T_1. Hence if $pV_1 = nRT_1$, $p\Delta V = nR\Delta T$.

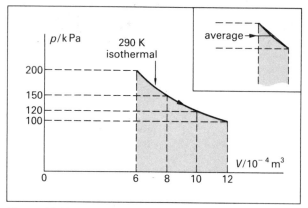

Figure 10.20 A reversible isothermal expansion at 290 K. The work done can be calculated by estimating the area beneath the curve.

We cannot use $\Delta Q = nC_V\Delta T$ or $\Delta Q = nC_p\Delta T$, since the process is neither constant-volume nor constant-pressure. We cannot even use $\Delta U = nC_V\Delta T$, since $\Delta T = 0$. We must calculate the work done ΔW by the gas as it expands. We know the way in which the pressure decreases as the volume increases: it obeys Boyle's law, i.e. $pV = nRT$ where $nRT = \text{constant} = 0.050\,\text{mol} \times 8.31\,\text{J K}^{-1}\,\text{mol}^{-1} \times 290\,\text{K}$. Substituting for $p_1 = 2.0 \times 10^5$ Pa and $p_2 = 1.0 \times 10^5$ Pa tells us that $V_1 = 6.0 \times 10^{-4}$ m³ and $V_2 = 12.0 \times 10^{-4}$ m³. Figure 10.20 shows this variation and reminds us that we need to find the area beneath the curve to find the work done. The area beneath the curve can be calculated exactly but the technique is beyond the scope of this book. We could plot the curve on graph paper, and count the squares beneath the curve, but we shall show a different method. In figure 10.20 the area has been split into three vertical slices, and we shall find the work done in each of these by multiplying the average pressure by the change in volume in each case. Since $pV = \text{constant}$ it is easy to calculate the values of p which correspond to the chosen V values of 8×10^{-4} m³ and 10×10^{-4} m³: these are 150 kPa and 120 kPa respectively. The calculation is performed in the table below:

$V/10^{-4}$ m³	$p/$kPa	change in volume $\Delta V/10^{-4}$ m³	average pressure $p/$kPa	work done $p\Delta V/$J
6.0	200⎫			
		...... 2.0	175	35
8.0	150⎬			
		...... 2.0	135	27
10.0	120⎭			
		...... 2.0	110	22
12.0	100⎭			

and hence the total work done is $(35 + 27 + 22)\,\text{J} = 84\,\text{J}$. This must be an over-estimate of the work done, since the simple averaging process treats the curve as if it was a series of three straight lines, and the area beneath these three lines would be greater than the area beneath the curve. The averaging process (e.g. $\frac{1}{2}(200 \times 150) = 175$) is only valid if the graph is straight. The inset in figure 10.20

should make this clear. The calculated value, however, is 83.5 J, so our method gives a good approximation.

Adiabatic processes

The second process is an *adiabatic* process, i.e. one in which there is *no heating* of the gas or by the gas. We have to assume that the cylinder insulates the gas perfectly from its surroundings: $\Delta Q = 0$, so that $\Delta W = -\Delta U$. In an expansion, where the gas does work so that ΔW is positive, ΔU will be negative, i.e. the temperature of the gas will fall. We can calculate ΔU (and therefore ΔW) if we can find ΔT, since, as always, $\Delta U = nC_V\Delta T$. (We cannot, of course, use $\Delta Q = nC_V\Delta T$ or $\Delta Q = nC_p\Delta T$: not only is this not a constant-volume or a constant-pressure process, but $\Delta Q = 0$!)

It can be shown that for a *reversible* adiabatic process

$$pV^\gamma = \text{constant}$$

Figure 10.21 shows a reversible adiabatic expansion, together with the isothermal curve for an expansion from the same initial pressure and volume to the same final volume. You will notice that the pressure falls to a lower value in the adiabatic expansion. Many adiabatic processes, however, are not reversible, and then the p-V curve lies somewhere between the curves $pV = \text{constant}$ and $pV^\gamma = \text{constant}$, but we do not know the values of the pressure which correspond to particular volumes and so we cannot plot the curve. The dashed lines in figure 10.21 do not represent actual expansions. Notice that at the end of a reversible adiabatic expansion the temperature is lower than it would have been after an isothermal expansion over the same change in volume: the temperature falls in an adiabatic expansion. This is to be expected, since the gas does work as it expands, and the insulation of the container

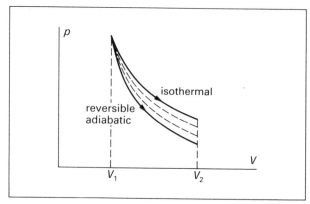

Figure 10.21 The p-V curves for an isothermal expansion, and for a reversible adiabatic expansion for an ideal monatomic gas. The dashed lines show *possible* curves for irreversible adiabatic expansions. Since the actual curves cannot be drawn (through lack of information about what the actual pressures are) the work done cannot be calculated.

prevents the surroundings from heating the gas. So it loses internal energy and its temperature falls. (This assumes that the gas does do some work as it expands: if it expands into a vacuum, no work is done, and there is no decrease in its internal energy, or fall in temperature.)

The work done in a reversible adiabatic expansion can, as usual, be found by estimating the area beneath the p-V curve, but there is a simple way of calculating it. Since $\Delta Q = 0$, in $\Delta Q = \Delta U + \Delta W$, $\Delta W = -\Delta U$, and $\Delta U = nC_V\Delta T$ (*always*, even if the process is not a constant-volume process). Therefore

$$\Delta W = -nC_V\Delta T$$

▶ Suppose 0.15 mol of an ideal monatomic gas is enclosed in a cylinder at a pressure of 250 kPa and a temperature of 320 K. It is allowed to expand adiabatically and reversibly until its pressure is 100 kPa. What is its final temperature, and how much work does the gas do? $R = 8.31$ J K^{-1} mol^{-1}. The initial volume V_1 is given by

$$V_1 = \frac{nRT_1}{p_1}$$

$$= \frac{(0.15 \text{ mol})(8.31 \text{ J K}^{-1} \text{ mol}^{-1})(320 \text{ K})}{2.5 \times 10^5 \text{ N m}^{-2}}$$

$$= 1.60 \times 10^{-3} \text{ m}^3$$

Using $p_1 V_1{}^\gamma = p_2 V_2{}^\gamma$, and remembering that $\gamma = 5/3$ for a monatomic ideal gas, we have

$$V_2 = V_1\left(\frac{p_1}{p_2}\right)^{\frac{1}{\gamma}}$$

$$= 1.60 \times 10^{-3} \text{ m}^3\left(\frac{250}{100}\right)^{\frac{3}{5}}$$

$$= 2.76 \times 10^{-3} \text{ m}^3$$

The p-V curve is drawn in figure 10.22. Such a curve should always be sketched before starting a question like this, and the information added as the solution proceeds. The amount n of gas is constant so

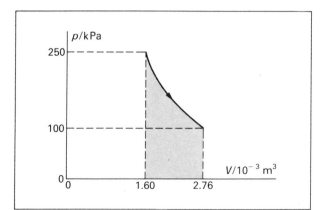

Figure 10.22 A reversible adiabatic expansion.

$$\frac{p_1 V_1}{T_1} = \frac{p_2 V_2}{T_2}$$

so that
$$T_2 = \frac{T_1 p_2 V_2}{p_1 V_1}$$

$$= \left(320 \text{ K}\right)\left(\frac{100}{250}\right)\left(\frac{2.76}{1.60}\right)$$

$$= 222 \text{ K}$$

As
$$\Delta W = -nC_V\Delta T$$
$$= -(0.15 \text{ mol})(\tfrac{3}{2} \times 8.31 \text{ J K}^{-1} \text{ mol}^{-1})$$
$$(222 - 320) \text{ K}$$

$$= 184 \text{ J}$$

The work done by the gas is 184 J. ◀

Reversibility

A process is *reversible* if at all stages its direction can be reversed by an infinitesimally small change in direction of the supply of energy. If any of the following features are observed in a process, it is *irreversible*:
(i) frictional or viscous forces which convert mechanical energy into internal energy,
(ii) spontaneous changes, which must imply non-equilibrium situations; these include
    (1) any heating process, i.e. energy transfer because of finite temperature differences
    (2) processes in which bodies accelerate, as a result of unbalanced forces.
You will realise that it is very unlikely that a process will, in practice, be reversible. A reversible process is an idealisation like a frictionless surface or a massless string. How close can we get?

An isothermal process is a reversible process. If we do work ΔW on some gas, and its temperature and internal energy do not change, it heats its surroundings by an amount ΔQ ($=\Delta W$). At any stage this process can be reversed: if the surroundings heat the gas, the gas does an equal amount of work on the surroundings. The problem is to make the process genuinely isothermal. If we push even a frictionless piston into a cylinder at a finite speed, there will be a temporary rise in the temperature of the gas and therefore a finite difference of temperature between the gas and its surroundings: that makes the process irreversible. There will also be turbulence in the gas, localised differences in pressure, and so the pressure of the gas cannot be specified: this also makes the process irreversible. If the process had been genuinely isothermal, it would have been reversible.

An adiabatic process, also, can be reversible. Now we need the material of the containing cylinder to be a perfect thermal insulator: if we compress the gas, its temperature will rise. This does not matter, provided the temperature difference between the gas and its surroundings does not cause energy to flow. If we do work ΔW on the gas, its internal energy will increase by an amount ΔU ($\Delta U = \Delta W$), and at any stage this process can be reversed: the increased pressure of the gas can push the piston back in the opposite direction. But in compressing the gas, or when the gas is allowed to expand, the piston must not move at a finite speed, or else, for the reasons mentioned above (turbulence, localised differences in pressure, localised differences in temperature) the process will be irreversible.

So to obtain good approximations, in practice, to reversibility, both isothermal and adiabatic processes have to be performed slowly compared with the speeds of the molecules. It is easier to get close to reversible adiabatic processes because pressure differences can even themselves out more quickly than can temperature differences. Notice that 'reversible' does not mean only that the working substance can, by some means or other, be returned to its original state. For example, some gas is initially at 300 K, and is quickly compressed by a piston which is not frictionless. The piston is then moved back to its original position, and after a time the temperature of the system returns to 300 K because the system has heated the surroundings. It should be clear that although the system is now in a state which is identical to its original state, the process has not been reversible.

11 Real gases and thermodynamics

11.1 Real gases

Figure 10.10 shows the behaviour of some real gases and compares this with the behaviour of an ideal gas. The real gases appear to behave in a simple manner, the value of pV/nT being linearly related to pressure, and the differences between them and the ideal gas are small. Indeed we have already seen (page 126) that the difference is less than 0.1 % at atmospheric pressure.

High pressures

But what happens at higher pressures? Figure 11.1 shows the behaviour of these gases at 273 K and pressures of up to 1000 times atmospheric pressure, and now we can see that their behaviour is much more complicated. Note that the vertical scale has had to be extended considerably to accommodate the very different values of pV/nT, which now range from about 7 J K^{-1} mol^{-1} to about 17 J K^{-1} mol^{-1} whereas previously they lay between 8.307 J K^{-1} mol^{-1} and 8.320 J K^{-1} mol^{-1}. Real gases deviate considerably from ideal gas behaviour when the pressure is high.

Low temperatures

We expect real gases to behave differently from an ideal gas at low temperatures too: we know that all gases can be liquefied, and we expect a real gas to behave non-ideally even before it has liquefied. Graphs of pV/nT for nitrogen at different temperatures (and very high pressures) are shown in figure 11.2. Here again there are differences between the behaviour of real gases and that of an ideal gas; even nitrogen behaves non-ideally in the range of temperatures shown. The graph shows a large decrease in the value of pV/nT at a constant pressure: if the temperature is constant this can mean only that the volume is decreasing – and what is actually happening is that the gas is liquefying. Nitrogen is not exceptional in behaving like this: the curves for all gases have a similar vertical portion if the temperature is low enough. This fact was first recognised by Thomas Andrews in a long series of experiments from 1863 to 1870. Before that time there had been much speculation as to whether the so-called permanent gases, like hydrogen, nitrogen and oxygen, could be liquefied at all. Pressures of up to 3000 times atmospheric pressure had been tried with these gases, but no liquefaction occurred. Andrews, working with carbon dioxide, performed experiments which produced the first comprehensive results, covering wide ranges of pressure, volume and temperature. He showed that carbon

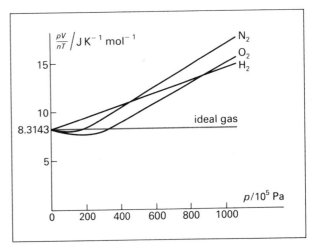

Figure 11.1 The variation of pV/nT with p for high pressures (the highest pressure shown on the pressure axis is 1000 times greater than atmospheric pressure) for some real gases at 273 K. Note the much greater range of the vertical scale, compared with figure 10.10: the behaviour or these gases is very different from that of an ideal gas.

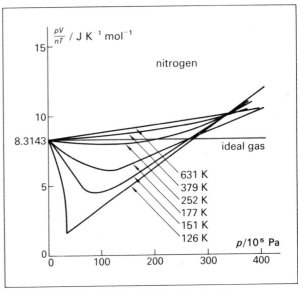

Figure 11.2 The variation of pV/nT with p for nitrogen at different temperatures. Its behaviour is very different from that of an ideal gas, particularly at low temperatures.

dixoide could be liquefied by compression if the temperature was low enough, and he argued that all gases could be liquefied if their temperatures were first reduced to a low enough value before they were compressed.

p-V-T surfaces

We shall illustrate Andrews's results presently, but first let us illustrate the variation of p, V and T for an ideal gas. It is usual to draw two-dimensional graphs for the variation of p with V, keeping T constant at particular values (as in figure 10.8, page 126) and also graphs of p against T (keeping V constant) and graphs of V against T (keeping p constant). All the information can, however, be represented, in three dimensions, by a single surface, as shown in figure 11.3. This helps to reinforce the fact that there is no freedom of choice about the value of the third quantity once the values of the first two have been chosen. For instance, if we have 5 moles of an ideal gas at a temperature of 400 K in a vessel of volume 0.10 m³, its pressure will necessarily be 166 kPa (using $pV = nRT$). Figure 11.3 also shows, by means of projections, the more familiar p-V, p-T and V-T graphs. On the p-V graph, for example, each line represents the variation on the surface for a particular value of T.

We should expect the p-V-T surface for a real gas (or, better, a real substance, since we shall need to consider the liquid and solid phases also) to be much more complicated, and figure 11.4, which is the p-V-T surface for carbon dioxide, certainly is. The surface is typical of the surfaces

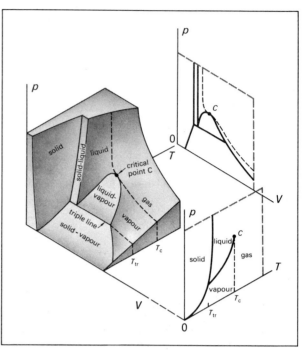

Figure 11.4 The p-V-T surface for a real substance (like carbon dioxide), together with the p-V and p-T projections in two dimensions. All substances which expand on melting have similar surfaces.

which can be drawn for any substance which expands on melting (as nearly all substances do: water is the common exception). The projections show the variation of p with V for constant temperature, and the variation of p with T for constant volume. These are the more important of the three possible projections. The reason for studying carbon dioxide, incidentally, is primarily historical: it was the substance on which Andrews performed his experiments. Similar surfaces (and projected curves) are obtained for all other substances, and in what follows you should think of the substance not as carbon dioxide particularly, but as any real substance.

Paths on the p-V-T surfaces

The surface contains a great deal of information, and we cannot afford the space to explore it in detail. We should at least make the following points:
(i) at the temperature T_1 the volume decreases as the pressure increases in much the same way as an ideal gas does (path ab in figure 11.5).
(ii) at the lower temperature T_2 the volume again begins to decrease (in an approximately ideal-gas manner) as the pressure is increased, but at a particular pressure the gas seems to collapse, and no further increase in pressure is needed to reduce the volume. The gas is liquefying. When the gas has completely liquefied very large increases of pressure are required to reduce the volume further (path

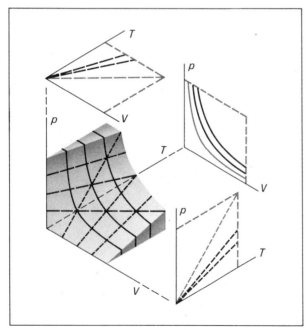

Figure 11.3 The p-V-T surface for an ideal gas, together with the p-T, V-T and p-V projections in two dimensions.

cdefgh), and eventually continually increasing pressure solidifies the substance.

(iii) there is a *critical temperature* T_c somewhere between T_1 and T_2. Above T_c the volume of a gas decreases continuously with increasing pressure; below T_c liquefaction occurs.

(iv) it is *conventional* to use the term *vapour* to describe a gas which is at a temperature below the critical temperature for that substance, but there is no obvious difference between a vapour just below T_c (e.g. at A in figure 11.5) and a gas just above T_c (e.g. at B). Nor is there any obvious difference between a liquid just below T_c (e.g. at C) and a gas just above T_c (e.g. at D). This is probably harder to accept; but at a point such as D the pressure is very high, and so the density of the 'gas' is very high, and the dense 'gas' is indistinguishable from a 'liquid'. The dashed line on figure 11.4 and figure 11.5 does not mark a boundary between obviously different phases.

(v) there is a *triple line* (which becomes the *triple point* on the *p-T* projection). This occurs at a particular temperature T_{tr} and a particular pressure p_{tr}; only at this temperature and pressure can the three phases co-exist *in equilibrium*.

(vi) below the triple-point temperature T_{tr} an increase in pressure solidifies the substance. It passes directly from the vapour phase to the solid phase (path *ijkl*). The reverse process is common in laboratories; carbon dioxide passes directly to the vapour phase when its temperature is allowed to rise at constant (atmospheric) pressure.

(vii) commonly the temperature of a substance in the solid phase is allowed to rise at constant pressure (e.g. when a cold substance is placed in a warm room). The changes which occur are typically shown by the paths *mnopqr* (when the pressure is greater than p_{tr}) in figure 11.6 and by *stuv* (when the pressure is less than p_{tr}). Water passes through the liquid phase when it changes from solid to vapour because its triple-point pressure is less than atmosphere pressure; carbon dioxide passes directly from the solid phase to the vapour phase because its triple-point pressure is greater than atmospheric pressure.

Van der Waal's equation

The surface shown in figure 11.3, for an ideal gas, is represented by the equation $pV = nRT$. It would obviously be impossible to find a single equation to represent the surface for a real substance, but there have been attempts to find an equation to represent parts of the surface. The best-known of these was put forward by J.D. Van der Waals in 1879: he suggested that two reasons why a gas or vapour might not behave like an ideal gas were:

(i) that at low temperatures and high pressures (and correspondingly low volumes) the volume of the molecules themselves would not be negligible compared with the volume of the vessel;

(ii) that the forces of attraction between the molecules would no longer be negligible, since the molecules would be much closer together.

Let us rewrite the equation $pV = nRT$ as

$$p\frac{V}{n} = RT$$

or $$pV_m = RT$$

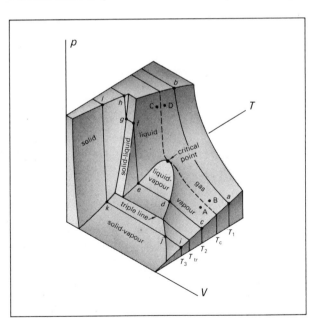

Figure 11.5 The paths *ab*, *cdefgh* and *ijkl* on the *p-V-T* surface for a real substance. These paths are all isothermal: one path is above T_c, one between T_c and T_{tr}, and the other is below T_{tr}.

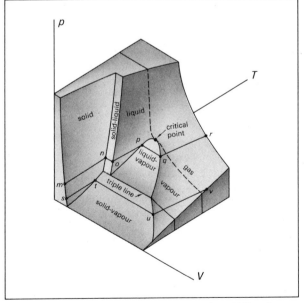

Figure 11.6 The paths *mnopqr* and *stuv* on the *p-V-T* surface for a real substance. They are constant-pressure paths: one is above p_{tr} and one is below p_{tr}.

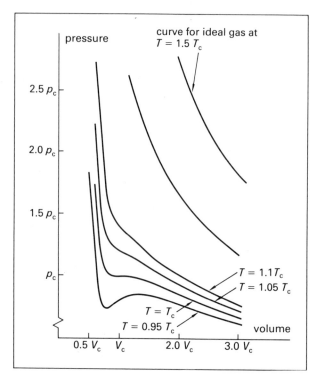

Figure 11.7 Isothermals predicted by the Van der Waals equation, together with one isothermal for an ideal gas.

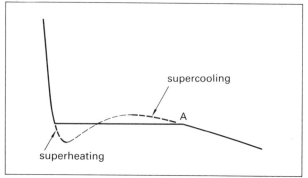

Figure 11.8 A modification to one of the Van der Waals isothermals; after modification it has the right shape for a real substance in the liquid-vapour region.

where V_m is the molar volume, i.e., the volume occupied by one mole of the substance. To allow for the fact that the whole of the volume of the vessel is not available to a molecule (because of the presence of the other molecules) let us replace V_m by $(V_m - b)$: b is called the *co-volume* of the molecules, and should be about four times the volume of the molecules themselves. The forces of attraction will reduce the pressure of the substance, and we could replace p by $(p + z)$ where z is a term which Van der Waals suggested was proportional to $1/V_m^2$. The full Van der Waals equation is

$$\left(p + \frac{a}{V_m^2}\right)(V_m - b) = RT$$

If it fits them well, it lends further support to our ideas about the way in which molecules behave. Figure 11.7 shows isothermal curves (i.e. graphs of the variation of p with V_m at constant temperatures) for different temperatures above, at and below the critical temperature T_c *as predicted by the equation*. Certainly the general shape of the curves agrees very well with the general shape of the curves plotted from the actual experiments on carbon dioxide, although the lowest curve shown (for $T = 0.95\ T_c$) has an upward-sloping portion which cannot be correct (since pressure and volume cannot both increase simultaneously). Figure 11.8 shows this curve re-drawn with a constant pressure line drawn across it (there are theoretical reasons for thinking that there should be equal areas above and below the line):

then we see that we have a very plausible curve. Moreover the dashed lines represent the existence of metastable states which actually occur: for example, vapour needs nuclei on which condensation may start, and if these nuclei are not present the substance follows the dashed line instead of beginning to condense at the point A. This is known as super-cooling: super-heating (represented by the other dashed line) also occurs.

We find, then, that the Van der Waals equation gives the right sort of shape of surface for a real substance in some regions: How well does it make quantitative predictions? One simple test is to consider the value of the quantity $p_c V_c/RT_c$ (known as the *critical coefficient*). The Van der Waals equation predicts that this should be 2.67 for all gases; for real gases the values are all higher than 2.67, though they are all of the same order of magnitude and vary surprisingly little. But then the Van der Waals assumptions are very simple. Considering their simplicity, the agreement is good, and we are therefore confident that two reasons for real gases not behaving like an ideal gas are those he suggested. All this confirms the model of atomic behaviour we have outlined in earlier chapters.

11.2 Vapour saturating a space

We now look in more detail at the liquid-vapour region in figures 11.4, 11.5 and 11.6. We have seen that two simple processes (i) doing work on the substance and allowing it to cool (path *de*) and (ii) heating the substance and allowing it to do work (path *pq*) both result in a similar movement (though in opposite directions) on the surface. In these movements both the pressure and the temperature remain constant. The only change is in the volume of the substance as it either

(i) liquefies, by any process (heating, or doing work on the substance) which takes the substance along a path parallel to *de* (or *qp*).

(ii) evaporates, which takes the substance along a path parallel to *pq* (or *ed*).

If we imagine the substance to be in a cylinder fitted with a piston we can illustrate the state of the substance at the beginning and end of these paths, and at some intermediate point, in figure 11.9. The volume can have different values but the pressure depends only on the temperature, and to each temperature there corresponds a particular pressure, known as the *saturation vapour pressure* (s.v.p.). For example, the s.v.p. of water at 390 K is 1.94 kPa, and the pressure in a vessel containing water in both the liquid and the vapour phases at this temperature is therefore 1.94 kPa, whatever the volume of the vessel and whatever the proportions of liquid and vapour. The s.v.p. increases roughly exponentially (see appendix B7) with ideal-gas scale temperature, but there is no simple relationship between the two quantities: this variation is shown by the line separating liquid and vapour on the *p-T* graph in figure 11.4.

Dependence of s.v.p. on temperature

Why does s.v.p. depend only on temperature? Consider a vessel containing liquid and vapour. As the pressure in the vessel is constant, the number of molecules of vapour must stay constant; but at the same time, molecules must continue to escape from the liquid at a steady rate. So molecules must be returning to the liquid at a rate equal to the rate of escape. It is not always the same molecules in the space above the liquid, but there is always the same *number* of molecules there. Suppose now that the temperature of the vessel is increased. More molecules in the liquid will have enough energy to escape from the liquid, and so the rate of escape

will increase. The number of molecules in the space will increase, and therefore the rate of returning will increase. The molecules are not 'forced' back into the liquid: it is just that the greater number now in the space means that the chance of a molecule moving downwards near the surface, and into the liquid, is increased because there are more molecules there. So the number of molecules in the space increases until the rate of returning is equal to the new, greater, rate of escape. The greater number of molecules above the liquid, and the fact that they are moving faster, means that the pressure is greater, so an increase in temperature produces a greater s.v.p.

The limit to the number of molecules in the space is not the sort of limit imposed by a notice in a lift which says 'only four persons allowed in this lift'. It is more like people going into a large shop. At the beginning of the day the number going in will be greater than the number coming out, but after a short time the number in the shop will be (roughly) constant, as the rates of going and coming become equal.

S.v.p. independent of volume

If instead of increasing the temperature we had reduced the volume of the vessel, there would temporarily have been a greater number of molecules per unit volume in the vessel (same number, greater density) and so the rate of return to the liquid would have been proportionally higher. Thus the rate of return would have been greater than the rate of escape (constant at constant temperature) and the number of molecules in the space would have decreased until the rate of returning was equal to the rate of escape. When this occurs, the density of molecules in the space must be the same as it was originally, and so the pressure must be the same as it was, although we have reduced the volume. If we had increased the volume, we could not have been sure that the space would have remained saturated with molecules, and therefore that the pressure would have remained constant. If the increase in volume were large enough the liquid might all have evaporated, and the arguments would not hold. As long as there is even a trace of liquid present and we have allowed enough time for evaporation, however, we can safely assume that the space is saturated and the pressure is the s.v.p.

It sometimes happens that a space contains a gas well above its critical temperature (e.g. air at room temperature and pressure) as well as a liquid and its vapour. We can predict what will happen by considering air and the other substance separately, treating the air as an ideal gas, and making use of the fact that the pressure of the mixture is the sum of the pressures which each would exert if it were there by itself (Dalton's law, page 129). An example will make this clearer.

▶ A cylinder fitted with a piston contains air and water vapour in a volume of 1.0×10^{-2} m³, and at a pressure of 100 kPa and a temperature of 360 K. The walls of the cylinder are damp, so that we can assume that the vapour is saturating the space. The piston is now moved inwards until the volume is 0.80×10^{-2} m³, and simultaneously the

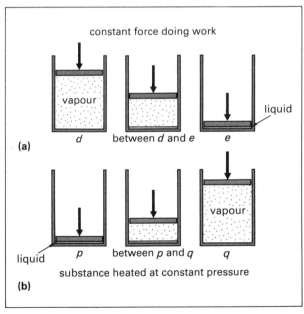

Figure 11.9 In (a) a constant force does work and liquefies the vapour; in (b) the substance is heated and does work against a constant pressure. The pressure and temperature remain constant throughout.

temperature is lowered to 340 K. What is the new pressure in the cylinder, if the s.v.p.s of water at 340 K and 360 K are 23 kPa and 53 kPa respectively?

Initially the pressure of the air in the cylinder is 100 kPa − 53 kPa = 47 kPa.

Using $\dfrac{p_1 V_1}{T_1} = \dfrac{p_2 V_2}{T_2}$

we have, for the air alone,

$$p_2 = (47 \text{ kPa}) \left(\frac{1.0 \times 10^{-2} \text{ m}^3}{0.80 \times 10^{-2} \text{ m}^3} \right) \left(\frac{340 \text{ K}}{360 \text{ K}} \right)$$

$$= 55 \text{ kPa}$$

Since both the change in volume and the change in temperature are in such a direction as to ensure that the space remains saturated, the pressure due to the water vapour is now 23 kPa, so that the total pressure in the cylinder is 55 kPa + 23 kPa = 78 kPa. Since the volume is smaller, and the temperature lower, some vapour will have condensed, but the volume of this water is negligible. ◀

Evaporation

Consider some liquid in an open vessel; it will be evaporating. The rate of evaporation can be increased by increasing the surface area (more molecules are then near the surface) or by blowing air across the surface (the air molecules will knock away some of the molecules which might have returned to the liquid) or by increasing the temperature (this will increase the number of molecules with enough energy to escape). Since it is the most energetic molecules which escape, the *average* kinetic energy of the remaining molecules will initially fall. This will continue until the temperature is sufficiently less than that of the surroundings for the surroundings to be able to heat the liquid at a rate exactly equal to the rate of loss of energy by evaporation (page 114). Hence we should expect the temperature of a liquid which is free to evaporate to be a little lower than that of its surroundings; how much lower will depend on the other factors mentioned earlier (e.g. the surface area, and the existence of draughts) and on how near the temperature of the liquid is to its boiling point. This effect can be demonstrated by bubbling air through a liquid: the liquid evaporates into the bubbles. With water the fall in temperature is less than 1 K, but with ether the fall in temperature may be 20 K or more. The latent heat needed for the evaporation of the liquid is being taken from the liquid itself.

Boiling

If the same liquid is now supplied with energy its temperature will begin to rise. At some stage we shall observe large bubbles forming and rising throughout the liquid, usually quite violently. These bubbles began as very small air bubbles which were saturated with the liquid's vapour; they were probably attached to the walls of the vessel. They grow because the liquid evaporates into them at a rate which increases with temperature: the pressure of the air in them soon becomes insignificant (because they now occupy a volume which may be 1000 times greater, while the temperature has not risen very much) and so the pressure is almost exactly the s.v.p. at the temperature of the liquid. The bubble cannot exist, however, unless the pressure in the bubble (the s.v.p. of the liquid) is at least equal to the pressure in the liquid at that depth. So *boiling* (defined as the continuous formation of large bubbles of vapour within the liquid) occurs at a particular temperature only – the temperature at which the s.v.p. of the liquid is equal to that above the liquid surface. When boiling begins the temperature of the liquid does not rise further. As with surface evaporation, the latent heat required for boiling is being taken from the liquid itself. At the boiling point, the evaporation into the bubbles removes energy from the liquid at a rate equal to that at which it is being supplied. When a liquid has been boiling for some time, all the air bubbles have been driven off. Boiling can then proceed only if there are *nuclei*, i.e. literally empty (though very small) spaces into which the liquid can evaporate. The existence of nuclei can be encouraged by adding pieces of broken porous pot: they roll about on the bottom of the vessel, and form nuclei as they roll. If there are few nuclei the temperature of the liquid rises above the temperature at which it normally boils at that pressure: this is known as *superheating*. Bubble chambers contain superheated liquid hydrogen: the ionisation caused by radiation entering the chamber produces a trail of nuclei on which a row of very small bubbles forms. The tracks of the particles are thus revealed in much the same way as in a cloud chamber (page 254).

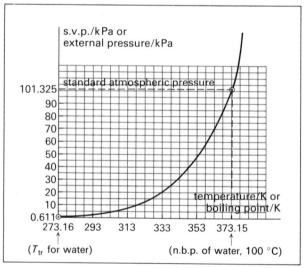

Figure 11.10 This graph shows *either* the variation of s.v.p. of a liquid with temperature *or* the variation of the liquid's boiling point with external pressure.

Effect of pressure on boiling point

Boiling will obviously occur at a relatively low temperature if the external pressure is low (e.g. on a mountain) and at a relatively high temperature if the external pressure is high (e.g. in a pressure cooker, where the intention is to prevent the water boiling until a temperature of perhaps 390 K is reached. The hotter water cooks food more quickly.). The *normal boiling point* (n.b.p.) of a liquid is the temperature at which it boils when the external pressure is standard atmospheric pressure. For example, the n.b.p. of water is 373 K at an external pressure of 101.3 kPa (because at 373 K the s.v.p. of water is 101.3 kPa).

The graph of the variation of boiling point with external pressure is inevitably exactly the same as the graph of the variation of s.v.p. with temperature (which, as we stated on page 143, is the boundary line between the liquid and vapour phases on a *p-T* graph). This graph is shown in figure 11.10: note the alternative labelling of the axes.

Measurement of s.v.p.

Barometer tube method

There are two simple ways in which the s.v.p. of a liquid can be measured. The first is suitable for small s.v.p.s (i.e. of less than standard atmospheric pressure). A vertical barometer tube is fitted with a tap at its upper end. With the tap open air is pumped out from the upper end, and mercury rises in the tube. When it has stopped rising the tap is closed (figure 11.11a). The liquid is then poured into the space above the tap and allowed to run into the barometer tube very slowly. As soon as any liquid reaches the evacuated space it evaporates, and the pressure of the vapour pushes the mercury down. When the mercury stops falling or, better, when a trace of liquid appears on top of the mercury column, the tap is closed, since then the space is saturated with the liquid's vapour. The s.v.p. of the liquid is then measured by the difference between the present level of the mercury and the original level, with allowance made for the effect of the liquid lying at the top of the mercury column.

As it stands this apparatus provides a measurement of s.v.p. at room temperature only. It can, however, be modified to measure s.v.p. at any required temperature. Two parallel barometer tubes are then used, jacketed by a water bath whose temperature can be varied. One of the tubes has liquid introduced into it as already described; the other is an ordinary barometer tube which provides a reference so that the difference in levels can easily be measured. A considerable difficulty in this experiment is the maintenance of a uniform, constant temperature.

Boiling-point method

The second method depends on the fact that a liquid boils at the temperature at which its s.v.p. is equal to the external

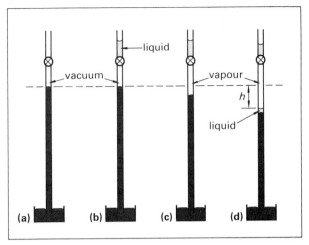

Figure 11.11 Measurement of the s.v.p. of a liquid at room temperature. *h*, the difference between the original and final levels of the mercury, measures the s.v.p.

pressure. It can therefore be used for s.v.p.s which are both less than and more than standard atmospheric pressure.

The liquid can be heated in a vessel fitted with a reflux condenser; see figure 11.12. The purpose of the condenser is merely to return evaporated liquid to the vessel, so that it neither needs to be replaced nor passes through the pump. The outlet tube from the condenser passes to a mercury manometer or some other pressure gauge and then through a large flask which serves as a reservoir to absorb fluctuations in pressure which may be produced by the pump. One thermometer measures the temperature of the liquid, and the other the temperature of the vapour: if there is no superheating these should be identical. A combination of

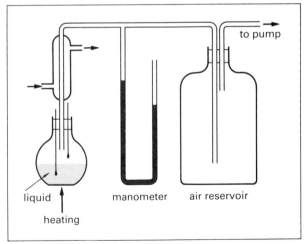

Figure 11.12 Measurement of s.v.p. The manometer indicates less than atmospheric pressure in the apparatus, so the liquid will boil at a temperature lower than the n.b.p. The pressure in the apparatus when the liquid is boiling is the s.v.p. of the liquid at that temperature.

pumping and heating is then employed until the liquid is boiling steadily. The temperature and pressure are then measured; this pressure is the s.v.p. of the liquid at that temperature. With suitable modifications to strengthen it, this type of apparatus has been used to measure s.v.p.s up to 30 times standard atmospheric pressure. It also demonstrates very effectively the variation of boiling point with external pressure: in a school laboratory it is not difficult to show water boiling at 20°C.

11.3 Heat engines

The term *heat engine* includes internal combustion engines, steam engines, refrigerators. It is obviously an extremely important topic for practical purposes, and we shall see that a study of heat engines also leads to some ideas of enormous theoretical importance.

All heat engines have a *working substance*. All heat engines must perform a *cyclic* process, i.e. they must be able to repeat their operation for any number of cycles. The working substance must therefore be returned to its original state at the end of each cycle: if the substance is disposed of at the end of each cycle (as in an internal combustion engine) then we must start the new cycle with fresh working substance in the same state as the working substance at the beginning of the previous cycle.

In all heat engines the working substance is heated, usually by contact with a *source* which we shall call the *hot reservoir*, but sometimes through a direct injection of energy, as in the internal-combustion engine; the working substance then expands and does work. Most heat engines therefore convert internal energy into mechanical energy (though some, e.g. the refrigerator, reverse this process). We shall see that at some stage all heat engines heat a *sink* which we shall call the *cold reservoir*; e.g. the exhaust of an internal combustion engine heats the surroundings. In that case the surroundings are the cold reservoir. The two reservoirs are assumed to have infinite heat capacity so that energy taken from them or given to them does not change their temperature.

Thermal efficiency

The thermal efficiency η of a heat engine is defined by the equation

$$\eta = \frac{\text{work done by engine}}{\text{energy absorbed by heating}}$$

although sometimes the efficiency is expressed as a percentage. Let us fix our ideas by considering the following example.

► A heat engine, with a monatomic gas ($C_V = \frac{3}{2}R$) as its working substance, performs the reversible cycle illustrated in the p-V diagram in figure 11.13. Find thermal efficiency.

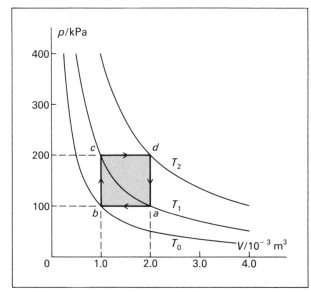

Figure 11.13 *abcd* represents a reversible process, and the shaded area represents the net work done by the gas. The curves are isothermal curves for temperatures T_0, T_1 and T_2. Along *ab* the gas is cooled at constant pressure, so its volume decreases. Along *bc* it is heated at constant volume, so its pressure increases. *cd* is the reverse of *ab*, and *da* is the reverse of *bc*.

There is no work done along the paths *bc* and *da*, since there is no change in volume. The work done ΔW along the paths *cd* and *ab* can be calculated from $\Delta W = p\Delta V$, since the work is done by constant-pressure processes:

$$\text{along } cd: \Delta W = p\Delta V$$
$$= (2.0 \times 10^5 \text{ Pa})(1.0 \times 10^{-3} \text{ m}^3)$$
$$= +200 \text{ J}$$

$$\text{along } ab: \Delta W = p\Delta V$$
$$= (1.0 \times 10^5 \text{ Pa})(-1.0 \times 10^{-3} \text{ m}^3)$$
$$= -100 \text{ J}$$

so the total work done by the gas (and therefore by the engine) in one cycle is 100 J.

We now have to calculate the energy absorbed by the gas by heating. The most methodical way is to set out a table; the values of ΔU are calculated from the relation

$$\Delta U_{ab} = \tfrac{3}{2}nR(T_0 - T_1) = \tfrac{3}{2}nRT_0 - \tfrac{3}{2}nRT_1$$

We can use $pV = nRT$ to show that

$$\tfrac{3}{2}nRT_0 = \tfrac{3}{2}(100 \times 10^3 \text{ Pa})(10^{-3} \text{ m}^3) = 150 \text{ J}$$

Similarly $\tfrac{3}{2}nRT_1 = 300 \text{ J}$

So $\Delta U_{ab} = -150 \text{ J}$

Check the others for yourself. ΔQ is calculated from $\Delta Q = \Delta U + \Delta W$ (the first law).

path	ΔW/J	ΔU/J	ΔQ/J
ab	-100	-150	-250
bc	0	$+150$	$+150$
cd	$+200$	$+300$	$+500$
da	0	-300	-300

We see that energy is *absorbed* by the gas along the paths bc and cd: a total of 650 J. The energy rejected to the surroundings along ab and da is wasted. So the thermal efficiency η is given by

$$\eta = \frac{\Delta W}{\Delta Q}$$

$$= \frac{100 \text{ J}}{650 \text{ J}}$$

$$= 0.15 \text{ or } 15\%$$

We see that the thermal efficiency is very low, and this takes no account of energy conversion by frictional forces, unwanted heating of the surroundings, etc. The overall efficiency will be even less than the thermal efficiency. ◀

What can we do to make an engine more efficient? To answer this question we must first return to the topic of reversibility. This was mentioned at the end of chapter 10, and before going further you should read that reference again. There we mention mechanical reversibility and thermal reversibility. The first of these is the easier to understand: a process will be reversible if there are *no dissipative forces* to convert mechanical energy into internal energy and *no finite unbalanced forces* which will produce accelerations and speed and more dissipative forces. For instance, consider a spring of stiffness 20 N m^{-1} and natural length 1 m hung up vertically with a load of 8 kg on it. It will stretch 4 m, so that its total length will now be 5 m, as shown in figure 11.14. Suppose it now just touches the floor. How much elastic potential energy E_p does the spring have stored in it? From page 23

$$E_p = \tfrac{1}{2}kx^2 = \tfrac{1}{2}(20 \text{ N m}^{-1})(4 \text{ m})^2 = 160 \text{ J}$$

How can we make use of that? If we unload the whole 8 kg at once, at floor level, the spring will merely contract violently and uselessly, and we shall get no benefit at all. Suppose we unload 4 kg at ground level: the other 4 kg will then move up to the 2 m level, and after some oscillations, will come to rest there. We can unload the 4 kg there, and we shall have extracted 80 J ($= 4 \text{ kg} \times 10 \text{ N kg}^{-1} \times 2 \text{ m}$) of gravitational potential energy. That is better. But suppose we unload 2 kg at ground level: the other 6 kg will reach the 1 m level where we can unload a further 2 kg. The remaining 4 kg will reach the 2 m level, and we can unload another 2 kg. The final 2 kg will reach the 3 m level, and we can unload that there. The gravitational p.e. we have now extracted

$$= (2 \text{ kg})(10 \text{ N kg}^{-1})(1 \text{ m}) + (2 \text{ kg})(10 \text{ N kg}^{-1})(2 \text{ m}) +$$
$$(2 \text{ kg})(10 \text{ N kg}^{-1})(3 \text{ m})$$
$$= 120 \text{ J}$$

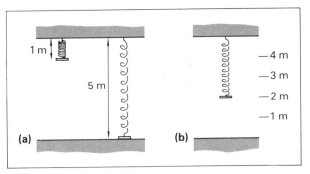

Figure 11.14 The natural length of the spring is 1 m: a load of 2 kg (weight ≈ 20 N) stretches it by 1 m, so a load of 8 kg pulls it down to the floor.

You can check for yourself that if we had unloaded the masses 1 kg at a time we should have extracted 140 J, and if they had been unloaded 0.5 kg at a time we should have extracted 150 J. Presumably if we could remove the masses in infinitesimally small amounts we could extract the full 160 J.

We have been describing here processes which have been becoming more nearly reversible: all of them have been irreversible because the oscillations of the spring have been converting mechanical energy into internal energy (through air resistance, and the internal friction in the metal of the spring). The smaller the lumps in which the mass was removed, the smaller were these oscillations, and the more reversible the process; when the mass is removed in infinitesimally small amounts the process is reversible and the conversion of elastic p.e. into gravitational p.e. has the maximum efficiency (100%).

The Carnot cycle

We can see that we must avoid *finite* unbalanced forces: we must remove the loads in infinitesimal amounts so that the pull of the spring is never a finite amount greater than the weight of the masses. This gives us a clue as to what we should aim for to make our heat engine thermally reversible: we must not have finite temperature differences between the working substance and the hot or cold reservoirs. We must not allow energy to move from a hot reservoir to a cold reservoir without doing any work, just as we could not allow the spring to contract and waste its energy. It is easy to see how this could be done, at least in theory. We can put the working substance of our heat engine in contact with the hot reservoir, and let the working substance expand *isothermally*. All we need to do is to allow the working substance to expand so that its temperature does not rise (only we must be careful not to allow the heating to be so rapid that we produce finite unbalanced forces on the piston!). Equally we can let the working substance heat the cold reservoir isothermally at the end of

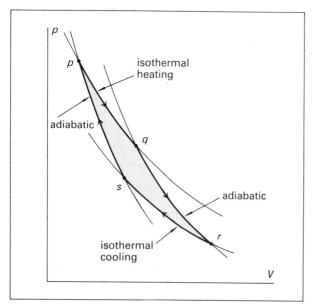

Figure 11.15 The Carnot cycle *pqrs* consists of two isothermal and two adiabatic curves. All the processes must be reversible.

the cycle. And we can arrange for the cycle to be completed by two reversible paths. One way of doing this is to use two reversible adiabatic processes; this cycle is known as a Carnot cycle, after Sadi Carnot, a French engineer, who proposed it in 1824. It is shown in figure 11.15.

In the Carnot cycle energy Q_h is absorbed only at a temperature T_h, and energy Q_c is rejected only at a temperature T_c, and since $\Delta U = 0$ for the whole cycle, the first law ($\Delta Q = \Delta U + \Delta W$) tells us that ΔW, the work done by the substance, is equal to $Q_h - Q_c$. So

$$\eta = \frac{Q_h - Q_c}{Q_h}$$

$$= 1 - \frac{Q_c}{Q_h}$$

For any reversible cycle in which energy is absorbed and rejected only at the extreme temperatures it can be shown that

$$\frac{Q_c}{Q_h} = \frac{T_c}{T_h}$$

so that $\eta = 1 - \dfrac{T_c}{T_h}$

for a Carnot cycle or any other reversible cycle in which the energy is wholly absorbed at one temperature T_h, and wholly rejected at another temperature T_c. For convenience we shall describe all such cycles as Carnot-efficiency cycles. We can see that the efficiency of such cycles depends only on the two temperatures T_h and T_c, and increases as the ratio T_h/T_c is made larger. The nature of the working substance is irrelevant (we shall prove this later); e.g. it need not be a gas. It could begin the cycle as a vapour, and

be liquefied during the compression. This would not affect the value of the efficiency, which depends only on T_h and T_c.

Other reversible cycles

Any cycle can be made reversible, and it can be approximated by a large number of small Carnot-efficiency cycles. Figure 11.16 shows a very rough approximation, in which just a few Carnot-efficiency cycles are used to approximate to a simple constant volume and constant pressure cycle. The approximation can be made as good as we wish, if a larger number of smaller Carnot-efficiency cycles are used. But the small Carnot-efficiency cycles will, generally, have a smaller temperature difference between their high and low temperatures than there is between the highest and lowest temperatures T_h and T_c in the cycle, so each of these small cycles will be less efficient than a cycle working between T_h and T_c. So this non-Carnot-efficiency, but reversible, cycle will be less efficient than a Carnot-efficiency cycle. *Between two particular temperatures the most efficient engine is a Carnot-efficiency engine* and this is its importance.

In all reversible engines the direction of operation can be reversed without changing the amount of work done or the amounts of energy absorbed or rejected by heating – just because the cycle is reversible. These amounts would not remain the same for an irreversible cycle. Any engine can be run backwards, in reverse, but only in a thermo-dynamically reversible engine do these quantities remain the same.

The conditions necessary for a process to be reversible seem very strict, and unlikely to be achieved in a real engine. For example, we have specified that the heating should be slow enough for the acceleration of a piston in an engine to be zero! In practice the speeds of the moving parts of engines are considerably less than the speeds of the gas molecules, and this means that thermal near-reversibility is achieved.

Real engines

Real engines are inefficient because
(i) the temperature difference between T_h and T_c is too

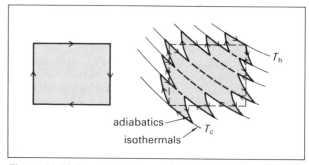

Figure 11.16 A cycle being approximated, very roughly, by a series of alternate isothermal and adiabatic steps which are parts of small Carnot cycles. In this way we can show that *any* cycle can be reversible.

small, so even an engine using a Carnot-efficiency cycle will not be very efficient.

(ii) even if the cycle is reversible it is unlikely to be a Carnot-efficiency cycle, and is therefore less efficient than a Carnot-efficiency cycle would be.

(iii) friction in the moving parts converts mechanical energy into internal energy.

Carnot-efficiency cycles, although potentially the most efficient (between two particular working temperatures) are obviously unattainable in practice.

Two common *internal combustion* engines use crude approximations to the Carnot-efficiency cycle called the *Otto cycle* and *Diesel cycle*; they have overall efficiencies of about 0.26 and 0.38 respectively when working at normal temperatures. A disadvantage of the Diesel engine is that its higher compression ratio means that it needs to be more robust, and is therefore more massive. The *Stirling cycle* engine is a hot-air engine which uses any means to heat the air; its theoretical efficiency is the same as that of a Carnot cycle, and its overall efficiency is about 0.45. Although the principle was first stated in 1816, it still has not been fully developed, so at present its cost is high because it is not mass-produced, and it has a small power/mass ratio. However it is quiet, and causes little pollution.

In steam engines and steam turbines water is evaporated and then further heated to produce *superheated steam* at temperatures of up to 850 K. The adiabatic expansion of this steam (which does work) cools and condenses the steam; this cycle (the *Rankine cycle*) is then repeated. The next example shows how the efficiency of steam engines has increased with the use of higher temperatures.

► Calculate the thermal efficiency of (i) an early steam engine which used steam at atmospheric pressure (and therefore 373 K), (ii) a steam engine of around 1820, which used steam at a pressure of about 5 atmospheres (boiling point of water at that pressure ≈ 150°C = 423 K), and (iii) a modern power station's steam turbine, which might work at a temperature of 850 K. Assume that all the engines use a Carnot-efficiency cycle, and exhaust their working substance at a temperature of 300 K.

(i) $\eta = 1 - \dfrac{300}{373} = 0.20$

(ii) $\eta = 1 - \dfrac{300}{423} = 0.29$

(iii) $\eta = 1 - \dfrac{300}{850} = 0.65$

We see the advantage of increasing the ratio T_h/T_c of the temperatures of the hot and cold reservoirs. The engines would not use a Carnot-efficiency cycle, and for this reason, and because the cycles are not reversible, and because of mechanical inefficiency, the overall efficiencies of these engines might be 0.02, 0.08 and 0.47 respectively, which shows how the mechanical efficiency of steam engines has improved in the last 200 years. ◄

Heat pumps and refrigerators

We shall go on to prove some other statements about a Carnot engine, but before that let us see what happens if a Carnot engine (or any other heat engine) is operated in reverse, i.e., so that a net amount of work is done on the working substance as shown in the flow diagram of figure 11.17b. (Figure 11.17a shows the flow diagram for a heat engine working forwards). But in this mode of operation (b) energy is being *removed* from the cold reservoir and *given* to the hot reservoir. The actual mechanism is not important, but one way in which this can be done is to have a substance with a high s.l.h. of vaporisation, and which can be easily liquefied. Figure 11.18 shows that a compressor C pumps the substance round the circuit. On the high-pressure side the vapour liquefies, and gives its latent heat to the surroundings. The liquid then passes through a valve into a low-pressure region, where it evaporates, taking the necessary latent heat from the cold reservoir. The mechanism is unimportant: an ideal-gas

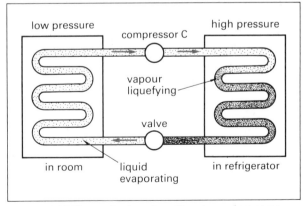

Figure 11.18 The mechanism of a refrigerator, or heat pump. Internal energy is being removed from the already-cold region, on the right, and is being delivered to the already-warm region, on the left. Work must be done (by the pump) to achieve this.

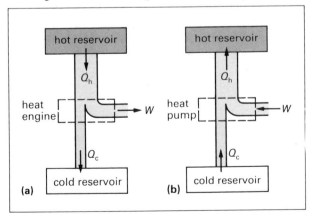

Figure 11.17 A flow diagram for (a) a heat engine, and (b) the same heat engine running in reverse as a heat pump.

refrigerator could be constructed, and it could perform the Carnot-efficiency cycle of figure 11.15 in reverse. The essential point is that energy is being taken from an already cold region and delivered to a warmer region. As we should expect, work has to be done to achieve this: the normal direction of flow of energy is from hot to cold, and the engine 'pumps up' energy from a cold reservoir to a hot reservoir. It can be put to use in two apparently different but actually identical ways.

(i) *The refrigerator.* Here the cold reservoir is the inside of the refrigerator, and the hot reservoir is the room in which the refrigerator is placed. The room becomes warmer and the inside of the refrigerator colder; essentially all that is happening is that energy is being moved from one place to another, but because the direction is abnormal, work must be done.

(ii) *The heat pump.* Here the cold reservoir is a river, or a well – or the coils are buried in the ground; the hot reservoir is the inside of a building. Just as a refrigerator heats a room, at the expense of the inside of the refrigerator, the heat pump heats the building, at the expense of the river, well, or ground. (The two different names merely reflect the different uses to which the same engine is put.) This is an excellent method of heating a house, since the energy from the surroundings is itself *free*, and the only running cost is the energy needed to drive the pump – and this energy may be several times less than Q_h, the energy delivered to the building. The owner of the building is using a little energy to make use of a lot of energy.

Coefficient of performance

The *coefficient of performance* (or figure of merit) η_r for a refrigerator or heat pump is defined by the equation

$$\eta_r = \frac{Q_h}{W}$$

As in any heat engine $W = Q_h - Q_c$, so

$$\eta_r = \frac{Q_h}{Q_h - Q_c}$$

$$= \frac{1}{1 - \dfrac{Q_c}{Q_h}}$$

For a Carnot-efficiency engine $\dfrac{Q_c}{Q_h} = \dfrac{T_c}{T_h}$

and so $\eta_r = \dfrac{1}{1 - \dfrac{T_c}{T_h}}$ for a Carnot-efficiency engine working in reverse

If the river temperature T_c is 5°C ($= 278$ K) and the temperature inside the house T_h is 20°C ($= 293$ K)

$$\eta_r = \frac{1}{1 - \dfrac{278}{293}}$$

$$\approx 20$$

i.e. the amount of energy delivered to the house is about 20 times the amount of work done. This calculation is for a Carnot-efficiency engine (and therefore for a reversible process of maximum efficiency) but values of η_r of 4 are quite possible for any cycle. The principle of the heat pump was known over 100 years ago, but they are not as common as they should be because until the early 1970's fossil-fuel chemical energy was cheap. Even now there are few commercially-available heat pumps (apart from refrigerators!) The capital cost is high because of the lack of development, and the main reason for lack of development is probably that people do not believe that they can get *free* energy.

11.4 The second law of thermodynamics

The *second law of thermodynamics* can be stated in many equivalent ways, each expressing a different facet of its meaning. There are so many different forms of it because it is of such significance. William Thomson (later Lord Kelvin) in 1851 stated it in this form:

no heat engine can perform a cyclic operation whose only result is to convert internal energy into mechanical energy.

For instance 1000 J of internal energy cannot be wholly converted into gravitational potential energy, although it is very easy to convert 1000 J of gravitational p.e. into 1000 J of internal energy. The flow diagram of figure 11.17a illustrates this: we must waste some energy. We cannot make use of all of it.

The second law was stated by Rudolph Clausius in 1850 in this form:

no refrigerator (or heat pump) can transfer internal energy from a cold reservoir to a hot reservoir without some external agent doing work.

Figure 11.17b illustrates this.

The second law, like the law of conservation of mass-energy, is one of the great fundamental laws of physics, to which no exceptions are known. You should be able to get an understanding of its significance from what follows, where we shall use it to make some statements about the efficiencies of heat engines.

Figure 11.19 shows a reversible engine R, the same reversible engine working as a heat pump P, and an irreversible engine I. Let us assume that I is more efficient than R, and that the efficiencies of I and R are 0.40 and 0.30 respectively. The energies (in arbitrary units) shown in figure 11.19 correspond with these efficiencies. Figure 11.19d shows I connected to P, so that 18 units of I's total output of 28 units is being used to drive P. Looking at the overall situation, we see that the remaining 10 units of I's output of work has being converted directly from internal energy,

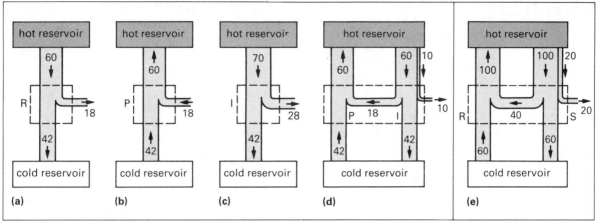

Figure 11.19 In (d) 18 units of I's total output of 28 units goes to drive P, thus providing a self-sufficient cycle in which both the hot and cold reservoirs show no net gain of internal energy. In addition, however, I manages to convert 10 units of internal energy directly to work, which violates the second law. In (e) a similar argument proves that S and R must be equally efficient: the situation shown, in which 20 units of internal energy are converted directly into work, is impossible.

which is a violation of Kelvin's statement of the second law. So the efficiencies of I and R cannot be 0.40 and 0.30. If we try different numbers we shall find that η_I must be less than η_R. With numbers we can only illustrate this, but with symbols we could prove that between two particular temperatures, *any reversible engine is more efficient than any irreversible engine.*

Suppose we have two Carnot-efficiency engines R and S, and let us assume that R has an ideal gas as its working substance, and S is a steam engine. Let us suppose that S has a higher efficiency than R: let their values be 0.50 and 0.40 respectively. Figure 11.19e shows S being used to drive R backwards. In a very similar argument to the one used above, you will see from the energies in the figure that the net result of this is to have S converting internal energy wholly to work, which contravenes the second law, so S cannot be more efficient than R. Interchanging the engines we can show that R cannot be more efficient than S – so R and S are equally efficient. *All Carnot-efficiency engines have the same efficiency, whatever their working substance.* Hence the Carnot-efficiency engine's thermal efficiency is (i) independent of its working substance, and (ii) greater than that of any other engine, reversible or irreversible, working between the same extreme temperatures. It provides a standard against which other engines can be judged and pinpoints the ratio of the reservoir temperatures as the key to obtaining high thermal efficiency. Until Carnot proposed his cycle, and developed the arguments about its efficiency, designers of steam engines relied on intuition and experiment; at a time when the development of steam engines was rapid, some theory was essential. We shall go on to look at one very important application of his work.

The (kelvin) thermodynamic scale of temperature

The temperatures used in deriving the equation

$$\frac{Q_h}{Q_c} = \frac{T_h}{T_c} \qquad \text{(for a Carnot cycle)}$$

were, of course, ideal-gas scale temperatures. We had no choice: the ideal-gas scale was the only defined temperature scale. Suppose the cold reservoir had been at the triple-point T_{tr}: then

$$T_h = T_{tr} \times \left(\frac{Q_h}{Q_{tr}}\right)$$

i.e., the *ideal-gas scale temperature* T_h is given by this equation. Suppose we now imagine that we have another Carnot-efficiency engine (any working substance will do) whose cold reservoir is at the triple point, but that we have no thermometer with which to measure the temperature of the hot reservoir. We make some measurements on the energy absorbed from the hot reservoir and rejected to the cold reservoir, and notice that the hotter the hot reservoir, the greater the quantity Q_h, and we decide to use this quantity to measure the temperature of the hot reservoir. Using T^* to represent temperatures measured in this way, we define T^* by the equation

$$T^* = T_{tr} \times \left(\frac{Q_h}{Q_{tr}}\right)$$

just as we defined temperature on the ideal-gas scale (on page 122). This *thermodynamic scale* (suggested by Kelvin) has been defined without reference to the properties of any particular substance (which was the objection to the ideal-gas scale, even though the objection was minimised by its being an *ideal* gas) and we see that for identical values of Q_h, $T^* = T$, i.e. the temperature measured on the thermodynamic scale is the same as the temperature measured on the ideal-gas scale. So any lingering doubts about the independence of the ideal-gas scale are removed: ideal-gas scale temperatures are identical to thermodynamic temperatures, which do not depend on the properties of any substance.

11.5 Entropy

We look again at the Carnot cycle shown in figure 11.15 and draw up a table of the values of T and ΔQ for the working substance: we have also tabulated the values of $\Delta Q_{subs}/T$.

path	T	ΔQ_{subs}	$\Delta Q_{subs}/T$
pq	T_h	$+Q_h$	$+Q_h/T_h$
qr		0	0
rs	T_c	$-Q_c$	$-Q_c/T_c$
sp		0	0
whole cycle		$Q_h - Q_c$	$+\dfrac{Q_h}{T_h} - \dfrac{Q_c}{T_c}$

But we have already seen (page 148) that for a Carnot-efficiency cycle

$$\frac{Q_c}{Q_h} = \frac{T_c}{T_h}$$

and it follows that

$$\frac{Q_c}{T_c} = \frac{Q_h}{T_h}$$

so the sum of the quantities $\Delta Q_{subs}/T$ for the whole (Carnot) cycle is zero. It is useful to be able to consider quantities which return to their original values after a process has occurred; another such quantity is the internal energy U of a substance. If in figure 11.15 the values of U at p and q are U_p and U_q, the change in U in moving from p to q is $U_q - U_p$, *whatever* the path taken from p to q – and for the complete cycle $\Delta U = 0$. So too for this cycle $\Delta Q_{subs}/T$ is zero, and we give this quantity the symbol ΔS, and call it the change in *entropy S* of the substance. We are therefore *defining* it by the equation

$$\Delta S = \frac{\Delta Q}{T}$$

where ΔS is the change in entropy of the substance which occurs when it gains energy ΔQ by *reversible* heating at a temperature T. For ΔQ we use the same sign convention as for the first law: e.g. $\Delta Q = -300$ J means that 300 J of energy are removed by heating the surroundings. This implies that a *gain* of energy by a body by heating will produce a *gain* in entropy by that body.

▶ (i) Calculate the change of entropy of 5.0 kg of ice (s.l.h. of melting $= 3.33 \times 10^5$ J kg^{-1}) which melts reversibly at 0°C.

The ice is reversibly heated by the surroundings, and its gain of energy ΔQ ($= ml$) at a constant temperature T ($= 273$ K) results in the ice gaining entropy ΔS_{subs} (using the suffix to emphasise that this is the entropy change for the substance) given by

$$\Delta S_{subs} = \frac{\Delta Q}{T}$$

$$= +\frac{(5.0 \text{ kg})(3.33 \times 10^5 \text{ J kg}^{-1})}{273 \text{ K}}$$

$$= +6.1 \times 10^3 \text{ J K}^{-1}$$

(ii) Calculate the change of entropy of 0.050 mol of an ideal gas when it is compressed isothermally (and therefore reversibly) at 290 K from a volume of 12×10^{-4} m³ to a volume of 6.0×10^{-4} m³.

The reverse of this situation was considered in the example on page 135. There we saw that the work done ΔW by the gas was $+83.5$ J.

Here, $\Delta U = 0$ (since $\Delta T = 0$) and so $\Delta Q = \Delta W = -83.5$ J

$$\Delta S_{subs} = \frac{\Delta Q}{T} = \frac{-83.5 \text{ J}}{290 \text{ K}} = -0.29 \text{ J K}^{-1}$$

(iii) Calculate the change of entropy in any reversible adiabatic expansion. In any reversible adiabatic expansion $\Delta Q = 0$, so $\Delta S_{subs} = 0$. This result applies equally to a reversible adiabatic compression.   ◀

Values of S

So far we have considered only changes in entropy. Usually the changes in entropy are all that matter, just as it is only the changes in potential energy which matter. However, just as we can choose an arbitrary zero for potential energy, so we can choose an arbitrary zero for the entropy of a substance, and it is usual to say that $S = 0$ when $T = 0$. It may help you to understand better what is meant by entropy if you look at table 11.1, which gives some values of entropy. You can see that the entropy of a substance increases whenever energy is given to it, whether the result is to raise its temperature or to melt or evaporate it. The entropy also depends on the mass: the energy required to bring a substance to a particular temperature is proportional to its mass. The variation in entropy for the different metals reflects the fact that they have different s.h.c.s: a metal with a low s.h.c., like lead, will not have needed as much energy to raise its temperature from 0 K to 273 K, so its entropy is relatively low.

substance	entropy $S/10^3$ J K^{-1}
1.0 kg of ice at 273 K	2.28
1.0 kg of water at 273 K	3.52
1.0 kg of water at 373 K	3.54
1.0 kg of water vapour at 373 K	10.9
2.0 kg of water at 273 K	7.04
1.0 kg of aluminium at 273 K	0.969
1.0 kg of copper at 273 K	0.488
1.0 kg of lead at 273 K	0.302

Table 11.1 Values of entropy. Notice (a) that there are large increases in entropy when ice melts and water evaporates, (b) that the changes are small when its temperature rises, (c) that the entropy of 2.0 kg of water is twice that of 1.0 kg of water, (d) that the entropies of the metals decrease in order of decreasing s.h.c.

There is zero change of entropy of the working substance in a Carnot-efficiency cycle, and in any other reversible cycle consisting, as it does, of a number of smaller Carnot-efficiency cycles (page 148).

Nor is there, when a reversible engine is working, any change in the entropy of the surroundings: the cold reservoir gains entropy Q_c/T_c, and the hot reservoir loses entropy Q_h/T_h, and we have seen that these amounts are equal, and opposite in sign. This can be extended to any reversible process: in part (i) of the last example, the surroundings lost as much energy as the ice gained, at the same temperature, so $\Delta S_{surr} = -6.1 \times 10^3$ J K^{-1}. If we use the suffix u to refer to the 'universe', i.e. the substance and the surroundings together,

$$\Delta S_u = \Delta S_{subs} + \Delta S_{surr}$$

and $\Delta S_u = 0$ in a reversible process

Before we can consider entropy changes in irreversible processes we must add something to our definition of change of entropy. We *define* the change of entropy ΔS in a complete cycle to be zero, *whether the process is reversible or irreversible*. This addition does not allow us to use the equation $\Delta S = \Delta Q/T$ to calculate ΔS for particular parts of the cycle: we are now saying only that for a substance taken round a complete cycle, reversibly or irreversibly, ΔS *for the substance* is zero. It does mean, however that we can calculate ΔS_{irrev} by considering the change of entropy when the substance is brought back reversibly to its original state, since ΔS for the whole process, out and in, is zero.

▶ Calculate the change of entropy of 0.050 mol of an ideal gas when it expands freely (i.e. into an evacuated space) in a thermally insulated vessel from a volume of 6.0×10^{-4} m^3 to a volume of 12×10^{-4} m^3. When the gas expands there are finite pressure differences, and the process is irreversible, so we cannot use $\Delta S = \Delta Q/T$ to calculate ΔS. But we have already considered the (reversible) isothermal compression of this gas (page 152) and there we saw that $\Delta S = -0.29$ J K^{-1} for a process which would return the gas to its original state after its free expansion. Since ΔS for the complete cycle (irreversible expansion followed by reversible compression to the original state) is zero, ΔS for the free expansion $= +0.29$ J K^{-1}. Note that the equation $\Delta S = \Delta Q/T$ would in any case have given us the wrong answer: $\Delta Q = 0$, so it predicts (wrongly) that $\Delta S = 0$. Here we have an increase in the entropy of a substance *without* an increase in its energy. The gas simply takes up more space. The *disorder* of the system has increased. The association of entropy with disorder is a very important idea to which we shall return. ◀

The entropy of the universe

Let us consider the melting of 0.10 kg of ice which is at 0°C when the surroundings are at 10°C. It happens spontaneously because there is a finite temperature difference, and because there is a finite temperature difference the process is irreversible. The s.l.h. of melting of ice is

3.33×10^5 J kg^{-1}. So the ice absorbs 3.33×10^4 J of energy at 273 K, so

$$\Delta S_{subs} = \frac{+33\,300 \text{ J}}{273 \text{ K}} = +122 \text{ J K}^{-1}$$

The surroundings lose this energy irreversibly at a temperature of 283 K; if the energy were restored reversibly to the surroundings (so that for the whole operation $\Delta S = 0$) ΔS_{surr} would be given by

$$\Delta S_{surr} = \frac{+33\,300 \text{ J}}{283 \text{ K}} = +118 \text{ J K}^{-1}$$

so when the surroundings lost this energy irreversibly $\Delta S_{surr} = -118$ J K^{-1}. For the system as a whole (the universe u)

$$\begin{aligned}\Delta S_u &= \Delta S_{subs} + \Delta S_{surr}\\ &= +122 \text{ J K}^{-1} - 118 \text{ J K}^{-1}\\ &= +4 \text{ J K}^{-1}\end{aligned}$$

In this *and all other* irreversible processes the entropy of the universe increases.

We have already seen that $\Delta S_u = 0$ for reversible processes, but these are idealisations. All real processes are irreversible, so the entropy of the universe is continually increasing. There are no processes in which the entropy of the universe decreases.

We take it for granted that the *energy* of an isolated system will be conserved; a consideration of the *entropy* is sometimes necessary before we can tell whether or not a process will happen. Consider the two processes shown in figure 11.20. In AB a brick falls to the floor. Its initial gravitational p.e. is converted to k.e. as it falls; when it hits the floor its k.e. is converted into the internal energy of the brick and the floor. There was no change of entropy for the brick while it was falling, but after it has stopped the brick and the floor are warmer. They have therefore gained entropy, and hence $\Delta S_u > 0$. The reverse happens in process BC, in which the internal energy of the brick and floor is converted into k.e. in the brick, and hence into gravitational p.e. Now the brick and floor *lose* entropy but there has been no gain of entropy by any other part of the universe. So $\Delta S_u < 0$, which is impossible, and this process cannot happen.

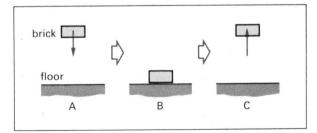

Figure 11.20 In both processes, AB and BC, energy is conserved, and so energy gives no guidance as to whether the processes will happen. But in BC the entropy of the universe would increase, and this is impossible.

You may be thinking that you did not need a knowledge of entropy to predict that AB happens and BC does not. But you can imagine that there are situations, particularly in chemistry, where the possibility of an event is not obvious – and where, for example, a slight change of temperature or pressure may turn an impossibility into a possibility. A knowledge of entropy is indeed all-important in the chemical industry. The example which follows returns to the melting of ice.

▶ Calculate the change of entropy which would occur if 0.10 kg of ice at 0°C melted while the surroundings were at −10°C. We already know that $\Delta S_{subs} = +122$ J K^{-1}. The energy needed to melt the ice would be taken from the surroundings while their temperature was 263 K, so

$$\Delta S_{surr} = \frac{-33\,300\ \text{J}}{263\ \text{K}} = -127\ \text{J K}^{-1}$$

So for this process $\Delta S_u = +122$ J K^{-1} − 127 J K^{-1} = −5 J K^{-1}, which is an impossibility. So ice does not melt when the surrounding temperature is −10°C. ◀

But there is a further point. Why do we not think that a brick will jump up off a warm floor? The molecules in the brick have the velocities of their own random vibrational energy; when the brick is falling they also have, superimposed on this, the downward velocity of the brick as a whole. After the fall the energy of the downward motion has become the energy of the increased random vibration of the molecules. The number of molecules in the brick is of the order of 10^{25}; to rise from the ground, these molecules would all, simultaneously, need to regain their upward velocities! It could happen, some day: it is only a matter of *probability* – but of course the number of ways in which the molecules in the floor and the brick could be moving is so vast that we are safe in saying that it will never happen. It would be a spontaneous change from disorder to order.

Orderliness is improbable: disorder is probable. So gases mix, buildings fall down, eggs break, a hot bath cools down, a room becomes more untidy – because these processes are probable, because there are many ways in which they can happen. There is only one way in which two gases can be unmixed, but many ways in which they can be mixed; only one way in which a particular house can be built, but many ways in which the rubble can be scattered; only one way in which an egg can exist, but many ways in which the broken egg can be smeared over its surroundings.

What can happen in many ways is, probably, what will happen. We could in fact have begun this section on entropy by defining it in terms of 'the number of ways' in which something can happen, or be arranged. A process such that 'the number of ways' increases (which is a probable process) will make the entropy, defined in this way, increase.

Another form of the second law

We are now using the idea of entropy to predict the possibility of physical and chemical changes, so we can see the significance of the concept of entropy. We could study

mechanics without the concept of momentum, but it would be much harder to make sense of it; the concept of entropy helps in the same way. The second law can be stated in the following way:

It is impossible for the entropy of the universe to decrease.

We can easily see that this is equivalent to the other two forms (due to Kelvin and Clausius) which we have already stated and which are illustrated in (a) and (b) in figure 11.21. These engines are assumed to be performing cyclic processes, so that in one complete cycle $\Delta S_{subs} = 0$, so $\Delta S_u = \Delta S_{surr}$.

In (a) an engine has taken in energy Q and converted it into work W, e.g. by lifting a load. The entropy of the load has not changed, but the entropy of the surroundings has *decreased* by an amount Q_h/T_h, i.e. $\Delta S_u < 0$, which is impossible. A heat engine must also deliver energy to the cold reservoir: if it is a reversible engine, the energy and associated entropy make $\Delta S_u = 0$. If it is irreversible it does less work, delivers more energy to the cold reservoir, and hence more entropy, and $\Delta S_u > 0$. In (b) a heat pump is absorbing energy Q at a temperature T_c and delivering the same amount Q to the hot reservoir at a temperature T_h.

$$\Delta S_{surr} = \frac{-Q}{T_c} + \frac{+Q}{T_h}$$

which is < 0, so $\Delta S_u < 0$, which is impossible. A heat pump must have $Q_c \leqslant Q_h$; only in this way can $\Delta S_u \geqslant 0$. So the Kelvin and Clausius statements of the second law are equivalent to this new statement $\Delta S_u \geqslant 0$.

▶ It is intended to design a steam engine which will work with a hot reservoir at 820 K and a cold reservoir at 300 K. From the consideration that $\Delta S_u \geqslant 0$, find the maximum efficiency under these conditions.

Suppose that the engine takes in 1.00 MJ from the hot reservoir: the change of entropy of the surroundings

$$\Delta S_{surr} = \frac{-1.00 \times 10^6\ \text{J}}{820\ \text{K}} = -1.22 \times 10^3\ \text{J K}^{-1}$$

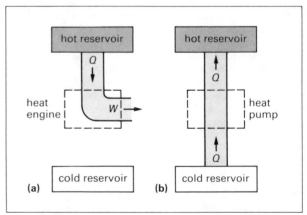

Figure 11.21 Two impossible processes: in both of them the entropy of the universe would decrease.

Suppose that of this 1 MJ, an amount Q_c is rejected to the cold reservoir: the change of entropy of the surroundings must be at least $+1.22 \times 10^3$ J K^{-1} in order that $\Delta S_u \geqslant 0$, and so

$$\frac{Q_c}{300 \text{ K}} \geqslant 1.22 \times 10^3 \text{ J K}^{-1}$$

or $\quad Q_c \quad \geqslant 3.66 \times 10^5$ J

Thus the work done by the engine

$$= 1.0 \times 10^6 \text{ J} - 3.66 \times 10^5 \text{ J} = 6.34 \times 10^5 \text{ J}.$$

The maximum efficiency would be

$$\frac{6.34 \times 10^5 \text{ J}}{1.0 \times 10^6 \text{ J}} = 0.63$$

This could be achieved with a Carnot-efficiency cycle, but of course in practice there will be both mechanical and thermal irreversibility, so the efficiency will drop below this figure. The overall efficiencies of (fossil fuel) power stations are up to 0.47, but the overall efficiency of a nuclear-fuel power station, where the working temperature is lower, might be only 0.30. So even the most efficient power station rejects to the surroundings (i.e. *wastes*) more than half the energy which it takes in.   ◀

The second law tells us that this is a one-way universe. The processes on which we depend cannot be reversed: the burning of coal and oil, the decay of radioactive materials, are spontaneous, irreversible processes which cause an increase in the entropy of the universe. The total *energy* remains the same, but after the process the energy is not as useful as it was. Occasionally there are local decreases in entropy, but according to the second law there must be a corresponding and larger increase in entropy elsewhere. The universe proceeds inexorably to a state of maximum entropy.

12 Centripetal forces and gravitation

12.1 Describing circular motion

When a line rotates, we measure its angular displacement from some agreed origin. The average *angular velocity* ω_{av} of the line is then defined by the equation

$$\omega_{av} = \frac{\Delta\theta}{\Delta t}$$

For motion with a constant angular velocity ω, e.g. a gramophone record, the angle θ rotated in a time t is given by

$$\theta = \omega t$$

The units of ω (the Greek letter omega) are radians per second, rad s^{-1}. (Appendix B4 explains radian measure.) It is however quite common to quote angles in degrees and angular speeds in revolutions per second. We must always use radians in calculations so it is useful to remember that

$$1° \equiv \frac{2\pi}{360} \text{ rad}$$

and $\quad 1 \text{ rev s}^{-1} = 2\pi \text{ rad s}^{-1}$

The time taken to complete one revolution is called the *period T* and is related to ω by

$$2\pi = \omega T \quad \text{or} \quad T = \frac{2\pi}{\omega}$$

The number of revolutions completed per second is called the rotational frequency n and is measured in rev s^{-1}.

Clearly $\quad n = \dfrac{1}{T}$

If a rigid object e.g. a gramophone record, completes a revolution, the line joining every point on it to the centre of rotation moves through an angle of 360° (2π rad), but the speeds at different points are not the same. In figure B4 (appendix B4) suppose the radius OAB moves to OPQ in a time t. The point A covers a distance $r\theta$, and B covers a distance $2r\theta$. (By definition the size of the angle θ in radians is given by $\theta = \text{arc}/r$.) The speed of A is given by

$$v_A = r\theta/t = r\omega$$

and for B

$$v_B = 2r\theta/t = 2r\omega$$

i.e. the speed v of any point on a rigid body is proportional to the distance r of that point from the axis of rotation.

$$v = r\omega \tag{1}$$

This equation does not hold for bodies such as gymnasts and acrobats who bend and stretch as they rotate.

➤ What are (i) the angular velocity and (ii) the speed of a point on the Earth's surface at a latitude of 60°N, e.g. the Shetland Isles or Leningrad? Take the Earth to be a sphere of radius 6400 km.
(i) All points on the Earth complete a circle once every 24 hours. Therefore Leningrad's angular velocity

$$\omega = \frac{2\pi}{T} = \frac{2\pi \text{ rad}}{24 \times 3600 \text{ s}} = 7.3 \times 10^{-5} \text{ rad s}^{-1}$$

and this is independent of the latitude.
(ii) A point at latitude 60° is moving in a circle of radius (6400 km) cos 60° ($= 3200$ km) about the Earth's axis of rotation. Therefore Leningrad's speed v is given by

$$\begin{aligned} v = r\omega &= (3.2 \times 10^6 \text{ m})(7.3 \times 10^{-5} \text{ rad s}^{-1}) \\ &= 230 \text{ m s}^{-1} \end{aligned}$$

We could of course have found v by calculating the distance ($2\pi r$) covered in a day and then calculating $2\pi r/T$. ◀

Measuring angular velocity

Measuring constant or steady angular velocities is straightforward. We make a mark on the rotating body and time how long it takes to make a counted number of revolutions. Alternatively, for high values of $n(=\omega/2\pi)$ we can use a calibrated flashing stroboscope to 'freeze' the rotating object and then read off the frequency f of the flashes: f is then equal to n. There is, however, a chance that we have selected a flashing rate $f = n/2, n/3$ etc, for then the rotating object will still appear to be frozen. To check this we should increase f gradually and measure the *highest f* which gives a *single* frozen view of the object. Where the object is like a bicycle wheel and appears identical when rotated through an angle θ which is less than 2π rad, it is best to mark it in some way, e.g. by sticking some white tape to one spoke or to one part of the wheel.

For varying values of ω the techniques outlined in figures 1.1 and 1.2 can be adapted. In particular where the object is a disc the tape can be wrapped on it (or off it) as the disc rotates (figure 13.8, page 172).

➤ The side of a record player deck has a series of equally-spaced vertical lines painted on it. A small mains-operated (50 Hz) neon light is arranged to illuminate the lines as the deck rotates. At a rotational frequency of $33\frac{1}{3}$ rev min^{-1} the lines appear to be stationary. Explain this effect and calculate the number of lines there are around the circumference of the deck.

The effect is a stroboscopic one. The neon light is 'on' 100 times every second when the voltage across it rises above a certain value. If the record deck rotates at such a speed that each vertical line moves to the position occupied by its neighbour in 0.010 s the lines will appear 'frozen'.

$$33\tfrac{1}{3} \text{ rev min}^{-1} = \frac{100}{3} \times \frac{1}{60} \frac{\text{rev}}{\text{s}}$$

so that in 0.010 s the deck rotates

$$\frac{100}{3} \times \frac{1}{60} \times 0.010 \text{ rev} = \frac{1}{180} \text{ rev}$$

Thus 180 vertical lines will produce the desired effect. You should be able to explain that 360 lines would also give rise to a stationary effect. ◀

Angular acceleration

The value of the instantaneous angular velocity $\omega \,(= d\theta/dt$, appendix B3) of a rotating object may not be constant. A fairground roundabout slows to a halt; an electric drill speeds up. For angular motion with uniform angular acceleration a, the equations for uniformly accelerated linear motion have equivalent angular forms. These are:

$$a = \frac{\omega - \omega_0}{t}$$

where ω_0 and ω are the initial and final values of the angular velocity. Rearranging

$$\omega = \omega_0 + at$$

The angular displacement θ can be found from

$$\theta = \left(\frac{\omega_0 + \omega}{2}\right) t$$

and, when $\omega_0 = 0$,

$$\theta = \frac{\omega t}{2} = \tfrac{1}{2}at^2$$

▶ A gramophone record takes 2.0 s to reach its (constant) angular velocity of 45 rev min^{-1} from rest. Find its angular acceleration, assuming that it is constant, and the number of revolutions it makes before it reaches this speed.

$$45 \text{ rev min}^{-1} = 45 \times \frac{2\pi}{60} \text{ rad s}^{-1} = 4.7 \text{ rad s}^{-1}$$

Thus $\quad a = \dfrac{\omega - \omega_0}{t} = \dfrac{4.7 \text{ rad s}^{-1}}{2.0 \text{ s}} = 2.4 \text{ rad s}^{-2}$

and $\quad \theta = \left(\dfrac{\omega_0 + \omega}{2}\right) t = \left(\dfrac{4.7 \text{ rad s}^{-1}}{2}\right)(2.0 \text{ s}) = 4.7 \text{ rad}$

$$= \frac{4.7}{2\pi} \text{ rev} = 0.75 \text{ revolutions} \qquad ◀$$

12.2 Centripetal forces

Figure 12.1 shows a particle which moves from P to Q in time Δt at a constant speed v. Its velocity changes from v_P to v_Q and the vector triangle (b) shows the change of velocity Δv, where (vector) $\Delta v = v_Q - v_P$. The average acceleration of the particle is thus $\Delta v/\Delta t$ in the direction of Δv. As Δt is made smaller the direction of Δv becomes more closely perpendicular to v_P, i.e. along PO. Therefore the instantaneous acceleration of the particle at any point on its circular path will be inward along a radius. It is said to be *centripetal*.

To establish the size of the average acceleration between P and Q refer to figure 12.1c where $v_P = v_Q = v$.

$$\Delta v = 2v \sin \frac{\Delta\theta}{2}$$

But provided $\Delta\theta/2$ is small, we can write   (appendix B5)

$$\sin \Delta\theta/2 \approx \Delta\theta/2 \quad \text{(radians)}$$

$$\therefore \ \Delta v \approx 2v\left(\frac{\Delta\theta}{2}\right) = v\Delta\theta$$

As $\quad v = r\omega = r\dfrac{\Delta\theta}{\Delta t}$

then $\quad \Delta t = \dfrac{r\Delta\theta}{v}$

so that acceleration $\quad \dfrac{\Delta v}{\Delta t} \approx \dfrac{v\Delta\theta}{r\Delta\theta/v} = \dfrac{v^2}{r}$

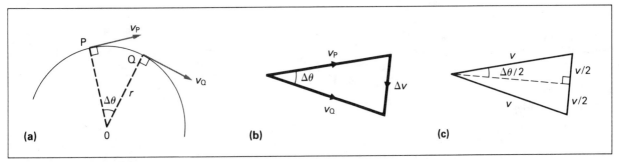

Figure 12.1 Changing velocity and circular motion: to v_P add Δv in order to get v_Q.

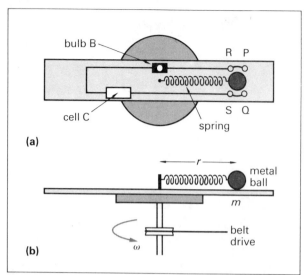

(a)

(b)

Figure 12.2 An apparatus for illustrating $F_{res} = mr\omega^2$. The bulb circuit is omitted in (b).

The instantaneous size of the acceleration a at P is the size of $\Delta v/\Delta t$ as Δt and $\Delta\theta$ tend to zero. It is therefore given by

$$a = \frac{v^2}{r} = r\omega^2 \quad \text{(as } v = r\omega\text{)}$$

i.e. a particle which is moving in a circle of radius r with a constant speed v has a centripetal acceleration which is of constant size v^2/r directed at any instant towards the centre of the circle.

In a laboratory centrifuge a particle suspended in a liquid at a distance of 0.10 m from the axis of rotation will, at 200 rad s^{-1} (about 30 rev/s) have an acceleration of

$$a = (0.10 \text{ m})(200 \text{ rad s}^{-1})^2 = 4000 \text{ m s}^{-2}$$

which is about $400g$!

A satellite circling at 200 km above the Earth's surface passes overhead about once every 88 minutes. Its acceleration, towards the centre of the Earth (which has a radius of 6400 km) is therefore

$$a = r\omega^2 = (6.6 \times 10^6 \text{ m}) \left(\frac{2\pi}{88 \times 60 \text{ s}}\right)^2$$

$$= 9.3 \text{ m s}^{-2}$$

which is only just less than g at the Earth's surface. A similar calculation for the Moon in its orbit round the Earth gives its acceleration as 0.0027 m s^{-2}. Both these results will prove significant when we consider satellites in section 12.6.

Measuring centripetal forces

Newton's second law states that a body of mass m moving in a circle of radius r with constant speed v requires a resultant centripetal force F_{res} of size mv^2/r or $mr\omega^2$. This can be demonstrated in a number of ways. Figure 12.2 is a sketch of a piece of apparatus which enables F_{res}, m, r and ω to be measured separately and for each to be varied. The bulb B lights when a metal ball contacts either RS or PQ, two pairs of metal pegs which can be set in a number of places on a wooden board. The board is fixed to a horizontal turntable which is driven at different speeds. The angular velocity $\omega = 2\pi n$ of the board is increased until the contact of the ball across RS is broken and the bulb goes out, and this speed is maintained while n is measured. If the speed rises too far the bulb comes on again through contact across PQ. The pull of the spring on the ball is the centripetal force F_{res} (if frictional forces are negligible). With this arrangement we can measure F_{res} for a range of values of the variables m, r and ω, and show that $F_{res} \approx mr\omega^2$. The experiment is not very precise and an agreement of better than $\pm 10\%$ should not be expected.

The pull of the spring on the ball is a *centripetal* force; the pull of the ball on the spring is a centrifugal force which is the same size (Newton's third law). The latter is seldom of any interest to us as it does not act on the object which is moving in a circle – the ball in this case.

Many types of force act as centripetal forces, e.g. the sideways frictional push of the ground on a car as it rounds a corner, the gravitational pull of the Earth on a satellite or the electromagnetic pull on an electron moving in a magnetic field. In many other cases the resolved part of a force, e.g. the pull of a string on a conker rotating in a horizontal circle, acts as the centripetal force.

▶ Some eggs are placed in an open basket and (safely) rotated in a vertical plane so that the eggs describe a circle of radius 1.0 m. What is the least speed at the top of the circle at which the eggs must move?

Figure 12.3b shows a free-body diagram for one egg at the moment it passes through the top of its circle. So long as there is contact between the egg and the basket the push P of the basket on the egg acts downward as shown. By Newton's second law and using $a = v^2/r$

$$m\frac{v^2}{r} = mg + P$$

Clearly the least value of v for which the egg stays in the basket is that for which $P = 0$, i.e. when the egg just loses contact with the basket.

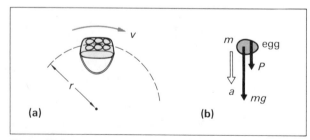

(a)   **(b)**

Figure 12.3 Whirling a basket of eggs in a vertical circle. If the speed at the top of the circle is above a certain value the eggs remain in the basket.

So then $m\dfrac{v_{min}^2}{r} = mg$

or $v_{min} = (rg)^{\frac{1}{2}} = \sqrt{(1.0\ \text{m})(9.8\ \text{m s}^{-2})}$

$\qquad\qquad = 3.1\ \text{m s}^{-1}$

This is equivalent to a (steady) rotational frequency of

$$n = \frac{\omega}{2\pi} = \frac{v}{2\pi r} = \frac{3.1\ \text{m s}^{-1}}{2\pi(1.0\ \text{m})}$$

$$\approx 0.5\ \text{s}^{-1}$$

At less than half a revolution per second the eggs will drop out of the basket at the top of the circle. ◀

▶ A car of mass 900 kg is driven round a circle of radius 150 m on a horizontal track at a steady speed of 20 m s^{-1}. What frictional force must be exerted by the ground on the car? Take $g = 10\ \text{m s}^{-2}$.

Figure 12.4 is a free-body diagram of the car showing the centripetal (sideways) frictional push S of the ground on the car. By Newton's second law

$$ma = S \quad \text{and as} \quad a = \frac{v^2}{r}$$

$$S = (900\ \text{kg})\frac{(20\ \text{m s}^{-1})^2}{(150\ \text{m})} = 2400\ \text{N}$$

If the car is moving at a steady speed there must be another frictional force, the forward push F of the ground, acting on the car. For instance if the resistive push of the air on the car is 3200 N backwards, then $F = 3200$ N forwards. The two frictional pushes F and S, act at right angles. Their sum is 4000 N, as

$$(3200\ \text{N})^2 + (2400\ \text{N})^2 = (4000\ \text{N})^2$$

and this total frictional push of the ground on the car acts in the direction shown in figure 12.5. Note that the size of the air resistance force does *not* affect the size of the centripetal force; in solving problems to find S, the forward push of the ground is often ignored. ◀

Weightlessness

The pull of the Earth on a body is called its *weight*. We

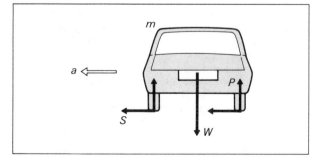

Figure 12.4 A free-body diagram for a car, seen from behind, making a left hand turn.

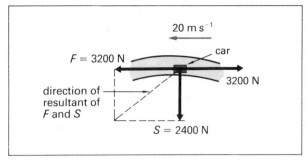

Figure 12.5 The horizontal forces acting on a car rounding a corner at 20 m s^{-1}.

can perceive our own weight only indirectly; it would be indeed surprising if we could feel the pull of the Earth on us just as it would be surprising if we could feel the air pressing against us. If a force of similar size to our weight acts on a small area of our bodies (e.g. the perpendicular contact push of the floor on the soles of our feet), we do feel it. If we temporarily feel no such supporting force the brain mistakenly thinks that we have no weight either. It is in this way that the idea of weightlessness arises. A person might *feel* weightless (and perhaps have a peculiar feeling in the stomach) when, for example, he

(i)   has jumped off a spring board or trampoline,

(ii) treads on a non-existent floor in the dark (perhaps on meeting an unexpected step down), or

(iii) travels in a car over a hump-backed bridge at high speed.

An astronaut is in a state of *free fall* for much longer than a diver or trampolinist. He feels no supporting force from the spacecraft around him when its engines are off and so feels weightless.

There is sometimes confusion between mass and weight; the confusion is perhaps the result of both words being used loosely before Newton's second law is properly understood and the equation $mg = W$ for free fall clearly established. Note, however, that

(i)   mass is a scalar, weight (being a force) is a vector;

(ii) mass is measured in kg, weight in N;

(iii) mass is invariable, being associated with *how much matter* there is in a body. Weight depends on how far the body is from the Earth or some other object which attracts it gravitationally; weight can, of course, be zero in deep space.

12.3 Newton's law of gravitation

The study of the motions of the Sun, Moon, planets and stars is as old as man himself; how can one live under the night sky and not be fascinated by it? (Today we live under roofs and move about in boxes so that modern man is relatively unaware of the sky.) Describing the motions of the heavens led to careful observations and to ingenious rules for predicting the positions of the Moon and planets. Johannes Kepler (1571–1630) eventually established the following:

(i) the planets describe *elliptical* orbits, with the Sun at one focus (1609).

(ii) the line drawn from the Sun to a planet sweeps out *equal areas* in equal time intervals (1609).

(iii) the squares of the planets' periods of revolution are proportional to the cubes of the semi-major axes of the ellipses in which they move around the Sun, i.e. $T^2 \propto r^3$ (1619).

The difficulties encountered in this sort of research can be appreciated by comparing the elliptical orbit of Mars with a circle. Take two drawing pins and stick them, with their heads touching, in a board covered with a piece of A4 paper. Make a loop of cotton around the pins and draw a large ellipse by keeping a pencil taut at the outside of the loop. This ellipse will be very close to the shape of Mars's orbit around the Sun, which would coincide with one of the pins. If a compass is now used to draw a circle with its centre between the drawing pins and of radius equal to the average size of the ellipse it is very difficult to tell the circle from the original ellipse. And Mars has the most elliptical path of the planets for which Kepler had information.

Can we find a mechanism which correctly predicts these rules? The work of Galileo Galilei (1564–1642) and Isaac Newton (1642–1727) provided the first answer. We must beware, however, of thinking that the role of the physicist is to aim to answer the question 'why?'. Galileo's and Newton's work does not explain why the planets come to move as they do. When we apply Newton's third law and find that it makes useful predictions, we do not know of a reason why forces obey this rule; it is simply found that they have the property of appearing in pairs. Similarly the law of gravitation (below) helps to solve problems, from explaining the tides to putting men on the Moon, but nowhere does it suggest a *mechanism* for gravitational forces: they exist, and they obey this rule.

Newton's law of gravitation (1687) states that every particle in the universe attracts every other particle with a force F, where

$$F \propto \frac{m_1 m_2}{r^2}$$

m_1 and m_2 are the masses of the two particles and r is their separation. Predictions made with this *inverse square law* can be verified experimentally, the most convincing correlation between theory and practice coming from predictions about planetary motions rather than laboratory experiments. For example, the planets Neptune and Pluto were both first seen after their very existence had been predicted by analysing the variations in the motions of planets near them. We can insert a constant and write

$$F = G \frac{m_1 m_2}{r^2}$$

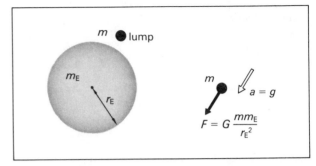

Figure 12.6 A freely falling body at the Earth's surface.

where $G = 6.673$ N m² kg⁻² and is called the universal constant of gravitation or 'big gee'. It can be measured experimentally (see below) but is the most difficult of all the fundamental constants to measure precisely. Newton was not able to measure G himself but, by guessing the mean density of the Earth he was able to deduce a rough value for it.

▶ If the mean density of the Earth is 5500 kg m⁻³ and its mean radius is 6400 km, show that taking g to be 9.8 m s⁻² at the Earth's surface leads to a value for G and find that value.

Applying Newton's second law to a lump of mass m at the Earth's surface (figure 12.6) we have

$$mg = F$$

But
$$F = G \frac{m m_E}{r_E{}^2}$$

$\therefore$
$$mg = G \frac{m m_E}{r_E{}^2} \quad \text{or} \quad G = \frac{g r_E{}^2}{m_E}$$

As $m_E = \frac{4}{3}\pi r_E{}^3 \rho$ for an Earth of uniform density ρ

$$G = \frac{3 g r_E{}^2}{4\pi r_E{}^3 \rho} = \frac{3g}{4\pi r_E \rho}$$

$$= \frac{3(9.8 \text{ m s}^{-2})}{4\pi (6.4 \times 10^6 \text{ m})(5500 \text{ kg m}^{-3})}$$

$$= 6.6 \times 10^{-11} \text{ m}^2 \text{ kg}^{-1} \text{ s}^{-2} \quad \text{or}$$

$$6.6 \times 10^{-11} \text{ N m}^2 \text{ kg}^{-2}$$

The method adopted in this example, of substituting numbers only when the final expression for G is established, is generally sensible, but particularly useful in problems on gravitation, where the numbers are often very large.

In this example we have assumed that the Earth attracts a particle which is above its surface as if all the mass of the Earth were concentrated at the Earth's centre. It can be shown that this assumption is justified and we shall use it for other spherically symmetric bodies, for example the lead and mercury spheres in the experiment below. ◀

Measuring *G*

Figure 12.7 shows an apparatus designed to enable G to

be found, albeit very roughly, in the school laboratory. Two small lead spheres are mounted about 100 mm apart on a springy beam to form an object like a dumb-bell; the beam is suspended by a very fine tungsten wire. The dumb-bell is shielded from convection currents by a plastic case and from mechanical vibration by having the support at the upper end of the fibre resting on some plastic foam. The beam carries a small mirror which is arranged to reflect light from a lamp to a scale. As electrical forces may be larger than the gravitational forces in this experiment we must ensure that none of the apparatus has any electric charge.

Two round bottomed flasks containing 'spheres' of mercury or fine lead shot (shown empty in the photograph) are now placed so as to twist the wire by exerting a pair of gravitational forces on the suspended dumb-bell. If the masses of the lead spheres and flasks are respectively m and M, then the two twisting forces are each of size GmM/r^2, and the twisting couple (or the sum of the moments of the forces – figure 12.7b) is $GmMd/r^2$. When the dumb-bell reaches an equilibrium position

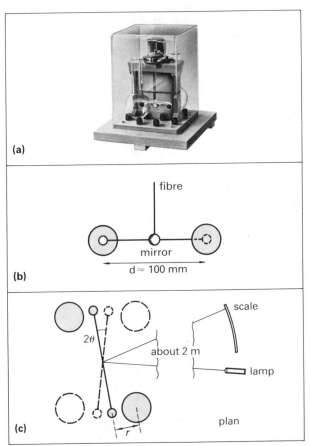

Figure 12.7 (a) A small torsion balance for measuring G. The flasks are filled with mercury or lead shot to produce spherical attracting masses. (b) and (c) show the principle of the method.

$$\frac{GmM}{r^2} d = k\theta$$

where θ is the angle through which the beam has rotated from the untwisted position and k is the torsional constant of the suspending wire. The angle θ is best found by positioning the flasks so as to turn the beam the other way and measuring 2θ (figure 12.7c). When a mirror rotates, the reflected beam rotates twice as much so that the movement of the light on the scale corresponds to an angle of rotation of 4θ. We can find k by removing the flasks and measuring the period of free oscillations of the beam. If (page 173) its moment of inertia is $I \; (\approx 2m(d/2)^2)$ then

$$T = 2\pi \sqrt{\frac{k}{I}}$$

and hence G can be found.

The period T is so long (of the order of 5 minutes) that much patience is needed with this apparatus, and a time and place when vibration from road traffic, etc. is as small as possible. To measure G more precisely no new principle is needed, improved control of temperature and mechanical vibration being most important.

▶ Estimate the size of the gravitational forces involved in a school experiment to measure G.

Refer to figure 12.7: the flasks look to have a capacity of about 50 cm³ capacity so that using mercury of density $\rho_{Hg} \approx 14\,000$ kg m⁻³ we get

$$m_{Hg} = (50 \times 10^{-6} \text{ m}^3)(1.4 \times 10^4 \text{ kg m}^{-3})$$
$$= 0.7 \text{ kg}$$

The beam is about 10 cm long; and the lead spheres have a diameter of about a centimetre, and so their volume is less than 1 cm³, say 0.5 cm³. (There is no point in using volume $= 4\pi r^3/3$ in an estimate of this type.) $\rho_{Pb} \approx 11\,000$ kg m⁻³ so we get

$$m_{Pb} = (0.5 \times 10^{-6} \text{ m}^3)(1.1 \times 10^4 \text{ kg m}^{-3})$$
$$= 6 \times 10^{-3} \text{ kg}$$

The separation of the attracting lead and flask is about 5 cm, so that

$$F = G \frac{m_1 m_2}{r^2}$$

$$= \frac{(6.7 \times 10^{-11} \text{ N m}^2 \text{ kg}^{-2})(6 \times 10^{-3} \text{ kg})(0.7 \text{ kg})}{(5 \times 10^{-2} \text{ m})^2}$$

$$\approx 10^{-10} \text{ N}$$

which is roughly equal to the weight of a particle of dust.   ◀

12.4 Gravitational fields

Newton's law of gravitation tells us that in the region near any lump of matter a mass will experience a gravitational force; the region is described as a *gravitational field* or a *g-field*. (To produce observably large forces, the lumps must be very large.) We measure the strength g of a gravitational field by the equation

$$F_g = mg$$

where F_g is the gravitational force experienced by a body of mass m; g is measured in N kg^{-1}. It is a vector quantity which has the same direction as the force. We have seen this equation before as an expression for the pull of the Earth on a body of mass m in a state of free fall; g was then measured in m s^{-2}. You should check that the units are identical.

Uniform fields

On or very near the Earth's surface the pull of the Earth on a body is constant in size and, over a limited area such as a town, is constant in direction. The gravitational field g is thus uniform within these limits. If we move away from the Earth's surface into 'space', g changes in size and, of course, it is different in direction above, for instance, London and Tokyo. We can represent a uniform g-field by a set of parallel lines of force just as we do for uniform E-fields and B-fields (Chapters 15 and 16). Figure 12.8 shows such a gravitational field. To demonstrate that it is uniform we need only take a test mass on a spring balance and show that the reading on the balance is independent of where we place it in the field; a trivial experiment but an important idea.

When a body is projected in a uniform g-field (it is then usually called a *projectile*) it follows a parabolic path. Figure 1.10 shows the path; projectiles are discussed on page 8.

Gravitational potential in a constant field

On page 27 the change in gravitational potential energy ΔE_g of a body moving in the Earth's g-field was defined as

$$\Delta E_g = \Delta(mgh)$$
$$= mg\,\Delta h$$

in a uniform field. We could call ΔE_g the gravitational potential energy difference of this body; $\Delta E_g/m$ is then called the *gravitational potential difference* or, if the context is clearly one involving gravity, simply the potential difference ΔV_g

$$\Delta V_g = \frac{\Delta E_g}{m} = g\,\Delta h$$

Thus $\quad g = \dfrac{\Delta V_g}{\Delta h}$

This is called the *potential gradient* of the g-field.

It is often useful to be able to consider surfaces on which all points have the same potential; some are shown in figure 12.8 for a uniform field. V_g has units of J kg^{-1} or

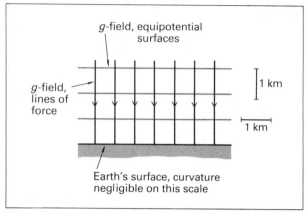

Figure 12.8 The uniform gravitational field, $g = 9.8$ N kg^{-1} (at the Earth's surface).

m^2 s^{-2}; you should satisfy yourself that these units are equivalent.

As in the electrical case only *differences* of potential usually concern us; but it is convenient to have a conventional zero for gravitational potential energy (g.p.e.). The Earth's surface is normally taken as this zero; thus a book of mass 2.0 kg which is on a shelf 1.5 m above the floor is described as having a g.p.e. of ($g = 10$ N kg^{-1}) 30 J, meaning that ΔE_g shelf-to-floor for the book is 30 J.

In this section three different ways of defining or using g have appeared. (The values given in the following are at the Earth's surface):
(i) the gravitational acceleration of a freely falling body; $g = 9.8$ m s^{-2}. This is the same for all masses at the chosen point.
(ii) the strength of the gravitational field; $g = 9.8$ N kg^{-1}. A body of mass 1.0 kg will experience a gravitational force of 9.8 N at the chosen point.
(iii) the gravitational potential gradient; $g = 9.8$ J kg^{-1} m^{-1}. This is the gradient of a graph of the gravitational potential against distance at the point (page 27).

The units m s^{-2}, N kg^{-1} and J kg^{-1} m^{-1} are all identical and the first definition of g is equivalent to the other two if the rotation of the earth is ignored.

Radial fields

The inverse square law of gravitation applies to *particles*. But provided the bodies we consider possess a spherical symmetry we can treat them, so far as they attract and are attracted by bodies which lie beyond their surfaces, as particles with all their mass concentrated at their centres. Thus above the Earth's surface (i.e. at more than $r_E = 6400$ km from its centre) the Earth attracts gravitationally just as would a particle of mass $m_E = 6.0 \times 10^{24}$ kg placed at the Earth's centre. Below the surface of a spherically symmetrical body the rule is that we can ignore that part of the mass of the sphere which lies at a greater distance

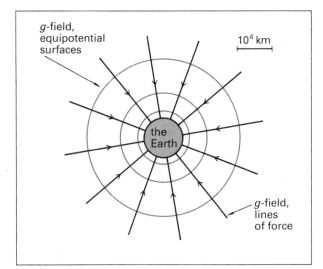

Figure 12.9 The radial gravitational field lines and spherical equipotential surfaces for the Earth, a typical planet.

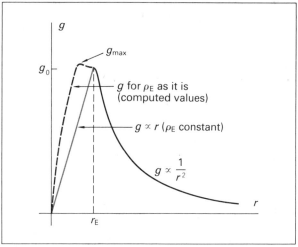

Figure 12.10 The size of the Earth's g-field. Note that as ρ_E is not constant, $g_{max} > g_0$ and occurs near $r = r_E/2$.

from the centre than the point considered; for the rest of the sphere the above applies. Both these results can be shown to follow from an inverse square law of attraction, such as Newton's law of gravitation.

By definition, a gravitational field g is measured by $g = F/m$. The pull of the Earth on the mass m a distance r from the Earth's centre ($r > r_E$) is given by $F = Gmm_E/r^2$. Thus

$$g = \frac{Gm_E}{r^2}$$

where $Gm_E = (6.7 \times 10^{11} \text{ N m}^2 \text{ kg}^{-2})(6.0 \times 10^{24} \text{ kg})$
$= 4.0 \times 10^{14} \text{ N m}^2 \text{ kg}^{-1}$

This gravitational field is not uniform; a diagram of lines of force for the Earth ($r > r_E$) shows a pattern of radial lines with arrows towards the Earth's surface (figure 12.9). For a given value of r, g is the same all round the Earth so this field has a spherical symmetry.

A graph of g against r, for $r > r_E$, looks like the curved part of figure 12.10.

The straight grey part of the figure tells us how g would vary for $0 \leqslant r \leqslant r_E$ for an Earth of uniform density ρ_E. Again $g = Gm/r^2$ but now $m \neq m_E$; instead

$$m = \tfrac{4}{3}\pi r^3 \rho_E$$

the mass of that part of the Earth of radius less than r.

Thus $g = \left(\dfrac{G}{r^2}\right)\left(\dfrac{4}{3}\pi r^3 \rho_E\right) = \dfrac{4}{3}\pi G\rho_E r$

and $g \propto r$ as $4\pi G\rho_E/3$ is a constant. The Earth's density is not, however, constant: at the surface $\rho \approx 3500 \text{ kg m}^{-3}$ and at the centre $\rho \approx 11\,000 \text{ kg m}^{-3}$. (This information comes from studies of earthquake waves.) The dotted line of figure 12.10 indicates the sort of values we expect for g at different depths for the Earth.

The maximum value g_0 of the gravitational field strength for a uniform planet occurs at its surface. For Mars $g_0 = 3.8 \text{ N kg}^{-1}$, for the Moon $g_0 = 1.7 \text{ N kg}^{-1}$ (and one can hit golf balls a long way!) and for Jupiter $g_0 = 25 \text{ N kg}^{-1}$.

Variation of g

If we go from sea level to the top of the Earth's highest mountain we find that g falls by about 0.3%; it is 1% below its sea level value at a height of about 32 km. The measured free fall acceleration g does, however, vary *at* sea level from 9.83 m s^{-2} at the poles to 9.78 m s^{-2} at the equator, a variation of 0.5%. There are two causes:
(i) 0.2% is the result of the Earth lacking symmetry; it is not a sphere but is like a sphere flattened slightly at the poles, and
(ii) 0.3% is the result of the Earth's present rotation (the 0.2%, caused by its lack of symmetry, is mainly the result of its rotation in the distant past).

On a local scale tiny variations of g, of the order of one part in 10^7 or less, can be detected with gravimeters, very sensitive spring balances used by geologists to help them to predict what lies below the Earth's surface at that point.
▶ Explain why the measured free fall acceleration of a body at the Earth's equator, (9.78 m s^{-2}), is less than that predicted by the local value of $g_0 = Gm_E/r_E^2 = 9.81$ m s^{-2}.

A body of mass m at rest on the Earth's surface at P, figure 12.11, is acted on by two forces
(i) the gravitational pull of the Earth, $F = mg_0$, and
(ii) the perpendicular contact push S of the ground.
The body is *not* in equilibrium under the action of these two forces; it is accelerating centripetally towards the centre O of the Earth, around which it rotates at one revolution per day (7.3×10^{-5} rad s^{-1}). Therefore applying Newton's second law

$$mr_E\omega^2 = mg_0 - S$$

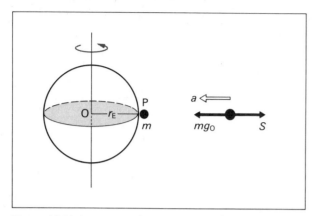

Figure 12.11 A mass m at the equator and a free-body diagram for it.

If we write $S = mg$, where g is the *measured* free fall acceleration at P, then we get

$$r_E\omega^2 = g_0 - g$$

or $g_0 - g = (6.4 \times 10^6 \text{ m})(7.3 \times 10^{-5} \text{ rad s}^{-2})^2$

$$= 0.03 \text{ m s}^{-2} \quad \text{(to one significant figure)}$$

This is the difference between the 9.81 m s^{-2} and 9.78 m s^{-2} given in the question.

At a latitude λ, the principle of the calculation is the same; we get $g_0 - g = r_E\omega^2 \cos^2 \lambda$. Clearly at $\lambda = 90°$, $g = g_0$. ◄

Gravitational potential in a radial field

The gravitational potential difference ΔV_g between two points a distance Δr apart in a uniform g-field is related to the field strength by the relation

$$\Delta V_g = g\Delta r \qquad \text{(page 162)}$$

This also holds for small values of Δr and ΔV_g in a non-uniform field. Over large distances we can calculate the gravitational potential difference as

$\Delta V_g =$ area between the graph-line of a g-r graph and the r-axis

The example on page 24 finds the area under a graph to show that the change in gravitational potential energy ΔE_g for a satellite of mass 5000 kg is 7.8×10^{10} J when it moves from 4000 km to 1000 km above the Earth's surface. The change in V_g is thus

$$\Delta V_g = \frac{\Delta E_g}{m} = \frac{7.8 \times 10^{10} \text{ J}}{5000 \text{ kg}} = 1.6 \times 10^7 \text{ J kg}^{-1}$$

It can be shown that the gravitational potential difference between any two points in the Earth's inverse square law field can be expressed as

$$\Delta V_g = Gm_E\left(\frac{1}{r_2} - \frac{1}{r_1}\right) \qquad (1)$$

where r_2 and r_1 are the distances from the *centre* of the Earth. Thus given $Gm_E = 4.0 \times 10^{14}$ N m^2 kg^{-1} and $r_E = 6400$ km we get for the above example

$$\Delta V_g = (4.0 \times 10^{14} \text{ N m}^2 \text{ kg}^{-1}) \times$$
$$\left(\frac{1}{7.4 \times 10^6 \text{ m}} - \frac{1}{10.4 \times 10^6 \text{ m}}\right)$$
$$= 1.6 \times 10^7 \text{ J kg}^{-1} \quad \text{as expected}$$

Equation (1) is the difference between two gravitational potentials V_g each of the form

$$V_g = -\frac{Gm_E}{r} \qquad (2)$$

V_g is such that the gradient of the graph of V_g against r (page 27) is equal at every point to g ($= Gm_E/r^2$). Thus the *gravitational potential* at points above the Earth's surface varies inversely with the distance r from the centre of the Earth. Figure 12.9 shows *equipotential surfaces* around the Earth. V_g falls to zero at infinity; and this replaces the convenient zero at $r = r_E$ which we often use when moving close to the Earth's surface. For other planets (or stars) we must replace m_E by m_P, the mass of the planet.

The gravitational potential V_g at a point in a gravitational field depends *only* on position. The gravitational potential energy (g.p.e.) of a body of mass m is given by $E_g = mV_g$ at the point and is independent of the path followed by the body in getting there. Thus to put a satellite into a geosynchronous orbit (page 166) we have to raise it from a point on the Earth's surface ($r_E = 6400$ km) to a height $r = 42000$ km; the energy needed for this will not depend on the path taken by the satellite; it might be 'parked' in a lower orbit for a time during its journey, but the energy needed to get it into its final orbit will be the same.

► What is the gravitational potential at the Earth's surface, and 100 km above the Earth's surface? Take $Gm_E = 4.0 \times 10^{14}$ N m^2 kg^{-1} and $r_E = 6400$ km.

At the Earth's surface

$$V_g = -\frac{4.0 \times 10^{14} \text{ N m}^2 \text{ kg}^{-1}}{6400 \times 10^3 \text{ m}} = -6.25 \times 10^7 \text{ J kg}^{-1}$$

and at $r = 6500$ km, i.e. 100 km up

$$V_g' = -\frac{4.0 \times 10^{14} \text{ N m}^2 \text{ kg}^{-1}}{6500 \times 10^3 \text{ m}} = -6.15 \times 10^7 \text{ J kg}^{-1}$$

You must add 0.10×10^7 J kg^{-1} to V_g to get V_g', i.e. V_g' has a larger value than V_g. The gravitational potential is thus a minimum at the Earth's surface and gets bigger as r gets bigger.

Figure 12.12 shows a way of helping you to visualise changes in V_g. The surface is called a potential 'well' and is so shaped that its height above the flat disc (at centre) for different values of r represents V_g at that place. The zero of potential is where the surface becomes flat, a very long way from the centre. ◄

► With what velocity must a body be projected from the Earth's surface if it is to 'escape' from the Earth's gravitational field? Comment on your calculation. Take $r_E = 6400$ km and $g_0 = 9.8$ m s^{-2}.

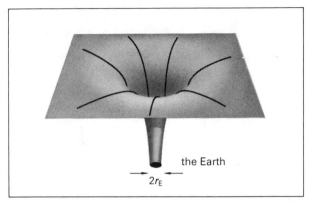

Figure 12.12 A potential well, an analogue for $V_g \propto 1/r$. A ball rolling on the curved surface of the well behaves like a free falling body in the Earth's gravitational field.

A body of mass m at the Earth's surface has a gravitational potential energy $-Gmm_E/r_E$ and at infinity it would have zero g.p.e. The minimum kinetic energy it must be given is therefore equal to Gmm_E/r_E. The potential well of figure 12.12 can be used to illustrate this energy conservation. Imagine a small ball bearing placed at the edge of the surface shown: it would roll, slowly at first, towards the centre losing g.p.e. and gaining k.e. Similarly if the ball were projected from the centre it would reach the edge if (returning to our example)

$$\tfrac{1}{2}mv_e{}^2 \geqslant \frac{Gmm_E}{r_E}$$

i.e. $\qquad v_e \geqslant \sqrt{2Gm_E/r_E}$

where v_e is called the *escape speed* of the body. As $Gm_E/r_E{}^2 = g_0$, we get

$$v_e \geqslant (2g_0 r_E)^{\frac{1}{2}}$$
$$\geqslant \sqrt{(2 \times 9.8 \text{ m s}^{-2})(6.4 \times 10^6 \text{ m})}$$
$$\geqslant 11 \text{ km s}^{-1} \quad (\text{more than } 40\,000 \text{ km h}^{-1})$$

Notice that v_e is indeed an escape *speed* as it does not matter in what direction the body is projected.
Note that
(i) The Earth's atmosphere has been ignored; such a simple expression is only valid once one is beyond the frictional drag of air.
(ii) It would be worth projecting the body in the sense in which the Earth rotates. Any body at the equator has a tangential speed of $r_E \omega \approx 0.5$ km s^{-1}.
(iii) The escape speed is independent of the mass of the body. However, the escape energy $\tfrac{1}{2}mv_e{}^2$ is proportional to m.

12.5 Satellites

A body which moves under the action of only one force,

the pull on it of the Earth or some other planet, star etc. is in a state of *free fall*. Objects which remain in free fall for any length of time are called *satellites*. Our observations used to be confined to natural satellites, often called moons, and to the planets as satellites of the Sun. Today we can consider artificial satellites which have for limited periods orbited the Sun, Venus, Mars, Earth and our Moon. Consider the four readily visible (Galilean) satellites of Jupiter shown in figure 12.13 as they might be seen through a small telescope on two different occasions. Table 12.1 lists them and gives their average orbital radius r, and period of revolution T. A graph of r^3 against T^2 is a straight line passing through the origin; see Kepler's third law (page 160).

Newton's mechanics can explain this result: consider one satellite of mass m circling Jupiter, mass m_J, at a distance r from its centre. Using Newton's second law and the law of gravitation we get

$$mr\omega^2 = G\frac{mm_J}{r^2}$$

or $\quad r^3\omega^2 = Gm_J$

As $\omega = 2\pi/T$, where T is the period of revolution

$$r^3\left(\frac{2\pi}{T}\right)^2 = Gm_J$$

$$\Rightarrow \qquad r^3 = \frac{Gm_J}{4\pi^2}T^2$$

i.e. $\qquad r^3 = kT^2$

	$r/10^8$ m	$T/10^5$ s
Io	4.2	1.5
Europa	6.7	3.1
Ganymede	10.7	6.2
Callisto	18.8	14.7

Table 12.1 The four easily observed satellites of Jupiter.

Figure 12.13 Two photographs of the readily visible moons of Jupiter. In the lower picture one of the four is hidden by the planet itself.

which is Kepler's law. We can now calculate the mass of Jupiter if we know G and similarly the mass of any planet with observable satellites. Using Io we find that

$$m_J = \frac{4\pi^2 r^3}{GT^2}$$

$$= \frac{4\pi^2 (4.2 \times 10^8 \text{ m})^3}{(6.7 \times 10^{-11} \text{ N m}^2 \text{ kg}^{-2})(1.5 \times 10^5 \text{ s})^2}$$

$$= 1.9 \times 10^{27} \text{ kg}$$

We could now deduce the free-fall acceleration or gravitational field strength at the surface of the planet ($g_0 = Gm/r_J^2$) given that $r_J = 7.2 \times 10^7$ m, and check the value quoted on page 163.

▶ Early Bird is a geosynchronous satellite. Explain as fully as you can what this means.

Geosynchronous means that the satellite has a period of revolution equal to that of the daily rotation of the Earth and is so positioned as to appear to remain above the same place on the Earth's surface. Clearly this place must lie on the equator and the satellite must rotate in the same sense and about the same axis as the Earth itself.

Using Newton's laws we get

$$mr\omega^2 = G\frac{mm_E}{r^2}$$

where r is the radius of the satellite's orbit and ω is equal to the angular velocity of the Earth.

$$\omega = 2\pi/(24 \times 3600) \text{ rad s}^{-1} = 7.3 \times 10^{-5} \text{ rad s}^{-1}$$

Therefore $r^3\omega^2 = Gm_E$,
and as $Gm_E = 4.0 \times 10^{14}$ N m^2 kg^{-1}, we get

$$r^3 = \frac{4.0 \times 10^{14} \text{ N m}^2 \text{ kg}^{-1}}{(7.3 \times 10^{-5} \text{ rad s}^{-1})^2}$$

$$= 7.5 \times 10^{22} \text{ m}^3$$

and $r = 4.2 \times 10^7$ m or 42 000 km

i.e. between 36 000 km and 35 000 km above the Earth's surface.

As both r and T are determined the satellite must have a speed given by

$$v = 2\pi r/T = 3.1 \times 10^3 \text{ m s}^{-1}$$

so that, in placing it in orbit, it must be lifted to a predetermined height and given this tangential speed for it to become geosynchronous.   ◀

Orbits

Figure 12.14 summarises the possible paths for satellites projected tangentially at the Earth's surface (or the surface of any planet) with speed v. If v is zero the path is a straight line vertically downwards (not shown); on the other hand an infinitely large v (if that were possible) would give rise to a straight-line tangential path. All other speeds give rise to paths which can be described as shown. Two interesting cases are

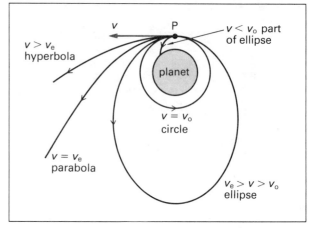

Figure 12.14 Satellite orbits: all are projected tangentially from P but with a variety of speeds v.

(i) $v = v_e = \sqrt{2g_0 r_E}$. This result was proved on page 165 and we saw there that with $g_0 = 9.8$ N kg^{-1}, $v_e \approx 11$ km s^{-1}. The path is a parabola.

(ii) $v = v_0$, the speed sufficient for the satellite to have a circular orbit near the Earth's surface. Since the centripetal acceleration must equal the free-fall acceleration

$$\frac{v_0^2}{r_E} = g_0$$

$$v_0 = \sqrt{g_0 r_E}$$

$$= \sqrt{(9.8 \text{ N kg}^{-1})(6.4 \times 10^6 \text{ m})}$$

$$\approx 8 \text{ km s}^{-1}$$

In this case the kinetic energy which needs to be given to the satellite is half that which would be needed to enable it to escape completely from the Earth's gravitational field. In practice a satellite cannot orbit close to the Earth's surface because of the air resistance force caused by the atmosphere: the first artificial satellite (Sputnik I, 1957) was in orbit for only three months because its minimum height in its (elliptical) orbit was only 230 km above the Earth's surface.

The tides

When we think of the Moon rotating around the Earth, we usually think of a 'stationary' Earth, with the centre of the Moon's orbit at the centre of the Earth. But the Earth is only 80 times more massive than the Moon, and they *both* rotate – about the centre of mass of the Earth-Moon system, which is at a point $\frac{3}{4}r_E$ from the centre of the Earth, towards the Moon. (This motion is, of course, in addition to the annual motion of the Earth around the Sun, and the daily spin of the Earth on its axis: this monthly rotation of the Earth and Moon about their common centre of mass is superimposed on the other two motions.) The water lying on the surface of the Earth nearest the Moon at any time will (because it is nearer the Moon) have a larger gravitational force on it than normal, and because the water is not rigid, will actually move towards the Moon: the water level will rise. The water on the surface of the Earth furthest from the Moon will have a smaller than normal gravitational force, which will not be large enough to pull the water in to its normal level: the water level will rise here too. So there will be two tidal bulges on opposite sides of the Earth. We might expect the bulges to lie on a line joining the Earth to the Moon, but the daily rotation of the Earth drags the water round so that the bulges lie ahead of that line, but fixed in position relative to the Earth and the Moon, so that as the Earth rotates different places on its surface experience, twice a day, the higher level of the water, which we call the *tides*.

▶ If the masses of the Earth and Moon, m_E and m_M, are such that $m_E/m_M = 81$ and the Earth-Moon separation is $d = 3.8 \times 10^8$ m, at what distance from the Earth is there a region where a spacecraft would experience zero resultant gravitational pull? (Ignore the Sun).

Suppose the neutral point is a distance r from the Earth, then for a spacecraft of mass m at the null point

$$G\frac{mm_E}{r^2} = G\frac{mm_M}{(d-r)^2}$$

$$\therefore \quad (d-r)^2 = \frac{m_M}{m_E}r^2$$

$$\text{or} \quad d-r = \left(\frac{m_M}{m_E}\right)^{\frac{1}{2}}r \quad \text{(ignoring the negative square root)}$$

This gives

$$r = \frac{d}{1 + \sqrt{(m_M/m_E)}}$$

$$= \frac{3.8 \times 10^8 \text{ m}}{1 + \sqrt{(1/81)}}$$

$$= 3.4 \times 10^8 \text{ m}$$

so that the null or *neutral point* occurs when the spacecraft has covered more than 90% of the distance if it is going from the Earth to the Moon.   ◀

13 Equilibrium and rotation

13.1 Moments and equilibrium

The effect of a force is often to cause rotation. The pull of a hand on a doorknob or the push of a foot on the starting pedal of a motorcycle both produce rotation about a fixed axis, the vertical door hinges and the horizontal engine axle respectively. Sometimes a force produces both rotation and translation as when a cue strikes a snooker ball or a hand throws a boomerang. To study the turning effect of a force in the laboratory it is only necessary to support a rod or ruler on a pivot and to exert known forces on it – a seesaw type of experiment. The smaller the push of a girl on the left of a real seesaw, e.g. 300 N, the further she must be from the centre in order to balance the greater push of a man on the right of the seesaw, perhaps 700 N. We say that the *moment* of the 300 N force *about the axis of rotation* increases as its distance from the axis increases. In pulling open a door the knob is at the edge of the door and not in the middle, thus increasing the moment of the pull. The greatest spin on a snooker ball is achieved when it is cued as much off-centre as possible.

We define the *moment M of a force* about a chosen axis by the equation

$$M = rF$$

where F is the force and r is the perpendicular distance from the chosen axis to the line of action of the force (figure 13.1). The unit of M is the N m. This looks the same as the unit of work and energy (the joule). But the *moment of a force* and *energy* are quite different concepts so we shall keep N m as the unit for moment and not abbreviate it to J.

We shall need to consider only forces which effectively act in one plane (known as coplanar forces) so that we can draw cross-sectional diagrams with the axis of rotation going into the page and appearing as a point: O in figure 13.1. We shall refer to the moment of a force about an axis

through O and perpendicular to the page simply as the moment of the force about O.

The principle of moments

For a body which is in equilibrium the sum of the moments, about any axis, of the external forces acting on the body is zero.

$$M = 0 \quad \text{for equilibrium}$$

To illustrate how we use the principle of moments in a practical calculation consider the gate shown in figure 13.2a. The top hinge H is in working order but the bottom one is missing. so that the gate post may be assumed to push horizontally against the bottom of the gate. The gate is of symmetrical design and has a weight of 2000 N. We can find the forces exerted by the gatepost on the gate as follows. First a free-body diagram of the gate is drawn (figure 13.2b). We do not know the direction of the pull of the upper hinge of the gate so two forces are best drawn, P and Q, which are the vertical and horizontal resolved parts of this pull.

The moment of P about H = zero
The moment of Q about H = zero
The moment of S about H = (1.1 m) S, anticlockwise
The moment of W about H = (0.5 m) (2000 N), clockwise

As the sum of the moments about H must be zero, we have, taking clockwise moments to be positive

$$1000 \text{ N m} - (1.1 \text{ m}) S = 0$$
$$\Rightarrow \quad S = 910 \text{ N}$$

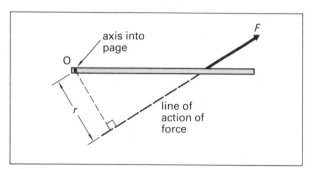

Figure 13.1 The moment of a force about O, $M = rF$.

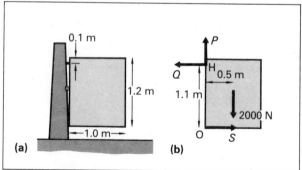

Figure 13.2 A gate with a broken lower hinge. (b) is a free-body diagram of the gate.

To find P we similarly calculate the moment of each force about O and then again use the principle of moments to equate their sum to zero. Or, as S is now known we can use the fact that the sum of the forces acting on the gate, resolved in *any* direction, must be zero for equilibrium. Thus

vertically $P - 2000 \text{ N} = 0$
horizontally $Q - 910 \text{ N} = 0$
so that $P = 2000 \text{ N}$
and $Q = 910 \text{ N}$

The size of the resultant pull R of the hinge on the gate is given by

$$R = \sqrt{(2000^2 + 910^2)} \text{ N}$$
$$= 2200 \text{ N}$$

and the angle R makes with the horizontal is θ where

$$\tan \theta = \frac{2000 \text{ N}}{910 \text{ N}} = 2.2$$

which gives $\theta = 66°$.

Centre of gravity

The above example assumes (as we have done on several occasions earlier in the text) that we know where to put the 2000 N force, the pull of the Earth on the gate. The Earth exerts a gravitational pull on every molecule of the gate, but we can replace all these by a single resultant force which acts through a point G called the *centre of gravity* of the gate. Its position is defined as being that point about which the sum of the moments of the pull of the Earth on the molecules of the body is zero however the body is placed in the Earth's gravitational field.

This definition enables us to calculate the position of G for bodies of simple shape, though a look at the symmetry of the situation is usually enough to tell us where G lies: thus for a uniform rod it is half-way along, for a uniform disc it is at the centre, etc. For non-uniform and/or complex bodies, e.g. a human being, it can be found only by experiment and may anyway change its position as when, for example, the person raises his arms or moves to a sitting position. For many bodies G can be located experimentally by balancing them on a narrow support or hanging them from a single string. G must lie respectively above or below the point of support when the body balances. Where necessary the process can be repeated to locate G in two or three dimensions.

Three-force equilibrium

When a body is in equilibrium under the action of three non-parallel forces then the *lines of action* of the three forces must pass through a single point. That this is so can be seen by taking moments about a point where two of the forces cross: their moments will each be zero. If the third force F has a moment rF about this point then the body can only be in equilibrium if $r = 0$; i.e. the third force must pass through this point. In figure 13.2b the resultant of P and Q,

called R in the text, must pass through the point of intersection of S and the 2000 N force. If all we are trying to find is the direction of R then it must act at θ to the horizontal, where $\tan \theta = 1.1$ m/0.5 m $= 2.2$ which gives $\theta = 66°$, the same value as was found above after calculating P and Q.

Structures

In analysing the equilibrium of rigid structures such as bridges it is the internal forces, the push or pull of one part of the structure on another, that concern us. This being the case we must take particular care in choosing the object or bodies for which we draw free-body diagrams. To illustrate this let us consider the forces acting on and in a simple arch such as that used as the basis of many stone bridges and in Norman doors and windows. Figure 13.3a shows a simple arch made of blocks (voussoirs). Once the arch has been built these need not be held together by mortar but during the building process they will have to be supported from below. To find the forces acting on a *half-arch* we need to know (i) where the push of the ground on the base stone effectively acts, (ii) the total gravitational pull of the Earth on the half-arch and (iii) the size of the push of the removed half of the arch on the top block of the half-arch, which will be horizontal for a symmetrical arch. We can proceed for a model half-arch as follows (figure 13.3b):
(i) slide a narrow plastic probe under the base stone. When the base stone does not rock over to touch the bench on either side of the probe we know that the whole push of the ground on the arch must be acting through the probe.
(ii) weigh the half-arch (for a real arch the weight would be found by calculation from the volume of stone to be used).
(iii) arrange a cradle as shown in (b). Measure the push of the cradle on the half-arch R using the pulley arrangement shown or with a spring balance.

Figure 13.3 shows a free-body diagram for the half-arch. It is the dissection of the model arch that enables us to measure the size of R. When analysing real structures we must mentally dissect them by considering free-body diagrams of half or part of the whole. Taking moments about A (the half-arch is in equilibrium)

$$yR - x\frac{W}{2} = 0$$

so that $R = \dfrac{xW}{2y}$

If we know x the point where R and $W/2$ intersect is determined. The line of action of the third force Q must pass through this point (see above) and so the angle Q makes with the horizontal is found from $\tan \varphi = y/x$. Resolving vertically we have

$$Q \sin \varphi - \frac{W}{2} = 0$$

and so Q can be found. The experimental and analytical techniques outlined for a half-arch can be extended to

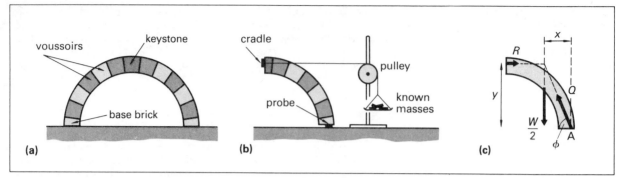

Figure 13.3 (a) A model arch with keystone plus (b) an experimental arrangement for analysing the forces on a half-arch, and (c) a free-body diagram for the half-arch.

discover the thrust line, the locus of the effective points of contact of adjacent voussoirs, for an unloaded or for a loaded arch. A technique, photoelastic stress analysis, for determining the stresses within the parts of solid structures is outlined on page 375.

▶ A man weighing 800 N stands at the top of a symmetrical stepladder. The sides of the ladder are light and their weight can be ignored. They rest on an effectively frictionless floor and are connected by a cord. The dimensions of the ladder are shown in figure 13.4a. Find the tension in the cord.

Since the cord is an internal part of the stepladder we cannot, by considering the whole ladder, discover anything about the pull of the cord on the sides of the ladder any more than we can discover the push of the bottom half of one side on the top half of the same side. Figure 13.4b shows how by drawing a free-body diagram of the left side FH *only* we can introduce the pull of the cord T on FH. P is the push of the floor on the side FH and must be 400 N, equal to half the weight of the man, as the ladder is symmetrical. X and Y are respectively the push of the right hand half of the ladder and the push of the man on the left hand half of the ladder.

We can now find T by taking moments about H for the body FH (which is in equilibrium):

$$(1.8 \text{ m})T = (0.8 \text{ m})(400 \text{ N})$$
$$\Rightarrow \quad T = 180 \text{ N}$$

If the base of the ladder is made wider this will increase T.   ◀

13.2 Couples and torques

Two coplanar forces of equal size and opposite directions which act on a body so as to twist it are said to form a *couple*. Suppose (figure 13.5a) that a book is held by one corner and supported in equilibrium in a vertical plane. If the weight of the book is P then the person holding it must exert an upward force of the same size P. The book is not, however, in equilibrium under the action of these two forces even though they add to produce no net vertical force. They do not act through the same point and so the sum of their moments is *not* zero. About G there is a total anticlockwise moment of $0 \times P + d \times P = dP$. About

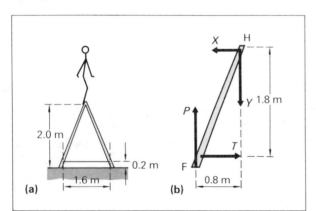

Figure 13.4 A stepladder problem, (b) is a free-body diagram of one half of the stepladder.

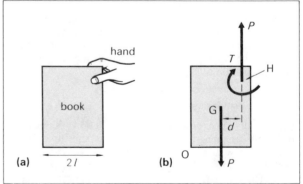

Figure 13.5 A man holds a book. For equilibrium he must exert a (clockwise) torque $T = dP$ as well as the upward force P.

O there is a total anticlockwise moment of $(d + l)P - lP = dP$, and about H (where the book is held) the total moment is $d \times P + 0 \times P = dP$, anticlockwise again. This moment depends only on the size of the force P and the distance d, i.e. the turning effect or couple (of dP anticlockwise) is the same about *any* axis perpendicular to the plane in which the two forces act.

In order to keep the book in equilibrium (figure 13.5b) the person holding it must grip it and produce a turning action or *torque T*, such that

$$T \text{ (clockwise)} + dP \text{ (anticlockwise)} = 0.$$

It is not possible to represent this torque (or similarly the effect of a disc brake) by two equal but opposite forces. A torque, which like a couple has units of N m, produces only a turning effect and in no way alters the translational motion of the centre of gravity of the body on which it acts. Other commonly met examples include the torque produced when gripping a doorhandle or screwdriver and the frictional torque on the bearings of any rotating shaft.

Torques and power

Electric motors and internal combustion engines – in fact almost any device man uses for producing mechanical energy – are designed to drive a shaft about a fixed axis. In dynamic terms their output is not a force which does work in moving along a line but a torque which does work in rotating through an angle. If the mechanical torque produced by a machine on a load is T, then the work done on the load in turning it through an angle θ is given by

$$W = T\theta \qquad (1)$$

where θ (in radians) $= 2\pi N$

where N is the number of revolutions through which the shaft is driven; For instance, a torque of 12 N m produced by an electric motor might turn through 300 rev in one minute. The work done

$$= (2\pi \times 300)(12 \text{ N m})$$
$$= 22\,600 \text{ J}$$

and the power is $\dfrac{22\,600 \text{ J}}{60 \text{ s}} = 380 \text{ W}$

Notice that the N m from the torque becomes the J for the energy. This example illustrates that the power at which a torque operates, the rate at which it does work, is given by

$$P = \frac{dW}{dt} = T\frac{d\theta}{dt} = T\omega \qquad (2)$$

Equations (1) and (2) are similar in form to those for work and power in translational motion:

$$W = Fs \quad \text{and} \quad P = Fv \qquad \text{(page 24)}$$

To measure the power at which a machine does work while turning at a steady rate of rotation n it is necessary to measure the torque. Figure 13.6 shows a simple *torque-meter* made by measuring the tensions on either side of

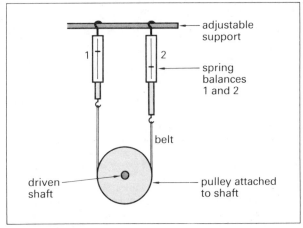

Figure 13.6 The principle of a band brake for torque measurement.

a belt which rubs against a rotating pulley. If the spring balances read F_1 and F_2 then, for a pulley of radius a, the applied torque is $a(F_1 - F_2)$, as the pulley, though moving, is in equilibrium. The power P is given by $P = \omega a(F_1 - F_2) = 2\pi na(F_1 - F_2)$. This is called the *brake power*, which is the useful power output. The machine turning the shaft will need to produce more than this to overcome the friction losses at the bearings. To measure n, we need a chalk mark on the pulley and a stop watch, though for high values of n a flashing stroboscope may be needed.

▶ A shaft is being rotated by a person pedalling with his feet on a fixed bicycle arrangement such as that found in a gymnasium. At two revolutions per second the tensions in the belt are $F_1 = 200$ N and $F_2 = 50$ N on a pulley of radius 0.20 m (see figure 13.6). What is the power delivered by the person to the shaft?

$$P = T\omega = 2\pi na(F_1 - F_2)$$
$$= 2\pi(2.0 \text{ s}^{-1})(0.20 \text{ m})(150 \text{ N})$$
$$= 380 \text{ W}$$

A fit man can keep up this power in pedalling for a minute or so; it is more than the power necessary for manpowered flight. ◀

Efficiency

The power output of an electric motor or car engine is always less than the power input. Energy is converted to 'wasted' internal energy.

The *efficiency* η of any energy converting device or transducer is defined by the equation

$$\eta = \frac{\text{energy output}}{\text{energy input}}$$

or $\quad \eta = \dfrac{\text{power output}}{\text{power input}}$

Efficiency is often expressed as a percentage. The list on

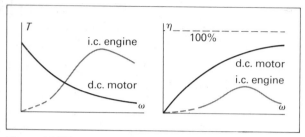

Figure 13.7 Torque T and efficiency η for a direct current electric motor and a petrol internal combustion engine as a function of angular velocity ω. The curves are only representational.

page 29 gives some typical values for η. Of course the efficiency of an electric motor or internal combustion engine varies with the conditions under which it is operated. An electric motor produces its highest torque T at low angular speeds ω, but T falls as ω rises. Nevertheless it is most efficient at high speeds and for a large motor η might rise to 85%. The engine of an ordinary car is only reasonably efficient, 25%, within a limited range of ω and at low speeds the torque T falls to zero – the engine 'stalls'. Thus a car requires a gear box to match a limited range of engine speeds to a much larger range of travelling speeds. Electrically driven vehicles (e.g. a milk float) do not need gear boxes. Figure 13.7 shows in a general way how T and η vary with ω for electric motors and internal combustion engines.

Gear trains and pulley systems do not transfer energy from one form to another as do the motors and engines mentioned above; their input and output energies are both mechanical. Their efficiency η is defined in the same way as above and losses are usually the result of frictional effects and of transferring mechanical energy (kinetic or potential) to their moving parts. For instance, in a pulley system we often lift a set of pulleys as well as the sack or barrel we want to lift. However much we reduce friction by lubrication and by good design (e.g. by using roller bearings) it is not possible to make $\eta > 100\%$. This mythical machine (which would give out more energy than was put into it) is often referred to as a 'perpetual motion' machine.

Many mechanical systems are deliberately made to have efficiencies of less than 50%. As a result they do not go into reverse when the driving force is removed and are said not to *overhaul*. The car jack in the example which follows is such a device – it would indeed be a nuisance if having raised the car the jack unwound as we crawled beneath it. ◄ A screw jack of the type used to lift a car has a handle which moves in a circle of radius 150 mm and a screw with a pitch of 2.5 mm. It is 20% efficient. If a man can conveniently exert a force of 60 N in turning the handle, what is the weight of the heaviest car he can lift?

If we consider one complete turn (2π rad) of the handle of the jack it raises the side of the car a distance equal to the pitch of the screw.

$$\text{energy input} = T\theta = 2\pi rF$$
$$= 2\pi(150 \times 10^{-3} \text{ m})(60 \text{ N})$$
$$\text{and} \quad \text{energy output} = Ps = P(2.5 \times 10^{-3} \text{ m})$$

where P is the push of the jack on the car. As $\eta = 20\%$ then

$$0.2 = \frac{(2.5 \times 10^{-3} \text{ m})P}{2\pi(150 \times 10^{-3} \text{ m})(60 \text{ N})}$$

$$\Rightarrow \quad P = 4500 \text{ N}$$

The man could just lift one side of a car which had a total weight of 9000 N, i.e. a mass of about 900 kg. ◄

13.3 Moment of inertia

When a force accelerates a body in a straight line the greater the mass of the body the more difficult it is to accelerate. We say that the body has inertia. It is much the same when you try to spin something, a bicycle wheel, perhaps, or a small roundabout in a childrens' playground. The more massive the object the more difficult it is to get going, but the way the mass is distributed now also counts. For example, with the roundabout it is harder to increase its angular velocity if there are people on it than when it is empty; but it is easier if the people are standing close to the centre, the (vertical) axis of rotation, than if they are all near the edge.

If we consider the kinetic energy of a rotating body its value again depends on the distribution of mass. It is possible to see why this is so for the roundabout, for if it is rotating at a fixed angular velocity a person near the centre has a lower speed (and hence lower kinetic energy) than a person of the same mass at the edge.

Figure 13.8 shows a laboratory turntable which can help us to study *rotational inertia*. Tickertape is first wound on to one of the steps of the table. The table can be speeded up by pulling on the tape with a measured force F. This force, together with a second force, the push of the axis on the turntable, forms a couple of size Fr. By having a number of steps we can vary r. Various experiments are now possible:

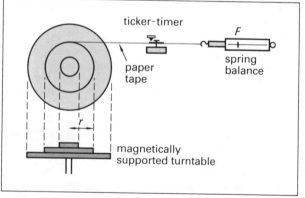

Figure 13.8 A low friction turntable arrangement for experiments on angular motion; alternatively an air suspension system can be used.

let us consider investigating how the angular velocity ω of the table depends on the work done by the couple. If we start with the table at rest then both $\omega(=v/r)$ and $\theta(=s/r)$ can be found by measuring distances s and speeds v from the ticker-tape. The work done by the couple is then

$$W = T\theta = Fr\theta$$

The experiment can be repeated for different constant values of F and for different values of the disc radius r (there are three shown in figure 13.8). The increase of kinetic energy of the turntable is equal to the work done (page 25). In parallel with linear motion, the kinetic energy E_k proves to be proportional to the square of the speed; i.e. a graph of Fr (the work done) against ω^2 is a straight line.

$$E_k = (\text{constant})\omega^2$$

By analogy with $\frac{1}{2}mv^2$ for translational k.e. we write the constant, the gradient of the graph, as $\frac{1}{2}I$ and call I, which is a measure of the rotational inertia of the turntable, its *moment of inertia*. The $\frac{1}{2}$ is shown below to stem directly from the $\frac{1}{2}$ of $\frac{1}{2}mv^2$.

In an experiment with a turntable that tells us that

$$Fr\theta = 0.250 \text{ N m when } \omega = 4.0 \text{ rad s}^{-1}$$

then $\frac{1}{2}I\omega^2 = 0.250$ N m

and $I = \dfrac{2 \times 0.250 \text{ N m}}{(4.0 \text{ s}^{-1})^2} = 0.031 \text{ N m s}^2$

$$= 0.031 \text{ kg m}^2$$

I is a constant for the body for rotation about one axis but a body will have *different* moments of inertia about different axes. The unit of I is the kg m². The experiment could be improved by using photographic techniques for θ and ω and a small mass on a string passing over a pulley to provide the constant accelerating force. The purpose of the experiment is mainly, however, to demonstrate that I is a fixed quantity for the rotation of a body about a given axis.

Calculating moments of inertia

If instead of investigating the kinetic energy of a rotating body experimentally we worked out the kinetic energy of each small bit of it and added these k.e.s. (k.e. is a scalar), then we would get

$$E_k = \tfrac{1}{2}m_1 v_1^2 + \tfrac{1}{2}m_2 v_2^2 + \ldots + \tfrac{1}{2}m_n v_n^2$$

For the rotation of a *rigid* body about a fixed axis

$$v_1 = r_1\omega,$$
$$v_2 = r_2\omega, \ldots v_n = r_n\omega$$
$$\Rightarrow \quad E_k = \tfrac{1}{2}m_1 r_1^2\omega^2 + \tfrac{1}{2}m_2 r_2^2\omega^2 + \ldots + \tfrac{1}{2}m_n r_n^2\omega^2$$
$$= \tfrac{1}{2}(\text{sum of all } mr^2 \text{ terms})\omega^2$$

so that $I = \text{sum of all } mr^2 \text{ terms}$

The moment of inertia I is thus related to the mass of the body and to the way that its mass is distributed about the

body	axis of rotation	I
solid ring or hoop of radius a	through centre, perpendicular to plane of ring	ma^2
flat disc of radius a	through centre, perpendicular to plane of disc	$\frac{1}{2}ma^2$
uniform solid sphere of radius a	through centre of sphere	$\frac{2}{5}ma^2$
uniform thin rod of length l	through centre, perpendicular to length	$\frac{1}{12}ml^2$

Table 13.1 Some useful moments of inertia.

axis of rotation. For bodies possessing a high degree of symmetry I can be calculated. Table 13.1 lists I for some simple shapes.

▶ It has been suggested that a car propulsion unit which could replace the internal combustion engine might consist of a solid cylinder (flywheel) of high-strength material rotating in a vacuum. The cylinder would be spun up to a very high speed at a garage in a matter of minutes and its kinetic energy used, through a system of gears, to drive the car. If a litre of petrol is equivalent to 6×10^7 J of energy, of which only about 25% is transformed to mechanical energy in running a car, estimate the properties of the flywheel which would be equivalent to 20 litres (over 4 gallons) of petrol.

Twenty litres of petrol provide $20 \times 0.25 \times 6 \times 10^7$ J $= 3 \times 10^8$ J of mechanical energy, so we need a flywheel for which

$$\tfrac{1}{2}I\omega^2_{\max} = 3 \times 10^8 \text{ J}$$

The size of the space under the car bonnet limits the radius a and height h of the cylinder to a maximum of about 0.3 m and 0.4 m respectively and hence limits I (for a cylinder $I = \frac{1}{2}ma^2 = \frac{1}{2}\pi a^2 hpa^2$). The strongest materials which can best withstand very high centripetal accelerations are of relatively low density (e.g. resin-bonded carbon fibre which has a density $\rho \approx 1800$ kg m^{-3}).

So $I \not> 0.5\pi (0.3 \text{ m})^4 (0.4 \text{ m})(1800 \text{ kg m}^{-3})$
$\not> 9 \text{ kg m}^2$

Hence $\tfrac{1}{2}(9 \text{ kg m}^2)\omega^2_{\max} = 3 \times 10^8$ J
$\Rightarrow \qquad\qquad \omega_{\max} \approx 8000 \text{ rad s}^{-1}$

which is over 1000 rev sec^{-1}, i.e. about 10 times faster than the maximum rate of rotation of a typical car engine. The centripetal acceleration at the rim of the cylinder is

$$r\omega^2 = (0.3 \text{ m})(8000 \text{ rad s}^{-1})^2 \approx 2 \times 10^7 \text{ m s}^{-2}$$

This is 2 million g, a colossal acceleration producing correspondingly high stresses ($\approx 2 \times 10^{10}$ N m^{-2}) in the cylinder at its rim. Modern matrix materials can be used with stresses as high as this. You might think that the flaw in this discussion is that having bought some energy it would gradually leak away while the car was parked. But it is expected that the leakage would be slight from day to day.

▶ A flywheel of moment of inertia 0.20 kg m² is rotating at 20 rad s⁻¹ about a horizontal axis. Its shaft 'catches' a piece of string to which is attached a body of mass 1.0 kg and the body is lifted 3.0 m before it hits the shaft. What is the final angular velocity of the flywheel?

We need to make several assumptions: let us ignore the kinetic energy of the 1.0 kg mass and friction effects; and let us take $g = 10$ m s⁻².

The gain in gravitational potential energy of the mass is

$$mg\Delta h = (1.0 \text{ kg})(10 \text{ m s}^{-2})(3.0 \text{ m})$$
$$= 30 \text{ J}$$

The initial kinetic energy of the flywheel

$$\tfrac{1}{2}I\omega_0{}^2 = \tfrac{1}{2}(0.20 \text{ kg m}^2)(20 \text{ rad s}^{-1})^2$$
$$= 40 \text{ J}$$

Therefore with the above assumptions the final kinetic energy of the flywheel $= (40 - 30)$ J $= 10$ J, and hence its final kinetic energy $\tfrac{1}{2}I\omega^2$ is given by

$$10 \text{ J} = \tfrac{1}{2}(0.20 \text{ kg m}^2)\omega^2$$
$$\Rightarrow \quad \omega = 10 \text{ rad s}^{-1}$$
◀

Torque and angular acceleration

For translational motion $ma = F_{res}$; for rotational motion

$$Ia = T_{res} \qquad (3)$$

where I is the moment of inertia about the given axis of rotation, a is the angular acceleration and T_{res} is the sum of the external torques about the axis. This is *Newton's second law for rotational motion* in its simple form. The apparatus of figure 13.8 can be used to stress the content of this relationship and if $T(= rF)$ and a are measured then I for the turntable can be found. The angular acceleration is measured from an analysis of the tape;

$$\text{for} \quad a = \frac{\omega - \omega_0}{t}$$

Equation (3) applies to rotation about a fixed axis; it can also be applied to a body whose centre of mass C is accelerating, e.g. a bus rounding a corner, but then a and T_{res} *must* be measured about an axis through C.

▶ A turntable roundabout in a children's playground has a radius of 2.0 m, a moment of inertia of 600 kg m² and is at rest. A boy starts it and accelerates it to an angular velocity of 1.5 rad s⁻¹ by exerting a tangential force of 100 N at its circumference while running round with the turntable. If the bearings of the turntable exert a steady frictional torque of 50 N m find
(i) the angular acceleration of the turntable, and (ii) the distance the boy has to run.

(i) Using $Ia = T_{res}$ we have

$$(600 \text{ kg m}^2)a = (100 \text{ N})(2.0 \text{ m}) - 50 \text{ N m}$$
$$\therefore \quad a = 0.25 \text{ rad s}^{-2}$$

(ii) The time taken to accelerate the turntable t is given by

$$0.25 \text{ rad s}^{-2} = \frac{1.5 \text{ rad s}^{-1} - 0}{t}$$

so that $\qquad\qquad t = 6.0$ s

The average angular velocity $= 0.75$ rad s⁻¹, so that the turntable turns through an angle θ given by

$$\theta(= \omega_{av}t) = (0.75 \text{ rad s}^{-1})(6.0 \text{ s}) = 4.5 \text{ rad}$$

and the boy runs a distance of

$$(4.5 \text{ rad})(2.0 \text{ m}) = 9.0 \text{ m}$$
◀

13.4 Angular momentum

Angular momentum is defined as the product $I\omega$ just as the product mv defines linear momentum. Its units are kg m² s⁻¹ or N m s. A single particle of mass m and velocity v has an angular momentum mvr about an axis at a perpendicular distance r from its line of motion. If a rigid body is both moving along a path and spinning, as does a planet or an electron bound to an atom, it possesses both (orbital) angular momentum mvr about its sun or nucleus and (spin) angular momentum $I\omega$ about its own centre of mass. These concepts have played a large part in the development of theories of the atom and of the origin of the solar system. The rate of change of angular momentum ($I\omega$) is equal to the resultant torque T_{res} acting on a body. Thus

$$\frac{\mathrm{d}(I\omega)}{\mathrm{d}t} = T_{res} \qquad \text{(appendix B3)}$$

or $\qquad \dfrac{\mathrm{d}(mvr)}{\mathrm{d}t} = T_{res}$

When the resultant torque T_{res} about the axis of rotation is zero then either (i) $I\omega$ is constant (e.g. for a spinning skater) or (ii) mvr is constant (e.g. for a planet moving round the sun). To illustrate case (ii) figure 13.9a shows an orbiting body which is pulled directly towards a central post (the force is in this case provided by a piece of elastic but it could be a gravitational force on a comet moving round the Sun). (b) shows a body which is pushed directly away from a central point (the force in this case is provided by the hill up which the body is rolling but it could be the electrical force on an a-particle moving near a nucleus). The *central* pull or push produces zero torque about the centre in both cases as the pull or push on the body has no moment about the centre. Thus mvr is constant or, as m is fixed, $v \propto 1/r$ (r is the perpendicular distance from the centre to the line of v). You can check this by taking measurements from the photographs. A more complete analysis shows that this implies that the area swept out by

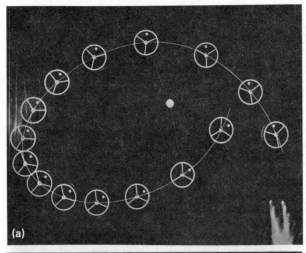

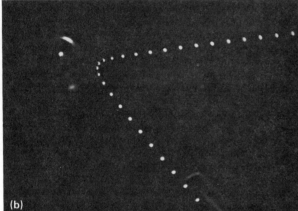

Figure 13.9 (a) A low friction puck attached by a piece of elastic to a central post and (b) a ball bearing rolling on a curved hill. Both puck and ball are illuminated stroboscopically and are seen from above.

the radius in a given time is the same at all parts of the orbit – Kepler's second law (page 160).

Demonstrating angular collisions

The turntable shown in figure 13.8 can be used to demonstrate the *conservation of angular momentum* for rigid bodies. It can also be used to measure the moment of inertia I of discs etc. which are dropped on to it while it rotates freely. The moment of inertia of the turntable I_t needs to be known. If ω_1 and ω_2 are the measured angular velocities of the turntable before and after such a collision then for angular momentum to be conserved,

$$I_t\omega_1 = (I_t + I)\omega_2$$

Figure 13.10 shows an alternative system for studying angular momentum conservation which enables two bars

to interact. They rotate independently about a vertical axis on low-friction bearings. Their angular velocity can be measured by timing complete or half revolutions with a stopwatch. The bars can be made to explode from rest (in a similar way to that shown in figure 2.1) or to collide, with one of them being initially at rest. In the latter situation they might be arranged to stick together (a wholly inelastic collision) or to bounce apart, perhaps magnetically (a perfectly elastic collision). Using different initial values of ω and using more than two rods (for which $I = ml^2/12$) it can be demonstrated that angular momentum is conserved in all types of angular collisions.

▶ A friction clutch consists of two facing discs of moment of inertia 0.80 kg m² and 0.60 kg m² respectively which rotate about the same axis. The first is rotating at 280 rad s⁻¹ while the second is stationary. Discuss what happens when the clutch is engaged if it takes 1.2 s to complete the linkage.

If we assume that there are no torques acting on the discs other than that of their interaction then angular momentum is conserved and we have for a final angular velocity ω

$$(0.80 \text{ kg m}^2)(280 \text{ rad s}^{-1}) = (0.80 + 0.60) \text{ kg m}^2 \times \omega$$

$$\Rightarrow \qquad \omega = 160 \text{ rad s}^{-1}$$

The assumption is a reasonable one provided the clutch mechanism operates quickly, as it does here. We can see that mechanical energy is converted in the interaction, for

initial kinetic energy $= \frac{1}{2}(0.80 \text{ kg m}^2)(280 \text{ rad s}^{-1})^2$
$\qquad\qquad\qquad\quad = 31\,000 \text{ J}$

final kinetic energy $= \frac{1}{2}(1.60 \text{ kg m}^2)(160 \text{ rad s}^{-1})^2$
$\qquad\qquad\qquad\quad = 20\,000 \text{ J}$

so that 11 000 J are converted to internal energy; the friction pads of the clutch become hot.

As the action takes 1.2 s then the average torque can be estimated. The angular acceleration of the disc which starts at rest is $(160/1.2)$ rad s⁻². The average torque T is thus

$$T = (0.60 \text{ kg m}^2) \left(\frac{160}{1.2} \text{ rad s}^{-2} \right)$$

$$= 80 \text{ N m}$$

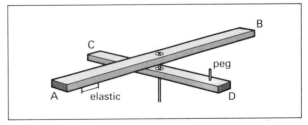

Figure 13.10 Two interacting bars. The bar CD is swung to make (in the case shown) a partly elastic collision with the bar AB.

Non-rigid bodies

A cat held upside down and released without rotation lands on its feet. At first sight this experiment seems to flout the rules for angular motion, for if the cat has zero angular momentum at the start of its fall how can it then rotate? However, the cat behaves as if it were *more than one body*: it rotates its hind legs in one sense about a horizontal axis and rotates its front legs and head in the opposite sense thus preserving (as it must) zero total angular momentum. It now alters the moment of inertia of the two parts of its body and then reverses the original movement. The net effect is that the cat achieves a rotation but at no time, including the time when it has completed the manoeuvre, does it have any angular momentum.

This is a complicated example of the behaviour of a non-rigid body. Simpler cases include that of a skater going into a spin. Figure 13.11 shows two positions for which the moments of inertia about a vertical axis are about 1.2 kg m² and 8.0 kg m² respectively. The ratio of the angular velocities obtainable is thus of the order of 6 or 7 to 1 in going from (b) to (a) since $I_a\omega_a = I_b\omega_b$. If, as here, $\omega_a > \omega_b$ then $\frac{1}{2}I_a\omega_a^2 > \frac{1}{2}I_b\omega_b^2$ and the skater gains energy. He must pull in his arms and leg in reaching the closed position and in so doing transforms chemical energy to kinetic energy (plus, of course, some internal energy).

When a person is in a state of free fall, e.g. a diver or

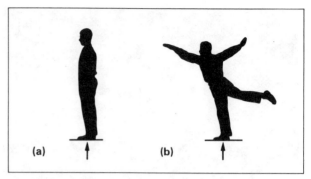

Figure 13.11 For rotation about a vertical axis the ratio of the moments of inertia of the man in the two positions shown is about 1:7.

trampolinist, his angular momentum about a horizontal axis through his centre of mass is constant. If he wishes to achieve a faster rotation, or a slower one, he tucks or opens his position and alters his moment of inertia – the greatest ratio of values of I he can achieve about this axis being about 3.5. Similarly a long jumper who finds himself rotating forwards can prevent this by whirling his arms or pedalling his legs. You may appreciate this better by tilting slowly forward and at the last minute trying to get back to the vertical without moving your feet.

14 Storing electric charge

14.1 Capacitors

An arrangement of two conductors close together, but insulated from one another, is called a *capacitor*. A capacitor may take many different forms – two sets of interleaved flat sheets, two insulated sheets of metal foil rolled up into a 'cigar', a pair of concentric tubes, etc. (see figure 14.3 below). But for the present we consider only the simplest arrangement, which consists of just a pair of parallel metal plates (figure 14.1). The symbol of a capacitor is based on this arrangement whatever the actual form of construction in a particular case.

A series circuit containing a capacitor is not a closed circuit, and there can be no continuous current in it. However, when the capacitor is first connected to the source of e.m.f., it is found that there is a momentary current in the circuit. We can demonstrate this with a commercial capacitor (10 μF) and microammeter using only a 6 V supply. But with a pair of plates as in figure 14.1 several mm apart we need to use a supply of 1000 V or more and a very sensitive meter e.g. a light-spot galvanometer. A very high resistance should be included in the circuit, as shown, as a protection in case the plates are accidentally touched together.

When the capacitor is first connected by joining the flying lead to the point A on the high-voltage supply, the meter swings as a small pulse of current passes through it. If now the meter is allowed to come to rest and the flying lead is taken out and joined to the point B, the meter deflects in the opposite direction. But the amount it swings

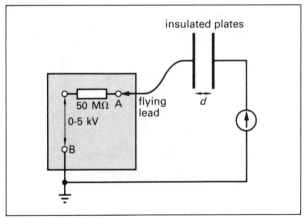

Figure 14.1 The momentary flow of charge in a 'circuit' containing a parallel-plate capacitor; with a commercial capacitor (say 10 μF) a 6 V supply would be sufficient.

is the same as before, and it is clear that the same charge has flowed in the opposite direction in the capacitor leads, thus discharging it.

A meter used like this to detect a momentary pulse of current is called a *ballistic galvanometer*. It has the useful property that the first swing θ of the coil caused by the pulse of current is proportional to the total charge Q passed during the pulse. Thus

$$Q = b\theta$$

where b is the calibration constant of the instrument. So we may use it to find the total charge Q passed in a short pulse of current.

The experiment of figure 14.1 may be repeated with different values of the p.d. V, and it is found that the charge Q that passes during charging or discharging is proportional to V. So

$$Q \propto V$$

For a given value of V we may vary the separation d of the plates; when d is increased, we find that the charge Q is less. The maximum value of Q is obtained when the plates are close together. This suggests the result

$$Q \propto \frac{1}{d}$$

By moving the plates sideways we can also vary the area of overlap A at constant separation; and this suggests that

$$Q \propto A$$

The insulator between the plates is also found to affect the charge Q that flows in this experiment. We usually think of an insulator as something that prevents an electric current, since there are no free charged particles in it to carry a current. However, the presence of *bound* electric charges in the insulator between the plates has the effect of modifying the *electric field* there (page 195), and causes a much greater charge to flow onto and off the plates on either side. This may be demonstrated by fixing a slab of any good insulating material (such as polythene) between the plates; the charge Q is then much greater than with air between them.

When a capacitor is connected to a source of e.m.f. as in the above experiments, there is thus a momentary flow of charge in the circuit, as shown in figure 14.2a. This leaves a surplus of positive charge on one plate of the capacitor and a surplus of negative charge on the other, as shown. The capacitor is then said to be storing a charge Q; provided the insulation between the plates is good enough, this charge ($+Q$ on one plate and $-Q$ on the other) will remain isolated upon them when the source of e.m.f.

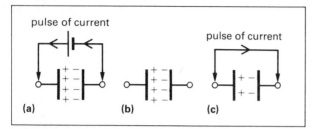

Figure 14.2 The flow of charge onto capacitor plates: (a) charging, (b) isolated, and charge stored, (c) discharging.

is disconnected (figure 14.2b). When the plates are subsequently connected together (figure 14.2c), a pulse of current passes in the circuit, carrying a charge Q from one plate to the other, so discharging the capacitor once more.

The simple arrangement of two parallel plates in figure 14.1 cannot store very much charge, probably less then 1 μC even with a p.d. of 5 kV. To increase the quantity of charge carried for a given p.d. we must increase the area of the plates and decrease their separation as much as possible, consistent with keeping the necessary strength of insulation between them. As we have already seen the use of a suitable insulator also increases the charge stored. Figure 14.3 displays a number of types of capacitor. A 'cigar-roll' form of construction is used with strips of plastic or paper insulation (a). Mica insulation requires a design consisting of a stack of metal plates (b). Ceramic insulation is most easily arranged in the form of a tube (c) with the metal electrodes deposited on its inside and outside surfaces.

Electrolytic capacitors make use of a quite different principle. In these the insulator is formed by electrolysis. When an electric current is passed through a solution of ammonium borate using aluminium electrodes, a very thin film of oxide forms on the anode; this is an example of a process known as anodic oxidation or *anodising*. The thickness of the film depends ultimately on the p.d. used. In the electrolytic capacitor the oxide layer is the insulator; the two 'plates' of the capacitor are the *anode* and the *electrolyte* (to which connection is made via the cathode). The thickness of the oxide layer may be no more than 10^{-7} m, and a very compact unit can thus be produced. The actual construction of the capacitor is similar to the cigar-roll type (a), but in this case the paper strips between the lengths of aluminium foil are soaked in the electrolyte.

In order to maintain the electrolytic deposit it is necessary for a small leakage current to pass through the capacitor in the right direction (typically between 0.1 mA and 1 mA). This limits its use to circumstances in which the applied p.d. is always in the same direction; one terminal is marked to indicate that this must be kept positive with respect to the other. The capacitor can be used with an alternating p.d., but only when this is superimposed on a steady p.d. sufficient to maintain the polarising current in the right sense in the capacitor. Another form of electrolytic capacitor uses tantalum instead of aluminium; this requires even smaller polarising currents (about 1 μA). For many purposes tantalum electrolytic capacitors can even replace plastic insulation capacitors (as well as being much more compact).

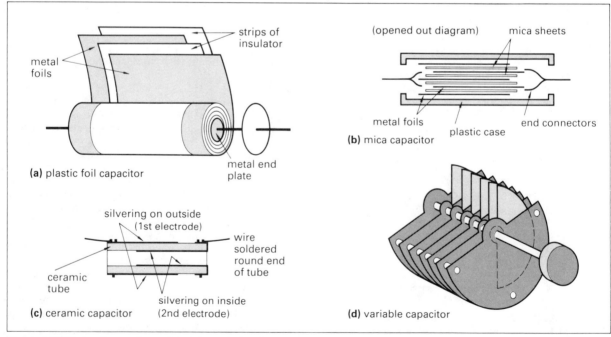

Figure 14.3 Various types of capacitor: (a) plastic foil (0.01 μF to 10 μF); (b) mica (100 pF to 0.01 μF); (c) ceramic (2 pF to 500 pF); (d) variable (50 pF to 500 pF maximum).

Figure 14.3d shows a *variable capacitor*. This consists of two sets of interleaved parallel plates (or vanes). One set is fixed to the frame, the other to the central shaft. Rotation of this varies the area of the interleaved parts of the vanes. Miniature variable capacitors are also made with sheets of plastic held loosely between the vanes. This enables the plates to be thinner and closer together than with air insulation alone.

14.2 Capacitance

When a potential difference V is connected across the plates of a capacitor, a charge Q flows onto the plates charging them up; i.e. a charge $+Q$ appears on one plate, and $-Q$ on the other. This charge Q is proportional to V. Therefore

$$Q = CV$$

This equation defines C, a constant which is called the *capacitance* of the capacitor. The unit of capacitance implied by this definition is the C V^{-1}; but it is convenient to have a special name for this much-used unit, and it is called the *farad* (F) after Michael Faraday (1791–1867) who first explored the nature of electric fields near charged conductors. The farad turns out to be a very large quantity, and capacitances are usually expressed in μF or even pF (appendix A2, table A2).

Figure 14.4 shows a method of measuring directly the capacitance of a capacitor. This is in fact a modification of the basic experiment of figure 14.1, but adapted for use with relatively large capacitances; we can now use p.d.s of 0 to 6 V, and a less sensitive ballistic galvanometer. With the two-way switch on contact X the capacitor is charged to the p.d. indicated by the voltmeter. The switch is then moved to contact Y, discharging the capacitor through the galvanometer. The maximum deflection on the first swing is noted, and so the charge delivered is calculated. If the calibration constant b of the instrument is not known, it may be found by first performing the experiment with a capacitor of known capacitance.

◄ In the above experiment the ballistic galvanometer has a calibration constant of 5.0×10^{-7} C div^{-1}. The volt-

meter reads 6.0 V. The maximum deflection of the meter when the capacitor is discharged through it is found to be 48 divisions. Calculate the capacitance C.

The charge Q that flows through the meter is given by

$$Q = (5.0 \times 10^{-7} \text{ C div}^{-1})(48 \text{ div})$$
$$= 2.4 \times 10^{-5} \text{ C}$$
$$\Rightarrow \quad C = \frac{2.4 \times 10^{-5} \text{ C}}{6.0 \text{ V}}$$
$$= 4.0 \times 10^{-6} \text{ C V}^{-1} = 4.0 \text{ μF} \quad \blacktriangleleft$$

The charging current

The current I in the wires joined to a pair of capacitor plates is the rate of flow of charge Q onto or off the plates. Thus

$$I = \frac{dQ}{dt} \qquad \text{(appendix B3)}$$

As the charge Q stored in the capacitor grows, so does the potential difference V between the plates. Since $Q = CV$, and C is a constant, we can write the above equation in terms of C and V as

$$I = C \frac{dV}{dt}$$

◄ A capacitor of capacitance 50 μF is being charged by a current of 45 μA. (i) What is the rate of growth of p.d. between its plates? (ii) If the capacitor continued to be charged at this rate, how long would it take for the p.d. to grow from 0 to 18 V?
(i) The rate of growth of p.d. (dV/dt) is given by

$$\frac{dV}{dt} = \frac{I}{C} = \frac{45 \times 10^{-6} \text{ A}}{50 \times 10^{-6} \text{ F}}$$
$$= 0.90 \text{ V s}^{-1}$$

(ii) If the p.d. grows at a steady rate of 0.90 V s^{-1}, the time t for it to grow from 0 to 18 V is given by

$$t = \frac{18 \text{ V}}{0.90 \text{ V s}^{-1}} = 20 \text{ s} \quad \blacktriangleleft$$

If the charging current I of a capacitor is kept constant, the p.d. across it rises at a steady rate. This may be observed with the apparatus in figure 14.5. The voltmeter used to measure the p.d. V needs to have a very high resistance so that it draws truly negligible current from the capacitor; a cathode-ray oscilloscope is suitable (used with 'direct' connection to the plates), or else an electrometer (page 182). First, both switches are closed, and the current I is adjusted to a convenient value with the rheostat. Then the switch B is opened, and a stop-clock is started at the same moment. The current that is being registered by the microammeter then starts charging the capacitor, causing the c.r.o or electrometer to deflect. As the p.d. across the capacitor grows, the current tends to fall. Continual adjustments are therefore made to the rheostat to keep the current constant. A graph of p.d. V against time t is a straight line (registered

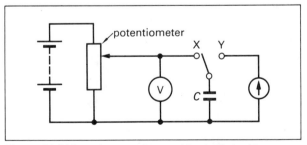

Figure 14.4 Measuring capacitance with a ballistic galvanometer (whose swing is proportional to the total charge that flows through it in a pulse of current).

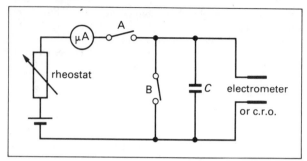

Figure 14.5 Charging a capacitor with a steady current.

directly on the screen if a c.r.o. with a very slow time-base is used). The rate of growth is best found by opening the switch A after a suitable time t, and taking the final reading of p.d. V. Then we have

$$I = C\frac{V}{t}$$

This provides us with a way of measuring C that avoids the difficulties of observing the momentary maximum reading of a ballistic galvanometer.

The vibrating-reed capacitance meter

In this instrument the capacitor under test is charged to the p.d. V of the supply (up to 25 V, say), and is then discharged through a microammeter. But this sequence of connections is performed many times a second by means of a metal strip (or reed) kept vibrating at a known frequency (figure 14.6). One way of doing this is to employ a device known as a *reed-switch*. In this kind of switch the contacts to be joined are steel strips fixed in the ends of a glass tube (figure 14.6b). When the magnetic strips are magnetized by current in a coil round the tube, they are drawn rapidly together. The action is rapid enough to allow the switch to be opened and closed 400 times a second or more. In the unmagnetized

condition the flexible strip springs back on another (non-magnetic) contact. The reed switch is mounted inside a coil which is joined to an a.c. supply of known frequency through a semiconductor diode (page 287). The switch then closes in the conducting half-cycle of the diode and springs back in the other half-cycle, when the diode is non-conducting. The number of discharges per second is equal to the frequency f of the a.c. supply.

If the pulses of current through the microammeter follow one another at high enough frequency the meter shows a steady deflection which records the average current I passing through it.

Thus I = rate of passage of charge

The charge Q stored in the capacitor at p.d. V is given by

$$Q = CV$$
$$\Rightarrow I = fCV$$

The method can be used for smaller values of capacitance than the previous methods. For instance, if we take $f = 400$ Hz and $V = 25$ V as maximum values, then $C = 1$ nF gives a current I of

$$I = (400 \text{ Hz})(10^{-9} \text{ F})(25 \text{ V})$$
$$= 10^{-5} \text{ A} = 10 \text{ μA}$$

With a sufficiently sensitive meter we can make precise measurements of capacitance down to 100 pF or less.

The resistances R_1 and R_2 in figure 14.6a are included in the charging and discharging circuits to avoid excessive pulses of current as the contacts close. But these resistances must be low enough to ensure that the capacitor is *fully* charged and discharged in each cycle of operations (see example on page 185 below).

Capacitors in parallel

The combined effective capacitance C_{par} of capacitors in parallel is equal to the sum of their separate capacitances. This may be shown as follows. In figure 14.7a the p.d. V is the same across the parallel capacitors C_1, C_2 and C_3, but the charges on them will be different. C_{par} is given by

$$Q = C_{par}V$$

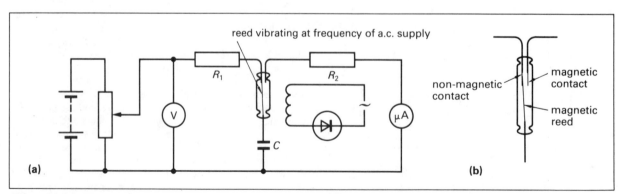

Figure 14.6 (a) The arrangement used in a vibrating-reed capacitance meter; (b) A reed switch.

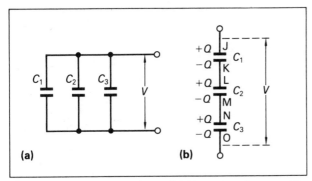

(a)   **(b)**

Figure 14.7 (a) Capacitors in parallel (the same p.d. across all). (b) Capacitors in series (the same charge on all).

where Q is the total charge on the capacitors, i.e. the sum of their separate charges.

So $Q = C_1 V + C_2 V + C_3 V$
$\qquad = V(C_1 + C_2 + C_3)$
$\Rightarrow \quad C_{\text{par}} = C_1 + C_2 + C_3$

Capacitors in series

We need to consider first what happens when the series capacitors in figure 14.7b are charged by applying a p.d. V between the points J and O. This will cause a charge $+Q$ to flow onto the plate J, and $-Q$ therefore onto the other plate K. The plates K and L and the connecting link between them form a single insulated conductor, whose total charge must remain zero throughout the process. Therefore if a charge $-Q$ appears on plate K a charge $+Q$ must appear on plate L of the next capacitor, and so on through the sequence of capacitors. Thus, with capacitors in series the *same charge* is stored in each capacitor, $+Q$ on one of its plates and $-Q$ on the other. The combined effective capacitance C_{ser} is given by

$$Q = C_{\text{ser}} V$$

where V is the total p.d. across the combination.

Now $V = \dfrac{Q}{C_1} + \dfrac{Q}{C_2} + \dfrac{Q}{C_3}$

$$= Q\left(\frac{1}{C_1} + \frac{1}{C_2} + \frac{1}{C_3}\right)$$

But $\dfrac{V}{Q} = \dfrac{1}{C_{\text{ser}}}$

$$\Rightarrow \quad \frac{1}{C_{\text{ser}}} = \frac{1}{C_1} + \frac{1}{C_2} + \frac{1}{C_3}$$

The above results may be tested very simply with our vibrating-reed capacitance meter, measuring the capacitances of a set of capacitors individually, and then trying them in various combinations in parallel and in series.

Alternating current in capacitors

If an alternating supply is joined to the plates of a capacitor, charge flows onto and then off the plates as the potential difference between them oscillates; and there may be a large alternating current in the connecting wires. This current can be large enough to light a powerful electric bulb in series with the supply and the capacitor (figure 14.8a). One of the functions of a capacitor in an electric circuit is to 'block' the passage of direct current while allowing an alternating current in the rest of the circuit (page 280).

The magnitude of the alternating current in a capacitor of capacitance C may be calculated as follows. The current I in the capacitor leads is given by

$$I = C\frac{dV}{dt}$$

If a sinusoidal alternating p.d. of peak value V_0 and frequency f is joined to the capacitor, we have

$$V = V_0 \sin 2\pi f t \qquad \text{(page 85)}$$

The maximum current occurs at the moments when the p.d. V is zero, i.e. when the p.d. is changing at the maximum rate. The peak value I_0 of the current is therefore given by

$$I_0 = C\left(\frac{dV}{dt}\right)_{\text{max}}$$

$$= C(2\pi f V_0) \qquad \text{(appendix B6)}$$

We may use this result to work out the current in figure 14.8a. For the a.c. mains supply, typical values are

$$V_{\text{rms}} = 240 \text{ V}$$

and so $V_0 = \sqrt{2} \times 240 \text{ V} = 340 \text{ V}$ \qquad (page 272)

Then $I_0 = 2\pi(50 \text{ Hz})(340 \text{ V})(10^{-5} \text{ F}) \approx 1 \text{ A}$

We may measure a capacitance C with a.c. ammeter and voltmeter, as in figure 14.8b. Such meters normally register r.m.s. values of current and p.d. (rather then peak values, page 272), but this does not matter; for we may re-arrange the above result to give

$$\frac{1}{2\pi f C} = \frac{V_0}{I_0} = \frac{V_{\text{rms}}}{I_{\text{rms}}}$$

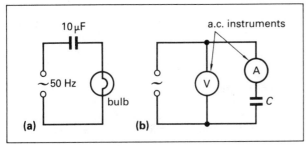

(a)   **(b)**

Figure 14.8 (a) An alternating current in a circuit containing a capacitor. (b) Measuring capacitance with a.c. ammeter and voltmeter.

14.3 Electrometers

For many purposes we need a voltmeter with an exception-ally high resistance. An ordinary moving-coil voltmeter is often quite inadequate, e.g. for measuring the p.d. across a 1 µF capacitor; for the current taken by the instrument would discharge the capacitor in milli seconds. For some purposes we can use a cathode-ray oscilloscope as a high resistance voltmeter, but even this takes a current, which although small is large enough to matter when we are dealing with very small quantities of electric charge. We then need voltmeters that take currents of 10^{-12} A or less; such instruments work on new principles and are called *electrometers*.

One form of electrometer makes use of electronic com-ponents (transistors, etc, page 293) to produce a *d.c. amplifier* or *measuring amplifier*, as it is called. The p.d. at the input terminals is amplified (figure 14.9a), and the output is connected direct to a moving-coil meter, which is calibrated to read the p.d. at the input (typically in the range 0 to 1 V). The instrument is thus primarily a very high resistance voltmeter (10^{12} Ω or more); but it may also be adapted to measure a very small current I by passing this through a known high resistance R joined across the input terminals (figure 14.9b). The amplifier and meter measure the p.d. V across this; and we have

$$I = \frac{V}{R}$$

For instance, if the resistance R is 1.00×10^{10} Ω, then the current range obtained is 0 to 10^{-10} A (if the input p.d. is to be between 0 and 1 V). With suitable values of R currents of 10^{-12} A or less may be measured without difficulty by this means.

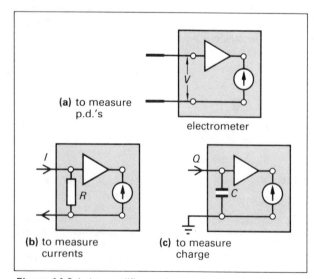

Figure 14.9 A d.c. amplifier used as an electrometer: (a) to measure potential difference; (b) to measure current; (c) to measure charge.

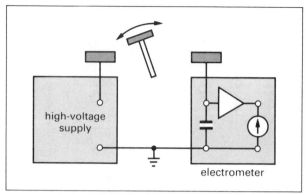

Figure 14.10 Spooning charge from a high-voltage source into an electrometer; the p.d. across the capacitor rises in equal steps.

By joining a known capacitance C across the input terminals of an electrometer (figure 14.9c) the instrument may also be adapted to give the charge Q delivered to the terminals. If the capacitor is initially uncharged, the p.d. V across it is then given by

$$Q = CV$$

For instance suppose $C = 0.01$ µF (10^{-8} F); then for input p.d.s up to 1 V the charge range of the instrument is 0 to 10^{-8} C. You will notice when you use the instrument in this way that the reading of charge on the meter falls very slowly as the capacitor discharges through the input resistance of the amplifier; but in practice the change is slow enough not to matter over the time required for taking a reading.

Used in this way an electrometer enables us to measure small quantities of charge collected on another insulated conductor or stored on the plates of a small capacitor. For instance, a small metal disc on an insulating handle (usually called a *proof-plane*) may be used to collect a sample of the charge on some other conductor, and then this may be measured by touching the proof-plane on one terminal of the capacitor across the electrometer input. In practice almost the entire charge on the proof-plane is given up to the capacitor when this is done. It is instructive to use an arrangement of this sort to take successive equal quantities of charge from a plate maintained at a high potential and to 'spoon' them into the electrometer capacitor (figure 14.10). It will be found that the reading of the instrument goes up in equal steps, confirming that the p.d. across the capacitor is directly proportional to the charge stored in it.

The same thing may be done to measure the capacitance C' of a small capacitor (if it is much less than the elec-trometer capacitance C). The capacitor is charged to a suitable p.d. V (say 100 V); then it is disconnected from the supply and joined across the input capacitor of the elec-trometer, and gives up almost its entire charge Q to it. Then we have

$$C' = \frac{Q}{V}$$

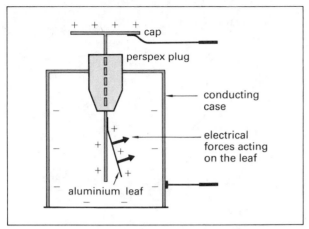

Figure 14.11 A leaf electrometer for measuring high potential differences.

The leaf electrometer

Before the advent of efficient electronic amplifiers the only way of designing electrometers was to make use of the force that acts between two charged conducting surfaces. When a potential difference is produced between two conductors, the force acts in such a way as to draw them together, since the charges on the two conductors are of opposite sign. This principle is made use of in the leaf electrometer (figure 14.11). It consists of a rectangular piece of fine aluminium leaf attached at its top edge to the side of a brass rod. This is fixed in an insulating plug in the top of a conducting box with glass windows. A small metal disc, called the *cap* of the electrometer, is usually mounted at the top of the brass rod. When a p.d. is produced between the cap and the case, the electrical forces draw the metal leaf out towards the case at an angle to the vertical, as shown. A scale may be mounted on one glass window to indicate the angle of deflection, or to show directly the p.d.; but often the instrument is used only for qualitative observations. Full-scale deflection requires a p.d. of about 1500 V. The leakage current through the insulating plug can well be less than 10^{-13} A; and the capacitance of the instrument is only a few pF, so that a charge of about 10^{-9} C can be sufficient to produce maximum deflection.

Whichever way round the p.d. is connected, the leaf is pulled towards the case. The leaf electrometer is therefore suitable for measuring large alternating p.d.s as well as steady p.d.s.

14.4 Charging and discharging

Energy storage

The existence of a potential difference between the plates of a capacitor implies that the charge stored there carries electrical potential energy. If the p.d. is V, and a small quantity of charge ΔQ is allowed to flow from one plate to the other, the energy converted from electrical form to other forms (chiefly internal energy in the connecting wire) is $V\Delta Q$. This movement of charge partly discharges the capacitor, and the p.d. between the plates falls slightly. A graph of p.d. against total stored charge Q is shown in figure 14.12. It can be seen that the loss of energy $V\Delta Q$ is represented by the area of the shaded strip under the graph. As further quantities of charge ΔQ are allowed to flow from one plate to the other, the areas of the strips under the graph get steadily less as the p.d. falls. The total energy W_E stored in the capacitor is represented by the total area between the graph line and the axis of Q.

$$\Rightarrow \quad W_E = \tfrac{1}{2}QV$$

Since $Q = CV$, we may substitute for Q or V in this expression to obtain the three alternative formulae:

$$W_E = \tfrac{1}{2}QV = \tfrac{1}{2}CV^2 = \tfrac{1}{2}\frac{Q^2}{C}$$

Notice that for a given capacitance C the energy stored is proportional to the square of the p.d. V. Doubling the p.d. across the capacitor quadruples the energy W_E stored. When the capacitor is discharged this energy may be changed into internal energy in a resistor, or into some other form of energy; for instance, if a large capacitor (1000 μF) is discharged through a small electric motor, the energy can be used to lift a weight, thereby transforming some of it to gravitational potential energy.

▶ A capacitor of capacitance 8.0 μF is charged to a p.d. of 400 V and then isolated from the supply. (i) What is the energy stored in it? (ii) If an identical capacitor (initially uncharged) is joined across it, what is the energy now stored in the pair of capacitors? Comment on the result.

(i) The initial energy W_E is given by

$$W_E = \tfrac{1}{2}CV^2$$
$$= \tfrac{1}{2}(8.0 \times 10^{-6}\text{ F})(400\text{ V})^2 = 0.64\text{ J}$$

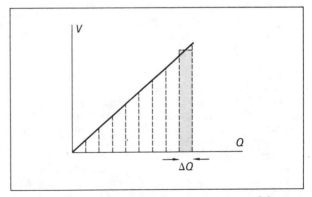

Figure 14.12 The relation between p.d. V and charge Q for a capacitor; the energy stored is given by the area between the line and the Q-axis.

(ii) The two capacitors are now joined in parallel, and their combined capacitance is therefore 16.0 μF. Half the charge in the first capacitor flows into the second one; but the total charge remains constant (assuming the insulation is good enough). The p.d. across the capacitors therefore falls to half the initial value, i.e. it falls to 200 V. The final energy W_E' is therefore given by

$$W_E' = \tfrac{1}{2}(16.0 \times 10^{-6} \text{ F})(200 \text{ V})^2$$
$$= 0.32 \text{ J}$$

Thus, half the energy (0.32 J) is apparently lost in this process of sharing the charge between the capacitors. Most of this energy is converted to internal energy in the conductors joining the two pairs of capacitor plates. Very often the flow of charge in these conductors is oscillatory – the charge flows back and forth between the two capacitors a few times before settling in its new equilibrium position; (this arises because of the inductance of the conductors, page 280). The frequency of such oscillations is probably many MHz. In these circumstances a small pulse of radio waves is emitted (page 368); a radio set nearby may produce a crackle when the capacitors are joined. But only a very small proportion of the original energy is radiated away in this manner.   ◀

Exponential decrease of p.d.

Suppose (as in the last example) that we have a capacitor of capacitance 8.0 μF charged to 400 V, and we then join a 2.0 MΩ resistor across it. The initial discharge current I is given by

$$I = \frac{400 \text{ V}}{2.0 \times 10^6 \text{ Ω}} = 200 \text{ μA}$$

This starts to discharge the capacitor; the rate of change of p.d. is given by

$$I = -C\frac{\mathrm{d}V}{\mathrm{d}t}$$

(a minus sign because the current is *discharging* it).

Hence $\dfrac{\mathrm{d}V}{\mathrm{d}t} = -\dfrac{200 \times 10^{-6} \text{ A}}{8.0 \times 10^{-6} \text{ F}}$

$$= -25 \text{ V s}^{-1}$$

If the current continued at 200 μA, the capacitor would be completely discharged in 16 s (since 16 s × 25 V s⁻¹ = 400 V). However, as the p.d. falls the current also decreases in the same proportion; when the p.d. is 300 V, the current is 150 μA; when the p.d. reaches 200 V, the current drops to 100 μA; and so on.

The p.d. V across a discharging capacitor therefore falls more and more slowly, as shown in figure 14.13a; we can show that this is an *exponential decay* process (appendix B7). For the current I through the resistor is given by

$$I = \frac{V}{R} = -C\frac{\mathrm{d}V}{\mathrm{d}t}$$

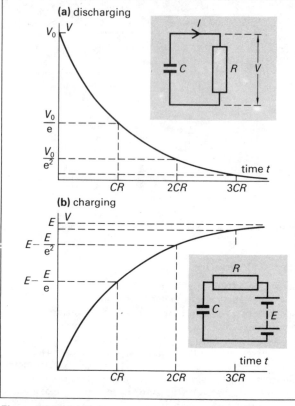

Figure 14.13 (a) The exponential decay of the potential difference across a capacitor. (b) The charging of a capacitor by a battery of e.m.f. E; in this case $(E - V)$ decays exponentially. In both cases the time constant is CR.

$$\Rightarrow \quad \frac{\mathrm{d}V}{\mathrm{d}t} = -\frac{1}{CR}V$$

Thus, the rate of change of p.d. ($\mathrm{d}V/\mathrm{d}t$) is proportional to V which is the condition for an exponential change (appendix B7). V is therefore given by

$$V = V_0 \mathrm{e}^{-t/CR}$$

where V_0 is the initial p.d. This is the equation of the curve in figure 14.13a. The *time constant* of the decay process is CR; i.e. when $t = CR$, $V = V_0/\mathrm{e} = 0.37\,V_0$. Thus, in the case discussed above, we have

$$CR = (8.0 \times 10^{-6} \text{ F})(2.0 \times 10^6 \text{ Ω})$$
$$= (8.0 \times 10^{-6} \text{ C V}^{-1})(2.0 \times 10^6 \text{ V A}^{-1})$$
$$= 16 \text{ C A}^{-1} = 16 \text{ s}$$

This means that in the first 16 s the p.d. falls from 400 V to 0.37×400 V (= 148 V). In the next 16 s the p.d. again decreases in the same radio to 0.37×148 V (= 55 V). The same happens in each successive interval CR. You can calculate that the p.d. falls to less than 1 V in a time $6CR$ (= 96 s).

A similar result occurs when a capacitor is *charged* by

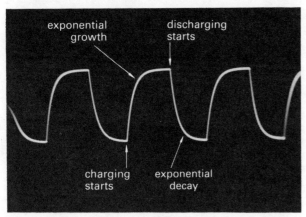

exponential growth

discharging starts

charging starts

exponential decay

Figure 14.14 An oscilloscope trace of the exponential charging and discharging of a capacitor in a vibrating-reed capacitance meter; the resistances must be small enough to ensure effectively complete charging and discharging.

connecting it to a source of e.m.f. In a high-resistance circuit the p.d. V across the capacitor rises exponentially towards its final value equal to the e.m.f. E of the source (figure 14.13b). In this case it is the difference $(E - V)$ that decays exponentially. So we have

$$E - V = Ee^{-t/CR}$$
$$\Rightarrow \qquad V = E(1 - e^{-t/CR})$$

In the vibrating-reed capacitance meter (figure 14.6, page 180) we have a capacitor alternately charged through a resistance R_1, and then discharged through a meter and another resistance R_2. Figure 14.14 shows an oscilloscope trace of the p.d. across the capacitor. Each time the reed switch changes over the p.d. changes exponentially towards the p.d. of the supply or towards zero.

▶ A vibrating-reed capacitance meter is being used to measure a capacitance of about 0.1 μF. The frequency of the alternating supply driving the reed switch is 50 Hz. Calculate the maximum value of R_1 or R_2 if the capacitor is to be effectively fully charged and discharged each time.

The period of oscillation = (1/50) s. The maximum time t available for charging or discharging is half a period. So $t = 0.01$ s. We have seen above that the p.d. falls to 1/400 of the supply p.d. in a time $6CR$. To achieve this degree of discharging we must have

$$t > 6CR$$

or $\quad R < \dfrac{t}{6C} = \dfrac{10^{-2}\ \text{s}}{6 \times 10^{-7}\ \text{F}} = 17\ \text{k}\Omega$

In practice we can protect the contacts of the reed switch sufficiently with resistances far less than this (say 300 Ω). ◀

15 Electric fields

15.1 Electrical forces

In the space between the plates of a charged capacitor strong electrical forces act on any charged particles present. This is obvious enough when the insulation breaks down; a solid insulator will have a hole punched in it by the hot spark that passes through it in such a case. The action of these electrical forces can be demonstrated less destructively by placing a flexible test strip in the gap between two plates charged to a high p.d. (figure 15.1a); almost any piece of thin foil on an insulating handle may be used, but a strip of thin metallised plastic is usually best. If this test strip is touched on the positive plate it picks up from this some positive charge, and is then seen to be bent towards the negative plate, so revealing the electrical force that acts on any positive charge in the gap. Similarly if the test strip is touched on the negative plate, picking up a sample of negative charge, it is deflected in the opposite direction, showing the electrical force that acts on a negative charge in the gap.

These forces are simply the electrical forces that push the positive charge in one direction in an electric circuit, and the negative charge in the other. In a metallic circuit there are free electrons able to move through the conductors, and an electric current (i.e. a flow of charge) is

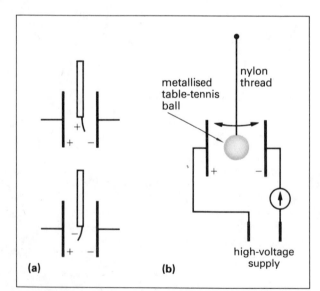

Figure 15.1 (a) Detecting the electric field between parallel plates with a flexible test strip. (b) A metallised table-tennis ball 'completes' the circuit between parallel plates.

produced. But in an insulator (such as air) there are no free charged particles, and no charge flows. However, the forces still act on any charged particle introduced into the space, even if it is not free to move. Any space in which charged particles are acted on by forces is said to contain an *electric field* or *E-field*.

The deflection of the test strip in figure 15.1a shows that there is an electric field between the plates of a charged capacitor (page 177). The field is produced by the charges on the surfaces of the plates. The behaviour of the test strip shows that:

> *like charges repel one another*
and *unlike charges attract one another*

We may demonstrate these rules directly with *two* test strips. These may be charged by touching each of them on one plate or the other of the charged parallel plates (figure 15.1a); we may then show the forces acting between them when they are far removed from the pair of plates. With like charges on the strips they bend away from one another, and with unlike charges they bend towards one another.

The nature of an electric current is revealed in a striking way if a light metallised sphere is suspended on an insulating thread between a pair of charged parallel plates (figure 15.1b). As long as the sphere remains uncharged it experiences no force. But as soon as it touches one of the plates (say the positive one) it acquires from it some positive charge and is pushed towards the other plate; here it delivers up this charge, receiving a negative charge instead. It then returns to the positive plate, and so the cycle is repeated indefinitely at a rate of several oscillations per second. A meter joined in series with the capacitor and the high-voltage supply registers a steady deflection. Presumably the current passes in a series of pulses in this case, but the meter response is too slow to do more than show the average current. If the plates are disconnected from the supply, leaving the sphere bouncing back and forth, it discharges the capacitor formed by the pair of plates just as leaky insulation would do; and the rate of discharge slowly decreases as the p.d. between the plates decreases.

Frictional or contact electricity

When the surfaces of insulating bodies are placed in contact, they are found to become electrically charged, i.e. they attract or repel each other. Two surfaces 'in contact' actually touch only over a small fraction of the area involved because of the surface irregularities present (page 19). To increase the surface charge produced we need to rub the surfaces vigorously together. Charges produced by surface

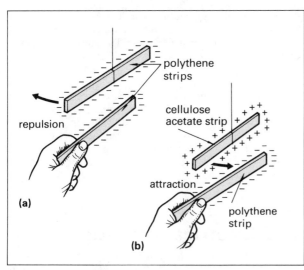

Figure 15.2 The forces between frictionally produced charges: (a) repulsion between like charges; (b) attraction between unlike charges.

contacts are therefore usually called *frictional charges*. Polythene, for instance, rubbed with a woollen duster acquires a *negative* charge; and two polythene strips so treated will be found to repel one another (figure 15.2a). Perspex and cellulose acetate (the substance used for the base of a photographic film) acquire a *positive* charge when rubbed with a woollen duster; and these will be attracted towards a polythene strip that has been rubbed with wool (figure 15.2b).

It is at once apparent from such experiments that there are just two kinds of electric charge, and that the two kinds are opposite, i.e. they cancel one another's effects. The choice of sign was originally an arbitrary matter; but, as we now know, it has led to the sign of the charge on an electron being called negative, and that on a proton positive. The sign of the charge on a rubbed insulator may be tested by suspending it between a pair of charged parallel plates, and observing the direction of the force acting on it.

The frictional charge that appears on the surface of an insulator is small by other standards, probably not more than 10^{-10} C on the polythene strips mentioned above. But the difference of potential between the strip and the duster may be 10^4 V or more; we often notice sparks passing between frictionally charged objects near one another – pulling off a nylon vest provides a familiar example!

Sometimes we find that unwanted frictional charges appear on parts of our apparatus producing disturbing effects. An insulating support may easily become charged through contact with a coat sleeve, and then the distribution of electric charge in the apparatus is disturbed by the presence of the frictional charge; also dust and other small pieces are attracted to it. Discharging of a charged insulator is most easily effected by using a flame (e.g. a match or a small gas burner flame). The flame contains large numbers of ions of both signs, and when it is placed in an electric field some of these are drawn out and pass quickly and invisibly through the air towards the sources of the field. Thus a negatively charged polythene strip held a few centimetres from a flame attracts positive ions out of it, and is discharged completely in about a second.

If a small flame is held near the pair of plates in figure 15.1b the current carried across the gap by the ions may be quite large, and can be registered by a meter joined in series.

Normally frictional charges are observed only on good insulators, since with a conductor or poor insulator any charge at once leaks away to the Earth. But if a conductor is carefully insulated from the Earth and rubbed with an insulator, charges are separated just as with a pair of insulators. Frictional processes in the exhaust of an aeroplane can lead to the separation of large charges; accidents have been caused by the *sudden* discharge and spark that occurs as the plane touches down. This is now prevented by adding substances to the rubber of the tyres to make them slightly conducting; the charge on the plane then leaks away *gently* during the first moments of contact with the ground.

Many substances that we ordinarily regard as insulators (at the low p.d.s of the batteries used in simple electric circuits) must be treated as conductors at the high p.d.s commonly used when we are studying electric fields. For instance, if the parallel plates of figure 15.1 are charged by connecting them for a moment to a high-voltage supply we may discharge them in a fraction of a second by joining the plates with a wooden ruler or a piece of paper, though in especially dry conditions these may behave as insulators. The only really adequate insulators are water-repellant substances such as nylon and other modern plastics. In most experiments on electric fields we can assume that the bench, the walls, the experimenter and the Earth are in good electrical contact. The only electric fields that exist will be between these objects and others that have been specially insulated and charged.

Lines of force

Electric fields may be described by means of lines of force in much the same way as magnetic and gravitational fields (page 162), the lines indicating the direction of the force that would act on any charged particle placed in the field. The arrow on a line of force is always drawn to show the direction of the force that would act on a *positive* charged particle. In a diagram the electric lines of force are therefore drawn starting from surfaces carrying positive charge and ending on those carrying negative charge. The lines thus give the direction in which current would pass (in the conventional sense) under the action of the field if a conducting substance occupied that region of space.

It is possible to plot the lines of force of an electric field in a rather similar way to that used with magnetic fields. It is found that a small particle or fibre tends to set itself

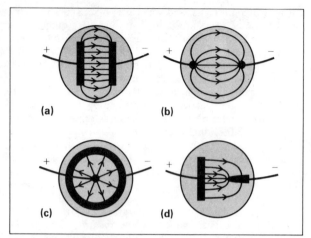

Figure 15.3 The lines of force of the electric field between various pairs of electrodes; in (a) there is a uniform field in the central region between the parallel plates, in (c) there is a radial field between the central rod and the cylindrical electrode.

with its longest axis parallel to a line of force. Particles of semolina (which are conductors) are effective for this purpose. This material will float on a layer of tetrachloromethane (carbon tetrachloride) in a small dish; the evaporation of this rather volatile liquid can then be inhibited by pouring a layer of castor oil on top of it, and the particles float at the interface of the two liquids (which are both good insulators). Pairs of electrodes of various shapes can be inserted in the liquid and connected to a source of a few thousand volts. The particles of semolina immediately reveal the pattern of lines of force. A similar demonstration may be performed with small clippings of nylon or other such fibre.

Figure 15.3 shows the patterns of lines of force found for a variety of arrangements of electrodes with the arrows inserted according to the usual rule. The simplest is that between a pair of parallel plates (a). Near the edges of the plates the field is non-uniform, varying in direction and magnitude from point to point; but in the central region between the plates the lines of force form a pattern of parallel lines across the gap between the plates; and here the field is a *uniform* one. By this we mean that the force acting on a small charge placed in the field is the same in magnitude and direction at all points. This may be demonstrated roughly with the aid of a flexible test strip as described above (figure 15.1a). If the charged test strip is moved about to different positions between the plates, the amount it bends under the action of the electrical forces is clearly the same at all points – except near the edge of the gap where the strength of the field falls off and is obviously no longer uniform. A more precise test of the uniformity of the field between parallel plates is provided by observing the movements of charged oil drops as in Millikan's experiment (page 192). The force acting on such a drop is found to be precisely the same in all positions in the central region between the plates.

The strength of the electric field, the E-field, at a point is measured by the force per unit electric charge that acts on a small positive charge placed there. Thus the force F_E acting on a small charge Q is given by

$$F_E = QE$$

(We have to assume that an electric charge Q introduced to measure the value of E is indeed small enough to produce negligible disturbance of the field.) The unit of electric field strength implied by the above definition is the $N\ C^{-1}$; but any other equivalent combination of units may be used. In fact it is usually convenient to use the $V\ m^{-1}$, as explained in the next section.

15.2 Electrical potential

Up to this point we have used the potential as a way of describing the electrical situation in a *circuit*. Current passes (in the conventional sense) from points at high potential in a circuit to points at low potential, and the size of current depends on the potential difference between these points. The p.d. V is the energy converted from electrical to other forms per unit electric charge that passes (page 81); thus if the charge is Q, we have

$$V = \frac{W}{Q}$$

where W is the energy converted. This conversion of energy takes place under the action of the electrical forces acting in the circuit.

The same idea is useful for describing an electric field in an insulating medium such as the air. The electrical forces act on any charged particle that happens to be present, and if it is a positive charge they act to drive it in the direction of the arrow along a line of force. Just as in an electric circuit the potential decreases as we pass along a line of force from the positive charge from which it starts towards the negative charge on which it ends; and the potential difference V between two points in the field is the energy converted by the forces of the electric field per unit charge that passes from one point to the other.

As with electric circuits only *differences* of potential really concern us; but it is often convenient to have a conventional *zero of potential* with reference to which other potentials are measured. The Earth is normally taken as this. In practice this means that the walls, the floor, the bench top and the experimenter are all at zero potential, unless special steps are taken to insulate one of them.

A knowledge of the potentials at two points enable us to calculate the work done by the electrical forces in moving a charged particle from one point to the other, without having to work out the fields strength at intermediate points. Suppose a particle of charge Q is moved from a point at a potential V_1 to another at a potential V_2. Then the work W done by the electrical forces is given by

$$W = Q(V_1 - V_2)$$

Calculate the speed v of an electron as it strikes the anode of a thermionic tube (page 236), if the p.d. between anode and cathode is 150 V; assume that the electron emerges from the cathode at negligible speed, and that its mass is 9.1×10^{-31} kg and charge -1.60×10^{-19} C.

The electron emerges from the cathode with high electrical potential energy. As the electrical forces accelerate it this potential energy is converted to kinetic energy. The work W done on the electron between cathode and anode is given by

$$W = (1.60 \times 10^{-19} \text{ C})(150 \text{ V})$$
$$= \text{kinetic energy gained} = \tfrac{1}{2}m_e v^2$$

where m_e is its mass. Re-arranging,

$$v^2 = \frac{2(1.60 \times 10^{-19} \text{ C})(150 \text{ J C}^{-1})}{(9.1 \times 10^{-31} \text{ kg})}$$
$$= 5.27 \times 10^{13} \text{ J kg}^{-1}$$

[The units of v^2 emerge here as J kg^{-1}. But

$$1 \text{ J kg}^{-1} = (1 \text{ N m}) \text{ kg}^{-1} = (1 \text{ kg m s}^{-2} \text{ m}) \text{ kg}^{-1}$$
$$= 1 \text{ m}^2 \text{ s}^{-2}]$$

$$\Rightarrow \quad v = 7.3 \times 10^6 \text{ m s}^{-1}$$

Potential gradient

A knowledge of the way in which the potential varies from point to point in a field enables us to calculate the value of the E-field. Thus figure 15.4 shows the field between two parallel plates, and (inset) a small part of the same field. Suppose a particle of charge Q is moved by the forces of the field between two points A and B a short distance d apart. It therefore loses electrical potential energy. In a vacuum this is changed into kinetic energy as the particle accelerates. But in air the viscous drag on the particle causes the potential energy to be converted to internal energy of the air. This is essentially the same process of energy conversion as occurs in a conducting wire with electrons moving through it. If the field strength is E, the force F_E acting on the particle is given by

$$F_E = QE$$

Since the field is uniform, this force remains constant as the particle moves. In moving it the distance d the work W done by the forces of the field is given by

$$W = F_E d = QEd$$

If the potentials at A and B are V_1 and V_2 respectively, we can also say

$$W = Q(V_1 - V_2)$$

$$\Rightarrow \quad E = \frac{V_1 - V_2}{d}$$

The quantity on the right in this equation is called the *potential gradient* of the field, and we can see that it is equal to the strength of the E-field. The unit of potential gradient is the V m^{-1}, and this is usually the most convenient unit for recording electric field strengths. You can check for

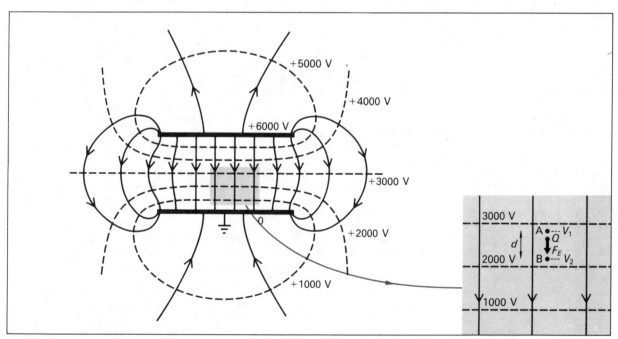

Figure 15.4 The equipotential surfaces (dotted) and lines of force between a pair of parallel plates between which there is a p.d. of 6 kV; inset we see how to calculate the work done on a small charge Q moved from A to B by the forces of the field.

yourself that the V m^{-1} is the same as the N C^{-1}.

It is worth giving some estimates of the size of the E-field in different circumstances. For instance a typical dry cell (e.m.f. 1.5 V) used in a torch is 60 mm long. The field in the space round the cell as it lies in a cupboard is not of course uniform; but if it was uniform from top to bottom, we should have

$$E = \frac{1.5 \text{ V}}{60 \times 10^{-3} \text{ m}}$$

$$= 25 \text{ V m}^{-1}$$

The electric field will actually be rather greater than this near the top where the outer case and central positive terminal are quite close together.

With a pair of parallel plates joined to a high-voltage supply, as in figure 15.1b (but with the sensitive meter removed), we can find what E-field is sufficient to disrupt molecules of air in the gap; when this happens some of the air is ionised and a spark passes between the plates. For instance, with the plates set 1.5 mm apart a spark passes when the p.d. of the supply is increased to 4.5 kV. Thus in the gap

$$E = \frac{4.5 \times 10^3 \text{ V}}{1.5 \times 10^{-3} \text{ m}}$$

$$= 3.0 \text{ MV m}^{-1}$$

At the surfaces of good insulators that have been charged by friction the E-field can well be close to this value; hence the small sparks observed in such circumstances.

In a thunder cloud the breakdown of the insulation of the air occurs at a much lower figure, about 1 MV m^{-1}; this is because of the presence of many water drops. However, the E-field needs to reach this level only in a small region of the storm cloud; once the discharge has started it extends rapidly to regions where the E-field is much less, perhaps only 0.01 MV m^{-1}. It is found that the net effect of the electrical processes occurring in a thunder storm is to transport positive charge from the Earth's surface to the higher conducting layers of the atmosphere, known as the *ionosphere*; in this region a proportion of the air molecules are permanently ionised by the ultra-violet radiation from the Sun. Because of thunder storms the ionosphere carries a permanent positive charge. The result is that in normal (fair-weather) conditions there is a downward electric field at the Earth's surface of about 300 V m^{-1}.

In an air-insulated capacitor (as in figure 14.3d, page 178) the E-field must be kept much less than 3 MV m^{-1}. But solid insulators have much greater insulation strengths; and very thin films of plastic or mica can be used without danger of breakdown. For instance, a certain capacitor has a plastic film 5.0×10^{-6} m thick between its layers of foil, and is rated for potential differences up to 300 V. The E-field in the film at the maximum safe p.d. is given by

$$E = \frac{300 \text{ V}}{5.0 \times 10^{-6} \text{ m}}$$

$$= 60 \text{ MV m}^{-1}$$

In an electrolytic capacitor (page 178) the figures are even more startling. The oxide layer that provides the insulation in these may be no more than 10^{-8} m thick; but the maximum E-field it can stand without breakdown can be as much as 1000 MV m^{-1}.

Equipotentials

We have two alternative ways of describing electric fields: (i) by means of lines of force and the electric field strength E; or (ii) by means of diagrams showing the electric potential at points in the field. Figure 15.4 shows both these forms of description for the field between a pair of parallel plates, one of which is earthed (zero potential), and the other at a potential of 6000 V. The dashed lines connect points at a common potential. Seen in three dimensions these lines would be parts of *surfaces* on each of which the potential has a given value. Such surfaces are referred to as *equipotential surfaces*. The direction of the line of force at any point is that in which the potential changes most rapidly with distance (i.e. in which the potential gradient is a maximum). This means that the lines of force are always at right-angles to the equipotential surfaces. In the central region of figure 15.4, where the field is uniform, the equipotentials are planes parallel to the plates, and the potential decreases uniformly through the air gap from one plate to the other.

We may demonstrate that the electrical force on a charged object depends on the potential gradient with the aid of the flexible test strip. The test strip is given a charge and supported between a pair of parallel plates connected to a high-voltage supply. If the p.d. V and separation d of the plates are now varied in such a way as to keep the potential gradient V/d constant, the amount the test strip bends remains constant (as long as the separation d is not so great that the field in the gap can no longer be considered uniform). Thus, if the p.d. V and separation d are both doubled, the test strip bends the same amount.

It is easier to measure how the *potential V* varies within an electric field than to measure the *field strength E* directly. We cannot of course use an ordinary voltmeter to find the potential at a point in the air. Even an electrometer or a cathode-ray oscilloscope takes a small current, and in an insulating medium there is no way in which charge can flow to whatever probe we place there to connect to the instrument. We may get round this difficulty by incorporating a source of ionisation in the probe. The easiest way of doing this is to make the probe out of a hypodermic syringe on the end of which a very small gas flame is maintained. The flame produces a copious supply of ions which move out into the surrounding air if there is any tendency for charge to flow onto or off the probe. In this way the probe reaches the same potential as there was before the probe was put there. A leaf electrometer is a suitable instrument for recording the potentials. Figure 15.5 shows how this arrangement can be used for studying the distribution of potential both between a pair of parallel plates and outside

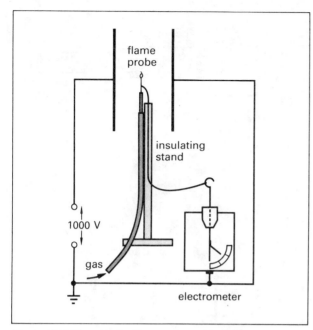

Figure 15.5 The use of a flame-probe and leaf electrometer to measure the potentials at points in an electric field.

the gap. By moving the probe in such a way as to keep the deflection of the electrometer constant we can trace out the shape of any required equipotential surface.

If we wish to investigate the pattern of equipotential surfaces round a complicated system of electrodes it is usually easier, however, to make a *scale model* of the electrode system. In many cases it will be sufficient to study the nature of a field in two dimensions only; we can then use a special sort of paper coated with a uniform poorly-conducting layer. The positions of the electrodes can be marked in on this paper with good-conducting paint, and a suitable low p.d. applied between them. Figure 15.6 shows a potentiometer technique (page 98) for marking the positions of the equipotential lines on the paper. For each potential selected by the switch S the conducting pencil P is moved to find positions at which no deflection of the meter occurs. A series of pencil dots is thus obtained outlining the position of each equipotential line. From a map of the equipotentials we can then proceed, if we wish, to draw in the lines of force so that they cross them at right-angles everywhere.

This kind of procedure is used in designing vacuum tubes. On account of the small mass of an electron its path in such a tube is, to a first approximation, along an electric line of force (though this is of course modified by the action of any magnetic fields present, page 209).

The electron-volt

The energy gained by a charged particle accelerated by an

electric field in a vacuum depends only on its charge and the difference of potential through which it is accelerated. For instance, a proton and an electron (which have the same size of electric charge) gain the same kinetic energy; but the proton, having a larger mass than the electron, reaches a smaller speed. In atomic and nuclear calculations we are not much concerned to know the speed; only the energy is important. It is therefore convenient in such applications to use a special unit of energy adapted for this purpose; it is called the *electron-volt* (eV). Thus an electron (or proton) accelerated between two points A and B, between which there is a potential difference of 100 V gains 100 eV of kinetic energy. An α-particle (with a charge $2e$) gains 200 eV of kinetic energy in moving between the same points; and an ion with a charge of $3e$ gains 300 eV; and so on. Since the electronic charge $e = 1.60 \times 10^{-19}$ C,

$$1 \text{ eV} = (1.60 \times 10^{-19} \text{ C})(1 \text{ V}) = 1.60 \times 10^{-19} \text{ J}$$

Some idea of the magnitude of the eV may be gained if we say that the energy of vibration of an atom of a solid at room temperature (say, an atom on the page you are reading) is about 0.03 eV ($= 4.8 \times 10^{-21}$ J). The energy converted in a chemical reaction is typically about 2 eV per reacting molecule. By contrast the energy converted in a nuclear process (e.g. radioactivity, page 266) is typically about 5 MeV. If we need to accelerate charged particles to produce nuclear reactions in the target they strike, we must use p.d.s of many millions of volts. For instance we might use a van de Graaff generator (page 199) to produce a p.d. of 4×10^6 V. If this is used to accelerate helium nuclei (α-particles), the kinetic energy E_k of a nucleus is given by

$$E_k = (2e)(4 \times 10^6 \text{ V}) = 8 \text{ MeV}$$

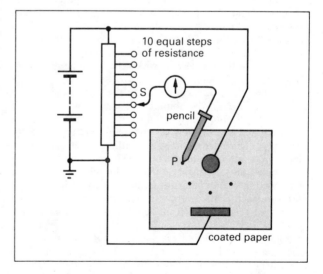

Figure 15.6 Plotting the equipotential lines between two electrodes on paper coated with a uniform resistance layer. (The model electrodes represent a metal rod parallel to a metal plate.)

15.3 Millikan's experiment

The principle of this experiment is to measure the electrical force acting on a charged oil drop in a known electric field, and so to find the charge it carries. The drops used are of microscopic size, and the charges they carry are many orders of magnitude smaller than the smallest charge that could be measured by any other means.

In 1909 the American physicist R.A. Millikan found that the charges on such oil drops are always integral multiples of a certain smallest quantity of charge e; i.e. the charges obtained were

$$\pm e, \ \pm 2e, \ \pm 3e, \ \ldots . \text{ and so on}$$

The charge was never (for instance) $0.6e$, $1.3e$, $4.7e$, or any other fractional multiple. The experiment demonstrates conclusively that electric charge exists in indivisible packets of size e. Millikan assumed that e was the electronic charge; and this is in agreement with the results of other methods of measuring e. Millikan's experiment has not proved eventually to be the most precise method available. But other methods do no more than measure the *average* charge of electrons. It is the special achievement of Millikan's experiment that it demonstrates that all electrons (and presumably protons too) carry the *same* charge.

A simplified form of Millikan's apparatus is shown in figure 15.7. The electric field is produced by a steady p.d. applied between two horizontal plates P and Q (a); these are held exactly parallel by an accurately-made spacer ring of insulating material. The ring has two windows let into it. The upper plate has a small hole in the centre through which oil drops are allowed to fall from a spray. Frictional effects in the nozzle of the spray result in at least some of the oil drops being charged. A beam of light is concentrated through one window at the side into the space between the plates; and a low-power microscope at the other window (b) is used to observe the drops by means of the light reflected from them. They are seen as sharp points of light against a relatively dark background. There is a transparent scale fixed in the eyepiece by which the distances the drops move can be measured.

When the p.d. is connected to the plates, the motion of some of the slowly falling drops is reversed because of the electrical forces acting on them; a suitable drop is selected, and by switching off and on alternately it may be held near the centre of the space until all other drops have landed on one plate or the other. Measurements are now conducted on this single oil drop. The oil must be of the type used in vacuum apparatus; this has a very low vapour pressure so that the evaporation of the drop is slow, and its weight remains practically constant for a considerable time.

First we have to find the weight W of the drop. With the electric field switched off (and the plates connected together) the drop is allowed to fall under the action of gravity. Using a stop-watch the time is measured for the spot of light to move between two selected divisions of the eyepiece scale; hence the speed v of the drop is obtained.

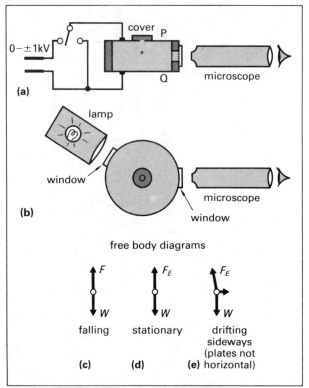

Figure 15.7 A simplified modern form of Millikan's apparatus for measuring the charge on an oil drop. Underneath are the free-body diagrams showing the forces acting on the oil drop: (c) when it is falling freely at its terminal speed, (d) when it is stationary in the electric field, (e) when it is drifting sideways through the plates not being horizontal.

For instance, in a particular case we find that a drop takes 15.0 s to fall 1.00 mm. So we have

$$v = \frac{1.00 \times 10^{-3} \text{ m}}{15.0 \text{ s}} = 6.67 \times 10^{-5} \text{ m s}^{-1}$$

The drop is so small that it reaches its terminal speed (page 67) almost at once; as the drop is then in equilibrium, the viscous force F of the air is equal in magnitude to the weight W (figure 15.7c). The viscous force F acting on a sphere of radius a moving with speed v through a fluid is given by

$$F = 6\pi a \eta v \qquad \text{(page 67)}$$

where η is the viscosity of the fluid (air in this case). The graph of figure 15.8 gives the values of η for air over the possible range of room temperatures. If ρ is the density of the oil, we have

$$W = (\text{volume of drop}) \times \rho g$$
$$= \tfrac{4}{3}\pi a^3 \rho g$$

Since the drop is moving at its terminal speed we have

$$F = W$$

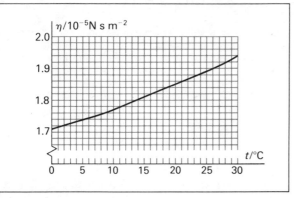

Figure 15.8 The variation of the viscosity η of air with temperature, needed for calculating the weight of the oil drops in Millikan's experiment.

Hence $\quad \frac{4}{3}\pi a^3 \rho g = 6\pi a \eta v$

$\Rightarrow \qquad a^2 = \left(\frac{9\eta}{2\rho g}\right) v \qquad\qquad (1)$

Hence a may be found, and so we may calculate W. For instance, with the drop considered above, if room temperature was 18°C, from the graph we take $\eta = 1.83 \times 10^{-5}\,\text{N s m}^{-2}$. Suppose the density of the oil was $0.920 \times 10^3\,\text{kg m}^{-3}$, substituting in equation (1), we obtain

$$a^2 = \frac{9(1.83 \times 10^{-5}\,\text{N s m}^{-2})(6.67 \times 10^{-5}\,\text{m s}^{-1})}{2(0.920 \times 10^3\,\text{kg m}^{-3})(9.81\,\text{m s}^{-2})}$$

$$= 60.9 \times 10^{-14}\,\text{m}^2$$

$\Rightarrow \quad a = 7.80 \times 10^{-7}\,\text{m}$

(This is about equal to the wavelength of red light – it is indeed a *small* drop!)

Hence $\quad W = \frac{4}{3}\pi(7.80 \times 10^{-7}\,\text{m})^3\,(0.920 \times 10^3\,\text{kg m}^{-3}) \times$
$\qquad\qquad (9.81\,\text{m s}^{-2})$

$$= 1.79 \times 10^{-14}\,\text{N}$$

We can now proceed to measure the charge Q on the drop. The field is switched on in such a direction as to oppose the weight of the drop. The p.d. V between the plates is then adjusted until the drop is held stationary. The drop is then again in equilibrium and the electrical force F_E is equal in magnitude to its weight (figure 15.7d). This will be possible only if the plates are exactly horizontal; otherwise the electrical force has a horizontal component (figure 15.7e), and the drop soon drifts out of the field of view. If the electric field strength is E when the drop is stationary, we have

$$F_E = W$$

But $\quad F_E = QE = Q\left(\frac{V}{d}\right)$

Hence $\quad \dfrac{QV}{d} = W$

$\Rightarrow \qquad Q = \dfrac{Wd}{V}$

If with the oil drop we have been considering the p.d. needed to hold the drop stationary was 670 V, with the top plate negative, and the plate separation d was 12.0 mm, then

$$Q = +\frac{(1.79 \times 10^{-14}\,\text{N})(12.0 \times 10^{-3}\,\text{m})}{(670\,\text{V})}$$

$$= +3.21 \times 10^{-19}\,\text{C}$$

The charge on the drop can then be changed by holding a radioactive source nearby (a β-source is most effective); this slightly ionizes the air between the plates. The drop will soon collide with one or more ions and its charge is then changed – and suddenly a new value of V is needed to hold the drop stationary. With our specimen drop the next p.d. needed might be 1340 V (exactly twice as much as before); but this time the top plate might need to be positive. The charge on the drop was therefore -1.60×10^{-19} C. Subsequent values of the p.d. and charge (after further use of the radioactive source) could be of both signs, but always close to the values:

p.d./V	$\pm Q/10^{-19}$ C
1340	1.60 $=\ e$
670	3.21 $= 2e$
445	4.80 $= 3e$
335	6.41 $= 4e$

and so on

The charge on any drop is always found to be an integral multiple of the smallest charge obtained above. This smallest charge is taken to be the electronic charge e. The presently accepted value is

$$e = 1.602 \times 10^{-19}\,\text{C}$$

It is of some interest to note that the electrical force on just *one* electron in a quite moderate electric field (about 10^5 V m^{-1}) is equal to the gravitational pull of the whole Earth on *all* the molecules in the drop; in our oil drop above there must have been more than 10^9 molecules, each containing many protons, electrons, etc. We shall consider later (page 200) the relative magnitudes of the electrical and gravitational forces with which a proton attracts an electron; but clearly Millikan's experiment shows that the gravitational force is *very* small.

15.4 Permittivity

The strength E of the electric field between the plates of a capacitor is equal to the potential gradient across the gap (page 189). Thus

$$E = \frac{V}{d}$$

where V is the p.d. between the plates, and d is their separation. We need now to discover how the field strength E depends on the charges on the surfaces of the plates that produce the field. In the last chapter (page 177) we saw how the charge Q carried on the plates depends on their area A and separation d, giving

$$Q \propto \frac{AV}{d}$$

Re-arranging, we can write this as

$$\frac{Q}{A} \propto \frac{V}{d}$$

The quantity Q/A is the charge per unit area on each plate, and is referred to as the surface density of charge on the plates; it is denoted by the symbol σ (Greek letter 'sigma'), and is measured in $C\ m^{-2}$. On the other side of the proportionality, V/d is equal to the electric field strength E. Therefore we can write this as

$$\sigma \propto E$$

Thus the electric field strength E is directly proportional to the charge density σ. It is not affected by the separation d of the plates as long as the charge upon them remains constant. We may confirm this with the flexible test strip used previously. The plates are connected for a moment to a suitable high-voltage source and are then *isolated* from it. If the insulation is good enough, the charge on the plates then remains constant. The test strip is charged and held as usual in the gap to indicate variations in field strength. When the separation d of the plates is now varied, the deflection of the test strip stays the same, as long as d is not so great that the field can no longer be considered uniform in the gap.

It follows that the potential difference between the plates must *increase* with the separation in this experiment, if the potential gradient is to remain constant. If d is doubled, the p.d. must double also if E is to remain unchanged. This may be shown directly by joining the plates to the terminals of a leaf electrometer. When the plates are charged and isolated, the p.d. is then seen to increase as they are moved apart; and it falls when they are brought together again. (If the plates are kept *connected* to a constant high-voltage supply, the p.d. between them stays constant; and the field *decreases* as they are moved apart, since $E = V/d$; in this case the charge on the plates decreases.)

We can therefore express the field between parallel plates by

$$\sigma = \varepsilon E \qquad (2)$$

where ε (Greek letter 'epsilon') is a property of the insulator in the gap between the plates known as its *permittivity*.

The parallel-plate capacitor

A knowledge of the permittivity ε for a given insulator

enables us to write down an expression for the capacitance C of a parallel-plate capacitor with this insulator between the plates. Thus, writing equation (2) in terms of Q and V, we have

$$\frac{Q}{A} = \frac{\varepsilon V}{d}$$

But $C = \dfrac{Q}{V}$

$$\Rightarrow \quad C = \frac{\varepsilon A}{d} \qquad (3)$$

From this expression we see that the unit of permittivity is the $F\ m^{-1}$. Its value may be found for any insulator by measuring the capacitance of a capacitor of known dimensions made with this insulator (see below).

Permittivity is one of the properties possessed by empty space (a vacuum); for an electric field can exist in a vacuum, and we can use a vacuum as the insulator between the plates of a capacitor. The permittivity of a vacuum is denoted by ε_0, and we find that

$$\varepsilon_0 = 8.85 \times 10^{-12}\ F\ m^{-1}$$

This universal constant is also referred to as the *electric constant*.

▶ Estimate the charge density σ on the facing surfaces of a pair of parallel plates 20 mm apart across which there is a p.d. of 5 kV. Assume that the permittivity of air is the same as that of a vacuum. If the plates are 0.20 m square, estimate the total charge Q on one plate.

The field strength E in the gap is given by

$$E = \frac{V}{d} = \frac{5 \times 10^3\ V}{20 \times 10^{-3}\ m}$$
$$= 2.5 \times 10^5\ V\ m^{-1}$$
$$\Rightarrow \quad \sigma = \varepsilon_0 E = (8.84 \times 10^{-12}\ F\ m^{-1})(2.5 \times 10^5\ V\ m^{-1})$$
$$= 2.2 \times 10^{-6}\ C\ m^{-2}$$

The charge density will in fact be rather greater than this at the edges of the plates where the field is not uniform. But assuming σ is constant over the surface of a plate we have

$$Q = (2.2 \times 10^{-6}\ C\ m^{-2})(0.20\ m)^2$$
$$= 8.8 \times 10^{-8}\ C \approx 10^{-7}\ C \qquad ◀$$

The permittivity of all material insulators is greater than ε_0. In other words a greater charge density σ is required to produce a given field strength E in the gap between the plates for a material insulator than for a vacuum. The ratio $\varepsilon/\varepsilon_0$ is called the *relative permittivity* of the insulator, and is denoted by ε_r. Thus

$$\varepsilon = \varepsilon_r \varepsilon_0 \qquad (4)$$

(An older name for ε_r is the *dielectric constant* of the insulator.) Relative permittivity is a dimensionless quantity, and is exactly 1 for a vacuum (by definition). Table 15.1 shows the values of ε_r for a number of materials. For gases ε_r is very little more than 1. But for liquid and solid insulators it lies between 2 and 10 mostly. For a few materials it is vastly greater than this; but in these substances the simple law (2) breaks down, and it is an over simplification to describe the behaviour of the material by a single constant.

▶ Calculate the number of plates and mica sheets required to produce a mica capacitor (constructed as in figure 14.3b, page 178) of capacitance 2.0 nF. The mica sheets are of thickness 5.0×10^{-5} m, and the overlapping parts of the plates measure 15 mm by 25 mm.

Taking the relative permittivity of mica as 6.0, the capacitance C of a single pair of plates is given by

$$C = \frac{\varepsilon_r \varepsilon_0 A}{d}$$

$$= \frac{(6.0 \times 8.85 \times 10^{-12}\ \text{F m}^{-1})(15 \times 10^{-3}\ \text{m} \times 25 \times 10^{-3}\ \text{m})}{5.0 \times 10^{-5}\ \text{m}}$$

$$= 4.0 \times 10^{-10}\ \text{F}$$

When the plates are stacked as in figure 14.3b all the capacitors (formed by a pair of parallel plates enclosing a mica sheet) are in parallel. The number of such capacitors required is

$$\frac{2.0 \times 10^{-9}\ \text{F}}{4.0 \times 10^{-10}\ \text{F}} = 5$$

We therefore need 5 mica sheets and 6 plates to enclose them. ◀

The measurement of ε_0

This is a matter of finding the capacitance C of a parallel-plate capacitor of known dimensions. Ideally the insulator between the plates should be a vacuum, but the difference between air and a vacuum will be undetectable with simple apparatus. The capacitor may be formed from two flat rectangular plates. The upper one rests on a set of small accurately ground insulating spacers on top of the lower plate. The capacitance we need to measure is that between these plates. But inevitably there are 'stray' capacitances included in the arrangement – between the non-earthed plate and earthed objects round about, and between the connecting wires of the apparatus; and we must allow for these.

The simplest way to measure C is with the vibrating-reed capacitance meter (page 180). The p.d. V and frequency of vibration f should be as large as the design of the apparatus allows. Even then the current I may be only a few µA (since the capacitance is only about 100 pF), and a sensitive meter must be used. The total capacitance is then given by

$$C = \frac{I}{fV}$$

But this includes the stray capacitance C_0 as well as the capacitance between the two plates that is given by equation (3) above. These two capacitances are in parallel, and we therefore write

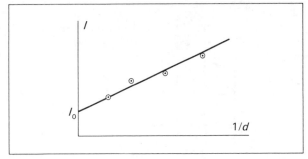

Figure 15.9 Measuring the permittivity of air in a parallel-plate capacitor; the plot of current I in the vibrating-reed capacitance meter against reciprocal of plate separation d.

$$\frac{I}{fV} = \frac{\varepsilon_0 A}{d} + C_0$$

To eliminate the stray capacitance C_0 from the calculations we repeat the measurements for a range of values of gap width d. A graph is then plotted of I against $1/d$, as in figure 15.9. The intercept on the axis of I gives the value of I_0, from which C_0 ($=I_0/fV$) may be found; and the gradient of the line is equal to $\varepsilon_0 Af V$ (appendix B2). Hence we may calculate ε_0. The straight line obtained in this experiment confirms that the capacitance is indeed inversely proportional to d.

Fields in insulators

In atomic terms we may explain the effect of an insulator on an electric field as follows. In an insulator the charges are *bound* in the sense that they cannot move away from the site in the crystal lattice to which they belong; but they are capable of very small amounts of movement under the action of an electric field, the positive charges (atomic nuclei) in one direction and the negative ones (electrons) in the other. Suppose we have a slab of insulator between two parallel plates (figure 15.10). When a p.d. is applied between them, the electric field acts on the bound charges

substance	ε_r
gases:	
air	1.000 6
hydrogen	1.000 3
helium	1.000 07
non-polar substances:	
polythene	2.4
perspex	3.3
glass	4 to 7.5
mica	6.0
paper	2.7
polar substances:	
water	80
ethanol	27
glycerine	43

Table 15.1 The relative permittivities of various dielectrics.

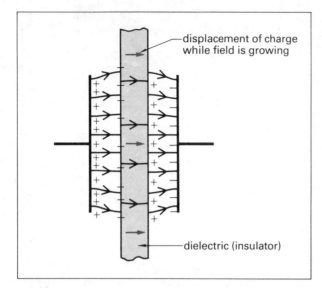

displacement of charge
while field is growing

dielectric (insulator)

Figure 15.10 The polarisation of a dielectric by the displacement of the bound charges within it, leading to a reduction of the field inside it.

in the insulator slightly displacing them. Thus there is a small pulse of electric current in the insulator while the electric field is changing; as soon as the field is steady, no further displacement of the charges takes place. The effect of this is to produce two layers of bound charges on the surfaces of the insulator; the field of these layers of charge partly cancels out the electric field in the slab. The bound charge on the surface of the insulator is always less than that on the neighbouring metal plate, so that the field in the insulator is never completely cancelled.

The effect of an insulating medium in the gap is thus always to *reduce* the field compared with its value in a vacuum; or, to put it the other way round, a greater density of charge σ is required on the plates to produce a given field strength E in the insulator. The permittivity ε of all insulating materials is therefore greater than ε_0. Although a steady current cannot exist in an insulator, an electric field can; in this connection an insulator is often referred to as a *dielectric*. When the charges in a dielectric are displaced by an electric field, we say that it is *polarised*.

The polarisation of a dielectric can take place by two distinct mechanisms. In many materials the molecules are initially uncharged, i.e. the effective centres of the electron clouds in them coincide with the centres of positive charge. When an electric field is applied, these centres of charge are strained slightly (in opposite directions) from their normal positions, and so the dielectric is polarised. But in some substances the effective centres of negative and positive charge do not coincide, and the molecule has a built-in polarisation; these are called *polar* molecules. (Water and ethanol are examples.) In the absence of an electric field the electric axes of polar molecules are orientated at random, and no polarisation of the material as a whole is observed.

But when a field is applied the molecular axes tend to line up with the field. This tendency is opposed by the thermal agitation of the molecules, which tends to keep them pointing at random; so the lining up is never more than partial. However, the relative permittivity of such materials is always much greater than those of non-polar substances (see table 15.1); also the permittivity of a polar substance is much greater at reduced temperatures.

It is worth contrasting the effect of a sheet of metal on an electric field with that of a sheet of insulator. In the insulator the field is *reduced* by the displacement of charges within it. But in the metal some of the charges are completely free to move; under the action of an electric field they continue in motion until they have taken up positions that reduce the field inside the metal everywhere to *zero*. Only while there is a current in a conductor can there be any electric field inside it. When a sheet of metal replaces the slab of insulator between the parallel plates of figure 15.10, the charges induced on the surfaces of the sheet are *equal* to those on the neighbouring plates, thus reducing the field inside the sheet to zero.

Electric flux

The electric field strength E between the plates of a capacitor in a vacuum is given by

$$\sigma = \varepsilon_0 E$$

(and for practical purposes the same expression applies for *air* between the plates). The right-hand side of the equation is called the *flux density* of the electric field. Thus

electric flux density $= \varepsilon_0 E$

In a non-uniform field E varies from point to point, and so therefore does the flux density. But at the surface of a charged conductor the same result is found to apply, namely that the flux density is equal to the surface density of charge σ on which it arises.

The picture of electric fields that we are developing is this. The lines of force of a field start on positive charge, move out through the surrounding space and end on negative charge somewhere else. Let us think about the way the field varies as we move along the lines of force from beginning to end. Consider a small area ΔA on the surface of a charged conductor (figure 15.11). We imagine all the lines of force that can be drawn from the points on the perimeter of the area ΔA; these mark out a kind of tube of variable cross-section through the insulator. Where the tube widens out the field is less concentrated and the flux density decreases; where it contracts the flux density increases. It can be shown that the product:

(electric flux density) $\times$ (area of cross-section)

is constant for the tube. It is called the electric *flux* Ψ of the field in the tube. At the positive end of the tube Ψ is given by

$$\Psi = \varepsilon_0 E \Delta A = \sigma \Delta A$$

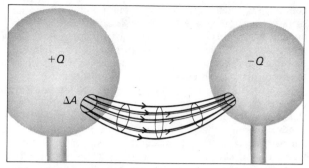

Figure 15.11 Electric flux Ψ starts on the positive charge (on the area ΔA), spreads out into the surrounding space, and ends on an equal negative charge on another body.

Now $\sigma\Delta A$ is the charge on the small area ΔA. The electric flux Ψ is thus equal to the electric charge on which it arises; and this is the flux through all cross-sections of the tube we have considered.

Electric flux always behaves like this. A body carrying a charge Q is the source of electric flux Ψ and always we have $\Psi = Q$. This flux is directed outwards into the space around the charged body, and there is the same total flux Ψ out through any imaginary surface surrounding the body; this flux ends on an equal and opposite charge $-Q$ on another body (or bodies) nearby. This result is known as *Gauss's theorem*, and gives us useful insights into the nature of electric fields around conductors. We shall see how to make use of this idea in actual cases below.

15.5 Fields near conductors

In the absence of any source of e.m.f. producing a continuous supply of electrical energy, the charges in a system of conductors reach equilibrium under the action of the electric field in a very short time (less than 10^{-12} s usually). When this has happened the field inside the material of a conductor must be everywhere zero. If not, the free charges in it will move under the action of the field until the distribution of charge is such as to make this so. Only if there is a current is the field inside a conductor non-zero. It follows that the potential inside the material of a conductor (with no current in it) is everywhere the same. Only when there is a current in a conductor can the potential vary through it, decreasing in the direction of the current.

By similar reasoning the field at the surface of a conductor must be everywhere perpendicular to the surface. If it was not, it would have a component parallel to the surface and would set in motion the free charges in the surface. Only where there is a current in the conductor does the field have a component parallel to the surface.

A hollow conductor

The charge on a hollow conductor is entirely on its *outside*

surface. There is no charge at all on the inside surface of the hollow space. Strictly speaking this applies only for a *completely* closed-in hollow space, but in practice it is sufficiently clearly demonstrated by using a deep metal can on an insulating support. The can is given a charge by raising it to a high potential. We now make a test of the charge on different parts of it with a proof-plane (figure 15.12a). When placed on a conducting surface the proof-plane becomes for a moment electrically part of the surface, and collects a sample of the charge (if any) upon it. The charge may be tested by touching the proof-plane on the terminal of an electrometer.

We find that no charge can be collected from the *inside* of the can, provided the proof-plane is not touched too near the lip of it. The charge resides entirely on the outside. (Warning: you need to make sure there is no frictional charge on the handle of the proof-plane for this to work, page 187.)

If there is no charge on the inside surface of a hollow space in a conductor, it follows that the electric field must be zero everywhere inside the hollow space. For an electric field can arise only from a charge; and if there is no charge inside, there is no place where electric flux could start or end. The point may be tested with a leaf electrometer. In general, if the electrometer is placed in an electric field, a difference of potential is created between cap and case, and the leaf deflects. For instance a frictionally charged plastic rod produces a large electric field all around it, and causes the leaf of the electrometer to deflect when the rod is brought anywhere near the cap. A leaf electrometer and cap is thus an indicator of the presence of an E-field. But if we place the electrometer in a closed metal box

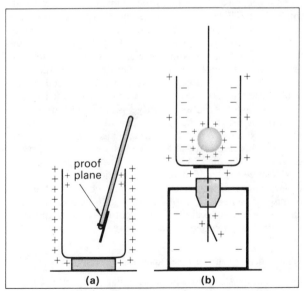

Figure 15.12 (a) No charge can be collected from the inside surface of a hollow charged conductor. (b) Demonstrating that the charge induced on the inside of the metal can is equal to the inducing charge.

(with windows of metal gauze through which the instrument can be observed), we find no deflection of the leaf in any circumstances. Large electric fields may be produced near the box; or it may be stood on an insulating slab and raised to a high potential. In no case do we observe any deflection. There cannot therefore be any electric field inside the hollow space.

The result is important in the design of high-voltage apparatus. A hollow metal compartment may be raised to a high potential, and any apparatus or observers inside it are not in any way affected. There is *no* E-field inside, although there are large E-fields outside. The ability of a closed metal screen to shield apparatus inside it from stray electric fields outside (and vice versa) is also made use of in high-frequency radio equipment. Many of the components (coils, etc) are mounted inside metal cans; without these the fields between different parts of the circuit would make its behaviour unpredictable. At high frequencies the cans also provide shielding from stray magnetic fields (page 225).

Induced charges

If an electric charge is insulated on some object, the electric flux that starts on it must end on an equal and opposite charge at some other place. Thus an insulated electric charge always causes an equal and opposite charge to flow onto other objects in its surroundings. The latter charges are referred to as *induced charges*; induced and inducing charges are always equal in size.

To test this prediction a can may be connected to a leaf electrometer (figure 15.12b). We then lower a charged metal object on an insulating support inside the can without touching it. This induces an opposite charge on the inside surface, and leaves a charge of the same sign as the original inducing charge on the outside surface; a proportion of this charge is shared with the electrometer. The flow of charge raises the potential of the electrometer cap and leaf, and we observe a deflection. If we move the charged object about inside the can, we observe no change in the deflection, showing that the induced charge is not affected by the position of the inducing charge, provided it is well inside the can. Now we *touch* the charged object on the inside of the can; this is the crucial test. We know from the previous experiment (figure 15.12a) that the total charge inside the can is now zero, and careful observation shows that the deflection of the electrometer *does not change at* the moment of making contact. It follows that the induced and inducing charges exactly cancel at the moment of contact, and that therefore they were equal and opposite.

Charge density and surface curvature

On a charged conductor of irregular shape the charge density varies from point to point of its surface. A simple test with a proof-plane and electrometer shows that the charge density is always greatest at the most highly curved convex parts of the surface. Thus the charge is found to concentrate chiefly near corners and points, where the surface curvature is greatest. Samples of charge may be taken from various parts of the charged conductor with the proof-plane and then transferred to the electrometer. We find that the deflection of the electrometer is greatest when the proof-plane has been touched on the most highly curved parts of the surface, less on any flat parts, and least on any concave part (almost zero in any deep concavity). The charge density also depends on the distribution of other conductors and charged objects in the neighbourhood. An earthed conductor (the hand) placed nearby greatly increases the charge density on the part of the surface nearest to it; and a charged object held close to the conductor may even produce a change of sign of the charge at some part of the surface, although the potential must be the same all over the conductor.

A consequence of this concentration of charge at the most curved parts of a conductor is that it is possible, without using very high p.d.s, to produce a very large field locally near a sharp point or fine wire, because the field strength E at the surface of the conductor is proportional to the charge density σ. Near a sharp point or fine wire it may even be sufficient to ionise the air there (which requires a field strength of 3 MV m^{-1}). Those ions that have the same sign of charge as the conductor are violently repelled from it, giving rise to an appreciable electric 'wind'. If we hold a lighted candle near a sharp needle raised to a high potential, the wind can be sufficient to blow the flame

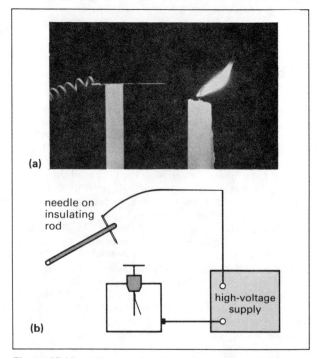

Figure 15.13 (a) The electric wind from a needle at a high potential. (b) The electrometer is charged by the electric wind from the needle.

horizontal (figure 15.13a). It carries away the charge on the needle into the surrounding air, and eventually to earthed objects nearby. This is called a *point discharge* or *corona discharge*.

Any object placed in the way of the 'wind' may collect some of the charge on its surface. We may show this by holding a sharp needle connected to a high-voltage source near the cap of an electrometer (figure 15.13b); we then observe a rapidly increasing deflection, as the charge blows onto the cap.

Corona discharge has to be allowed for in the design of high-voltage equipment. All parts at a high potential must be gently curved or covered with smoothly curved screens. All sharp corners or protruding ends of wire must be eliminated. This feature may be noticed in the high-voltage parts of a television set. Similarly, the wires used for high-voltage power transmission must not be too thin; otherwise the power loss into the air through corona discharge becomes significant.

The van de Graaff generator

This is a form of electrical machine that is much used to provide the large p.d.s needed for atomic particle accelerators (page 248); the largest models give p.d.s. up to 10 MV. Figure 15.14 shows the principle of the machine. Positive charge is sprayed by corona discharge from a row of points K onto a moving belt of insulating material (usually rubberised silk). This is carried up into the hollow conductor C and induces a negative charge on the inside of it. This negative charge is sprayed by the comb L onto the belt,

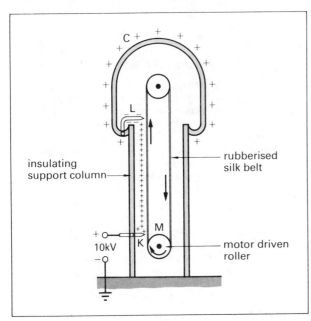

Figure 15.14 A van de Graaff generator, in which charge sprayed on the belt from the comb K is carried to the upper dome, raising it to a high potential.

neutralising the positive charge. The net result is the continuous transfer of positive charge from the points K to the belt and so to the upper conductor. A considerable tension is set up in the belt owing to the repulsion between the charge on the upper conductor and the charge being carried up on the belt. The electrical energy is supplied in the form of work done against this repulsion by the motor the drives the lower roller M. The maximum potential reached depends on the quality of the insulation round the machine.

In the simplified type of van de Graaff machine often used in schools the row of points K is earthed, and the action depends on the frictional charge developed on the inside of the belt by its contact with one roller (which is made of insulating material). This charge induces an opposite charge on the row of points K, and this is then sprayed onto the outside of the belt and carried up to the hollow conductor C. But even these simple machines may produce p.d.s up to 0.5 MV in the best conditions.

15.6 Radial fields

An isolated charged particle

The electric field of an isolated particle must have spherical symmetry. The total electric flux Ψ coming from the particle is equal to the charge Q that it carries. Imagine a spherical surface of radius r concentric with the particle, and suppose that the strength of the field at the distance r (in air) from the particle is E. Then the electric flux Ψ through the spherical surface is given by

$$\Psi = \varepsilon_0 E (4\pi r^2)$$

But $\quad \Psi = Q$

$$\Rightarrow \quad E = \frac{Q}{4\pi\varepsilon_0 r^2}$$

If a second particle with charge Q' is placed in this field at the distance r from the first one, the force F which each exerts on the other is given by

$$F = Q'E = \frac{QQ'}{4\pi\varepsilon_0 r^2}$$

The force acting between two particles thus varies inversely with the square of their distance apart; if the distance is doubled, the force falls to a quarter as much, and so on. This result is known as the *inverse square law* for electric charges.

We may check the law approximately by measuring the force F which two charged metallised pith balls exert on each other at various distances apart. One of these is fixed on an insulating support, while the other is suspended from a pair of long nylon threads (figure 15.15a). If the pith balls carry charges of the *same* sign, the repulsion

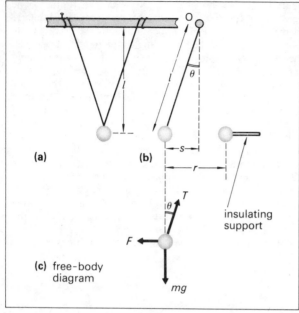

(a)

(b)

insulating
support

**(c) free-body
diagram**

mg

Figure 15.15 Testing the inverse square law for the force F between small charged particles; the free-body diagram shows the forces acting on the suspended pith ball.

between them deflects the suspended pith ball; the sideways displacement of this is best measured by observing its shadow cast on a screen by a small light source. If the displacement is small compared with the length l of the support system, the force F is proportional to the *horizontal* movement s of the pith ball (figure 15.15b). In figure 15.15c the three forces acting on the suspended pith ball are shown; T is the pull of the thread on the pith ball. Resolving the forces horizontally and vertically, we have

$$F = T \sin \theta = T \frac{s}{l}$$

and $mg = T \cos \theta \approx T$ (if θ is small)

dividing $\dfrac{F}{mg} \approx \dfrac{s}{l}$

i.e. $F \propto s$ (if θ is small)

To check the law we therefore need to show that the horizontal movement s is inversely proportional to the square of the distance apart r of the pith balls; and both these distances we find from the positions of the shadows on the screen. Leakage of charge from the pith balls produces uncertainties; but with care we can show that a graph of s against $1/r^2$ is approximately a straight line.

The best tests of the law are indirect ones. It can be shown that Gauss's theorem, and all the results deduced from it, apply only if the inverse square law is true. For instance the absence of any field inside a hollow closed conductor is a very sensitive test of the inverse square law.

The hydrogen atom

In a hydrogen atom the average distance apart of the proton and the electron is 5.3×10^{-11} m. It is interesting to work out the strength of the E-field of the proton at this distance from it. The charge of the proton is the electronic charge e ($= 1.60 \times 10^{-19}$ C). Therefore

$$E = \frac{1.60 \times 10^{-19} \text{ C}}{4\pi (8.85 \times 10^{-12} \text{ F m}^{-1})(5.3 \times 10^{-11} \text{ m})^2}$$

$$= 5.1 \times 10^{11} \text{ V m}^{-1}$$

This is an enormous field strength; compare it with the maximum E-field in air (3×10^6 V m^{-1}) or in the very strongest insulators (about 10^9 V m^{-1}).

The force F acting on the electron in this field is given by

$$F = (1.60 \times 10^{-19} \text{ C})(5.1 \times 10^{11} \text{ V m}^{-1})$$

$$= 8.2 \times 10^{-8} \text{ N}$$

This may not seem a very large force. But it is vastly greater than the gravitational force with which the proton attracts the electron (page 160). You can work out that the gravitational force at this separation of the particles is only 4.1×10^{-47} N, which is less than the electrical force by a factor of 2×10^{39}.

Bearing in mind the small mass of an electron (9.1×10^{-31} kg), the electrical force pulling it in a hydrogen atom is very large. The acceleration a of the electron in its orbit is given by Newton's second law as

$$a = \frac{F}{m_e}$$

$$= \frac{8.2 \times 10^{-8} \text{ N}}{9.1 \times 10^{-31} \text{ kg}}$$

$$= 9.0 \times 10^{22} \text{ m s}^{-2} \approx 10^{22} \text{ g}$$

The electric potential near a point charge

In this case it is convenient to take our zero of potential at an *infinite distance* from the isolated point charge Q. In practical terms this means that we are assuming that the earthed walls and floor of the laboratory are at a sufficient distance for their effects to be ignored.

The potential V at a distance r (in air) from a particle with a charge Q may be derived from the expression for the electric field strength near the charge:

$$E = \frac{Q}{4\pi\varepsilon_0 r^2}$$

It is necessary to find an expression for the potential V such that the potential gradient is equal at every point to E as given above. This may be done using calculus; but we shall just quote the result here:

$$V = \frac{Q}{4\pi\varepsilon_0 r} \quad \text{(compare the calculation on page 164)}$$

Thus the potential varies inversely with the distance r from the point charge. It falls to zero at an infinite distance from the point.

The potential V at a distance of 5.3×10^{-11} m from a proton is given by

$$V = \frac{1.60 \times 10^{-19} \text{ C}}{4\pi (8.85 \times 10^{-12} \text{ F m}^{-1})(5.3 \times 10^{-11} \text{ m})}$$

$$= 27 \text{ V}$$

An electron moving from a large distance away to its normal average distance in a hydrogen atom therefore loses 27 eV of electrical potential energy. It can be shown that exactly *half* this energy is converted to the kinetic energy of the electron in its orbit, so that the net loss of energy if the electron is to remain in its normal orbit must be 13.5 eV. Conversely, this amount of energy must be given to a hydrogen atom to detach the electron from it; in other words the first ionisation potential of hydrogen is 13.5 V (page 246).

A conducting sphere

Again, the field has spherical symmetry; and the total electric flux arising from the sphere is equal to the charge Q it carries. *Outside the sphere* the E-field is exactly the same as though the charge Q carried on its surface was concentrated at its centre. The field strength E at a distance r from the centre of the sphere (in air) is therefore given by

$$E = \frac{Q}{4\pi \varepsilon_0 r^2}$$

and the potential V is given by

$$V = \frac{Q}{4\pi \varepsilon_0 r}$$

(again taking the zero of potential at an infinite distance). *Inside the sphere* (a hollow space) the E-field is zero. If the sphere is of radius r_s, then at its surface the field E_s and potential V_s are given by

$$E_s = \frac{Q}{4\pi \varepsilon_0 r_s^2}$$

$$V_s = \frac{Q}{4\pi \varepsilon_0 r_s}$$

The potential has the same value as this throughout the interior of the sphere, since there is no E-field inside the sphere, and therefore no potential gradient. Figure 15.16 shows how E and V vary with r near a charged conducting sphere.

We may investigate the variation of V with r for a conducting sphere with the aid of a flame-probe (page 191). The sphere (e.g. a metallised football) needs to be suspended on insulating threads in the centre of a room, well away from walls, ceiling, floor, bench top, etc. It is joined by a fine wire (producing only a small field itself) to one terminal of a high-voltage source, the other terminal of this being earthed. The flame-probe is then joined by a long, well-insulated lead to a calibrated leaf electrometer. The potential

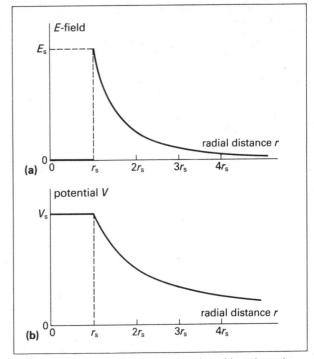

Figure 15.16 The electric field inside and outside a charged sphere: (a) the E-field plotted against r; (b) the potential V plotted against r. Inside the sphere the E-field is zero and the potential constant.

may be measured reliably with the electrometer and flame-probe from close to the sphere up to about 2 m from it, provided the experimenter keeps his own person well away from the probe and is very careful about the insulation of the wires going from probe to electrometer.

The capacitance of an isolated sphere

A conducting sphere carrying a charge $+Q$ necessarily induces a charge $-Q$ on its surroundings. We have assumed above that this induced charge $-Q$ is at a great distance, and have ignored any effect caused by it, but its existence is always implied. When we talk about an isolated object like this as a capacitor, we are implying that the other 'plate' of the capacitor is another conductor completely surrounding the sphere and at a very great distance from it. The potential V of the isolated sphere is therefore the p.d. across the capacitor formed by the sphere and the earthed conductor at 'infinity' that forms the other electrode of the system – in practice the walls, floor and ceiling of the laboratory.

Since $\quad V_s = \dfrac{Q}{4\pi \varepsilon_0 r_s}$

the capacitance C (in air) of the sphere is given by

$$C = \frac{Q}{V} = 4\pi \varepsilon_0 r_s$$

For instance suppose the metallised football used above was of diameter 0.20 m (i.e. of radius 0.10 m). Then we have

$$C = 4\pi(8.85 \times 10^{-12}\ \text{F m}^{-1})(0.10\ \text{m})$$
$$= 1.1 \times 10^{-11}\ \text{F} = 11\ \text{pF}$$

Charged to a potential of 1000 V, the charge Q stored on the sphere would be given by

$$Q = CV = 1.1 \times 10^{-8}\ \text{C}$$

This is well within the range of charges that may be measured with an electrometer (page 182). It is therefore possible to use an electrometer to check the expression for the capacitance of an isolated sphere for a range of spherical objects of differing radii, so confirming the above theoretical analysis.

16 Magnetic forces

16.1 Magnets

A magnet influences the space around it so that forces act on any other magnet placed nearby. Any such region of space in which a magnet would be acted on by forces is said to contain a *magnetic field*. The magnetic properties of a magnet appear to be concentrated in certain parts of it only. Thus if a simple bar magnet is plunged into a bowl of iron filings scarcely any stick to the central parts of the bar, while many filings cluster round its ends.

Likewise the forces on a pivoted magnet when another magnet is brought near appear to act chiefly at its ends. The forces on the two ends are in opposite directions. If one end is *attracted* to the magnet brought up to it, the other is *repelled*. A compass is just a very small pivoted magnet, used to indicate the direction of the magnetic forces. When it is placed in a *uniform* magnetic field the forces on it form a *couple* (i.e. a pair of equal, opposite and parallel forces (figure 16.1). This couple tends to turn the compass until its *axis* is parallel to the forces and to the field. In a non-uniform field the pivoted magnet not only turns but is pulled in some direction as well, since the forces on its ends differ in size or line of action.

The apparent concentration of the magnetic properties of a bar magnet at its ends is an illusion. In reality all the material in the magnet contributes to the strength of the magnet. This may be seen if a strip of magnetized material is broken in two (figure 16.2). Before the break is made no iron filings will cling to the central part of the strip, but only to its ends. But the filings cling to the new ends formed by the break just as much as to the original ends. This shows that the central parts of the magnet were magnetised

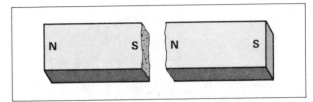

Figure 16.2 The magnetisation of the centre of a magnetised strip is revealed as soon as the strip is broken.

before, but this was not apparent outside the magnet until the break was made. We can imagine the breaking up of a magnet continued in this way to the limit, until the specimen is broken down into its constituent atoms. We would expect to find that the individual atom is magnetised and behaves as though it were a small magnet. This may in fact be demonstrated directly with a narrow beam of atoms in a vacuum. If the beam is passed through a highly non-uniform magnetic field, a resultant force acts on each atom (in addition to any couple rotating it), and the beam is deflected out of a straight line.

A coil of wire carrying an electric current behaves in a similar way to a magnet. Placed in a magnetic field forces act on it that tend to turn it so that its axis is parallel to the field; and in a non-uniform field there is a resultant force acting on it tending to move it as a whole in some direction. Likewise a coil of wire with a current in it produces a magnetic field similar to that of a magnet of the same shape.

The modern picture of the atom with its circulating systems of electrons therefore leads us to expect that magnetic effects will arise in the individual atom. We now understand that the magnetic properties of a magnet are simply the summed effect of the magnetic behaviour of the charged particles in the atoms of which it is composed. To a large extent the orbits of electrons in an atom form equal and opposite pairs cancelling out their magnetic effects; but with many atoms there is some residual magnetic effect, caused chiefly by the *spin* of the electrons. Even so, with most substances the effects observed are very small (page 232). But with *iron*, *nickel*, *cobalt* and a few others interactions between atoms produce intense magnetic properties. These are called *ferromagnetic* materials, and it is from these that magnets may be made.

The Earth's field

The Earth itself has a weak magnetic field. This is caused by electric currents circulating within its core (figure 16.3a). The currents are probably generated by convection in the liquid core maintained by the conversion of nuclear energy

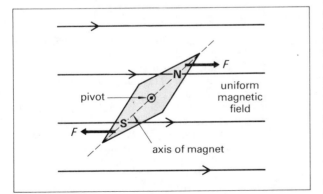

Figure 16.1 A magnet in a uniform field experiences a couple (a pair of equal, opposite and parallel forces); this tends to turn its axis in line with the field.

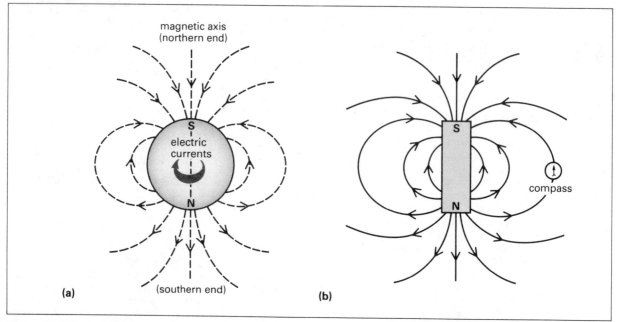

Figure 16.3 Magnetic fields: (a) of the Earth; (b) of a bar magnet. The Earth has a magnetic S pole at the northern end of its magnetic axis.

in radioactive substances in the Earth's interior. The direction of the Earth's field varies from place to place on its surface. It is vertical at the ends of the Earth's magnetic axis, and horizontal at the 'magnetic equator'. The *magnetic meridian* at a point is the vertical plane in which the Earth's magnetic field acts at the point. A freely suspended magnet (e.g. a compass) therefore tends to align itself so that its magnetic axis is in the magnetic meridian. This provides a simple way of distinguishing and labelling the two ends of a magnet (*poles* they are usually called, by analogy with the geographical terms): the north-seeking pole or N pole of a magnet is the one that is drawn towards the northern end of the Earth's magnetic axis; and the other end of the magnet is called its south-seeking pole or S pole. Simple experiments with pairs of magnets show that *like poles repel each other, while unlike poles attract each other.* (Notice that this implies that in the magnetised core of the Earth the *northern* end is a magnetic S pole.) The magnetic axis of the Earth is slowly changing its position. The direction of magnetisation of the rocks of the Earth's crust provide a kind of magnetic history of the Earth; when a rock first forms it is weakly magnetised in the direction of the local field at that time. This magnetisation is subsequently retained, and provides a record of the crustal movements of the Earth, as well as indicating changes that have occurred in the magnetisation of the Earth as a whole. The record shows that the Earth's magnetic field has reversed at irregular intervals (of half a million years or so) throughout the Earth's history. At present the northern end of the Earth's magnetic axis is located in northern Canada.

The *direction* of a magnetic field at any point is defined as the direction in which the N pole of a small freely pivoted magnet would point when placed there. Magnetic fields may be described pictorially by means of *lines of force* (in much the same way as electric fields, page 187). This kind of pattern is most easily revealed by sprinkling iron filings on a piece of card placed in a field. Figure 16.3b shows a sketch of the field of a simple bar magnet derived from an iron filings pattern of this sort. The directions of the arrows on the lines must of course be determined by a compass in the field. The arrows are therefore drawn emerging from the N pole of a magnet and entering the S pole.

The field of a current

The production of a magnetic field by an electric current was discovered by the Danish scientist Hans Christian Oersted in 1819, who detected the deflection of a compass placed near a conductor when an electric current was switched on and off in it. A descriptive picture of the fields produced by circuits of various shapes may be obtained by observing lines of force in the usual way with iron filings on a card. The diagrams for three cases are shown in figure 16.4. Notice in particular that the lines of force near a straight wire (a) are circles concentric with the wire; the field definitely has no component parallel to the wire or radially away from it. The direction of the arrow on the lines of force may be determined as usual with a small compass placed near the wire. This direction may be described by a *right-hand screw rule*:

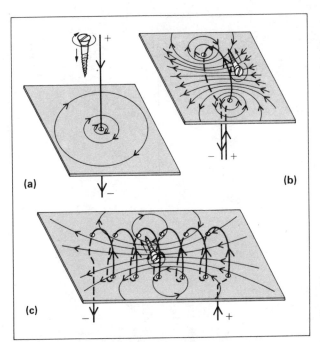

Figure 16.4 The magnetic field of an electric current: (a) a straight wire; (b) a plane coil; (c) a solenoid. The direction of the field is given in each case by the right-hand screw rule.

The direction of the lines of force round a current is that in which a right-hand screw would be turned in order to advance it in the direction of the current.

The same rule may be applied to describe the direction of the field of a *coil*. The imaginary right-hand screw may be placed, as shown, at the position of any section of the wire of the coil; this gives the direction of the field near this section, and enables us to predict the sense of the arrows on the lines of force through and near the coil.

In figure 16.4c with the current in the direction shown the arrows on the lines of force emerge from the left-hand end of the tubular coil of wire (or *solenoid*) and enter its right-hand end. The left-hand end of the solenoid therefore behaves like the N pole of a bar magnet, and the right-hand end like the S pole. Likewise the plane coil of figure 16.4b produces a field resembling that of a thin disc-shaped magnet with a N pole on the left-hand face and a S pole on the right.

16.2 The force on a current

When a straight length of wire carrying a current is placed at right-angles to a magnetic field, it is found that a force acts on it whose line of action is at right-angles to both the field and the current. The direction in which the force acts is traditionally described by *Fleming's left-hand rule*:

If the thumb and first two fingers of the left hand are put mutually at right-angles, and the First finger is pointed in the direction of the Field while the seCond finger is in the direction of the Current, then the thumb gives the direction of the force (figure 16.5).

This basic magnetic effect may be demonstrated with a simple *current balance* consisting of a rectangle of copper wire supported on a pair of razor blade knife edges. A magnetic field may be applied locally to one end of the rectangle by means of a suitable U-shaped magnet (figure 16.6a). When a current is passed through the copper wire via the knife edges, the balance is deflected up or down according to the directions of current and field, in agreement with Fleming's rule. The magnetic field may alternatively be produced by means of a current in a coil placed near the end of the balance, and the same rule is found to be obeyed. In every case the deflecting force is a maximum when the field is arranged to be at right-angles to the current, and it falls to zero if the field is parallel to the current (figure 16.6b).

The force may be measured by balancing it against the weight of a small rider on the end of the balance. If two or three identical U-shaped magnets are placed side by side under the end of the balance it may be shown that, for a given current, the force F is proportional to the length l of wire in the field. In a given field the force F is inevitably proportional to the current I in the wires of the balance, since precisely the same magnetic effect is used in the moving-coil meter that we employ to measure the current! However, the result is worth checking as a practical exercise. The force F also of course depends on the strength of the magnetic field, and may therefore be used as a means of measuring it. The quantity so obtained is called the *flux density* of the magnetic field, and is denoted by the symbol B. (The reason for using the name *flux density* will become apparent in the next chapter, page 225). We can therefore write

$$F = BIl$$

and use this equation to define B. The unit of magnetic flux density is the $N\ A^{-1}\ m^{-1}$; but it is convenient to have a separate name for this unit, and it is called the *tesla* (T), after Nikola Tesla (1856–1943), who in the 1890's pioneered

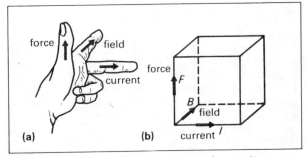

Figure 16.5 Fleming's left-hand rule for the force acting on an electric current in a magnetic field.

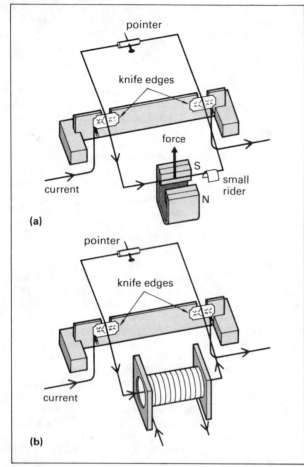

(a)

(b)

Figure 16.6 A current balance for measuring the force on a current-carrying conductor; the insulating rod that carries the pointer breaks the wire loop at that point, so that all the current passes through the other end of the balance. In (a) the field is at right-angles to the current (maximum force); in (b) the field is parallel to the current (no force).

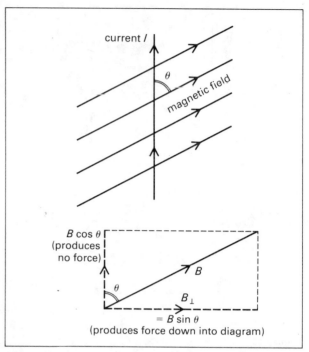

Figure 16.7 Only the component (or resolved part) of the field perpendicular to the current produces any force.

in America the development of electrical machinery and the use of a.c. systems.

Thus $1\text{ T} = 1\text{ N A}^{-1}\text{ m}^{-1}$

If a wire is placed parallel to the field the force on it is zero. In the general case the force can be regarded as being caused by the *component* or *resolved part* of the B-field *perpendicular* to the current; we denote this by the symbol $B_\perp$ ($= B \sin \theta$ in figure 16.7). The component of the B-field parallel to the current produces no force. The flux density B of a magnetic field is thus a *vector* quantity; flux densities can be summed, or resolved into components, like any other vector (page 3). Thus in figure 16.7 the force F is given by

$$F = B_\perp Il = BIl \sin \theta$$

Using a simple current balance like that in figure 16.6 we can now measure the flux density B of the various magnet and coil systems that we have used.

► Estimate the strength of the B-field in the gap between a pair of magnadur magnets on a steel yoke (as in figure 16.6a) from the following data: the length l of wire between the faces of the magnets is 50 mm; the mass of the small rider placed on the end of the balance is 0.35 g; and the current I then required to bring the balance level once more is 2.2 A.

The force F is equal in size to the pull of the Earth on the rider (i.e. its weight); taking $g = 10\text{ N kg}^{-1}$, we have

$$F = mg = (0.35 \times 10^{-3}\text{ kg})(10\text{ N kg}^{-1})$$

Hence $B = \dfrac{F}{Il}$

$$= \frac{(0.35 \times 10^{-3}\text{ kg})(10\text{ N kg}^{-1})}{(2.2\text{ A})(50 \times 10^{-3}\text{ m})}$$

$$= 3.2 \times 10^{-2}\text{ T} \approx 30\text{ mT}$$

The strength of the field depends on the width of the gap between the pole pieces of such a magnet and steel yoke system; the narrower the gap, the greater the field. The largest magnets with a very narrow gap can achieve a flux density of about 0.1 T (100 mT). ◄

With an *electromagnet* (an iron core with current-carrying coils wound on it) we can do rather better, achieving flux densities up to about 1 T; massive electromagnets with water-cooled coils can perhaps produce about 2 T. The maximum field attainable (at great expense) in physics laboratories is about 20 T. Such fields are, however, commonplace on the surfaces of some kinds of stars, and in the interiors of many kinds of atoms.

By contrast the flux density of the magnetic field of the Earth is about 50 μT, too small to measure with the simple current balance of figure 16.6. In Great Britain the Earth's field is inclined downwards at about 70° to the horizontal. A current balance arranged as in figure 16.6 (with the end wire lying east-west) would respond only to the *horizontal component* of the field, which is

(50 μT) cos 70° = 17 μT

The force on the end wire of the balance would then probably be less than the weight of even a 1 mg rider!

The torque acting on a coil in a magnetic field

In a uniform *B*-field opposite sides of a rectangular current-

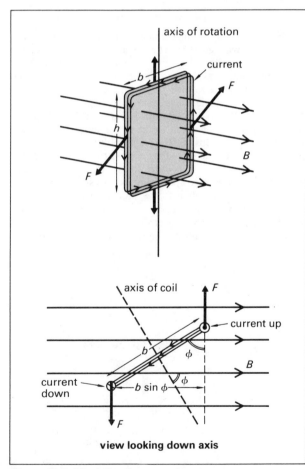

Figure 16.8 Calculating the torque *T* acting on a coil in a magnetic field.

carrying coil experience forces which are equal in size but opposite in direction; thus a couple acts on it tending to turn it to a position in which all the forces are in the plane of the coil. In this position the plane of the coil is at right-angles to the magnetic field and has its axis in line with it. The torque *T* of the couple may be calculated as follows.

We shall confine ourselves to the case of a rectangular coil; suppose it measures *b* by *h*, and has *N* turns each carrying a current *I*. It is free to rotate about an axis in its own plane, this axis being at right-angles to a field of flux density *B* (figure 16.8). The forces acting on the top and bottom of the coil (through which the axis of rotation passes) are parallel to this axis and produce no turning moment. The total force *F* acting on each of the other two sides is given by

$$F = N(BIh)$$

These two sides remain at right-angles to the field as the coil rotates so that the force has the same magnitude in all positions. Let the angle between the *axis* of the coil and the field be ϕ. The perpendicular distance between the lines of action of the two forces is $b \sin \phi$. Therefore the torque *T* of the couple is given by

$$T = Fb \sin \phi = N(BIh)b \sin \phi$$
$$\Rightarrow \quad T = BANI \sin \phi$$

where *A* is the area of the coil ($= hb$). By an extension of this reasoning the expression may be proved quite generally for a plane coil of area *A* and of any shape. The product *ANI* is called the *electromagnetic moment* of the current-carrying coil, and is given the symbol *m*. Thus

$$T = mB \sin \phi$$
$$\text{where} \quad m = ANI$$

The moving-coil meter

In a moving-coil meter of the type commonly used in a laboratory current is passed through a pivoted coil in the field of a permanent magnet. The instrument in effect measures the torque acting on the pivoted coil caused by the current in it, and so gives an indication of the current. A meter of this kind is therefore a form of current balance using a constant magnetic field. Two forms of construction are in use:

(i) The pivoted-coil instrument

In this the coil is wound on a light metal frame pivoted in jewelled bearings so that it is free to rotate between the poles of a permanent magnet (figure 16.9). The motion of the coil is restrained by two hair-springs, one at each end. These hair-springs are also used to conduct the current into and out of the coil. When there is a current in the coil, it rotates until the restoring torque of the hair-springs is equal and opposite to the deflecting torque caused by the current. Attached to the coil is a light balanced pointer,

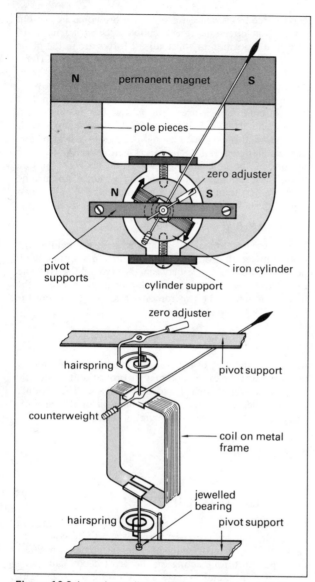

The effect of this is to produce a field of almost constant magnitude, but acting radially, i.e. the lines of force in the gap are along the radii of the central cylinder. Thus as the coil rotates its plane is always parallel to the field in the gap, and the deflecting torque acting on the coil does not vary as it rotates; it depends only on the current in the coil. The restoring torque of the hair-springs is proportional to the deflection. Therefore the equilibrium deflection is proportional to the current, and the instrument has a uniform scale.

To bring the coil quickly to rest when a current is passed through it some sort of damping of the movement has to be provided. Without this the pointer would oscillate back and forth for an excessive time, and measurements would be very tedious to make. Damping is provided by the light metal frame on which the coil is wound. This can be regarded as an additional short-circuited coil of one turn of very low resistance. When the coil moves, currents are induced in the frame which is quickly brought to rest (page 221).

(ii) The suspended-coil instrument or light-spot galvanometer

In the most sensitive instruments the slight friction that occurs even with good jewelled bearings would make the movement irregular. It is then necessary to dispense altogether with bearings; instead the coil is suspended between two taut phosphor-bronze strips (figure 16.10). These are maintained in sufficient tension by the spring strips at top and bottom to keep the coil centred between the pole pieces of the magnet. Connection to the coil is made through the two suspension strips. Instead of a pointer a small plane mirror is fixed to the coil. This is used to cast an image of an illuminated cross-wire onto a scale (figure 16.11). The whole meter and optical system is fixed in a single containing box so that once aligned the optical system needs no further adjustment. The torsion head carrying the top suspension strip can be rotated to bring the undeflected position of the beam of light to the required point on the scale.

Figure 16.9 A moving-coil meter of the pivoted type. (a) The magnet and coil system. (b) The coil, pivot and hair-spring assembly; the connections to the coil are made through the hair-springs and pivot supports.

which moves over a suitable scale. This is the kind of instrument that forms the basis of ordinary ammeters and voltmeters (page 91); the scale is then of course calibrated in amperes or volts, as the case may be.

It is convenient to design the instrument so that its scale is uniform, i.e. so that the deflection is proportional to the current. This is achieved by careful design of the magnetic system. The magnet is fitted with concave iron pole pieces, and an iron cylinder is fixed centrally between them so that it partly fills the space inside the coil. This forms a narrow cylindrical gap in which the sides of the coil can move.

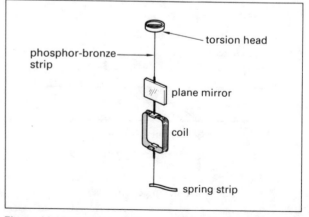

Figure 16.10 The coil and mirror assembly of a taut-suspension type of meter.

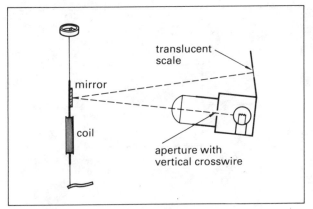

Figure 16.11 The arrangement of lamp and scale in the box of a mirror galvanometer.

In this type of instrument the coil is wound on a non-conducting frame that produces no damping. However, the shunts normally joined across the coil provide a low resistance path for currents induced in the coil itself, and this causes sufficient damping in the same way. The instrument can, however, be used in a high resistance circuit without any shunts, and the coil then oscillates freely when the current in it changes. This is required when the instrument is to be used as a *ballistic galvanometer* for measuring the charge that flows in short pulses of current (page 177). When not in use the coil of a suspended instrument is short-circuited so that it does not oscillate violently when being carried about; this provides a certain amount of protection against mechanical damage.

Sensitivity

Suppose we have a meter coil of N turns each of area A carrying a current I. The deflecting torque T acting on the coil is given by

$$T = BANI \qquad \text{(since } \sin \phi = 1\text{)}$$

The restoring torque T_k of the springs is proportional to the deflection θ of the coil (Hooke's law, page 53). So we write

$$T_k = k\theta$$

where k is called the *suspension constant*; it is the torque produced per unit deflection. The coil comes to rest in an equilibrium position such that $T = T_k$, i.e. at a deflection θ given by

$$k\theta = BANI$$

The *current-sensitivity* s of the instrument is defined by

$$s = \frac{\theta}{I}$$

hence $\quad s = \dfrac{BAN}{k}$

To achieve high current-sensitivity the instrument must be

designed with a coil of large area and many turns; very fine wire must then be used, and this increases the resistance of the coil. The magnetic field in the gap must also be as large as possible, and the suspension constant k must be small.

With the pivoted-coil type of instrument the manufacturers usually quote the current required to give full-scale deflection (f.s.d.), since this is what we need to calculate the resistances of shunts, etc., to make the instrument into an ammeter or voltmeter with a given scale range (page 91). Typical modern instruments range from f.s.d. 1 mA (with a coil resistance of about 100 Ω) to f.s.d. 50 μA (with a coil resistance of about 2000 Ω). However, with the suspended-coil type of instrument manufacturers often quote a *calibration constant*, which is defined as I/θ (the reciprocal of the current sensitivity). The scale of these instruments is normally marked out in millimetre divisions, and the calibration constant is quoted in μA div^{-1}. Any given scale reading can then be converted to a current reading simply by multiplying by the calibration constant. A typical modern light-spot galvanometer has a calibration constant of 4.0×10^{-8} A div^{-1}, and a coil resistance of 14 Ω. Used as a ballistic galvanometer for charge measurement its calibration constant is about 1.3×10^{-8} C div^{-1}. The most sensitive instruments available are about 20 times as sensitive as this, and have coil resistances of 1500 Ω or more. The smallest current that can be detected with such an instrument is about 10^{-9} A, and the smallest charge in a short pulse of current is about 10^{-9} C.

16.3 The force on a moving charge

The magnetic force acting on a wire carrying a current is the summed effect of the forces acting on the individual electrons moving through it. It is therefore possible to work out the force acting on a single charged particle moving in a magnetic field of flux density B. This may be done as follows.

The current I in a conductor is the rate of passage of electric charge past any given point in it. Thus if N electrons each of charge $-e$ flow past the point Y in time t (figure 16.12), then the current I is given by

$$I = \frac{Ne}{t}$$

If the average speed of the electrons is v, then the length l of the conductor occupied by the N electrons concerned is given by

$$l = vt$$

The force F on this length of conductor is given by

$$F = B_\perp Il \qquad (B_\perp = B \sin \theta)$$

$$= B_\perp \frac{Ne}{t} vt \quad = B_\perp Nev$$

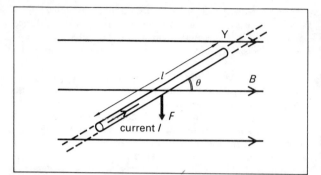

Figure 16.12 The force on a conductor in a magnetic field is the summed effect of the forces on the charged particles in it.

The B-field force F_B acting on each particle is therefore given by

$$F_B = B_\perp ev = Bev \sin \theta$$

The force is at right-angles to the magnetic field and to the path of the particle in a direction given by Fleming's left-hand rule (not forgetting that the current in the conventional sense is opposite to the motion of an electron).

The effect of this force can be observed directly in the deflection of beams of charged particles in magnetic fields, e.g. electron beams in cathode-ray tubes (page 239). Since the magnetic force acting on a charged particle is at right-angles to its path, it does no work on the charged particle (page 22), and thus causes no change of speed – only a change of direction. If the motion is exactly at right-angles to a uniform field, the path is turned into a circle. In the special case of motion along a line of force, the force reduces to zero ($\theta = 0$). In general, with the motion inclined to the field, the path is a *helix* round the lines of force.

The Hall effect

When a conductor carrying an electric current is placed in a magnetic field, it is found that a potential difference is developed between the sides of the conductor in a direction perpendicular to the field. This is called the *Hall effect*. Thus in figure 16.13a, if a current I is passed through the slab of conductor between the edges P and Q, and a magnetic field B is applied at right-angles to the faces of the slab, there is a potential difference V_H between the points R and S on the sides.

The effect is readily explained in terms of the forces acting on the moving charges in the conductor. In figure 16.13a with the directions of current and field shown, according to Fleming's left-hand rule the B-field force on the charged particles acts across the slab from S towards R; the charged particles therefore drift sideways across the slab towards R, creating the p.d. V_H between R and S. For a conventional current from P to Q the direction of

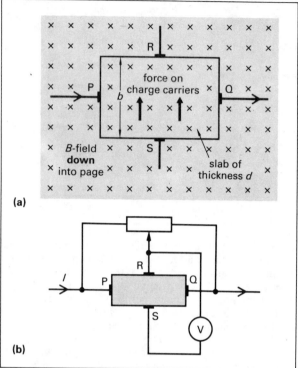

(a)

(b)

Figure 16.13 (a) The Hall effect; with the field shown the magnetic forces on the charge carriers move them towards the side R of the slab, producing the Hall p.d. V_H between R and S. (b) The potentiometer is used to reduce the p.d. between R and S precisely to zero in the absence of a magnetic field.

the *force* is the same whichever sign the charged particles in the material happen to have. The sign of the Hall p.d. may thus be used as a direct test of the sign of the charged particles that carry the current in any given material. If R becomes *negative* with respect to S, then the charge carriers must be the negative electrons; but if R becomes *positive*, then the conduction must be by means of positive holes. In other words the Hall effect provides the means of finding out whether the electrical conduction in a material is predominantly n-type or p-type (page 77).

When the moving charge carriers drift sideways across the slab of conductor, they cause an electric field E given by

$$E = \frac{V_H}{b} \qquad \text{(page 82)}$$

where b is the width of the slab. The sideways separation of charge ceases when the magnetic and electrical forces (sideways) on them are equal in magnitude and opposite in direction. This enables us to work out the value of the Hall p.d. The electrical force F_E and magnetic force F_B on each particle are given by

$$F_E = eE = \frac{eV_H}{b}$$

and $F_B = Bev$

where v is the average speed of the charge carriers flowing between P and Q. We have already shown that

$$v = \frac{I}{nAe}$$ (page 78)

where n is the number of charge carriers per unit volume and A is the cross-sectional area through which the current passes ($= bd$ in figure 16.13a).

Hence $F_B = \frac{BI}{nbd}$

Equating the two forces F_E and F_B, we have

$$\frac{eV_H}{b} = \frac{BI}{nbd}$$

$$\Rightarrow \quad V_H = \frac{1}{ne}\frac{BI}{d}$$ (1)

This equation predicts that for a given slab the Hall p.d. is proportional to the product BI. This is readily confirmed by experiment. For a given material and given values of B and I, V_H is inversely proportional to the thickness d of the slab; it is greatest for a very thin specimen. With metals the Hall effect is very small; with the electromagnets available in schools V_H is unlikely to exceed 1 mV even with very thin specimens. But in semiconductors the volume density n of charge carriers is much less. This means that for a given current the speed v of the charge carriers is greater, so that the magnetic force F_B is greater, yielding a larger value of V_H (as the equation shows). The value of n in a semiconductor depends on the amount of impurity it contains (page 77). But a typical value for germanium (lightly doped with impurity) is 1.0×10^{20} m^{-3}. Suppose we have a specimen 0.20 mm thick through which we send a current of 100 mA. If we now bring up a small magnet that produces a magnetic flux density near its poles of 5.0 mT, then the Hall p.d. V_H is given by

$$V_H = \frac{1}{(1.0 \times 10^{20}\text{ m}^{-3})(1.6 \times 10^{-19}\text{ C})}$$
$$\times \frac{(5.0 \times 10^{-3}\text{ T})(100 \times 10^{-3}\text{ A})}{(0.20 \times 10^{-3}\text{ m})}$$

$$= 0.16\text{ V}$$

The measurement of V_H for known values of the other quantities in equation (1) provides us with a means of measuring the volume density n of charge carriers in the specimen. For metals there is generally about one free electron per atom, which gives a value of n of about 10^{29} m^{-3}.

The chief practical difficulty in demonstrating the Hall effect (even with semiconductors) is in the positioning of the Hall electrodes R and S (figure 16.13a). We need to ensure that, in the absence of a magnetic field, R and S are at exactly the same potential. Figure 16.13b shows one way of doing this. With no magnetic field the potentiometer

is adjusted to give zero reading on a meter joined between R and S. The Hall p.d. is then read directly from the meter when a magnetic field is applied.

The Hall probe

The Hall effect provides a simple and direct means of comparing *B*-fields. If a constant current is maintained through a suitable slice of germanium, the Hall p.d. generated between its sides is proportional to the component of the *B*-field perpendicular to the slice. For this purpose the germanium slice is usually mounted in the end of a rod through which the necessary leads pass. This arrangement is known as a *Hall probe*. To find the *direction* of the resultant field at a point we have to turn the probe until the Hall p.d. is a maximum; the direction of the field is then at right-angles to the germanium slice. In general the Hall probe gives the resolved part of the *B*-field at right-angles to the slice.

In experiments with a Hall probe the Earth's field is inevitably superimposed on those of the magnets or coils under investigation, and allowance may have to be made for this. But in practice we use it only for measurement on fields appreciably greater than 10^{-4} T, to which the Earth's field is a negligible addition.

16.4 *B*-fields of coils and wires

Solenoids

If we move a Hall probe slowly into a solenoid carrying a steady current, we find that the *B*-field increases at first, and then reaches a steady value which is constant through the whole inner region of the solenoid. Furthermore the field in this inner region is constant over the whole cross-section. The simplest way of obtaining a uniform magnetic field is therefore to use the central region of a long solenoid. If we place the Hall probe at an end face of the solenoid we shall find that the *B*-field here is exactly half that in the uniform central region, though the field at the end is obviously non-uniform, as we can see from the line of force diagram (figure 16.4c, page 205). Outside the solenoid tube the field is almost zero except near the ends. Evidently the fields of the various turns of the solenoid almost competely cancel one another on the outside; and the longer the solenoid, the more exactly is this found to be so.

With the aid of a moving-coil meter we can readily check that the field in a solenoid (or in any other sort of coil for that matter) is proportional to the current I in it. A more surprising result is that the field does not depend on the cross-sectional area or shape of the solenoid. It is useful for this purpose to have a set of solenoids of different cross-sections and shapes. These are joined in series, so that they all carry the same current. With the aid of the Hall probe we can then show that the flux density for a given current depends only on the number of turns per unit length (N/l), where N is the number of turns and l the length of the solenoid tube. (But this holds only if the solenoids are long compared with their diameters; otherwise the fields at their centres are not truly uniform and are somewhat less than for a 'long' solenoid.) These experiments show that the B-field in a long solenoid is given by

$$B \propto \frac{N}{l} I$$

The field also depends on the material filling the solenoid. But if the space in and around it is empty (it makes virtually no difference whether it is a vacuum or air), then we can write

$$B = \frac{\mu_0 N I}{l}$$

where μ_0 is a constant, which is a property of empty space, called the *magnetic constant*. It is also referred to as the *permeability* of a vacuum (or of free space). It is the constant that specifies the magnetic forces that act between given electric currents, just as the electric constant ε_0 specifies the forces that act between given electric charges (page 194), and the gravitational constant G specifies the forces that act between given masses (page 160). The above equation shows that the unit of the magnetic constant μ_0 is the T m A^{-1}. But since we have

$$1\,\text{T} = 1\,\text{N A}^{-1}\,\text{m}^{-1} \qquad \text{(page 206)}$$

we can express the unit of μ_0 more significantly as N A^{-2}. Its value is fixed as soon as we select the ampere as our unit of current. By using a simple current balance to measure the B-field inside a long solenoid carrying a current measured by a moving-coil ammeter, we may find the value of μ_0 implied by our choice of the ampere as the unit of current. In this way we find

$$\mu_0 = 1.26 \times 10^{-6}\,\text{N A}^{-2}$$

We shall see below how the exact value of μ_0 may be *calculated from the definition of the ampere*.

A flat circular coil

Sometimes it is more convenient to produce a magnetic field with a flat circular coil (as in figure 16.4b). We quote here the expression for the B-field *at its centre*. (The mathematics involved in deducing this is beyond the scope of this book.) If the coil is of N turns each of radius r, then

$$B = \frac{\mu_0 N I}{2r}$$

A straight wire

We may use the Hall probe to investigate the magnetic field produced by a long straight wire. We have to make sure that the return leads from the ends of the wire are sufficiently distant from the apparatus not to affect the

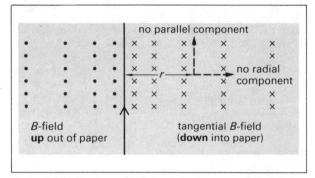

Figure 16.14 The magnetic field of a straight wire is entirely tangential; there is no radial or parallel component.

field significantly. Also there needs to be a rather large current in the wire (several amperes) to give a measurable field at even 50 mm from it.

We have already seen that the lines of force of such a field are concentric circles round the wire (figure 16.4a, page 205). It may be confirmed with the Hall probe that the magnetic field indeed has no component parallel to the wire or radially away from it (figure 16.14). But turning the probe so that its plane is parallel to the wire (i.e. with the lines of force perpendicular to it), we may find how the field varies with the distance r from the wire. We find that B is *inversely proportional* to r. Doubling the distance r halves the field, etc. At a given position the flux density is also proportional to I. Thus

$$B = \frac{kI}{r}$$

where k is a constant whose value may be determined from the actual measurements in this experiment. We thus find

$$k = 2.0 \times 10^{-7}\,\text{N A}^{-2}$$

This is not of course the same as μ_0, the magnetic constant defined in terms of the field of a solenoid; but it is related to it. It may be shown (by mathematics beyond the scope of this book) that $k = \mu_0/2\pi$. The field at a distance r from a long straight wire is therefore given by

$$B = \frac{\mu_0 I}{2\pi r}$$

Coaxial cables

It is no surprise to discover that practically no magnetic field is produced by a pair of conductors very close together carrying equal and opposite currents (like a typical electric light flex). Actually there is a very small field *close* to a light flex. But with a coaxial cable of the kind often used for the lead-in wires of a television aerial there is definitely *no* magnetic field outside the cable. A coaxial cable consists of an insulated inner copper wire that carries the current in one direction, surrounded by an outer tube of fine braided wires that carry the return current. Outside the

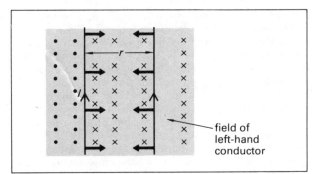

Figure 16.15 Calculating the force between two parallel conductors; we work out the force on the right-hand conductor in the field of the left-hand one.

cable the field of the current in the outer cylinder exactly cancels the field of the current in the inner conductor. The only magnetic field is in the inner space between the conductors.

The magnetic constant μ_0

The modern definition of the ampere (appendix A2) envisages an arrangement of two long, straight, parallel conductors. We can now derive an expression for the forces acting between such a pair of conductors. Suppose they are a distance r apart, and each carries a current I.

In figure 16.15 the magnetic field produced by the left-hand conductor at the position of the right-hand one is *down* into the paper. The flux density B of this field is given by

$$B = \frac{\mu_0 I}{2\pi r}$$

By Fleming's left-hand rule, the force on the right-hand conductor is to the left – i.e. the two conductors *attract* each other. (If the currents are in opposite directions they repel each other, page 76.) The force F acting on a length l of the right-hand conductor caused by the current in the *whole* length of the left-hand one is given by

$$F = BIl = \frac{\mu_0 I^2 l}{2\pi r}$$

In the definition of the ampere the following figures are specified:

$$l = r(= 1 \text{ m})$$
$$F = 2 \times 10^{-7} \text{ N}$$
$$I = 1 \text{ A}$$

Substituting, we obtain the value of μ_0 as

$$\mu_0 = 4\pi \times 10^{-7} \text{ N A}^{-2}$$

and so this is the exact value implied in the definition of the ampere. It is often convenient to express μ_0 in terms of

the unit of inductance, the *henry* (H), explained in the next chapter (page 227); and we can then write

$$\mu_0 = 4\pi \times 10^{-7} \text{ H m}^{-1}$$

16.5 Induced e.m.f.s

When a conductor is made to move across a magnetic field, a potential difference is generated between its ends. If we consider the forces acting at the atomic level, such an effect is indeed to be expected; for we have seen how a magnetic force acts on any charged particle moving through a magnetic field. A moving conductor contains equal numbers of charged protons and electrons, on all of which the magnetic field therefore acts. However, only the electrons are free to move inside the conductor. The result of the motion is therefore an accumulation of negative charge at one end of the conductor, leaving a surplus of positive charge at the other.

Thus, with the direction of the field and the motion indicated in figure 16.16 the application of Fleming's left-hand rule shows that the force acting on an electron (with negative charge) moving with the wire will be from A to B; the end B therefore becomes negative, leaving A positive. Such a moving conductor behaves as a source of e.m.f.; like an electric cell it causes separation of charge between its ends, and the charge separated has electrical potential energy. If the conductor is connected in a complete circuit, it causes a current in it, and supplies electrical energy to it. The energy is derived from the work done by the mechanical forces maintaining the motion of the conductor against the magnetic force $F(= BIl)$ acting on it while there is a current in the circuit.

The effect is known as *electromagnetic induction*, and the e.m.f.s generated by such means are called *induced e.m.f.s*. Electromagnetic induction was discovered by Michael

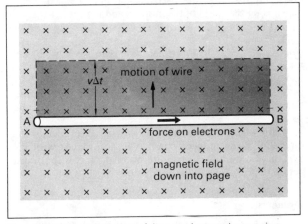

Figure 16.6 The induced e.m.f. in a moving conductor arises from the forces acting on the charged particles in it.

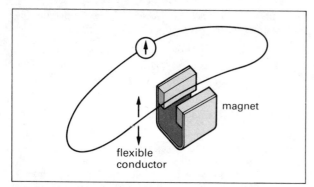

Figure 16.17 The e.m.f. induced in a circuit when part of it is moved across the lines of force of a magnetic field.

Faraday in a brilliant sequence of experiments – difficult to design in his day because of the insensitive measuring instruments available to him. However, with modern sensitive meters the effect is a very obvious one. In figure 16.17 if the flexible conductor is moved up and down between the poles of the permanent magnet the meter joined to its ends registers a small current. The current continues only while the conductor is moving. The same happens if the magnet is moved while the conductor is held still; *relative motion* of conductor and source of magnetic field is required to produce the induced e.m.f. Also the motion must 'cut across' the lines of force of the magnetic field. The maximum e.m.f. is produced when the motion is at right-angles to the field (up and down in figure 16.17); and it falls to zero for motion in the same direction as the field (i.e. horizontal motion in figure 16.17, across the gap from one pole to the other of the magnet).

In such experiments allowance has to be made for the relatively slow response of a sensitive meter. When a current through the meter has ceased, the deflection takes time to fall to zero. Only if the movements of the conductor are slow in relation to the response time can definite indications of what is happening be obtained. If care is taken in this way, we discover that the induced e.m.f. depends on the speed of the motion, and the results suggest

$$\text{e.m.f.} \propto \text{speed}$$

The simplest way to investigate this is to see how long it takes to move the wire through the field in such a way as to keep the meter deflection constant. If we move the wire faster, in such a way as to *double* the chosen deflection, it takes *half* as long to move the wire the same distance, and so on.

In any of these experiments it is easily checked that the direction of the induced e.m.f. is in accordance with the explanation we have given in terms of the forces that must act on the electrons in the conductor.

Dynamos

It is worth examining at this point the parts of a small d.c. dynamo. The moving part (called the *armature*) consists of a set of wires mounted in slots round an iron cylinder (figure 16.18) so that when this is turning the wires cut across the *B*-field produced by a suitable electromagnet. The wires are connected to the external circuit through sliding contacts at one end in such a way that all those wires that are cutting through the field at any instant are contributing an e.m.f. in the same sense in the circuit.

A simple dynamo of this sort may be used to test more exactly whether induced e.m.f.s are indeed proportional to the speed of the moving conductors. A voltmeter is connected to the armature terminals, and arrangements are made to measure the rate at which the armature is rotated. A graph of e.m.f. against speed of rotation will then be found to be a straight line through the origin (appendix B2). We may also use a simple dynamo to test how the induced e.m.f. depends on the flux density *B* of the magnetic field. Over the range of currents up to the maximum current for which the dynamo is designed the flux density produced by the electromagnet is approximately proportional to the current in the field coil. Using the dynamo at constant speed of rotation we may then show that

$$\text{e.m.f.} \propto B$$

About the simplest possible form of d.c. dynamo is *Faraday's rotating disc* (figure 16.19). It consists of a metal disc that may be rotated at a steady speed in a magnetic field. Any given radius (such as OP) cuts through the field at a steady rate; therefore a steady e.m.f. is generated in the circuit formed by the meter and the radius momentarily between the sliding contacts A and B. Again we may demonstrate that the induced e.m.f. is proportional to the speed. If the disc is mounted inside a solenoid supplied with an adjustable current, we may also demonstrate that the e.m.f. is proportional to *B*.

Alternators

If we have a coil rotating in a magnetic field, an e.m.f. is induced in it that varies from one position of the coil to another, and reverses its direction twice in every revolution;

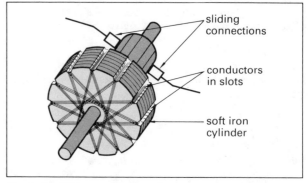

Figure 16.18 The armature (moving part) of a d.c. dynamo; the e.m.f. is the sum of the p.d.s induced in all those wires in the slots that are cutting across the lines of force at any instant.

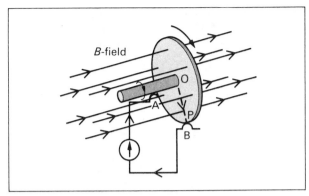

Figure 16.19 Faraday's rotating disc dynamo; the e.m.f. is produced by the radius OP cutting through the lines of force of the field.

in other words there is an alternating current in the circuit of which the coil is a part. This may be investigated with a large rectangular coil rotated in the Earth's field. It may be connected by long flexible leads to a sensitive meter (figure 16.20). The e.m.f. induced in the two halves of **AB** cancel out (by symmetry); and the same applies to **CD**. We can concentrate our attention therefore on the e.m.f.s induced in the other two sides **BC** and **DA** which are parallel to the axis of rotation. The induced e.m.f. is a maximum when the plane of the coil is in the magnetic meridian, since at this moment the sides **BC** and **DA** are cutting across the lines of force of the magnetic field at the greatest speed; since they are moving in opposite directions, the e.m.f.s induced in them act *in the same sense round the coil*. When, however, the plane of the coil is at right-angles to the field the sides **BC** and **DA** are moving momentarily in the direction of the field, and no e.m.f. is generated. If the coil is turned slowly by hand, the way in which the e.m.f. varies may be seen from the deflection of the meter; in fact we are thus generating a very low frequency alternating current.

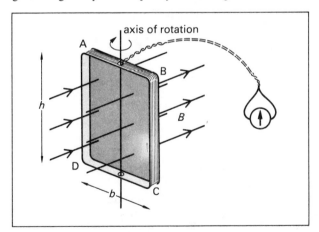

Figure 16.20 A coil rotating in a magnetic field produces an alternating e.m.f., which performs one oscillation per revolution of the coil.

If slip ring connections are provided for the coil, the speed of rotation may be increased, and the alternating e.m.f. may then be observed by connecting the coil to an oscilloscope. The frequency is equal to the number of revolutions per second of the coil. At steady speeds of rotation a *sinusoidal* waveform is produced (page 85); the peak value of the e.m.f. is again found to be proportional to the speed. This is the principle of the a.c. generator or *alternator*. However, in practice it is more satisfactory to keep the coil fixed and to rotate the magnetic field in relation to it (figure 16.21); the slip ring connections then have to carry only the relatively small current in the rotating electromagnet. The coil is wound in two parts in series on an iron core; between these the electromagnet rotates. The electromagnet is supplied with current from a separate d.c. supply through the slip rings and graphite blocks (called *brushes*) that are held against them. The rotating electromagnet is called the *rotor* of the alternator, while the fixed system of coils with their core is called the *stator*.

In small alternators (e.g. cycle dynamos) a rotating *permanent* magnet is used; and no sliding connections are then needed.

Calculations

We can derive an expression for the magnitude of the

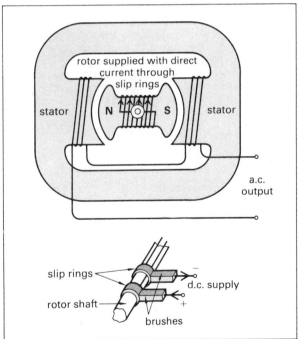

Figure 16.21 A simple alternator, consisting of a rotating electromagnet (the rotor) between the two halves of a fixed coil (the stator). The rotor is supplied with current through a pair of slip rings and brushes.

induced e.m.f. E in a moving conductor from the previous result for the force on a charged particle moving in a magnetic field. Consider a conductor of length l at right-angles to a magnetic field of flux density B and moving across it at a steady speed v (as in figure 16.16). The conductor is not connected in a circuit; so the p.d. across it is equal to the e.m.f. E. When the conductor is first set in motion the magnetic forces cause a movement of the electrons in one direction along it, thereby setting up the p.d. between its ends. This continues until the magnetic forces are exactly balanced by the electrical forces tending to restore the electrons to their original uniform distribution. The electrons are moving at the same speed v as the wire, and the B-field force F_B acting on one of them is given by

$$F_B = Bev$$

The E-field force acting on an electron is given by

$$F_E = e \times \text{(potential gradient)} \qquad \text{(page 189)}$$

$$= e\frac{E}{l}$$

The electrons reach equilibrium when $F_E = F_B$, i.e. when

$$\frac{eE}{l} = Bev$$

Hence $E = Blv$ $\qquad\qquad\qquad\qquad$ (2)

(which is in agreement with the previous experimental results).

In the more general case we must remember that the above calculation requires that the value of B used is the component $B_\perp$ of the flux density perpendicular to the plane in which the conductor is moving.

The product lv in equation (2) is equal to the rate at which the wire sweeps out area A in the plane at right-angles to $B_\perp$. For in figure 16.16 in time Δt the wire moves a distance $v\Delta t$; and the area ΔA swept out is given by

$$\Delta A = lv\Delta t$$

$$\frac{\Delta A}{\Delta t} = lv$$

In the usual notation (appendix B3) we can therefore express equation (2) in the alternative form

$$E = B_\perp \frac{dA}{dt} \qquad\qquad\qquad\qquad (3)$$

This way of looking at the calculation is very useful when we are dealing with a conductor of irregular shape; we do not need to consider the e.m.f.s induced in the separate parts of it, provided we can write down the rate at which the conductor as a whole sweeps out area. Thus, in figure 16.16 a kinked wire would give rise to the same e.m.f. as the straight one if the distance between its ends was the same; for, if it moves through the field at the same speed, it will sweep out area at the same rate lv. The point may be tested with the arrangement of figure 16.17 if a deliberately kinked wire is used in place of the straight one.

▶ Calculate the e.m.f. E induced between the wing-tips of an aeroplane of wing span 30 m flying horizontally at 250 m s^{-1}, if the vertical component of the Earth's B-field is 4.0×10^{-5} T.

The area swept out by the wing per second

$$= (30 \text{ m})(250 \text{ m s}^{-1})$$

$$\Rightarrow \quad V = (4.0 \times 10^{-5} \text{ T})(30 \text{ m})(250 \text{ m s}^{-1}) = 0.30 \text{ V}$$

You can check for yourself that in the northern hemisphere the port (left) wing-tip will be positive. (You should also check that the T m^2 s^{-1} is the same as the V.) We shall consider below (page 217) what must be done to measure such an e.m.f. It is important to realise that it *cannot* be done with a voltmeter moving with the plane. ◀

The same method can be used for calculating the e.m.f. induced between centre and rim of a rotating disc, such as Faraday's disc dynamo in figure 16.19. Here the conductor that is sweeping out area is a radius such as OP. If the disc is of radius r and makes n revolutions per unit time, the rate of sweeping out area by any given radius is $\pi r^2 n$. Therefore the e.m.f. E is given by

$$E = \pi r^2 n B$$

This result may be checked by mounting a Faraday disc inside a long solenoid, in which we may *calculate* the value of B (figure 16.22). When we have proved the above result, we may use a Faraday disc as a flux density meter; but it is not as convenient as a Hall probe.

▶ A horizontal record player turntable (made of aluminium) of diameter 0.30 m rotates at $33\frac{1}{3}$ revolutions per minute in a uniform vertical B-field of 1.0×10^{-2} T. Calculate the e.m.f. E induced between the centre and rim of the turntable.

A given radius sweeps out an area of $\pi(0.15 \text{ m})^2$ per revolution.

$$\text{The number of revolutions per second} = \frac{100}{3 \times 60} \text{ s}^{-1}$$

Hence the rate of sweeping out area by a given radius

$$= \pi(0.15 \text{ m})^2 \left(\frac{100}{3 \times 60} \text{ s}^{-1} \right)$$

$$\Rightarrow \quad V = \pi(0.15 \text{ m})^2 \left(\frac{100}{3 \times 60} \text{ s}^{-1} \right)(1.0 \times 10^{-2} \text{ T})$$

$$= 3.9 \times 10^{-4} \text{ V}$$

If the magnetic field is directed vertically *downwards*, with the usual direction of rotation of record players (clockwise) you can check that the rim will be positive. ◀

The Lorenz rotating disc apparatus

This is the method used for the absolute measurement of resistance in the standardising laboratories. It consists of a Faraday disc mounted inside a long solenoid, as shown in figure 16.22. The apparatus is turned so that the plane of the disc is in the magnetic meridian. The e.m.f. induced is then caused by the field of the solenoid only. Sliding con-

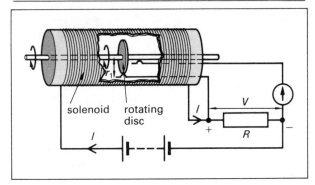

Figure 16.22 The Lorenz rotating disc apparatus for the absolute measurement of a very small resistance.

tacts are made to bear on the rim and axle, as shown. If the distances of these contacts from the central axis are r_1 and r_2, and the disc performs n revolutions per unit time, then the rate of sweeping out area $(\mathrm{d}A/\mathrm{d}t)$ by the part of a radius of the disc between the two contacts is given by

$$\frac{\mathrm{d}A}{\mathrm{d}t} = n\pi(r_1{}^2 - r_2{}^2)$$

The B-field in the solenoid is given by

$$B = \frac{\mu_0 NI}{l}$$

The induced e.m.f. is equal to the rate of cutting magnetic flux by the part of the radius concerned. This e.m.f. is made equal to the potential difference V across a very small resistance R; since R is in series with the solenoid, the same current I therefore flows in it. The adjustment is brought about by varying the speed of rotation of the disc until the meter gives no deflection. Then we have

$$E = \frac{\mu_0 NI}{l}\frac{\mathrm{d}A}{\mathrm{d}t} = IR$$

$$\Rightarrow \quad R = \frac{\mu_0 N}{l}\frac{\mathrm{d}A}{\mathrm{d}t}$$

The resistance R is thus found in terms of the dimensions of the apparatus and the speed of rotation n; the constant μ_0 is already fixed (by the definition of the ampere) as $4\pi \times 10^{-7}$ N A^{-2}. This is called an *absolute* method of measurement, since it does not depend on any other electrical measurement; the only measurements required are those of lengths and times. The resistances that may be measured by this method are not larger than a few milliohm; larger resistances must then be found by comparison with these (page 101).

A rotating coil

The peak value E_0 of the e.m.f. generated in a rotating coil, such as that in figure 16.20 (page 215), may be calculated as follows. The peak e.m.f. is generated at the moment

when the plane of the coil is parallel to the field, since at these moments the two sides BC and DA are moving at right-angles to the field. If these sides are of length h, the peak p.d. E_1 induced between the ends of a single conductor in one of them is given by

$$E_1 = Bhv$$

where v is the speed. If the breadth of the coil is b, the sides BC and DA are moving in circular paths of diameter b, and we have

$$v = \pi bn$$

where n is the number of revolutions per unit time. If the coil is of N turns, the total number of such conductors is $2N$, and the combined peak e.m.f. E_0 is the sum of the e.m.f.s induced in them all. Hence

$$\begin{aligned}E_0 &= 2NBhv \\ &= 2NBh(\pi bn) \\ &= 2\pi nBAN\end{aligned}$$

where $A(=bh)$ is the area of the coil. This result may in fact be proved for a coil of area A of any shape. The frequency f of the e.m.f. is equal to the number of revolutions of the coil per unit time, and so

$$f = n$$

▶ Calculate the peak e.m.f. induced in a circular coil of 100 turns of radius 0.10 m rotating at 50 revolutions per second about a diameter which is at right-angles to a B-field of 0.20 T.

Using the above result the peak e.m.f. E_0 is given by

$$\begin{aligned}E_0 &= 2\pi nBAN \\ &= 2\pi(50\ \mathrm{s}^{-1})(0.20\ \mathrm{T}) \times \pi(0.10\ \mathrm{m})^2 \times 100 \\ &= 200\ \mathrm{V}\end{aligned}$$

Since the coil makes 50 revolutions per second, the frequency of the alternating e.m.f. generated is 50 Hz. ◀

Frames of reference

It is worth considering how the e.m.f. induced between the ends of a moving conductor is to be detected. Consider for instance the wing frame of an aeroplane flying through the Earth's magnetic field. This must be regarded as an *open-circuited* source of e.m.f., since there is no closed path for any current to follow. Suppose an observer in the plane decides to investigate the induced e.m.f. by connecting a moving-coil voltmeter to the ends of the wing frame. The voltmeter and its connecting leads are then moving through the magnetic field with the plane and sweep out area at the same rate as the wing frame. Therefore an equal e.m.f. is induced in them. We now have two equal e.m.f.s in parallel, and no current flows in the loop joining them. The voltmeter therefore registers zero; if he has no other information, the observer in the plane must conclude that there is no induced e.m.f., and therefore that the magnetic field is *moving along with the plane*.

Another observer based on the ground analyses the situation rather differently. His laboratory observations show that the magnetic field is stationary in *his* frame of reference; he calculates that an e.m.f. must be induced in the wing of the plane as it moves through the field. To settle the point he contrives to make momentary connection between the wing-tips and his measuring instruments on the ground; and he duly detects the predicted p.d. The observer in the plane agrees that a p.d. should be detected in this experiment, but accounts for it in a different way. He says that the magnetic field is stationary in his frame of reference (based in the plane), and that the voltmeter on the ground and its connecting leads are receding rapidly through the field; the p.d. is therefore induced in the ground-based part of the circuit. Both observers agree (as they must if the laws of physics are valid) about the effect – namely the voltmeter reading; but they disagree about the site of the e.m.f. in the circuit. Each believes that the e.m.f. arises in the part of the circuit that is moving with respect to his frame of reference. And a third observer in a train moving with respect to both the others regards the e.m.f. as arising partly in one section of the circuit and partly in the other – but he too agrees about the voltmeter reading.

This is in fact a very simple application of the principle of special relativity, put forward by Albert Einstein in 1905. This states that there is no preferred observer whom we can assert to be 'at rest', and whose observations have any sort of priority over those of other observers based in other frames of reference. Two observers may analyse a given situation in different ways – but the results they predict will be the same. There is therefore no conceivable way of telling which observer has got the right analysis. The final result does not depend on what frame of reference we choose (provided it is a non-accelerating one). What What matters is that we should always be clear what frame of reference we are working in; and calculations of induced e.m.f.s must always state or imply what this is – e.g. a frame of reference fixed in the Earth, or perhaps a space ship.

16.6 Energy conversions

In any of the examples of electromagnetic induction we have considered so far you should be able to show that the *direction* of the induced e.m.f. is such that the resulting current *opposes* the movement that caused it. This result is known as *Lenz's law*. The opposition to the movement occurs only if there is a complete circuit so that a current actually passes. This opposition is immediately obvious when a small dynamo is turned by hand. As long as no current is taken from the dynamo, the only opposition we can feel is the ordinary frictional resistance. But as soon as a current is taken, an extra opposition to the motion is apparent, and increases with the current.

Lenz's law is in fact a necessary consequence of the principle of conservation of energy (page 28). If the e.m.f. acted in the opposite direction to that which opposes the

motion, the result would be the production of electrical and mechanical energy from nothing. The current in the circuit in fact produces magnetic forces that oppose the applied forces. Work then has to be done by the applied forces to turn the armature; and this work re-appears as electrical energy in the armature circuit.

We may use Lenz's law to predict the direction of an induced e.m.f. (Alternatively, we may predict the direction by considering the magnetic forces acting on the charged particles in the moving conductor, as we have done previously.) For instance, in figure 16.23 if the conductor AB is made to slide to the right in contact with the metal rails PQ and RS it moves through the magnetic field (at right-angles to the diagram), and an e.m.f. is induced in it. This e.m.f. drives a current anti-clockwise in the circuit as shown, since this is the direction that will cause the magnetic force F_B to be to the left, i.e. opposing the motion to the right that caused it.

We may use reasoning based on the conservation of energy as an alternative way of deriving the expression for the e.m.f. E induced in a moving conductor. Thus in figure 16.23 suppose the conductor AB is being propelled at a steady speed v by a force F along the rails PQ and RS (which we assume to be frictionless); let the flux density at right-angles to the plane of motion be $B_\perp$. We shall assume that the induced current I produces a negligible B-field compared with $B_\perp$, and that the resistance of the circuit is high enough for this to be so. The opposing magnetic force F_B is given by

$$F_B = B_\perp Il$$

where l is the distance between the rails. If the wire is moving at a *steady* speed, it is in equilibrium, so that

$$F = F_B$$

The rate of doing work (i.e. the power input)

$$= F \times \text{(distance moved per second)}$$
$$= Fv = B_\perp Ilv$$

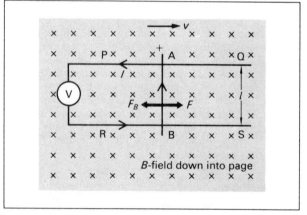

Figure 16.23 The rate of conversion of energy in a circuit containing a conductor cutting through a magnetic field.

The work done re-appears as electrical energy in the circuit, and is there changed into internal energy. The rate of conversion of electrical energy to internal energy is EI.

Hence $\quad EI = B_\perp Ilv$
$\Rightarrow \quad\quad E = B_\perp lv$

in agreement with the previous result.

The d.c. machine

A coil carrying a current in a magnetic field experiences a torque that tends to turn it until its plane is at right-angles to the field. To produce continuous rotation some arrangement must be used that reverses the connections to the coil whenever it passes through this equilibrium position. In a practical motor the moving part (the *armature*) consists of a set of coils sunk in slots spaced round an iron core (figure 16.24). The coils are joined to a set of sliding contacts arranged round the armature shaft (called a *commutator*); current is led into and out of the commutator segments by a pair of graphite blocks (called *brushes*) that bear upon them. The commutator and its brushes are so arranged that the connections to each coil are automatically reversed at the right positions in each revolution. The coils are interconnected in such a way that at least two-thirds of them contribute to the torque of the motor at any given moment. The armature rotates between the poles of an electromagnet whose coils draw current from the same d.c. supply that feeds the armature.

You will at once notice that the construction of a d.c. motor is identical with that of a d.c. dynamo (compare figure 16.24 and figure 16.18, page 214). Any d.c. machine may be used equally as a dynamo or as a motor. The difference between the two is only a question of function. When mechanical energy is being supplied to the machine, and is being changed into electrical energy, we call it a *dynamo*; and when electrical energy is being converted into mechanical energy, we call it a *motor*.

Closer thought shows that the d.c. machine must always behave *simultaneously* both as dynamo and motor. Any current in the armature gives rise to a torque. Thus, when current is taken from a dynamo, a torque arises, as in a motor; by Lenz's law this *opposes* the rotation. Likewise, whenever the armature of a motor rotates, an e.m.f. is induced in it, just as in a dynamo; by an extension of Lenz's law the e.m.f. acts *in opposition* to the applied p.d.; (it is therefore often called a *back e.m.f.*). The fact is that a rotating coil carrying a current in a magnetic field both generates an e.m.f. and experiences a torque. The difference between its functions as dynamo and motor is in the *direction of the current*.

(i) *As a dynamo*: the current is in the direction of the induced e.m.f., which is of course responsible for driving the current through the armature and round the circuit. The product of e.m.f. E and current I is the electrical power generated, which is equal to the mechanical power *input* needed to maintain the rotation against the opposing electromagnetic torque. Because of the internal resistance R_{int} of the armature the p.d. V at the terminals of the dynamo is less than E by an amount IR_{int} (page 94). Thus

$$E = V + IR_{int}$$

Multiplying by I we have

$$EI = VI + I^2 R_{int}$$

and this equation says:

total power generated = electrical power output
+ rate of heating armature

(ii) *As a motor*: the current is driven through the armature by the applied p.d. in opposition to the induced e.m.f. The product of induced e.m.f. E and current I is now equal to the mechanical power *output*. In this case the applied p.d. V is greater than E by an amount IR_{int}. Hence

$$V = E + IR_{int}$$

Multiplying by I we have

$$VI = EI + I^2 R_{int}$$

and this equation says:

electrical power supplied = mechanical power output
+ rate of heating armature

These relationships are revealed in the energy flow diagram of figure 16.25.

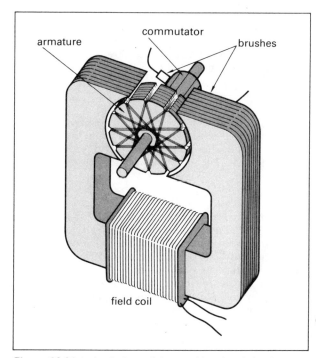

armature
commutator
brushes
field coil

Figure 16.24 A simple form of d.c. machine; for clarity the bearings that support the armature axle are not shown.

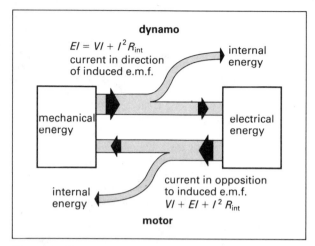

Figure 16.25 The d.c. machine is a reversible device; whether it acts as a dynamo transforming mechanical energy to electrical, or as a motor transforming electrical energy to mechanical depends on the direction of the current in its armature.

The d.c. machine is thus a reversible energy converter in much the same way as a rechargeable electric cell (such as a lead-acid cell, page 95). A lead-acid cell is a device for the interconversion of electrical and *chemical* energy; whereas the d.c. machine is for the interconversion of electrical and *mechanical* energy. Their functions in an electric circuit are essentially the same. A practical example will help to make this clear for you.

Consider a small electric delivery van (e.g. a milk float), worked from a 100 V lead-acid battery, climbing over the brow of a hill. As it ascends the hill it travels fairly slowly; the induced e.m.f. generated in its motor is proportional to the speed, and in this case is less than the battery e.m.f.; suppose it is 95 V. The difference between the two e.m.f.s, namely 5 V, drives a large current through the armature; this current provides the torque to keep the float moving up the hill. The net result is the conversion of the chemical energy of the battery into gravitational potential energy of the float. Suppose the current at this stage is 40 A. The electrical power P supplied (i.e. the rate of conversion of chemical energy to electrical energy) is given by

$$P = (100 \text{ V})(40 \text{ A}) = 4000 \text{ W}$$

The rate of conversion of electrical energy to mechanical P_m is given by

$$P_m = (95 \text{ V})(40 \text{ A}) = 3800 \text{ W}$$

and the rate of conversion of electrical energy to internal energy of the circuit P_R by

$$P_R = (5 \text{ V})(40 \text{ A}) = 200 \text{ W}$$

(You can check incidentally that the resistance of the circuit must be 0.125 Ω.)

As the milk float approaches the brow of the hill, and the slope becomes less, the torque required decreases, and the vehicle accelerates. The increased speed raises the

induced e.m.f. generated in the armature to a figure very near the battery e.m.f.; it might be 99 V. The difference between the two, now 1 V only, drives a current one-fifth as great (8 A) through the motor; but this is sufficient to maintain the speed on the smaller gradient and to supply the smaller power needed to overcome friction. As the float starts to descend the other side of the hill, it speeds up further, until the induced e.m.f. is *equal* to the battery e.m.f. At this point the loss of gravitational potential energy is just sufficiently rapid to make up for the frictional loss; no current passes, and the motor is idling. When the speed increases still further, the induced e.m.f. becomes more that 100 V, and the d.c. machine, now behaving as a dynamo, drives current in the reverse direction through the battery. The current in the machine now produces a torque which opposes the motion (acting therefore like a brake). The rate of doing work against this torque is the rate of conversion of mechanical energy to electrical (and then mostly to chemical energy in the battery). Much of the gravitational potential energy acquired by the van as it ascended the hill is now being changed back into chemical energy.

Starting a motor

At the moment of starting the induced e.m.f. is zero, and it increases to its final value only as the motor gathers speed. If no other resistances are included in the circuit, the starting current I' is given by

$$V = I'R$$

and I' can well reach a large and damaging value. Consider for instance the motor of our van above, operating on a 100 V d.c. supply in a circuit of total resistance 0.125 Ω. In this case at starting

$$I' = \frac{100 \text{ V}}{0.125 \text{ }\Omega} = 800 \text{ A}!$$

When starting such a motor it is therefore necessary to join a suitable resistance in series with the armature. Suppose that in the present example we must limit the current to 50 A. The total resistance required must then be

$$\frac{100 \text{ V}}{50 \text{ A}} = 2.0 \text{ }\Omega$$

As the motor speed increases the induced e.m.f. rises from zero and the starting resistance is progressively switched out of the circuit; near full speed the induced e.m.f. alone is sufficient to limit the current to a safe value, and the starting resistance is switched out of the circuit altogether (figure 16.26). In a low-power motor the armature resistance is normally sufficient to limit the current to a safe value without any additional resistance.

Reversing an electric motor

The direction of rotation of a d.c. motor is not altered by reversing the polarity of the supply; this merely reverses

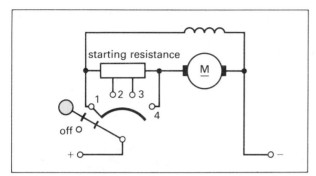

Figure 16.26 The starting resistance used with a large electric motor; as the motor gathers speed, the resistance is progressively switched out.

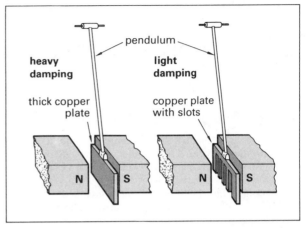

Figure 16.27 The damping of the swing of a pendulum by eddy currents induced in the copper plate; the effect is lessened by slots that reduce the size of the internal loops in which the eddy currents can flow.

both the armature current and field-magnet current, leaving the torque still in the same direction. To reverse the rotation we need rather to reverse the connections to one only of the coils – say, the field coil. However, very small motors, such as those used in toy trains, are made with *permanent magnets*. Reversing may then be effected by simply changing the connections of the d.c. supply.

Eddy currents

When a conductor moves in a magnetic field an e.m.f. is induced in it. So far we have been studying the effects of induced e.m.f.s acting in circuits consisting of loops of wire, etc. But if the conductor is a lump of metal, significant e.m.f.s can act round closed paths *inside* the lump of metal itself. Although these e.m.f.s are not usually very large, the resistance of the current paths is so low that large currents may pass. These induced currents circulating inside a piece of metal are known as *eddy currents*. By Lenz's law they must flow in such directions as to *oppose* the motion. They can indeed act as a very effective brake on the motion of the body, the mechanical energy being transformed into the electrical energy of the eddy currents, which in turn is transformed into internal energy, thus warming the metal.

The effect may be demonstrated by swinging a pendulum with a thick copper bob between the poles of an electromagnet (figure 16.27). When the electromagnet is switched

on large eddy currents are generated in the bob, and there is a marked braking effect. The eddy currents can be reduced in this case by using a bob with a series of slots cut in it so that the currents are now free to circulate only inside the relatively narrow teeth left between the slots. Very little braking effect can then be detected.

Eddy currents arise also in a stationary piece of metal if the source of a magnetic field is moving in relation to it. Thus, if a thick brass plate is held just below a suspended magnet performing torsional oscillations in the Earth's field, the movement of the magnet is seen to be heavily damped on account of the interactions between the magnet and the eddy currents induced in the plate.

Eddy currents must be minimised in iron-cored apparatus, (e.g. motors and dynamos), otherwise considerable losses of energy would occur on this account. In these cases the iron parts of the apparatus are therefore built up out of a stack of thin iron sheets, called *laminations*, as you can see in figure 16.24. The laminations are insulated from one another, sometimes by a thin layer of paper stuck on one side of each lamination.

17 Changing magnetic fields

17.1 More induced e.m.f.s

In the previous chapter the induced e.m.f.s we considered were produced by moving a conductor through a magnetic field. An e.m.f. may also be induced in a conductor by *changing* the magnetic field it is in *without any relative movement of the conductor and the source of the magnetic field.* Thus, if a wire is clamped between the poles of an electromagnet and joined to the terminals of a sensitive meter (figure 17.1), a deflection of the meter is observed when the current in the coils of the electromagnet is changing. This effect cannot be explained in terms of the magnetic force acting on a moving charge, since the conductor is stationary in the changing magnetic field; it is an entirely new phenomenon. However, it has many features in common with the simpler form of electromagnetic induction (with moving conductors); and, as we shall see, it is possible to analyse both processes in the same way.

As before, the induced e.m.f. E is found to be proportional to a *rate of change*, in this case the rate of change of the B-field. So we can write

$$E \propto \frac{dB}{dt}$$

Thus, if the current in the electromagnet is increased or decreased at a steady rate, the meter shows a constant reading which is proportional to the speed at which the change is brought about. Furthermore the induced e.m.f. is exactly the same whether the change in the B-field round the conductor is brought about by reducing the current in the electromagnet to zero or by moving the conductor out of the field.

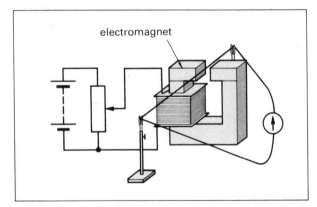

Figure 17.1 An e.m.f. is induced in a fixed conductor in a changing magnetic field.

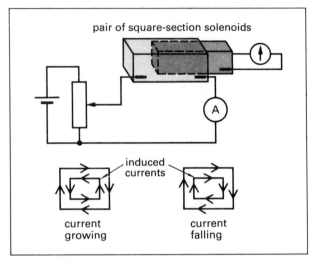

Figure 17.2 A pair of solenoids, one inside the other; an e.m.f. may be induced in the inner one, either by relative movement of the solenoids, or by changing the current in the outer one.
When the current is growing the induced e.m.f. is in the opposite sense round the inner solenoid; and when the current is falling it is in the same sense.

The point may be tested more exactly by using a pair of solenoids that may be placed one inside the other. This increases the effect by the use of many turns of wire without the complications introduced by the presence of iron cores. A variable potential difference may be supplied to the outer solenoid (in series with an ammeter), while the inner one is connected to a sensitive meter (figure 17.2). An e.m.f. may now be induced in the inner coil either by moving it slowly out of the outer one or by slowly reducing the current in the outer coil to zero. The changes may be made smoothly so as to keep the meter reading at very nearly the same value for both cases. It is then found that the same time is taken to reduce the field in the inner solenoid to zero *whichever process is used.* And, as before, the meter reading is doubled by halving the time taken over the change, whichever process is used – showing that the e.m.f. is indeed proportional to the rate of change of the field.

The direction of the induced e.m.f. may again be described by Lenz's law; i.e. the induced current *opposes* the change that caused it. Thus, when the current is reduced to zero in the turns of the outer solenoid, current flows in the *same* sense round the turns of the inner one, so tending to prevent the reduction of magnetic field that caused the induced e.m.f. When the current in the outer solenoid is once more increased, the current in the turns of the inner one flows in the opposite sense, tending to *prevent* the magnetic field increasing inside it (figure 17.2). The solenoids also exert forces on each other while this is going on, tending to prevent the changes. Such forces will generally be too small to detect, but their existence can be used to help us predict which way induced currents are going to flow in a given case – and this is the function of Lenz's law.

The same pair of laws may therefore be used to describe the induced e.m.f.s both in the above case of changing currents in stationary circuits and in the case of relative motion between a circuit and the source of magnetic field, considered previously. The laws take the following forms:
(i) the induced e.m.f. is proportional to the rate of change causing it (*Faraday's law*).
(ii) the induced current opposes the change causing it (*Lenz's law*).

Magnetically linked circuits

In the induction of e.m.f.s between stationary circuits electrical energy is transferred from one circuit to the other through the agency of the magnetic field. The process is called *mutual induction*. It can occur only when the current is changing. In d.c. circuits, therefore, effects caused by mutual induction arise only at the moments of switching on or off. But if an *alternating* current is caused to pass in one of the coils, an alternating e.m.f. of the same frequency is induced in the other. Thus in a.c. circuits energy is transferred from one circuit to the other at every oscillation of the current through the linking action of the magnetic field.

At high frequencies the rate of change of field is very rapid, and the process is effective even with air-cored coils. But at low frequencies the transfer of energy from one coil to another can only be really effective if the B-field is increased by winding the coils on a closed loop of iron. This is the basis of the design of the transformer (page 230). For the moment we shall avoid the complications introduced by iron cores and consider coils with air cores only. We may use for this purpose the pair of solenoids, one inside the other, that we have employed previously. One coil, called the *primary*, is joined to a suitable low-voltage a.c. supply. The meter in series with the other coil, called the *secondary*, must be replaced with a suitable a.c. meter; a cathode-ray oscilloscope is best, since it enables us to see what is happening at different points in the cycle of variation of current (figure 17.3a). If a double-beam oscilloscope is available we may use one beam to show the e.m.f. induced in the secondary coil, while the other beam shows the current in the primary. The oscilloscope deflection depends on the *p.d.* between the deflector plates. To show the *current* in the primary coil we must therefore join a resistor in series with it, and connect the oscilloscope across this.

If the current supplied to the primary varies sinusoidally (page 85), the e.m.f. induced in the secondary is also

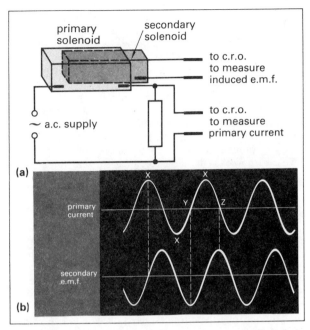

Figure 17.3 Mutual induction between two magnetically linked coils. The oscilloscope trace (b) shows how the induced e.m.f. in the secondary is 90° out of phase with the current in the primary.

sinusoidal, but it is related to the primary current as shown in the double-beam oscillscope trace of figure 17.3b. At moments such as X the magnetic field produced by the current in the primary solenoid is at its maximum value in one direction or the other, but is momentarily unchanging; at this moment the e.m.f. induced in the secondary solenoid is zero. However at moments such as Y or Z the magnetic field is momentarily zero, but is changing at the maximum rate in one direction or the other. The induced e.m.f. is therefore a maximum or minimum. We can describe this by saying that the e.m.f. induced in the secondary lags a quarter of a cycle behind the variations of current in the primary.

If the frequency of the primary current is varied, while preserving the same peak current (and the sinusoidal waveform), we find that the peak value of the induced e.m.f. is proportional to the frequency. This may be tested with a signal generator (an electronic device producing a sinusoidal output p.d. variable over a wide range of frequencies, page 295). At 50 Hz the e.m.f.s induced in the secondary may be too small to be detectable. But in the range from 1 kHz to 100 kHz useful oscilloscope deflections may be obtained.

The a.c. search coil

By using very fine wire we can make a small secondary coil of many turns to take the place of the secondary solenoid above. Such a coil may be used to explore how the field varies from one part to another of the primary solenoid. This is called a *search coil*. With it we can show, for instance, that the field is practically constant over the whole cross-section of a solenoid – a result we have already studied (page 211), but of great practical importance.

The e.m.f. E induced in a search coil is proportional to the number of turns N it contains, since the same e.m.f. is induced in each turn of the coil. This is easily confirmed with a set of coils of the same area of cross-section, but with varying numbers of turns. By using coils of different sizes we can show that the induced e.m.f. is proportional to the area of cross-section A of the coils, but does not depend on the shape of the cross-section. Thus in a given oscillating field

$$E \propto NA$$

The induced e.m.f. depends only on the component (or resolved part) $B_\perp$ of the magnetic field perpendicular to the plane of the search coil. Thus if the search coil is placed in a solenoid, the induced e.m.f. is a maximum when the axis of the coil is parallel to that of the solenoid, but falls off to zero for a coil turned at right-angles to this position.

Summarising these results, the induced e.m.f. E is given by

$$E = kNA\frac{dB}{dt} \tag{1}$$

In fact you can show experimentally that the constant k is exactly 1 by measuring all the other quantities in equation (1) in any particular case. (There are also theoretical reasons for expecting k to be 1, but we cannot go into these in this book.) However you should check for yourself that k is a *dimensionless* quantity by proving that

$$1 \text{ T m}^2 \text{ s}^{-1} = 1 \text{ V}$$

To do this you need to remember that $1 \text{ T} = 1 \text{ N A}^{-1} \text{ m}^{-1}$ (page 206), and that $1 \text{ V} = 1 \text{ J C}^{-1}$ (page 80).

The e.m.f. E induced in a coil of N turns of area A in a changing B field can therefore be calculated from the relation

$$E = NA\frac{dB}{dt} \tag{2}$$

► A search coil of 5000 turns of average area $1.00 \times 10^{-4} \text{ m}^2$ is mounted coaxially at the centre of a solenoid of length 0.300 m uniformly wound with 380 turns of wire. An oscillator joined across the solenoid supplies a sinusoidal current of peak value 0.400 A and of frequency 800 Hz. Calculate the peak e.m.f. E_0 induced in the search coil.

The peak value B_0 of the flux density in the solenoid is given by

$$B_0 = \frac{\mu_0 NI}{l} \tag{page 212}$$

$$= \frac{(4\pi \times 10^{-7} \text{ N A}^{-2})(380)(0.400 \text{ A})}{(0.300 \text{ m})}$$

$$= 6.37 \times 10^{-4} \text{ T}$$

The variation of B at frequency f is represented by

$$B = B_0 \sin 2\pi ft \tag{page 85}$$

Hence the maximum rate of change of B is given by

$$\left(\frac{dB}{dt}\right)_{max} = 2\pi fB_0 \tag{appendix B6}$$

$$= 2\pi(800 \text{ Hz})(6.37 \times 10^{-4} \text{ T}) = 3.20 \text{ T s}^{-1}$$

Thus according to equation (2) we have

$$E_0 = (5000)(1.00 \times 10^{-4} \text{ m}^2)(3.20 \text{ T s}^{-1})$$
$$= (1.60 \text{ T m}^2 \text{ s}^{-1}) = 1.60 \text{ V} \qquad \blacktriangleleft$$

If we need only to find out how an alternating B-field *varies* from one point to another we can do so without detailed calculations. For we have

$$E_0 \propto B_0$$

For instance, in the above experiment, if the search coil is moved level with the end of the solenoid, we find that E_0 falls to 0.80 V, showing that the B-field here is exactly half its value in the central region of the solenoid.

Using a search coil and oscilloscope with alternating current we may repeat many of the investigations of fields discussed in the last chapter where we used a Hall probe; for instance, we can study how the field varies near a straight wire or a flat circular coil. Only the alternating component of the field is detected with an a.c. search coil; therefore the steady field of the Earth or of other equipment nearby does not affect the measurements. The method can well be 100 times as sensitive as the Hall probe if sufficiently high frequencies are used.

When we need to use the search coil actually to *measure* the peak value B_0 of the field, we have

$$\left(\frac{dB}{dt}\right)_{max} = 2\pi fB_0 \tag{appendix B6}$$

and the peak e.m.f. E_0 is then given by

$$E_0 = NA\left(\frac{dB}{dt}\right)_{max} = NA(2\pi fB_0)$$

$$= 2\pi NAfB_0$$

Eddy currents

In the last chapter we saw how currents could be induced inside a solid piece of metal moving or rotating in a magnetic field (page 221). Eddy currents also occur in a stationary piece of metal in a changing magnetic field. The effect is very obvious if a solid lump of any metal is placed in a solenoid carrying a large alternating current; the lump rapidly becomes hot owing to the eddy currents circulating in it.

This conversion of electrical energy to internal energy needs to be prevented as far as possible in any iron cores of coils that carry alternating currents. This is achieved by making them of laminations, just as with the iron parts of motors and dynamos (page 221). The higher the frequency, the more the cores have to be subdivided to keep the rate of energy conversion by eddy currents down to acceptable limits. If thin laminations are not sufficient, a bundle of iron wires may be used, or even a core packed with iron dust. Another way of reducing this effect is to use magnetic substances that have very high resistivity. Such, for instance, are the iron compounds known as *ferrites*. Many of these may be classed as electrical insulators, though their magnetic properties are similar to those of iron. Ferrites are widely used to make magnetic cores for coils in radio receivers.

According to Lenz's law eddy currents must flow in such a way as to *reduce* the alternating magnetic field that causes them. At high frequencies this process is so efficient that we may use a *non-magnetic* metal sheet as a screen for magnetic fields. This may be investigated with a search coil held near the end of a solenoid with an alternating current in it. If an aluminium sheet is placed between the solenoid and the search coil, the induced e.m.f. in the coil is greatly reduced; and at frequencies of 10 kHz and over an aluminium sheet of 0.5 mm thickness is sufficient to provide almost perfect screening. In electronic apparatus aluminium cans are normally placed around coils and other components that carry high-frequency currents. These prevent alternating magnetic fields passing through the cans and inducing unwanted e.m.f.s in other parts of the circuit. (The same cans are also effective in screening the components from stray *electric* fields, (page 198.) However, this magnetic screening is not effective at low frequencies, when other methods must be used (page 234).

17.2 Magnetic flux

We now bring together the two methods of calculation we have used for the induced e.m.f.s in the two types of electromagnetic induction.

(i) *E.m.f.s induced in moving conductors.* For a conductor sweeping out area A at a steady rate we have

$$E = B_\perp \frac{dA}{dt}$$ (page 216)

If we bear in mind that the moving conductor must in practice form part of a complete circuit (if we are to detect the e.m.f.), then A can be regarded as the area of the *circuit*, which is changing at the rate dA/dt as the conductor moves. $B_\perp$ is the component of the flux density perpendicular to the plane in which the area is changing.

(ii) *E.m.f.s induced by changing currents.* For each turn of a coil of area A in a changing B-field we have

$$E = A \frac{dB_\perp}{dt}$$

$B_\perp$ is the component of the flux density perpendicular to the plane of the coil. In both of these cases the induced e.m.f. E is equal to the rate of change of the product $(AB_\perp)$; for we can write

in (i) $E = B_\perp \dfrac{dA}{dt} = \dfrac{d(AB_\perp)}{dt}$

since $B_\perp$ is constant; and

in (ii) $E = A \dfrac{dB_\perp}{dt} = \dfrac{d(AB_\perp)}{dt}$

since A is constant. The product $(AB_\perp)$ is called the *flux Φ* of the magnetic field through the area A (figure 17.4). Thus

$$\Phi = AB_\perp$$

The unit of magnetic flux is therefore the T m²; but it is convenient to have a special name for this unit, and it is called the *weber* (Wb), after Wilhelm Weber, who in 1846 set out a unified theory of electric and magnetic forces and effects. Thus

$$1 \text{ Wb} = 1 \text{ T m}^2$$

The reason for using the name *flux density* for the quantity B should now be clear. For

flux density = flux per unit area

and $1 \text{ T} = 1 \text{ Wb m}^{-2}$

The word *flux* itself perhaps needs some explanation. In developing the mathematical theory of magnetic fields the nineteenth century scientists were always making comparisons with the theory of the flow of fluids; they found it a helpful aid to the imagination. And so many of the words used in describing these fields have come down to us from the terms used in the theory of fluids. *Flux* is one of these. We talk about flux 'emerging from' a solenoid, or 'passing through' a coil. But it must be understood that this is only figurative language. The position is a stationary one, and nothing actually flows along the lines of force!

We can now summarise the laws of electromagnetic induction in the single statement

$$E = -\frac{d\Phi}{dt}$$ (3)

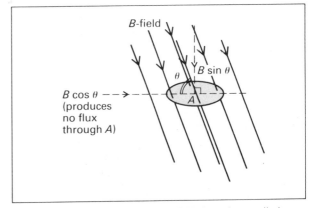

Figure 17.4 The flux of a magnetic field through a small plane area A.

This is sometimes known as *Neumann's law*. The minus sign is needed to express Lenz's law, since the induced e.m.f.s always act in such a way as to drive a current that would *oppose* the flux changes. Equation (3) reveals another relationship between the units that is useful to bear in mind:

$$1 \text{ Wb} = 1 \text{ V s}$$

(Compare the relations between the units referred to on page 224.)

Although we have defined the concept of flux only for areas over which the field is uniform, we may use Neumann's law to measure flux in other cases also, e.g. to find the flux through a large coil (over which the field may well not be uniform). For instance, if we can arrange for the flux to alternate at frequency f, we can use the coil as an a.c. search coil and so find the peak value Φ_0 of the flux. If the coil is of N turns, the *flux linkage* $N\Phi$ of the field with the coil is N times the flux through any one turn of it; and the total induced e.m.f. E is N times the e.m.f. induced in one turn of the coil. So for a sinusoidally varying B-field we can write

$$N\Phi = N\Phi_0 \sin 2\pi f t$$

so that $\quad E_0 = N\left(\dfrac{d\Phi}{dt}\right)_{\max} = N(2\pi f \Phi_0) \quad$ (appendix B6)

$$\Rightarrow \qquad \Phi_0 = \frac{E_0}{2\pi f N} \qquad\qquad (4)$$

►   Two flat coils are placed on top of one another on a table. There is a sinusoidal alternating current of frequency 400 Hz in the lower coil; and an alternating e.m.f. of peak value 6.8 V is induced in the top coil (which has 100 turns). (i) Calculate the peak value of the flux through the top coil. (ii) If the top coil is 35 mm square, calculate the peak value B_0 of the flux density over its cross-section.
(i)   Using equation (4) above, we obtain

$$\Phi_0 = \frac{6.8 \text{ V}}{2\pi(400 \text{ Hz})(100)}$$

$$= 2.7 \times 10^{-5} \text{ Wb} = 27 \text{ μWb}$$

(ii) The flux density B is probably by no means uniform over the cross-section of the coil. But the average value of B_0 is given by

$$B_0 = \frac{2.7 \times 10^{-5} \text{ Wb}}{(35 \times 10^{-3} \text{ m})^2}$$

$$= 2.2 \times 10^{-2} \text{ T} = 22 \text{ mT} \qquad \blacktriangleleft$$

The continuity of magnetic flux

Pictures of magnetic fields by means of lines of force (such as in figure 16.4, page 205) suggest that the flux of a magnetic field always exists in closed loops, and detailed measurements show that this is exactly the case. Whenever flux enters a space, such as the inside of a solenoid, just as much leaves it at some other point. We describe this by saying that magnetic flux is *continuous*. In this respect magnetic flux has the same property as electric current; whatever electric current enters any region must flow out of it again at some other point – electric current also is *continuous*. But of course magnetic flux does not consist of a flow of anything.

The continuity of magnetic flux applies not only with the fields produced by electric currents but also with the fields of magnets. Magnetic flux does not 'sprout' from the ends of a magnet, but exists in closed loops – down the inside of the magnet from S pole to N pole, out of the N pole into the air, and back again through the air into the S pole – in fact just like the field of a solenoid. Although we cannot put a probe into the inside of a magnet, we can do experiments to find the total flux passing through the body of the magnet and so show the existence of the internal B-field of the magnet. For instance, we may use a close-fitting search coil, and move the magnet into or out of this, at the same time measuring the induced e.m.f.

►   A close-fitting search coil of 80 turns is placed round the centre of a bar magnet. The magnet is removed from the search coil in such a way that a meter in series registers a steady e.m.f. of 1.2 mV; this process takes 5.0 s. (i) Calculate the flux Φ through the centre of the bar magnet. (ii) If the magnet is of rectangular cross-section, 25 mm × 10 mm, calculate the average flux density B in the central cross-section.
(i)   The induced e.m.f. E is given by

$$E = \frac{\Delta(N\Phi)}{\Delta t}$$

As the magnet is removed from the coil, the flux linkage changes steadily from its initial value $N\Phi$ to zero. So we can write

$$1.2 \times 10^{-3} \text{ V} = \frac{N\Phi}{5.0 \text{ s}}$$

$$\Rightarrow \qquad N\Phi = 6.0 \times 10^{-3} \text{ Wb}$$

Hence the flux Φ through the centre of the magnet is given by

$$\Phi = \frac{6.0 \times 10^{-3} \text{ Wb}}{80}$$

$$= 7.5 \times 10^{-5} \text{ Wb or 75 μWb}$$

Such an experiment may be extended to find the flux through other sections of the bar magnet. It will be found that less then half the total flux emerges through the end faces; most of it emerges from the sides of the magnet near its poles.
(ii) The average flux density B through the centre of the magnet is the flux per unit area of cross-section at this point.

$$\Rightarrow \qquad B = \frac{7.5 \times 10^{-5} \text{ Wb}}{(25 \times 10^{-3} \text{ m})(10 \times 10^{-3} \text{ m})}$$

$$= 0.30 \text{ T or 300 mT} \qquad \blacktriangleleft$$

17.3 Inductance

In the process of *mutual induction* the current changing in one coil causes an induced e.m.f. in a second coil. But e.m.f.s are also induced in a single coil *by changes in its own current*, since this also produces magnetic flux through the coil. This process is known as *self-induction*. From Lenz's law we expect such induced e.m.f.s to act in opposition to the changes of current that cause them.

In air-cored coils (and in some iron-cored coils too) the flux linkage $N\Phi$ of the magnetic field with a single coil is proportional to the current I in it. So we can write

$$N\Phi = LI$$

where L is a constant known as the *inductance* of the coil; it depends on the size, number of turns and geometry of the coil. The unit of inductance is therefore the Wb A^{-1}; but it is convenient to have a special name for this unit, and it is called the *henry* (H) after the American electrical engineer, Joseph Henry, who in 1832 observed the effects of self-induction in electric circuits. Thus

$$1\ \text{H} = 1\ \text{Wb A}^{-1}$$

You can check for yourself the following connections with other units:

$$1\ \text{H} = 1\ \text{V s A}^{-1} = 1\ \Omega\,\text{s}$$

From Neumann's law, equation (3) above, the induced e.m.f. E produced by self-induction in a single coil is given by

$$E = -\frac{\mathrm{d}(N\Phi)}{\mathrm{d}t} = -\frac{\mathrm{d}(LI)}{\mathrm{d}t} = -L\frac{\mathrm{d}I}{\mathrm{d}t}$$

since L is constant. Thus the induced e.m.f. is proportional to the rate of change of current.

An inductive coil (or *inductor*) inevitably has resistance R as well as inductance L (unless it is superconducting). It is useful to think of a coil as though it consisted of a 'pure' inductance (without resistance) in series with a 'pure' resistance (without inductance), as in figure 17.5. But we must remember that L and R are actually inseparable properties of the whole coil. If the current in such a coil is to be I, and this is to grow at a rate $\mathrm{d}I/\mathrm{d}t$, we must apply a potential difference V across it made up of two components.
(i) A p.d. equal in size but opposing the induced e.m.f.

$$= L\frac{\mathrm{d}I}{\mathrm{d}t}$$

This is required to maintain the *growth* of current in the inductance.
(ii) The p.d. required to maintain the current in the resistance

$$= IR$$

The total p.d. V across the inductive coil is therefore given by

$$V = L\frac{\mathrm{d}I}{\mathrm{d}t} + IR \qquad (5)$$

When the current has reached its final steady value, the first term on the right of this equation vanishes, and we have, as for any resistance,

$$V_{\text{final}} = IR$$

But this term of the equation is zero at the moment of switching on, when $I = 0$, and before the current has had time to grow appreciably. The growth of current at this moment is controlled entirely by the inductance of the coil, and we have

$$V_{\text{initial}} = L\frac{\mathrm{d}I}{\mathrm{d}t}$$

In other words, at the moment of switching on the e.m.f. induced in the coil is exactly equal to the applied p.d., and this enables the current to grow at the rate given by this equation. The way this works out is best understood by considering a numerical example.

▶ A coil of inductance 0.10 H and resistance 4.0 Ω is connected to a 12 V car battery (of negligible resistance). Calculate (i) the final steady current in the coil; (ii) the initial rate of growth of current; (iii) the current after 1.0 ms.

(i) The final current I is given by

$$I = \frac{V}{R}$$

$$= \frac{12\ \text{V}}{4.0\ \Omega} = 3.0\ \text{A}$$

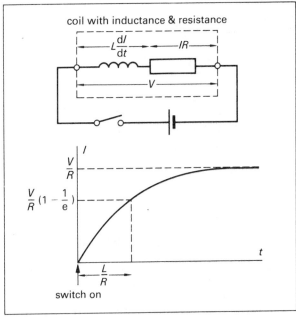

Figure 17.5 The exponential growth of current in an inductive coil; the time constant is L/R.

(ii) The initial rate of growth of current dI/dt is given by

$$\frac{dI}{dt} = \frac{V}{L}$$

$$= \frac{12\ V}{0.10\ H} = 120\ A\ s^{-1}$$

(iii) Assuming dI/dt is nearly constant for a short time interval Δt after switching on, the increase in current ΔI is given by

$$\frac{\Delta I}{\Delta t} = 120\ A\ s^{-1}$$

If $\Delta t = 1.0$ ms, we have

$$\Delta I = (120\ A\ s^{-1})(1.0 \times 10^{-3}\ s)$$
$$= 0.12\ A$$

Equation (5) above, which describes the growth of current, may be re-arranged as

$$\frac{L}{R}\frac{dI}{dt} = \frac{V}{R} - I$$

This is related to the form discussed in appendix B7; there is a linear relation between the rate of change dI/dt and the current I itself. This means (though we shall not prove it here) that the current in an inductor grows *exponentially* towards its final value V/R (figure 17.5) (In other words, if you examine the graph in figure 17.5, the difference between the final current (V/R) and momentary current I *decays* exponentially, until eventually the difference is zero, and the current has reached its final steady value.) The quantity L/R is the *time constant* of the combination. In other words, the current grows to within $1/e$ ($= 0.37$) of its final value in the time L/R; when $t = L/R$, the current I is given by

$$I = \frac{V}{R}(1 - 0.37) = 0.63\frac{V}{R}$$

In the example considered above, the time constant is

$$\frac{L}{R} = \frac{0.10\ H}{4.0\ \Omega}$$

$$= 0.025\ s = 25\ ms$$

In this time interval the current grows from zero to the value

$$0.63 \times 3.0\ A = 1.9\ A$$

In d.c. circuits the effects of self-induction are apparent only at switching on or off – while the current is rising to its maximum value or falling again to zero. At switching on the induced e.m.f. acts in opposition to the applied potential difference, so that it delays the growth of current (figure 17.6). If we arrange to observe this growth of current on a c.r.o., we can measure the gradient dI/dt just after switching on (when the induced e.m.f. is equal to the applied p.d.), and so find the inductance of the coil. With

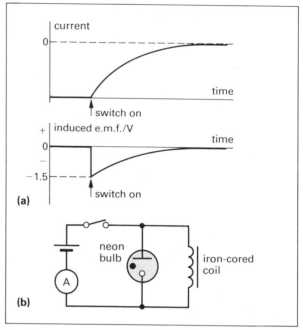

Figure 17.6 (a) At switching on the current grows initially at such a rate that the induced e.m.f. in the coil is exactly equal to the applied potential difference. At switching off a very large e.m.f. is produced, sufficient to light the neon bulb (>100 V).

a very large iron-cored coil the growth of current may well take several seconds, and then we can even use a moving-coil ammeter to give the value of dI/dt.

At switching off the behaviour of an inductive circuit like this is much more dramatic. When the switch is opened, the current is obliged to drop almost instantly to zero, and the flux linked with the coil changes extremely rapidly. A very large pulse of e.m.f. is thus induced in it, which acts so as to tend to keep the current going. The peak value of this e.m.f. is often many hundreds of times greater than the original applied p.d., sufficient in figure 17.6 to light the neon lamp (which requires at least 100 V). If two fingers (of *one* hand) are placed on the switch terminals, an appreciable shock may be felt even when only a single 1.5 V cell is used in the circuit. (You should not try this with more than one cell; it could be dangerous.)

This effect poses problems for the designer of switches. If the contacts are separated too slowly at switching off, the e.m.f. of self-induction may be sufficient to cause an arc between them. In extreme cases the arc can melt the switch arms. In a domestic light switch it is sufficient to use a spring-loaded switch arm, which results in rapid separation of the contacts. In large switch systems jets of compressed air are made to blow out the arc that forms as the contacts separate; or else the contacts are mounted in a tank of non-inflammable oil under pressure, which quickly quenches the arc.

In a.c. circuits the e.m.f.s of self induction must act in such a way as to oppose the applied p.d. at every instant.

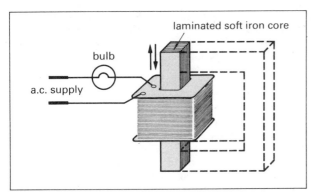

Figure 17.7 The inductance may be varied by moving the iron core in or out of it; this controls the brightness of the bulb.

The value of the current depends not only on the resistance of the circuit but also on its inductance. This may be demonstrated with a coil of many turns with a removable (laminated) iron core. The coil is joined in series with an electric bulb to the a.c. mains (figure 17.7). With the core removed the inductance of the coil is fairly small, and (at a frequency of 50 Hz) the bulb brightness is scarcely affected. But when the iron core is introduced, the inductance is increased, and the current falls to a small value. If it can be arranged so that a closed loop of laminated iron is formed through the coil (as suggested by the dotted lines) the effect is even more marked. An inductive coil used in this way is called a *choke*; an alternating current can be controlled by this means without much conversion of electrical energy to internal energy, such as would occur if a resistor was used instead.

In a later chapter (page 277) we shall analyse exactly how the alternating current *I* in an inductor and the p.d. *V* across it are related. But it is useful at this point to work out the connection between the peak values V_0 and I_0 for a 'pure' inductance. In practice it is not quite as unrealistic to talk about a 'pure' inductance as you might think. For if we use a high enough frequency the reversals of p.d. occur at intervals that are short compared with the time constant L/R of the coil; the current then never rises to the level at which the second term on the right in equation (5) is anything but negligible compared with the first. This would apply, for instance, with the coil considered above at a frequency of 500 Hz or more; the reversals of p.d. then occur at intervals of 1 ms, which is small compared with the time constant of 25 ms. So at high enough frequencies we can write for *any* coil

$$V = L\frac{dI}{dt}$$

If the current *I* is sinusoidal, we have

$$I = I_0 \sin 2\pi f t$$

Hence $\quad V_0 = L\left(\dfrac{dI}{dt}\right)_{max} = L(2\pi f I_0) \qquad$ (appendix B6)

This provides us with a simple means of measuring the inductance *L*, using just an a.c. ammeter and voltmeter to measure *V* and *I*, and an a.c. supply of sufficiently high frequency. Such meters normally register r.m.s. values of current and p.d. (rather than peak values, page 272); but this does not matter, as we can re-arrange the above result to give

$$2\pi f L = \frac{V_0}{I_0} = \frac{V_{rms}}{I_{rms}}$$

Magnetic energy

Because of the e.m.f. of self-induction that acts when the current in a coil increases, electrical energy must be supplied in setting up the current against this induced e.m.f. Multiplying equation (5) above by *I* we have

$$VI = LI\frac{dI}{dt} + I^2 R$$

and this equation says:

power supplied = rate of supplying magnetic energy
+ rate of supplying internal energy

The magnetic energy supplied is stored in the *B*-field of the coil. It grows as the current grows, and is returned to the circuit again as the current falls. Thus

$$\text{rate of supplying magnetic energy} = LI\frac{dI}{dt}$$

If the current grows by an amount ΔI in a small time interval Δt, the magnetic energy ΔW_B supplied in this interval is given by

$$\Delta W_B = \left(LI\frac{\Delta I}{\Delta t}\right)\Delta t = LI\Delta I$$

A graph of LI against I is of the form shown in figure 17.8; the gradient of the line is *L*. The increase of magnetic energy

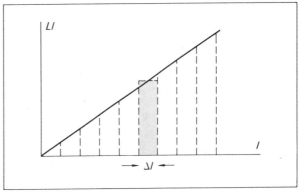

Figure 17.8 A graph of *LI* against *I* for an inductor; the energy stored is given by the area between the graph line and the *I*-axis.

$LI\Delta I$ is represented in figure 17.8 by the area of the shaded strip under the graph. If the current is increased in a series of equal steps ΔI, further quantities of energy $LI\Delta I$ are stored which increase steadily as the current I grows. The total energy W_B stored in the inductor is represented by the total area under the graph of LI against I. Hence

$$W_B = \tfrac{1}{2}(LI)I = \tfrac{1}{2}LI^2$$

Notice that for a given inductance L the energy stored is proportional to the *square* of the current I. Doubling the current in the coil quadruples the energy W_B stored.

In the example considered above, when the current has grown to its final value of 3.0 A, we have

$$W_B = \tfrac{1}{2}LI^2$$
$$= \tfrac{1}{2}(0.10 \text{ H})(3.0 \text{ A})^2$$
$$= 0.45 \text{ J}$$

It is this energy that appears in the spark at the switch contacts or the flash of the neon bulb in the experiment of figure 17.6.

Calculating inductance

To calculate the inductance L of a given coil we need to know the flux linkage $N\Phi$ when the current in it is I; then

$$N\Phi = LI$$

We shall consider only one important case, that of a *long solenoid*. Suppose the solenoid is of cross-sectional area A, contains a total of N turns, and is of length l. If we ignore the reduction of field that takes place near the ends (justified only if the end region is a *small* part of the total length), we can write

$$B\text{-field in the core} = \frac{\mu_0 NI}{l}$$

So flux Φ through the core $= \dfrac{\mu_0 ANI}{l}$

and flux linkage $N\Phi$ with the whole coil $= \dfrac{\mu_0 AN^2 I}{l}$

$$\Rightarrow \quad L = \frac{N\Phi}{I} = \frac{\mu_0 AN^2}{I}$$

▶ A solenoid used for the study of magnetic fields is of square cross-section, 70 mm by 70 mm, and of length 0.30 m; it is uniformly wound with 380 turns of wire. Estimate the inductance of the solenoid.

According to the above result

$$L = \frac{(4\pi \times 10^{-7} \text{ N A}^{-2})(70 \times 10^{-3} \text{ m})^2(380)^2}{0.30 \text{ m}}$$

$$= 3.0 \text{ mH}$$

However the flux through the end turns of the solenoid is only half that through the centre turns; and we expect the measured inductance of such a solenoid to be appreciably less than this figure (probably about 2.5 mH). ◀

Two results follow from all such calculations.
(i) For coils of the same dimensions the inductance is proportional to N^2. Thus a solenoid with *half* as many turns as the above, but of identical length and cross-section will have a *quarter* the inductance.
(ii) For geometrically similar coils, the inductance is proportional to the linear dimensions. (Suppose we have two coils, one twice the diameter, length, etc, of the other. The larger coil has *four* times the cross-sectional area A of the other, and *twice* the length l; therefore A/l is twice as big for this coil.)

These two results apply not only for long solenoids but also for coils for other shapes as well (for which the above formula giving the inductance of a long solenoid does not apply).

17.4 The transformer

A low-frequency transformer consists of two coils, a primary and a secondary, wound on a closed iron core, which must be laminated to minimise eddy currents (figure 17.9). This design ensures that the maximum magnetic flux is produced in the core for a given current. It also ensures that the magnetic coupling between the coils is almost complete – i.e. nearly all the flux threads equally through all the turns of both coils.

The a.c. supply is joined to the primary coil, which has a large number of turns, enough to ensure that the current in it is controlled entirely by its inductance; its resistance has negligible effect. The alternating flux produced in the core induces alternating e.m.f.s E_p and E_s of the same frequency in the primary and secondary coils. In the primary E_p is the induced e.m.f. of self-induction; it is nearly equal to the supply p.d. V_p at every instant. Thus

$$E_p \approx V_p$$

The same flux in the core threads equally through all the turns of both coils. Therefore the e.m.f. induced per turn is the same throughout the transformer, so that

$$\frac{E_s}{N_s} = \frac{E_p}{N_p}$$

where N_s and N_p are the numbers of turns in the secondary and primary respectively.

$$\Rightarrow \quad \frac{E_s}{E_p} = \frac{N_s}{N_p}$$

As long as there is no current in the secondary circuit, the output p.d. V_s at the terminals of the secondary coil is equal to E_s. When current is taken, the p.d. V_s falls because of the secondary coil resistance; but provided the current is not too great, the difference is negligible, and we can write

$$E_s \approx V_s$$

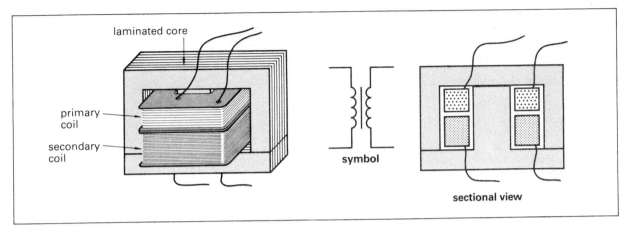

Figure 17.9 The structure of a transformer.

and so, provided the transformer is not overloaded, we have

$$\frac{V_s}{V_p} \approx \frac{N_s}{N_p} \qquad (6)$$

By suitable choice of the *turns ratio* (N_s/N_p) we can thus transform an alternating supply at one p.d. to a supply of the same frequency at any other p.d. For instance, a transformer to run on a 12 V a.c. supply has a primary coil of 96 turns; to give an output p.d. of 50 V the number of turns N_s required in the secondary coil is given by

$$N_s = \frac{50 \text{ V}}{12 \text{ V}} \times 96 \text{ turns} = 400 \text{ turns}$$

In practice (because of the approximations we have made) a few more turns than this will be needed to give the desired output p.d. of 50 V.

As long as there is no current in the secondary circuit, the current in the primary circuit is *very small* (since its inductance is so large). When a current I_s is taken from the secondary, by Lenz's law its effect must be to *reduce* the magnitude of the alternating flux in the core that caused it. This causes a drop in the primary induced e.m.f. E_p. which allows the primary current I_p to grow. The primary current I_p tends to *maintain* the alternating flux in the core, while the secondary current I_s tends to *reduce* it. The primary and secondary currents therefore flow in opposite senses round the core (or, as we say, they are in antiphase, page 274). In fact I_p and I_s have almost equal and opposite effects in producing the alternating flux in the core; the small difference between their effects is enough to maintain the alternating flux in the core at almost its initial value.

The flux produced by a coil in a given core is proportional to the product of the current and the number of turns. If the coils produce nearly equal and opposite effects, we can write

$$N_s I_s \approx N_p I_p$$
$$\Rightarrow \quad \frac{I_s}{I_p} \approx \frac{N_p}{N_s} \qquad (7)$$

This is the *reciprocal* of the turns ratio. If the p.d. is transformed up in a given ratio, then the current is transformed down in the same ratio (approximately), and vice versa. For instance, with the transformer considered above, if an ammeter in series with the primary shows a reading of 2.5 A, the current I_s in the secondary will be given by

$$\frac{I_s}{2.5 \text{ A}} = \frac{1}{\text{turns ratio}} = \frac{96}{400}$$
$$\Rightarrow \quad I_s = \frac{96}{400} \times 2.5 \text{ A} = 0.60 \text{ A}$$

In practice (because of the approximations we have made) the current in the secondary will be a little less than this.

Multiplying together both sides of equations (6) and (7) above, we get

$$\frac{V_s I_s}{V_p I_p} \approx 1$$
$$\Rightarrow \quad V_s I_s \approx V_p I_p$$

i.e. power output of secondary $\approx$ power input of primary. With the transformer just considered

power output of secondary $= (50 \text{ V})(0.60 \text{ A}) = 30 \text{ W}$
power input of primary $= (12 \text{ V})(2.5 \text{ A}) = 30 \text{ W}$

In practice, however,

$$V_s I_s < V_p I_p$$

since some of the energy input is converted to internal energy in the transformer. This happens in three different ways.

(i) *Resistive heating* – internal energy produced in the coils by the currents I_p and I_s.

(ii) *Eddy current heating* – internal energy produced in the iron core by the eddy currents circulating inside the individual laminations.

(iii) *Magnetisation heating* – internal energy produced in the iron core by the continual reversals of its magnetisation (page 234).

But the transformer is in fact an exceedingly efficient device. The large ones used in power supply systems handling hundreds of kilowatts are usually more than 99 % efficient. However, in saying this we need to bear in mind that even 1 % of 500 kW is 5 kW (the same as the output of five 1 kW electric fires). Large transformers therefore require elaborate cooling systems, with cooling fluid arranged to circulate through the core and coils.

Transformers are often wound with several secondary coils, electrically insulated from one another. The mains transformer of an audio amplifier is usually made this way. In addition to the main winding supplying power for the circuit there may be several additional windings; these provide power for indicator lamps, etc. The resistive heating in a coil of a transformer is proportional to its resistance and to the square of the current ($P = I^2R$). To minimise this wastage of energy, high-current low-voltage windings are made of thick wire; while the low-current high-voltage coils can well be of relatively fine wire – but of course such coils require many more turns, and probably take up as much space as the low-voltage coils.

17.5 Magnetic materials

Magnetic forces arise between charged particles in motion (page 75). All atoms are therefore inherently magnetic because of the circulating charged particles they contain. To a large extent the orbital of electrons exist in pairs whose magnetic effects cancel out; but with many atoms there is some residual magnetic effect, caused chiefly by the *spin* of the electrons (page 78). All substances are to some extent magnetic; but the only materials whose magnetic properties are sufficiently marked to produce substantial effects are *ferromagnetics* (like iron), considered below. Other materials are by comparison only weakly magnetic; but we shall deal with these first. They fall into two classes, depending on the presence or absence of a residual magnetic effect in their atoms.

Paramagnetics

These are substances in which the electron orbitals and spins of the individual atoms or ions are not completely balanced; the atoms or ions therefore behave like small magnets. Placed in a magnetic field the atoms or ions tend to align themselves with the field. However, at ordinary temperatures the thermal agitation of the atoms is sufficient to cause their magnetic axes to be arranged almost entirely at random even when strong magnetic fields are applied in an attempt to align them in one direction. Magnetic fields of more than 10^3 T would be required to align most of the atoms – quite unattainable in the laboratory. Only at temperatures close to the absolute zero is it possible to achieve such an effect. In the ordinary way the degree of magnetisation of a paramagnetic specimen *increases* the B-field inside it by less than one part in 1000.

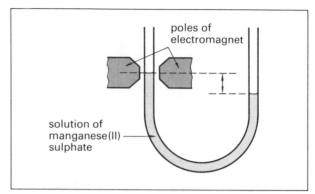

Figure 17.10 The movement of a paramagnetic liquid in a magnetic field.

The forces acting on a paramagnetic specimen in a magnetic field are similar to those acting on a piece of iron, though much smaller. Thus it is *attracted* towards the stronger parts of a non-uniform field. This can be demonstrated with a paramagnetic liquid, such as a concentrated solution of manganese(II) sulphate, contained in a U-tube. If one meniscus is placed between the poles of a powerful electromagnet, as shown in figure 17.10, a difference of level of about 15 mm is produced by a field of 0.5 T. Experiments of this sort enable us to find the electromagnetic moment of the individual ions of the solution – in this case the manganese(II) ion.

Diamagnetics

In these substances the electron orbitals and spins are exactly balanced under normal conditions. But when a magnetic field is applied, the motions of the electrons are modified slightly in such a way as to *reduce* the B-field in the material. This is in fact an example of the process of *electromagnetic induction* taking place on an atomic scale, and the reduction in B-field is in agreement with Lenz's law. The effect is very small; inside a diamagnetic specimen the B-field is *decreased* by less than one part in 10^5 by the 'magnetisation' of the specimen, and it is nearly independent of the temperature. Presumably the diamagnetic effect takes place in all materials, including paramagnetics, but is normally swamped by the much greater paramagnetic effect.

A diamagnetic material is *repelled* from the stronger parts of a non-uniform field; but the effect is difficult to demonstrate. If the experiment of figure 17.10 is repeated with a diamagnetic liquid, such as water, there is a just perceptible *fall* in the left-hand meniscus when a field of 0.5 T is switched on; but the meniscus needs to be watched through a low power microscope to see this.

Ferromagnetics

In these substances the electron orbitals and spins are unbalanced, just as in a paramagnetic substance; and the magnetism of the individual atoms is of the same order of magnitude. But their characteristic property is that the interactions between neighbouring atoms are such as to align them with their magnetic axes parallel to one another; and this is so even when no external magnetic field is applied. On an atomic level a ferromagnetic material thus has an inherent local magnetisation. Thermal agitation tends to disrupt this, of course, and at high enough temperatures the local alignment of the atoms breaks down, and the material ceases to exhibit its characteristic ferromagnetic properties. The temperature at which this happens is known as the *Curie temperature*. For iron this is 1040 K, though for some ferromagnetic alloys of nickel it is below 373 K. Above the Curie temperature ferromagnetic materials become paramagnetic. But below the Curie temperature the B-field inside a ferromagnetic specimen can be increased by a factor of anything up to 10^5 times by the local magnetisation of the material – and the effect is therefore an obvious and familiar one. Only iron and a few other substances of closely related atomic structure exhibit ferromagnetic properties.

Within any one crystal of a ferromagnetic substance there is thus a tendency for all the atoms to be lined up in one direction. The equilibrium structure of the crystal is one in which it divides up into a series of regions called *domains* (figure 17.11). In any one domain all the atoms are aligned with their magnetic axes in one direction, but the direction varies from one domain to another. The domain boundaries are shown up in this photograph by placing the specimen in a liquid carrying a suspension of fine magnetic particles. These particles concentrate along the boundaries between domains, where the magnetic field is highly non-uniform. The arrows drawn in on the photograph mark the direction of magnetisation of each domain. We must emphasise that this photograph is of a *single* crystal (which in this case has 5 magnetic domains).

In a freshly formed crystal the magnetic fields of the domains cancel out, and virtually no external field can be detected; in this condition we say that the specimen is *unmagnetised*, although on the atomic level there is a high degree of local magnetisation. If now it is placed in a magnetic field, each atom experiences a torque tending to align its magnetic axis with the field. For small fields the result is a general shifting of the domain boundaries in such a way that the proportion of atoms lined up with the field increases (as shown in the succeeding pictures of figure 17.11). The specimen now produces a magnetic field adding to the field that magnetised it. To some extent this shifting of the domain boundaries is reversible; when the magnetising field is removed, the boundaries return more or less to their original positions, and much of the magnetisation disappears. In larger applied fields the axes of whole domains rotate abruptly one after another, until the point is reached at which *all* the atoms in the specimen are aligned with the field. No further magnetisation is then of course possible. These latter changes are mostly irreversible, so that the material retains much of its magnetisation when the field is removed – and so we are left with a *magnet*.

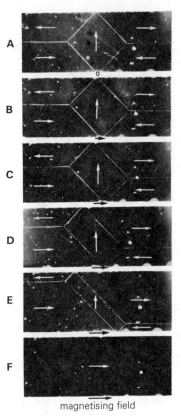

magnetising field

Figure 17.11 The domain structure of a crystal of iron. The domain boundaries are shown up by immersing the crystal in a weak suspension of magnetic particles, which tend to congregate along the boundaries where the field is non-uniform. The magnetising field increases through the sequence of pictures.

It is possible to obtain single large crystals of iron and other ferromagnetic materials. But generally these substances solidify in polycrystalline form (i.e. myriads of microscopic crystals orientated at random throughout the specimen, page 44). Each crystal divides separately into domains, and responds more or less independently to an applied field. The observed properties of the specimen are the summed effects of this random arrangement of crystals.

The magnetic properties of iron (or any other ferromagnetic material) may be represented graphically as in figure 17.12 (which is actually for mild steel). Starting with an unmagnetised specimen a magnetic field is applied and increased slowly from zero. At first, magnetisation is produced by the shifting of the domain boundaries (O to A on the curve). A larger field causes sudden alterations of domain axes, and the magnetisation then increases more rapidly (A to B on the curve). But beyond a certain strength of magnetic field virtually all the domains are aligned, and no further increase of magnetisation is possible; the magnetic material is then said to be *saturated* (C). (The flux density in a saturated piece of pure iron is about 2 T; for steels it is about 1.5 T.) When the magnetising field is

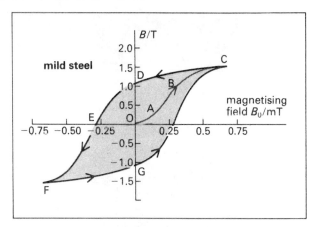

Figure 17.12 A hysteresis loop showing the way in which the magnetisation of a loop of mild steel varies with the magnetising field.

reduced again, the magnetisation does not return to zero, but remains (D) not far below its saturation value; and an appreciable reverse field has to be applied before it is much reduced again (E). A large enough reverse field changes over the alignment of the domains, and a rapid change of magnetisation then occurs until the specimen is saturated in the reverse direction (F). A similar sequence is followed when the field is once more reversed to produce saturation in the original direction (FGC).

Thus the magnetisation of a specimen does not depend in a simple way on the magnetising field, but appears to lag behind the changes in the field. This phenomenon is known as *magnetic hysteresis*; and a magnetisation curve, like that in figure 17.12 obtained by varying the magnetising field cyclically, is called a *hysteresis loop*. When a material is magnetised, magnetic energy is stored in it (derived from the electrical energy available in the surrounding solenoid). Taking a piece of material through a cycle of magnetisation like this causes some of the magnetic energy in the material to be converted to internal energy, which raises the temperature of the material. Magnetisation heating has to be allowed for in the design of equipment (such as transformers) with magnetic cores. For a typical material used in a transformer core the energy converted per cycle of magnetisation is about 2×10^{-2} J per kilogram of material. Thus in a small a.c. mains transformer (for 50 Hz), whose core contains 0.5 kg of iron, the power conversion on this account is

$$(2 \times 10^{-2} \text{ J kg}^{-1})(0.5 \text{ kg})(50 \text{ Hz}) = 0.5 \text{ W}$$

The hysteresis loops obtained for various ferromagnetic substances are all generally similar to figure 17.12; but the horizontal scales of the graphs show large variations; the field required to reverse the magnetisation (point E on the graph) varies from about 4 μT for an alloy called *mumetal* to more than 0.1 T for some of the materials used to make magnets.

In the early investigations of magnetism the extremes of

magnetic properties were represented by *soft iron* and *hard steel*. Soft iron is very easily magnetised and demagnetised; hard steels require large magnetic fields to do this. Soft iron was therefore used for the cores of electromagnets and other apparatus; hard steels were used for permanent magnets. Modern magnetic materials show much more extreme properties than soft iron and hard steel, but the words 'soft' and 'hard' are still used (in a magnetic sense only) to describe the properties of these new materials. Thus *mumetal* (74% nickel, 20% iron, 5% copper, 1% manganese) is one of the modern 'soft' materials; *physically* it is a hard substance, but *magnetically* it is very soft. Its magnetisation can be reversed even in the Earth's field. Mumetal is one of the best materials to use for making screens to shield pieces of apparatus from magnetic fields. (However, high-frequency alternating fields are best screened by using sheets of aluminium, page 225). Typical of the modern 'hard' magnetic materials are the alloys in the *alnico* range, and the compounds like *magnadur*; these are used for making magnets, and require fields of 0.1 T or more to magnetise or demagnetise them.

Permeability

The effect of filling the core of a coil or solenoid with iron is to increase the flux density B produced in it for a given current. Because of hysteresis the flux Φ in a core of real (soft) magnetic material is not in general proportional to the magnetising current I in the coil (figure 17.13). The result of this is that electrical components with iron cores, such as transformers and chokes, always introduce some distortion into the waveforms that the circuits are handling. This is very obvious if the input and output waveforms of a transformer are compared, particularly if it is deliberately run for a short time at higher input p.d.s than it is designed for (so that the core is saturated at the peaks of current).

However, in this discussion we shall confine our attention

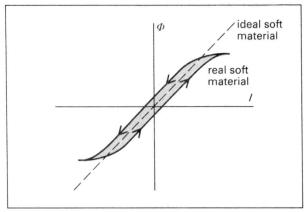

Figure 17.13 The variation of flux Φ with magnetising current I for a coil on a loop of soft magnetic material. An ideal soft material would show no hysteresis or saturation effects.

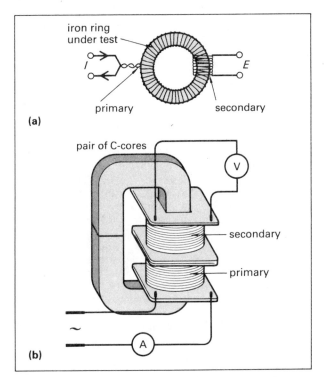

(a)

(b)

Figure 17.14 Measuring relative permeability: (a) for an iron ring uniformly wound with a primary coil; (b) for a pair of C-cores linking primary and secondary coils.

to magnetic cores consisting of *ideal* soft magnetic materials, in which the B-field is exactly proportional to the magnetising current I, and the hysteresis loop is imagined to be collapsed into a straight line (figure 17.13). At best this is an approximation to the real situation; but it is quite a good approximation in many cases, and is easily analysed. Also we shall deal only with *closed loops* of ferromagnetic material. Even a small air gap in a magnetic core markedly reduces the B-field in it, and we shall avoid such complications in this book.

Consider a ring-shaped coil of N turns uniformly wound on a closed loop of an ideal magnetic material (figure 17.14a). We can think of the coil as a ring-shaped solenoid, so that in the absence of the magnetic material the flux density B_0 inside the ring would be given by

$$B_0 = \frac{\mu_0 N I}{l}$$

where l is the average length round the lines of force inside the ring. The effect of filling the coil with the magnetic material is to increase the B-field in it by a constant factor. The flux density B in the material is thus given by

$$B = \mu_r B_0$$

where μ_r is a constant called the *relative permeability* of the material. Notice that μ_r is a dimensionless quantity. For a

vacuum $\mu_r = 1$, by definition. For air, and for any paramagnetic or diamagnetic substance μ_r is also very little different from 1. But most soft ferromagnetic materials have relative permeabilities over 10^3, and in some materials values up to 10^5 are obtained.

To measure the relative permeability μ_r we have to find the flux density B in the ring for a given magnetising current I (from which we can calculate B_0). This may be done by winding a secondary coil (i.e. a search coil) round part of the ring and measuring the e.m.f. induced in this when an alternating current of known frequency is supplied to the primary coil (i.e. the solenoid). The calculation of B from the e.m.f. E induced in the secondary coil is done in the same way as with other a.c. search coils (page 224). We have to make sure that the specimen we are studying is in such a form that eddy currents are not induced in it significantly. This means that it must be *laminated* or in the form of a bundle of wires. In fact for this experiment we need to make up what amounts to a transformer. It makes little practical difference what shape the core is, as long as it contains no air gaps; and the result is little affected by having the coils bunched into convenient units, as in figure 17.14b, rather than being spread out uniformly round the core. The primary current I and secondary e.m.f. E can be measured with a.c. ammeter and voltmeter, as shown. Alternatively we may use an oscilloscope to measure the peak values, as in figure 17.3, page 223.

Iron-cored inductances

The inductance L of a coil of N turns wound on a *closed loop* of magnetic material is simply μ_r times the inductance of the equivalent air-cored coil. Again it makes little difference whether the coil is spread out round the core as in figure 17.14a or bunched together as in (b). Quoting the result on page 230, we therefore have

$$L = \frac{\mu_r \mu_0 A N^2}{l}$$

where again l is the average length round the lines of force inside the core. In this case the iron core ensures that nearly all the flux remains concentrated in the core; we can therefore assume that the same flux links all the turns of the coil. This was a very poor assumption to make with the air-cored coil considered earlier, but here it is a good assumption.

As an example let us work out the inductance of the primary coil (of 96 turns) of the transformer considered on page 231. Suppose it has a square-section core (30 mm by 30 mm) of average length (round the lines of force) 0.40 m, and that its relative permeability is 3.0×10^4. Hence

$$L = \frac{(3.0 \times 10^4)(4\pi \times 10^{-7} \text{ H m}^{-1})(30 \times 10^{-3} \text{ m})^2 (96)^2}{0.40 \text{ m}}$$

$$= 0.78 \text{ H}$$

This sort of calculation is valid only provided the current never rises to the point at which the core is saturated.

18 Electrons and quanta

18.1 The properties of electrons

When a piece of metal is heated to a high temperature, electrons emerge from its surface. This is known as the *thermionic effect*. We may demonstrate it with the apparatus in figure 18.1. A tungsten wire K is enclosed in an evacuated bulb; it can be heated by passing an electric current through it. Opposite the wire is a metal plate A. When the wire is cold, we find there is no current through the vacuum between K and A. But when the wire is heated, there is a current across the gap between the two electrodes. The direction of this current shows that the charge emitted is negative. Thus, if A is made positive with respect to K, a current is registered on the meter; in this case the electric field in the gap acts on the negative charge of the electrons to attract them towards A and so complete the circuit. But if the potential difference between A and K is reversed, no current passes; the electric field now draws any negative charge back again into K, and these is no complete circuit.

The two electrodes of an electron tube are labelled in the same way as in electrolysis (page 79). The electrode from which the negative charge emerges (K) is called the *cathode*; that towards which this charge travels (A) is called the *anode*. However, unlike the phenomenon of electrolysis, no chemical change occurs in either cathode or anode; nor does any gas remaining in the bulb affect the phenomenon, provided the pressure is below about 10^{-3} Pa (when the mean free path is so large that the electrons rarely strike gas molecules).

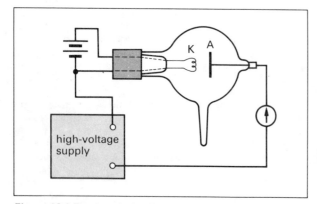

Figure 18.1 The thermionic effect, in which electrons are emitted by the heated cathode K and collected by the anode A. At high p.d.s some of the electrons strike the glass bulb beyond the anode, producing fluorescence.

When the p.d. between anode and cathode is large enough, the glass of the bulb opposite the cathode *fluoresces* with a pale green colour; this is made very much brighter by coating the inside of the bulb with a special fluorescent material (as on the inside of the front of a television tube). The anode seems to cast a shadow in the middle of this fluorescence, and we gain the clear impression that the effect is caused by rays travelling in straight lines out from the cathode. This is even more striking if the anode is cut in some distinctive pattern (such as a Maltese cross), when a shadow of the same shape appears in the fluorescence on the opposite wall of the tube. In the early development of the subject these rays were inevitably called *cathode rays*, and the name has stuck for beams of electrons formed in this way.

Although cathode-ray beams travel in approximately straight lines in the simple situation of figure 18.1, they are readily deflected into curved paths if additional electric or magnetic fields are applied; and this enables us to confirm that the cathode rays are indeed beams of negatively charged particles. For this purpose we need a narrow beam like that produced in the *cathode-ray tube* shown in figure 18.2a. In this the anode A is a disc with a horizontal slit in it; a narrow stream of cathode rays therefore fans out in a horizontal plane beyond the anode. This stream strikes a vertical sheet of fluorescent material obliquely, so that its path through the tube is made visible where it intersects the sheet.

In experiments with cathode rays it is usual to use the tubes with the anode earthed. This ensures that the region beyond the anode, in which the deflections are studied, is close to Earth potential; the electric fields in the tube are not then much affected by movements of the observer's hand nearby. (But such apparatus must always be approached with caution, as the p.d.s used can be lethal.)

When a potential difference is applied between the two horizontal metal plates (figure 18.2), the electric field pulls the cathode rays towards the positive plate and away from the negative one, and the beam is deflected accordingly. Over most the space between the pair of deflector plates the *E*-field is nearly uniform. The path of an electron in such a field is therefore very similar to that of a projectile in a uniform gravitational field (page 8). The horizontal velocity *v* is unaltered by the force. But in the vertical direction the electrons experience a uniform acceleration (figure 18.2b). The resulting path is then part of a *parabola*.

The cathode rays may also be passed through a magnetic field, produced by a magnet or by a current in suitably placed coils. If the beam carries a stream of *negative* charge, we must regard it as a current directed (in the conventional

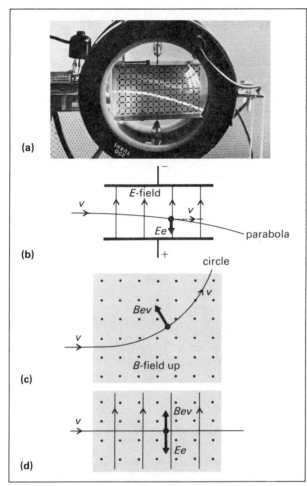

Figure 18.2 (a) A cathode-ray tube used for studying the deflection of cathode rays (electrons) in electric and magnetic fields. (b) The parabolic path of cathode rays in a uniform electric field. (c) The circular path of cathode rays in a uniform magnetic field. (d) The path is straight when the electric and magnetic forces on the particles are balanced.

supplied by the high-voltage source is the rate of converting electrical energy in the tube, and is given by

$$P = V_a I$$

where V_a is the p.d. between anode and cathode, and I is the beam current through the tube. In figure 18.2 typical values of these quantities might be $V_a = 5.0$ kV and $I = 100$ μA. giving

$$P = (5.0 \times 10^3 \text{ V})(100 \times 10^{-6} \text{ A})$$
$$= 0.5 \text{ W} = 0.5 \text{ J s}^{-1}$$

This is the rate at which energy arrives at the end wall of the tube as kinetic energy of the electrons, and is there converted to internal energy, etc. If the p.d. V_a between anode and cathode is increased, the production of internal energy, light and X-radiation goes up; at much higher p.d.s (more than 6 kV) the X-radiation also becomes more penetrating and is then dangerous. (Modern television tubes using p.d.s of more than 20 kV have their front walls made of a special thick glass that gives sufficient protection.) Some of the energy of the beam is always given up in ionising the small amount of gas left in the tube. If a cathode-ray tube is operated at a gas pressure of about 10^{-1} Pa, the ionisation makes the path of the beam visible in a well-darkened room. This provides another means of studying the deflection produced (as in figure 18.3).

Specific charge

The measurement of the deflections of cathode rays in electric and magnetic fields enables us to calculate the speed v of the electrons and the ratio of their charge e to mass m_e; the latter quantity (e/m_e) is known as the *specific charge* of the electrons. One of the simplest ways of doing this is to arrange for the forces of the E-field and B-field to act in opposition on the electrons. In figure 18.2 this may be done by applying a horizontal B-field directed towards the reader, in addition to the vertical E-field between the deflector plates. By adjusting the strength of one or both fields the forces can be balanced so that the beam is undeflected (at any rate in the centre where the fields are uniform). The force F_E of the E-field on a particle of charge e is given by

$$F_E = eE \qquad \text{(page 188)}$$

and the magnetic force F_B of the B-field is given by

$$F_B = Bev \qquad \text{(page 210)}$$

When the beam is undeflected, $F_B = F_E$ (figure 18.2d). Hence

$$Bev = eE$$

$$\Rightarrow \quad v = \frac{E}{B}$$

Thus v can be calculated if we measure E and B. The

sense) *towards* the cathode. Thus the beam in figure 18.2c may be deflected *upwards* by applying a B-field at right-angles to the plane of the diagram and directed *towards* the reader. The direction of the force is then seen to be in agreement with Fleming's left-hand rule (page 205). The magnetic force is always at right-angles to the beam and therefore acts as a centripetal force (figure 18.2c), deflecting the beam into a *circular* path (page 157).

It is evident that a beam of cathode rays carries energy. As the electrons accelerate between cathode and anode, their initial electrical potential energy is converted to kinetic energy. When the beam strikes the end wall of the tube, this kinetic energy is converted mostly into internal energy, and the end wall of the tube becomes warm; there is also some light produced and a very little X-radiation. The power P

Figure 18.3 A fine-beam tube, in which the path of the electrons is made visible by the ionisation they produce in the small amount of gas in the tube; the beam shown is in a uniform magnetic field.

E-field is the potential gradient between the two plates and so

$$E = \frac{V_\text{p}}{d}$$

where d is the separation of the plates, and V_p the p.d. between them. Thus in figure 18.2 the plates are 45 mm apart and the p.d. between them is 3.6 kV,

$$E = \frac{3.6 \times 10^3 \text{ V}}{45 \times 10^{-3} \text{ m}} = 8.0 \times 10^4 \text{ V m}^{-1}$$

The B-field may be produced by a current I in a pair of vertical coils outside the tube on either side of the beam (as in the photograph in figure 18.2. B is then proportional to the current I, and we can write

$$B = kI$$

where k is a constant for the pair of coils. We can find k by measuring B for any chosen value of I; this can be done with a Hall probe (page 211) or with a simple current balance (page 206). We can then calculate B for any other value of I. With the arrangement in figure 18.2 we might find that a current of 0.67 A giving a B-field of 1.9 mT was able to give an undeflected beam. This gives

$$v = \frac{8.0 \times 10^4 \text{ V m}^{-1}}{1.9 \times 10^{-3} \text{ T}}$$

$$= 4.2 \times 10^7 \text{ m s}^{-1}$$

which is about one-seventh of the speed of light! Since the cathode rays are observed to remain in a single concentrated beam in this experiment, it is clear that all the particles must have the *same speed*.

Once the speed v is known, e/m_e may be calculated from the value of the accelerating p.d. V_a between anode and cathode. The kinetic energy gained by a particle accelerating through this p.d. is given by

$$\tfrac{1}{2} m_\text{e} v^2 = eV_\text{a} \qquad \text{(page 25)}$$

$$\Rightarrow \quad \frac{e}{m_\text{e}} = \frac{v^2}{2V_\text{a}}$$

In figure 18.2 the accelerating p.d. was 5.0 kV, so that we have

$$\frac{e}{m_\text{e}} = \frac{(4.2 \times 10^7 \text{ m s}^{-1})^2}{2(5.0 \times 10^3 \text{ V})}$$

$$= 1.8 \times 10^{11} \text{ C kg}^{-1}$$

The currently accepted value of the specific charge of the electron is

$$\frac{e}{m_\text{e}} = 1.759 \times 10^{11} \text{ C kg}^{-1}$$

An alternative way of finding e/m_e is to switch off the E-field between the deflector plates, and allow the B-field to act alone. The force F_B now acts as a centripetal force accelerating the electrons at right angles to their path, and drawing the beam into a circular path of radius r, and we have

$$F_B = Bev = \frac{m_\text{e} v^2}{r}$$

$$\Rightarrow \quad r = \frac{m_\text{e} v}{Be} = \frac{v}{B(e/m_\text{e})}$$

The radius of curvature r may be measured with the tube in figure 18.2; for instance, with the B-field of 1.9 mT acting alone and the other quantities as before, the radius of curvature of the path is given by

$$r = \frac{(4.2 \times 10^7 \text{ m s}^{-1})}{(1.9 \times 10^{-3} \text{ T})(1.8 \times 10^{11} \text{ C kg}^{-1})}$$

$$= 0.12 \text{ m}$$

But such a measurement is not very precise. It is better to use a *fine-beam tube* (figure 18.3), in which the path of the beam is made visible by ionisation of the gas left in the tube. Each of the expressions we have deduced can be checked experimentally by performing the measurements with a range of different magnetic fields, currents and p.d.s. For instance the last result may be tested by verifying that the radius of curvature r is inversely proportional to the current I in the coils for a fixed accelerating p.d.

The deflection produced by a given magnetic field is always the same for the entire beam; this shows that all the particles of the beam are identical. If it were not so, a 'spectrum' of deflections would be produced for the different values of e/m_e represented. There is clearly only one sort of electron in the cathode-ray beam. Furthermore it can be shown that e/m_e is the same for the electrons coming from all kinds of cathode. This suggests that the electron is indeed a universal constituent of matter.

The proton

The specific charge of the hydrogen ion (i.e. the proton) can be found from electrolytic experiments (page 80), which give

$$\frac{e}{m_p} = 9.58 \times 10^7 \text{ C kg}^{-1}$$

where m_p denotes the mass of the proton. (It carries a charge of the same size e, but of opposite sign.) Dividing e/m_e by e/m_p we get

$$\frac{m_p}{m_e} = \frac{1.759 \times 10^{11} \text{ C kg}^{-1}}{9.58 \times 10^7 \text{ C kg}^{-1}} = 1836$$

The mass of the electron m_e is thus much less than the mass of even the smallest positive ion; and the mass of the electrons in any piece of matter is a minute proportion of the total mass. The value of m_e may be calculated from the known values of e/m_e and e (page 193); hence

$$m_e = \frac{e}{e/m_e} = \frac{1.602 \times 10^{-19} \text{ C}}{1.759 \times 10^{11} \text{ C kg}^{-1}}$$
$$= 9.11 \times 10^{-31} \text{ kg}$$

18.2 The cathode-ray oscilloscope

The electron beam of a cathode-ray tube is readily deflected by electric or magnetic fields and can therefore be used to study rapid variations in the potential differences or currents by which these fields are produced. An instrument employing this principle is called a *cathode-ray oscilloscope* (c.r.o.) The construction of a typical cathode-ray tube for this purpose is shown in figure 18.4. Electrons are emitted by the heated filament F. The anodes A_1, A_2, A_3 are kept at positive potentials with respect to the filament. The electron beam is therefore accelerated down the axis of the tube (which is highly evacuated). The shapes and potentials of the anodes are chosen so that the electric fields between them converge the beam into a fine spot on the fluorescent screen S. The filament F is surrounded by a cylindrical electrode G (called the *grid* of the cathode-ray tube); this is kept at a negative potential with respect to F. This potential controls the proportion of the emitted electrons that reach the hole in the first anode A_1, and so controls the brightness of the spot of light on the screen. The side walls beyond the last anode are usually coated with a conducting layer of graphite, which is electrically continuous with a conducting film on the inside of the fluorescent screen. These are connected to the final anode, and provide a return path for the electrons to complete the anode circuit. The current in the electron beam is typically about 100 μA.

The tube has two pairs of deflector plates, X_1 and X_2 for horizontal deflections, Y_1 and Y_2 for vertical. One of each pair of plates is connected to the final anode A_3. The deflection of the spot of light is proportional to the p.d. between a pair of deflector plates. The time of transit of the electrons through the tube is so small that the spot faithfully follows the variations of X and Y p.d.s up to very high frequencies of oscillation. The beam may also be deflected magnetically using currents in coils mounted round the neck of the tube. But this method is normally used only in television tubes.

Measurements

The cathode-ray oscilloscope may be used to study the waveforms of alternating potential differences, and to measure their frequency and amplitude. For this purpose the X-plates are connected to an auxiliary circuit that generates a *saw-toothed* p.d. (figure 18.5); such an arrangement is called a *time base*. The waveform to be investigated is joined to the Y-plates. During the interval from A to B the spot of light is drawn at a steady speed in the X-direction across the screen, while in the Y-direction it follows the p.d. being studied. The trace is therefore a graph of this

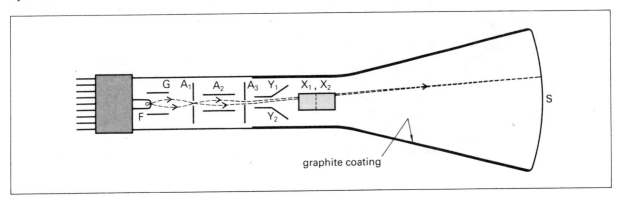

Figure 18.4 A cathode-ray tube for an oscilloscope.

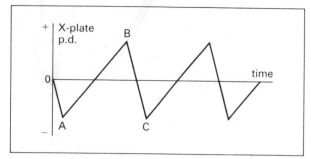

Figure 18.5 The saw-toothed waveform required for the time base of a cathode-ray oscilloscope.

p.d. against time. During the interval from B to C the spot flies back quickly to its starting point.

The horizontal time scale provided by the time base is arranged to be switched to a series of exact values marked '100 ms/div, 10 ms/div, ...' (figure 18.6). It is then a simple matter to read the time interval between any two points of a trace by means of a graticule in front of the tube. Thus in figure 18.6 we see that one complete oscillation occupies 2.5 div of the tube width, while the time base switch is set at 1 ms/div. The period T of one oscillation is therefore given by

$$T = (2.5 \text{ div})(1 \text{ ms/div}) = 2.5 \text{ ms}$$

The frequency f is the reciprocal of this; hence

$$f = \frac{1}{T} = \frac{1}{2.5 \times 10^{-3} \text{ s}} = 400 \text{ Hz}$$

If the pattern observed is to be a steady one, it is necessary for the second and subsequent traces to be exactly superimposed on the first. This will occur only if the frequency of of the time base is an exact submultiple of the frequency of the p.d. under test; this p.d. will then complete exactly

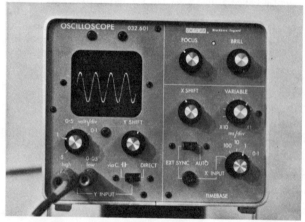

Figure 18.6 The front panel of a modern c.r.o. The settings of the time base and Y-amplifier controls enable the frequency and peak p.d. of the waveform displayed to be determined (as shown in the text).

a fixed number of oscillations in the time of one oscillation of the time base. This conditions is achieved by using another control to vary the point B at which the fly-back is triggered.

It is usually necessary to amplify the waveform to be examined before it is applied to the Y-plates, since a peak p.d. of at least 50 V is likely to be needed to give an adequate height of trace. The amplification is usually arranged to be switched to a series of exact values, chosen to give a convenient range of sensitivities of the oscilloscope when used as a voltmeter. These will be marked for instance '5 V/div, 1 V/div. . . .' Thus in figure 18.6 the total height of the trace from crest to trough is 4.6 div; and the *amplitude* (i.e. the distance from the centre line to crest or trough) is half this, namely 2.3 div. Since the vertical sensitivity scale is set at 1 V/div, the peak p.d. V_0 of the trace is given by

$$V_0 = (2.3 \text{ div})(1 \text{ V/div}) = 2.3 \text{ V}$$

In this case the trace is a sinusoidal one (page 272), so that we can write

$$V_{rms} = \frac{V_0}{\sqrt{2}} = 1.6 \text{ V}$$

where V_{rms} is the r.m.s. value of the alternating p.d.

The c.r.o. is primarily a very high resistance voltmeter (more than 10 MΩ at the input terminals of the internal Y-amplifier). We can adapt it to measure *currents* (and to study their waveforms) by connecting the Y-input across a resistance through which the current in question is made to pass; the p.d. across the resistance is then at every moment proportional to the current ($V = RI$). (See for instance the application in figure 17.3, page 223.) Usually an oscilloscope is used with alternating p.d.s. The Y-amplifier is then equipped with a blocking capacitor (page 280) at its input, so that the trace is not affected by any steady potential difference superimposed on the alternating p.d. But if we are to use the instrument to measure a steady p.d. (or a very slowly varying one) the blocking capacitor must be switched out (with the switch marked 'via C – direct' in figure 18.6).

Oscilloscopes are also made with *double-beam* cathode-ray tubes. Such a tube produces two electron beams; these pass between the same pair of X-deflector plates, but separate pairs of Y-deflector plates (and associated amplifiers). The two beams thus move together horizontally at the same speed, but can be used to observe two alternating p.d.s at the same time. This is useful for studying phase differences in a.c. circuits (as in figure 20.8, page 276).

18.3 The photoelectric effect

The emission of electrons from a metal may also be caused by illuminating its surface with light of sufficiently short wavelength; for most metals ultra-violet radiation is needed.

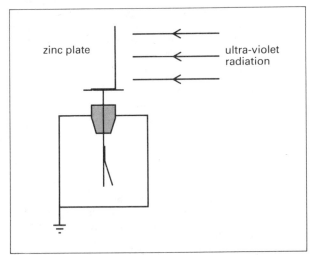

Figure 18.7 Demonstrating the photoelectric effect with a leaf electrometer; the electrometer discharges only when it is first given a *negative* charge.

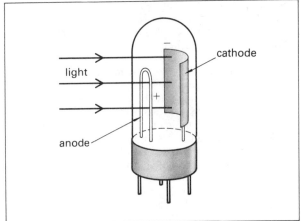

Figure 18.8 A vacuum photocell; the current between cathode and anode is proportional to the illumination of the cathode.

This phenomenon is called the *photoelectric effect*. It may be demonstrated with the apparatus shown in figure 18.7. A freshly cleaned zinc plate is mounted on the cap of an electrometer so that it can be illuminated with ultra-violet radiation from a mercury-vapour lamp. The zinc plate and electrometer are now given a negative charge. As soon as the radiation is directed at the plate the deflection starts to fall; negative charge is evidently being lost from the illuminated surface. If the beam of radiation is intercepted with a sheet of ordinary glass (which absorbs the ultra-violet) the effect stops; while with an ultra-violet filter (opaque to visible light) the discharging of the electrometer continues unaffected.

If, however, the electrometer is given a *positive* charge, no fall in the deflection occurs however bright the illumination of the zinc plate. In this case the electrons emitted are at once drawn back to the plate.

For a given metal emission of electrons occurs only for light of wavelength less than a certain value. This cut-off wavelength varies from one type of surface to another; for most metals it is in the ultra-violet region of the spectrum. But for the alkali metals (sodium, potassium, caesium, etc) the cut-off is in the visible region; for specially treated surfaces it is even in the infra-red. However, the alkali metals can be used only in a vacuum tube or in one containing an inert gas.

Figure 18.8 shows a common design of vacuum photocell that makes use of this effect to control an electric current by means of light. The cathode is a curved plate coated with the photosensitive material. An alloy of antimony and caesium is satisfactory for the visible spectrum. For the light from a tungsten filament lamp, a special surface consisting of a layer of caesium deposited on silver oxide is often used, since it responds well in the infra-red as well as in the visible spectrum. The anode is a loop of wire, which is sufficient to collect all the electrons emitted provided a p.d. of more than about 20 V is maintained across the tube. The current measured by a meter in series with the photocell is then found to be proportional to the illumination of the cathode.

Other types of photocell make use of the action of light on semiconductor materials. One type is the *photodiode* (page 286), in which the energy of the light creates extra electron-hole pairs near the *p-n* junction of the diode; this enables a current to pass through the photodiode even in the reverse (normally non-conducting) direction. Another type is the *photoconductive cell*; in this cell the electron-hole pairs created by the light cause the resistance of the surface layer of semiconductor to fall (page 286).

The quantum theory

When the photoelectric effect was discovered (towards the end of the 19th century) it had already been established that light was an electromagnetic wave motion. Experiment and theory had shown that the energy of the light was carried by oscillations of the electric and magnetic fields in the space through which it passed (page 369). The wavelengths of light of different colours had been measured by interference effects (page 361), which can arise only in wave processes.

However the photoelectric effect failed completely to fit in with the accepted wave theory. It is true that emission of electrons from an illuminated metal might be expected on an electromagnetic wave theory. The oscillating fields would be expected to cause oscillations of the charged particles in the metal, and an electron might in this way be given sufficient energy to escape from the surface. But the way in which the emission depends on the illumination of the surface is quite contrary to expectations. On the wave theory we expect to find that no emission of electrons occurs for very low intensities of light, since the electromagnetic field should not in this case be sufficiently strong. But in fact there is no threshold illumination for the photoelectric effect; some emission of electrons occurs no matter how small the intensity of the light. Again, the wave theory leads us to expect the *energy* of the emitted electrons to depend on the intensity of illumination of the cathode. But experiment shows that the energy of the electrons is not affected at all by the illumination; it depends only on the *frequency* of the light wave. The only thing the *illumination* decides is the number of electrons emitted per second (i.e. the current); we can check with any photocell (including the photodiode) that the current through it is proportional to the illumination.

These details are easily explained if we regard a beam of light as a stream of particles. The intensity of the beam is now to be measured by the number of particles arriving per second, and we expect this to be proportional to the number of electrons emitted per second. The energy of the electrons does not now depend on the intensity of the beam, but rather on the amount of energy carried by each light 'particle'. It appears that the blue (and ultra-violet) 'particles' in the spectrum carry more energy than the red ones that are generally unable to knock electrons out of a metal.

The only objection to this is that a particle theory cannot possibly explain the wave effects, such as interference and diffraction, which are so readily demonstrated for beams of light (page 352). It appears that in some experiments we must regard light as a stream of particles, and in others as a wave motion. Or should we regard it as both things simultaneously – a sort of wave-particle that defies the imagination? This is indeed the paradoxical situation with which nature presents us.

It was left to Albert Einstein (after the work of Max Planck) to follow through this paradox to its logical conclusion. Without denying the wave nature of light he proposed that light was emitted and absorbed in indivisible packets of energy, called *quanta*. The theory was therefore known as the *quantum theory*. The energy W of each quantum of light Einstein took to be proportional to the frequency f of the light wave. Thus

$$W = hf$$

where h is a universal constant, called the *Planck constant*. The quantum of energy is assumed to be emitted by a single electron in the source, and to continue as an indivisible packet of energy until it is absorbed, when it gives up all the energy to a single electron. If the electron is in the surface of a metal, it may receive sufficient energy to eject it from the surface, as in the photoelectric effect. Einstein supposed that in order to extract an electron from a given metal surface a certain minimum energy would have to be transferred to it. This would be equivalent to moving the electron through a potential difference Φ, which he called the *work function* of the metal; the minimum energy to be given to an electron by the light to enable it to escape is then $e\Phi$. The energy might well need to be more than this

for some electrons (i.e. those far below the metal surface); but it could not be less. The *cut-off frequency* f_0, below which there is no emission of electrons, is therefore given by

$$hf_0 = e\Phi$$

When light of higher frequency f is used, the surplus energy is given to the escaping electron as kinetic energy. If v_{max} is the *maximum* speed of the emitted electrons, then

$$\tfrac{1}{2}m_e v_{max}^2 = hf - e\Phi = h(f - f_0)$$

We can measure the kinetic energy of the electrons by making the anode of a vacuum photocell slightly *negative* with respect to the cathode (figure 18.9). A retarding force then acts on the emitted electrons; but some of them have sufficient kinetic energy to overcome this, and a small current continues to pass. In other words, the electrons crossing from cathode to anode are being slowed down by the electric field, gaining electrical potential energy at the expense of their kinetic energy. We can measure their kinetic energy by increasing the reverse p.d. between anode and cathode (i.e. with anode negative) to a value V_s at which the current is just stopped altogether. We then have

$$\text{maximum k.e. of the electrons} = eV_s = \tfrac{1}{2}m_e v_{max}^2$$

$$\Rightarrow \quad eV_s = h(f - f_0) \qquad (1)$$

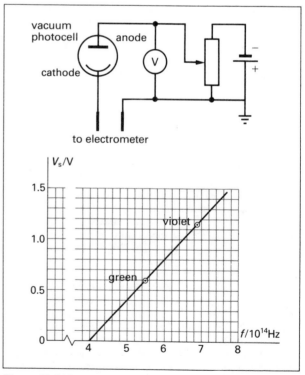

Figure 18.9 Measuring the stopping potential difference V_s for photoelectrons emitted by light of different frequencies.

In this way we can find the stopping potential V_s for monochromatic light of a selected range of frequencies. The graph of figure 18.9 shows the kind of result obtained with a vacuum photocell with a cathode coated with potassium. (The simplified apparatus usually available in schools allows this experiment to be done only for a very limited range of frequencies. But with more elaborate equipment Einstein's theory can be tested for a full range of frequencies.) The graph of V_s against f is a straight line, as predicted by equation (1) (appendix B2); the intercept on the axis of f (where $V_s = 0$) gives the cut-off frequency f_0 for the metal. The gradient of the line is h/e. Since we already know the electronic charge e, we may use the experiment to measure the Planck constant h.

The two experimental points marked on the graph of figure 18.9 are those that might be obtained with the violet and green lines in the spectrum of a mercury vapour lamp (frequencies 6.88×10^{14} Hz and 5.49×10^{14} Hz). In this cases the values of V_s for these two frequencies are 1.15 V and 0.60 V, respectively. This gives

$$\frac{h}{e} = \frac{1.15 \text{ V} - 0.60 \text{ V}}{(6.88 - 5.49) \times 10^{14} \text{ Hz}}$$

$$\Rightarrow \quad h = \frac{(0.55 \text{ V})(1.60 \times 10^{-19} \text{ C})}{(1.39 \times 10^{14} \text{ Hz})}$$

$$= 6.3 \times 10^{-34} \text{ J s}$$

The currently accepted value is

$$h = 6.626 \times 10^{-34} \text{ J s}$$

The graph of figure 18.9 gives the cut-off frequency for potassium as 4.0×10^{14} Hz. From this the work function Φ is given by

$$\Phi = \frac{hf_0}{e} = \frac{(6.63 \times 10^{-34} \text{ J s})(4.0 \times 10^{14} \text{ Hz})}{(1.60 \times 10^{-19} \text{ C})}$$

$$= 1.7 \text{ V}$$

(But when allowance is made for small p.d.s that arise at the junctions of different metals in the circuit the work function of potassium is in fact slightly greater than this.) A serious practical difficulty in this experiment is the contamination of the anode with potassium (evaporated onto it from the cathode); this causes an unwanted emission of electrons from the *anode* when the light falls on it. This current can easily mask the rather small current in the right direction through the tube (perhaps only 10^{-12} A), and makes it difficult to determine the true stopping potential V_s.

Photons

The quantum of energy in the Einstein theory has all the properties of a definite particle; it is often referred to as a *photon*. At the same time we do not deny the wave picture of light, since the energy of the photon is expressed in terms of the *frequency* of the wave, which is still supposed to consist of oscillations in the electric and magnetic fields in space. At first sight this appears to be nonsense. But closer thought shows that this is only because of our preconceptions of what particles and waves should be like. If someone talks of a particle, our thoughts turn to billiard balls, bullets, etc; and, if waves are mentioned, we think of ripples on a pond and the like. If we are asked to make a record of the path of a billiard ball on a table, we might take a motion picture of the table. But there is no way of doing the same thing for a photon. Suppose, for instance, we have a point source of light on the axis of a lens, which forms an image of the source on a screen the other side of it. There is no doubt that a photon emitted by the source will fall on the screen in the image spot. But what path has it followed in between? This is a question which, in the very nature of things, cannot be answered. To decide whether the photon passed through a given point we should need to put something there (a photocell, say) to detect it. But this would absorb and destroy the photon, so that it never completes the journey at all. In other words, there is no conceivable way of determining the path followed by an individual photon between the points where it is emitted and absorbed. *It has not got a path* in the sense that a bullet has. To describe the way in which the light is propagated through space we turn to the wave picture and the oscillating electric and magnetic fields in which the light energy moves.

We can pass from the wave picture back to the particle picture of light by regarding the *intensity of the wave* at a given point as a measure of the *probability* of the photon appearing there. We cannot predict the path of a photon with certainty; but we can state the probability of its following a given path. Whether or not it actually goes that way can be stated only *after* the event. Looked at in this way a wave is seen to be the only possible way of describing the travelling of a particle such as a photon, which has no clearly defined path. The wave and particle pictures of light are complementary to one another. Between them they enable us to give an exact description of the behaviour of light. While it is travelling from one place to another we study the behaviour of the wave. But it is emitted and absorbed by interaction with single electrons, and then we consider the behaviour of the particles, the photons. What the wave tells us is the probability of the particle being at a given point.

Practical sources of light emit vast numbers of photons. For instance, the energy W of a quantum of sodium light (frequency 5.1×10^{14} Hz) is given by

$$W = (6.6 \times 10^{-34} \text{ J s})(5.1 \times 10^{14} \text{ Hz})$$

$$= 3.4 \times 10^{-19} \text{ J}$$

A small sodium vapour lamp may emit about 10 W of light energy of this frequency. The number of photons emitted per second is therefore

$$\frac{10 \text{ J s}^{-1}}{3.4 \times 10^{-19} \text{ J}} \approx 3 \times 10^{19} \text{ s}^{-1}$$

Even the weakest beam of light that we could use in an ordinary experiment will deliver an enormous number of photons per second; and because of the large numbers the statistical predictions of probability that the wave theory gives us then become virtual certainties. The intensity of the wave at a given point still gives the probability of photons turning up there; but this is now a 'certain' prediction of the number of photons arriving per second, which is the intensity of the light. A sharply defined *ray* of light is then a meaningful thing to talk about and to study. But we must not forget that 'the path of a single photon' is an essentially meaningless phrase, which will only mislead us if we use it.

Electron waves

The description of light as a wave motion can be regarded as a way of expressing our uncertainty about the path followed by an individual photon. There is a similar uncertainty about the movements of an electron. Any attempt to determine the exact path of the electron is doomed to failure, because a photon that tells us where the electron is must have have bounced off it and so *changed* its path. Inside an atom, in particular, there is no possible way of determining the actual position at any instant of an electron in its orbit; any experiment to do this would change the orbit of the electron or perhaps knock it out of the atom altogether.

When we are dealing with vast numbers of electrons (as in a cathode-ray beam) we can make accurate predictions on a statistical basis. But for a single electron all we can state is the *probability* of its following a given path.

In 1923 Prince Louis de Broglie made the suggestion that the behaviour of an electron might be more satisfactorily described by means of an associated wave. The intensity of the wave at any point he took as a measure of the probability of the electron being there. He suggested that the wavelength λ of the electron *wave* would be connected with the momentum $(m_e v)$ of the *particle* by

$$\lambda = \frac{h}{m_e v}$$

Let us see what this would mean for the wavelength of an electron in a typical cathode-ray beam. If the p.d. between anode and cathode in the tube 5000 V, the kinetic energy $(\frac{1}{2}m_e v^2)$ of an electron is given by

$$\frac{1}{2}m_e v^2 = eV$$

$$\Rightarrow \quad \text{momentum} = m_e v = \sqrt{2em_e V}$$

The de Broglie wavelength λ is therefore given by

$$\lambda = \frac{h}{\sqrt{(2em_e V)}}$$

Using the values we have found for h, e and m_e, we have

$$\lambda = \frac{(6.6 \times 10^{-34} \text{ J s})}{\sqrt{\{2(1.6 \times 10^{-19} \text{ C})(9.1 \times 10^{-31} \text{ kg})(5000 \text{ V})\}}}$$

$$= 1.7 \times 10^{-11} \text{ m}$$

This wavelength is of the same order of magnitude as for X-rays (page 251). The two sorts of wave are of course quite different; X-rays are electromagnetic waves, which electron waves are not; X-rays are very penetrating, while electron beams are absorbed by the thinnest layers of matter. The wavelengths, however, of the two sorts of wave are similar, and therefore we may expect diffraction and interference effects on the same kind of *scale* in the two cases. If X-rays are diffracted by the regularly spaced atomic layers of a crystal, so too should electron waves.

The physical reality of the de Broglie waves is indeed readily confirmed by showing that cathode rays suffer diffraction when passing through a thin foil. Such a specimen of foil is polycrystalline, and the effects observed are similar to those obtained with the X-ray powder camera described on page 364; figure 18.10 shows a pair of such diffraction pictures obtained with the same aluminium foil (a) using X-rays, (b) using electrons of the same wavelength. The electron wavelengths derived from such experiments agree with those predicted by de Broglie.

Figure 18.11a shows a cathode-ray tube modified to demonstrate *electron diffraction*. The hole in the anode is covered with the thin specimen (e.g. a film of graphite); if the electrons are to pass through this, it must be thin enough to see through! On the screen at the end of the tube, as well as the undeflected central spot of light, we observe a pattern of concentric circles formed by the diffracted electrons. The diameter of a circle is proportional to the

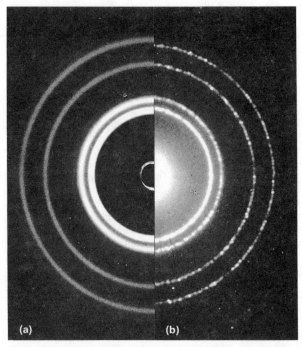

(a)   (b)

Figure 18.10 Photographs of diffraction patterns obtained with the same aluminium foil, (a) using X-rays, and (b) using electrons of the same wavelength.

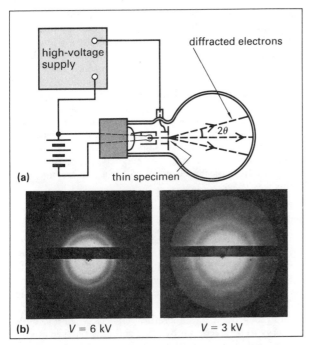

Figure 18.11 A cathode-ray tube adapted for studying the diffraction of electrons by a very thin specimen of graphite. The electron wavelength is inversely propotional to $\sqrt{V}$, so that the ring diameters in the right-hand picture are $\sqrt{2}$ times greater than those in the left-hand picture.

electron wavelength; and, using the expression derived in the course of the previous calculation, we can write

$$\text{diameter of circle} \propto \lambda \propto \frac{1}{\sqrt{V}}$$

This is a result we can check for a suitable range of wavelengths. Thus, if we halve the p.d. V, the diameter of a given circle is multiplied by $\sqrt{2}$. Figure 18.11b shows photographs of diffraction patterns obtained with the graphite film at two different values of V. The diffraction patterns are similar in both; but the right-hand one was obtained with half the accelerating p.d. of the left-hand one. (The angle 2θ is explained in figure 27.24, page 364.)

18.4 Energy levels

A chemical element in the form of a gas or vapour may be made to emit light by supplying energy to its atoms – either by heating it, as in a flame, or by passing electrons of sufficient energy through it. The spectrum emitted consists of a number of sharply defined lines, whose frequencies are characteristic of the element concerned (page 242). In 1913 Niels Bohr put forward a revolutionary extension of the quantum theory to account for atomic spectra of this sort. The existence of sharply defined line spectra suggested that a given atom was capable of emitting or absorbing energy only in certain definite amounts. He therefore supposed that an electron in an atom could exist only in sharply defined states of fixed energy (W_1, W_2, etc); the emission of light then accompanies the change from one such state to another, according to the usual quantum relation

$$\text{energy emitted} = W_1 - W_2 = hf$$

The problem was to find the correct way of describing these fixed states so that the behaviour of an atom could be predicted.

In 1911 Ernest Rutherford had put forward the model of the atom with which we are now familiar – a massive, but very small, positively charged nucleus around which circulate the appropriate number of negatively charged electrons to make the atom as a whole electrically neutral. Bohr accepted this model, and suggested that the fixed states of the electrons that he was looking for consisted of particular 'allowed' orbits. In order to get a good fit with the observed spectrum of hydrogen (figure 28.20, page 378) he proposed that the allowed orbits of an electron were those in which its *angular momentum* (page 174) was an integral multiple of $h/2\pi$. Thus

$$\text{angular momentum} = n\frac{h}{2\pi}$$

where n is an integer (i.e. a whole number). All other orbits he supposed to be impossible. The emission and absorption of radiation then occurs when the electron changes from one allowed orbit to another.

In general the analysis of the orbits for an atom containing many electrons is a problem too complex to be contemplated. However, when there is only a single circulating electron, the working is fairly simple (though beyond the scope of this book). Bohr made detailed predictions therefore for hydrogen, deuterium, and for heavier atoms ionised to the point at which only one electron remains near the nucleus. The predicted frequencies agree with experiment to within 1 part in 40000. This striking result left little doubt of the fundamental correctness of Bohr's ideas.

Figure 18.12 is a chart showing the energy levels of the single electron in a hydrogen atom for different values of n. The groups of transitions represented by the arrows each correspond to well-known series of lines in the hydrogen spectrum. The *Lyman* series arises from transitions to the ground state ($n = 1$), and involves large energy changes (10.2 eV is the smallest). The quanta emitted are of high energy and therefore of short wavelength; and the Lyman series is thus entirely in the ultra-violet part of the spectrum. The *Balmer* series is the only part of the hydrogen spectrum in the visible range. The quantum of lowest energy (and therefore of greatest wavelength) in this series arises from transitions from the energy level at $n = 3$ to the level at $n = 2$. From the chart in figure 18.12 the energy W of this quantum is given by

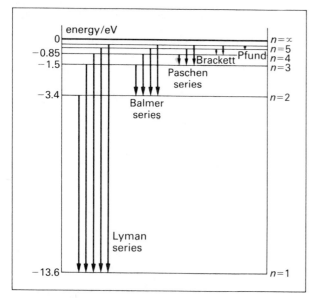

Figure 18.12 The energy level chart and the transitions that can occur in a hydrogen atom. The energy levels are spread out horizontally so as to display the transitions clearly.

$$W = 3.4 \text{ eV} - 1.5 \text{ eV} = 1.9 \text{ eV}$$
$$= (1.6 \times 10^{-19} \text{ C})(1.9 \text{ V}) = 3.0 \times 10^{-19} \text{ J}$$

The frequency f is given by

$$f = \frac{W}{h} = \frac{3.0 \times 10^{-19} \text{ J}}{6.6 \times 10^{-34} \text{ J s}}$$
$$= 4.5 \times 10^{14} \text{ Hz}$$

and so the wavelength λ is given by

$$\lambda = \frac{c}{f} = \frac{3.0 \times 10^8 \text{ m s}^{-1}}{4.5 \times 10^{14} \text{ Hz}}$$
$$= 6.6 \times 10^{-7} \text{ m}$$

This is in the red part of the spectrum (page 378).

Although Bohr's theory cannot be worked through for anything more complicated than a hydrogen atom the idea that the electrons in any atom have a limited number of allowed orbits and corresponding energy levels has turned out to be a universally consistent one. Even the most complex spectrum can be successfully analysed in terms of transitions between a relatively small number of atomic energy levels. Thus, in the light of Bohr's theory, the study of an apparently chaotic atomic spectrum becomes a means of measuring the energy levels of the atom. And energy level charts similar to figure 18.12 can now be constructed for all the elements.

In modern practice we try to avoid talking about an electron in an atom as though it was a concentrated particle; we therefore avoid the word 'orbit' in this connection. The reality is the stable electron wave pattern around the nucleus; this is referred to as the electron *orbital*. The position of the electron within the orbital can only be stated in terms of probabilities; and it is correct to regard its

charge and mass as being 'smeared out' over the orbital in proportion to the intensity of the wave at each point.

The charge distribution in an atom is thus a stationary one. But the wave pattern still represents an electron in motion around the nucleus; and the orbital may have angular momentum (like a planet, page 174) and produce a magnetic field. In addition each electron has internal angular momentum (also like a planet), which is called its *spin*, and this causes it to produce a magnetic field quite apart from its motion around the nucleus. Generally the electron orbitals in an atom exist in pairs of opposite spin whose magnetic effects cancel out. The magnetic properties of an atom arise from the few electrons within it that are unpaired in this way (page 232).

Ionisation of a gas

In the undisturbed state all the atoms in a gas are at their lowest possible energy level, the *ground state* of the atom, as it is called. For hydrogen (figure 18.12) the ground state is 13.6 eV below the arbitrary zero, which is taken to be the state in which the electron is just on the point of being removed from the atom ($n = \infty$). To ionise a hydrogen atom in its ground state therefore requires at least 13.6 eV of energy to be given to it by some colliding particle (atom, electron, photon, etc); thus, if the colliding particle is another electron, it must be accelerated through a potential difference of at least 13.6 V to gain enough energy to ionise a hydrogen atom that it strikes. This is called the *ionisation potential* of the gas.

We may demonstrate this, and measure the ionisation potential, by using a special gas-filled thermionic tube (represented in figure 18.13a). In this tube the cathode K is heated by an insulated heater wire embedded in it. The space between cathode K and anode A contains a third (perforated) electrode G, called a *grid*. In this experiment the anode is maintained several volts *negative* with respect to the cathode. This prevents any electrons from reaching it; and initially the meter in series with it reads zero. However, as the potential V of the grid is raised, the point is eventually reached at which the electrons accelerated towards the grid gain sufficient energy to ionise those atoms with which they collide. The positive ions are then accelerated towards the anode, and the anode current I suddenly starts to rise. The grid potential at which this happens is the ionisation potential (figure 18.13b). At this point the gas in the tube may be observed to glow, as some of the atoms recapture electrons and return to their ground states emitting the appropriate quanta of energy as light. The ionisation potential of helium is 24.6 V; other gases have lower ionisation potentials, e.g. argon (15.7 V), hydrogen (13.6 V), xenon (12.1 V).

Excitation of a gas

If you examine the energy level diagram of hydrogen in figure 18.12, you will see that the energy level next above the ground state is the one at −3.4 eV. This means that

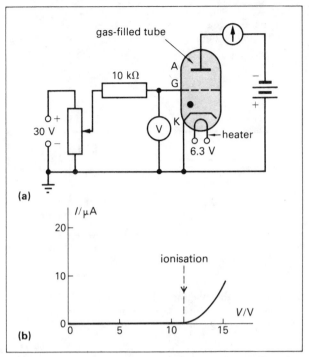

Figure 18.13 Finding the ionisation potential of a gas. There is no current in the anode circuit until the grid reaches the ionisation potential, so that positive ions are being produced in the tube.

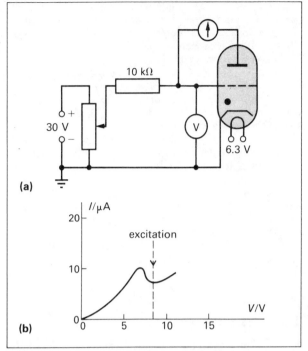

Figure 18.14 Measuring the excitation potential of a gas. There is a sudden drop in the anode current when the grid reaches the excitation potential, since at this point many electrons are able to give up all their energy to gas atoms.

the smallest quantity of energy that can be imparted to a hydrogen atom in its ground state is

$$(13.6 \text{ eV} - 3.4 \text{ eV}) = 10.2 \text{ eV}$$

An electron bearing less than this amount of energy cannot impart *any of it* to any hydrogen atom with which it collides; or, as we say, it cannot *excite* the hydrogen atom. In fact a collision between an electron and a gas atom under these circumstances is a *perfectly elastic* collision (page 32). Furthermore the gas atom is so much more massive than the electron that it gains a negligible amount of kinetic energy from the electron in such an elastic collision – it behaves as a virtually immovable object to the electron. In other words the electron rebounds in such a collision with a hydrogen atom with its energy unaffected.

Only if the potential difference through which the electron has been accelerated is more than 10.2 V (for hydrogen), can it give energy to the gas atom; this is called the *excitation potential* of the gas. This behaviour may be demonstrated by using the gas-filled tube as in figure 18.14a. For this purpose the anode is joined to the grid through a sensitive meter. As the potential of the grid and anode are raised from zero, the current slowly grows. Initially the collisions in the gas are of the perfectly elastic kind, and the electron energy is unaffected by the collisions; some of the electrons are collected by the grid, but many pass through it to be collected at the anode, causing the observed anode current.

When the potential of the grid reaches the excitation potential of the gas, however, some of the electrons are able to excite the gas atoms near the grid above their ground states; in the process these electrons lose nearly all their energy. Because of the current through the resistance of the meter the anode is in fact very slightly negative (about 0.2 V) with respect to the grid. Consequently those electrons which now have less than 0.2 eV of energy are unable to reach the anode, and the anode current falls. Theoretically this fall should be quite sudden, starting at the excitation potential, but not all the electrons emitted by the cathode have the same energy, and there are a number of other complicating factors.

The result (figure 18.14b) shows, however, that an electron beam can lose energy to gas atoms in its path even when the electrons have insufficient energy to ionise the gas. It thus demonstrates the existence of an energy level within the atom. More refined experiments are able to show many other excitation potentials corresponding with the higher energy levels of the atom.

Gas discharge tubes

An electric current can be maintained in a gas at low pressure if the p.d. between the electrodes is sufficiently high. This is called a *gas discharge*. Electrons in the tube are sufficiently accelerated by the high p.d. to excite the gas atoms when they collide with them; these then emit quanta of light energy as they return to their ground states. This process is used in fluorescent tubes for lighting. These tubes are filled with mercury vapour, which emits light of a blue-green colour – quite unacceptable by itself for interior illumination. The mercury vapour spectrum also includes a considerable amount of ultra-violet radiation (page 379); this is used to correct the colour of the light emitted by using suitable fluorescent powders on the inside wall of the tube. The ultra-violet radiation is absorbed by the fluorescent coating and re-emitted as visible light, chiefly in the red part of the spectrum in which mercury light is deficient.

There is a tendency for the current in a gas to grow indefinitely once it has started. Fluorescent tubes are generally run on an alternating supply, and are then best controlled by a *choke* joined in series (page 229), which automatically limits the current. The gas-filled tubes in the experiments of figures 18.13 and 18.14 need suitable resistors in series with the supply, otherwise the currents might become excessive and damage the tubes.

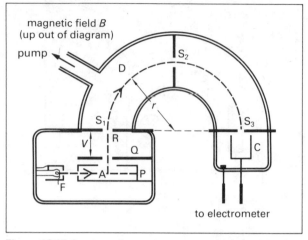

Figure 18.15 One type of mass spectrometer. As the accelerating p.d. *V* is varied ions of different mass are able to reach the collecting electrode C.

18.5 Ion beams

In an ionised gas there are likely to be many different kinds of ions present, particularly if the gas contains any complex molecules. As well as individual atoms (which may have not only one, but perhaps two or more electrons removed) there may be fragments of molecules carrying charge of either sign; also some atoms (e.g. oxygen) have a tendency to pick up free electrons in their paths forming *negative* ions. All these ions, together with any free electrons present, move under the action of any electric or magnetic fields in which the gas is placed. By the combined use of such fields it is possible to separate out the different kinds of ions and to focus them, one kind at a time, onto a detecting electrode connected to an electrometer. An instrument that does this is called a *mass spectrometer*; it enables the relative abundances of the different ions present in a gas mixture to be measured.

The principle of one type of mass spectrometer is shown in figure 18.15. The material to be examined is introduced as a gas or vapour into the space A. An electron beam is directed across this space from the filament F, and ionises some of the atoms there. The positive ions are drawn out of A by making the electrode Q slightly negative with respect to P, and are then accelerated by the much larger *E*-field between Q and R. Thus a narrow beam of ions emerges from the slit S_1 into the space D. If the p.d. between Q and R is *V*, the speed *v* acquired by ions of charge *e* and mass m_i is given by

$$\tfrac{1}{2}m_i v^2 = eV$$

$$\Rightarrow \qquad v^2 = \frac{2eV}{m_i}$$

In the space D the particles move in semi-circular paths under the influence of a *B*-field directed at right-angles to the plane of the diagram. The slit system S_1-S_2-S_3 is arranged to select particles whose orbits are of one particular radius *r*, which is given by

$$r = \frac{m_i v}{Be} \qquad\qquad \text{(page 238)}$$

Eliminating *v* between these equations, the specific charge of the particles selected by the system is given by

$$\frac{e}{m_i} = \frac{2V}{B^2 r^2}$$

Any ions with this specific charge are collected by the electrode C, and a small current is registered by the electrometer. To make an analysis of the ions present in the apparatus, the ion current is measured as the p.d. *V* is varied, the magnetic field being kept constant. One type of ion after another is thus brought into the collector electrode C, and the relative sizes of the peaks of current indicate the proportions of isotopes and complex ions present.

Particle accelerators

Most of our knowledge of the composition of the nucleus has come from collision experiments, in which we study the effects of firing high-speed particles at a piece of matter. Nature provides us with a limited range of such projectiles from radioactive substances, which spontaneously emit particles of various kinds (page 252). Another source of particles of much higher energies is provided by *cosmic rays* that enter the Earth's atmosphere from outer space; but to make use of these we need to carry our equipment in balloons or satellites high above the Earth's atmosphere. In addition to both these useful sources of high-speed particles, we clearly need controlled sources of chosen particles for use in the laboratory.

The most obvious way of accelerating a charged particle is to use a very high potential difference between two electrode in an evacuated tube. The van de Graaff machine was developed for this purpose (page 199). The p.d. reached with this is limited by the quality of the insulation that can reasonably be put around the high potential electrode, somewhat less than 10 MV in practice.

The cyclotron

In this machine the path of the particles being accelerated is wrapped into a spiral by using a magnetic field. We can work out the period of revolution T of a particle in a given B-field as follows. In one revolution of radius r the particle travels a distance $2\pi r$ round the circumference at speed v; hence

$$T = \frac{2\pi r}{v}$$

But we have

$$r = \frac{m_i v}{Be} \qquad \text{(page 238)}$$

$$\Rightarrow \quad T = \frac{2\pi m_i}{Be}$$

You will notice that T does not depend on the radius r of the orbit or the speed v. All particles of mass m_i and charge e take the same time to complete an orbit in a given B-field. A fast particle moves in a circle of larger radius, but takes the same time for one revolution as a slow particle. Thus for protons in a B-field of 0.50 T, we have $m_i = 1.67 \times 10^{-27}$ kg, and $e = 1.60 \times 10^{-19}$ C; and so

$$T = \frac{2\pi (1.67 \times 10^{-27} \text{ kg})}{(0.50 \text{ T})(1.60 \times 10^{-19} \text{ C})}$$

$$= 1.31 \times 10^{-7} \text{ s}$$

The frequency f of revolution (i.e. the number of revolutions per second) is given by

$$f = \frac{1}{T} = \frac{1}{1.31 \times 10^{-7} \text{ s}} = 7.62 \text{ MHz}$$

In the cyclotron the particles are accelerated by applying an alternating p.d. of exactly the above frequency f to a pair of D-shaped electrodes that form a box within which the particles move. The whole system is in an evacuated chamber between the poles of a very large magnet (figure 18.16). The positive ions are injected near the centre, and are accelerated by the E-field between the D's each time they cross the gap between them; this always happens at the moments when the E-field between the D's reaches its peak value in the direction that will accelerate the particles. Inside the D's there is no electric field, and the particles move in semicircular orbits. Each successive semicircular orbit has a larger radius than the previous one, but the period of revolution remains constant; the particles thus stay synchronised with the accelerating p.d.. After per-

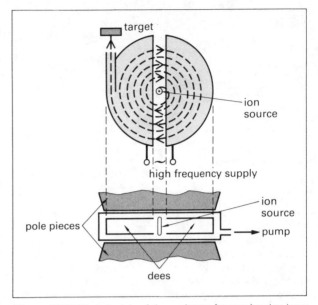

Figure 18.16 The principle of the cyclotron for accelerating ions to high energies. The magnetic field wraps the paths of the particles into a spiral.

forming about 100 revolutions the particles reach the edge of the system and enter a subsidiary electric field that deflects them out of the circle to strike the target in which the collision experiment occurs.

The energy attained by the particles is limited by the relativistic increase of mass. For instance, for protons (hydrogen ions) at 20 MeV the mass is about 2% greater than their rest mass (page 263), and beyond this point the synchronism between the particle orbits and the alternating p.d. breaks down. However, by lowering the frequency during the accelerating process, synchronism can be maintained for several thousand revolutions of the particles; and pulses of protons of energies up to about 600 MeV have been produced by this means. The maximum energy is then decided by the strength of the magnet and the radius of its pole pieces (which have a mass of hundreds of tonnes in the largest machines).

18.6 X-rays

In 1895 Wilhelm Röntgen discovered that a very penetrating radiation was being emitted by a cathode-ray rube across which there was a high p.d. The radiation produced fluorescence on screens several metres away, and fogged photographic plates nearby, even when these were securely wrapped in dark paper. Cathode rays themselves could not even penetrate the walls of the tube; and it was therefore clear that a new kind of radiation was involved, to which Röntgen gave the name *X-rays*.

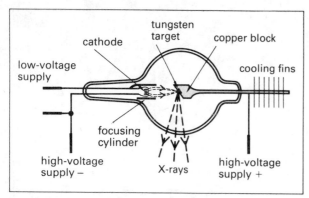

Figure 18.17 A high-vacuum X-ray tube.

X-rays are produced whenever electrons strike a piece of matter and so are brought to rest. In a cathode-ray tube they are therefore emitted at the anode and from all parts of the glass wall struck by the beam of electrons. In order to produce a localised source of X-rays it is necessary to focus the cathode rays into a small spot on the *target*. Figure 18.17 shows the construction of a modern X-ray tube. A high vacuum is used (less than 10^{-4} Pa), and the beam of electrons is focused by means of an electrode system similar to that used in the electron gun of a cathode-ray tube. The efficiency of X-ray production is increased by using a target material of high atomic number. But at best only about 0.2% of the energy of the cathode-ray beam is converted into X-rays; the rest is changed into internal energy. The target is therefore made of tungsten, which has a high atomic number and also a high melting point; this is set in a substantial copper bar with cooling fins. Potential differences up to 100 kV are commonly employed, and for special purposes 1 MV or more. The *intensity* of the X-ray beam depends on the number of electrons striking the target per second (i.e. it depends on the current through the tube). This is controlled by varying the temperature of the cathode by means of the heater current passed through it.

Safety

X-rays are to some extent absorbed by all matter through which they pass; but their penetrating power is very great. They are therefore to be treated with the greatest possible respect; and adequate, well-tested safety precautions must always be observed. Their effects on the human body are only partly understood; but they are known to produce deep-seated burns, destruction of living cells and unpredictable chemical changes. Most insidious of all is their tendency to produce serious diseases, sometimes many years after the time of irradiation. They can also produce genetic changes which may become apparent only in subsequent generations. All unnecessary exposure to X-rays should be rigorously avoided. P.d.s. above 6 kV should never be used across any cathode-ray tube or gas discharge tube without adequate lead shielding.

Medical uses

When X-rays pass through matter, their immediate effect is to produce ionisation. Electrons may be ejected from all levels of the atoms, and the energy given to them is then dispersed through the rest of the matter by collisions. By this means the energy of an X-ray beam is mostly converted into internal energy. For a given thickness of screen the absorption is proportional to the total number of electrons in the material per unit volume. It is therefore greatest for dense substances containing elements of high atomic number.

The X-ray photography of the human body is made possible by this effect. The bones contain much denser material than the soft organs, and therefore absorb the X-rays passing through them to a much greater extent. An X-ray picture is simply a *shadow photograph*, which is made sharp enough by using a very small source of X-rays and placing the photographic film as close as possible behind the part of the body under investigation. When the film is developed, all the tissues show up relatively pale against the darkened background; and the bones show up very clearly. Apart from these, the differences of density between the parts of the body are slight, and it is not normally possible to distinguish the details of the soft organs – stomach, liver, brain, blood vessels, etc. To photograph such parts doctors have to introduce into the organ a harmless compound containing a dense element. For instance, in order to observe the stomach and intestinal tract, the patient is given a *barium meal*, consisting of a paste of barium sulphate and water. Each part of the alimentary canal in turn then shows up clearly as the barium meal reaches it.

When X-ray photography must be used for medical purposes, care is always taken to ensure that the dose is far below that which is known to be harmful. Sometimes, however, the destructive power of X-radiation is deliberately used for therapeutic purposes. Rapidly dividing cells are more readily damaged by X-rays than stable ones. (So care is taken to avoid taking X-ray pictures of a growing foetus.) But a carefully controlled dose can be used to destroy a growing tumour without doing irreparable harm to the surrounding tissue. Similar techniques are also used with the radiations from radioactive substances (page 257).

The properties of X-rays

These clearly point to their being electromagnetic waves, essentially of the same kind as radio waves, light, etc, but of much shorter wavelength even than ultra-violet radiation.
(i) They are not deflected by electric or magnetic fields. We can therefore dismiss the possibility of their being charged particles.
(ii) They are produced by the deceleration of electrons; and theory leads us to expect the production of electromagnetic waves in such a case (page 369).
(iii) They affect photographic emulsions just like light and

ultra-violet radiation; and they ionise the air through which they pass just like ultra-violet radiation. However, other kinds of radiations (page 256) also do this.

(iv) They show diffraction effects when they pass through a very finely ruled grating (page 359); and similar effects occur with the regularly spaced layers of atoms of a crystal (page 362). This clearly establishes the wave nature of X-rays, and shows that the wavelengths used are mostly in the range 10^{-11} m to 10^{-10} m, about $1/10\,000$ of the wavelength of visible light. With modern X-ray equipment the range available is from 10^{-8} m to 10^{-15} m, overlapping ultra-violet radiation on one side and γ-radiation on the other, and conclusively demonstrating that X-rays form part of the electromagnetic spectrum (page 366).

X-ray photons

The quantum theory of electromagnetic radiation (page 242) predicts that X-radiation of frequency f should be emitted as packets of energy W given by

$$W = hf$$

where h is the Planck constant. Now the electrons in an X-ray tube are all accelerated through the same p.d. V, and therefore all reach the target with practically the same energy W_{max} given by

$$W_{max} = eV$$

This is also the maximum energy of an X-ray photon; the energy may be *less* than this, since the electron may not give up the whole of its energy in one collision, but it cannot be more. We therefore expect to find a *maximum* X-ray frequency f_{max} given by

$$hf_{max} = eV$$

This means that for a given p.d. V across the tube the spectrum of X-ray wavelengths emitted has a *minimum* value λ_{min} given by

$$f_{max}\lambda_{min} = c$$

where c is the speed of electromagnetic radiation. Thus

$$\lambda_{min} = \frac{c}{f_{max}} = \frac{hc}{eV}$$

For instance, for a tube p.d. of 25 kV we have

$$\lambda_{min} = \frac{(6.6 \times 10^{-34}\ \text{J s})(3.0 \times 10^{8}\ \text{m s}^{-1})}{(1.6 \times 10^{-19}\ \text{C})(25 \times 10^{3}\ \text{V})}$$

$$= 5.0 \times 10^{-11}\ \text{m}$$

Furthermore λ_{min} is inversely proportional to V, so that at 50 kV the minimum wavelength in the spectrum is halved to 2.5×10^{-11} m; and so on.

This is in good agreement with the experimental curves obtained with an X-ray tube over a range of different p.d.s (figure 18.18). The striking feature of these curves is the sharply defined *minimum wavelength* for a given accelerating p.d.. This shows that the quantum theory applies to X-ray photons as well as to the photons of light. In fact, measurement of λ_{min} by the methods of X-ray spectrometry (page 363) gives the best available estimate of the Planck constant h.

In addition to the continuous spectrum of X-ray wavelengths depicted by the curves of figure 18.18, an X-ray tube also emits certain sharply defined wavelengths, which are characteristic of the element used in the target of the tube (figure 28.26, page 380). Just as the optical spectrum of an element gives information about the energy levels of the outermost electrons of the atomic structure, so the X-ray *line spectrum* gives information about the energy levels of the innermost electrons of the atom. The incident beam of electrons may have enough energy to eject electrons even from these levels. Transitions then occur deep within the atom as it returns to its ground state; and the surplus energy is radiated away, as usual, as a photon. Because the innermost electrons of the atom are very tightly bound, the differences between their energy levels are considerable. The photon emitted in such a transition is therefore of high frequency and short wavelength – i.e. in the X-ray region of the spectrum. The higher the atomic number of the target element, the larger the charge on the nucleus and the greater the energy level differences. The X-ray line spectra of elements of high atomic number are therefore of shorter wavelength than those of elements at the lower end of the periodic table; this matter is considered further in chapter 28 (page 380).

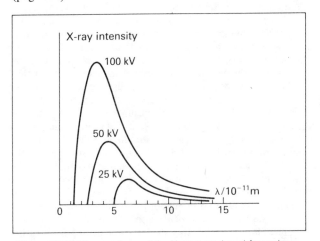

Figure 18.18 The spectrum of the X-rays produced for various values of the p.d. across the tube; the cut-off wavelength is inversely proportional to the p.d.

19 Probing the nucleus

19.1 Radioactivity

In 1896 the French scientist Henri Becquerel discovered a hitherto unknown type of radiation which was emitted spontaneously by any sample of the element *uranium* or its compounds; the radiation was detected in the first instance by its ability to blacken a photographic plate on which it fell. The phenomenon was called *radioactivity*.

At first it appeared that the new radiation was similar to the X-radiation discovered by Röntgen (page 249). Thus, it was very penetrating; also it was strongly absorbed by a sheet of metal. A metal screen placed between the uranium and the photographic plate cast a 'shadow' on the plate, just as with X-rays. However, fuller investigation showed that the effect was a complex one, in which several different kinds of radiation were involved; some of these were found to have properties very different from X-rays.

Becquerel's discovery led to an intense search for other radioactive materials. Within a few years a long list of radioactive elements had been discovered; notable among these discoveries was that of *radium* by Madame Curie and her husband; this element is a million times more radioactive per unit mass than uranium. Many of the new radioactive substances turned out to be isotopes of already known elements, and nearly all of them were of large atomic mass. In recent years the list has been enormously extended following the discovery of techniques for making artificial isotopes not found in nature; most of these are radioactive. It is now possible to find radioactive isotopes of every element in the periodic table, light elements as well as heavy ones. The same techniques have also given rise to new radioactive elements with atomic masses greater than that of uranium, the heaviest element normally found in nature.

Radium is still one of the most useful radioactive sources for laboratory purposes, since a small quantity of it strongly emits radiations of several different kinds simultaneously. But it is sometimes convenient to have a source that provides a single kind of radiation only. For this purpose certain artificial radioactive isotopes are particularly suitable; those usually encountered in school laboratories are isotopes of *americium, strontium, cobalt,* and *caesium.*

The chief effect produced by the radiations from radioactive substances is the ionisation of the matter through which they pass. This may be demonstrated by placing a radioactive source near a charged leaf electrometer (page 183), when the deflection at once starts to decrease. A more detailed investigation of the ionising effect of the radiation may be conducted using an ionisation chamber connected to an electrometer of the measuring-amplifier sort (page 182), as in figure 19.1. The ionisation chamber is

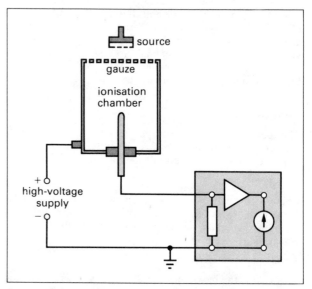

Figure 19.1 Measuring the ionisation current produced by a radium source. The ionisation current falls to about 1% of its initial value when the source is lifted more than 60 mm above the gauze.

provided with a gauze cover through which the radiation can enter. The outer metal cylinder of the chamber forms one electrode and the insulated central rod the other. The 'circuit' is completed by the movement of ions in the air inside the chamber. The ionisation current is a measure of the rate at which ions are formed in the air by any radiation passing through the chamber.

α-radiation and β-radiation

The most strongly ionising component in the radiations emitted by radium has a very short range – only a few centimetres in air, and proportionately less in a denser substance. When a radium source is held a short distance above the gauze cover of the ionisation chamber, a steady ionisation current is recorded; this can be caused only by that part of the radiation which has passed through the gauze and produced ionisation *inside*. When the source is raised, the current decreases, and finally falls to a very small value when the radium is about 60 mm from the gauze. If the radium is placed once more close to the gauze, it will be found that even a single thick sheet of paper is sufficient to absorb the radiation almost completely.

Using thinner absorbing sheets, a foil of aluminium

10^{-2} mm thick reduces the current to about a quarter of its initial value. With three of four such foils the current falls to about 1% of its original value. However, it then becomes apparent that the radium is also emitting a second kind of radiation, which is both *more* penetrating and also produces *less* ionisation. The thickness of absorbing foil under the source may be doubled or trebled without much affecting the remaining 1% of ionisation current; and a sheet of aluminium 1 mm thick is barely sufficient to halve it. These two kinds of radiation are called *α-radiation* and *β-radiation*. The α-radiation from radium is stopped by a few centimetres of air or a thick film of solid material; but within this short range it causes much ionisation. To stop the β-radiation from radium requires a thickness of about 0.5 mm of lead or several mm of aluminium; its range is about 10 times that of the α-radiation, and the ionisation it produces in a given distance is about 1/10 as much.

A graph of ionisation current I against effective thickness of stopping foil for a radium source is of the form shown in figure 19.2a. Since the current covers a large range of values, it is convenient to plot this sort of graph using a logarithmic scale of current. The stopping power of a slab of material is found to be proportional to its mass per unit area of cross-section, regardless of what the material is; it is therefore convenient to express the *absorbing thickness* of the foil in units of kg m^{-2}. At the point on the graph where there is a total absorbing thickness of 0.09 kg m^{-2} there is a pronounced discontinuity. At this point the α-radiation has been completely absorbed, and the small remaining current is caused by the β-radiation only. The absorption of the β-radiation from radium may be studied in the same way, though we must now use a range of very much thicker foils. The ionisation current caused by the β-radiation is approximately halved for an absorbing thickness of about 3 kg m^{-2}; the absorption is complete when the absorbing thickness approaches 7 kg m^{-2} (figure 19.2b).

γ-radiation

Using even thicker absorbing screens than these and a very

sensitive electrometer, it is possible to show that a radium source emits yet a third kind of radiation, called *γ-radiation*. This is even more penetrating than β-radiation, and produces a much lower density of ionisation, about 1/1000 of that produced by β-radiation in the same distance. With the radium sources available in schools the ionisation current produced by the γ-radiation will not be much more than 10^{-13} A. However, there are other means of detecting the radiations, as we shall see below, and these avoid the need for using very sensitive measuring instruments.

To cut down the γ-radiation to even half its original intensity requires an absorbing thickness of about 100 kg m^{-2} (about 10 mm of lead); this is about 30 times as much as is required to produce the same reduction for β-radiation.

Safety precautions

The effects on living organisms of the radiations from radioactive substances are in many ways similar to those of X-rays (page 250). They can cause burns and destruction of living cells, and are capable of initiating serious diseases and even genetic changes. Also the effects are to some extent cumulative, so that a very small daily dose of radiation may add up in the course of a year to a total dose sufficient to cause irreparable harm. Radioactive sources must therefore be treated with the greatest possible care – particularly because, unlike an X-ray tube, they are never 'switched off', and it is all too easy to forget their presence.

The damage produced by α-rays is intense, but their range is so short that they would not penetrate even the surface layers of skin. There is therefore little danger from α-active materials provided they are not absorbed into the system through the stomach or lungs. In particular, radium sources must be carefully sealed to prevent any escape of the α-active gas *radon* that is generated continually by radium. There is less damage produced by β-rays, but their penetration is rather greater. However, a layer of wood or

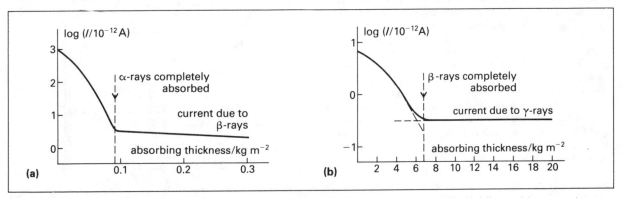

Figure 19.2 Ionisation current (on a logarithmic scale) plotted against absorbing thickness. In (a) thin foils are able to stop the α-radiation. In (b) much thicker plates are needed to stop the β-radiation. (On a logarithmic scale we do not attempt to start the axis at the point marked '0', since this is merely the point where $I/10^{-12}$ A = 1.)

perspex 5 mm thick gives complete protection against external β-radiation. The main hazard from β-active materials is again the possibility of absorbing them into the system through the stomach or lungs. The chief danger in handling radioactive materials is from γ-rays. Even a layer of lead 10 mm thick is only sufficient to absorb about 50% of the γ-radiation incident on it, and so it is difficult and expensive to provide adequate screening. All γ-sources must therefore be as weak as possible, and must be kept at a sufficient distance from people in the laboratory to reduce the radiation dose well below the acceptable limit, allowing for whatever lead screening is in use.

It is usual to insist on the following rules being observed in a school laboratory in which experiments of the types described in this chapter are performed.
(i) All sources must be weak (less than 10μCi, page 257), and must be sealed in foil to prevent the escape of radioactive material.
(ii) All sources must be stored in a special cupboard in a little-frequented part of the laboratory; they must be kept and transported in a suitable lead container.
(iii) Sources must be handled only with forceps, and should always be held well away from the body.
(iv) No eating, drinking or smoking must take place in a laboratory where radioactive materials are in use; also licking of labels, etc, should be avoided.

Having said all this, it is only fair to point out that by observing these rules the dangers are reduced to quite negligible proportions.

19.2 Particle detectors

The spark counter

The ionisation produced in air by α-radiation is sufficient to trigger off a spark discharge in an electric field that would not otherwise be strong enough. One form of spark counter consists of a grid of fine wires stretched about 1 mm above the surface of a metal plate, and insulated from it (figure 19.3). The potential difference between wires and plate is adjusted to a value slightly less than that required to produce sparking. When a radium source is held close to the wire, sparks are seen between the two electrodes. The sparks occur at irregular intervals and appear to be distributed at random on the grid of wires. They are clearly caused by the α-radiation only, since no sparks occur when the source is moved to more than 60 mm from the wire or when a sheet of paper is placed to intercept the α-radiation. With this type of spark counter β-radiation does not give sufficient ionisation to start a spark.

By watching the spark counter we get the impression that the α-radiation does not arrive in a steady stream, but rather as a random 'rain' of particles. We thus have the means of *counting* the individual α-particles (as we shall call them) that arrive near the wire. We can use the spark

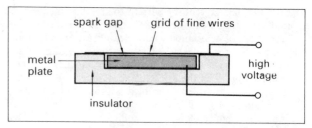

Figure 19.3 A simple spark counter for detecting α-particles.

counter to test which of our sources emit α-particles and which do not. Thus americium is an α-emitting source, while strontium and cobalt sources are not.

The cloud chamber

In this device the tracks of the α-particles are made visible, thus providing an even more vivid demonstration that they are indeed particles emitted at random by the source. A common form of cloud chamber is shown in figure 19.4. The walls and top of the chamber are made of a transparent plastic. Round the top of the walls is fixed a strip of felt which is kept soaked with a suitable volatile liquid, such as ethanol. The air in the chamber is therefore saturated with the ethanol vapour. The floor of the chamber consists of a metal plate, which is cooled by a layer of 'dry ice' (solid carbon dioxide) packed beneath it. The ethanol condenses on the plate, and there is thus a steady diffusion of vapour from the top to the bottom of the chamber. A short distance above the metal plate the vapour passes through a narrow region in which it becomes supercooled (page 142). In this condition it will condense on any ions that happen to be present. A string of droplets therefore forms along any track in which ionisation occurs, and in

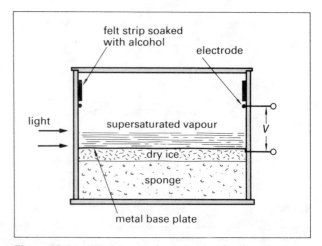

Figure 19.4 A diffusion-type cloud chamber for observing particle tracks.

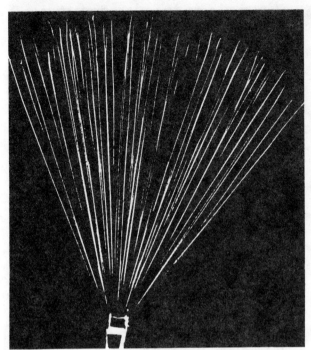

Figure 19.5 A fan of α-particle tracks in a cloud chamber. The tracks are almost straight and of equal length; heavy ionisation is produced along them.

this way the paths of any ionising particles are made visible. To see the tracks clearly we need to illuminate brightly the active region in the chamber, and the base-plate needs to be painted black to cut out background reflections. The chamber must be levelled to avoid convection currents that make the tracks drift sideways. Also the ions must be swept out of the chamber by an electric field so that the space is continually being cleared for the detection of fresh ionising particles. The clearing field may by provided by joining a suitable potential difference V between the base-plate and another electrode mounted near the top of the chamber. Alternatively, the plastic top may be rubbed with a duster, when the static charge should produce a sufficient clearing field.

When a suitable radioactive source is mounted in the chamber, the tracks of α-particles are strikingly shown up (figure 19.5). Their range in air and the manner in which they are stopped by thin foils shows that the tracks are indeed those of α-particles such as we have already detected by other means. Detailed measurements show that each α-particle produces about 3000 ion pairs per mm of its path in air at s.t.p. Thus the total number of ion pairs in the 60 mm track of an α-particle is about 180 000. The energy given to each air molecule ionized by the α-particle cannot be less than its ionisation energy, and is likely to be more; suppose it is 30 eV. This means that the total energy carried by an α-particle must be about

$$180\,000 \times 30 \text{ eV} = 5.4 \text{ MeV}$$

The α-particle must come from an individual atom; yet the energy it carries is about a million times greater than the energy converted when an atom takes part in a chemical reaction (page 263). Clearly in radioactive materials we are dealing with a quite new source of energy.

The paths of the α-particles are seen to be almost exactly straight, though small deflections of about 1° are fairly common, particularly towards the ends of the tracks. The patient observer, who watches a chamber for some time, may be rewarded by seeing an α-particle deflected through a much larger angle (90° or more). The study of the large-angle deflections of α-particles provided Rutherford with the evidence he needed to propound the nuclear theory of the atom (page 261).

It is also possible to use a cloud chamber to observe β-radiation. Again we find a series of clearly marked tracks, and it seems that β-radiation also must be regarded as particles. The ionisation along the track of a β-particle is much less dense than for an α-particle, and the path is only thinly marked out by a line of droplets; it vanishes within a fraction of a second of being formed. But with a magnifying glass focused on the right region of the chamber an alert observer will manage to see the occasional track. The paths of the particles are very far from straight (figure 19.6); they seem to suffer frequent small deflections and occasional large deflections of 90° or more. The deflections become more and more frequent as the speed of a particle falls; and the end of the track is usually very tortuous.

A beam of γ-rays shows up in a cloud chamber in the

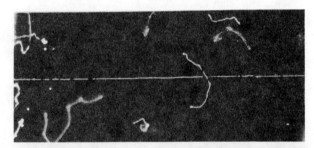

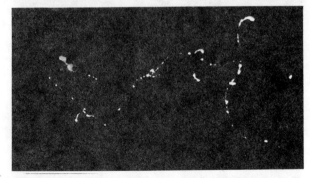

Figure 19.6 Typical β-particle tracks. The top picture shows the track of a fast β-particle, almost straight and very thinly marked by droplets. The lower picture shows slow β-particle tracks, tortuous and frequently deflected.

Figure 19.7 Cloud chamber tracks produced by a beam of X-rays entering the chamber from the right. The tracks start in the path of the beam and are identical with those of slow β-particles. Similar tracks are caused by γ-rays.

same manner as a beam of X-rays (figure 19.7). In this case there is no clear line of droplets marking the path of the beam; but several short tracks are observed, resembling those of β-particles, each track originating in the path of the beam. This is quite different from the tracks of α-particles or β-particles, which only rarely show 'whiskers' of the kind that characterise the paths of γ-rays. It is difficult to observe these tracks in the conditions of a school laboratory.

Photographic film

The tracks of all three kinds of radiation may also be studied in photographic films. The emulsion reduces the range of α-particles to about 0.02 mm, and even β-particle tracks are only about 1 mm in length. A microscope must therefore be used to observe them. Special nuclear emulsions are used for this purpose which are thicker than normal and have a higher density of silver bromide grains. The method has the advantage of automatically making a permanent record of the events studied. (See figure 3.17b, page 33.)

The Geiger-Müller tube

This instrument is probably the most versatile and useful of the devices available for detecting radiation from radioactive substances. It is essentially a form of discharge tube containing gas at a pressure of 10^4 Pa; it is operated at a p.d. somewhat less than that which would produce a continuous discharge through the gas. A typical design is shown in figure 19.8a. The anode consists of a fine rod, which runs along the axis of the cylindrical cathode. A large electric field is therefore produced near the surface of the rod (page 198); and in this region any free electrons are sufficiently accelerated to cause further ionisation. In this way the ionisation builds up quickly, and a small amount of initial ionisation produces a considerable

'avalanche' of electrons. Thus any ionisation of the gas in the tube (even a single ion-electron pair) can be sufficient to trigger off an appreciable pulse of current.

The GM tube (as it is usually called) is connected in the circuit shown in figure 19.8b. When an ionising particle enters the tube, the resulting pulse of current causes a corresponding pulse in the potential difference across the resistance R in series with it. This is amplified and registered by a suitable detecting device. For instance the output of the amplifier could be joined to a pair of headphones or a loudspeaker. Each pulse is then detected as an audible click. At low count rates it is possible to note the individual clicks; but at higher speeds they merge into a continuous crackle.

The exact number of pulses occurring in a given time interval may be recorded by using an electronic counting circuit (page 297); the apparatus designed for this purpose is called a *scaler*. There is another type of electronic circuit that records the *average rate* at which pulses are delivered by a GM tube, the figure being represented by a moving-coil meter incorporated in the circuit. Such an apparatus is known as a *ratemeter*. Because of the random nature of the emission of particles there is always a certain amount of wandering of the pointer, caused by the statistical fluctuations of the count rate.

The behaviour of a GM tube provides clear evidence that β-radiation and γ-radiation also are emitted by a radioactive source in an irregular and random manner, rather than as a steady stream of radiation. Just as with α-radiation it is clear that we are dealing with *particles* ejected at random by the source.

It is possible to design a GM tube to detect all three kinds of radiation; but to detect α-particles the mica window through which they enter must be made very thin. Even supported by a metal grill this is very easily damaged. To detect β-particles a rather thicker window can be used. It is usually made of mica or glass, and is still thin enough to be treated with great care.

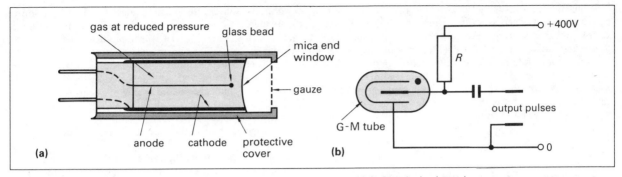

Figure 19.8 (a) An end-window Geiger-Müller tube. (b) The circuit into which the tube is plugged.

To detect γ-radiation we can dispense with a window altogether, since γ-radiation readily penetrates through any part of the tube. Indeed the radiation is so weakly ionising that it has only a small chance of producing any ionisation in the tube at all. Only about 1 % of the particles passing through the tube are actually counted in this case, even with a tube of large volume.

The solid-state particle detector

A semiconductor diode, with a reverse p.d. joined across it, may also be used to detect ionising radiations (page 286). The passage of an ionising particle through the *p-n* junction produces many extra electron-hole pairs, and causes a pulse of current through the diode. The pulses may then be recorded with a scaler or ratemeter, as with a GM tube.

However, unlike a GM tube the solid-state detector produces pulses whose amplitude depends on the number of electron-hole pairs caused by the particle. It is thus able to discriminate between the different types of particle. If the output is connected to a cathode-ray oscilloscope, the variation of pulse size is very obvious. The largest pulses are of course produced by α-particles; but to detect these the thickness of material on one side of the junction in the diode must be reduced to a minimum. It provides a small, convenient type of detector, operating on a low-voltage supply.

Activity

In using a radioactive source we are not usually concerned with the mass of active material present. What matters is the rate at which it emits particles; this is called its *activity*. A traditional unit of activity is the curie (Ci). This is defined as an activity of 3.7×10^{10} s^{-1}. (This figure was chosen as being the activity of 1.00 g of the isotope of radium first discovered by the Curies.) Thus an americium source marked '0.1 μCi' is emitting 3.7×10^3 α-particles per second (in all directions). But half of these are emitted backwards and absorbed in the source support; and many others are stopped by the protective grid in front or by the foil in which the active material is embedded.

A measurement of the activity of a source (allowing for the particles lost by absorption) enables us to estimate the mass of radioactive material it contains. Thus a pure radium source with an activity of 18.5×10^4 s^{-1} contains a mass *m* of radium given by

$$m = \frac{18.5 \times 10^4 \text{ s}^{-1}}{3.7 \times 10^{10} \text{ s}^{-1}} \times 1.00 \text{ g}$$

$$= 5.0 \text{ μg}$$

In measurements with a counter (either GM tube or solid-state type) and scaler we must always make sure that we count a sufficiently large number of particles to allow for the random fluctuations in the emission rate. For instance, if the count rate in some experiment averages 10 000 per minute, the actual numbers recorded in a series of one-minute counts might be:

9954, 10 108, 10 082, 9911, 9945

These differences are not caused by errors, but are a genuine record of the fluctuations in the numbers of particles received. A total count of at least 10 000 is needed if we wish to reduce the uncertainty on this account to less than 1 %.

In school laboratories 10 μCi is the greatest activity allowed for any one source. With larger sources safety precautions must be much more strict than those we have described. For medical purposes radium sources of up to 10 Ci are sometimes used; and γ-emitting sources of cobalt-60 of more than 2000 Ci have been used for deep therapy. Such sources can give lethal doses of radiation in quite a short time.

The background count

Even without any special radioactive source in the neighbourhood some background radiation is recorded by a GM tube (or any other particle detector). This is caused partly by radioactive impurities inevitably present in the apparatus and its surroundings, and partly by *cosmic radiation* entering the Earth's atmosphere from outer space. The background count rate of a GM tube usually amounts to between 20 and 50 per minute, though with lead shielding it can be reduced to about 10 per minute.

When we are measuring the emission rate from a radio-active source, it is necessary first to record the background count, and then subtract this from all subsequent readings. To obtain satisfactory measurements we must ensure that the total count rate is substantially greater than the background count rate. For instance, if the emission rate to be measured is only about 5 particles per minute, it would be difficult to detect this with certainty against a background count rate of 50 per minute.

19.3 Identifying the particles

β-particles

These are readily deflected by a magnetic field (figure 19.9). A β-emitting source (strontium) is mounted at the end of a tube whose walls are thick enough to stop the β-radiation completely. A GM tube is supported in a clamp so that it can be moved to various positions along an arc, as shown. The β-radiation is virtually confined to a narrow beam near the axis of the tube. Outside the beam the count rate falls to a low value about equal to the background count. When a magnetic field is applied at the mouth of the tube (at right-angles to the plane of the diagram), the beam is deflected to one side in the plane of the paper; the count rate falls to a low value in the centre and the β-particles may now be detected by moving the tube to the correct position on one side.

The direction of the deflection shows that the β-particles are *negatively* charged. In figure 19.9, if the magnetic field is *down* into the paper, the beam is deflected towards the *lower* edge of the diagram, as shown. Fleming's left-hand rule (page 205) then shows that the beam is behaving like an electric current directed (in the conventional sense)

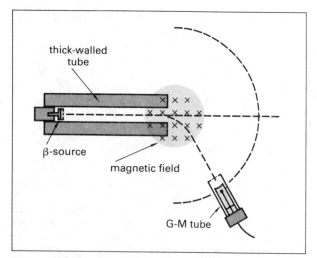

thick-walled
tube

β-source

magnetic field

G-M tube

Figure 19.9 The magnetic deflection of β-rays shows that they consist of a stream of negatively charged particles.

towards the source; the particles emerging from the source must therefore be negatively charged. Quite moderate magnetic fields are found to produce large deflections. A flux density of 10^{-2} T (produced, for instance, by a pair of magnadur magnets) is quite sufficient for this demonstration.

The β-particles may also be deflected by an electric field, though this is harder to demonstrate in a school laboratory. Again the deflection shows that the particles are negatively charged; i.e. passed between a pair of deflector plates with a large potential difference between them, the deflection is towards the positive plate.

In these experiments there is not much point in making precise measurements, since the β-particles are being slowed down and scattered by collisions with air molecules along their paths; and without vacuum apparatus little useful information is gained. But it is possible to measure the speed and specific charges of β-particles by the same kinds of methods as those used in studying cathode rays (page 237). The results show that:
(i) the β-particles have all possible speeds from zero up to a definite maximum for a given source.
(ii) for the β-particles with *low* speeds the specific charge is identical with that of the electrons in cathode rays. It seems therefore that β-particles are *electrons* emitted with speeds that range from zero up to two-thirds the speed of light. Their energies range from zero up to several MeV.
(iii) for the β-particles with *high* speeds the specific charge is less than that of low-speed electrons. This reduction is in agreement with the prediction of the theory of relativity that the mass of a particle should increase with speed (page 34).

It is now understood that in the process of β-emission a second particle is emitted at the same time as the electron. This accounts for the variable kinetic energy of the β-particle. In all other atomic and nuclear processes energy is lost or absorbed only in definite packets (quanta) corresponding with the changes in energy levels that can occur within the atom. It would be surprising if the process of β-emission was an exception to this. The second particle has no charge and is called a *neutrino*. The total energy carried away by the electron and neutrino in β-emission is constant for a given radioactive isotope; but it may be divided between the two particles in any way whatever. Like the electron the neutrino has spin and therefore carries away angular momentum as well as energy.

The neutrino has very little interaction with other forms of matter – so little that it can pass right through the Earth with only a slight chance of interacting with any of the particles in its path. Nevertheless its interaction with matter has been directly observed in the intense streams of radiation emerging from a nuclear reactor (page 266).

α-particles

When radioactivity was first investigated, it was concluded that α-rays could not be deviated by electric or magnetic fields. Flux densities sufficient to produce large deflections of cathode rays and β-radiation could give no detectable

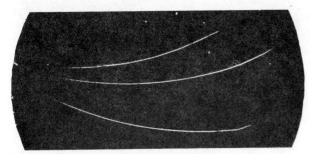

Figure 19.10 An α-particle is not readily deflected by a magnetic field. For this cloud chamber photograph a field of more than 4 T was used (directed down into the picture.)

deflection of the α-radiation. However, Rutherford and others eventually succeeded in devising sufficiently sensitive techniques for observing the minute deflections that do actually occur; and thus managed to show that α-particles are relatively massive and carry a *positive* charge. Figure 19.10 shows a cloud chamber photograph of α-particle tracks in a *B*-field of more than 4 T, taken at Cambridge by the Russian scientist Peter Kapitza.

A simple experiment of this kind can be performed in a school laboratory using the apparatus shown in figure 19.11. It serves at least to show how small is the deflection of α-particles by even a large magnetic field, and it gives the sign of the charge they carry. Some reduction of pressure is needed in the system so as to increase the range of the α-particles sufficiently for the path length required. The particles are detected by a solid-state particle detector, which can be used with care in a vacuum system provided the changes of pressure to which it is subjected are not too sudden.

In the absence of the magnetic field the flexible tube is bent into a shallow arc just sufficient to cut off the α-particles from the detector. The magnetic field is now applied at right-angles to the curved part of the tubing using a large U-shaped magnet. If the field is strong enough and in the right sense, the detector will once more register the arrival of α-particles. In figure 19.11 the field needs to be *down* into the diagram; and Fleming's left-hand rule then shows that the beam of particles must carry *positive* charge.

Detailed experiments show that:
(i) with some sources *all* the α-particles are emitted with the same energy. Other sources produce α-particles which have one or other of a few particular energies. We describe this by saying that the α-particle energies form a *line spectrum* (page 378).
(ii) the specific charge of the α-particles from all sources is 4.8×10^7 C kg^{-1}. This is just half the specific charge of the hydrogen ion (page 80). The α-particle must therefore be either a singly charged ion of mass number 2 (i.e. a deuterium ion), or else a doubly charged ion of mass number 4 (i.e. a helium ion), etc.

The matter was settled by the celebrated experiment of Rutherford and Royds depicted in figure 19.12. A small quantity of the radioactive gas *radon* was placed in the central tube; this had sufficiently thin walls (0.01 mm) to allow some of the α-particles emitted by the radon to pass through into the evacuated outer compartment B. Once they have come to rest the α-particles will acquire electrons from the material of the walls, and will form atoms (of gas to be identified). Any gas which collected in the space B could be compressed up into the discharge tube at the top by raising the level of the mercury. After six days enough gas had collected to enable an electric discharge to be passed between the two electrodes. The light from this was examined with a spectrometer, and the usual spectrum of helium was revealed. The α-particle must therefore be a doubly ionised helium atom, or *helium nucleus*.

Collision tracks

A further confirmation of Rutherford's result is obtained by studying the tracks of α-particles in a cloud chamber filled with helium instead of air. An occasional collision is observed of an α-particle with a helium atom, in which the struck atom is ionised and projected forwards, while the

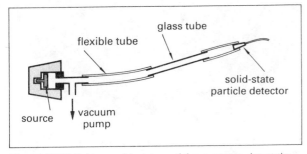

Figure 19.11 Using a solid-state particle counter to detect the slight magnetic deflection of α-particles travelling in air at reduced pressure.

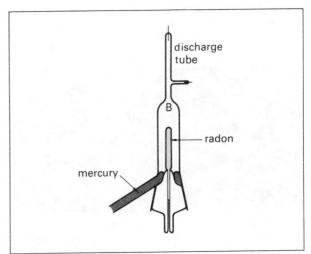

Figure 19.12 Rutherford and Royds' experiment to show that α-particles are helium nuclei.

Figure 19.13 An α-particle collision in a cloud chamber filled with helium; the angle between the paths of the particles is a *right-angle*. The plane of collision is not quite square to the camera. But comparison with the picture from a second camera on another axis shows that the angle between the paths is indeed a right-angle. This can happen only if the two particles are of identical mass.

incident α-particle is deflected, giving a Y-shaped appearance to the tracks (figure 19.13). The tracks in the two arms of the Y both appear to be of identical density compared with one another and with other α-particle tracks. Furthermore the angle between the arms of the Y is found to be 90°. The dynamics of colliding particles shows that this angle occurs only with particles of *equal* mass (page 33).

Collision tracks may also be observed in cloud chambers filled with other gases. For instance with hydrogen, collisions occur between α-particles and hydrogen nuclei (protons). The tracks of the knocked-on protons are longer and are more thinly marked by droplets than α-particle tracks; and measurement of the angles involved shows in this case that the masses of the colliding particles are in the ratio 4:1, in agreement with Rutherford's result (figure 19.14).

In a school cloud chamber the tracks of protons knocked on by α-particle collisions can be observed by covering an α-source with a sheet of polythene. This material is rich in hydrogen atoms, presenting plenty of scope for the type of collision we are seeking. The source is placed near one

edge of the cloud chamber completely enclosed in a container, one side of which is formed by the polythene sheet. This must be just thick enough to stop the α-particles completely. Proton tracks will then occasionally be observed coming from the polythene. If a pin-hole is made in the polythene, an occasional α-particle track will also be seen emerging from this; this enables us to compare the length and density of the α-particle and proton tracks. Figure 19.15 shows a cloud chamber photograph obtained with an α-source which is partly covered with a thin film of paraffin wax (which, like polythene, is rich in hydrogen atoms).

Figure 19.16 shows a pair of photographs obtained with α-particles in a cloud chamber filled with oxygen; the oxygen nuclei are four times as massive as the α-particles, as is confirmed by measurements of the angles between the emerging tracks (which in this case are *more than* a right-angle).

The nucleus

It is known that atoms are of such a size that in a piece of solid material they are tightly packed together (page 43). When therefore an α-particle traverses a thin piece of metal foil, it must actually pass right through the interior of a large number of atoms. Any changes of direction that occur in the process give some indication of the nature of the forces that act on a charged particle *inside* the atom. The tracks of α-particles in a cloud chamber are very nearly

Figure 19.15 A cloud chamber photograph of α-tracks; the left-hand half of the source has been covered with a film of paraffin wax (which is rich in hydrogen). One proton track can be seen emerging from the wax, knocked out by an α-particle.

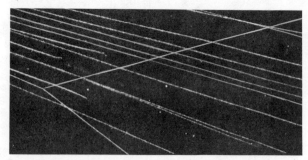

Figure 19.14 An α-particle collision with a proton in a cloud chamber filled with hydrogen. The α-particle is deflected and the proton is projected forward. The angle between their paths is less than a right-angle, since the proton is less massive than the α-particle.

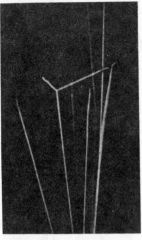

Figure 19.16 In these pictures the α-particle collisions are with nuclei of oxygen. Since these are more massive than the α-particles, the angles between the emerging tracks are more than a right angle, and the α-particle may be scattered back, as in the right-hand picture.

straight; but close inspection shows that they suffer frequent small deflections of 1° or less. These deflections, small though they are, cannot be accounted for by collisions with the electrons in the atoms. The α-particle is 7000 times more massive than the electron, and the maximum possible deflection in one such collision is less than 0.01°. The deflections observed are therefore evidence of the inter-action of the α-particle with the positively charged part of the atom.

In 1909 acting under Rutherford's guidance, Geiger and Marsden embarked on a detailed investigation of the deflections of α-particles in a metal foil. They then found that a small proportion of the particles (about 1 in 8000) were scattered through angles of more than 90°; in fact they appeared to be scattered *back* from the metal foil. Rutherford pointed out that this was like finding a shell from a large field gun bouncing back from a piece of tissue paper!

Assuming that the force acting on an α-particle was electrical in nature, it was clear that the electric field in at least some part of the atom was an exceedingly intense one. Rutherford therefore suggested that the positive charge of the atom and most of its mass was concentrated in a very small central nucleus. The electric field close to the nucleus could then be of sufficient size to account for the deflections observed. The results of Geiger and Marsden's measure-ments agreed in every detail with the predictions of Rutherford's theory; and there was therefore little doubt that this *nuclear model* of the atom was correct.

One of the predictions of the theory was to give the proportion of scattered particles in terms of the *charge* on the central nucleus. The α-particle scattering experiments therefore enabled this charge to be measured. Expressed in electronic charge units (*e*) the charge of the nucleus was found to be equal to the *atomic number* of the element concerned (page 36). This number had been used hitherto to describe the position of an element in the periodic table, which indicated its chemical properties and type of spectrum. Rutherford had now shown that the atomic number was to be identified with the *charge on the nucleus*. This charge controls the number of electrons in the atom, and in this way fixes its chemical properties and spectrum.

γ-radiation

There is clear evidence to show that the properties of γ-radiation and X-radiation are identical in every respect. The list of properties of X-radiation given on page 250 apply equally to γ-radiation. We conclude that γ-rays, like X-rays, are electromagnetic waves of very short wavelength. While γ-rays are generally of shorter wavelength than X-rays, there is in fact a considerable overlap between the X-ray and γ-ray parts of the spectrum (page 366). The difference between γ-rays and X-rays is one of origin rather than of nature; γ-rays arise in the nucleus, whereas X-rays are generated in the inner parts of the electron structure outside the nucleus.

From a given γ-source only certain sharply defined wave-lengths are observed, which are characteristic of the type of nucleus concerned. In other words, like the visible and X-ray spectra of the *atom*, the γ-radiation from the *nucleus* forms a *line spectrum*. This implies that the particles in the nucleus, like the electrons in the atom, can exist only in certain 'allowed' states, each of well-defined energy. As with optical and X-ray spectra the energy W and frequency f of the radiation are connected by

$$W = hf$$

where h is the Planck constant (page 242). The quantities of energy liberated in nuclear processes are generally far larger (a few MeV) than those arising from re-arrangements of the electrons in the outer parts of the atom (a few eV); a γ-ray photon is therefore of very high frequency (and so of very short wavelength). The energy carried by one γ-ray photon is quite sufficient for it to be detected indi-vidually by a GM tube (although some may pass right through the tube without causing any ionisation). The behaviour of a GM tube in a beam of γ-rays is therefore a convincing demonstration of the particulate (photon) nature of electromagnetic radiation (page 243).

19.4 Nuclear transformations

With the model of the nuclear atom clearly established, we are in a position to see what happens in the atom when radioactive emission takes place. The source of the particles is the nucleus itself, whose charge and mass must therefore be changed in the process. If a nucleus ejects an α-particle, which carries a charge 2*e*, its atomic number *decreases* by 2; at the same time its nucleon number *decreases* by 4, since that is the nucleon number of helium. Since the charge on the nucleus is less by $2e$, the atom of which it is part now requires 2 fewer electrons in its outer electron structure. These surplus electrons are disposed of to the surroundings; and we then have an atom of some different element (page 37).

Likewise, the emission of the negatively charged β-particle *increases* the positive charge on the nucleus by one electronic charge, and so raises its atomic number by one; the atom must now gain an electron from the surroundings to complete its outer electron structure. However, the mass of the β-particle (and the neutrino that goes with it) is very small compared with the mass of a nucleon – much less than the small difference between the *relative atomic mass* of a substance and the nearest whole number (which is its *nucleon number*). The emission of a β-particle therefore leaves the nucleon number of the atom *unchanged*.

Thus radioactive elements are transmuted spontaneously into other elements according to the rules outlined above. We shall consider how this works out by analysing what happens in a radium source. The atomic number of radium is 88, and the nucleon number of its principal isotope is 226. Using the usual chemical symbol for radium, this information can be expressed symbolically:

$$^{226}_{88}Ra$$

The subscript indicates the atomic number, and the superscript the nucleon number. Since an α-particle is a helium nucleus, it can be described in the same notation by

$$^{4}_{2}He$$

Strictly speaking the information carried by the subscript is redundant; if the chemical element is known, a copy of the periodic table tells us also the atomic number. To specify a nucleus it is only necessary to state the element and its nucleon number. Sometimes therefore the above isotopes are simply described as radium-226 and helium-4.

When a nucleus of radium-226 emits an α-particle, its atomic number falls to 86 and its nucleon number to 222. It therefore becomes a nucleus of the *gas radon*.

$$^{226}_{88}Ra \rightarrow {}^{222}_{86}Rn + {}^{4}_{2}He$$
$$(\alpha\text{-particle})$$

The evolution of radon from a quantity of radium is readily demonstrated, though the experiment is not safe to attempt in a school laboratory. Radon itself is a radioactive element which emits an α-particle, and so changes into an isotope of *polonium*.

$$^{222}_{86}Rn \rightarrow {}^{214}_{82}Pb + {}^{4}_{2}He$$

Polonium in its turn emits an α-particle and decays to form a radioactive isotope of *lead*.

$$^{218}_{84}Po \rightarrow {}^{214}_{82}Pb + {}^{4}_{2}He$$

This lead isotope is a β-emitting substance, which therefore increases its atomic number by one when it decays, forming bismuth-214.

$$^{214}_{82}Pb \rightarrow {}^{214}_{83}Bi + {}^{0}_{-1}e + \nu$$
$$\text{(electron)   (neutrino)}$$

Thus the original radium-226 is the start of a sequence of radioactive changes, which involves 8 intermediate isotopes and terminates eventually in the stable isotope lead-206. A sample of originally pure radium, left for a long enough time, is found to contain all of these elements. Several of them are present in sufficient quantity to be detected by chemical means, though some of them last such a short time before decaying into another element that their presence can be inferred only from the radiations they emit.

The nature of the radiations observed from a radium source depends on whether the first decay product, radon, remains trapped in it or not. In practice care is taken to seal a laboratory source for safety reasons; a radium source usually consists of a piece of foil of a radium-silver alloy. Almost all the radon then remains trapped in the foil. What we call a radium source is really a mixture of isotopes of nine elements, including the parent radium. Between them they emit particles of all three kinds of radiation of many different energies. A freshly purified sample of radium emits α-particles of one range only, together with some γ-radiation. But if it is kept with its decay products it soon produces β-particles also, and in time its α-emission and γ-emission increase by a factor of about 4.

γ-radiation

Since this consists of electromagnetic radiation, it does not alter the charge of the nucleus from which it comes and affects its mass very little. It therefore leaves the composition of the nucleus unaltered. A γ-ray photon can only *follow* the emission of either an α-particle or a β-particle. When one of these particles emerges, it may leave the resulting nucleus in an excited state. Shortly afterwards the nucleus emits the surplus energy as a γ-ray photon, as it reverts to its ground state.

Mass and energy

One of the results of the special theory of relativity is that energy and mass are equivalent (page 34). The theory predicts that a body gaining energy ΔE thereby increases in mass by an amount Δm given by

$$\Delta E = c^2 \Delta m$$

where c is the speed of light. The quantities of energy handled in everyday life are far too small to involve detectable changes of mass. Thus 1 kg of water absorbs 4.2×10^5 J of energy to raise its temperature from 0°C to 100°C. The theory of relativity therefore predicts that its mass will increase accordingly by an amount Δm, given by

$$\Delta m = \frac{4.2 \times 10^5 \text{ J}}{(3.0 \times 10^8 \text{ m s}^{-1})^2}$$

$$\approx 5 \times 10^{-12} \text{ kg}$$

In chemical reactions also the changes of energy are of this order of magnitude, involving at the most a few eV of energy per atom; and the associated changes of mass are again quite undetectable.

However, in nuclear processes the quantities of energy released are of the order of a few MeV, and are sufficient to produce significant changes in the masses of the particles concerned. The mass equivalent Δm of 1.00 MeV of energy may be calculated as follows:

$$1.00 \text{ MeV} = 1.60 \times 10^{-13} \text{ J} \qquad \text{(page 191)}$$

Hence $\quad \Delta m = \dfrac{\Delta E}{c^2}$

$$= \frac{1.60 \times 10^{-13} \text{ J}}{(3.00 \times 10^8 \text{ m s}^{-1})^2}$$

$$= 1.78 \times 10^{-30} \text{ kg}$$

Thus 1.00 MeV of energy (of any form) is equivalent to 1.78×10^{-30} kg of mass. (This is nearly *twice* the mass of an electron.) A 1 MeV γ-ray carries this mass away from the nucleus that emits it. In the same way a 1 MeV β-particle has a total mass *three* times the rest mass of a stationary electron (its rest mass + the mass of its kinetic energy). Thus the emission of a β-particle not only carries away from a nucleus the rest mass of an electron, but also the mass equivalent of its kinetic energy.

The processes of radioactive decay always involve appreciable reductions in the total rest mass of the nuclei and particles concerned, though the changes are still small compared with the mass of a nucleon, so that the relative atomic masses of all atoms remain close to whole numbers. The changes in mass can readily be measured by modern techniques of mass spectrometry (page 248).

Consider for instance the α-decay of radium-226. Its relative atomic mass is 226.0254; i.e. the actual mass m_{Ra} of one atom of this isotope is given by

$$m_{\text{Ra}} = 226.0254 \; m_{\text{u}}$$

where m_{u} is the unified atomic mass constant ($= 1.66 \times 10^{-27}$ kg, page 37). Similarly the masses of one atom of the decay product radon-222 and of helium-4 are given by

$$m_{\text{Rn}} = 222.0176 \; m_{\text{u}}$$
$$m_{\text{He}} = 4.0026 \; m_{\text{u}}$$

so that $\quad m_{\text{Rn}} + m_{\text{He}} = 226.0202 \; m_{\text{u}}$

This is less than the mass of the parent radium atom by $0.0053 \; m_{\text{u}}$, a quantity known as the *mass defect* of the reaction. This mass is conserved as kinetic energy of the emitted particle. So, for the α-emission from radium-226 we have

$$\text{surplus energy} = \frac{0.0052 \times 1.66 \times 10^{-27} \text{ kg}}{1.78 \times 10^{-30} \text{ kg/MeV}}$$

$$= 4.85 \text{ MeV}$$

Experiments show that the α-particle carries away most of this energy, the remainder being emitted as a γ-ray photon shortly afterwards.

In every case of radioactive decay the large quantities of energy released are found to be provided at the expense of the masses of the nuclei and particles involved, as predicted by the theory of relativity. Indeed a spontaneous nuclear change cannot take place unless the total mass of the daughter products is *less* than the mass of the parent nucleus, so that a surplus of energy can be made available to provide kinetic energy for the decay products. In general, the greater the mass defect of any possible nuclear change, the more probable the change becomes, and the shorter-lived is the nucleus concerned. In other words a radioactive element emitting a particle of very high energy will generally be very short lived.

The positron

In 1930 Paul Dirac produced a theory of the electron, predicting that there should be a positive electron as well as the familiar negative one; and a few years later the new particle, called the *positron*, was discovered. It was found that a γ-ray photon of very high energy passing near an atomic nucleus may give rise to an electron-positron *pair* (figure 19.17). The discovery of pair production provided one of the most convincing demonstrations of the equivalence of mass and energy. The minimum energy that might produce an electron-positron pair is $2m_{\text{e}}c^2$, where m_{e} is the mass of an electron (or positron). This is equal to 1.02 MeV. Any γ-ray photon above this energy is capable of causing pair production, and at very high energies pair production accounts for most of the absorption of γ-rays by matter.

A positron has a short life, since it soon annihilates itself in combination with an electron. The mass-energy is conserved by the production of *two* γ-ray photons; each of these is of energy 0.51 MeV.

Dirac's theory is found to apply more widely than just to electrons and positrons. There is complete symmetry in the table of fundamental particles. To every particle there is an *anti-particle* of opposite charge (if any), but otherwise of similar properties.

Figure 19.17 Pair production of electrons and positrons in a cloud chamber. The group of particles has arisen from the γ-rays in a cosmic ray shower. Since the chamber is in a magnetic field, the negative electrons curve in one direction and the positrons in the other.

Artificial transmutations

In elements of low atomic number the charge on the nucleus is relatively small, and the electrical repulsion between this and an incident α-particle is much less than for a heavy nucleus. There then exists the chance that an α-particle may actually penetrate the nucleus and cause a nuclear transmutation.

The first effect of this kind was discovered in nitrogen. It was found that the passage of α-particles through nitrogen caused the production of a few protons, apparently knocked out of the nitrogen nuclei by the α-particles. A reaction of the following kind is apparently occurring:

$$^{14}_{7}\text{N} + ^{4}_{2}\text{He} \rightarrow ? + ^{1}_{1}\text{H}$$
$$\text{(proton)}$$

The need to balance atomic numbers and nucleon numbers before and after the reaction leads us to expect a nucleus of oxygen-17 ($^{17}_{8}\text{O}$) to be formed in this process; this is an isotope that is known to form a small part of naturally occurring oxygen.

Cloud chamber photographs with nitrogen in the chamber show that about one α-particle track in 50 000 is branched as shown in figure 19.18. The α-particle track comes to an end at the fork. Of the two tracks emerging from this point, one is more thinly marked than an α-particle track, and is that of a proton; the other is short and thick, and is taken to be that of the oxygen nucleus which has been produced.

Many other examples have been subsequently discovered of the transmutation of nuclei by bombardment with α-particles, as well as by other projectiles such as high-speed protons and deuterons (deuterium nuclei). By using particle accelerators (page 248) these particles may be given high energies and used to bombard a target. By these and other means the list of known radioactive isotopes has been enormously extended. In most cases the nuclei produced turn out to be radioactive isotopes not found in nature. Such for instance is the isotope phosphorus-30, formed in the α-bombardment of aluminium. This isotope exhibits a positron from of β-disintegration, decaying to silicon-30 by emission of a positron (and neutrino):

$$^{30}_{15}\text{P} \rightarrow ^{30}_{14}\text{Si} + ^{0}_{1}\text{e} + \nu$$
$$\text{(positron)} \quad \text{(neutrino)}$$

19.5 The neutron

Up to this point in the development of the subject the composition of the nucleus had remained something of a mystery. The *proton* was clearly one ingredient, and presumably accounted for the positive charge of the nucleus. Since, however, its relative atomic mass is usually more than double its atomic number, it must contain other particles besides protons. It was suggested by theorists that the undetected ingredient was an uncharged particle of

Figure 19.18 A nuclear transmutation produced by an α-particle in a nitrogen nucleus. The long thin track is that of a proton emerging from the struck nucleus; the α-particle is absorbed in the process forming a nucleus of oxygen-17, which leaves a short thick track.

about the same mass as the proton; this was referred to as the *neutron*.

The neutron was eventually observed in 1932 by James Chadwick in the bombardment of beryllium with α-particles. A type of radiation emerged from this that at first sight appeared similar to γ-radiation. It could not consist of charged particles, since these would produce much greater ionisation in the detector. But when Chadwick inserted sheets of paraffin wax (or any other material containing a large amount of hydrogen) in the path of the beam, a big increase in the ionisation current was noted. This was found to be caused by protons knocked out of the paraffin wax by the new radiation. It seemed to have no interaction with the electrons in the matter through which it passed, but only with the atomic nuclei. The effect was duly detected in cloud chamber photographs, with hydrogen in the chamber (figure 19.19). The tracks visible are those of protons that have been projected forwards by collisions with neutrons; the neutrons themselves, being uncharged, cause no ionisation and leave no tracks.

By careful measurement of the energies of the nuclei struck by the neutrons Chadwick was able to estimate that the rest mass of the neutron was about equal to that of the

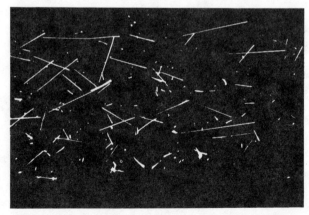

Figure 19.19 The passage of a beam of neutrons through a cloud chamber filled with hydrogen. The neutrons themselves leave no tracks. What we see are the tracks of protons knocked on by the impact of neutrons.

proton. It is now known that the neutron is in fact slightly more massive.

Mass of neutron $= 1.008\,67\;m_u$
Mass of proton $= 1.007\,28\;m_u$

Since the neutron is more massive than the proton, there exists the possibility that it might decay into a proton, the surplus mass being conserved as kinetic energy of the decay products. This very process has been observed in neutrons outside the nucleus. In the free state the neutron itself is a *radioactive particle*, which decays into a proton by emission of a β^--particle (together with a neutrino); the half-life of this decay process is about 12 minutes (page 267).

The discovery of the neutron settled for the time being the problem of the composition of the nucleus. The number of protons in the nucleus is equal to its atomic number, and the additional mass is made up of neutrons. On this basis the α-particle must consist of just 2 protons and 2 neutrons. The composition of other nuclei can be worked out similarly (see table on page 37). In the case of light nuclei the neutron number N is always approximately equal to the atomic number Z (which is the proton number of the nucleus). But with more massive nuclei the proportion of neutrons goes up, until in the largest nuclei there are nearly twice as many neutrons as protons.

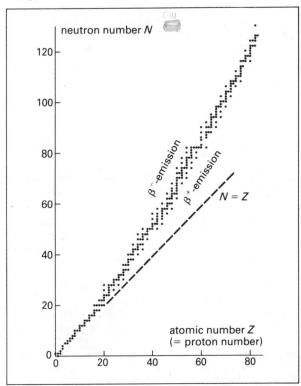

Figure 19.20 A plot of neutron number N against proton number Z for the stable nuclei. Nuclei above the region of stability have surplus neutrons and tend to decay by β-emission; those below by β^+-emission.

β-decay

Figure 19.20 shows a plot of neutron number N against atomic number Z; each dot on this represents a known stable nucleus. For most elements there are several dots above each value of Z, representing the different possible stable isotopes of the element concerned. Unstable isotopes are not shown in figure 19.20; they would lie mostly either above or below the fuzzy line formed by the plot. When they decay by radioactive emission, they do so in such a way as to bring the resulting nucleus nearer to the region of stability close to the line. Thus, nuclei above the line are those that have a surplus of neutrons; these will tend to decay by β^--emission, which increases the proton number by one at the expense of the neutron number. Nuclei below the line have a surplus of protons, and tend to decay by β^+-emission, which has the opposite effect. Phosphorus-30 is one of these (see above).

The process of β-decay thus consists in the transformation of a neutron into a proton (in β^--emission), or else of a proton into a neutron (in β^+-emission). The process is accompanied by the creation of an electron or positron to carry away (with a neutrino) the surplus energy and angular momentum. In any nuclear process the total number of *nucleons* (protons + neutrons) remains unaltered, if we include any carried away by α-particles, etc. This is the reason for balancing the nucleon numbers before and after a nuclear reaction. The masses of the nucleons change slightly in a reaction, but the total number of nucleons does not.

Neutron reactions

By using beams of neutrons many new kinds of nuclear reaction can be produced. A *charged* particle entering a nucleus needs to have sufficient energy to overcome the intense electrical repulsion. Protons and α-particles must therefore have large energies to do this. But the neutron, being uncharged, can pass readily *through* a nucleus, whatever its energy; and it may in the process be absorbed by it.

One such reaction provides one of the most convenient modern methods of detecting neutrons;

$$^{10}_{5}B + \underset{\text{(neutron)}}{^{1}_{0}n} \rightarrow ^{7}_{3}Li + ^{4}_{2}He$$

The boron-10 nucleus absorbs neutrons very readily, and then disintegrates into a lithium nucleus and an α-particle, both of which are heavily ionising particles. The ionisation produced by neutrons themselves in most gases is so slight that an ordinary GM tube scarcely responds to them at all. But by filling the tube with boron trifluoride gas a high proportion of the incident neutrons can be detected.

Safety

Because of their high penetrating power neutrons constitute a serious health hazard. The most effective substances for stopping them are those that contain as many light nuclei as possible per unit volume. Protective screens are best made of compounds of hydrogen and other light nuclei (e.g. water or paraffin wax) rather than lead. (By the same reasoning the human body is dangerously effective as an absorber of neutron radiation.)

Nuclear energy

Matter in the nucleus of an atom is potentially the source of prodigious quantities of energy. This is apparent even in the ordinary processes of radioactive decay. For instance in the decay of one atom of radium-226 (page 255) the kinetic energy produced per nucleus is about 5 MeV. This is provided at the expense of part of the rest mass of the original nucleus; there is a decrease of about 1 part in 40 000 in the rest mass of the particles, and this re-appears as the kinetic energy of the decay products, (and is then converted to internal energy of the material in which the particles are stopped). This may not sound very much. But suppose we have 1 kg of radium-226. The total energy ΔE converted from nuclear energy to internal energy when all the atoms of this have decayed is given by

$$\Delta E = c^2 \Delta m$$

$$= (3.00 \times 10^8 \text{ m s}^{-1})^2 \times \frac{1 \text{ kg}}{40\,000}$$

$$\approx 2 \times 10^{12} \text{ J}$$

The total energy converted as the radium goes through its complete decay sequence is 8 times as much – equivalent to the energy output of a 100 MW power station for 2 days. We do not notice the presence of this energy in the radium, since in the ordinary way its release is spread out over several thousand years; and in any case the samples of radium in laboratory radioactive sources are kept very small.

If energy is to be released in a nuclear reaction, the rest masses of the products of the reaction must be less than the rest mass of the original ingredients. The total number of nucleons cannot change in such a reaction, but the *mass per nucleon* decreases. Atomic masses can be measured with great precision with a mass spectrometer (page 248); the average mass per nucleon is found by dividing the atomic mass by the nucleon number. Expressed in terms of the unified atomic mass unit (m_u) the mass per nucleon is always close to 1; but it varies slightly over the range of mass numbers as shown in figure 19.21. The points for the free proton and free neutron come at the extreme left ($A = 1$). The nucleons in a compound nucleus are always less massive than these; and as we pass to larger mass numbers the mass per nucleon falls sharply at first, and then levels out. However, the important feature of the curve is the shallow minimum at about $A = 60$. In atoms of this size the nucleons have the minimum possible mass, and therefore minimum energy. Beyond this mass number the mass (or energy) of the nucleons steadily increases towards the higher atomic masses; uranium comes at the extreme right of the plot.

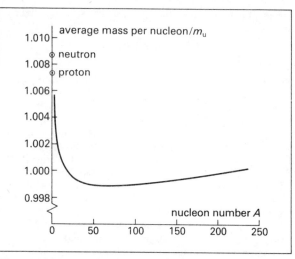

Figure 19.21 The variation of the average mass per nucleon with the number of particles in the nucleus (the mass number A). It has a minimum value at about $A = 60$.

There are therefore two possible ways in which the nuclear energy of matter can be tapped. Either we seek to break down the more massive nuclei ($A = 230$ to 250) into smaller fragments i.e. nearer to the minimum at $A = 60$; or we endeavour to fuse together the lighter nuclei into more massive units. Both these processes are possible.

Fission

When uranium is bombarded with neutrons, it is found that some of the nuclei absorb neutrons and forthwith break up into two fragments, a process known as *nuclear fission*. The two fragments are nuclei of elements of medium atomic mass, nearer therefore to the minimum energy condition for nucleons. The energy released in the fission of one uranium nucleus is an enormous 200 MeV. In 1939 it was discovered that in addition to the two main fragments several neutrons are emitted at the same time. This provides the possibility of starting a chain reaction in a piece of uranium. The fission of one nucleus can produce neutrons to cause the disintegration of several more nuclei; these in turn can trigger the fission of an even larger number; and so on. In a sufficiently large lump of uranium this can happen spontaneously. In this way a substantial fraction of the energy locked up in the uranium nuclei can be released.

In the *atomic bomb* (properly called a nuclear bomb) this energy is converted in an uncontrolled way – with devastating results. But in a *nuclear reactor* the rate of reaction is controlled so that the internal energy produced from nuclear energy can be carried away steadily by circulating fluids, and used to drive the turbines of a power station.

Another of the important uses of nuclear reactions has been the production of radioisotopes. Many of these are formed as the fission products of the uranium, and are extracted when the fuel rods are changed. Caesium-137 and strontium-90 are examples. Others may be produced by irradiation with neutrons. To do this a suitable substance is placed in the centre of the core of the reactor, where the flux of neutrons is very high.

A serious practical problem in the operation of a nuclear reactor is the safe disposal of the large quantities of radioactive waste formed; some of this will retain its activity for thousands of years, during which time it must not be allowed to contaminate the environment.

Fusion

In this type of reaction light nuclei are fused together into larger units. By using high-energy beams of protons, deuterons, etc, many such reactions have been discovered. In each case the condition for the reaction to proceed is that the two reacting particles should approach one another at sufficient speed and therefore with sufficient energy to overcome the electrical repulsion between them. To achieve this on a large scale it is necessary only to raise the temperature of the ingredients to the point at which some of the particles have the required energy. For this reason these are known as *thermonuclear reactions*; unfortunately temperatures of 10^8 K are needed to bring them about.

The only places in which such reactions normally proceed is the interiors of the sun and stars. A variety of thermonuclear reactions take place there whose net effect is the conversion of hydrogen into helium. The formation of each helium nucleus liberates 27 MeV of energy; more than 0.8% of the rest mass of the original particles is converted to internal energy in the process. The sun converts about 4×10^6 tonnes of matter per second in this way; and there seems no reason why it should not continue in this profligate style for the next 10^{10} years; Attempts to harness thermonuclear reactions to produce power in a controlled way on Earth have not so far succeeded. If and when they do, a source of cheap and abundant energy will be available for the human race (page 30).

19.6 Radioactive decay

For many radioactive substances the decay rate is so slow that there is no perceptible reduction in the material in the lifetime of one observer. But in all cases which can be investigated in a reasonable time the decay is found to follow an exponential pattern (appendix B7). Thus, if N is the number of atoms of a radioactive isotope left in a sample after time t, we have

$$N = N_0 e^{-\lambda t} \qquad (1)$$

where N_0 is the number of atoms initially present (when $t = 0$), and λ is the *decay constant* (appendix B7). The time taken for the number of atoms of the isotope to fall to *half* its initial value is called the *half-life* $T_{\frac{1}{2}}$ of the isotope. You can check that

$$e^{-0.693} = \tfrac{1}{2}$$

Hence $N = \tfrac{1}{2}N_0$, when $\lambda t = 0.693$

$$\Rightarrow \qquad T_{\frac{1}{2}} = \frac{0.693}{\lambda}$$

The half-lives of known radioactive substances cover an enormous range – from less than 1 μs to more than 10^{15} years. The direct measurement of long half-lives is unthinkable. The half-life of radium-226 is 1620 years; but even this is trifling compared with that of uranium-238 (4.5×10^9 years). Even so, the amount of radium-226 in a given sample would have decreased by only 4% during the time since this element was discovered by the Curies; and this cannot be made the basis for a measurement of its half-life.

However, if we can estimate the number of atoms N in a sample, we can derive the half-life from a measurement of its activity (page 257). The activity A of a sample is the rate at which it emits particles; this is equal to the rate of decrease in the number N of atoms present. Thus

$$A = -\frac{dN}{dt} \qquad \text{(appendix B3)}$$

In appendix B7 it is shown how equation (1) above can also be expressed in terms of dN/dt as:

$$\frac{dN}{dt} = -\lambda N$$

Hence $A = \lambda N$

and so $\lambda = \dfrac{A}{N}$

Knowing λ we can then work out $T_{\frac{1}{2}}$.

▶ A speck of dust contains 1 μg of plutonium-239 (an α-emitting substance). This is found to have an activity of 2300 s^{-1}. Calculate the half-life of plutonium.

The Avogadro constant (6.02×10^{23} mol^{-1}) is the number of atoms per mole of any element (page 38). One mole of plutonium atoms has a mass of 239 g. Therefore 1 μg (i.e. 10^{-9} kg) of plutonium contains a number N of atoms given by

$$N = \frac{10^{-9} \text{ kg} \times 6.02 \times 10^{23} \text{ mol}^{-1}}{(0.239 \text{ kg mol}^{-1})}$$

$$= 2.52 \times 10^{15} \text{ atoms}$$

Hence $\lambda = \dfrac{A}{N} = \dfrac{2300 \text{ s}^{-1}}{2.52 \times 10^{15}}$

$$= 9.13 \times 10^{-13} \text{ s}^{-1}$$

$$\Rightarrow \qquad T_{\frac{1}{2}} = \frac{0.693}{\lambda} = \frac{0.693}{9.13 \times 10^{-13} \text{ s}^{-1}}$$

$$= 7.59 \times 10^{11} \text{ s}$$

and in years, $T_{\frac{1}{2}} = \dfrac{7.59 \times 10^{11} \text{ s}}{(3600 \times 24 \times 365) \text{ s yr}^{-1}}$

$$= 24\,000 \text{ years}$$

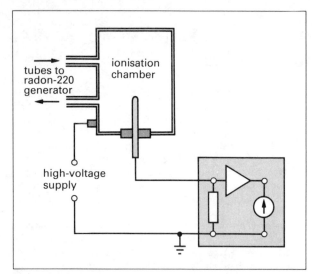

Figure 19.22 Measuring the half-life of radon-220 with an ionisation chamber and electrometer.

Notice how small the activity A of the speck is. To compare it with other radioactive sources we can express its activity in Ci. Thus

$$\frac{A}{1 \text{ Ci}} = \frac{2300 \text{ s}^{-1}}{3.7 \times 10^{10} \text{ s}^{-1}} \qquad \text{(page 257)}$$

$$\Rightarrow \quad A \quad \approx 6 \times 10^{-2} \text{ μCi}$$

Nevertheless even so small an activity as this is now regarded as a serious health hazard with a long-lived isotope like this. Absorbed into the lungs it emits 2300 α-particles per second for the rest of the person's life, and is more than likely to initiate dangerous alterations in the tissue surrounding it.

The *direct* measurement of half-life is possible only for short-lived radioactive isotopes. Many of these are too

dangerous for us to experiment with, because being short-lived, they have high activities. However, radon-220, formed as one of the decay products of thorium-232, can be handled safely in a school laboratory. About 25 g of thorium hydroxide forms a convenient source of this isotope. The thorium reaches equilibrium with its decay products (page 269) in about 12 years, and may then be used to produce radon-220. The 'vintage' thorium hydroxide is kept in a polythene bottle that can be connected to an ionisation chamber by flexible tubing in such a way that the radon is confined within the closed circuit of tubing. Inside the polythene bottle the thorium hydroxide is held in a dust-proof bag that acts as a filter to prevent escape of the fine powder. Radon-220 is formed continually, and may be transferred to the ionisation chamber by simply squeezing the bottle (figure 19.22).

The mass of radon-220 transferred from the bottle to the ionisation chamber is probably less that 10^{-20} kg, quite undetectable by chemical means. However, such a sample contains 3×10^4 atoms, and these will decay initially at a rate of about 400 per second; each α-particle produces about 1.5×10^5 ion pairs, so that the total charge passing through the ionisation chamber per second is about

$$(1.5 \times 10^5)(400 \text{ s}^{-1})(1.6 \times 10^{-19} \text{ C}) \approx 1.0 \times 10^{-11} \text{ C s}^{-1}$$
$$= 1.0 \times 10^{-11} \text{ A}$$

This is an ionisation current I that may be measured with an ordinary electrometer. The current I is proportional to the activity A of the sample, which in turn is porportional to the number of atoms of radon-220 remaining in it. We therefore expect to find I decaying exponentially in the same manner as N. Thus

$$I = I_0 e^{-\lambda t}$$

where I_0 is the initial current (when $t = 0$). A typical experimental result is shown in figure 19.23a. But as usual, the best way to test an exponential decay is to plot the equivalent logarithmic graph (appendix B7), which is

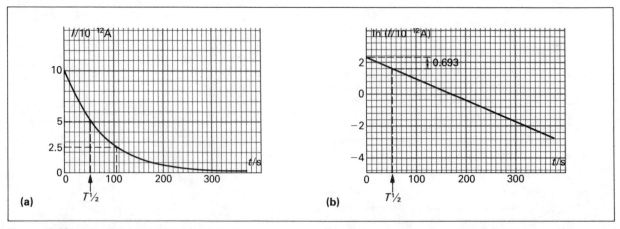

(a)

(b)

Figure 19.23 (a) The exponential decay of ionisation current; the current is halved in each successive interval of 52 s. (b) The same results plotted on a logarithmic scale of currents; the gradient of the line is $-\lambda$.

$$\ln I = \ln I_0 - \lambda t$$

as in figure 19.23b; this is a straight line whose gradient is $-\lambda$. From this graph we can take readings at $t = 0$ and $t = 300$ s, to give

$$\text{gradient} = \frac{(-1.70 - 2.30)}{(300 \text{ s} - 0)}$$

$$= -1.33 \times 10^{-2} \text{ s}^{-1}$$

$$\Rightarrow \qquad \lambda = +1.33 \times 10^{-2} \text{ s}^{-1}$$

Hence $\quad T_{\frac{1}{2}} = \dfrac{0.693}{\lambda} = \dfrac{0.693}{1.33 \times 10^{-2} \text{ s}^{-1}}$

$$= 52 \text{ s}$$

The graph in figure 19.23a shows the significance of this. Thus after the first time interval of 52 s the ionisation current has fallen to half its initial value; and half the radioactive sample has decayed. In a second interval of 52 s a half of what remains decays, leaving 1/4 of the original sample. After a total time $3T_{\frac{1}{4}}$ only 1/8 of the sample is left; and after about $7T_{\frac{1}{4}}$ less than 1% of it remains.

It is revealing to express the decay constant λ as a fraction; thus for radon-220

$$\lambda = 1.33 \times 10^{-2} \text{ s}^{-1} = \frac{1}{75} \text{ s}^{-1}$$

But $\quad \lambda = \dfrac{A}{N} = \dfrac{\text{no. of atoms decaying per second}}{\text{total no. of atoms}}$

so that λ can be thought of as the *fraction* of the total number of atoms that decay per second; with radon-220, 1 atom in 75 decays (on average) every second. It is impossible to say *which* atoms will decay; but we can say that the *probability* of a given atom decaying in the next second is 1/75. With long-lived substances the probabilities of decay of a given atom are fantastically small. For instance, the example of the radioactive speck of dust discussed above shows that for plutonium

$$\lambda = 9.15 \times 10^{-13} \text{ s}^{-1} \approx \frac{1}{10^{12}} \text{ s}^{-1}$$

The decay of a given plutonium atom in the next second is thus such an unlikely event that in other circumstances we should ignore it completely. It is only the enormous number of atoms in the sample of plutonium that make this minute probability something with detectable effects.

Radioactive tracers

The amount of radon-220 used in the above experiment would be regarded, by any other standards, as undetectably small. However, the number of atoms in the sample is large enough to provide a measureable ionisation current. Radioactive samples used in this way are called *radioactive tracers*. They have become an indispensable tool of research in many fields. By using a GM tube it is possible to follow a radioactive isotope through a physical or chemical process, and its presence can be detected by the radiations it emits;

but only a minute quantity of the isotope need be used, as long as it is sufficient to ensure that the count rate is significantly greater than the background count.

For instance, it is possible by this means to unravel some of the complex sequences of chemical reactions that take place in living organisms. The organism may be 'fed' with a compound in which a small proportion of some element (such as hydrogen-1) has been replaced with a radioactive isotope of the same element hydrogen-3, known as *tritium*. The subsequent growth of activity in other compounds present in the organism reveals the pattern of chemical reactions into which the original 'food' enters.

Radioactive equilibrium

The radioactive decay of radium is part of a series of disintegrations, ending eventually with a stable isotope of lead. Several such radioactive series are known. Indeed the series that follows radium is itself only part of a much longer series starting with uranium-238. The complete uranium-238 series is displayed in table 19.1.

The half-life of radium is very short compared with the age of the Earth, so that any radium originally present in the Earth must long since have decayed away. The radium we now have is that which has arisen as a daughter product of uranium and has not yet decayed into radon, etc. Radium and the other members of the uranium-238 series are therefore found naturally in uranium-bearing ores. Given sufficient time the amount of each isotope in the ore builds up to the point at which its rate of decay is equal to the rate at which it is formed from the decay of the previous member of the series. When this condition is reached, the proportions of the radioactive isotopes in the rock do not change with time, and we say that the uranium-238 is in *radioactive equilibrium* with its daughter products. The half-life of the parent uranium-238 is so long that its rate of disintegration is almost constant, and this is equal to the disintegration rate of each member of the series in equilibrium with it. It is also the rate at which the stable end-product (lead-206) is being formed. The net result is therefore the conversion of uranium-238 to lead-206 at a steady rate, while the amounts of the intermediate members of the series remain constant.

For a given radioactive series the activity A is therefore the same for each isotope of the series in the ore. Thus

$$A = \lambda_1 N_1 = \lambda_2 N_2 = \lambda_3 N_3 = \ldots$$

Therefore $\quad N \propto \dfrac{1}{\lambda} \propto T_{\frac{1}{2}}$

for a sample of the ore. The mass m of any isotope is proportional to the product of the relative atomic mass A_r and the number of atoms N. Hence

$$m \propto A_r N \propto A_r T_{\frac{1}{2}}$$

for the given sample.

isotope	symbol	emissions	half-life
uranium-238	$^{238}_{92}$U	α, γ	4.5×10^9 y
↓			
thorium -234	$^{234}_{90}$Th	β, γ	24 d
↓			
protoactinium -234	$^{234}_{91}$Pa	β, γ	1.2 min
↓			
uranium-234	$^{234}_{92}$U	α, γ	2.5×10^5 y
↓			
thorium-230	$^{230}_{90}$Th	α, γ	8.0×10^4 y
↓			
radium-226	$^{226}_{88}$Ra	α, γ	1620 y
↓			
radon-222	$^{222}_{86}$Rn	α	3.8 d
↓			
polonium-218*	$^{218}_{84}$Po	α	3.1 min
↓			
lead-214	$^{214}_{82}$Pb	β, γ	27 min
↓			
bismuth-214*	$^{214}_{83}$Bi	β, γ	20 min
↓			
polonium-214	$^{214}_{84}$Po	α	1.6×10^{-4} s
↓			
lead-210	$^{210}_{82}$Pb	β, γ	19 y
↓			
bismuth-210*	$^{210}_{83}$Bi	β	5.0 d
↓			
polonium-210	$^{210}_{84}$Po	α	138 d
↓			
lead-206	$^{206}_{82}$Pb	Stable	—

* In these isotopes alternative modes of decay occur in a small proportion of cases. For example polonium-218 may decay by β-emission to astatine-218; this then decays by α-emission to bismuth-214.

Table 19.1 The uranium series.

► A sample of ore is found to contain 1.0 kg of uranium-238. Estimate the mass of radium in the ore.

Assuming the ore to be in radioactive equilibrium, the mass m of radium in the ore is given by

$$\frac{m}{1 \text{ kg}} = \frac{(226)(1620 \text{ y})}{(238)(4.5 \times 10^9 \text{ y})}$$

$$\Rightarrow \quad m \quad = 0.34 \text{ mg}$$

You can see from this the problem the Curies faced when they sought to extract radium from its ore. To obtain significant amounts of radium they had to handle vast quantities of uranium-bearing minerals. ◄

Radioactive dating

Radioactive substances with long half-lives provide us with us with a means of measuring time intervals on a geological scale. If a radioactive isotope and its decay products remain trapped in a rock, the measurement of the proportion of the parent isotope that has decayed enables the age of the rock to be determined. In a uranium-bearing rock the net result is the production of lead-206 from uranium-238. The measurement of the ratio of the amounts of these two isotopes in the rock enables the age of the rock (since it first solidified) to be established – always assuming there was no lead-206 in the rock to start with.

Sometimes it is possible to make use of lighter radioactive elements. For instance about 30% of the element rubidium exists as a radioactive isotope rubidium-87. This decays by β^--emission into strontium-87, which is a stable isotope. The half-life of the decay process is 4.3×10^{10} years. The measurement of the relative concentrations of these isotopes in a rock provides another means of estimating its age.

Carbon-14 dating

The action of cosmic rays in the Earth's atmosphere causes a steady production of the radioactive isotope carbon-14 from nitrogen. Because of this a very small proportion of the carbon in the atmosphere is in this radioactive form. There is a continual interchange of carbon (in the form of carbon dioxide) between the atmosphere and all living creatures, whose cells therefore contain the same small proportion of carbon-14, about 1 part in 10^{12}.

However, when the creature dies, the interchange of carbon with the atmosphere ceases, and the carbon-14 content of the remains starts to decrease. By measuring the activity of the carbon-14 in a specimen (such as wood from a building or parchment from a manuscript) we can find the total amount of carbon-14 still remaining in it; and so we can find its age. The half-life of carbon-14 is 5730 years, which is ideal for archaeological purposes. The technique has enabled ages from 500 years up to about 10 000 years to be found, in some cases with uncertainties of less than 100 years.

20 Alternating currents

20.1 Measurements

There are several different ways in which we can express the *size* of an alternating current.

(i) *The peak value* of the current is its maximum value (in either direction). For instance, if the current has a sinusoidal waveform:

$$I = I_0 \sin 2\pi ft \qquad \text{(page 85)}$$

then I_0 is the peak value.

(ii) *The mean value* or *half-cycle average* of the current is its average value taken over half a cycle while the current is in one direction only. (The average current over the whole cycle is of course zero.)

(iii) *The root-mean-square (r.m.s.) value* I_{rms} of the current is the square-root of the mean value of the square of the current taken over a whole cycle. Thus

$$I_{rms} = \sqrt{(\text{mean value of } I^2)}$$

At first sight this may seem an unnecessarily complicated way of specifying an alternating current; but it is much the most useful way in practice. The reason for this may be seen as follows. The power P converted to internal energy in a resistance is proportional to the square of the current; at any instant it is given by

$$P = I^2 R$$

With alternating current, P and I both vary from moment to moment. The mean or average power $\bar{P}$ converted in the resistance is given by

$$\bar{P} = (\text{mean value of } I^2)\, R$$
$$= I_{rms}^2\, R$$

The power supplied by the alternating current is the same as that supplied by a steady direct current of I_{rms} passing through the resistance R. The r.m.s. current is the most convenient way of specifying the alternating current since it enables us to make calculations of power using the same methods as in d.c. situations. The above reasoning, incidentally, does not depend on the alternating current have a sinusoidal waveform; it applies equally to square waves, saw-toothed waves, or the irregular waveforms to be found in the amplifying circuits of a record player.

Alternating *potential differences* may be analysed in the same way: *the peak value*, *the mean value*, and *the r.m.s. value* of an alternating p.d. all have their uses in technical calculations; but the r.m.s. value is generally the most useful for the same reason as before. Thus the power P being converted at any instant in a resistance R when the p.d. across it is V is given by

$$P = \frac{V^2}{R}$$

$$\text{Hence} \quad \bar{P} = \frac{(\text{the mean value of } V^2)}{R}$$

$$= \frac{V_{rms}^2}{R}$$

$$\text{since} \quad V_{rms} = \sqrt{(\text{mean value of } V^2)}$$

Again, a steady p.d. of V_{rms} connected across the resistance R would supply energy to it at the same mean rate, and the result does not depend on the waveform of the p.d. being sinusoidal. It follows that light bulbs and heating elements designed for a given d.c. supply will work correctly on alternating supplies of the same r.m.s. value.

In many a.c. circuits capacitative and inductive effects are negligible, and the currents are controlled entirely by the resistances of the circuit. In this case the current I in the resistance R and the p.d. V across it are proportional to one another at every instant, and we have

$$\frac{V}{I} = R$$

The current and p.d. are therefore *in phase*, i.e. the peaks of current and p.d. occur at the same moments. Figure 20.1 shows how V, I and P vary with time for a sinusoidal waveform. The power P fluctuates sinusoidally between zero and P_0 twice in every oscillation of p.d. and current; its mean value $\bar{P}$ is given by

$$\bar{P} = V_{rms} I_{rms} = I_{rms}^2 R = \frac{V_{rms}^2}{R} \qquad (1)$$

The practical result of all this is then very simple: provided we use r.m.s. values of current and p.d., the analysis of the currents in networks of resistors and the calculations of power converted in them can be performed in exactly the same way as with direct currents.

However we must realise that in general the behaviour of a.c. circuits is not as simple as this. When capacitative and inductive effects are appreciable, the currents are no longer controlled by the resistances only, nor are the currents and p.d.s necessarily in phase, i.e. the peaks of current and p.d. may occur at different moments. The details of these effects are discussed later in this chapter.

The relation between the peak and r.m.s. values of a current or of a p.d. depends on the waveform. For a sinusoidal waveform we have

$$V = V_0 \sin 2\pi ft$$
$$\text{and} \quad I = I_0 \sin 2\pi ft$$

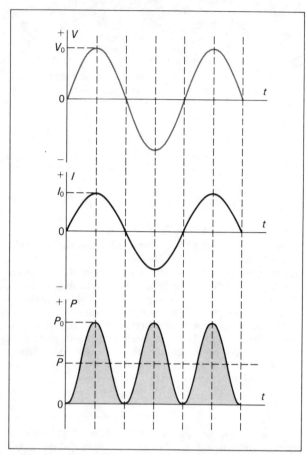

Figure 20.1 Parallel graphs of p.d. V, current I and power P for a resistor in which an alternating current is passing. The p.d. and current are in phase. The power fluctuates between P_0 and zero at twice the frequency of the current.

Figure 20.1 shows that in this case the mean power $\bar{P}$ is given by

$$\bar{P} = \tfrac{1}{2}P_0$$

Now P_0 is given by one of the three expressions

$$P_0 = V_0 I_0 = I_0^2 R = \frac{V_0^2}{R}$$

$$\Rightarrow \quad \bar{P} = \tfrac{1}{2}V_0 I_0 = \tfrac{1}{2}I_0^2 R = \tfrac{1}{2}\frac{V_0^2}{R}$$

Comparing this with equation (1) above we have

$$I_{\text{rms}}^2 = \tfrac{1}{2}I_0^2$$

Hence $I_{\text{rms}} = \dfrac{I_0}{\sqrt{2}}$

Similarly $V_{\text{rms}} = \dfrac{V_0}{\sqrt{2}}$

When we describe the a.c. mains as a 240 V supply, we are referring to the r.m.s. value of the potential difference between the live and neutral terminals of the socket. The peak value V_0 is therefore given by

$$V_0 = \sqrt{2} \times V_{\text{rms}}$$

$$= \sqrt{2}\,(240\ \text{V}) \approx 340\ \text{V}$$

The insulation between the conductors of the mains must therefore be sufficient to withstand 340 V, rather than 240 V.

Mean values

It may be shown (by mathematics beyond the scope of this book) that for a sinusoidal waveform the mean current $\bar{I}$ and mean p.d. $\bar{V}$ are related to the peak values as follows:

$$\bar{I} = \frac{2}{\pi}I_0 = 0.637\ I_0$$

$$\bar{V} = \frac{2}{\pi}V_0 = 0.637\ V_0$$

Meters for a.c.

It is important to know what aspect of an alternating quantity a given meter measures. As long as we are dealing with sinusoidal waveforms it makes no difference whether an instrument has its scale marked to give peak, mean or r.m.s. values (as long as we know which), since in this case the three quantities bear known ratios to one another (given above). In fact a.c. instruments are normally calibrated to read r.m.s. values, on the assumption that the ratios between the quantities are those for a sinusoidal waveform. However, when using a.c. meters with non-sinusoidal waveforms we must bear in mind that the readings may not be recording true r.m.s. values; in such cases we can use the meter only for purposes of comparison.

Rectifier instruments

A universal multi-range instrument, which incorporates a.c. as well as d.c. ranges, is based on a moving-coil meter; all the necessary shunts, multipliers, switches, etc. are included in the one case with the meter. The a.c. ranges are provided by using four semiconductor diodes, connected as in figure 20.2a to form a full-wave bridge rectifier (page 288). During one half-cycle diodes P and S conduct, and during the next half-cycle diodes Q and R. In both half-cycles the current is in the same direction through the meter. The arrangement is impracticable for small p.d.s since a semiconductor diode does not conduct significantly for p.d.s below a certain figure (0.2 V for germanium, 0.5 V for silicon). For measuring low p.d.s and for measuring currents it is usual to make use of a step-up transformer (page 230) to ensure that the applied p.d. is large enough to give effective rectification.

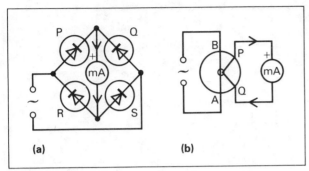

Figure 20.2 (a) A full-wave bridge rectifier as used in a multi-range meter for a.c. measurements. (b) The hot wire AB and junction of dissimilar metals (P and Q) in a thermocouple instrument.

Being based on a moving-coil meter a rectifier instrument is more sensitive than other kinds of a.c. meter. The meter reading is proportional to the half-cycle average or mean p.d. However, the a.c. ranges are calibrated to read r.m.s. values on the assumption that the waveform is sinusoidal.

By joining a storage capacitor across the meter in figure 20.2a the arrangement becomes like the rectifier circuits discussed on page 288, (figure 21.8). Used like this the instrument gives the *peak value* of the applied p.d. whatever the waveform may be.

Thermocouple instruments

The current to be measured is passed through a fine wire AB (figure 20.2b), raising its temperature. Attached to its centre are the ends of two other fine wires P and Q of dissimilar metals; this forms one junction of a thermocouple, the other junction being outside the unit at the temperature of the surroundings. The thermocouple produces a small e.m.f. that is recorded by a moving-coil meter joined to P and Q. The hot junction is enclosed in a small evacuated bulb, and so is not affected by convection currents.

Since the instrument uses the heating effect of the alternating current, its readings depend on the mean-square value of the current. In other words it may be calibrated by passing d.c. through the wire AB with an ammeter in series. Its readings then indicate r.m.s. values when used for a.c. It is particularly suitable for high frequency measurements, since the short wire AB has negligible inductance and produces little effect in a high frequency circuit.

Moving-iron instruments

Figure 20.3 shows one form of this kind of instrument. Two iron rods A and B are mounted parallel to the axis of the solenoid S (seen endwise on in the figure). A is fixed to the solenoid, and B is attached to the pointer. Whichever way the current flows in the solenoid, A is magnetised in the same sense as B; and the two iron rods therefore repel one another. The restoring torque is provided by hair-springs in the same way as in a moving-coil meter. The scale is

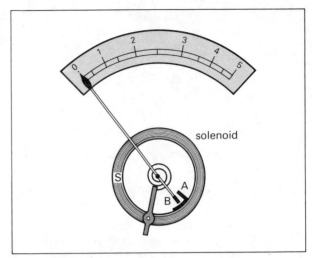

Figure 20.3 A moving-iron meter.

non-uniform, since the deflecting force is not proportional to the current. This can sometimes be an advantage, since we can arrange the scale so that the part of it to be used is highly expanded, making the instrument more sensitive in the required range.

The cathode-ray oscilloscope

This instrument has the obvious merit that it enables the waveform of an alternating p.d. to be studied as well its size. The use of this instrument has already been described on page 239.

20.2 Phase differences

When we have two alternating quantities whose peaks do not occur at the same moments we say that there is a *phase difference* between them. The way in which we analyse phase differences is best understood with the aid of a geometrical model that represents a sinusoidal alternating quantity (figure 20.4). OP is a rotating line, which we imagine to make f revolutions per second at a steady speed about one end O. The line OP in this model is referred to as a *phasor*. If the phasor OP makes an angle θ with the line OA, then (calculating the angle in radians, appendix B4) we have

$$\theta = 2\pi f t$$

where t is the time measured from the point where $\theta = 0$. The projections of the phasor OP on the lines OB and OA respectively are

$$ON = OP \sin 2\pi f t$$
$$\text{and} \quad OM = OP \cos 2\pi f t$$

We say that $OP \sin 2\pi f t$ and $OP \cos 2\pi f t$ are the *resolved parts* or *components* of the phasor OP in the directions OB

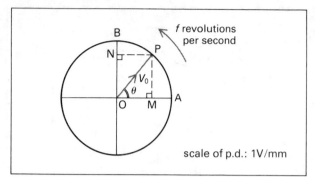

Figure 20.4 The geometrical model of a rotating line (called a *phasor*) to represent a sinusoidal alternating quantity.

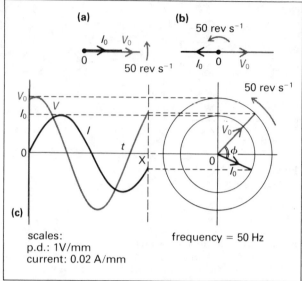

Figure 20.5 (a) A phasor diagram representing an alternating p.d. and current that are in phase. (b) The diagram for p.d. and current in anti-phase. (c) The general case in which there is phase difference ϕ between p.d. and current; the phasors are shown at the moment X in the acompanying graphs.

$$I = I_0 \sin 2\pi ft$$

$$V = V_0 \sin(2\pi ft + \phi)$$

and OA. If the *length* of the phasor OP represents the *peak value* of the alternating quantity, then the component of OP in a suitably chosen direction (such as ON) gives its instantaneous value. For instance, in figure 20.4 OP is 15.0 mm long and represents an alternating p.d. of peak value 15.0 V. Its instantaneous value may be represented by the component ON, and is 11.5 V.

The angle θ in this model is called the *phase* of the current or p.d. to be represented by the model. (The zero of phase is chosen arbitrarily, in this case at the moment $t = 0$ when P is at A.) The full description of a sinusoidal alternating current or p.d. requires us to state
(i) its frequency
(ii) its peak value
(iii) its phase
The phasor model provides a simple means of showing the relations between peak values and phases for several alternating quantities, since we can draw two or more phasors on the one diagram. For instance, in a resistor the current I and p.d. V are *in phase* (i.e. $\theta_V = \theta_I$); this is represented by two phasors coinciding in direction and rotating together (figure 20.5a). We have to choose two separate scales in this case, one for current and one for p.d., so that in general the two phasors will be of different lengths. For instance, in figure 20.5 the scale of p.d.s is again 1.0 V/mm, and the p.d.s shown are of peak value 15 V. The scale of currents, however, is 0.02 A/mm, and the currents shown are of peak value 0.20 A. Two quantities *in antiphase* are represented by two phasors pointing in diametrically opposite directions ($\theta_V - \theta_I = 180°$), as in figure 20.5b.

Figure 20.5c shows how in the general case we may represent the current I in some circuit component and the p.d. V across it; these are of the same frequency but there is a phase difference between them of ϕ. The phasors are drawn for the moment X in the accompanying graphs of V and I against time. As time advances both phasors rotate anticlockwise at the same steady speed (50 revolutions per second in this case, since the frequency is 50 Hz). As the phasors rotate, the angle between them remains constant; this angle is the phase difference ϕ; in this case $\phi = 70°$. Algebraically, I and V are given by

The three-phase system

The public electricity supply makes use of three separate circuits simultaneously; this enables considerable economies to be made. In a *three-phase alternator* (figure 20.6) one rotor is used to generate alternating current in three separate sets of stator coils spaced on pole pieces round the inside of the stator. (Compare this with the *single-phase alternator* shown in figure 16.21, page 215.) Each set of coils in the 3-phase alternator is connected to a separate distribution line; but the three share a common return wire (the *neutral* wire in the domestic supply).

The phases of the currents in the three circuits are equally spaced, i.e. the phase differences between them are 120°. This makes the action of the alternator smoother. But it also leads to economies in transmission, since only a very small current need flow in the common return wire. Indeed, if the currents in the three circuits are of equal size, the current in the common wire is actually zero, and in principle it might be dispensed with altogether. Figure 20.7 shows how this comes about. In the current/time graphs the sum of the three currents at any instant is seen to be zero; for instance, when one current is at its maximum value, the other two are each momentarily of half this size in the opposite direction, etc. In practice the three currents may not be exactly equal; but only a small resultant current

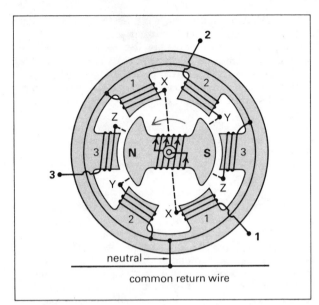

Figure 20.6 The arrangement of stator coils in a 2-pole, 3-phase alternator; alternating current is generated simultaneously in the three circuits.

need pass in the common return wire, which can therefore be of smaller gauge than the main cables. Three cables (and the small neutral wire) can therefore be used where six of the same size would be needed in the equivalent single-phase system.

Large industrial installations are normally supplied with all three phases, and these make use of special three-phase motors and other equipment. Like many other generators the three-phase alternator of figure 20.6 may be operated as a motor. If a three-phase supply is joined to the three sets of stator coils, the resultant magnetic field they produce is

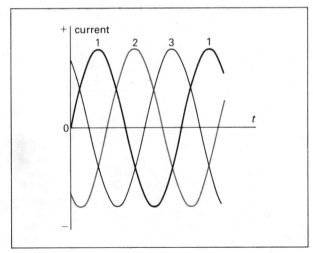

Figure 20.7 Balanced currents in a 3-phase system; the current in the common return wire is zero at every instant.

of constant size but rotating at a steady speed (50 times a second with the 50 Hz mains). Direct current from an auxiliary supply is passed through the rotor coils, as usual. An additional small motor is used to bring the speed of the rotor up to that of the field; the magnetic forces then keep it rotating at this speed, provided the load is not too great. The motor is therefore a *synchronous* one whose motion keeps in step with the oscillations of the supply.

The domestic consumer is supplied with only one phase (say, cable 1 and the neutral), but by arranging the number of houses on each phase to be approximately equal in any locality, the three currents remain in practice roughly balanced.

There are a number of specialised types of motor designed for use on a single-phase alternating supply (e.g. synchronous motors for electric clocks and record players). We shall not describe these in this book. However you should understand that an ordinary d.c. motor (page 219) may also be used with alternating current. Thus, if the direction of the currents through both the armature and the field windings of an ordinary d.c. motor are reversed at the same moments, the torque always remains in the same direction (page 220). To achieve this we must join the coils of the field magnet *in series* with the armature; the same current must then flow in both. The magnetic field and the armature current are then necessarily in phase and reverse at the same moments.

20.3 Reactive circuits

Circuits in which the currents are entirely controlled by capacitative or inductive effects are referred to as *reactive circuits*. Unlike purely resistive circuits, the alternating currents and p.d.s in reactive circuits are not in phase, and phasor diagrams must be used to analyse them. Practical a.c. circuits are inevitably partly reactive and partly resistive in their behaviour; but to start with we shall confine our attention to ideal reactive circuits, in which the effects of resistance are negligble.

Capacitance

When an alternating p.d. is joined across the plates of a capacitor, charge must continually flow onto and off the plates, as the potential difference between them changes. This means that there is an alternating current in the wires connected to the capacitor, although there is no continuous conducting path by which a direct current could pass. In this sense we often speak of the alternating current 'through' a capacitor, but this must be understood as a convenient piece of scientific jargon (page 181).

As with a resistor, the r.m.s. current in a capacitor is directly proportional to the r.m.s. potential difference applied. But with the capacitor there is a phase difference of 90° between current and p.d.; i.e. the current reaches its maximum value at the instants when the applied p.d. is momentarily zero. This is shown in the double-beam oscilloscope trace in figure 20.8a. One pair of deflector plates (Y_1 and earth) is joined across the the very small resistor R in series with the capacitor C (figure 20.8b), and this beam records the current I (page 240). The other pair of deflector plates (Y_2 and earth) is joined across the supply (which is almost the same as joining it across the capacitor, since the resistance R is so small); this beam therefore records the p.d. V across C.

We can understand what is happening by the following reasoning. The charge on the plates of a capacitor is directly proportional to the p.d. between them at every instant ($Q = CV$). At the moments marked X in figure 20.8a the applied p.d. V is a maximum, but is momentarily unchanging; therefore the charge on the plates is momentarily constant, and the current I (which is the rate of flow of charge) is zero. Now consider an instant such as Y, when the applied p.d. V is momentarily zero, but is increasing at the maximum rate; the charge on the plates is then also increasing at the maximum rate, and so the current is a maximum in the *positive* sense. Likewise, at an instant such as Z, the applied p.d. is again zero, but is now decreasing

at the maximum rate; the current is again a maximum but in the *negative* sense. Thus the current and p.d. are 90° out of phase as shown in the oscilloscope traces. It will be noticed that each peak of current occurs *less far along the time axis* (i.e. earlier) than the corresponding peak of p.d. so the peaks of current are 90° *ahead* of the peaks of p.d. This is described by saying that *the current leads the p.d. by 90°*

Figure 20.8c shows the phasor diagram representing this situation. We do not need to represent the rotation of the phasors on the diagram; it is assumed, as a matter of convention, that the phasors are rotating *anticlockwise*, so that I is ahead of V. Also in such a diagram we label the phasors "I" and "V" (rather than "I_0" and "V_0"), since we take the whole phasor model to represent the sinusoidally oscillating quantities I and V.

If we increase the frequency f of the supply while keeping the size of the applied p.d. constant, the same charge as before has to flow onto and off the plates, but it now has to do this in a proportionately shorter time. The current I is therefore proportional to the frequency. Increasing the capacitance C, while keeping the frequency and the size of the applied p.d. constant, causes a larger charge to flow onto and off the plates in the same time, and so again increases the current. In an earlier chapter we showed that for sinusoidal waveforms the relation between p.d. and current in a capacitor is

$$\frac{1}{2\pi f C} = \frac{V_0}{I_0} = \frac{V_{rms}}{I_{rms}} \qquad \text{(page 181)}$$

$$\Rightarrow \quad I_{rms} = (2\pi f C) V_{rms}$$

which is consistent with the above descriptive reasoning.

The quantity $1/2\pi f C$ is known as the *reactance* of the capacitor. For example, a 1.0 μF capacitor at 5.0 kHz has a reactance X given by

$$X = \frac{1}{2\pi (5.0 \times 10^3 \text{ Hz})(1.0 \times 10^{-6} \text{ F})}$$

$$= 32 \, \Omega$$

Connected to a 3.0 V sinusoidal supply at 5.0 kHz the current in it is given by

$$I_{rms} = \frac{3.0 \text{ V}}{32 \, \Omega}$$

$$= 94 \text{ mA}$$

At a frequency of 10 kHz the current would be *twice* as much (188 mA), and so on.

Reactance has the same units as resistance (i.e. both are given by a p.d. divided by a current), and are therefore measured in *ohms*. But the two things should not be confused. In a resistance the current and p.d. are in phase, and electrical energy is converted to internal energy. In a reactance the current and p.d. are 90° out of phase, and there is no net conversion of energy. But, as for resistance, we can write

$$V = XI$$

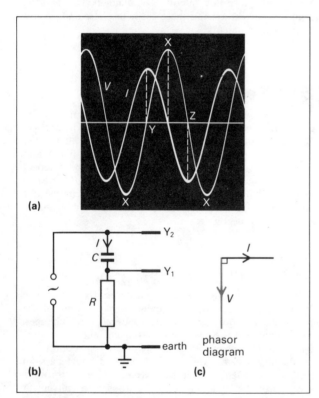

(a)

(b)   phasor diagram   (c)

Figure 20.8 (a) Double-beam oscilloscope traces of current I and p.d. V for a capacitor. (b) The arrangement used; R must be a very small resistance. (c) The phasor diagram representing I and V.

Inductance

An alternating current in an inductive coil (or *inductor*) produces an alternating flux through its core. This gives rise to an induced e.m.f. in the coil, and by Lenz's law this must act in opposition to the applied p.d. at every instant; for an inductor with negligible resistance the induced e.m.f. is *exactly* equal in size to the applied p.d. (page 229). The behaviour of an inductor in an a.c. circuit resembles that of a capacitor in that the current *I* and p.d. *V* are again 90° out of phase. However, with an inductor the current *lags* 90° behind the applied p.d. This is shown in the double-beam oscilloscope traces in figure 20.9a; the oscilloscope is connected as in (b) (where again *R* must be a very small resistance). This may be understood by the following reasoning.

The e.m.f. induced in the coil is proportional to the rate of change of current. At the moments marked X in figure 20.9a the current *I* (and with it the flux linkage) is at a maximum value (in one direction or the other), but is momentarily unchanging; therefore the induced e.m.f. is zero at these moments. Now consider an instant such as Y, when the current *I* is momentarily zero but is increasing

at the maximum rate; the induced e.m.f. is therefore a maximum, but in the *negative* sense, since by Lenz's law it must act so as to tend to hinder the change of current. Likewise, at an instant such as Z, the current *I* is again zero, but is now decreasing at the maximum rate; the induced e.m.f. is therefore a maximum in the *positive* sense. The applied p.d. *V* is equal in size at every instant to the induced e.m.f., so that the relation between *V* and *I* is as shown in the oscilloscope traces of figure 20.9. It can be seen that the current and p.d. are 90° out of phase. Also the peaks of current occur further along the time axis than the corresponding peaks of p.d., and thus occur *after* them. This is described by saying that *the current lags 90° behind the p.d.* This situation is represented by the phasor diagram in figure 20.9c.

In this case an increase in the frequency *f* of the supply allows a shorter time for the changes of current to take place. But to produce the same induced e.m.f. as before requires the same rate of change of current. The r.m.s. current must therefore fall, and is inversely proportional to the frequency. Increasing the inductance *L* of the inductor leads to an increase in the induced e.m.f. unless the current decreases in proportion. Therefore for constant size of applied p.d. the current varies inversely with the inductance.

In an earlier chapter we showed that with sinusoidal waveforms the relation between p.d. and current for an inductor is

$$2\pi f L = \frac{V_0}{I_0} = \frac{V_{rms}}{I_{rms}} \qquad \text{(page 229)}$$

$$\Rightarrow \quad I_{rms} = \frac{1}{(2\pi f L)} V_{rms}$$

in agreement with the above descriptive reasoning. The quantity $(2\pi f L)$ is known as the (inductive) *reactance* of the inductor. Like capacitive reactance it is measured in *ohms*. For example, a 1.0 mH inductor at 5.0 kHz has a reactance *X* given by

$$X = 2\pi (5.0 \times 10^3 \text{ Hz})(1.0 \times 10^{-3} \text{ H}) = 31 \ \Omega$$

Connected to a 3.0 V sinusoidal supply at 5.0 kHz, the current in it is given by

$$I_{rms} = \frac{3.0 \ V}{31 \ \Omega} = 97 \text{ mA}$$

At a frequency of 10 kHz the current would be *half* as much (48 mA). Thus, for an inductance

$$X = 2\pi f L$$

and, as before, $\quad V = XI$

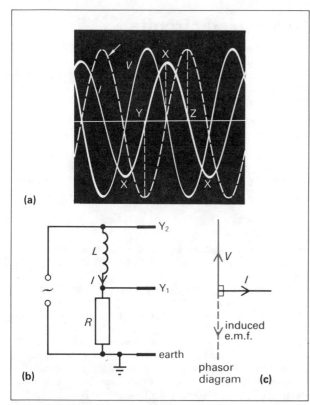

(a)

(b)

phasor diagram **(c)**

Figure 20.9 (a) Double-beam oscilloscope traces of current *I* and p.d. *V* for an inductor; the induced e.m.f. has also been traced on the picture. (b) The arrangement used; *R* must be a very small resistance. (c) The phasor diagram representing *I* and *V*.

Energy flow

Both capacitors and inductors may be used to control an alternating current without the conversion of electrical energy to internal energy. When there is difference of potential V between the plates of a capacitor, energy W_E is stored in the electric field between its plates; this is given by

$$W_E = \tfrac{1}{2}CV^2 \qquad \text{(page 183)}$$

Likewise, when there is a current I in an inductor, energy W_B is stored in the magnetic field of the coil; this is given by

$$W_B = \tfrac{1}{2}LI^2 \qquad \text{(page 229)}$$

In both cases the energy is stored *reversibly*, and is converted again to electrical energy in the supply when the p.d. or current falls. This can happen only when current and p.d. are 90° out of phase. Although a large current may be passing at a large p.d. and considerable quantities of energy may be flowing into and out of the apparatus, with a phase difference of 90° the net power taken from the supply is zero.

The details of the flow of energy in this case are shown in figure 20.10, which shows the variations of current and p.d. for an ideal inductor. During the first quarter-cycle from A to B both V and I are positive, and the power P supplied ($P = VI$) is also positive; this means that energy has been taken from the supply and is now stored in the

magnetic field of the inductor. During the next quarter-cycle from B to C the current I is still positive but V is negative; thus the power supplied is negative, and the energy that was supplied to the inductor during the first quarter-cycle ie returned again to the supply. This process recurs during the next half-cycle; from C to D energy is supplied to the inductor (V and I both negative, and therefore P positive), and this returns to the source of supply from D to E. The energy stored in the magnetic field reaches a maximum at the same moments as the current – namely, at the end of the interval AB, and again at the end of the interval CD. At these moments (B and D) the magnetic field is a maximum, and so therefore is the stored magnetic energy W_B.

A similar argument holds for the flow of energy into and out of a capacitor; the graphs of V and I are now as in figure 20.8a. In this case the energy stored in the electric field of the capacitor reaches a maximum at the same moments as the p.d. V, (i.e. when the electric field between the plates is a maximum), and so the energy is stored electrically (W_E).

20.4 Mixed circuits

In the general case of a circuit containing resistance as well as capacitance or inductance the phase difference between current and p.d. lies somewhere between 0 and 90°; the current leads the p.d. in a capacitative circuit and lags in an inductive circuit.

Impedance

As long as the only components making up a circuit are resistances, capacitances and inductances the r.m.s. current and r.m.s. p.d. are proportional to one another at a given frequency. Thus

$$\frac{V_{\text{rms}}}{I_{\text{rms}}} = Z$$

where Z is a constant known as the *impedance* of the arrangement; in general Z depends on the frequency employed.

The impedance of any part of a circuit component may therefore be measured by placing an a.c. ammeter in series with it and an a.c. voltmeter across it – much the same as the basic method for finding d.c. resistance (page 92), except that an a.c. supply is used instead of a battery.

In general, impedance Z has its resistive and reactive components; like resistance and reactance it is measured in *ohms*. In the special case of a pure inductor or capacitor (with negligible resistance) the impedance is equal to the reactance X; and for a pure resistor the impedance is equal to the resistance R.

Figure 20.11 shows the phasor diagrams for two simple combinations of circuit elements, (a) for inductance and resistance in series, (b) for capacitance and resistance in

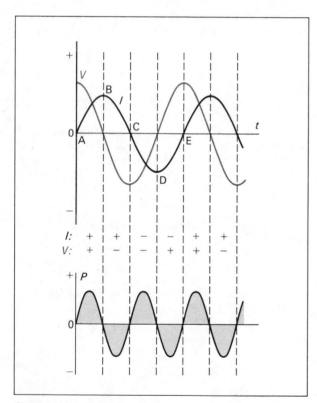

Figure 20.10 The reversible flow of energy into and out of an inductor. The energy stored in the magnetic field reaches its maximum value at the moments B and D when the current is a maximum.

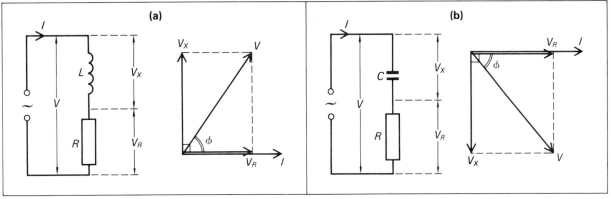

Figure 20.11 (a) Inductor and resistor in series, (b) Capacitor and resistor in series. The corresponding phasor diagram is shown alongside in each case.

series. With two components in series the current I is the same in both; *we therefore draw the phasor representing this first*. The p.d. V_R across the resistance is in phase with I, and the phasor representing this is therefore drawn parallel to I, and we have

$$V_R = RI$$

The p.d. V_X across the reactance is 90° out of phase with I, and is therefore drawn at right-angles to it. V_X *leads* I by 90° for the inductance, and is therefore drawn 90° further round (anticlockwise) on the diagram; while for the capacitance V_X lags behind I. In both cases

$$V_X = XI$$

The resultant p.d. V is the phasor sum of V_R and V_X; and so

$$V = \sqrt{(V_R{}^2 + V_X{}^2)}$$
$$= I\sqrt{(R^2 + X^2)}$$

The impedance Z in both cases is therefore given by

$$Z = \sqrt{(R^2 + X^2)}$$

The phase angle ϕ between V and I is given by

$$\tan \phi = \frac{X}{R}$$

► The impedance of a coil is 50 Ω at 500 Hz. Its resistance, measured with d.c. instruments, is 30 Ω. Calculate (i) its reactance at 500 Hz, (ii) its inductance, (iii) its impedance at 50 Hz, (iv) its impedance at 5 kHz.
(i) The connection between Z, R and X is

$$Z^2 = R^2 + X^2$$

Hence $X = \sqrt{(Z^2 - R^2)}$
$$= \sqrt{(50\,\Omega)^2 - (30\,\Omega)^2}$$
$$= 40\,\Omega$$

(ii) For an inductance, $X = 2\pi f L$

Hence $L = \dfrac{X}{2\pi f}$

$$= \frac{40\,\Omega}{2\pi(500\text{ Hz})} = 13 \text{ mH}$$

(iii) If the frequency is changed from 500 Hz to 50 Hz, the inductance L stays the same, but the reactance X becomes 1/10 as much, namely 4.0 Ω. X^2 is now negligible compared with R^2, and we can treat the coil as virtually a pure resistance. Hence its impedance Z is given by

$$Z = 30\,\Omega$$

(If you work it out, without this approximation, from $Z^2 = R^2 + X^2$, you will find it comes to 30.3 Ω; the difference is not significant.)
(iv) At 10 times the frequency (5 kHz) the reactance X becomes 10 times greater, namely 400 Ω. R^2 is now negligible compared with X^2, and we can ignore the resistance of the coil, and treat it as virtually a pure inductance. Hence the impedance Z is given by

$$Z = 400\,\Omega$$

(If you do the exact calculation, you will get 401 Ω; again the difference is not significant.)

This example reveals an interesting feature of impedance calculations. At 500 Hz the resistance and reactance of the coil were of comparable magnitude (30 Ω and 40 Ω). But, because X depends on the frequency, it does not require a very large change of frequency before R^2 and X^2 become of quite different orders of magnitude, and we can then ignore one or the other in the calculation. ◄

Filters

A choke (page 229) is an inductive coil used as a frequency filter. At zero frequency (d.c.) its impedance is equal to its resistance. But the above reasoning shows that at high frequencies we can forget about its resistance; and its impedance Z is then equal to $2\pi f L$, which increases with frequency. At high frequencies it has the property of offering high impedance to alternating currents and low resistance to direct currents. The higher the frequency, the more effective the filtering.

The capacitor-resistor combination of figure 20.11b also has filtering properties, but of the opposite kind. The reactance X of the capacitor is given by

$$X = \frac{1}{2\pi f C}$$

At low frequencies the reactance dominates, and the impedance varies inversely with the frequency. At zero frequency (d.c.) the impedance is of course infinite – there is no conducting path through the capacitor. But at high frequencies X becomes negligible, and the impedance of the combination is equal to the resistance R. This arrangement can be used for separating an alternating p.d. from a steady p.d. on which it is superimposed. The current through the resistance can only be a.c.; and at high enough frequency the p.d. across R is equal to the alternating component of the applied p.d. The potential difference across the capacitor is then equal to the steady (d.c.) component. A capacitor used in this way is called a *blocking capacitor* (page 181). A familiar example of this is at the Y-input of an oscilloscope (page 240); a switch enables us to use either direct coupling or capacitor coupling (which filters out any d.c. component of the waveform).

▶ A capacitance of 100 μF is joined in series with a resistance of 50 Ω. Calculate the impedance of the combination at 50 Hz and the phase angle between current and p.d.

The reactance X of the capacitor is given by

$$X = \frac{1}{2\pi f C} = \frac{1}{2\pi (50 \text{ Hz})(100 \times 10^{-6} \text{ F})}$$

$$= 32 \, \Omega$$

And so the impedance Z is given by

$$Z = \sqrt{(50 \, \Omega)^2 + (32 \, \Omega)^2}$$

$$= 59 \, \Omega$$

The phase angle ϕ is given by

$$\tan \phi = \frac{X}{R} = \frac{32 \, \Omega}{50 \, \Omega} = 0.64$$

$$\Rightarrow \qquad \phi = 32°$$

Notice that even at the low frequency of 50 Hz the impedance of this combination is very little more than the resistance (50 Ω); in other words the p.d. across the resistance R is equal to 50/59 of the applied p.d. (i.e. 85% of it). At higher frequencies the fraction is even higher, and we can then say that the p.d. across R is equal to the *alternating* p.d. applied to the combination, but any steady (d.c.) potential difference applied with it exists only across the capacitor. ◀

20.5 Electrical oscillations

When a charged capacitor is discharged through an inductor (figure 20.12), the interaction of inductance and capacitance causes the process to be an *oscillatory* one. The resistance of the inductor (represented by R in figure 20.12) causes *damping* of the oscillations. This may be investigated by connecting an oscilloscope across the capacitor; it will be found that the p.d. oscillates as shown in the figure. If a large inductance (10 H) is used and a large capacitance (10 000 μF), quite a leisurely oscillation of the spot of light will be observed. However the resistance of such a coil is likely to be so large that only a few oscillations will occur before the energy of the capacitor is all converted to internal energy. With smaller inductance and capacitance the frequency of the oscillations is much greater; also the resistance of the coil is now less significant, so that more oscillations occur before the spot of light comes to rest.

The reason for the discharge being oscillatory can be analysed as follows. At the moment X (figure 20.12) the switch is closed, but the inductance L delays the growth of current; and so the rate of discharge is small initially and only gradually increases. At Y the capacitor is fully discharged, but the rate of discharge (i.e. the current) is a maximum; the effect of the inductance is now to maintain the current, thus charging the capacitor in the reverse sense. As the p.d. grows in the reverse direction, the current slowly decreases until the point Z, when the p.d. across the capacitor is again momentarily steady. This process repeats itself many times over, giving rise to the oscillations we observe. These are of decreasing amplitude, since at each pulse of current some of the energy originally stored in the capacitor is converted to internal energy in the resistance of the circuit. The greater the resistance, the more heavily damped are the oscillations. Above a certain value of R *all* the initial energy is converted to internal energy by the

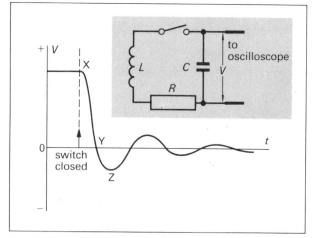

Figure 20.12 The oscillatory discharge of a capacitor C joined to an inductor L. The resistance R of the circuit affects the damping of the oscillations; when R is large enough no oscillations occur at all.

time the current has fallen to zero; and there is then no oscillation at all.

Another cause of energy loss from an oscillatory circuit is the emission of electromagnetic waves (page 368). It is found that radio waves are emitted from any circuit in which there is an alternating current. If we select the inductance and capacitance of figure 20.12 so that the frequency of the oscillations is in the range of a radio set nearby, a loud crackle will be heard on the radio each time the capacitor is discharged. However, the loss of energy from an oscillatory circuit from this cause is almost always much less then the internal energy converted in the resistance.

Resonance

As in other oscillatory systems (in mechanics, sound, etc) an *LC* loop as in figure 20.12 has a natural frequency at which electric current in it will oscillate. If it is *driven* by a small alternating e.m.f. at this frequency, the amplitude of oscillation builds up to a large value. One way of doing this is shown in figure 20.13a; in this we arrange another small coil to be magnetically linked with *L*. An alternating current in the small coil then produces an alternating e.m.f. of the same frequency in *L*. This is an example of *resonance* in an electric circuit (page 306). The resonant frequency *f* is easily worked out as follows.

If the resistance *R* is small, there is the same p.d. *V* at every instant across the inductance and capacitance; and being joined in a series loop there must be the same current *I* in both. This is possible only if the reactances of the two components are equal. Hence at resonance

$$2\pi f L = \frac{1}{2\pi f C}$$

$$\Rightarrow \quad 4\pi^2 f^2 LC = 1$$

Like any other resonant system the *LC* loop acts as a store of oscillatory energy. Electrical energy is continually lost from the system as internal energy (and radio waves); the amplitude of oscillation grows to the point at which the rate of loss of energy in these ways is equal to the rate at which energy is supplied by the driving e.m.f.

It is worth considering how the energy is stored in such a system. The variations of p.d. *V* and current *I* in a resonant *LC* loop are shown in figure 20.13b. At moments such as X the current is zero and the p.d. a maximum. These are the moments when *all* the energy is stored in the electric field of the capacitor; for the energy in a capacitor $(\frac{1}{2}CV^2)$ depends on the p.d. across it. At moments such as Y the p.d. is zero and the current a maximum. These are the moments when *all* the energy is stored in the magnetic field of the inductor; for the energy of an inductor $(\frac{1}{2}LI^2)$ depends on with the current in it. The variations of electrical energy W_E and magnetic energy W_B are shown in figure 20.13c. At points intermediate between X and Y the energy is stored partly in the capacitor and partly in the inductor.

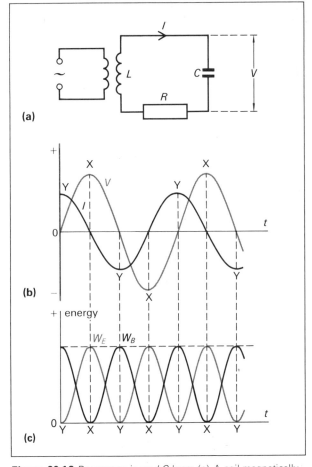

Figure 20.13 Resonance in an *LC* loop: (a) A coil magnetically linked to *L* can be used to drive the loop at its resonant frequency. (b) The variations of p.d. *V* and current *I*. (c) The energy stored in the loop stays constant, sometimes wholly as magnetic energy W_B, and at other times wholly as electrical energy W_E.

The total energy stored in the *LC* system is constant, and simply passes back and forth between the electric and magnetic fields.

This may be compared with the similar behaviour of an oscillating mechanical system such as the pendulum of a clock. In this case the total energy is again constant, and is being continually changed from gravitational potential energy to kinetic energy and back again. The clock mechanism has only to supply the loss of energy (as internal energy) caused by the slight damping forces acting on the pendulum. There is in fact a close analogy between the behaviour of oscillating mechanical systems and oscillatory electric circuits (page 307). The inductor of the electric circuit fulfils the same function as the mass (inertia) in a mechanical system. The capacitor acts as a store of potential energy (electrical), just as a spring does (or whatever else is responsible for the storage of potential energy) in a

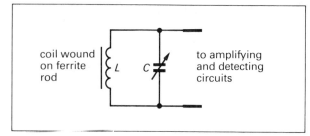

Figure 20.14 The resonant aerial circuit of a small radio.

mechanical system. The resistance R causes the loss of oscillatory energy from the electrical system, just like the damping forces in a mechanical system. In both kinds of oscillatory system, to maintain oscillations of constant amplitude, energy must be supplied, by an e.m.f. for the circuit or a force for the pendulum, varying at the resonant frequency of the system.

The tuning circuit of a radio

A resonant loop may be used as a filter to select one particular frequency from a mixture of many others. Figure 20.14 shows the aerial circuit of a small radio. The aerial consists of a ferrite rod with a coil wound round it. The rod is magnetised by the oscillating B-fields of the electromagnetic waves moving past it (page 370); e.m.f.s of many frequencies are therefore induced in the coil L of the resonant LC loop. At most of these frequencies no resonance occurs and the currents developed in the loop remain small. But if an e.m.f. is induced at the resonant frequency, the amplitude of the current in the loop increases, and an appreciable alternating p.d. is produced across it. The output of the circuit therefore effectively contains only those frequencies very close to the resonant one.

▶ The aerial circuit of a radio set is equipped with a tuning coil of inductance 1.8 mH. What tuning capacitor must be used with this to tune to the BBC long wave station (200 kHz)?

The resonant frequency f is given by

$$4\pi^2 f^2 LC = 1$$

and so

$$C = \frac{1}{4\pi^2 f^2 L}$$

$$= \frac{1}{4\pi^2 \,(200 \times 10^3 \text{ Hz})^2 \,(1.8 \times 10^{-3} \text{ H})}$$

$$= 3.5 \times 10^{-10} \text{ F} = 350 \text{ pF} \quad \blacktriangleleft$$

Resonant lengths of wire

When an e.m.f. is first generated in an electric circuit, it takes time for the electrical disturbance to be propagated round it. The charged particles in the conductors of the circuit drift at quite slow speeds (page 78); but the associated electric and magnetic fields are propagated round the circuit at about the speed of light, and this ensures that the charged particles are set in motion almost simultaneously throughout the circuit. With direct currents and with low-frequency a.c. the transmission of the disturbance round a circuit is so rapid that we are not usually aware of the finite time taken. But at very high frequencies this time may become comparable with the period of oscillation.

We then have a situation in which the e.m.f. reverses before the original disturbance has travelled very far along the wire. Figure 20.15a shows what happens when a twin cable is connected to a high frequency oscillator. At the instant shown the e.m.f. of the oscillator is at its peak (positive) value of 3.0 V; this has charged the capacitor formed by the parts of the cable near A and produced a pulse of current in the cable at A of peak value 10 mA. This pulse of p.d. and current then moves off along the cable, while the e.m.f. of the oscillator falls to zero and then reverses, so producing an opposite (negative) pulse of p.d. and current. After one complete cycle of oscillation the first pulse has reached the point A' on the cable, and the subsequent negative pulse has reached the point C, while a fresh positive pulse is starting at A. At the intermediate points B and D the p.d. and current are momentarily zero.

It may seem strange to find a situation in which the current is simultaneously in opposite directions at points close together (like A and C) in the same conductor. But we can see that the system of currents shown causes the whole pattern to move forward along the cable. Thus charge is flowing towards B in the upper conductor and away from it in the lower one; this charges up the capacitor

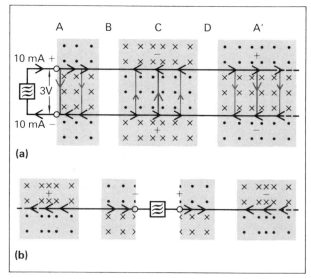

Figure 20.15 (a) The waves of p.d. and current that travel along a parallel wire cable joined to a high-frequency oscillator; associated waves of electric and magnetic fields travel in the space alongside. (b) Waves of p.d. and current (and the associated electromagnetic waves) travelling out from the centre of a dipole connected to a high-frequency oscillator.

formed by the part of the cable near B, so transferring the pulse of p.d. at A to B. At the same time the negative pulse at C is being transferred to D. In fact the high-frequency e.m.f. is sending *waves* of p.d. and current along the cable; these are accompanied by waves of electric and magnetic fields in the space surrounding the conductors, as shown. The lines of force of the electric field cross from one conductor to the other; and the strength of the *E*-field at any point is proportional to the p.d. between them. The lines of force of the magnetic field form closed loops round the wires, and are therefore at right-angles to the plane of the diagram; the strength of the *B*-field at any point is proportional to the current in the neighbouring section of the cable. The transmission of energy along a twin cable by an alternating current is thus accompanied by the passage of *electromagnetic waves* through the air (or other insulator) surrounding the cable. If the resistance of the cable is low enough, the current waves in the cable and the electromagnetic waves in the surrounding space travel at the speed *c* given by

$$c = 3.00 \times 10^8 \text{ m s}^{-1} \qquad \text{(page 365)}$$

The wavelength of the p.d. and current waves on the wire is the same as the wavelength of the electromagnetic waves alongside; wavelength λ and frequency f are connected by

$$c = f\lambda \qquad \text{(page 314)}$$

E.g. for the BBC long wave station at 200 kHz the wavelength λ is given by

$$\lambda = \frac{3.00 \times 10^8 \text{ m s}^{-1}}{200 \times 10^3 \text{ Hz}}$$

$$= 1500 \text{ m}$$

A similar result occurs if the pair of conductors joined to the oscillator is arranged in some other way; for instance, they can run in opposite directions away from the oscillator (figure 20.15b). As before, waves of potential and current travel along the wires, while the associated electromagnetic waves travel in the air alongside. At the ends of such a pair of wires the waves of p.d. and current are reflected and travel back towards their source. We then have a system of stationary waves, rather like the stationary waves that occur on a vibrating string (page 315), or in a vibrating air column (page 324). As in these other cases, the two waves travelling on the wires in opposite directions superpose to produce an interference pattern, giving *nodes* (points of minimum oscillation) at intervals half a wavelength apart along the wire (page 315). If we have a wire open at both ends, whose length is equal to a whole number of half-wavelengths, then a large amplitude stationary wave system can be set up on it, and the wire is *resonant* to the frequency of the wave. For instance, a wire 750 m long is the shortest wire than can be resonant to the frequency of the BBC long wave station at 200 kHz. Such a wire can carry a large alternating current at 200 kHz at its centre (an antinode of current), while there is a node of current at each end (where the current *must* be zero).

The transmitting aerial of a radio station usually consists of one or more resonant lengths of wire like this. In practice with long wave transmitters a much shorter aerial than this is used; it is made resonant to the required frequency by incorporating inductors and capacitors in the aerial system. But at shorter wavelengths, it is usual to employ a true resonant *dipole* aerial (of length $\lambda/2$), as in figure 20.16. Such an aerial emits electromagnetic waves efficiently into the surrounding space. A twin cable connects the centre of the dipole to a high-frequency oscillator, which supplies the energy to make up the energy radiated away (as well as the energy converted to internal energy in the wires). Some transmitting systems (e.g. in police cars and taxi cabs) employ a *quarter-wave* aerial (i.e. half a dipole). In this case the second terminal of the oscillator is connected to the metal frame of the vehicle.

A receiving aerial for television or v.h.f. radio is usually based on a similar resonant dipole, connected by a twin cable to the receiver. Such an aerial usually includes also one or more other metal rods parallel to the dipole; these act as reflectors, etc, so that the aerial responds mainly to radiation from only one direction (page 371). Electromagnetic waves in the space around the dipole generate alternating currents in it, and these build up to large amplitudes in the range of frequencies near the resonant frequency of the dipole; resonant *LC* loops in the receiver are then used to select the oscillations of the particular transmission required.

The electric constant ε_0

The speed c of electromagnetic waves, whether in empty space or along the surface of a wire is controlled by the electric and magnetic constants ε_0 and μ_0, (the permittivity and permeability of a vacuum, pages 194 and 213). It may be shown that

$$c^2 = \frac{1}{\varepsilon_0\mu_0}$$

It is interesting to see how the right-hand side of this equation comes to have the dimensions of (speed)^2. The

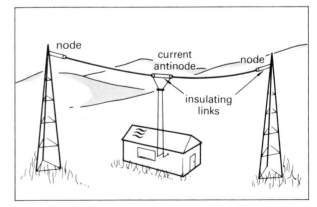

Figure 20.16 A resonant dipole transmitting aerial.

constants ε_0 and μ_0 are expressed in units of $\mathrm{F\ m^{-1}}$ and $\mathrm{H\ m^{-1}}$, respectively.

Now $1\ \mathrm{F\ m^{-1}} = 1\ \mathrm{C\ V^{-1}\ m^{-1}}$ (page 179)
and $1\ \mathrm{H\ m^{-1}} = 1\ \mathrm{V\ s\ A^{-1}\ m^{-1}}$ (page 227)
$$= 1\ \mathrm{V\ s(C\ s^{-1})^{-1}\ m^{-1}}$$
$$= 1\ \mathrm{V\ C^{-1}\ s^2\ m^{-1}}$$

Hence $1/(\varepsilon_0\mu_0)$ is to be expressed in units of

$$\frac{1}{(\mathrm{C\ V^{-1}\ m^{-1}})(\mathrm{V\ C^{-1}\ s^2\ m^{-1}})} \quad \text{or} \quad (\mathrm{m\ s^{-1}})^2$$

The magnetic constant μ_0 is chosen in the course of defining the ampere (page 213); ε_0 is then given by the theoretical relation

$$\varepsilon_0 = \frac{1}{\mu_0 c^2}$$

as the speed c of the electromagnetic waves is known.

Substituting

$$\mu_0 = 4\pi \times 10^{-7}\ \mathrm{H\ m^{-1}} \text{ and } c = 2.998 \times 10^8\ \mathrm{m\ s^{-1}}$$

we have

$$\varepsilon_0 = \frac{1}{(4\pi \times 10^{-7}\ \mathrm{H\ m^{-1}})(2.998 \times 10^8\ \mathrm{m\ s^{-1}})^2}$$

$$= 8.85 \times 10^{-12}\ \mathrm{F\ m^{-1}}$$

Conversely, the result of an experiment for measuring ε_0, such as that described on page 195, may be used to calculate the speed c of electromagnetic waves. The fact that *light* is found to travel at precisely *this* speed is strong confirmation that light consists of electromagnetic waves (page 365).

21 Electronic devices

21.1 Junction diodes

There are many devices that employ the special characteristics of junctions between two sorts of semiconductor. The simplest of these has just a single junction between *p*-type and *n*-type semiconductor, and is referred to as a *diode*. It has the property of low resistance for current in one direction through the junction, but high resistance for current in the reverse direction (page 90).

A diode is made by forming regions of *p*-type and *n*-type semiconductor *within the same crystal* of silicon or germanium. In the *n*-type region there are many free electrons; each such electron comes from an originally neutral atom, which remains fixed in the crystal lattice as a positive ion (while the electron moves freely through the lattice). Similarly, in the *p*-type region there are free positive holes (page 77), and each of these leaves behind a negative ion fixed in the crystal lattice. Both free electrons and holes carry a share of the internal energy of the crystal, and therefore move around at random in it – rather like an electron (or hole) 'gas' confined within the crystal structure.

At the junction some electrons diffuse across from the *n*-type side in the course of their random movements, and some holes from the *p*-type side (figure 21.1). Recombination of these electrons and holes produces on either side of the boundary a narrow *depletion layer* (or *barrier layer*) in which hardly any free charge carriers are left; this layer is therefore of high resistance. However, the depletion layer still contains the fixed ions left behind by the electrons and holes, positive ions on the *n*-type side of the junction (the right-hand side in figure 21.1), and negative ions on the *p*-type side. An electric field is thus created in the depletion layer, as shown, which leaves the *n*-type region at a positive

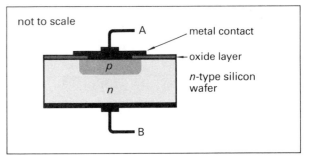

Figure 21.2 The construction of a silicon junction diode.

potential with respect to the *p*-type region. The diffusion of electrons and holes across the boundary continues until the *E*-field in the depletion layer is sufficient to prevent further movement of the charge carriers. Although there is an internal p.d. between the two regions of the semiconductor, no p.d. can be measured between the wires A and B joined at either end because of exactly balancing p.d.s that develop between the two metal contacts and the crystal (no depletion layers form at these points).

If now a p.d. is applied to the diode so that A becomes positive with respect to B, the electric field in the depletion layer is reduced, and charge can flow across the junction. But a p.d. applied the other way round increases the field in the depletion layer, and the flow of charge across the junction is further inhibited. The only current is then that caused by the very small number of electron-hole pairs created at random in the depletion layer by the thermal vibrations of the crystal. The junction thus conducts current readily in one direction (the *forward* direction), but has a high resistance in the *reverse* direction.

One form of silicon diode is shown in figure 21.2. A wafer of *n*-type silicon is soldered onto a metal base. The surface of the wafer is given a protective oxide coating by heating in steam at 1400 K. A gap is then etched in the oxide layer, as shown; and through this the appropriate impurity is diffused at high temperature to convert a layer at the top of the wafer into *p*-type silicon. Aluminium is then evaporated onto the surface to provide a contact pad to which a connecting lead is soldered. The whole unit is hermetically sealed in a light-proof capsule.

The electrical characteristics of a silicon diode are shown in figure 21.3. In the forward direction the current increases approximately exponentially with p.d. Very little current passes until the forward p.d. rises above about 0.5 V; it then increases very rapidly. In the reverse direction the current is usually about 10^{-8} A up to the point at which

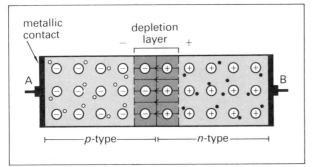

Figure 21.1 The rectifying action of a *p-n* junction. Diffusion of electrons and holes across the junction produces a narrow depletion layer free of charge carriers in which there is an electric field from the *n*-type to the *p*-type side.

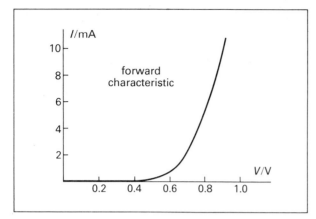

Figure 21.3 The forward characteristic of a silicon diode. In the reverse direction the current is only about 10^{-8} A up to near the breakdown voltage.

the insulation of the depletion layer breaks down (which can be anything up to 1000 V for silicon). In a germanium diode only 0.2 V is required before conduction starts, but the peak reverse p.d. cannot be more than about 100 V.

Silicon and germanium diodes for handling high currents have to be equipped with efficient means for conducting away the internal energy converted within them. If the temperature of a germanium diode is allowed to rise much above 100°C, irreversible changes take place in the crystal structure of the junction, and the diode is destroyed. It is therefore necessary to bolt a high-current diode to a *heat sink* consisting of a metal plate with cooling fins. The temperature of a silicon diode can be allowed to rise to about 200°C, and the problem of preventing undue temperature rises is not as acute as with germanium.

The photodiode

When a potential difference is applied to a diode in the reverse direction, the small reverse current arises from the spontaneous formation of electron-hole pairs in the depletion layer. If the depletion layer is illuminated with light of sufficiently high frequency, the energy of the photons (page 241) can be sufficient to create extra electron-hole pairs. An increase in the reverse current thus occurs which is proportional to the flux of light falling on the sensitive area. A diode adapted to be used in this way is called a *photodiode*. Another use of the same principle is in the *solid-state particle detector* (page 257). This also is a diode with a reverse p.d. across it; but in this case the extra electron-hole pairs are formed by high-energy particles such as α-particles passing through the depletion layer.

The photoconductive cell

Another type of photocell is the *light-dependent resistor*. This consists of a thin layer of semiconductor in polycrystalline form (often cadmium sulphide) to which connection

is made with metal electrodes. Light falling on the layer of semiconductor creates extra electron-hole pairs in it, and so decreases its resistance. Figure 21.4a shows the form of construction used. A pair of comb-like metal electrodes are evaporated onto the surface of the cadmium sulphide layer. This keeps the conducting path through the semiconductor layer short while providing a large surface area to be illuminated. The resistance of a typical cell of this type varies from about 10 MΩ in the dark to less than 100 Ω in bright daylight.

Figure 21.4b shows how we use a photoconductive cell to control a large current by joining a magnetic relay in series with it. The relay (c) consists of a coil of many turns wound on an iron core. When the core is magnetised by a small current (perhaps 10 mA) in the coil, it attracts a hinged iron plate, which carries the switch arm. In this way the

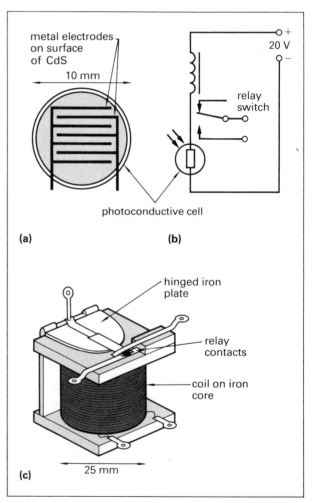

Figure 21.4 (a) The pattern of metal electrodes on the surface of a CdS photoconductive cell. (b) The use of the cell with a relay to control another circuit. (c) The construction of a magnetic relay.

light can be used to switch on or off some device that takes a very much greater current than that in the photoconductive cell.

The light-emitting diode (l.e.d.)

When the potential difference across a *p-n* junction diode is in the forward direction, recombination of electrons and holes is taking place continually in the neighbourhood of the junction. Electrons move across the junction from the *n*-type side and then recombine with holes in the *p*-type material; the same happens with holes moving across the junction from the *p*-type side. Each such recombination releases a quantum of energy. Many of these quanta are simply absorbed as internal energy in the material. But in some materials (such as gallium arsenide phosphide) many of the quanta of energy are emitted as light. To make use of this light the *p-n* junction is formed very near the surface of the material, and connection to one side of the diode is made through a very thin (transparent) layer on the surface. With a forward p.d. of about 2 V and a current of about 10 mA an l.e.d. forms a convenient light source of defined shape. They are commonly used in the displays of calculators. Each digit is formed by activating a selection from seven bar-shaped l.e.d.s; an eighth l.e.d. provides a decimal point when required.

The zener diode

When an increasing potential difference is applied across a silicon diode in the reverse direction, there is a sudden rise in current when the p.d. reaches the point at which the insulation of the barrier layer breaks down (figure 21.5a). At this point the diode is likely to suffer damage unless the current is limited by a resistance in series. But if this is provided, the p.d. across the diode remains nearly constant for a large range of currents. By careful adjustment of the impurity concentration near the junction it is possible to design diodes whose *zener p.d.* is anywhere between 2.5 V and 200 V. Zener diodes may be used as voltage reference devices for stabilising a varying potential difference to a predetermined value.

Figure 21.5b shows a simple arrangement for stabilising the supply p.d. of a small radio; notice that the zener diode is connected to the battery in the reverse sense. The series resistance *R* is chosen so that, with a fresh battery and no current taken by the radio (through the supply terminals), the maximum safe current at the zener p.d. passes through the diode. When a current is taken by the radio, the current through the diode falls by the same amount, so that the drop in p.d. across *R* remains almost constant. In this way variations in the current taken by the radio affect the p.d. between the terminals very little. The arrangement is also stabilised to some extent against variations in the battery p.d., provided this remains greater than the zener p.d. and the current taken is not too great.

Rectifier circuits

The ability of a diode to conduct current in only one direction enables it to be used as a rectifier for producing d.c. from an alternating supply. A diode joined in series with some device to an a.c. supply conducts only on those parts of the cycle when the p.d. across it is in the forward direction.

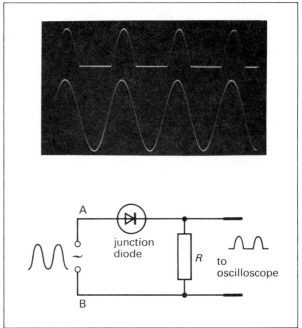

Figure 21.6 A diode in an a.c. circuit with a resistive load *R*. The output p.d. across the load is uni-directional but pulsating. The double-beam oscilloscope traces show this compared with the supply p.d. on the second beam.

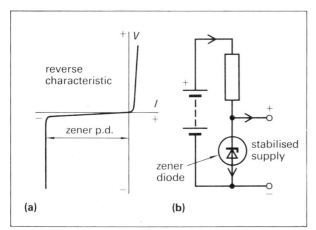

Figure 21.5 (a) The characteristic of a zener diode (which is always used in the reverse direction). (b) A simple arrangement for stabilising a battery supply for a small radio using a zener diode.

The current through the device is then uni-directional, but pulsating, as shown in the double-beam oscilloscope trace of figure 21.6.

To produce an approximately steady output p.d. we use the arrangement shown in figure 21.7. The connection of the *storage* or *reservoir capacitor* in parallel with the load radically alters the behaviour of the circuit. (The word 'load' in this connection means that to which the rectifier is supplying power in the form of direct current, e.g. a radio or tape recorder; this is represented in figure 21.7 by a resistor.) During the first positive half-cycle the diode conducts, charging the capacitor with the upper plate positive. At the moment X (on the oscilloscope trace) the p.d. across the capacitor is equal to the *peak* p.d. of the supply. After this the potential of the point A in the circuit starts to fall, and the diode ceases to conduct. The capacitor and load are now effectively disconnected from the supply. The capacitor therefore starts to discharge through the load; this continues until the point A once more reaches a higher potential than C. There is then a forward p.d. across the diode, which starts conducting again. A pulse of current therefore passes through it between Y and X', recharging the capacitor to the peak p.d. of the supply. The average p.d. across the load is thus slightly less than the peak p.d. of the supply, and there is a slight a.c. *ripple* (saw-toothed) superimposed on it, as shown in the double-beam oscilloscope traces of figure 21.7.

Suppose we have a 12 V (r.m.s.) a.c. supply. The peak p.d. V_0 is given by

$$V_0 = \sqrt{2} \times 12\ \text{V} = 17\ \text{V} \qquad \text{(page 272)}$$

and this is the p.d. to which the capacitor charges. Suppose the current I taken by the load is 10 mA. The rate of discharge $(\mathrm{d}V/\mathrm{d}t)$ of the capacitor is given by

$$I = C\frac{\mathrm{d}V}{\mathrm{d}t} \qquad \text{(page 179)}$$

The time of discharge Δt is *nearly* one period (0.020 s for a frequency of 50 Hz). The corresponding drop in p.d. ΔV is therefore given by

$$\Delta V = \frac{I\Delta t}{C} = \frac{(10 \times 10^{-3}\ \text{A})(0.020\ \text{s})}{(100 \times 10^{-6}\ \text{F})}$$
$$= 2.0\ \text{V}$$

Between X and Y the output p.d. thus drops from 17 V to 15 V; the average output p.d. is 16 V, and there is a saw-toothed ripple of peak value 1 V superimposed on it. The resistance R of the load in this case is given by

$$R = \frac{16\ \text{V}}{10 \times 10^{-3}\ \text{A}} = 1600\ \Omega$$

Because of the small size and cheapness of semiconductor diodes it is often preferable to use a full-wave type of rectifier circuit employing four diodes joined in a bridge network (figure 21.8). For this purpose sets of 4 diodes are available encapsulated as a single unit with 4 terminals. On one half-cycle current passes through diodes P and S, charging the storage capacitor to the peak p.d. of the supply;

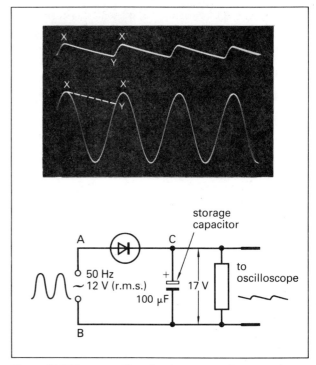

Figure 21.7 The connection of a storage capacitor across the load gives an approximately steady output p.d., with a saw-toothed a.c. ripple superimposed on it; the double-beam oscilloscope enables us to compare this with the supply waveform.

in the other half-cycle P and S remain non-conducting and current passes instead through Q and R. Again the average output p.d. is nearly equal to the peak p.d. of the supply, and there is an a.c. ripple (at 100 Hz now) superimposed on it. To reduce the a.c. component of the output we can use a resistor and zener diode, as explained above. The p.d. across the series resistance of 400 Ω with a current of 10 mA taken through it is 4.0 V. The output p.d. is therefore now 12 V; but the a.c. ripple is likely to be less than 0.02 V.

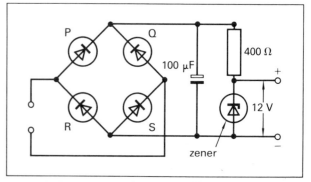

Figure 21.8 A full-wave bridge type rectifier with storage capacitor and zener diode smoothing to reduce the ripple to negligible proportions.

21.2 Transistors

Transistors are semiconductor devices with *three* terminals. They operate in such a way that the current through the device between terminals 1 and 3 (figure 21.9) is controlled by the p.d. between terminals 1 and 2 (between which only a very small current passes). The power input to the transistor to provide this control p.d. is therefore very small, but the power output in the main circuit joined to terminals 1 and 3 can be much larger. The energy converted in this circuit derives from the source of e.m.f. it contains, but the rate at which it is converted is controlled by the p.d. between terminals 1 and 2. The transistor is thus an amplifying device. This is quite different from what happens in a step-up transformer. In this, by using a high turns ratio we can make the output p.d. many times greater than the input p.d. But the power output can never be more than the power input. With the transistor the power output (derived from the battery) can be thousands of times greater than the power input. We shall describe two basic types of transistor.

The field-effect transistor (f.e.t.)

In this type of transistor the current is confined to a thin channel of semiconductor connecting two electrodes called the *source* s and *drain* d (figure 21.10). Only one kind of charge carrier is involved, electrons in an *n*-channel f.e.t., and holes in a *p*-channel f.e.t. However, the concentration of charge carriers in the channel is controlled by an electric field produced by applying a p.d. between an insulated electrode (called the *gate* g) and the bulk semiconductor under the channel. Figure 21.10a shows the internal structure of an *n*-channel f.e.t.; but you must realise that the diagram is not drawn to scale. The channel is produced by diffusion of *n*-type impurity into the surface of a chip of *p*-type silicon. The surface is then oxidised, and the aluminium gate electrode g is evaporated onto this. The oxide layer is only about 10^{-7} m thick, but it is sufficient to insulate the gate from the channel very effectively; the resistance of the oxide layer is usually more than 10^{12} Ω.

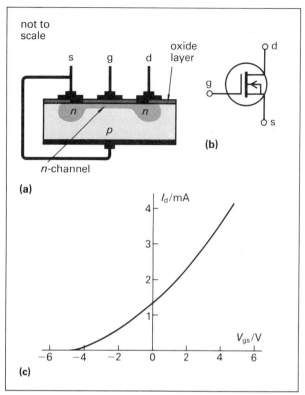

(a)

(b)

(c)

Figure 21.10 An *n*-channel field effect transistor (mosfet): (a) the internal structure; (b) the circuit symbol; (c) the characteristic of the device, showing drain current I_d plotted against the control p.d. V_{gs} between gate and source.

Connection to the source s and drain d is made by aluminium contact pads, as shown; holes are etched through the oxide layer at these points, so that the pads make contact with the semiconductor beneath.

This type of f.e.t. is often described as a *mosfet*, the first three letters being the initials of the structure-sequence (metal-oxide-semiconductor) from gate to channel. The circuit symbol of an *n*-channel mosfet is shown in figure 21.10b; the lead joined to the bulk *p*-type material is represented in the symbol by the central line (with an arrow on it). This is normally joined inside the transistor case to the source, as shown. (The symbol of a *p*-channel mosfet has the central arrow pointing the other way.)

The *n*-channel mosfet is operated with the drain at a positive potential with respect to the source. No current therefore passes through the *p*-type part of the chip, since there is a reverse p.d. across the junction between this and the drain. The only current is along the channel. Suppose now that the gate is made negative (with respect to the source and bulk semiconductor); this attracts positive charge into the *n*-type channel below the oxide layer, and reduces the concentration of free electrons in the channel. The drain current I_d therefore decreases. Conversely, with the gate positive with respect to the source, extra electrons

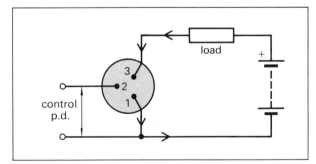

Figure 21.9 In transistor devices a p.d. applied between leads 1 and 2 controls the main current through the device between leads 1 and 3 in series with the load.

are attracted into the channel, causing the current I_d to increase. The variation of I_d with the p.d. V_{gs} between gate and source is shown in figure 21.10c. The *cut-off p.d.* at which I_d falls to zero is about -4 V.

The extraordinary insulation of the gate electrode in mosfet devices poses a practical problem that we must always be aware of in handling them. It is very easy inadvertently to raise the potential of the gate electrode to the point at which the insulation of the oxide layer breaks down (only about 50 V). Someone walking across a nylon carpet or on plastic tiles can well reach a potential of 10 000 V – as you find to your cost when you then touch an earthed pipe! Even a student sitting on a wooden stool on a dry day may reach a high enough potential by frictional contacts to risk damaging a mosfet near which he brings his hand. We therefore make sure that there is always a conducting path between gate and source; and when we are joining the mosfet in a circuit we make sure that the observer and his soldering iron are both securely earthed.

Figure 21.11 shows how we may use a mosfet to amplify a small alternating p.d. applied between gate and source. (The 10 MΩ resistor ensures that the gate remains close to the source potential even when there is nothing joined to the input terminals.) In this case the input p.d. is represented as having a sinusoidal waveform of peak value 1 V. To produce an amplified output p.d. we join a resistor R_d (called the *drain load resistor*) in the drain circuit, as shown. Variations in the drain current I_d then produce corresponding variations in the p.d. across R_d. It is best to see

how this works out in actual figures; in this discussion we shall take the source of the mosfet as being at zero potential.

Initially (at $t = 0$) the input potential V_{in} is zero, and from the transistor characteristic in figure 21.10c we see that

$$I_d = 1.3 \text{ mA}$$

Hence, p.d. across $R_d = I_d R_d$
$$= (1.3 \times 10^{-3} \text{ A})(10 \times 10^3 \text{ Ω})$$
$$= 13 \text{ V}$$

The top end of R_d remains at the potential ($+25$ V) of the positive terminal of the d.c. supply. The potential V_d of the drain is therefore 13 V less than this, namely 12 V. When the input potential has risen to its peak value of $+1$ V, the drain current I_d increases to 1.7 mA. The p.d. across R_d therefore increases, and the potential of the drain falls to a new value given by

$$V_d = 25 \text{ V} - (1.7 \times 10^{-3} \text{ A})(10 \times 10^3 \text{ Ω})$$
$$= 8 \text{ V}$$

Similarly, when the input p.d. falls to its maximum negative value of -1 V, the drain current falls to 0.9 mA, and we have

$$V_d = 25 \text{ V} - (0.9 \times 10^{-3} \text{ A})(10 \times 10^3 \text{ Ω})$$
$$= 16 \text{ V}$$

The drain potential thus varies sinusoidally between 8 V and 16 V; at this point in the circuit we have an alternating p.d. of peak value 4 V superimposed on a steady p.d. of 12 V. If we want to have only the alternating component at the output terminals, we must use a *blocking capacitor* (page 280) as shown. The p.d. across this then settles down to the steady value of 12 V, and the output p.d. V_{out} has a sinusoidal waveform varying between -4 V and $+4$ V. In this example the change in the output p.d. ΔV_{out} is 4 times the change in the input p.d. ΔV_{in}, and is in the opposite sense; when V_{in} *increases*, V_{out} *falls*. A single-stage amplifier like this is therefore an *inverting amplifier* – the output is in antiphase with the input. We describe this by saying that its *gain* is -4. In symbols

$$\text{gain} = \frac{\Delta V_{out}}{\Delta V_{in}} = -4$$

To obtain distortionless amplification the characteristic of the mosfet (figure 21.10c) needs to be linear over the range of input p.d.s used. This is approximately so in our example; but we can see from the characteristic that it would not be so if input p.d.s any greater than ± 1 V were used.

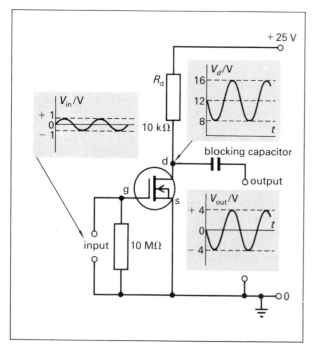

Figure 21.11 A simple amplifier employing an *n*-channel mosfet. An alternating p.d. of peak value 1 V gives an output p.d. of peak value 4 V in antiphase with the input.

The junction transistor

This consists of a thin layer of *n*-type or *p*-type semiconductor (usually silicon) sandwiched between two layers of semiconductor of opposite type, shown diagrammatically in figure 21.12(a and b). The circuit symbols for the two types are shown beneath (c and d). (The symbols derive from the construction of an early form of transistor, consisting of

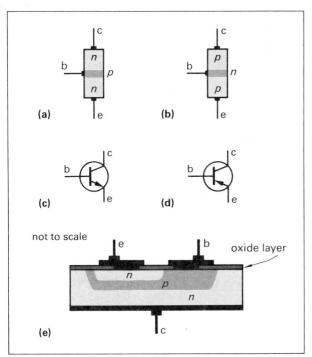

Figure 21.12 The junction transistor: (a) an *n-p-n* transistor; (b) a *p-n-p* transistor; the corresponding circuit symbols are shown below (c) and (d). (e) the internal structure of an *n-p-n* transistor.

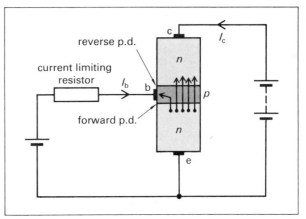

Figure 21.13 Transistor action. Most of the electrons crossing the emitter-base junction pass right through the thin base layer to the collector.

two fine wires touching a wafer of semiconductor.) The three layers are not just in electrical contact but must be formed within a single crystal. Figure 21.12e shows the internal structure of a typical *n-p-n* silicon transistor. The sequence of operations in forming it is just an extended version of the process used in other semiconductor devices we have considered (oxidation, localised etching, diffusion of impurity, etc). Finally aluminium contact pads are evaporated onto the crystal as shown, and connecting leads are soldered to these. The completed unit is enclosed in a hermetically-sealed, light-proof container.

The *n-p-n* transistor can be regarded as two *p-n* junction diodes joined back to back. It is operated with the p.d. across one junction (that between e and b) in the forward (conducting) direction, and the other in the reverse direction (figure 21.13). The layer e (called the *emitter*) is a strongly *n*-type region, so that the current across the lower junction is almost entirely by the movement of electrons from e to b. The electrons entering the central region (called the *base*) move across it; and, if it were thick, they would soon be neutralised by recombining with the holes there. But by making the base layer very thin ($< 10^{-6}$ m) a high proportion of the electrons reach the far side of it. They then enter the electric field at the upper junction in figure 21.13 and are quickly swept into the third region c (called the *collector*). In practice over 95 % of the electrons from the emitter reach the collector; and usually the proportion is more

than 99 %. The remainder recombine with positive holes in the base layer, and this produces a small current in the base lead.

In the *p-n-p* type of transistor the conduction process is similar, but takes place instead by movement of *holes* from emitter to collector, a few of which recombine with *electrons* in the base. The battery connections must of course then be the other way round.

In a typical small silicon transistor a current of about 5 mA may pass through the emitter-base junction when the p.d. across it is 0.7 V. Only about 1 % of this current (0.05 mA) passes to the base lead, leaving 4.95 mA to pass through the base to the collector lead. The ratio in which the current divides between collector and base leads is a fixed property of a given transistor; in the example just quoted the ratio is 99 : 1.

The current through the emitter-base junction is controlled by the p.d. V_{be} between base and emitter, as with any junction diode (figure 21.14a). No current passes at all if V_{be} is less than about 0.5 V, and over most of the operating range the value of V_{be} is close to 0.7 V, and the total current through the junction ($I_b + I_c$) changes very rapidly with V_{be}. The collector current I_c is always about 100 times (in our example) greater than the base current I_b; this ratio varies slightly with the collector-emitter p.d. V_{ce}, as shown in figure 21.14b, but it remains almost constant as long as V_{ce} is greater than about 0.5 V. Thus

provided $V_{ce} > 0.5$ V
$$I_c = \beta I_b$$

where β is a constant known as the *current amplification factor* of the transistor.* For the particular transistor considered above, $\beta \approx 100$.

You can see that the useful operating range of V_{be} is very small (between 0.5 V and 0.75 V). A common way of

---

*The current amplification factor β is often represented in technical literature by the symbol h_{fe}.

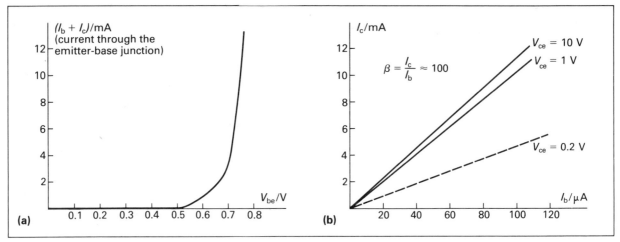

Figure 21.14 (a) The current through the emitter-base junction is controlled by the p.d. V_{be} across it.
(b) Provided $V_{ce} > 0.5$ V, $I_c = \beta I_b$.

using the transistor is with a current limiting resistor R_b in series with the base (figure 21.15); this ensures that the base current I_b is less than the maximum allowed for the transistor. To produce amplification of an input p.d. V_{in} applied as shown, we need to join a *collector load resistor* R_c in the collector circuit. Variations in the collector current I_c then produce corresponding variations in the p.d. across R_c. You can investigate the amplifying properties of this circuit with a variable input produced by a battery and potential divider, as shown in figure 21.15. V_{in} and V_{out} can be measured with two voltmeters. But if we want to measure V_{be} we must use a voltmeter of very high resistance (e.g. an electrometer, page 182).

Suppose we start with the input potentiometer near to the middle of its range; we might then find

$$V_{in} = 0.70 \text{ V}$$
and $$V_{be} = 0.60 \text{ V}$$

The p.d. across R_b is then 0.10 V, and the base current I_b (which is also the current in R_b) must be given by

$$I_b = \frac{0.10 \text{ V}}{10 \times 10^3 \text{ }\Omega}$$

$$= 10 \text{ }\mu\text{A}$$

Hence $I_c = \beta I_b$

$$= 100 \times 10 \text{ }\mu\text{A} = 1.0 \text{ mA}$$

This current passes through R_c, one end of which remains at the potential (5.0 V) of the positive terminal of the d.c. supply.

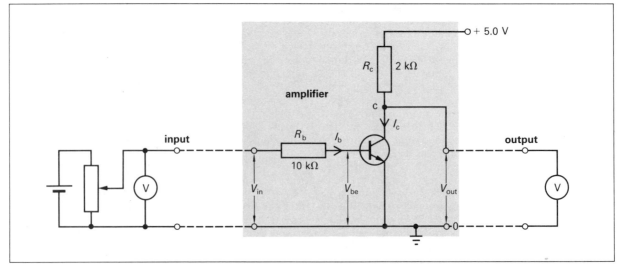

Figure 21.15 Testing a simple *n-p-n* transistor amplifier. A variable input p.d. may be applied with the potentiometer at the left, and the output p.d. is measured with the voltmeter at the right.

Hence   p.d. across $R_c = I_c R_c$
$$= (1.0 \times 10^{-3}\text{ A})(2.0 \times 10^3\ \Omega)$$
$$= 2.0\text{ V}$$

The potential of the collector c is therefore 2.0 V less than the positive d.c. supply potential (at 5.0 V), so that the reading of the output voltmeter V_{out} is 3.0 V
Let us suppose that the input potential is now increased so that we find

$$V_{in} = 0.90\text{ V}\quad\text{and}\quad V_{be} = 0.70\text{ V}$$

Then by similar reasoning we have

$$I_b = 20\ \mu\text{A}$$
$$I_c = 2.0\text{ mA}$$
$$\text{p.d. across } R_c = 4.0\text{ V}$$

and so V_{out} *falls* to 1.0 V. Thus an *increase* in the input p.d., causes a *decrease* in the output p.d.; and this is therefore also an *inverting amplifier*. Using the figures above

when $\Delta V_{in} = 0.90\text{ V} - 0.70\text{ V}$
$$= 0.20\text{ V}$$

then $\Delta V_{out} = 1.0\text{ V} - 3.0\text{ V}$
$$= -2.0\text{ V}$$

Hence   gain $= \dfrac{\Delta V_{out}}{\Delta V_{in}} = \dfrac{-2.0\text{ V}}{0.20\text{ V}} = -10$

There is only a limited range of values of V_{in} for which this amplifier can function without distortion. If V_{in} is less than 0.5 V, the transistor is *cut off* ($I_c = 0$), and we then have $V_{out} = 5.0$ V. If on the other hand V_{in} rises above about 1.00 V in this circuit, I_b rises to 25 μA, I_c rises to 2.5 mA, and the p.d. across R_c becomes 5.0 V, so that $V_{out} = 0$. The transistor is then said to be *bottomed*. In fact the collector potential never falls below about 0.3 V, since at this point normal transistor action ceases. Thus the useful swing of input p.d.s is only from 0.50 V to 1.00 V (i.e. 0.50 V), which with a gain of -10 gives the maximum possible swing of output p.d.s of nearly 5 V. Figure 21.16 shows how V_{out} varies with V_{in} for our simple transistor amplifier. Over most of the range of possible values of V_{in}

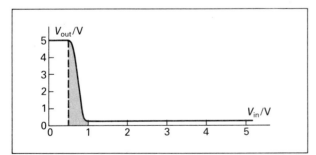

Figure 21.16 The output-input characteristic of a single-stage transistor amplifier; linear amplification occurs for only a limited range of input p.d.s. The greater the gain of the amplifier, the more limited the range.

the transistor is either cut off or bottomed. Only over a small part of the range is amplification possible. The same kind of result is obtained with any kind of amplifier; the higher the gain, the more limited the range of input p.d.s for which amplification occurs.

If the alternating input p.d. of an amplifier exceeds the allowable part of the range, *clipping* of the top and bottom of the waveform occurs as the transistor cuts off or bottoms. The amplifier is then said to be overloaded, and the resulting distortion will be obvious both in the sound from the loud-speaker output and in the waveform produced, if this is examined with an oscilloscope. Incidentally a grossly overloaded amplifier provides a simple means of generating a *square waveform* from a sinusoidal one. The sinusoidal input is so heavily clipped top and bottom that the resulting waveform is virtually a square one.

A complete amplifier will usually consist of several interlinked amplifying stages. The simplest way is to join them in cascade, the output of one stage being supplied to the input of the next. If we want just to amplify the alternating component of the input, then we can use blocking capacitors between the stages (as in figure 21.11 above). With an odd number of such stages the output is in antiphase with the input, and we have an inverting amplifier; with an even number of stages output and input are in phase, and we have a non-inverting amplifier.

The amplifier used in an *electrometer* (page 182) must have direct connection between the stages, as we need to use it with steady p.d.s. The first stage uses a mosfet, since this gives us the very high input resistance that we require. The voltage gain of the amplifier is substantially *less* than unity (a 1 V input gives full scale deflection of the meter, which requires perhaps 0.2 V). But the current gain may be 10^{10} or more.

Integrated circuits (i.c.s)

In manufacturing transistors and diodes the remarkable cheapness of these devices comes from our ability to make a very large number of them simultaneously. For instance about 7000 silicon transistors can be made on a single chip of silicon 50 mm in diameter. These are then sawn apart and individually mounted. However, it is also possible to use the same processes to fabricate complete circuits of transistors, diodes, resistors, etc., within a single chip of silicon. We are not limited to making these in a single layer. We can produce new layers of silicon on top of already formed layers by condensation from the vapour phase. This is called *epitaxial growth*. By a succession of such depositions and diffusions of impurities in carefully masked patterns many thousands of circuit components can be formed within a single chip less than 10 mm square (as in figure 21.17). The fabricating process may take more than a month; but again, many units can be made simultaneously, and the cost per unit is not great. Such a device is called an *integrated circuit* (i.c.). Many different types are now marketed, ranging from a complete radio (except for an

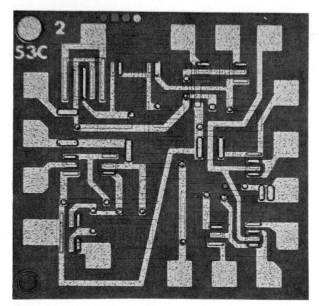

Figure 21.17 A photograph taken through a microscope of an integrated circuit containing many transistors, diodes, resistors, etc.

LC tuning circuit, page 281 to the complete operating heart of a small calculator. The art of using electronic devices has now become a matter of learning how to put together the i.c.s available so as to make circuits that do the jobs we require of them.

The range of i.c.s available includes a number of different types of amplifier. It matters very little to us what exactly the chip of silicon contains; but we need to understand how to use it to form the kind of amplifier we want. Figure 21.18a represents a type of i.c. called an *operational amplifier*. (The triangle is the general symbol for any amplifier.) The amplifier has two inputs. One of these (marked with a minus sign) is an inverting input; thus a small potential V_1 applied to this causes a greatly amplified output potential of *opposite* sign. The other input is a non-inverting one. The output potential V_{out} is therefore proportional to the difference of potential between the two inputs ($V_2 - V_1$). However, the gain of the amplifier is so large ($> 10^5$) that ($V_2 - V_1$) cannot be very large if the operational amplifier is to remain within the allowable operating range. For instance, if the maximum swing of output p.d. is 10 V (set by the supply p.d.), then the maximum value of input p.d. is

$$V_2 - V_1 = \frac{10 \text{ V}}{10^5}$$

$$= 10^{-4} \text{ V}$$

The operational amplifier is designed to have a very high input resistance ($> 10^{12} \, \Omega$), so that there is always negligible current through its input terminals.

Figure 21.18b shows in principle how we might use an operational amplifier to perform the same kind of function

as the two simple amplifiers we have discussed already (figures 21.11 and 21.15). The non-inverting input is joined to earth, and the output potential V_{out} is then (ideally) proportional to the input potential V_{in}. However, in practice an operational amplifier has not got a very linear response (any more than our earlier simple amplifiers); also a voltage gain of 10^5 is much greater than we are likely to require for most purposes. What we need to do is to find how to sacrifice some of the gain so as to achieve greater linearity of response. This may be done by equipping the amplifier with feedback paths so that part of the amplified output is fed back to the input; this feedback must of course be in antiphase with the input, so that it partly cancels it out. This is called *negative feedback*. The effect of negative feedback is to make the behaviour of the amplifier less dependent on the properties of the i.c.; instead, the gain now depends on the values of the resistances used in the feedback network. Since the response of a resistor is linear ($V \propto I$), the response of the whole amplifier becomes linear too, and the gain is exactly predictable.

Figure 21.18c shows the simplest way of doing this. The output is connected to the inverting input through the

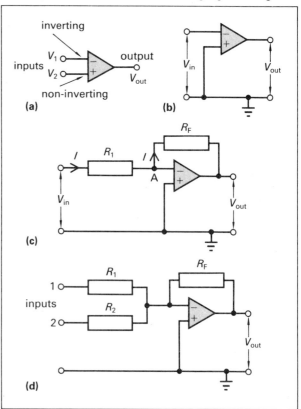

Figure 21.18 (a) The symbol for an operational amplifier (power supply connections are omitted). (b) A basic amplifier. (c) The use of feedback through the resistance R_F to give a linear amplifier of precisely defined gain. (d) A mixer amplifier for combining two input p.d.s in any desired proportions.

feedback resistor R_F; and another resistor R_1 is joined in series with the input. To understand this circuit we need to remember that the potential at the inverting input (at point A) cannot vary much from zero (less than 10^{-4} V from it in the example considered above). The p.d. across R_1 is therefore almost exactly equal to V_{in}. Likewise R_F is joined between the output terminal and A (which is nearly at zero potential); and therefore the p.d. across R_F is almost exactly equal to V_{out}. Only a negligible current enters the inverting input of the i.c. (because of its very high input resistance), and therefore the same current I flows in R_1 and R_F. Hence

$$I = \frac{V_{in}}{R_1} = \frac{-V_{out}}{R_F}$$

(A minus sign is used because V_{out} and V_{in} are of opposite sign.)

Therefore $gain = \dfrac{V_{out}}{V_{in}} = -\dfrac{R_F}{R_1}$

Thus if $R_F = 100$ kΩ, and $R_1 = 10$ kΩ, then the gain is precisely -10; and this is independent of the properties of the i.c. For instance, if the gain of the i.c. depends to some extent on its temperature (which is likely), the gain of the whole amplifier is still equal to $-R_F/R_1$; and if the two resistors are of the same kind, their *ratio* will not vary at all with temperature.

The same feedback system may be adapted to provide a *summing amplifier*, which enables us to mix together two inputs in any desired proportions (figure 21.18d). By the same reasoning the current through R_F is now equal to the sum of the currents through R_1 and R_2. Hence

$$\frac{-V_{out}}{R_F} = \frac{V_1}{R_1} + \frac{V_2}{R_2}$$

If we make $R_F = R_1 = R_2$, then $V_{out} = -(V_1 + V_2)$ We can use this type of circuit as a *mixer* to combine alternating p.d.s from a number of different inputs (microphones, pick-ups, etc). In this case R_1, R_2, etc, are made variable so that the balance of the different inputs can be adjusted at will.

Oscillators

An amplifier can become a self-maintained generator of oscillations if some of the output is fed back to the input in the right phase. Indeed quite unwanted oscillations can easily be generated in this way. In a public address system it sometimes happens that the sound fed back from the loudspeakers to the microphone is of greater intensity than the original sound to be amplified. The electrical oscillations then build up to a very large amplitude, and a piercing howl develops at some frequency to which the system is resonant. This is an example of acoustic feedback. But the same effect may occur if there is an internal electrical connection between output and input. If there is sufficient

feedback *in phase* with the input (*positive* feedback), the circuit is unstable and the amplitude of its oscillations increases.

Amplifiers with controlled positive feedback provide oscillators that are useful for many purposes. For instance, an electronic organ has a number of such circuits tuned to give the frequencies of the various notes of the keyboard. One way to control the frequency of an oscillator is to include in the circuit a resonant combination of inductance and capacitance (page 281). This kind of oscillator works well at radio frequencies (page 372), but in the lower audio frequency range the inductances required need to be rather large.

It is better then to control the frequency by means of capacitors and variable resistors. There is always a phase shift through an amplifier with its network of capacitors and resistors; this varies over the frequency range of the amplifier. The circuit can oscillate at the frequency for which the net phase shift through the amplifier and its feedback path is 360°; the feedback is then *in phase* with the input. The *loop gain* through the amplifier and back to the input must of course be sufficient; and if it is *just sufficient* to make up the energy losses of the circuit, then the oscillations are sinusoidal. With too much gain the transistors are bound to be cut off or bottomed at some part of the oscillation, and the waveform is then more complicated. Most commercial audio frequency oscillators consist of phase-shift circuits of capacitors and resistors built into amplifiers with some form of automatic gain control.

21.3 Sound reproduction

The movements of air in a sound wave are extremely small. A microphone will commonly be expected to produce a measurable p.d. with air displacements of only 10^{-10} m (less than the diameter of one atom). The corresponding pressure changes amount to about 10^{-3} Pa (not much more than the total pressure in the 'vacuum' of a cathode-ray tube). The power available to vibrate the diaphragm of the microphone may not be more than 10^{-12} W. It is not surprising that the alternating p.d. produced is very small, requiring considerable amplification before it can be used to operate a loudspeaker or recording head.

Good sound reproduction equipment is expected to respond uniformly to all frequencies to which the ear is sensitive (from 20 Hz to about 20 kHz), usually known as the *audio frequency* range. Amplifiers can now be made that fulfil the most exacting requirements almost to perfection. The weak links in the equipment are the transducers used at the inputs and outputs of an amplifier (microphones, pick-ups, etc, and loudspeakers).

The carbon microphone

This consists of a small box loosely packed with carbon granules (figure 21.19a). An electric current from a small

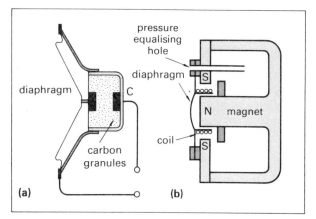

Figure 21.19 (a) A carbon microphone, as used in telephones. (b) A moving-coil or dynamic microphone of high quality.

battery is driven through the granules between the carbon block C at the back and a carbon button attached to the light aluminium diaphragm at the front. When the granules are compressed, the resistance of the unit decreases, and the current in the circuit rises. Thus the oscillations of the diaphragm produced by the pressure changes in the incident sound wave cause corresponding oscillations of the current.

The microphone and battery are joined to the primary of a high-ratio step-up transformer, so that a large alternating p.d. is produced at the output terminals. Although the p.d. generated in this way is controlled by the sound wave, its power is derived from the battery. The carbon microphone is thus an amplifying device, and is the only kind of microphone that can be used without additional amplification. It is therefore very suitable for telephones. It is not suitable, however, for high-quality sound reproduction because of the background hiss produced by the loose contacts between the carbon granules.

The crystal microphone and pick-up

In certain types of crystal, such as *quartz* and *Rochelle salt*, mechanical stresses produce electrical polarisation (page 196). Equal and opposite bound charges then appear on opposing faces of the crystal, and if there are two metal electrodes in contact with them, a p.d. is produced between them. This is called the *piezo-electric* effect. With large stresses the p.d. produced in this way can be several thousand volts, sufficient to produce a spark between two metal points joined to the electrodes. This is used in piezo-electric gas lighters and cigarette lighters.

A section from a single crystal of Rochelle salt can be used as a microphone by arranging it so that the incident sound wave causes the appropriate stress in it; the alternating p.d.s are typically in the range from 10 mV to 100 mV. The same principle is used in the crystal pick-up; in this case the stresses of the crystal are caused by the movements of the stylus vibrating in the groove of the record.

The moving-coil or dynamic microphone

In this type of microphone a small coil is attached directly to the diaphragm. It is arranged to move in the strong radial field in an annular gap between the pole pieces of a permanent magnet (figure 21.19b). When the diaphragm oscillates, the coil cuts the magnetic flux, and an alternating e.m.f. is generated in it. The construction is very similar to that of a moving-coil loudspeaker (figure 24.4, page 320). In fact a loudspeaker may be used as a microphone, though its bulk is usually a bit inconvenient for this purpose! The output of such a microphone is typically in the range from 1 mV to 10 mV, less than a crystal microphone; but it is capable of giving sound reproduction of much higher quality.

21.4 Switching circuits

A transistor amplifier may be used in the same way as a relay switch, in which the input current causes the switch contacts to close, so controlling a much larger current in another circuit (page 286). Thus in figure 21.20a, when the potential of A is zero, the transistor is cut off, and the output B will be at the potential of the positive supply rail, namely 5 V. Similarly when A is at 5 V the transistor is bottomed, and the output B is at a potential close to zero (strictly 0.3 V). In this simple arrangement the output potential thus takes one of two values depending on the input potential applied at A. When A is at 5 V, B is at zero potential, and vice versa. The virtue of using a transistor to achieve this result is that the input potential can be provided by another electronic circuit, and so the 'position' of the switch can depend on potentials derived from other points in the apparatus. Furthermore, electronic switches of this kind can be reversed millions of times a second, and so enable interlinked switching operations to be carried out with great rapidity. It is this function of a transistor that has made possible the construction of high-speed digital computers.

Any process of computation consists in performing a sequence of operations on the data of the problem; at each stage in the computation the nature of the operation to be performed is decided partly by the pre-determined programme and partly by the outcome of earlier stages in the process. We therefore need switches with multiple inputs, whose outputs are determined in specified ways by the condition of their inputs. These are known as *gates*, and mostly they are elaborations of the simple transistor switch of figure 21.20a.

Consider the gate in figure 21.20b. If one (or both) of its inputs A_1, A_2 is switched 'on' (i.e. joined to the positive supply rail), this is sufficient to raise the potential of the base of the transistor to the point at which it conducts and bottoms; B is then nearly at zero potential (the 'off' position). But when both inputs are 'off' (at zero potential), the transistor is cut off, and the output B is at 5 V. The

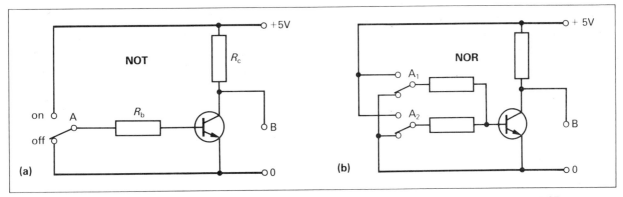

Figure 21.20 (a) A transistor used as a switch; the potential at B is the opposite of that at A, so that this is called a NOT gate. (b) A two-input NOR gate; B is 'on' only when neither A_1 nor A_2 are 'on'.

output B is 'on' only when neither input A_1 nor input A_2 is 'on'. The gate therefore performs a function equivalent to the logical connective NOR, and it is described as a NOR gate. A NOR gate can be constructed with any number of inputs (2, 3, 4, . . .) by adding further resistors alongside the two in figure 21.20b. The simple switch in (a) can be described as a NOR gate with only one input, or we can call it a NOT gate; sometimes it is known as a *negater*, since its output is always the opposite of its input.

The two-input NOR gate can be converted to a two-input OR gate by connecting a NOT gate to its output (figure 21.21a). In (b) is a simplified diagram of the same arrangement, in which the positive and negative supply rails are not drawn, and the amplifier is represented by the usual triangular symbol.

We can make an AND gate by using a NOR gate and connecting NOT gates to each of its inputs (c). In this combination the output B is 'off' if either of the inputs is 'off' (at zero potential). Only if input A_1 and input A_2 are 'on' will the output B be 'on' also. This is precisely the function required of an AND gate. All the logical processes required for the operation of a digital computer or a pocket calculator can be provided by circuits like the ones we have described.

A binary scale counter

Arithmetical operations in a digital computer are normally carried out using a binary scale of arithmetic. This is because only two digits (0 and 1) are required in this system, and these can be represented by the two possible 'positions' of a switch or gate. A binary counting circuit can be made by joining two NOT gates in cascade, with the output of the second gate connected back to the input of the first (figure 21.22). When one transistor is bottomed, it holds the other cut off, and vice versa. The combination is stable in either of its two states. One of these can represent the digit 0 and the other the digit 1. It is known as a *bistable* or *flip-flop* circuit.

To make it perform its counting function we must arrange for it to change over from one state to the other when a suitable electrical pulse is fed into it. In figure 21.22 the shaded parts of the circuit are concerned with steering the input pulses so that they switch off whichever transistor is conducting, so causing the circuit to change over to its other stable state. We need not concern ourselves with exactly how this is achieved. But you can trace through the action of the circuit, if you wish. (Bear in mind that the bottomed transistor causes the diode joined to its base to be *just* conducting; this diode therefore behaves as a low

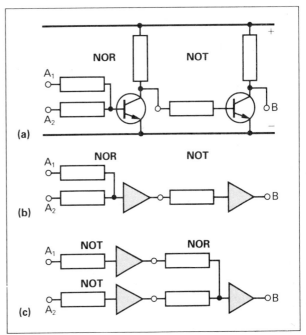

Figure 21.21 (a) A two-input OR gate. (b) A simplified diagram of the same OR gate, omitting the d.c. supply rails and using the 'amplifier' symbol. (c) A two-input AND gate made from a NOR gate and two NOT gates.

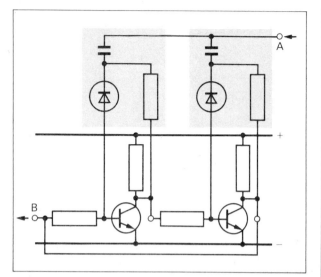

Figure 21.22 One stage of a binary counter; it consists of two NOT gates joined in a closed loop. The shaded parts of the circuit are designed to steer the input pulses to switch off the conducting transistor, so changing over the state of the gate.

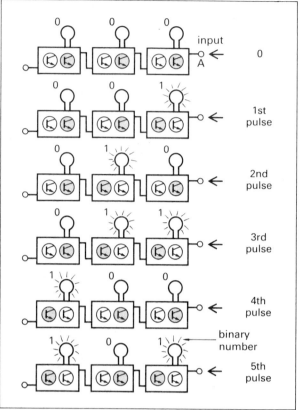

Figure 21.23 A simple binary counter of 3 bistable circuits in cascade. The indicator lamps show when the left-hand transistor of each circuit is conducting, representing the digit 1.

resistance, while the other diode remains non-conducting and of high resistance.) When the input A goes from 0 to 5 V, nothing happens; but when it goes back again from 5 V to 0, the change is communicated to the base of the conducting transistor, switching it off and changing over the state of the circuit.

The circuit thus changes its state every time the input A is taken from 5 V to 0, and on each occasion the output terminal B changes its potential from zero to 5 V or the other way. If a square wave generator is joined to A, the bistable circuit changes over once in each cycle of oscillation. The output at B is a similar square wave, but is of *half* the frequency. This output may be applied to the input of another similar bistable circuit; the latter will then change over its state on every other zero-going pulse applied to A. A third bistable will change over once in every four such pulses, and so on.

A complete binary scale counter is made by joining up a sufficient number of such units; figure 21.23 shows a counter with just three units. Suppose we decide that the digit 0 is to be represented by the right-hand transistor conducting in each case, and 1 by the opposite state. It is convenient to arrange for an indicator light to show when the digit 1 is being represented. The initial condition of the counter will then be as in the top diagram, with all the indicator lights out; the conducting transistor is shown shaded in each case. If we now feed in (say) 5 zero-going pulses at

the input A, the counter will pass through the states represented in the ensuing diagrams.

A digital computer contains many such counters, the condition of each one representing a number involved in the computation. Arithmetical operations of all kinds may be effected by performing the right sequence of logical processes on the data represented by the potentials in the counters. For this purpose gates such as those described above are used.

A counter can be regarded as a temporary *memory* for the number that has been fed into it. Other methods for storing the pulses that represent binary numbers are used when a longer access time to the memory is permissible. For instance, the pulses may be recorded on magnetic tape; the technique is similar to that used in an ordinary tape recorder.

22 Oscillatory motion

22.1 Describing oscillations

A swinging pendulum bob, a bouncing rubber ball and a lump on the end of a spring are all examples of objects which vibrate or oscillate. The motions are repetitive: to-and-fro or up-and-down. Both the pendulum and the lump oscillate in such a way that the time taken for a complete cycle T is constant although the energy of the oscillation is gradually being transformed to internal energy. The bouncing of the rubber ball is quite different; T gets smaller and smaller as the bounce gets smaller. The great majority of oscillating systems are like the pendulum and the lump-on-a-spring; their motion is *isochronous*, i.e. the periodic time is independent of the size of the oscillation.

If the periodic time or *period* of an oscillation (a complete to-and-fro) is T and it repeats itself with a frequency f then

$$f = \frac{1}{T}$$

For example a ruler clamped to the bench may vibrate 17 times per second: $f = 17$ Hz and $T = (1/17)$ s $= 0.059$ s.

Measuring time

All clocks use a repetitive or periodic vibration or oscillation. Mechanical clocks may employ a pendulum with a frequency of a fraction of a hertz, a balance wheel for which f may be a few hertz or a tiny quartz crystal which oscillates at tens of kilohertz (the whole crystal pulses in and out elastically). To find the size of f in these cases we must calibrate the clocks against an agreed standard clock. This is now chosen to be a clock using the oscillations of the caesium-133 atom. (appendix A2)

We study oscillations partly because there is a need for accurate timekeeping but mainly because of the phenomenon of *resonance*, which is relevant to all branches of physics. We walk by swinging our legs and we learn to use our muscles to maintain the rhythm of our pendulum-like legs. A car door rattles violently at certain frequencies only. We shall deal with these and other examples of resonance later in this chapter.

Simple harmonic motion

Figure 22.1a shows an air puck which is tethered to two fixed posts by long light springs. The puck is a small hovercraft with a battery powered motor in it; it glides over a carefully levelled sheet of plate glass, but any other low

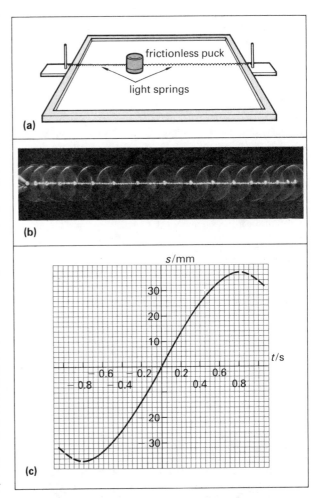

Figure 22.1 An experiment to analyse the motion of an oscillator. The scale for displacements in the graph is taken from the print in (b).

friction device would do. (b) shows the result of photographing stroboscopically (page 2) half a complete oscillation of the air puck from its release on the extreme left of its motion to its coming momentarily to rest at the extreme right. The half period shown took just over 1.5 s.

If we plot a graph of displacement s measured on the print (the real distances were about 15 times larger) against time t for the puck we obtain graph (c) (the time interval between exposures was 0.10 s, so the time for half a period $(T/2)$ was about 1.6 s and $T \approx 3.2$ s). We chose the zero for *s* at the centre of the oscillation, counting positive to the left; and the zero for *t* at *s* = 0 for symmetry. The *s-t* graph looks like part of a sine curve.

We can similarly plot *s-t* graphs for an air track glider attached to springs, for a lump oscillating vertically on a spring or for a pendulum (where we have an *angular* displacement θ (page 156)). Using a ticker-tape technique we can analyse the motion of a dynamics trolley tethered by springs to fixed posts or of a massive pendulum on the end of a very long suspension. Such graphs all have the characteristic *sinusoidal* shape shown in figure 22.1c. A particularly neat way of drawing an *s-t* curve experimentally is to move a camera steadily in a direction perpendicular to the line of an oscillation. Figure 22.2 shows the result of doing this for a lump oscillating up and down on a spring.

From the graph of figure 22.1c we can deduce both the velocity-time and the acceleration-time graph for the oscillator. The method is outlined in section 1.3: we draw tangents to the *s-t* curve (as $v = \mathrm{d}s/\mathrm{d}t$) to find values of *v*, and tangents to the *v-t* curve (as $a = \mathrm{d}v/\mathrm{d}t$) to find values of *a*. We find that the resulting graphs are themselves sinusoidal curves; the positions of their peaks and troughs are related as shown in figure 22.3 which continues the curves for a period or so. They all have the same period *T* (and frequency *f*).

At moments such as X the oscillator (let us call it a puck) is at its maximum (positive) displacement s_0 from the centre and is momentarily at rest, so that $v = 0$. However this is the point at which *v* is changing at its maximum rate (from positive to negative values), thus *a* is at a (negative) maximum. Similarly at moments such as Y, $T/4$ later, the puck has zero displacement but its displacement is changing at its maximum rate (from positive to negative values), i.e. *v* is at a (negative) maximum $-v_0$. At this point *v* is momentarily not changing; so that $a = 0$. You should continue the argument for moments such as Z or at $t = 0$, and be quite clear how the curves of figure 22.3 are related.

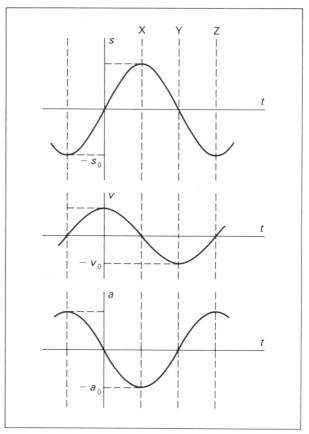

Figure 22.3 Motion graphs for a linear simple harmonic oscillator. The amplitudes are related thus: $v_0 = 2\pi f s_0$ and $a_0 = 2\pi f v_0$. Note the phase relationships between *s*, *v* and *a*.

Equations for s.h.m.

The equation for the *s-t* graph of figure 22.3a is

$$s = s_0 \sin 2\pi f t \qquad (1)$$

The maximum value of *s* is s_0 in either direction. This is known as the *amplitude* of the oscillation. An oscillatory motion for which *s* is related to *t* in this way is called *simple harmonic motion* or s.h.m. With $f = 1/T$ we can see that the term $2\pi f t$ will be 2π rad (or 360°) when $t = T$, and the curve then repeats itself and does so every period. We notice that the *a-t* graph is upside down compared to the *s-t* graph (i.e. troughs of *a* coincide with peaks of *s*), and we can write

$$a = -(\text{constant})s \qquad (2)$$

at every point, where the constant has units s^{-2}.

Since $s = s_0 \sin 2\pi f t$

the maximum rate of change of *s* (which is v_0) is given by

$$v_0 = \left(\frac{\mathrm{d}s}{\mathrm{d}t}\right)_{\text{max}} = 2\pi f s_0 \qquad (\text{appendix B6})$$

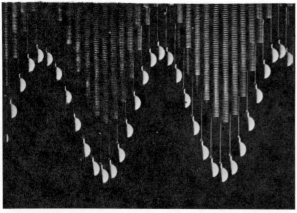

Figure 22.2 A lump oscillating on a spring photographed with a moving camera. The result is effectively a displacement-time graph for the lump.

The *v-t* curve is also sinusoidal (a sine curve but displaced a quarter of an oscillation along the axis). Thus the maximum rate of change of *v* (which is a_0) is given by

$$a_0 = \left(\frac{dv}{dt}\right)_{max} = 2\pi f v_0 \qquad \text{(appendix B6)}$$

Thus $a_0 = (2\pi f)^2 s_0$

and the constant in (2) above is $(2\pi f)^2$, so that

we have $a = -(2\pi f)^2 s$ (3)

Equations (1) and (3) are equivalent statements of the same simple harmonic behaviour.

Returning to the particular case illustrated in figure 22.1 we see that as $T = 3.2$ s then

$$f = \frac{1}{T} = 0.31 \text{ s}^{-1}$$

and $(2\pi f)^2 = (2\pi \times 0.31 \text{ s}^{-1})^2 = 3.8 \text{ s}^{-2}$

We can also find the maximum speed v_0 $(= 2\pi f s_0)$ of the puck:

$$v_0 = 2\pi f s_0 = 2\pi(0.31 \text{ s}^{-1})(36 \text{ mm}) = 70 \text{ mm s}^{-1}$$

which should be the gradient of the *s-t* graph at $t = 0$. By drawing a tangent as the curve in figure 22.1c crosses the axis you can confirm this. (Remember that the true speed is greater than this by the factor, about 15, by which the print is a reduction of the real motion of the air puck).

▶ A punchbag oscillates with simple harmonic motion of amplitude 0.60 m and frequency 2.0 Hz. What is (i) the maximum acceleration of the punchbag and (ii) its acceleration when it is 0.30 m from the centre?
(i) If the motion is s.h.m., then $a = -(2\pi f)^2 s$ and $a_0 = (2\pi f)^2 s_0$.

For $f = 2.0$ Hz, $2\pi f = 12.6 \text{ s}^{-1}$, so that

$$a_0 = (12.6 \text{ s}^{-1})^2 (0.60 \text{ m})$$
$$= 95 \text{ m s}^{-2}$$

This maximum acceleration occurs when the punchbag is at its furthest from the centre of its motion, at which point it is momentarily at rest.
(ii) As $a = -(2\pi f)^2 s$ for any displacement, the size of acceleration when $s = 0.30$ m is given by

$$a = (12.6 \text{ s}^{-1})^2 (0.30 \text{ m})$$
$$= 48 \text{ m s}^{-2}$$

or (as we could have realised immediately as $s = s_0/2$) just half the maximum acceleration. ◀
▶ The tide on a mid-ocean island rises and falls with simple harmonic motion. If the period *T* of the motion is 12 h 25 min and the tide is high at 12 noon when will it next be a quarter of the way out?

The height of the tide can be expressed as *h* where

$$h = h_0 \sin 2\pi f t = h_0 \sin 2\pi \frac{t}{T}$$

The tide is half-way out at $h = 0$, and a quarter of the way out at $h/h_0 = 1/2$

i.e. when $\sin \frac{2\pi}{T} t = 0.5$

$$\Rightarrow \qquad \frac{2\pi}{T} t = \frac{\pi}{3} \text{ rad} \quad \text{or} \quad 60°$$

Therefore the tide will be half-way out at $t = T/6$ i.e. at 1406 hours and not at 1508 as we might have guessed. ◀
Equations (1) and (3) above are the important relations for s.h.m. You will notice that (1) contains the amplitude s_0 of the oscillation *s*, but that (3) does not. Oscillatory motions which are s.h.m. have the same period T $(= 1/f)$ no matter what the amplitude is, i.e., they are isochronous motions.

S.h.m. and circular motion

Figure 22.4 shows a model which can be used to demonstrate the equivalence of s.h.m. and the projection of circular motion along a diameter. Once the frequencies are adjusted to be the same the two shadows are found to move together throughout their motion. This equivalence is utilized again on page 273 where the phasor method of treating sinusoidally alternating electrical quantities is described.

Angular harmonic motion

Many oscillating or vibrating bodies, e.g. pendulums and balance wheels, involve angular displacements; they are rotating or twisting motions about a fixed axis. The equations for these motions are analogous to equations (1) and (3) above, namely

$$\theta = \theta_0 \sin 2\pi f t$$
and $a = -(2\pi f)^2 \theta$

where a (alpha) is the angular acceleration of the body when it is rotated an angle θ from its equilibrium position.

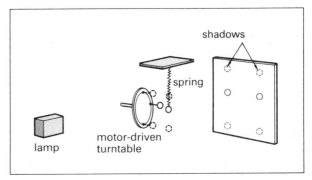

Figure 22.4 The equivalence of s.h.m. and projected circular motion. The shadows of the mass oscillating on the spring and the ball rotating on the turntable move together.

22.2 Simple harmonic oscillators

We want to discover under what conditions a body will move with an acceleration which is proportional to its displacement from a given point O, i.e. with $a \propto -s$. The minus sign indicates that the acceleration slows the body down as it moves away from O and then speeds it up as it approaches O again.

Referring to figure 22.1a and the air puck attached to the springs we see that the further the puck is from the centre the greater the resultant force F on it. If the springs obey *Hooke's law* (page 53), then F is proportional to the displacement s of the puck from its equilibrium position O (and the pull of the springs on the puck is such as to return the puck to O). With stiffer springs F will be larger for a given value of s, and the acceleration will be greater there, so reducing the period T of the oscillation. With a puck of greater mass the same force will produce a lower acceleration thus increasing T.

For a Hooke's law spring system of stiffness k

$$F = -ks \qquad \text{(page 23)}$$

By Newton's second law, the acceleration a of a puck of mass m acted on by a resultant force F is given by

$$a = \frac{F}{m} \qquad \text{(page 14)}$$

so that, if the resultant force on the puck is equal to the push or pull of the springs on it we can write

$$a = -\frac{k}{m} s$$

i.e. $a = -(\text{constant})s$

and the motion of the puck is simple harmonic motion. Comparing this with $a = -(2\pi f)^2 s$, which defines s.h.m., we see that

$$(2\pi f)^2 = \frac{k}{m}$$

and so $f = \frac{1}{2\pi}\sqrt{\frac{k}{m}}$

The period $T\,(=1/f)$ is thus given by

$$T = 2\pi\sqrt{\frac{m}{k}}$$

This argument deriving T by considering the forces acting on the puck shows that the period is independent of the amplitude of the motion s_0, and hence independent of the maximum velocity v_0 and acceleration a_0 of the puck. If $k = 3.7$ N m^{-1} and $m = 0.96$ kg we shall have an oscillation for which $T = 3.2$ s (like that in figure 22.1) for *any* value of s_0 (e.g. for $s_0 = 36$ mm or $s_0 = 17$ mm). Of course, other pairs of values of m and k could also produce a period of 3.2 s.

The answer to the question 'under what conditions will

a body oscillate with s.h.m.?' is thus something like, 'when the resultant force acting on it tends to restore the body to its equilibrium position and is proportional to its displacement from that position'.

Then $T = 2\pi\sqrt{\dfrac{m}{k}}$

$$= 2\pi\left(\frac{\text{mass of oscillating body}}{\text{net restoring force per unit displacement}}\right)^{\frac{1}{2}}$$

Any mechanically oscillating system must have these properties of *inertia* and of *elasticity* (or an equivalent mechanism for producing a restoring force).

A vertical mass-spring oscillator

When a mass m oscillates up and down on a spring its weight mg acts along the line of the spring and so is relevant in discussing the mechanics of its oscillation. We expect the motion to be s.h.m. (figure 22.2 illustrates this) but to prove it consider the mass attached to a light spring of stiffness k. Figure 22.5 shows (a) the spring before the mass is attached; (b) the spring after the mass is attached and a free-body diagram for the mass hanging in equilibrium with the spring extended a distance e; (c) the spring and mass at a moment during an oscillation and a free-body diagram for the mass then.

From (b) we get $T_0 = mg$, and, if the spring obeys Hooke's law $T_0 = ke$, so that

$$mg = ke$$

In (c) we have marked the displacement s *from the equilibrium* position of the particle and put the acceleration in the *positive s* direction. Using Newton's second law we have

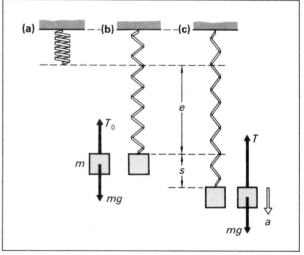

Figure 22.5 (b) A mass stretches a spring a distance e when it hangs in equilibrium, (c) shows the mass accelerating after being pulled down and released. A free-body diagram of the mass accompanies situations (b) and (c).

$$ma = mg - T$$
where $\quad T = k(e + s)$
so that $\quad ma = mg - k(e + s)$
$$= mg - ke - ks$$

But $mg = ke$, from above,

so that $\quad a = -\dfrac{k}{m} s$

This equation represents simple harmonic motion. The period T is therefore given by $T = 2\pi\sqrt{m/k}$, and will be the same no matter what the local value of the gravitational field g. Thus if $k = 420$ N m^{-1} and we measure T to be 0.356 s, then m *must* be 1.35 kg, on Earth or on the Moon. We could use the oscillations to measure m even in an orbiting (free-fall) laboratory.

The usefulness of $T = 2\pi\sqrt{(m/k)}$

When we lean on one corner of a car it might oscillate with a period of about 1 s; as the mass involved is at least a quarter of the mass of the car, say 200 kg, then the equation tells us that $k \approx 8 \times 10^3$ N m^{-1} or 8 N mm^{-1}, where k is the stiffness of the suspension at that corner.

A second example comes from the atomic world: if the ionic bond stiffness between adjacent ions in NaCl ≈ 150 N m^{-1} and the average mass of Na$^+$ and Cl$^-$ ions is close to 5×10^{-26} kg, then there should be a natural oscillation of frequency

$$f = \frac{1}{T} = \frac{1}{2\pi}\sqrt{\frac{k}{m}} = \frac{1}{2\pi}\sqrt{\frac{150 \text{ N m}^{-1}}{5 \times 10^{-26} \text{ kg}}}$$
$$\approx 9 \times 10^{12} \text{ Hz}$$

Electromagnetic waves have frequencies of this order in the infra-red – see page 366, and can induce oscillations in this range of frequencies in such atomic systems.

Pendulums

Simple pendulum

Suppose that the period T of a simple pendulum (a small bob of mass m on the end of a thin thread of length l) depends on m, l and the free-fall acceleration g, i.e.

$$T = km^x l^y g^z \qquad \text{(k a dimensionless constant)}$$

As $\quad [g] = LT^{-2}$ we can write $[T] = M^x L^y (LT^{-2})^z$

which gives $\quad x = 0 \quad y = \frac{1}{2}$ and $z = -\frac{1}{2}$ (page 68)

Thus a *dimensional analysis* suggests that

$$T = k\sqrt{\frac{l}{g}}$$

A quick experiment will confirm that T is independent of m and a graph of T against $\sqrt{l}$ turns out to be a straight line through the origin.

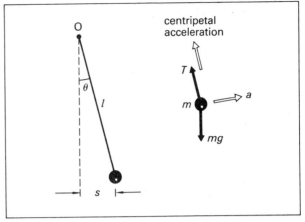

Figure 22.6 Analysing the motion of a simple pendulum.

To derive the expression and to find k consider a free-body diagram for the pendulum at a position displaced an angle θ from its equilibrium position (figure 22.6). Applying Newton's second law at right angles to the string (T has no resolved part in this direction)

$$ma = -mg \sin \theta$$

As $\sin \theta = s/l$ then

$$a = -\frac{g}{l} s$$

This is in the form of the defining equation for simple harmonic motion but it is only s.h.m. if s and a are along a line and this is only approximately true here for small values of θ, i.e. for *small amplitude* oscillations. With this restriction then,

$$T = 2\pi\sqrt{\frac{l}{g}}$$

Values of T calculated from this equation are within 1% of the experimental values if the amplitude θ_0 of the motion is below about 15°.

▶ How long is a 'seconds pendulum', that is, one which passes through its lowest point once every second?

The period of this pendulum is 2.00 s, so that in a laboratory where $g = 9.80$ m s^{-2} the length l of the pendulum is given by

$$l = \frac{T^2 g}{4\pi^2} = \frac{(2.00 \text{ s})^2 (9.80 \text{ m s}^{-2})}{4\pi^2} = 0.993 \text{ m}$$

i.e. the pendulum is just under 1 metre long. ◀

Measuring g

The equation $T = 2\pi\sqrt{l/g}$ for the period of a simple pendulum suggests a method for measuring g in the laboratory. Rearranging it we have

$$l = \frac{g}{4\pi^2} T^2$$

so that a graph of l against T^2 will have a gradient of $g/4\pi^2$. In performing the experiment we must keep the amplitude of the swing down to less than 15° or so. Several values of T for each value of l should be taken and a very firm support used for the pendulum. With care we might find the value of g with an uncertainty of about $\pm 1\%$.

We can improve the above experiment if we use a more rigid pendulum than a thread and bob; for instance a metal rule with a hole near one end for the pivot. You should consult books on practical physics for details of how such pendulums can be used to measure g to better than 0.01%. The variation of g at different places on the Earth's surface can be compared using these rigid or *compound pendulums* but standard determinations of g now use the free-fall technique mentioned on page 7.

Torsional pendulum

On page 161 an oscillating system consisting of a dumb-bell supported on a torsion wire was used to determine G, the universal constant of gravitation. For such a torsion pendulum we need to consider the dumb-bell as a rigid body with moment of inertia I (page 173). The torque T is provided by the twisting suspension and, if the wire obeys Hooke's law for angular deformation, $T = -c'\theta$

Newton's second law gives, for rotation

$$I a = T_{\text{res}} \qquad \text{(page 174)}$$

so that $\quad a = -\dfrac{c'}{I}\theta$

which represents angular harmonic motion (page 301) of period $T = 2\pi\sqrt{I/c'}$. Note that we do not need to make an approximation in deriving this equation as we did for the simple pendulum.

22.3 Oscillators and energy

Figure 22.7 shows a possible way of describing energy flow for a mechanical oscillator in which energy transfer from kinetic energy to potential energy (either elastic or gravitational) is wholly reversible. The motion is said to be *undamped*. At any time the sum of the kinetic and potential energies is constant.

$$\tfrac{1}{2}mv^2 + (mgh \text{ or } \tfrac{1}{2}ks^2) = \text{constant}$$

E.g. in figure 22.1 the mass of the puck is 0.96 kg and its maximum speed (allowing for the scaling factor of about 15) is

$$2\pi f s_0 \times 15 \approx 1 \text{ m s}^{-1}$$

so that the constant would be

$$\tfrac{1}{2}mv_0^2 = \tfrac{1}{2}(0.96 \text{ kg})(1 \text{ m s}^{-1})^2 = 0.5 \text{ J}$$

We can describe how this energy is shared between kinetic

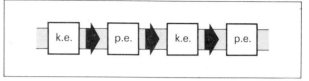

Figure 22.7 A way of describing energy transfer in an undamped oscillator.

and potential either (i) as time t varies or (ii) as displacement s varies.

Figure 22.8 shows how the *energy* varies with *time* for a simple harmonic oscillator. The kinetic energy of the oscillator, $E_k = mv^2/2$, involves the square of the velocity v. We know now that in s.h.m. v varies sinusoidally with time; v^2 also varies sinusoidally with time, but has a maximum value twice during each oscillation. (See the discussion for alternating current I and power $\propto I^2$ on page 272.) The frequency of the curve for E_k against time in figure 22.8 is therefore twice that of the oscillation it describes. The shape of the potential energy curve, E_p against time, is immediately predictable as the total mechanical energy is constant.

Figure 3.10 on page 28 shows the relationship $E_p = \tfrac{1}{2}kx^2$ (displacement x) for a particular mass-spring oscillator. The graph is a parabola ($y \propto x^2$) with maximum elastic potential energy at maximum displacement. As the total energy remains constant, 0.64 J in this case, the graph of *kinetic* energy against displacement is the parabola inverted with its peak at $x = 0$, when $E_k = 0.64$ J, and its sides reaching the energy axis ($E_k = 0$) at the maximum displacements ± 0.16 m. Similar graphs hold for all simple harmonic oscillators. For a simple pendulum E_p is *gravitational* potential energy.

Damped oscillations

Real oscillations always die away. They are said to be *damped*. Figure 22.9 shows two 'time-traces', i.e., displacement-time graphs for a damped oscillator. These were made by attaching a fine brush to the oscillator and moving a strip of paper steadily past it. For (b) the damping force was increased by attaching a piece of stiff card to the

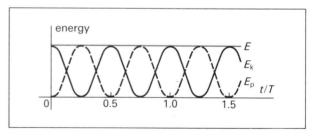

Figure 22.8 The total energy of an undamped mechanical oscillator $E\,(= E_k + E_p)$ is constant though both the kinetic energy E_k and the potential energy E_p vary sinusoidally at twice the frequency of the oscillator. Note that the time axis is expressed as a fraction of the period T.

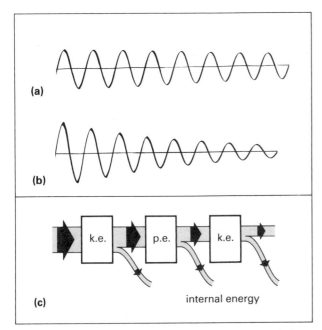

Figure 22.9 Two 'time traces' made by a fine brush attached to a damped oscillator. Notice how the period within each trace remains constant even though the amplitude decreases; (c) shows one method of representing the energy conversions as the oscillations die away.

oscillating body: the mechanical energy of the oscillator was thus converted to internal energy of the surroundings more rapidly. (c) is an energy flow diagram describing this process.

Damping forces may be either *viscous* forces in a fluid, (e.g. the air damping mentioned above) or *internal* forces in the oscillating system which prevent it being perfectly elastic. The internal energy transferred to the air or to part of the system results in their temperature rising a little.

Sometimes the damping of a mechanical system is a nuisance, sometimes it is desirable. A small child is disappointed by the damping on a swing but if a car continued to oscillate after hitting a bump the effect would be most uncomfortable. A large bell would ring for much longer if its oscillations were perfectly elastic, as the radiation of useful sound energy usually accounts for only a small fraction of the damping.

A system is said to be lightly damped if it undergoes a very large number of oscillations before coming to rest. Many mechanical systems such as car springs exhibit heavy damping and, once displaced, slowly approach their equilibrium position without oscillating at all. Some are fitted with shock absorbers, which often consist of a piston moving in a cylinder full of very viscous oil. Moving-coil meters are heavily damped in such a way that the pointer or needle reaches the value to be read as quickly as possible. Such *critical damping* is achieved electromagnetically in meters (page 208), the energy being converted to internal energy by induced currents.

22.4 Resonance

When a washing machine spin-drier is started the casing of the machine sometimes vibrates alarmingly at certain stages during the speeding-up process. The vibrations can be so violent that the motor fails to accelerate the tub through one of these stages. If we stop the machine and redistribute the clothes evenly the problem is often overcome. If a car is stuck in mud on a wet field and cannot be pushed out of the holes its wheels have made, the method we use to free it is to give a push, let it roll back, give it another push, and so on, timing the pushes so as to build up its oscillation until it rolls clear.

Or consider a child sitting on a garden swing that undergoes damped oscillations of period 4.0 s, i.e. has a frequency of 0.25 Hz, after being given a good push. About half the mechanical energy of the swing might be lost in only 5 oscillations. Suppose the pusher tries to push and pull the chains back and forth just below their point of suspension. Working at 0.25 Hz he would gradually be able to build up the amplitude of the swing's oscillation. The maximum amplitude of oscillation is reached when all the mechanical energy he feeds in during *one* cycle is converted to internal energy during the *same* cycle. If, however he tried to make the swing oscillate at some quite different frequency, the amplitude of the swing would be very low.

A 'swing' experiment of this type can be performed in the laboratory with a pendulum. For instance, if a simple pendulum with a natural frequency of 0.5 Hz is pushed and pulled at 5 Hz close to the point of suspension the amplitude is small – but there is an oscillation, and its frequency is 5 Hz. It is not a free oscillation but a forced one.

Forced oscillations

The two experiments which follow each demonstrate the properties of forced oscillations and *resonance*; each brings out some points best but there is some common ground. Figure 22.10a shows an apparatus which enables an experiment of the swing type to be performed with a mass-spring system. A mass supported by a vertical spring and an eccentric connection to a disc driven by a motor provides the *driving force*. The motor drive will probably need to be geared down considerably in order to provide a slow, periodic driving force which will not be affected by the movements of the vibrating mass. To the base of the mass is attached a thin rod which grips a piece of metal sheet. The degree of damping as the sheet oscillates up and down in a liquid bath can be varied by altering the liquid, e.g. water for medium damping and glycerol for heavy damping. To get light damping a smaller sheet can be used.

Keeping the amplitude of the driving force steady the amplitude s_0 of the *forced oscillations* of the mass is measured at a number of values of the driving frequency f. The system must be given time to settle down at each new frequency before the value of s_0 is measured. Figure 22.10b shows the results of such an experiment for medium and

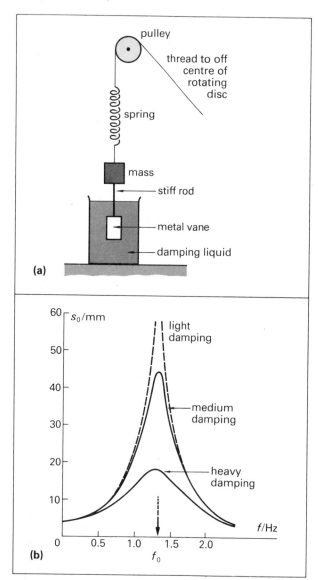

(a)

(b)

Figure 22.10 (a) An arrangement for investigating forced oscillations with different amounts of damping; (b) shows amplitude resonance at $f = f_0$.

heavy damping together with a dashed curve which indicates what would happen with very light damping. The peak is called a *resonance peak* and, for light damping, occurs when the driving frequency is equal to the natural frequency f_0 of the mass-spring system. In the case shown $f_0 \approx 1.3$ Hz, i.e. if the thread was fixed at the pulley and the mass was given a vertical displacement it would undergo damped oscillations of frequency 1.3 Hz.

The shapes of these curves are quite general and illustrate the phenomenon of *resonance*; namely, that

a system with a natural frequency f_0, if made to oscillate at a frequency f will oscillate with a large amplitude when $f \approx f_0$.

Here is the key to our earlier experiment with the swing and the explanation of why the washing machine casing vibrates violently at certain frequencies only. For the washing machine there is more than one natural frequency, as is nearly always the case with any but the simplest system. At resonance there is a *maximum transfer of energy* from the driving mechanism to the driven oscillator.

Barton's pendulums

Instead of studying one driven oscillatory body as the frequency of the driving force is varied we can, with a very simple system of pendulums, study the effect of a single driving frequency on a number of oscillators of different natural frequencies. Figure 22.11a shows such a system. P is a massive driver pendulum which forces the cord ABC to oscillate at a fixed frequency f_d (which depends on the length OP). It is set in motion and given an occasional push if necessary. Attached to AB, and thus forced to oscillate at f_d, are a number of very light pendulums of varying length. After some initial confusion their motions are like those shown in the photograph on the left of figure 22.11b. The air damping is quite heavy on the light (e.g. polystyrene) spheres but with two metal washers attached to each sphere the damping is (relatively) less and the photograph on the right is obtained. The demonstration shows that all the driven pendulums have the same frequency f_d but that their amplitudes vary. The sphere which oscillates with the greatest amplitude is found to be the one which would oscillate at f_d if it were given a push and set swinging freely, i.e. the one whose natural frequency f_0 is equal to f_d. The photographs also show that the driven pendulums are not in phase. By looking along the line of the pendulums we see that they all lag behind the forcing pendulum and at resonance the phase lag is exactly 90°, a quarter of an oscillation. The shortest pendulum is almost *in phase* with the driver P and the longest is almost *in antiphase* with it.

Examples of resonance

Mechanical systems

Musical instruments all depend on the resonant vibrations of strings or air columns or, less often, of rods and skins. More is said about them in sections 23.5 and 24.3. The driving force is generally a random one; stick-slip excitation for a violin etc., edge-eddy excitation for a recorder etc. The instrument itself selects the resonant driving frequency or frequencies because it accepts energy most effectively at these frequencies.

Rotating machinery undergoes resonant (rattling!) vibrations at certain frequencies, almost always more than one. These are generally a nuisance and have to be damped but

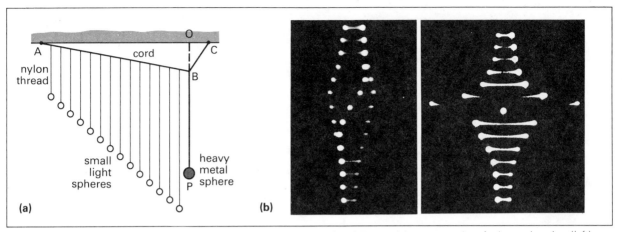

Figure 22.11 Barton's pendulums. (b) shows time exposure photographs of the pendulum seen end on for heavy damping (left), and medium damping (right). Superimposed on the time exposures are instantaneous flash photographs which show the relative phases of the driven pendulums.

better still removed by balancing the rotating parts so as to prevent vibrations at the resonant frequencies.

Large structures such as bridges and buildings (and even human beings) have resonant frequencies at which a small driving force may set up large-amplitude oscillations. The most spectacular example was the Tacoma Narrows Bridge collapse of 1940. The bridge, a large suspension bridge, obtained enough energy from a steady wind of 19 m s^{-1} (eddies providing the driving force) to reach an amplitude of oscillation beyond that which the structure could support. Factory chimneys, power transmission cables and cooling towers can also be set oscillating by winds. The resonant properties of the human stomach (in its cradle of supporting muscles) are at the roots of sickness induced by travel, e.g. seasickness.

Electrical systems

There is a close analogy between the behaviour of a mass-spring oscillator with damping and an inductor-capacitor circuit containing resistance. Electrical oscillations are dealt with fully in section 20.5. A resonant circuit is used to select the frequency we want when tuning a radio.

When electromagnetic waves act to produce oscillations in matter the resonating atoms or molecules absorb energy more effectively at their resonant frequencies than at other frequencies. This is the cause of absorption spectra, particularly at visible and infra-red frequencies and is considered briefly in section 28.5.

Chladni vibrations

The resonances in plates and membranes can be demonstrated using a technique first investigated by E.F.F. Chladni in 1878. The resonant condition can best be described in the language of stationary waves (page 315). The nodes and antinodes can be located by sprinkling dry sand on to the plate or skin and then setting it into oscillation. At resonance the sand is jogged away from the antinodes and settles at the nodes. The two-dimensional patterns so formed give us evidence about the vibrations; this technique is used in designing mechanical structures from car doors to turbine blades. Clamping the plate at different places causes the pattern to change. Figure 22.12 shows two resonant modes of the same plate vibrating in very different ways.

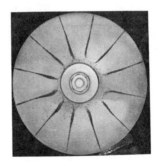
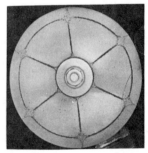

Figure 22.12 Sand patterns on a vibrating steel disc revealing patterns of nodes and antinodes.

23 Describing waves

23.1 What are waves?

Waves are the means by which we communicate; sound and light are both wave motions as is radio communication. *Waves transmit energy from place to place*; that is why we can send messages by providing stimuli for other people's ears and eyes. It is the radiant wave energy from the Sun on which life on Earth depends. Light and radio waves can travel through a vacuum and are two different kinds of *electromagnetic waves*; they are dealt with separately in chapter 28. Table 23.1 lists some different types of *mechanical waves*, i.e. waves which transmit mechanical energy through solids, liquids and gases by vibrations or disturbances in the medium. You should be able to add to the table – particularly to the second and third columns. The important thing to realise is that although there are many types of waves, they all have features in common; what we find in one wave system is likely to be present in some form in all others.

Let us start by considering waves on ropes or ripples on ponds where the properties and behaviour of the waves are a part of our everyday experience. These wave motions both leave the medium through which they pass *undisturbed*,

i.e. after a pulse on a rope or a ripple on a pond has passed, the rope and the water return to their original condition. This is true for all wave motions. Rope waves can move only along the line of the rope, i.e. in one dimension; water waves spread over the surface of a pond, in two dimensions. Sound waves from (say) a loudspeaker spread out in three dimensions through the surrounding air; they are dealt with in the next chapter.

To represent the behaviour of waves we usually draw graphs to show the displacements at points in the medium in which the wave is travelling. Figure 23.1a shows a rope carrying a wave pulse from left to right. (b) is a sequence of

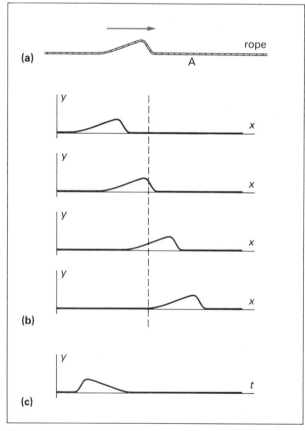

Figure 23.1 (a) A diagram of a wave pulse moving to the right along a rope; (b) shows four wave profiles drawn at regularly spaced times and (c) shows a displacement-time graph for one point A on the rope.

type of wave	cause	use (brackets indicate nuisance)
sound waves in air	vibrating solid bodies, vibrating air columns, the human larynx	(undesirable noise) desired communication, wind instruments
ultrasonic waves in fluids	oscillating quartz crystals, bats	drilling, sonar, echo location
seismic waves in Earth's crust	explosions, earthquakes	geophysical prospecting, (earthquakes), nuclear arms control
waves on ropes, etc.	plucked or bowed strings, winds on cables	stringed instruments, (galloping suspension bridges)
water surface waves	winds at sea, tidal effects, vibrating surface objects	possible energy source, (bores), ripple tanks
waves on sheets and rods	rotating machinery, drumsticks	(car panel rattles), (turbine blade failures), percussion instruments

Table 23.1 Mechanical waves (see also electromagnetic waves, page 366): many more examples could be added to the columns giving causes and uses.

displacement-position or *y-x* graphs drawn at regular time intervals. Each *y-x* graph shows at an instant in time the displacement *y* (sideways here) of each point on the rope; it is called a *wave profile* and looks very like the initial wave pulse, but we must remember that each wave profile is a graph. If we concentrate our attention on one point on the rope, e.g. A in figure 23.1a, then its displacement varies with time as shown in (c); this is called a *y-t* graph.

Transverse and longitudinal waves

Waves on ropes and strings are called *transverse* waves because their displacements are perpendicular to the direction in which the wave energy is travelling. Displacements can, however, be parallel to the wave direction as in sound waves where the vibration of a loudspeaker cone is an in-and-out one: such waves are called *longitudinal* waves. They are still represented graphically by *y-x* and *y-t* graphs. A longitudinal wave pulse on a wide-coiled spring (a slinky) is shown in figure 23.2b together with a *y-x* graph showing the longitudinal displacements *y* from the undisturbed position. In some waves the displacements are neither wholly transverse nor wholly longitudinal; e.g. a seagull floating on the sea is seen to move both up-and-down (transversely) and forwards-and-backwards (longitudinally) as a wave passes it and we feel these motions ourselves when bathing in the sea.

Transverse waves can involve particle displacements in more than one plane; you can vibrate the end of a rope up-and-down or side-to-side. Or you can do both at the same time. When movement of the source is confined to one plane the wave is said to be *plane polarised*.

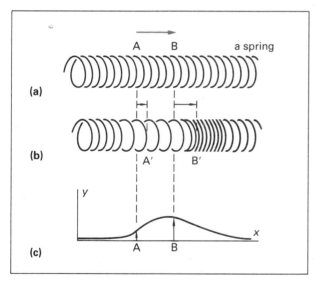

Figure 23.2 A diagram of a longitudinal wave pulse moving to the right along a slinky. (c) A *y-x* graph for the spring carrying the wave pulse in (b). Displacements to the right have been considered positive.

Wave energy

The energy transfer in a mechanical wave occurs continuously from one part of the medium carrying the wave to the next part. The energy is stored in any small region of the medium either as kinetic energy (k.e.) or potential energy (p.e.) or both. Thus when the front of a wave pulse such as that shown in figure 23.1a or 23.2a reaches A, the k.e. of the piece of rope or spring around A is increased. This k.e. is converted to p.e. as work is done in displacing the piece of rope or spring from its equilibrium position. The p.e. is then converted again to k.e. as work is done on the next piece of the rope or spring, accelerating it. These processes continue and the wave energy is propagated along the rope or spring. An energy flow diagram might look something like figure 22.7 with the boxes representing motion along the rope or spring rather than to-and-fro in a single oscillator. The net effect is the transfer of energy by the wave with no net movement of the carrying medium.

When the wave energy is gradually converted to internal energy in the medium or the surroundings, the amplitude of the wave is reduced (or *attenuated*); but the transfer is usually very efficient.

23.2 Wave speeds

The speed of a mechanical wave depends on the *elasticity* and the *inertia* of the medium through which it is travelling. Think of a pair of climbing ropes lying on a smooth slab; if one has three times the mass per unit length of the other then a wave pulse, a flick, sent along them both will travel along the heavier one more slowly. Likewise if two similar ropes are hanging over a cliff, one with a free end and the other supporting a fallen climber, then a wave pulse will travel down the taut one much more quickly than the other.

Figure 23.3a shows how a series of spring-linked trolleys can be used to produce a conveniently slow transverse wave pulse. Masses can be added to the trolleys to slow the pulse down even more and more spring-links can be added to speed it up; so its elastic and inertial properties can be altered independently. Figure 22.3c shows how a similar arrangement can be used for longitudinal waves. Such systems (and a coiled spring) illustrate one important feature of many wave motions, that the speed of the wave does *not* depend on the shape of the pulse, i.e. the way in which the end trolley is moved to create it, nor on the changing size of the pulse as it is attenuated. (Incidentally, if a long coiled spring is used we inevitably alter its mass per unit length when we alter its tension, its elastic property, by stretching it.)

Transverse pulses

Suppose we guess that the speed c_t of a transverse wave pulse along a rope or equivalent system depends only on the tension *T* in the rope and its mass per unit length μ,

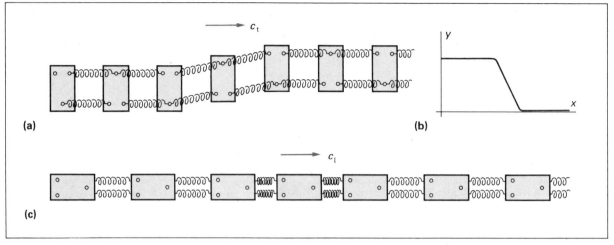

Figure 23.3 Investigating the speeds of mechanical wave with spring-linked trolleys; (b) shows a y-x graph for the pulse shown in (a).

i.e. $c_t = kT^x\mu^y$ (k a dimensionless constant)

T will change slightly as the wave pulse passes as the length of the rope must momentarily increase; let us assume that such changes are very small so that we can talk about *the* tension. Our guess then means that as $[c_t] = LT^{-1}$, $[T] =$ MLT^{-2} and $[\mu] = ML^{-1}$, we can write

$$LT^{-1} = (MLT^{-2})^x (ML^{-1})^y \qquad \text{(page 68)}$$

which gives x = 1/2, y = −1/2, so we have $c_t = k\sqrt{(T/\mu)}$. We can show experimentally that k = 1,

i.e. $c_t = \sqrt{\dfrac{T}{\mu}}$

Note that in figure 23.3a, the length involved is that of the springs only; it does not include the length of the trolleys. A whip tapers towards its end, so its mass per unit length μ alters and a pulse will increase in speed as it passes down the whip from the handle.

▶ A transverse wave is seen to travel along a power cable at about 50 m s⁻¹. If 100 m of the cable has a mass of 60 kg, what is the tension in it?

$$\mu = \frac{60 \text{ kg}}{100 \text{ m}} = 0.60 \text{ kg m}^{-1}$$

$$c_t = \sqrt{\frac{T}{\mu}} \Rightarrow T = \mu c_t^2$$

and so $T = (0.60 \text{ kg m}^{-1})(50 \text{ m s}^{-1})^2$

$$= 1500 \text{ kg m s}^{-2} \text{ or } 1500 \text{ N}$$

As the tension in such a cable varies from point to point the tension we have calculated is only an average value. ◀

Longitudinal pulses

The elastic property on which the speed depends is the size of the restoring forces which result from a longitudinal displacement in the system. In figure 23.3c suppose that each link has a spring constant k; that is, if adjacent trolleys are pulled or pushed a distance s from their equilibrium position, the restoring force is $F = ks$. For trolleys each of mass m_0, and with a distance l between trolley centres, it can be shown that the wave speed c_1 is given by

$$c_1 = \sqrt{\frac{kl}{m_0/l}} = l\sqrt{\frac{k}{m_0}}$$

You should check that the units are consistent.

In general we can express the speed of a mechanical wave as

$$c = \left(\frac{\text{elastic property of medium}}{\text{inertial property of medium}}\right)^{\frac{1}{2}}$$

provided we interpret the elastic and inertial properties correctly. The expressions for c_t and c_1 above are both of this form. Expressions for the speed of sound in air and of longitudinal waves in solids are also like it and so is that for the waves we see in a ripple tank.

Longitudinal waves in solids

Longitudinal mechanical waves travel in solid materials (they are sometimes called sound waves in the solid). We can measure their speed along a solid metal rod using the apparatus shown in figure 23.4. The signal generator is set to $f \approx 20$ kHz and at maximum voltage output. The cathode ray oscilloscope is arranged so that a trace appears when the hammer is placed in contact with the rod but disappears when the contact is broken. The experiment is now simple: we hit the rod with the hammer and count the number n of oscillations which appear momentarily on the screen. The

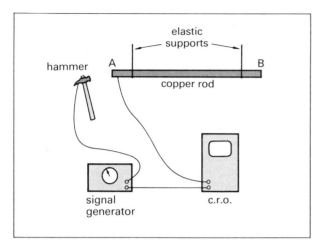

Figure 23.4 An experiment to measure the time of contact when a massive hammer hits a suspended metal rod. The time of contact is equal to the time taken for a longitudinal wave pulse to travel from A to B and back.

time of contact t is thus $nT(=n/f)$; e.g. if 12 full oscillations appear at 20 kHz

$$t = \frac{12}{20 \times 10^3 \text{ Hz}} = 6 \times 10^{-4} \text{ s}$$

When the hammer first makes contact with the rod it starts a *compression* pulse at A and as this wave pulse moves towards B the hammer remains in contact with the rod. The wave pulse is reflected at B as a *rarefaction* pulse (page 313) which when it reaches A breaks the contact between the rod and the hammer. If the length of the rod AB is d, then $c = 2d/t = 2df/n$. It is easy to take several readings of n for one rod and a simple matter to repeat the experiment for rods of different materials.

Suppose we consider the metal rod, e.g. copper, in the above experiment as lots of parallel rows of atoms, each of mass m_0 with interatomic spring constant k and separation l; we could argue that this relationship holds for real solids as well as idealised systems such as figure 23.3c. For the solid rod, if the Young modulus of the material is E and its density ρ, then from the definitions of k and E ($F = kx$ and $F/A = Ex/l$), we have

$$k = \frac{EA}{l}$$

For a single row of atoms the cross-sectional area $A = l^2$, so that

$$k = \frac{El^2}{l} = El$$

A single atom occupies a volume l^3, so that

$$m_0 = \rho l^3$$

These two equations relate the microscopic properties of the atoms, m_0, l and k to the macroscopic properties of the rod, E and ρ.

Substituting, therefore in $c_1 = l\sqrt{\dfrac{k}{m_0}}$

we get

$$c_1 = l\sqrt{\frac{El}{\rho l^3}} = \sqrt{\frac{E}{\rho}}$$

This equation tells us how fast longitudinal waves travel in a rod. Note that the speed is again of the form

$$c = \left(\frac{\text{elastic property}}{\text{inertial property}}\right)^{\frac{1}{2}}$$

For copper $E = 1.30 \times 10^{11}$ N m^{-2} and $\rho = 8.93 \times 10^3$ kg m^{-3}, so that

$$c_1 = \left(\frac{1.30 \times 10^{11} \text{ N m}^{-2}}{8.93 \times 10^3 \text{ kg m}^{-3}}\right)^{\frac{1}{2}}$$

$$= 3.8 \times 10^3 \text{ m s}^{-1}$$

Seismic waves

Seismic, or earthquake, waves result from a fracture or sudden deformation of the Earth's crust. The energy released is transmitted from the source by both longitudinal and transverse waves. The longitudinal waves, called the P (or push) waves, travel faster than the transverse waves, called the S (or shake) waves. The speeds of the P and S waves, near the Earth's surface, are about 7500 m s^{-1} and 4000 m s^{-1} respectively. Both speeds increase with depth for about the first 3000 km but beyond this depth the Earth's core behaves like a liquid in which the P waves can, but the S waves cannot, travel. The variation of speed with depth means that earthquake or seismic waves refract, so that their wave energy moves in curved paths through the body of the Earth. There are of the order of ten earthquakes each year which release more than 10^{18} J of energy each, and innumerable smaller earthquakes. They are unevenly distributed over the Earth, being concentrated along lines where a pair of the plates forming the Earth's outer mantle are sliding over one another. Figure 23.5 shows a simplified seismograph trace resulting from an earthquake.

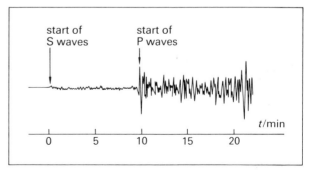

Figure 23.5 A diagrammatic seismograph record. The time interval between the arrival of the P and S waves indicates that the earthquake occurred about 5000 km from the recording station.

Evidence from P and S waves monitored at different stations can tell us about the internal composition of the Earth. Man-made seismic waves in the Earth's surface are now used in prospecting for oil and other useful deposits.

▶ A recording station observes that there is an interval of 200 s between the arrival of the P and S waves from an earthquake. At what distance from the station did the earthquake occur? We shall use the speeds given above and assume that the P and S waves have travelled along the same path to the station.

Suppose the P wave takes a time t to travel from the earthquake to the station; then the S waves will have taken $t + 200$ s. They travel the same distance so that

$$(7500 \text{ m s}^{-1})t = (4000 \text{ m s}^{-1})(t + 200 \text{ s})$$

$$s = (7500 \text{ m s}^{-1})(230 \text{ s})$$
$$= 1.7 \times 10^6 \text{ m or } 1700 \text{ km} \qquad ◀$$

Water waves

Water waves are neither transverse not longitudinal: the motion of a particle of water in the path of a wave travelling on very deep water will be circular. The plane of the circle is vertical and the motion of the particle is forward on the crest and backwards in the trough. As a circular *wavefront* spreads out from its source the wave energy may be dissipated as internal energy in the water; even if the wave energy is conserved the energy arriving a distance r from the source will be proportional to $1/r$ as the length of the wavefront (over which the energy is distributed) is proportional to r.

Types of water wave

(i) Small-wavelength water waves ($\lambda \approx 10$ mm) are called *ripples*. They depend on the surface tension γ of the water surface for their elastic property and are familiar in the laboratory in ripple tanks. Typical speeds in ripple tanks are 0.2 m s^{-1} to 0.3 m s^{-1}.
(ii) Ocean waves or *gravity* waves rely on the Earth's gravitational field g acting on the displaced water for their elastic property. Their speed c in deep water is given by

$$c = \sqrt{\frac{g\lambda}{2\pi}}$$

See also the example on page 314.

In shallow water (e.g. a river bed or a water trough) their speed is equal to $\sqrt{gh}$, and so depends on the depth of the water h. Ripple tank waves of a few cm wavelength can be used to demonstrate the result of changing the wave speed when h changes, i.e. they can be used to study refraction.

23.3 The principle of super-position

We can hear a person talking even though other sound waves are crossing the line between our ear and the speaker; we can see across someone else's line of sight; we can receive a radio programme even though the path from our aerial to the transmitting station is crossed by many other radio and TV carrying waves. Waves do not seem to bump into one another (unless the displacement of one of the waves is exceptionally large) and it is this property, that of waves of passing through one another, that is both surprising and most important. Figure 23.6 shows two sets of photographs of wave pulses on (a) a spring and (b) a string as they 'cross'. As the wave pulses cross they are said to *superpose* and

> the displacement of any point on the string or spring is the sum of the displacements caused by each disturbance at that instant.

This is called the *principle of superposition* and is the key to understanding many wave phenomena. It applies to all types of wave motion, both electrical and mechanical. The photographs of figure 23.6 are equivalent to two sets of

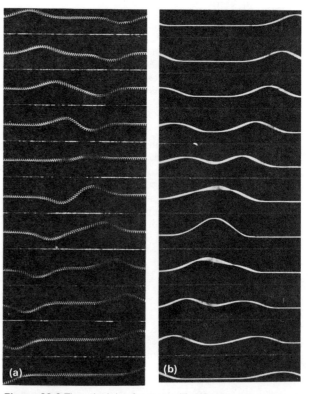

Figure 23.6 The principle of superposition for wave pulses, (a) on a spring, for displacements both positive and negative, and (b) on a string, with all the displacements positive.

y-x graphs taken at successive intervals of time. When drawing graphs to illustrate the principle we shall use idealised pulses of easily recognisable shape. For instance figure 23.7b shows the result of superposing two wave pulses at different moments as they cross. You should try to draw similar sets of y-x graphs for other wave pulses, perhaps for those shown in the photographs above, adding the grey lines yourself.

The behaviour of a single point in the path of more than one wave disturbance can be described by drawing a y-t graph for the point. For instance in the case of the point P in figure 23.7a the displacement-time graph is shown as (c). You should try drawing y-t graphs for points in the path of the waves shown in the photographs.

The energy of the wave pulses in figure 23.6a when the spring has zero displacement at all points seems to have disappeared. But it is all stored as kinetic energy in the spring, part of which is moving up and the other part moving down, as is shown by the blurring of the spring on

	fixed end	free end
transverse waves	the reflected pulse is inverted	the reflected pulse is not inverted
longitudinal waves	a compression pulse is reflected as a compression pulse	a compression pulse is reflected as a rarefaction pulse

Table 23.2 Reflecting waves at fixed and free boundaries.

the photograph where the pulses superpose. Conversely the lack of blurring in (b) where the two wave pulses coincide suggests that the spring is momentarily at rest and that here all the energy is stored as elastic potential energy.

Reflection

Table 23.2 summarises the effects of reflecting simple mechanical wave pulses. They can be demonstrated either with ropes and slinkies or using the trolley and spring models (figure 23.3). To explain the effects involves a careful application of Newton's laws of motion. E.g. in the interaction between the end of a rope and a fixed wall, the wall, because of Newton's third law, generates a reflected pulse by the action of the incident pulse upon it. The problem is like the problem of locomotion (page 17), where in order for a train or person to accelerate forward, the ground must be pushed back by the wheel or the foot. Near the fixed wall the incident and reflected pulses will superpose to give a composite shape before the interaction or reflection is complete. Figure 23.8 shows this at one instant during the reflection of a transverse wave pulse. Of course with a fixed wall the result of the superposition at any instant must produce zero displacement at the wall itself.

Where the end of the rope or spring is neither fixed nor free we have the possibility of some of the wave being reflected and some transmitted into a new medium, for instance into an attached rope or spring. The transmitted wave is now said to be *refracted* if its speed in the new medium is different from the speed of the incident wave.

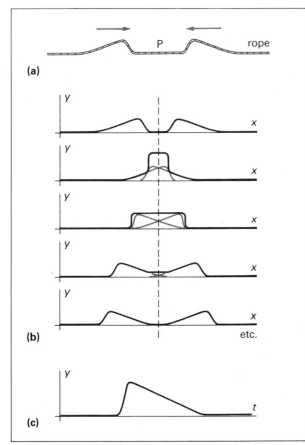

Figure 23.7 (a) is a diagram of two idealised pulses approaching along a rope; (b) shows five wave profiles at different times during the resulting superposition and (c) shows a y-t graph for one point P on the rope.

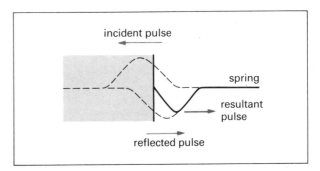

Figure 23.8 Reflection and the principle of superposition.

23.4 Repetitive waves

Many important wave effects involve periodic waves, that is, not single wave pulses but continuous *wavetrains* for which a *y-x* or a *y-t* graph shows regularly repeating features. The simplest sort of repetitive wave form, in terms of which all others may be analysed, is a sinusoidal one. Let us first, however define a number of important terms for any repetitive wave (figure 23.9).

The distance between corresponding points on neighbouring waves is called the *wavelength* of the wave and is denoted by λ. The *period T* of the wave is the period of oscillation of each point on the wave. Its frequency f is the number of oscillations each point undergoes per second.

Clearly $f = \dfrac{1}{T}$

The wave pattern moves forward one complete wavelength λ in the time T which any point takes to complete one oscillation. Thus the wave speed c is given by

$$c = \frac{\lambda}{T}$$

or $c = f\lambda$

which is a general relation for all periodic waves.
▶ The following information refers to deep-water waves produced by storms at sea.
(i) wavelength 16 m, frequency 0.31 Hz, and
(ii) wavelength 2.0 m, frequency 0.88 Hz.
Calculate the wave speeds and comment on the results.
(i) $c_1 = f_1\lambda_1 = (0.31\ \text{Hz})(16\ \text{m}) = 5.0\ \text{m s}^{-1}$
(ii) $c_2 = f_2\lambda_2 = (0.88\ \text{Hz})(2.0\ \text{m}) = 1.8\ \text{m s}^{-1}$
The speeds are quite different. One result will be that the longer wavelength waves (a swell) will reach a shore 100 km from the storm between 8 and 9 hours before the shorter wave-length waves (a choppy sea). ◀

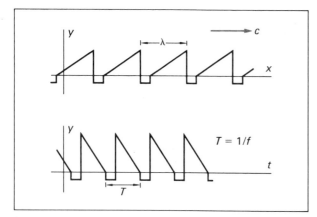

Figure 23.9 Displacement-position and displacement-time graphs for a non-sinusoidal periodic wave.

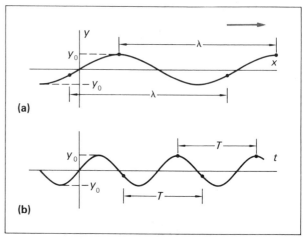

Figure 23.10 Sinusoidal waves; (a) $y = y_0 \sin \dfrac{2\pi}{\lambda}x$ and (b) $y = y_0 \sin \dfrac{2\pi}{T}t$.

Sinusoidal waves: phase

For a sinusoidal wave we can define c, λ, T and f as before (figure 23.10) and we can also define the amplitude y_0 of the wave as the maximum displacement of points in the path of the wave. Points which are separated by a whole number of wavelengths will be oscillating in step with one another, i.e. they will reach their maximum displacements at the same moments; such points in the path of the wave are said to be *in phase*.

As the curve (a) repeats itself in a distance λ and is a sine curve, we can write its equation as

(i) $y = y_0 \sin 2\pi \dfrac{x}{\lambda}$

while the curve (b) repeats itself in a time T and so

(ii) $y = y_0 \sin 2\pi \dfrac{t}{T} = y_0 \sin 2\pi ft$

Two points at x_1 and x_2 in the path of one sinusoidal wave have a *phase difference* given by

$$\text{phase difference} = \frac{2\pi}{\lambda}(x_1 - x_2)$$

If $(x_1 - x_2)$ is equal to a whole number of wavelengths $(= \lambda, 2\lambda, 3\lambda$ etc.) the phase difference will be 2π rad (360°), 4π rad, 6π rad etc., and in these circumstances the two points are *in phase*. If the phase difference is π rad (180°) or 3π rad etc. the two points are said to be wholly out of phase or in *antiphase*, and this occurs when $(x_1 - x_2) = \lambda/2, 3\lambda/2$, etc.

Sinusoidal waves: energy

By holding on to the end of a shaking rope we can feel that the rate at which mechanical wave energy arrives (the rate at which the rope does work on you) depends on both the frequency and the amplitude of the wave. The energy of a simple harmonic oscillator was shown (page 304) to be $\frac{1}{2}mv_0^2 = \frac{1}{2}m(2\pi fy_0)^2$, i.e. the energy is proportional to $f^2y_0^2$. The more rapidly the wave profile moves forward (its speed c), the more rapidly its motion is transferred to any point in its path, so that the rate at which a wave transfers energy is thus proportional to $cf^2y_0^2$.

When two waves arrive at the same point we can talk about the *phase difference* ε between the two waves at that point. Unless the waves have the same frequency the phase difference will vary with time. When the frequencies are the same the result of superposing the waves depends on the phase difference but the resultant oscillation at the point, the resultant y-t graph, is still sinusoidal as is shown by the example below.

For two sinusoidal waves of the same frequency f and the same amplitude y_0, each transporting energy at a rate P, at a given point we have

(i) if $\varepsilon = 0$:   oscillations in phase, resultant oscillation of frequency f and amplitude $2y_0$, energy $4P$.

(ii) if $\varepsilon = 180°$: oscillations in antiphase, no resultant oscillation, energy zero.

► What is the result of superposing two waves of the same frequency but different amplitude at a place where the waves arrive with a steady phase difference of 60°?

The answer is shown in figure 23.11, using y-t graphs. The dots on the graph show where it was easiest to add the two displacements caused by the incident (grey) waves. The resulting displacement is seen to be sinusoidal and of the same frequency as the superposing waves. ◄

23.5 Stationary waves

Sinusoidal waves on strings and springs can be produced in the laboratory using the sinusoidal output from a signal generator to drive a vibrator – a sort of robust moving-coil

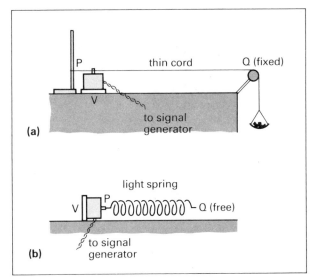

Figure 23.12 Demonstrating stationary waves; (a) transverse and (b) longitudinal. The frequency of the vibrator V is adjusted to find the natural resonant frequency of the cord or the spring.

loudspeaker (page 320) without the conical diaphragm. Figure 23.12 shows (a) how a transverse wave can be produced on a thin rubber cord, and (b) how a longitudinal wave can be produced on a light spring; V is the vibrator in both diagrams.

Transverse waves

The tension in the cord can be adjusted by placing small masses in the hanging tray. For a fixed tension, the frequency f of the vibrator V is slowly raised from its lowest value. The transverse waves from the source V travel along the cord and are reflected at the fixed ends P and Q. We thus have a system in which waves travelling along the cord in one direction are superposed on those reflected at the ends. Looking at the cord during the experiment there are some values of f which produce a fixed pattern of superposition in which some places show zero resultant oscillation (where the waves are in antiphase all the time) and some places which show large amplitudes of oscillation (where the waves are always in phase). There are waves travelling in both directions along the cord but no net transfer of energy between adjacent parts of it occurs. The energy from the vibrator is being fed into the oscillating rope at a steady rate. Once a stationary wave has been established, this mechanical energy is converted to internal energy in the surroundings. The observed pattern is called a *stationary wave pattern* or, more simply, stationary waves.

Nodes and antinodes

Figure 23.13 is a series of displacement-position graphs showing the resultant wave profile at successive intervals of time for a stationary wave. The graphs are drawn every

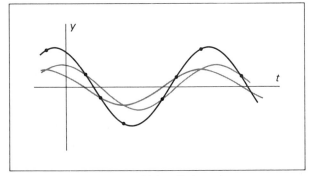

Figure 23.11 Adding sinusoidal oscillations; the two grey waves are most easily added at the points indicated by the dots. See also the phasor treatment on page 279.

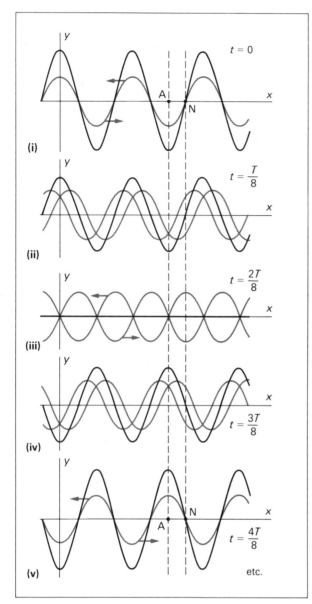

Figure 23.13 The formation of stationary waves; *y-t* graphs for N and A are shown in the next diagram, figure 23.14.

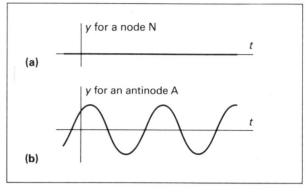

Figure 23.14 The *y-t* graphs for a node and an antinode. The frequency of vibration of A is that of the superposing waves.

which a stationary wave pattern fits onto PQ. Energy is transferred efficiently from V to the cord and large amplitude oscillations are built up when the driving frequency f_d has these values. In this sense the experiments of figure 23.12 are *resonance* experiments, and the frequencies at which the stationary wave patterns appear are the *natural frequencies* of the cord.

Using a flashing stroboscope to illuminate the stationary wave pattern on the rope we can see that all bits of the rope between adjacent nodes move up together and then down together, i.e. they are in phase. They are however in antiphase with the next half-wavelength loop so the rope moves with a sort of flip-flap motion: it is vibrating. A slinky or a row of linked trolleys such as those of figure 23.3a can produce a readily visible stationary wave pattern by fixing the ends and oscillating a piece of the slinky or one of the trolleys by hand.

Longitudinal waves

Figure 23.12b shows an experimental arrangement for demonstrating stationary waves on a spring. The frequency of V is again raised slowly from its lowest value. The stationary wave patterns or resonances are obvious to the eye as a blurring of the spring at antinodes for certain values of f_d. The distance between adjacent nodes or antinodes is again $\lambda/2$ and the *y-x* graphs of figure 23.13 describe what is happening. In this case Q, the free end of the spring, will always be an antinode. A stroboscopic view shows up the relative phase differences on the spring and the experiment can be reproduced on a more massive scale with trolleys linked end-to-end as in figure 23.3c. It is worth noting that the spring pulses in and out on both sides of a node, providing the maximum variation of compression and rarefaction there. *A displacement node is an antinode of compression variation.*

Longitudinal stationary waves can be set up in a solid rod, for instance, by hitting one end of it with a hammer (page 311). The frequency *f* at which the rod 'rings' can be measured with a microphone and frequency meter. As each end of the rod is a displacement antinode, the rod must be

eighth of a cycle, i.e. at time intervals of $T/8$. Points like N have zero displacement at all times; they are called nodal points or *nodes*. Neighbouring nodes are seen to be *half a wavelength $\lambda/2$ apart*. Similarly there are points like A with the largest amplitude of oscillation, antinodal points or *antinodes*, which are also $\lambda/2$ apart. Figure 23.14 shows separate *y-t* graphs for N and A.

The cord in the demonstration of stationary waves has two fixed ends which must both be nodes. There are therefore only certain wavelengths (and frequencies *f*) for

$\lambda/2$ long and hence the speed of the waves in the rod can be found.

Vibrating strings

The length PQ of the cord in figure 23.12a must be a whole number n of half-wavelengths if we are to get the sort of stationary wave pattern explained in the previous section.

Thus we have $PQ = l = n\dfrac{\lambda}{2}$

and as $c = f\lambda$, there is a definite value of f associated with a given wavelength

$$f = \frac{c}{\lambda} = \frac{c}{2l/n} = \frac{nc}{2l}$$

For a stationary wave pattern we must have $f = c/2l$, $2c/2l$, $3c/2l$ etc. only, as shown by the experiment.

The speed of a transverse wave on a cord or string is given by (page 310) $c = (T/\mu)^{\frac{1}{2}}$
so that under given conditions of tension T and mass per unit length μ, the frequencies which produce a stationary wave pattern are

$$f = \frac{n}{2l}\sqrt{\frac{T}{\mu}} \qquad (1)$$

Each value of n (1, 2, 3, etc.,) corresponds to a certain *mode* of vibration. The *fundamental* frequency (or first harmonic) occurs when the cord is half a wavelength long. The other resonant frequencies, called overtones or *harmonics*, are illustrated in figure 23.15.

The sharpness of any resonance is controlled by the amount of damping in the system (page 305) and although the damping is here not high there will be a noticeable oscillation at the antinodes for values of f_d on both sides of the values deduced above. The amplitude at the antinodes builds up as a resonant frequency is approached and dies down gradually after it is passed.

Test of $c_t = \sqrt{T/\mu}$

The vibrator and cord shown in figure 23.12 can be used to test this relationship for transverse waves, e.g. by plotting c_t against $\sqrt{T}$ (for a constant μ) and seeing if we get a linear graph passing through the origin. We measure c_t indirectly by measuring f and λ; $f(= f_d)$ is read from the dial of the signal generator; if the calibrations are unreliable they can be checked by comparing the output of the signal generator with a set of standard tuning forks. λ is measured from the stationary wave pattern by measuring the distance from one node to another about one metre away, and counting the loops between these points (then λ is found with an uncertainty of less than $\pm 1\%$). We deduce T from the mass of the hanging tray and its contents. For high tensions the cord might be replaced by a metal wire and

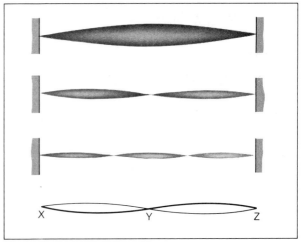

Figure 23.15 The vibration of a string in its first three resonant modes. The bottom diagram shows two instantaneous positions of the string for the middle mode. The string moves up along XY as it moves down along YZ with a 'flip-flap' motion.

the hanging tray by a calibrated newton balance directly connected to the wire. Such a wire, mounted above a sounding box, forms a simple one stringed musical instrument called a *sonometer* or monochord. It is similarly possible to test the formula for the speed of longitudinal waves on a spring using stationary wave patterns.

▶ A steel wire of length 0.80 m is stretched between two fixed points, and its fundamental frequency is found to be 150 Hz. Find the speed of the waves on the wire. If the ultimate tensile stress of steel is 1.6×10^9 N m^{-2}, what is the highest fundamental frequency which can be produced on this wire by increasing the tension? Density of steel $= 8.0 \times 10^3$ kg m^{-3}.

The wavelength will be twice the length of the wire, i.e. $\lambda = 1.60$ m, so that from $c = f\lambda$, we get the wave speed

$$c_t = (150 \text{ Hz})(1.60 \text{ m}) = 240 \text{ m s}^{-1}$$

As $\quad c_t = \sqrt{\dfrac{T}{\mu}} = \sqrt{\dfrac{T \times A}{\mu \times A}}$

$$= \sqrt{\frac{T}{A} \div \frac{\mu}{A}} = \left(\frac{\text{tensile stress}}{\text{density}}\right)^{\frac{1}{2}}$$

where A is the cross-sectional area of the wire; then the highest possible wave speed is given by

$$c_{max} = \left(\frac{\text{ultimate tensile stress}}{\text{density}}\right)^{\frac{1}{2}}$$

$$= \left(\frac{1.6 \times 10^9 \text{ N m}^{-2}}{8.0 \times 10^3 \text{ kg m}^{-3}}\right)^{\frac{1}{2}} = 4.5 \times 10^2 \text{ m s}^{-1}$$

For the fundamental frequency the wavelength must still be 1.60 m, and so the highest possible fundamental frequency with the new tension is

$$f_{max} = \frac{c_{max}}{\lambda}$$

$$= \frac{4.5 \times 10^2 \text{ m s}^{-1}}{1.60 \text{ m}} = 280 \text{ Hz}$$

Stringed instruments

The violin, 'cello, guitar and piano are all musical instruments which depend on the resonant vibration of stretched strings. In the piano a complete set of strings of fixed length and tension are available. But in the others the player makes do with a few strings by altering their effective length with his fingers: in a guitar the lengths to be used are predetermined by the frets; but for violins, etc., the length is continuously variable. The strings can be bowed or plucked. After the initial impulse the string quickly settles to vibrating only at the frequencies of its harmonics, given by equation (1) above. The relative amplitudes of these harmonics gives the instrument its *quality*, or tone, which enables us to distinguish the same notes played on the different instruments. Musically a doubling of frequency is said to be an increase in pitch of one octave; thus the musical progression from C to the next C above it is physically equivalent to a frequency ratio $f_0 : 2f_0$.

24 Sound

24.1 The nature of sound

Sound is a longitudinal mechanical wave motion. It covers the audible part of the spectrum of such waves, roughly the frequency range 20 Hz to 20 kHz. High frequency sounds are said to have a high *pitch*. The energy is normally transmitted through air, but longitudinal waves of this frequency in liquids and solids are also called sound; indeed it is through the bones and liquids of the middle ear that the sound is transmitted to the ear-brain mechanism. Sound can be produced by any vibrating body which transfers its energy of oscillation to the air around it, e.g. a loudspeaker, a guitar or the larynx. The wave energy propagates as a series of *compressions and rarefactions*, i.e. as a series of small pressure variations of amplitude p_0 above and below atmospheric pressure P. The size of p_0 at the ear when listening to normal conversation may be $\approx 10^{-2}$ Pa, a very small fraction of P ($\approx 10^5$ Pa). The displacements of air molecules in a sound wave are extremely small, but the ear is able to detect waves in which the movements are much less than the diameter of an atom ($\approx 10^{-10}$ m).

To get a mental picture of how a sound wave propagates consider figure 24.1. At standard atmospheric temperature and pressure a typical molecule in air makes about 10^{10} collisions per second with other molecules. When the air is pushed to the right by the loudspeaker it is compressed a little. The molecules now make more collisions per second in this denser air and there is a net movement of the molecules away from the region of compression. As the speeds of the molecules are of the order of 500 m s^{-1} (page 128) we might expect sound to travel at this speed in air. In practice the speed of propagation in air is found to be about 340 m s^{-1} which is of the same order of magnitude. For a

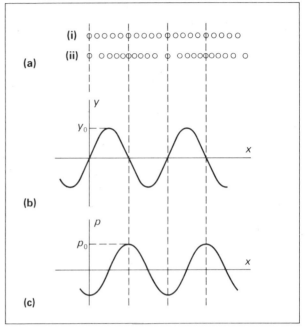

Figure 24.2 In (a) (i) the molecules are shown undisplaced; (b) and (c) are graphs which describe the displaced molecules in (a) (ii).

periodic wave there is no net movement of the molecules away from the source, because when the loudspeaker moves to the left they move back into the region of rarefaction and so on. A sound wave in a gas can thus be thought of as a series of periodic variations in the pressure (and density) of the gas superimposed on the random motion of the gas molecules.

The techniques described in section 23.1 for longitudinal mechanical waves can all be used for representing sound. In drawing graphs we sometimes plot the particle displacement y and sometimes the pressure variation p. For a sinusoidal wave figure 24.2 illustrates that y and p are 90° ($\pi/2$ rad) out of phase. Thus at places where the displacement is a maximum the *acoustic pressure*, that is the excess pressure above or below atmospheric is zero, and so on.

Diffraction

A person talking, or a record player, can often be heard when the source of sound cannot be seen, even when no reflection of the sound waves is possible. The wave energy

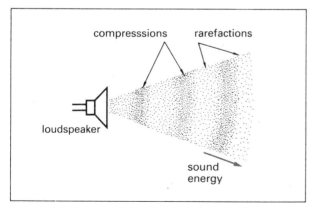

Figure 24.1 A representation of a sound wave in air.

spills round objects and through gaps by a process known as *diffraction*. This is dealt with for all waves including sound in section 25.4. To establish in a qualitative way whether or not much energy is going to spread sideways on passing through a gap, e.g. a door of width d, we need to compare d with the wavelength λ of the wave. If $\lambda \approx d$ then the sound spreads out very effectively beyond the door while if $\lambda \ll d$ the sound spreads out very little; figure 25.14. shows the effect for waves in a ripple tank.

▶ A loudspeaker system in a large hall consists of a number of rectangular speakers, each measuring 1.0 m by 0.1 m and set about 3 m above the ground. Explain why they are this shape and how they should be fixed, i.e. should the long side be vertical or horizontal? The speed of sound in air is about 340 m s^{-1}.

The sound from the rectangular speaker acts like sound coming through a door, the amount of spreading depending on the dimensions of the speaker and the wavelength of the sound. For a sound wave of frequency 1500 Hz, $c = f\lambda$ gives a wavelength of about 0.20 m. Figure 24.3 shows how the wave would spread from the speaker. In a large hall we would not want to lose too much sound (energy) upwards nor would we wish to send from the loudspeaker only to those listeners sitting directly in front of it. Therefore the diagram shows the speaker correctly mounted with its long side vertical. The width of the speaker is now less than the wavelength of the sound, so the sound spreads out horizontally, but the height of the speaker is greater than the wavelength of the sound, so not much sound spreads upwards or downwards. We chose 1500 Hz, which is a fairly high frequency; at (say) 300 Hz there will be more vertical and horizontal spreading of the sound. ◀

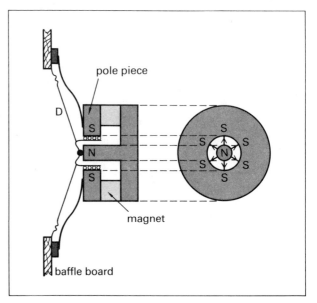

Figure 24.4 A moving-coil loudspeaker. The diaphgram D oscillates in and out when there is a varying current in the coils between the magnet's poles.

The moving-coil loudspeaker

Figure 24.4 describes the design of a simple moving-coil speaker. The moving part consists of a stiff paper diaphragm D supported at the circumference by a flexible ring of cloth. At the centre it is attached to a coil which lies in the gap between the pole pieces of a permanent magnet. The projected part of the diagram shows the shape of the magnetic field. An alternating current from a microphone is amplified and the current passed through the coil. The electromagnetic forces on the coil (try the left-hand motor rule, page 205) move the coil forwards and backwards in the gap, carrying the diaphragm with it like a piston. The diaphragm forces the air to oscillate and sound waves are thus generated.

Two sound waves are in fact produced – at the front and back of the cone; and these are bound to be in antiphase, a compression at the front coinciding with a rarefaction at the back. The cabinet or baffleboard is designed to reduce the possibility of these two waves reaching the listener at the same time where after superposition at the ear they would give little or no received signal. The cabinet must also be designed so as not to introduce unwanted resonances and to damp any resonance effects in the diaphragm itself.

The loudspeaker's useful function is to convert electrical energy to sound energy. The reverse process can be achieved with a moving-coil microphone (page 296) which has essentially the same construction as figure 24.4.

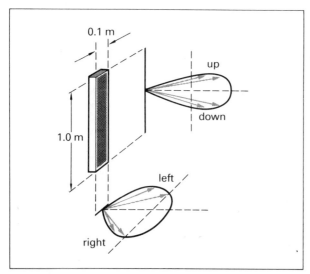

Figure 24.3 A method of representing the sound output from a long thin loudspeaker system. Diffraction enables the sound to spread widely in the horizontal plane but limits the loss of sound upwards and downwards.

Beats

Sound waves obey the *principle of superposition* and an experiment with two loudspeakers driven by the same signal generator is described on page 352. Suppose we connect two different signal generators to the two loud-speakers and set their scales at the same frequency. It is unlikely that their calibrations are exactly the same: suppose that they differ by 0.1 % at 2000 Hz, i.e. $f_1 - f_2 = 0.1\%$ of 2000 Hz = 2 Hz. Assuming that the speakers are equally loud we hear a sound of frequency about 2000 Hz which rises and falls (pulses) twice every second, i.e. at 2 Hz. This is called a *beat* phenomenon. It is easy to adjust one of the signal generators so as to 'speed up' the beats or to slow them down to zero. In the latter case $f_1 = f_2$ to within very precise limits as the ear can detect a slow beat which rises and falls only once every ten seconds.

Figure 24.5 shows oscilloscope traces (the time-base is the same in all three cases) for a sinusoidal wave at 700 Hz, at 600 Hz and then for the two waves superposed at the microphone. The beat phenomenon is clear in the bottom trace and, in this case, has a pulsing or beat frequency of (700 − 600) Hz = 100 Hz, that is the bottom trace bulges 100 times per second. In general

$$\text{beat frequency} = f_1 - f_2$$

where f_1 is the larger of the two frequencies. Any two sinusoidal sound sources can produce beats at a microphone or at the ear, though it is only when the intensities are similar that the effect is noticeable. With small values of $f_1 - f_2$, e.g. up to 10 Hz, we can use beats to measure f_2 when f_1 is known, by listening and counting. A piano tuner makes use of this technique.

Coupled oscillations

A beat phenomenon can occur when two oscillating systems are loosely connected. A ticker-timer sometimes beats when the natural frequency of vibration of the metal tongue is nearly 50 Hz, the frequency of the alternating magnetic field in the solenoid. We can demonstrate beats between two simple pendulums of roughly equal length which are hung from a string stretched between two points on the same level. The string loosely connects one with the other

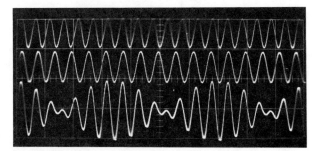

Figure 24.5 The beat phenomenon. The lower trace shows the result of superposing the upper two traces.

and energy can be slowly transferred between them. If one is set swinging it gradually slows while the second pendulum builds up to a maximum amplitude and then the process is reversed. A displacement-time graph for either pendulum would look something like the bottom trace in figure 24.5.

► Measure the beat frequency in figure 24.5. Does it support the result 'beat frequency $= f_1 - f_2$'? The two sinusoidal traces are of frequency 700 Hz and 600 Hz.

The upper trace must be the 700 Hz trace as it undergoes more oscillations on the figure. The timebase on the c.r.o. is thus such that there are 14 of these oscillations in 7.8 divisions of the screen

i.e. $7.8 \text{ div} \equiv \dfrac{14}{700}\,\text{s} = 0.020\,\text{s}$

A complete cycle of the beat trace occupies 4.0 div and thus takes a time

$$\left(\frac{4.0 \text{ div}}{7.8 \text{ div}}\right)(0.020\,\text{s}) = 0.0102\,\text{s}$$

so that the beat frequency

$$f = \frac{1}{0.0102\,\text{s}} = 98\,\text{Hz}$$

which is as close to (700 − 600) Hz = 100 Hz as we could expect having had to take measurements from the photograph. ◄

24.2 The speed of sound

The speed of sound c in air at 288 K (15°C) is 340 m s^{-1}. An estimate of this speed can be made by a simple clapping experiment using reflected sound (*echo*), but for a precise method we should measure f and λ in the laboratory and then calculate c. Figure 24.6 shows one possible arrangement. The signal generator produces a sinusoidal input potential difference at a small speaker L. This p.d. is also fed to the X-plates of a cathode ray oscilloscope the time-base of which is switched off. A microphone M receives the sound wave and after amplification its output potential difference is fed to the Y-plates of the c.r.o. When the two alternating p.d.s to the X- and Y-plates are in phase a slanting line appears on the c.r.o. screen from top right to bottom left ($LM = n\lambda$). If L is now moved away from M the p.d.s are no longer in phase, and an ellipse grows from the line on the c.r.o. screen. This becomes a line from top left to bottom right when the p.d.s are in antiphase, grows to an ellipse again but returns to the original slanting line after the loudspeaker has moved through one wavelength when the alternating p.d.s fed to the X- and Y-plates are again in phase ($LM = n\lambda + \lambda$).

For $\lambda \approx 0.5$ m a movement of M of ± 2 mm is enough to turn the line into a thin slanting ellipse. For $\lambda \approx 0.1$ m it is necessary to move L through several wavelengths to achieve similar precision. The frequency f must be read

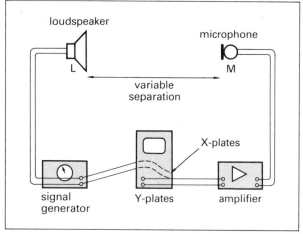

Figure 24.6 An experimental set up to measure λ and f for a sound wave and hence to deduce the speed of sound in free air.

from the signal generator and any error here affects the value of c, which is equal to the product $f\lambda$. The signal generator calibrations can be checked using a set of standard tuning forks. With this experimental arrangement we can quickly measure c for a variety of wavelengths. In the audible range c is found to be independent of λ (and thus also of f). A sensitive test of this result is that a church bell heard a long way off across a valley sounds the same as it does close to the church. With this apparatus we can also test qualitatively the variation of c with temperature: the air between M and L can be warmed with a bunsen and L moved to restore the trace on the c.r.o. screen.

The experiment measures the speed of sound in free air. Its speed in a pipe is a little less and can be measured by stationary wave techniques (page 324).

Factors affecting c

The speed of a mechanical wave through a medium (sound cannot, of course, be transmitted through a vacuum) depends upon the inertia and the elasticity of the medium. For sound waves in a gas the elastic property, the springiness of the gas, is represented by its pressure p and the inertia of the gas is represented by its density ρ. It can be shown that

$$c = \sqrt{\frac{\gamma p}{\rho}} \qquad (1)$$

where γ is a number which varies from gas to gas (page 135). For air in the usual range of temperatures and pressures γ is constant at 1.40. The expected value for the speed of sound in air at s.t.p. is therefore given by

$$c^2 = \frac{(1.40)(1.01 \times 10^5 \text{ N m}^{-2})}{(1.29 \text{ kg m}^{-3})}$$

which gives $c = 331 \text{ m s}^{-1}$

Adding water vapour to the air increases the speed slightly because ρ then decreases but the effect is very small.

For an ideal gas

$$\frac{p}{\rho} = \frac{RT}{M}$$

so that, from (1) $c = \sqrt{\dfrac{\gamma RT}{M}}$ (2)

This equation tells us that the speed of sound
(i) does not depend on the pressure of the gas. Of course, if p changes in $c = \sqrt{\gamma p/\rho}$, then so does ρ.
(ii) depends on the temperature; $c \propto \sqrt{T}$.
(iii) depends on the nature of the gas; $c \propto \sqrt{\gamma/M}$.

Atmospheric refraction

At dusk on a clear evening the ground radiates energy to space and cools rapidly, leaving colder air near the ground. As $c \propto \sqrt{T}$ sound waves are *refracted* as in figure 24.7, so an observer at O hears a sound from S more easily than during the day. You should try and construct the equivalent day-time diagram when the air is warmer near the ground.

The Doppler effect

When a low-flying aeroplane passes close to an observer on the ground the pitch of the sound heard changes from a higher to a lower pitch as the plane passes overhead. This is called the *Doppler effect* in sound and an analogous effect can be produced with all types of waves. Consider sound waves of frequency f_s. Clearly f_s is independent of the motion of the source or of the listener (whom we shall call the observer). Suppose that the frequency of waves received by the observer is f_o. There are two dissimilar cases.

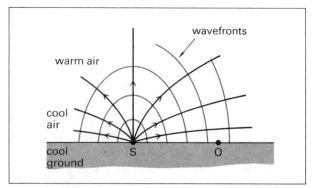

Figure 24.7 Wavefronts distorted by variations in temperature as sound waves travel faster in warm air than in cold air. The sound energy travels perpendicular to the wavefronts and the observer at O receives more wave energy than he would if all the rays from S were straight lines.

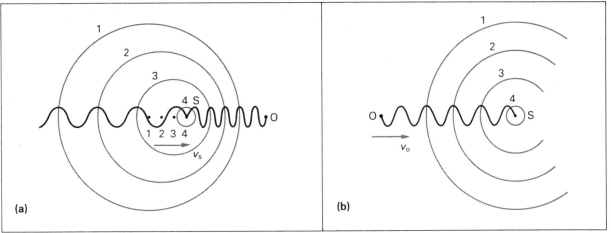

Figure 24.8 The Doppler effect for (a) a moving source and (b) a moving observer. The wavefronts are numbered 1, 2, 3, 4 in the order in which they originate at the source S.

(a) Moving source

Suppose an observer O is at rest in the medium in which the waves are travelling and that the source S is moving directly towards him at a speed v_s. The speed of sound c is constant in still air and does not depend on the motion of S. Figure 24.8a illustrates that O receives wavefronts which appear to him to be closer together than $\lambda_s = c/f_s$. This is equivalent to an *increase* of frequency. During one cycle of the source (a time interval of $1/f_s$) S moves a distance v_s/f_s towards O. Thus the apparent wavelength observed by O is λ_o, where

$$\lambda_o = \lambda_s - \frac{v_s}{f_s}$$

But the observed frequency is f_o, giving $\lambda_o = c/f_o$,

therefore $\dfrac{c}{f_o} = \dfrac{c}{f_s} - \dfrac{v_s}{f_s}$,

which gives $f_o = f_s\left(\dfrac{1}{1 - v_s/c}\right)$ (3)

Similar reasoning for a source moving away from the observer reverses the sign in the bracketed term. With the source moving away from the observer the term $(1 + v_s/c)$ increases as v_s increases and the observed frequency falls. With S moving towards O, however, the term $(1 - v_s/c)$ approaches zero when $v_s \to c$ and the observed frequency rises towards infinity. If $v_s' \gg c$ the analysis for equation (3) has no meaning for S moving towards Q; in this case (see below) shock waves are produced.

(b) Moving observer

In figure 24.8b we see that as the observer moves towards the source at a speed v_s he is receiving or crossing wavefronts more often than if he were stationary. The observed frequency f_o is greater than the source frequency f_s. In one second moving at v_o he will meet an extra v_o/λ_s wavefronts.

Therefore $f_o = f_s + v_o/\lambda_s$

As $c = f_s\lambda_s$, this gives

$$f_o = f_s\left(1 + \frac{v_o}{c}\right) \qquad (4)$$

Similar reasoning for an observer moving away from the source reverses the sign in the bracketed term. At $v_o = c$ no wave motion is observed; e.g., a successful surf rider does not experience the up and down motion of the water wave.

Of course we could have both source and observer moving relative to the medium carrying the waves or the medium itself may be moving, e.g. a wind or a flowing river which will affect the speed of the waves relative to the observer. The principles to be applied in predicting the new frequency are always the same as those illustrated above.

► What does a stationary observer hear when a horn of frequency 400 Hz attached to a sports car moving down the straight of a race track at a speed of 50 m s⁻¹ passes close to him? Take the speed of sound in air to be 340 m s⁻¹.

The frequencies heard will be given by equation (3) above:

$$f = 400 \text{ Hz} \times \cfrac{1}{1 \pm \left(\cfrac{50 \text{ m s}^{-1}}{340 \text{ m s}^{-1}}\right)}$$

$$= 470 \text{ Hz or } 350 \text{ Hz}$$

for the approaching and receding car respectively. Note that the change of frequency is +70 Hz for approach but only −50 Hz for recession. There is not the symmetry in the situation which we might at first have expected. ◄

Shock waves

Suppose we redrew figure 24.8a for the case where $v_s = c$. All the wavefronts would be coincident in the forward direction. In energy terms this means that the wave energy propagating in the forward direction would be concentrated and in the case of an aeroplane or a boat the resulting accumulation of air or water would greatly impede the forward motion. In the case of the aeroplane we are describing the situation where the plane is passing through the 'sound barrier'.

For $v_s > c$ the circular wavefronts of figure 24.8a lie behind the source and overlap each other so that they form a cone with S at its apex. It is easy to show that the semi-angle a of the cone is given by $\sin a = c/v_s$. This expression can be used to calculate v_s when c is known; the cone formed by the waves is readily visible behind a motor boat and can be made visible for an aeroplane model in a wind tunnel. The energy is concentrated along the edge of the cone and forms *bow waves* for water waves or a *shock wave* for sound. In aircraft engineering the ratio v_s/c is called the Mach number after E. Mach (1838–1916). A speed of Mach 2.0 does not necessarily mean a speed of 2×340 m s^{-1}, for the aeroplane may be travelling at a high altitude where c is less than 340 m s^{-1}; e.g. it might be 280 m s^{-1} at 10 km giving Mach 2 as only 560 m s^{-1}.

24.3 Vibrating air columns

Sound waves travelling in air from a source at one end of a pipe or tube will be reflected at the other end. A *stationary wave pattern* is thus produced by the resulting superposition of the two waves. Much of this section depends upon the general discussion of stationary waves (page 315) and is similar to the work on vibrating strings (page 317).

Figure 24.9 shows an experimental arrangement for studying the resonant oscillations of a (damp) air column. By adjusting the rate of delivery of water from the tap the water level can be made to rise or fall, quickly or slowly, or to remain in one place. Two types of experiment are possible.

(i) We can keep the frequency f of the signal generator fixed (at 1000 Hz for example) and alter the length L of the air column. It is best to move the water level quite rapidly from top to bottom of the tube. At certain levels a sharp increase in loudness of the sound is heard. For these values of L the sound wave reflected from the water surface is setting up a stationary wave pattern by superposition with the wave travelling down the tube. If the precise positions of the water level for two *adjacent* resonances are located then, as they are separated by $\lambda/2$, we can use $c = f\lambda$ to find the speed of sound in the gas in the tube. As it is possible to change the gas in the tube the relationship $c = \sqrt{\gamma RT/M}$ for the speed of sound waves in gases other than air can be tested.

(ii) We can fix the water level and vary f the frequency of the signal generator. Again there is a sharp increase in loudness at certain frequencies. This experiment is best seen as a *resonance* experiment; the signal generator acts as the driver making the air in the tube oscillate at its forcing

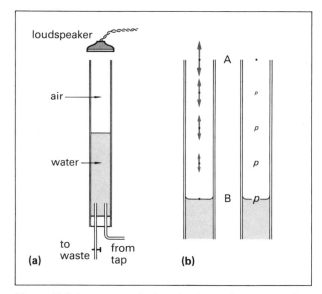

Figure 24.9 An apparatus for investigating stationary waves in an air column. (b) Shows ways of illustrating the displacement (left) and pressure variation (right) at points in the tube when it is resonating in its fundamental mode.

frequency f. As f passes through the lowest natural frequency f_0 of the air in the tube a graph of amplitude of the sound heard against f would look like that of figure 22.10b (page 306). Figure 24.9b shows the behaviour of the air in the tube when $f = f_0$. The air is said to be vibrating in its *fundamental mode*. B must be a place at which the longitudinal displacement y of the air is always zero; it is called a displacement node. The pressure variations (represented by p on the diagram) are then greatest at B; it is a pressure antinode. At A the pressure variations are very small since the pressure at the end of an open tube must be atmospheric pressure P. A is therefore close to a pressure node and to a displacement antinode, as y and p are 90° out of phase. Figure 23.14 (page 316) shows y-t graphs for a node and an antinode on a vibrating string; the same graphs hold for displacements in sound waves. As f is increased other resonant modes can be found. They are called overtones or *harmonics*. In each case there must be a displacement node at the water surface and a displacement antinode close to the top of the open tube.

To demonstrate such a pattern of nodes and antinodes directly we can use a dry horizontal glass tube with some very fine pollen or tiny expanded polystyrene spheres sprinkled along its length. A loudspeaker at one end can be made to set up resonant vibrations of the air in the tube, i.e. a stationary wave pattern, at certain frequencies. The powder is violently agitated at (displacement) antinodes but is quite still at the nodes so it tends to move away from the antinodes and settle at the nodes. It is a simple matter to measure the distance between antinodes; this is equal to $\lambda/2$.

Closed tubes

Consider a pipe or tube of length L which is *closed at one end*. The lowest frequency at which it will resonate is called its fundamental frequency f_0 and if the wavelength of this sound is λ then $L = \lambda/4$ as for this case there is a node at one end and an antinode at the other; see figure 24.10a. Thus if $L = 0.50$ m, $\lambda = 2.0$ m and $f_0 = 340$ m s^{-1}/2.0 m = 170 Hz. The next frequency at which it will resonate is $3f_0 = 510$ Hz as this has a wavelength one third that of (a) and thus 'fits' the tube as in (b). It is called the third harmonic; f_0 is the first harmonic but there is no resonance for $2f_0$, the second harmonic. You should draw further diagrams for the fifth harmonic $5f_0$, etc. One wind instrument which consists of just such a closed tube is the clarinet.

Open tubes

Consider a pipe or tube of length L which is *open at both ends*. The fundamental frequency (the first harmonic) is now different from f_0 as the tube has a displacement antinode at each end (c) giving $L = \lambda/2$ and $f_0' = c/2L$. Using the same values for L as above, 0.50 m, this gives $f_0' = 340$ m s^{-1}/1.0 m = 340 Hz, so that $f_0' = 2f_0$. The second harmonic (d) is $2f_0' = 680$ Hz. Table 24.1 summarises some of this information. One wind instrument which consists of an open tube is the flute.

▶ A glass tube is mounted vertically as in figure 24.9. A tuning fork known to be correctly marked 384 Hz is held above the tube and when the water level is 203 mm below the top of the tube a sharp increase in loudness is heard. The sound is loud again when the water is lowered to a level 635 mm below the top of the tube. Find the speed of sound in the (damp) air in this tube.

closed (at one end)	f_0		$3f_0$		$5f_0$	
	$L = \dfrac{\lambda}{4}$		$L = \dfrac{3\lambda}{4}$		$L = \dfrac{5\lambda}{4}$	
open (at both ends)	f_0'		$2f_0'$		$3f_0'$	
	$L = \dfrac{\lambda}{2}$		$L = \lambda$		$L = \dfrac{3\lambda}{2}$	

Table 24.1 Frequencies and wavelengths (all the λ are different) for overtones in pipes or tubes of length L. Note that $f_0' = 2f_0$.

The distance between the loud levels is $(635 - 203)$ mm $= 432$ mm and so the wavelength of the sound in the tube is given by

$$\frac{\lambda}{2} = 432 \text{ mm}$$

so that $\lambda = 864$ mm

Thus $c = f\lambda = (384 \text{ Hz})(0.864 \text{ m})$

$\qquad\qquad = 332 \text{ m s}^{-1}$

Notice that this is not the value we would have calculated had we assumed that the distance from the top of the tube to the first loud water level was $\lambda/4$. This is because our assumption that the pressure node (and displacement antinode) at the open end of the tube lie exactly level with the lip of the tube is wrong. In practice they lie just beyond the end of the tube, normally a distance of just over $d/4$ from the lip of a tube of diameter d. In the above case $\lambda/4 = 216$ mm and so the 'end correction' is $(216 - 203)$ mm $= 13$ mm. The tube probably had a diameter of about 50 mm. ◀

Wind instruments

As with stringed instruments (page 318) the character or *quality* of the sound from a wind instrument depends on the number and relative amplitudes of the overtones present. Open and closed tubes produce notes of different qualities as the resonant frequencies are different. For instance an open tube with a fundamental frequency of 170 Hz would have harmonics at 340 Hz, 510 Hz, 680 Hz etc., i.e. different from the closed tube above. The waveform, a p-t graph, of different wind instruments can be

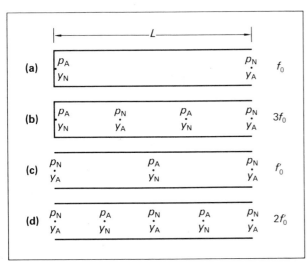

Figure 24.10 Overtones for a tube of length L; (a) and (b) show a closed tube, (c) and (d) show an open tube. The positions of nodes and antinodes for both displacement y and pressure p are indicated.

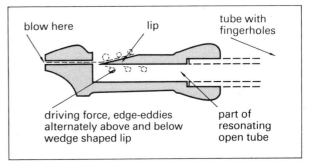

Figure 24.11 A section through the mouthpiece of a recorder.

observed by playing them in front of a microphone which is connected to a c.r.o.

To excite the air in the tube of a wind instrument we can use edge-eddy excitation, where the air is blown at a thin wedge and the eddies so formed set up vibrations in the tube; the recorder uses this method (figure 24.11). Other methods include the use of reeds or your lips. Most wind instruments have a set of holes which effectively varies the length of the tube. The trombone is the exception, having a tube of continuously variable length. It is also possible to 'overblow' wind instruments to excite only the higher frequencies.

24.4 Sound and hearing

The ear-brain system is a most remarkable device; you can listen to a pop record (all the sound from which is transmitted by a single loudspeaker) and yet hear the individual players, singers and background noises separately. In this sense the ear acts as a sort of sound spectrometer or frequency analyser.

The ear is not however as good at distinguishing the direction from which the sound comes. To judge direction we use both ears. At high frequencies we note the different intensities of the signals picked up by each ear; at low frequencies we note the different times of arrival of the signal at each ear. The brain can detect time differences of as little as 30 μs. You should compare both the ear's ability to analyse a complex sound into its components and its inability to judge direction with the equivalent properties of the eye-brain system.

Sound intensity

The ear responds to sound over a range of excess or acoustic pressure amplitudes p_0 from a maximum of about 50 Pa over a wide range of frequencies, down to about 10^{-5} Pa close to 3000 Hz. Figure 24.12 indicates these limits which are called, respectively, the threshold of feeling and the threshold of audibility. Both p_0 and f are represented on logarithmic scales.

The intensity I of a sound represents the rate of flow of sound energy per unit area. It is thus measured in W m^{-2} and can be shown to be proportional to $(fp_0)^2$. As we can detect pressure amplitudes p_0 over a range representing a ratio maximum/minimum $> 10^6$ at a given frequency, then we can detect intensities with a ratio maximum/minimum of $> 10^{12}$. To test this we could assume that the sound from a small loud-speaker radiates according to an inverse square law: stand, say, 2 m from the speaker and turn up the volume until the sound is painful, then move to just over 6 m ($3^2 \approx 10$) from the loud speaker. Now move in to 2 m again at the same time turning the volume down in such a way that the intensity seems to be constant. By repeating this until you can no longer detect any sound you could estimate the ratio of the maximum to minimum audible sound intensities.

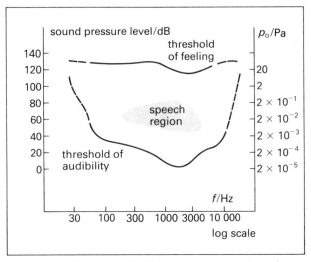

Figure 24.12 The range of p_0 for a normal ear plus the region of p_0 and f for normal speech.

The decibel (dB)

A pressure amplitude of 1 Pa represents a fairly loud noise when heard against a quiet background; if p_0 were to increase to 2 Pa we would detect an increase in loudness. An increase in p_0 from 10 Pa to 11 Pa would, however, be undetectable; from 10 Pa it would be necessary to raise p_0 to 20 Pa before the same increase in loudness was achieved as from 1 Pa to 2 Pa. The same is true of other parts of the nervous system; a pin-prick on the finger can hurt a lot if one is sitting quietly doing nothing but would be unnoticeable if someone were violently punching your upper arm. The nervous response, a subjective measure of the loudness of a sound, is thus best measured on a *logarithmic scale* based on the ratio of the pressure amplitude compared to a standard value $p_0 = 2 \times 10^{-5}$ Pa. The sound pressure level (or SPL) of a sound is defined in this way and is given the unit *decibel* (dB). The scales at left and right of figure 24.12 show how SPL is related to p_0.

Some sounds are more unpleasant than others. High-frequency whines or squeaks are usually judged to be more 'noisy' than low pitched rumbles of the same intensity. To legislate for noise levels in the world around us a new scale

change in noise level	± 2 dB(A) a barely perceptible change
	− 5 dB(A) normal hearing loss, age 45–55
	+ 10 dB(A) a doubling of the noise level
	− 20 dB(A) serious hearing loss
	− 60 dB(A) effect of a good car silencer
noise level	30 dB(A) a whisper
	60 dB(A) normal conservation
	80 dB(A) legal limit for car (1978) measured 10 m from (uphill) road
	95 dB(A) a dangerously noisy factory
	110 dB(A) a very large thunderclap

Table 24.2 Some representative noise levels measured on the A scale.

is defined: it is called the A scale with units dB(A) and is produced by weighting the measuring instrument to accentuate the higher frequencies. In other respects it is just like the decibel scale: Table 24.2 gives some representative environmental values for noise measured on the A scale.

Acoustics

The reduction of 60 dB(A) in the loudness of a car engine noise mentioned in Table 24.2 is an example of acoustic design to control unwanted sound. Such design, as well as the design of buildings to provide good audibility (e.g. in concert halls) is part of an area of study called acoustics. The car exhaust uses two principles to reduce the car engine noise. Firstly the sound can be 'muffled'. This involves the use of fibreglass to absorb the sound and convert its energy to internal energy. Secondly the sound energy can be transferred to resonating air cavities; the principle involved here is that the maximum power transfer from an oscillator (the air in the exhaust) to a driven system (the air in the cavity) occurs at resonance. For instance, if we have a bottle of air which resonates at 140 Hz when we blow over its top, then this bottle will be very effective at absorbing sound at and near this frequency from a random, noisy signal. In the car exhaust the main tube is perforated and surrounded by fibreglass padding while the tube is broken by a number of empty cavities, some of which may involve the pipe in bending back on itself. The cavities should be of different sizes to be most effective in reducing the noise over a range of frequencies.

The noise from a car engine cannot be completely removed by the exhaust silencer. To reduce the noise nuisance we can erect barriers alongside the road but the sound will tend to diffract round them. Another approach is to 'sound insulate' the houses or offices alongside the road. 250 mm of brick wall will reduce the sound level by about 50 dB(A) but an ordinary glass window by only 15 dB(A). Double glazing with sheets of glass about 100 mm apart and tightly sealed with absorbent materials will improve the reduction affected by the window to 30 dB(A). Such windows are widely used in aeroplanes and on high-speed trains.

The acoustics of rooms

If the listener in a concert hall is not to hear echoes and yet the hall is not to be acoustically 'dead', it is important to consider both its dimensions and the absorptive pro-

nature of surface	frequency/Hz		
	125	500	2000
brick wall	0.05	0.02	0.05
carpet on underlay	0.07	0.5	0.6
acoustic tile	0.2	0.7	0.9

Table 24.3 Some sound absorption coefficients a.

perties of the surfaces. Table 24.3 above gives a short list of absorption coefficients a, where a is the fraction of the incident sound energy which is absorbed. Note that a varies with frequency; the perforated acoustic tiles often seen covering the ceiling of a small lecture theatre or classroom are not a very good absorbers at low frequencies, say ≈ 200 Hz.

The time taken for a sound to die away in a room is an important measure of its acoustic properties. To study this, an abrupt sound (e.g. a pistol shot) is made and the time taken for the sound intensity to fall by a factor of 10^6, which is by 60 dB, is measured. This time period is defined as the *reverberation time* T_r of the room. For a small hall full of people $T_r \approx 2$ s; empty it might have twice this value. T_r increases with the volume V of the room and decreases with the area A of the absorbing surfaces

$$T_r = \frac{kV}{aA}$$

where $k (= 0.17$ s m^{-1}) is an empirical constant.

► A bare-walled bathroom measures 3 m by 4 m by 2 m (high). Estimate its reverberation time for a frequency of 500 Hz.

Let us assume that all the wall, floor and ceiling is like brick with $a = 0.02$, so that $aA = 0.02 \times$ total area $= 0.02 \times 52$ m^2 (for bare tiles a might well be less). The volume of the room is 24 m^3.

$$\Rightarrow \quad T_r = \frac{(0.17 \text{ s m}^{-1})(24 \text{ m}^3)}{0.02 \times 52 \text{ m}^2}$$

$$= 3.9 \text{ s}$$

Of course the floor cannot be wholly bare (there is the bath itself) so perhaps 3 to 4 seconds is a fair estimate. There is one other factor that needs to be considered; does the room resonate at 500 Hz? The wavelength of sound of this frequency is given by $\lambda = (340$ m s$^{-1})/(500$ Hz$) = 0.68$ m, so that $\lambda/2 = 0.34$ m or just less if the air is damp. The bathroom is thus an integral number of half wavelengths in all its dimensions and will probably resonate at 500 Hz, making its acoustic properties at this frequency more difficult to estimate. ◄

25 Wavefronts and rays

25.1 Introduction

What happens to the energy from wave sources can be conveniently described in two ways: either by drawing wavefronts or by drawing rays.

A *wavefront* is an imaginary line which we draw joining a set of points at which the wave displacements all have the same phase. Though this sounds complicated it is in practice a simple idea as demonstrated in a ripple tank where the crest lines are themselves the lines of the wavefronts. Along a crest line each particle of water has its maximum upward displacement and is about to move downwards. Wavefronts carry energy away from the source, in two dimensions on the surface of water. To understand how waves behave we need to appreciate a characteristic of waves first suggested by the Dutchman Christian Huygens in 1678 and known as *Huygens's construction*, namely, that every point on a wavefront may be considered to behave as a point source of waves sending energy in the direction of the wave's propagation. To find the new wavefront we have to use the principle of superposition and add the contributions from the sources on the original wavefront; this is usually a very complex operation. The diagram in figure 25.1 shows a series of dots along the wavefront which is just about to reach the barrier. Each dot acts as a point source of waves and the arcs show these *secondary waves* in the forward direction. At Q the secondary waves superpose to give a new plane wavefront while at P the result of the superposition is to extend the wave into the shadow region. The dashed line thus shows the predicted position of the new wavefront, and this is confirmed by the photograph.

A *ray* is an imaginary line which we draw to represent the path taken by the energy from a wave source. This idea is very readily appreciated if we consider light. Light rays can be followed either by passing the light through a smoky atmosphere (e.g. in a smoke box) or by using light streaks such as those illustrated in figure 25.2. Neither the bright beams in the smoke nor the streaks on the flat surface are strictly rays, but they do show where the light energy is going and hence help us to draw rays to describe what is happening. Bundles of rays are usually drawn emerging from a point source of light and the energy flux through a given area perpendicular to the rays is proportional to the number of rays passing through it. In this way ray diagrams form a convenient description of more than just the direction of travel of the energy.

Wave energy

Figure 25.3 shows both wavefronts and rays from a point source S. The diagram shows only two dimensions but S will, of course, radiate wave energy in all directions; it may be a radio star, a fire alarm, a light filament or some cobalt-60 (a γ-source). The principle of conservation of energy

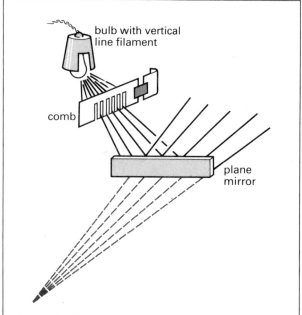

Figure 25.2 Light streaks produced by shadows from the comb slits on a flat white base. The light streaks show the paths of the rays.

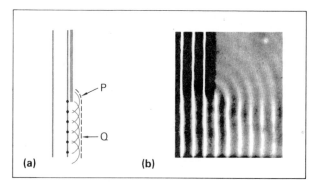

(a)   **(b)**

Figure 25.1 Wavefronts and wave propagation, (a) illustrates Huygens's construction.

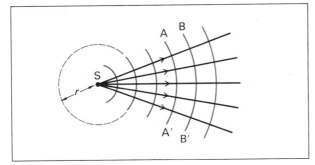

Figure 25.3 Wavefronts and rays from a point source. Here as elsewhere the rays are perpendicular to the wavefronts.

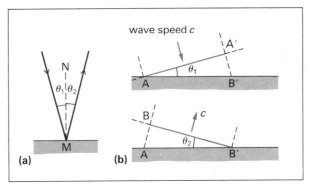

Figure 25.4 Reflection: rays and wavefronts. In (b) the wave diagram is drawn before and after reflection for clarity.

tells us that the energy flux through any of the spheres which have their centres at S is the same (assuming there is no absorption, etc.). Thus if S emits at a power (energy per unit time) P, then at a distance r the *intensity* of the radiation is I where

$$I = \frac{\text{power}}{\text{area}} = \frac{P}{4\pi r^2}$$

i.e. $I \propto \dfrac{1}{r^2}$

and in this case the intensity obeys an *inverse square law*. A very long way away from a source the energy flux is effectively all in one direction so that, for instance, light from the Sun does not decrease as it crosses a room.

Wave energy is sometimes transmitted in only two dimensions, e.g. over the surface of water. Here the relation between intensity I and distance r from a point source is $I \propto 1/r$. The square of the amplitude of a wave y_0^2 is proportional to the intensity I (page 314), so that $y_0 \propto 1/\sqrt{r}$ in a ripple tank and $y_0 \propto 1/r$ for waves spreading from a point source in three dimensions.

In figure 25.3 and all similar diagrams the rays and wavefronts are perpendicular. It is not therefore *necessary* to draw both; in studying light we usually consider only the rays and in ripple tank studies only the wavefronts. The time of travel of a disturbance from one wavefront, e.g. AA′, along a ray to another wavefront BB′, is the same for all points on AA′. This statement, sometimes known as *Malus's theorem*, will help us to use diagrams of wavefronts to make quantitative statements about wave behaviour from a knowledge of how the wavefronts propagate.

25.2 Reflection

The reflection of light is an everyday phenomenon and the *law of reflection* for light, namely: the angle of incidence is equal to the angle of reflection ($\theta_1 = \theta_2$ in figure 25.4) is readily demonstrated using light streaks. The incident ray, the perpendicular MN at the point of incidence and the reflected ray all lie in one plane.

The reflection of a wavefront, which can be studied in a ripple tank, is shown in (b). A plane wavefront such as AA′ approaching a flat reflecting barrier can be seen to reflect to form another plane wavefront BB′. If the time interval between AA′ and BB′ is Δt, then we have

$$AB = c\Delta t$$
and $A'B' = c\Delta t$

Triangles AA′B and ABB′ are therefore identical, so that

$$\theta_1 = \theta_2$$

One consequence of the law is that if a plane mirror is rotated through an angle a, a beam of light reflected from it is rotated through an angle $2a$. This result is readily demonstrated with light streaks and is used in 'spot' galvanometers where the mirror rotates with the coil (page 208); it has the effect of doubling the sensitivity of the instrument.

A particle model?

The result $\theta_1 = \theta_2$ can be predicted from Newton's laws of motion by considering light to consist of a stream of tiny particles each of mass m.

Suppose each particle (of mass m) moves at a speed v along the path of the incident ray. As it approaches the plane reflecting surface it experiences a force which must act perpendicular to the surface and so the linear momentum of the particle *parallel* to the surface remains unchanged,

i.e. $mv \sin \theta_1 = mv \sin \theta_2$

assuming that the particle's energy and hence speed is the same after the collision as before. Clearly therefore, $\sin \theta_1 = \sin \theta_2$ or $\theta_1 = \theta_2$, and the particle moves off along the path of the reflected ray. Both a consideration of wavefronts and a simple particle theory can thus predict the law of reflection.

Microwaves

Light is part of the electromagnetic spectrum and so we would expect it to behave as other waves do (such as water ripples). Other electromagnetic waves can also be shown to reflect. Microwaves of wavelength about 30 mm can be produced by the oscillation of electrons in small metal rods (dipole aerials – page 283). With a transmitter and receiver (the receiver also contains a small dipole) such as that represented diagramatically in the figure on page 332 the reflection of microwaves can be studied, for instance there will be a partial reflection from the face of the perspex prism on which the microwaves are incident. The received signal can be fed either (i) to a sensitive 100 μA meter or a mirror galvanometer or (ii) to an amplifier and loudspeaker if the transmitted signal has been modulated (switched on and off at a low frequency, e.g. 100 Hz). If we try to produce a narrow beam of microwaves using a slit as for light streaks the waves spread out beyond the slit and make any careful test of the law of reflection difficult.

It is more interesting to test which materials *do* reflect the microwaves. Electrical conductors, including the human body, reflect almost all the microwave energy, poor conductors such as wood are partial reflectors and electrical insulators such as perspex only weak reflectors. We might also note that good electrical insulators are often transparent to light, i.e. perspex and glass are poor reflectors of light, while metal surfaces are used as mirrors. A metal 'mirror' for microwaves can be made from wire gauze, as their wavelength (30 mm) is much greater than the gaps in the metal surface. A mirror for light, which has a wavelength 5×10^{-7} m in air, must however be a polished layer of the metal: irregularities in the surface must have only very small dimensions.

Phase change

When waves are reflected there is sometimes an abrupt change of phase. Wave pulses on ropes are reflected 'upside down' at a fixed end (page 313). Light and microwaves can also both suffer a phase change of 180° in the variations of electric field forming the wave; light when it is reflected from a medium in which it would travel more slowly and microwaves when reflected from a metal screen. Similarly sound waves in air undergo a phase change of 180° when reflected at a solid boundary.

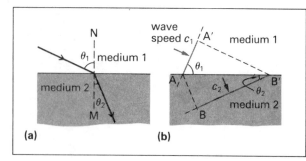

(a)   **(b)**

Figure 25.5 Refraction: rays and wavefronts. The result of the wave speed changing is to alter the direction of the rays except when $\theta_1 = 0$.

25.3 Refraction

Refraction occurs when a wave passes from one medium to another. Unless otherwise stated we shall deal with light of a single colour only (monochromatic light), that is, with waves of a single frequency. At a boundary where refraction occurs there will always be some reflected energy; we will ignore this unless it is of particular importance.

Snell's law – rays

Experiments with ray streaks passing from air to glass and from glass to air show that as θ_1 varies so does θ_2 (figure 25.5a) and that

$$\frac{\sin \theta_1}{\sin \theta_2} = \text{constant for two media}$$

This result is called *Snell's law*. If the direction of the energy propagation is reversed we still label as θ_2 the angle measured to the perpendicular *in* medium 2 etc. The incident ray, the perpendicular at the point of incidence and the refracted ray all lie in one plane.

The constant is usually written as n_2/n_1,

i.e. $$\frac{\sin \theta_1}{\sin \theta_2} = \frac{n_2}{n_1}$$

or $n_1 \sin \theta_1 = n_2 \sin \theta_2$ (1)

where n_1 and n_2 are called the refractive indices of the two media; n has no units. By definition we take the refractive index of a vacuum to be 1, and so *the refractive index* of a medium n is defined by the equation

$$\frac{\sin \theta_{\text{vac}}}{\sin \theta_{\text{med}}} = n$$

Values of n for two types of glass at different wavelengths (colour) are shown graphically on page 338. The refractive index of air under normal laboratory conditions is 1.0003; we usually take it to be 1 and measure n for other substances (such as glass) by studying refraction from air to glass or from glass to air with semicircular or rectangular glass blocks.

Where a ray is refracted successively through a series of *parallel* layers of different materials or through a material whose refractive index varies continuously with height, we can apply equation (1) for refraction at any convenient level, i.e.

$n_1 \sin \theta_1 = n_2 \sin \theta_2 = \ldots = n_8 \sin \theta_8 = n_9 \sin \theta_9 \ldots$ etc.,

or $n \sin \theta = \text{constant}$ (for a given ray)

Thus in Figure 25.6a pouring oil onto the surface of the water does not alter the angle φ at which the ray finally emerges into the air (it must anyway be parallel to the incident ray). In (b) the ratio of the refractive indices of the air at two different temperatures can be estimated by

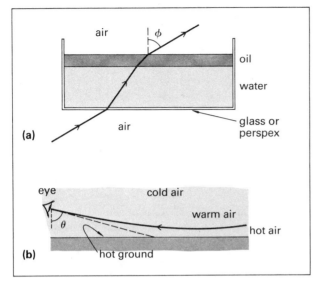

(a)

(b)

Figure 25.6 (a) φ is independent of the thicknesses of the water and oil layers for a fixed direction of the incident ray. (b) The observer finds he can't see the ground for θ greater than a certain value; he sees a reflection of the sky. This is called a mirage.

measuring θ, the angle at which the mirage is seen to start. For example, if $\theta = (90 - 0.5)° = 89.5°$, we get

$$n_{cold} \sin 89.5° = n_{hot} \sin 90°$$

so that $\dfrac{n_{cold}}{n_{hot}} = \dfrac{1}{\sin 89.5°} \approx 1.000\,04$

Snell's law – wavefronts

To demonstrate refraction of waves in a ripple tank involves arranging for the depth of the water to change abruptly – see figure 25.7a. The waves change speed (they slow down) as they go from deep to shallow water and so their wavelength changes. It is this *change of speed* that we call *refraction*. If the wavefronts meet the refracting boundary obliquely, as in figure 25.7b, the direction of energy propagation changes. The bending of waves at the seaside as they approach the shore is an everyday example of this.

In figure 25.5b a plane wavefront AA′ travelling at speed c_1 in medium 1 refracts to form another plane wavefront BB′ travelling at speed c_2 in medium 2. If the time interval between AA′ and BB′ is Δt, then we have

$$A'B' = c_1 \Delta t \text{ and } AB = c_2 \Delta t$$

but $A'B' = AB \sin \theta_1 \text{ and } AB = AB' \sin \theta_2$

so that $\dfrac{\sin \theta_1}{\sin \theta_2} = \dfrac{c_1}{c_2}$ (2)

which is a constant for any two media.

The *refractive index n* of a medium for light can thus be written

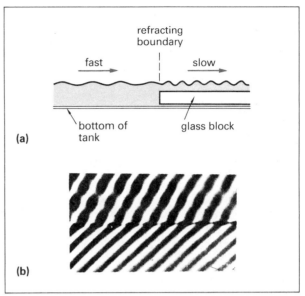

(a)

(b)

Figure 25.7 Demonstrating refraction in a ripple tank, (a) shows the practical arrangement and (b) is a photograph of the shadow cast by the refracting ripples.

$$n = \frac{\text{speed of light in a vacuum}}{\text{speed of light in the medium}}$$

The speed of light in a vacuum is 3.0×10^8 m s^{-1}; so if the measured refractive index for a block of glass is 1.5, we can deduce that light travels in the glass at 2.0×10^8 m s^{-1}. As $c = f\lambda$, and the frequencies of the light before and after refraction must be the same, then the ratio of the wavelengths will also be 1.5.

A particle model?

Snell's law can be predicted from Newton's laws of motion by considering light to consist of a stream of tiny particles, as can the law of reflection (page 329). Suppose each particle (of mass m) moves in medium 1 at a speed v_1 along the path of the incident ray in figure 25.5a. As it approachs the plane refracting surface it experiences a force which must act perpendicular to the surface; so the linear momentum of the particle *parallel* to the surface remains unchanged. If its speed in medium 2 was v_2 then

$$mv_1 \sin \theta_1 = mv_2 \sin \theta_2$$

we are assuming that the change of speed $v_2 - v_1$ is independent of θ_1, which it will be as the work done by the attractive and repulsive forces is constant and so, therefore, will be the particles' change in kinetic energy.

Snell's law is thus predicted to be

$$\frac{\sin \theta_1}{\sin \theta_2} = \frac{v_2}{v_1} = \text{constant}$$

but note that this implies that if the particle is refracted *towards* the perpendicular its speed will *increase*. The opposite prediction is made by the wavefront approach; see equation (2). Measurements of the speed of light show that the wavefront relationship is the correct one. The Newtonian particle model for light is thus inadequate.

Microwaves

Microwaves travel at less than 3.0×10^8 m s^{-1} in wax or paraffin or other dielectric materials. Figure 25.8 shows a refraction experiment for 30 mm waves passing through a prism. The incident wave is not plane but could be made more so by the use of a wax lens with the transmitter at its principal focus. Another lens used on the other side could produce a focused image for the receiver. The focal length of the lenses should first be roughly measured – a value of 300 mm is typical. The refractive index of the paraffin can then be measured by setting up the situation where the wave is symmetrically refracted through the prism. Figure 25.9 shows the waves passing through part of a prism of angle A. The angles in the diagram are correct only for the symmetrical passage of a wave. Thus, comparing with figure 25.5b and using equation (2) above, we have for refraction at the first surface

$$\frac{\sin \theta_1}{\sin \theta_2} = \frac{\sin \left(\frac{A}{2} + \frac{D}{2} \right)}{\sin \frac{A}{2}} = \frac{c_1}{c_2}$$

but $\dfrac{c_1}{c_2} = n$

$$\Rightarrow \quad \sin \frac{A + D}{2} = n \sin \frac{A}{2}$$

where A is the angle of the prism and D is the total deviation of the wave in passing through the prism, i.e. the angle between the incident and the emergent wavefronts. For a 60° prism, if D is found to be 40°, the ratio of the speeds is sin 50°/sin 30° = 1.5 to two significant figures; i.e. the refractive index n of paraffin for waves of this frequency

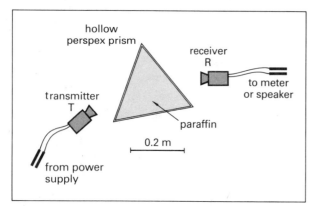

Figure 25.8 A refraction experiment using 30 mm electro-magnetic waves (microwaves).

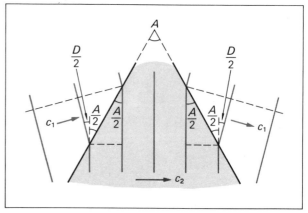

Figure 25.9 A plane wave is deviated through an angle $D = D/2 + D/2$ as it passes symmetrically through a prism.

(about 10^{10} Hz) is 1.5. For much lower frequencies, tending to zero, the refractive index can be shown to be equal to the square root of the relative permittivity of the material.

▶ A thin plane membrane is used to separate some gas at 288 K from the same gas at 392 K. A plane sound wave travels in the cooler gas and is incident on the membrane at an angle of 20°. Find the angle of refraction.

The speed of sound in a gas

$$c = \sqrt{\frac{\gamma p}{\rho}} = \sqrt{\frac{\gamma RT}{M}} \qquad \text{(page 322)}$$

Therefore for a particular gas at two temperatures T_1 and T_2

$$\frac{c_1}{c_2} = \frac{\sin \theta_1}{\sin \theta_2} = \sqrt{\frac{T_1}{T_2}}$$

In this example therefore

$$\sin \theta_2 = \sin \theta_1 \times \sqrt{\frac{T_2}{T_1}}$$

$$= \sin 20° \times \sqrt{\frac{392 \text{ K}}{288 \text{ K}}}$$

$$= 0.40$$

Thus $\theta_2 = 23.5°$, and the wavefront is *deviated* through only 3.5°. The higher temperature in this example (119°C) will not occur in the open air but the refraction of sound caused by temperature gradients in the atmosphere does occur. The clarity with which one can hear a distant noise on a still, clear evening after a hot day is evidence of this. The air near the ground cools first and provides the necessary temperature gradient (see figure 24.7 on page 322). ◀

Total reflection

With the apparatus of figure 25.6a suppose a narrow beam of light is sent into a tank of water (the oil being removed) through the side of the tank. As the angle of incidence is increased the beam, which at first is partly reflected and partly refracted, suddenly loses its refracted part and *all* the energy is reflected. Just before this happens the refracted beam is refracted along the surface of the water, i.e. $\theta_{air} = 90°$. Figure 25.10 shows three rays to illustrate this phenomenon which is called *total reflection* or sometimes total internal reflection. We have to consider light of only one colour as the different wavelengths (in air) reach the 'cut-off' at slightly different angles of incidence θ_c when white light is used. Applying Snell's law at Q we have

$$n_w \sin \theta_c = n_a \sin 90°$$

where n_w is the refractive index of water and n_a ($= 1$) that of air. Therefore

$$n_w = \frac{1}{\sin \theta_c}$$

θ_c is called the *critical angle* for water and will be 49° if $n_w = 1.33$.

Measuring the critical angle forms the basis of a number of methods of measuring refractive index, particularly for liquids or fats. They have been developed in the food industry as a sensitive quality control technique, e.g. for milk and margarine, and a smear of the material is all that is needed. Devices which produce a direct reading of n from such measurements of the critical angle are called *refractometers*.

In deriving Snell's law for wavefronts we took $c_2 < c_1$ in figure 25.5b and so the wavefront was refracted as shown. If $c_2 > c_1$ then AB ($= c_2 \Delta t$) would be greater than $A'B'$ ($= c_1 \Delta t$) and $\theta_2 > \theta_1$. Suppose AB were greater than AB'? It would not then be possible to draw a new wavefront in medium 2; i.e. refraction is only possible if

$$c_2 \Delta t \leqslant \frac{c_1 \Delta t}{\sin \theta_1}$$

the equals sign occurring when $\theta_1 = \theta_c$.
Thus *all* the wave energy must be reflected when

$$\frac{c_2}{c_1} > \frac{1}{\sin \theta_1}$$

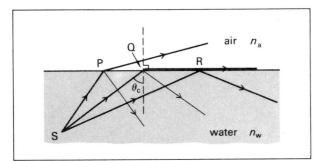

Figure 25.10 At P and Q refraction and reflection occurs but at R there is only reflection. When the refracted ray is parallel to the air-water surface the incident ray makes an angle θ_c where $\sin \theta_c = n_a/n_w$ with the normal.

which, as $\sin \theta$ is always less than 1, shows us that total reflection only occurs when $c_2 > c_1$, i.e. when the wave is travelling in a medium in which it travels slowly and meets the boundary of a medium in which it travels more quickly.

Examples of total reflection

Glass blocks exhibit total reflection for light and can be cut to encourage such internal reflections. Some transparent crystals, *precious stones* often with high values of n (for diamond $n = 2.4$), are cut in this way to produce the sparkle we associate with them. *Bubbles* rising through water or attached to an immersed object shine because light is totally reflected at their surface giving a mirror-like appearance. The ray in figure 25.6b is totally reflected near the hot ground giving rise to what the observer sees as imaginary lakes or pools of water. This is the classic *mirage*, now readily visible on a hot day on any road. The phenomenon is also utilised in the design of a *light pipe*: if a beam of light can be introduced into a narrow transparent fibre then it may meet the wall of the fibre obliquely and be totally reflected again and again; i.e. it may travel along inside the fibre even if the fibre turns corners. Light pipes can be used to illuminate otherwise inaccessible places and, by using a bunch of fibres, to transmit a picture, e.g. from the inside of a lung.

Sound pipes, or *speaking tubes* as they are usually called, are used on ships. As sound waves travel faster in a solid such as iron than in air, a sound wave meeting the wall of a metal tube can be totally reflected. Sound waves can also be totally reflected by temperature gradients in the upper atmosphere. A large explosion can by this process sometimes be heard a long distance from its source, the result being very similar to the total reflection of radio waves in the ionosphere – page 372.

Prisms and light

A special case of refraction at plane surfaces occurs when two surfaces make a prism of refractive index n and refracting angle A. When a narrow beam of light, e.g. from the ray streak apparatus, is incident on the first face of the prism, it is deviated in its passage through the prism by an angle D. It was shown on page 332 that when a microwave passes symmetrically through the prism,

$$n \sin \frac{A}{2} = \sin \frac{A + D}{2}$$

and this result holds for all waves, including light. On logical grounds the symmetrical case must be a case where D is a maximum or a minimum and it can be readily shown to be a *minimum* by experiment. In the above formula D should therefore really be written D_{min}. With white light only one wavelength can pass symmetrically through the prism at a time; figure 25.11 shows how yellow-green light (an 'average' colour) is deviated. The angular spread δ of

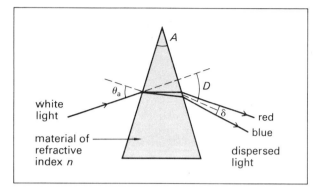

Figure 25.11 Refraction of white light through a glass prism. If A and θ_a are small then for yellow-green $D = (n - 1)A$.

the extreme wavelenghts is called the *angle of dispersion* (figure 25.11). It is larger for prisms of large A and for those for which a graph of refractive index against wavelength (figure 25.19 on page 338) has a large slope.

Thin prisms

For thin prisms A is small and, using the approximation

$$\sin \theta \approx \theta \qquad \text{(appendix B5)}$$

the above equation becomes

$$n\frac{A}{2} = \frac{A + D}{2}$$

which leads to $D = (n - 1)A$

It can be shown that D is now independent of the original angle of incidence providing that this is small. The result is important to the theory of lenses and can be stated as follows: for thin prisms all rays making small angles of incidence have the same deviation. By thin and small we mean A and $D < 0.1$ rad i.e. less than about 6°.

▶ If the refractive indices of a particular flint glass are 1.638 and 1.621 for blue and red light respectively, what is that distance between the red and blue ends of a spectrum produced on a wall 10 m from a thin prism of refracting angle 0.10 rad made of this glass?

Here the two angles of deviation will be:

$$D_{\text{blue}} = (1.638 - 1) \times 0.10 \text{ rad} = 0.0638 \text{ rad}$$
$$\text{and} \quad D_{\text{red}} = (1.621 - 1) \times 0.10 \text{ rad} = 0.0621 \text{ rad}$$

so that the angle of dispersion is

$$= (0.0638 - 0.0621) \text{ rad} = 0.0017 \text{ rad}$$

Suppose that the separation of the blue and red on the wall is x,

then $0.0017 = \dfrac{x}{10 \text{ m}}$ (appendix B5)

and $x = 0.017$ m or 17 mm

This is not a very long spectrum but shows that where long distances are employed, as in telescopes, the problems produced by colouring effects in the image (chromatic aberration) are serious.

Measuring n

Methods which depend on applying Snell's law directly to two angles measured with a protractor in a *ray streak* experiment are very imprecise and any one determination of n for glass by this method may have an uncertainty of as much as $\pm 10\%$. To improve this a number of readings are taken, not of θ_a and θ_g but of their sines calculated directly from right-angle triangles, so that it is lengths and not angles which we measure. Further, a graph of $\sin \theta_a$ against $\sin \theta_g$ will average out uncertainties in individual readings and enable n to be found from its slope. This is a case where a proper approach to a simple experiment can reduce the uncertainty considerably to (say) $\pm 2\%$ giving, e.g. $n = 1.52 \pm 0.03$.

Measuring *real and apparent depth* can lead to a value for n (page 338). It can be used for both liquids and solids and, with a travelling microscope a narrow depth of field gives a quick and precise result. To achieve an uncertainty of about $\pm 1\%$ the distance measured should be more than 100 mm and what is very important, the judging of when the microscope is properly focused should be a no-parallax method (page 338) between the cross wires of the eyepiece and the pollen or scratch which is being viewed.

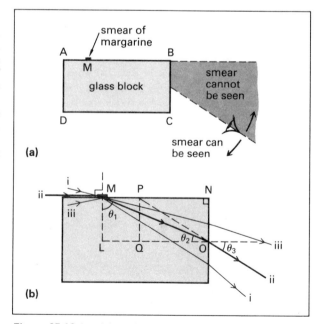

Figure 25.12 A quick method of finding the refractive index of oils and fats.

The *spectrometer* (page 348) is used for very precise measurements of n, for instance in determining the optical properties of a new type of glass.

► A narrow smear of margarine is made across one side of a rectangular block of glass at M – figure 25.12a. (The smear runs perpendicular to the plane of the paper). By looking into the block through the side BC the smear can be seen with the eye in the shaded area above the dotted line shown but when the eye is moved below the dotted line the smear disappears and only a shiny surface can be seen at M. Explain these observations and show that the refractive index of the margarine is equal to OL/OP.

Light, (e.g. the ray marked i in figure 25.12) enters the glass at M by refracting through the margarine and leaves the block below O on the side BC, enabling an eye viewing from there to see that there is a smear of margarine. When the eye is above O it receives a ray, e.g. that marked iii, which has been totally reflected inside the block at M and thus sees a mirror-like surface at M. At the cut-off between these two the ray marked ii enters the glass block from the margarine with an angle of incidence of 90°. Applying Snell's law for this critical ray we get

at M $n_m \sin 90° = n_g \sin \theta_1$

at O $n_g \sin \theta_2 = n_a \sin \theta_3$

where n_m is the refractive index of margarine, etc. Taking $n_a = 1$ and eliminating n_g gives

$$n_m = \frac{\sin \theta_1 \times \sin \theta_3}{\sin \theta_2}$$

But we can see in the figure that

$$\sin \theta_1 = \frac{OL}{OM}, \sin \theta_2 = \frac{ML}{OM} \text{ and } \sin \theta_3 = \frac{PQ}{OP}$$

so that $n_m = \left(\frac{OL}{OM}\right)\left(\frac{PQ}{OP}\right)\left(\frac{OM}{ML}\right)$

which as $ML = PQ$

gives $n_m = \frac{OL}{OP}$ ◄

25.4 Diffraction

Waves can bend round corners. Medium-wave radio signals can be detected over the horizon or behind a hill, sound waves carry conversations over walls or round open doors. The common feature in these changes in the direction of energy propagation is that where part of a wavefront is blocked or limited the remaining part spreads into the 'shadow' area. The phenomenon is called the *diffraction of waves*. The resulting behaviour of the wave can be predicted using Huygens's construction and the principle of superposition. This is done for the case of a slit in section 27.4 (page 357), but is generally a very difficult operation.

The ripple tank is an ideal tool for demonstrating the diffraction of water waves with wavelengths of the order of 20 mm. Figure 25.13 shows the result of plane waves striking a barrier in which there is a gap. The diffraction is seen to depend on the size of the gap – or, more correctly, on the ratio of the width d of the gap to the wavelength λ. We can see that a greater proportion of the wave energy spreads to the shadow area when λ/d is large than when it is small. A closer look reveals that the region beyond the barrier contains a complex wave pattern and the sketch in the figure represents this pattern diagrammatically for the right hand photograph. Obviously water waves could not propagate beyond the gap *without* spreading sideways for this would involve a 'wall' of water with vertical edges. The water thus falls outwards and in so doing the energy of the wave is diffracted. When the frequency of the wave is high (and thus the wavelength low relative to the width

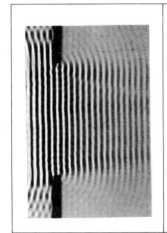

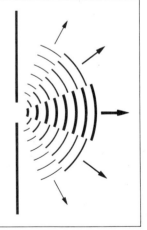

Figure 25.13 Photographs showing the diffraction of plane water waves at a gap. The wavelength is the same in each case. The sketch on the right is a diagrammatic representation of the right hand photograph. Note the phase differences between the fan-like portions.

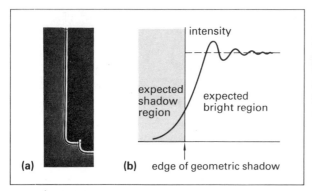

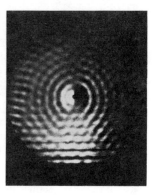

Figure 25.14 (a) The diffraction of light past the edge of a razor blade. (b) A graph showing the (rather surprising) intensity variation close to the edge of the geometrical shadow.

Figure 25.15 Diffraction and scattering: when the gap or obstacle is smaller than the wavelength of the wave it acts rather like a point source.

of the gap) there is less time for the water particles to fall sideways during their vertical oscillations and so in this case the diffraction is less noticeable. In the sketch of figure 25.13 the direction of travel of the wave energy is indicated by arrows the size of which is designed to illustrate roughly how much energy travels in each fan-like region.

Light waves bend round corners too. The wavelength is now very small ($\approx 5 \times 10^{-7}$ m) and so the gap through which we attempt to demonstrate a diffraction pattern similar to those above for water waves must also be very small. Suppose two razor blades are mounted, e.g. with bulldog clips on a piece of wood in which there is a hole, so that they can be slid together to provide a narrow slit. If now the slit is held vertically up to the eye and a vertical lamp filament a few metres away is viewed through it a pattern of light and dark bands can be seen. Moreover if the width of the slit varies from top to bottom the variation of the pattern observed through different parts of the slit illustrates how the energy distribution depends on the width (see also page 358). Instead of saying that because light has a very small wavelength we need very narrow slits to demonstrate its diffraction, we can reverse the logic and say that because light exhibits diffraction effects just as water waves do, then light must be wavelike – but of very small wavelength.

Diffraction occurs whenever a wavefront is limited in some way. If instead of a gap in a barrier there had been an obstacle in the path of the waves, they would diffract so that energy would travel round the edges into the shadow area. The photograph on page 328 showed an example of diffraction of this sort, for water waves. An equivalent diffraction pattern (sometimes called a *Fresnel* diffraction pattern) for light is seen in figure 25.14a. In (b) we see that the intensity I varies in a very complex way away from the blade's edge where we might expect a uniform intensity. We might describe the effect as a fuzzy edged shadow with a number of increasingly indistinct halos on the side away from the darkness.

Scattering

When we make (i) the breadth of the slit or (ii) the breadth of an obstacle in the path of the waves small compared with the wavelength, the wave is said to be scattered. In both cases the slit or obstacle act as a point source (Huygens's construction) and the secondary waves are (i) semicircles or (ii) circles, as there are no wave disturbances from other points on the original wavefront to take into account. The new wavefront is, of course, of very small amplitude (figure 25.15).

Scattering by small particles in the atmosphere gives rise to the blue colour of the sky. This is because for light of wavelength λ, the amount of scattering is proportional to λ^{-4}, so that as $\lambda_{red} \approx 2\lambda_{blue}$, blue light is scattered 16 times as effectively as red. On the other hand when we look *at* the Sun low on the horizon we receive light which has had a lot of the blue scattered by dust etc. in the air, and the Sun appears to be red. On the Moon the sky is black and the Sun looks white at all times.

25.5 Images

Wavefronts from a point source O spread out from O and some of them may enter a region which alters their shape in some way, e.g. by reflection or refraction. The wavefronts may now converge on a point I or appear to be diverging from a point I other than O – in both cases we call I the *image point*. If the waves are light waves then we would probably describe the formation of an image using rays; the rays first diverge from O, some of them are intercepted by an *optical system*, e.g. a mirror or a lens, and after passing through the system they converge on a point I, or appear to diverge from a point I, other than O. Figure 25.16 shows how an eye sees (i) a *real* image and (ii) a *virtual* image: the diagrams are ray diagrams with the wavefronts drawn as grey lines (we shall usually omit the

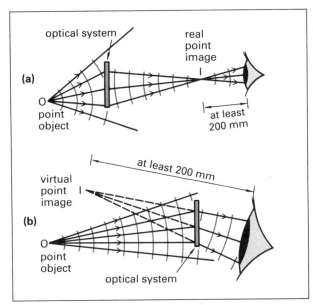

Figure 25.16 The formation of real and virtual images. Note that in (b) the eye could be placed right up against the optical system providing the image is formed more then 200 mm from it.

wavefronts unless they help to make a particular point). In both cases the eye receives a bundle of rays diverging from the image and brings them to a focus on the light-sensitive retina at the back of the eye. For a real image the energy is concentrated at I as the wave passes through the image position. This means that, for light, real images can be caught on a screen or recorded on a photographic plate. There is no energy at I for a virtual image.

Perfect images

The conditions for producing a perfect image, with *all* the rays passing through or appearing to come from a single point, are very stringent. When we further require that, for an extended object, the array of points which make up the object should form the same pattern of points (perhaps enlarged) in the image it is surprising that perfect images can ever be produced. Perfection *is* achieved in the image produced by a plane mirror. Figure 25.2 on page 328 shows light streaks illustrating how the image is formed. The optical image is three – dimensional with each point of the object producing a virtual image as far behind the mirror as the point is in front. The reflection of a circular wave at a straight barrier in a ripple tank can be used to help understand the formation of this perfect virtual image for it is easy to find a point behind the barrier which, if made a source of waves provides wavefronts *at* the barrier with exactly the same curvature as the reflected wavefront moving away from the other side. The reflected wavefronts are parts of perfect circles and this leads to our perfect image. Most images suffer imperfections called aberrations which can be roughly classified as geometrical

aberrations and chromatic aberrations. A further reason for imperfect images is that only part of the wavefront passes through the system and thus there are diffraction effects (page 359).

Spherical aberration

Using ray streaks we can demonstrate *longitudinal spherical aberration* (figure 25.17); the effect for mirrors is daily visible on the surface of the tea in any cup lit by a single bulb. Another way to demonstrate the effect for a lens is to use a pea bulb as the object and a short focal length lens and screen. Longitudinal spherical aberration is most noticeable where the optical system accepts a widely divergent beam of rays the edges of which are then strongly deviated. The effect is unavoidable in microscope objectives and strong magnifying glasses. The name exactly describes the wave situation; for an incident spherical wave emerges as a non-spherical one which cannot converge to (or appear to converge from) a single point. To reduce the resulting blurring of the image one solution is to use a system of lenses or mirrors to share the deviation of the rays among a number of refracting or reflecting surfaces. Another solution is to reduce if possible the aperture of the lens or mirror used. The image formed on the retina of the eye suffers from spherical aberration but the brain does a remarkable scanning job and tidies up what is seen in much the same way as poor quality photographs from space probes are improved by specially programmed computers.

Real and apparent depth

Figure 25.18 will help us to understand the nature of

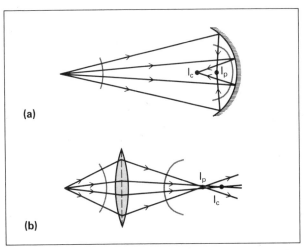

Figure 25.17 Spherical aberration. The wavefront is no longer circular after (a) reflection and (b) refraction. Rays passing through the edges of the system meet at I_p, not at I_c, where the central rays meet.

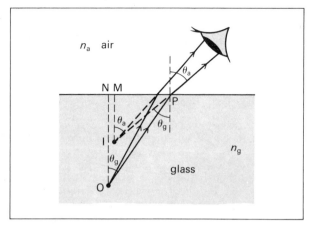

Figure 25.18 Real and apparent depth. If θ_g is small then $n_g/n_a = ON/IM$.

spherical aberration further. Applying Snell's law to the refraction at P, we get

$$n_a \sin \theta_a = n_g \sin \theta_g$$

But $\sin \theta_a = MP/IP$ and $\sin \theta_g = NP/OP$, so that putting $n_a = 1$,

$$n_g = \left(\frac{OP}{NP}\right)\left(\frac{MP}{IP}\right) \ exactly$$

The eye sees an image at I, so that

$$\frac{\text{real depth}}{\text{apparent depth}} = \frac{ON}{IM} \approx n_g$$

if the angles θ_a and θ_g are both small, for then $MP \approx NP$, $IP \approx IM$ and $OP \approx ON$. As the pupil of the eye is small the approximations above are valid when the eye is looking vertically downward into a glass block or a bowl of water.

Chromatic aberration

The refraction in figure 25.17 shows the path of a ray for a wave of only one frequency, i.e. one wavelength in air. But n_g varies with the frequency f for light. If we look obliquely at the lines on the bottom of a swimming pool we see that they have coloured edges; I_{red} is in a slightly different place from I_{blue}. Figure 25.19 show how n_g varies with wavelength in air for two types of glass. These curves are typical of most refracting materials. The refractive index increases as the wavelength decreases and the slope of the graph is larger for *optically denser* materials, i.e. materials with a higher average value of n. To eliminate the colouring of images would seem to be impossible for it would need a glass for which the slope of the graph line was zero. Lenses are more powerful in the blue region than in the red, and so there will be a *longitudinal chromatic aberration* in the formation of all images. A pea bulb,

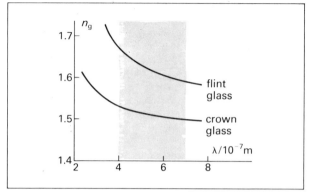

Figure 25.19 The dispersive properties of glass; the shaded area shows the visible region.

filters, a screen and a long focal length lens can be used to demonstrate the effect. Some lowering of the resulting blur of the image at one place can be achieved by using pairs of lenses, *achromatic combinations*, e.g. a converging crown glass component and a diverging flint glass component. The flint glass can be weaker than the crown glass lens yet produce chromatic effects of the same size and the other way around. So we can eliminate aberration for two chosen colours, e.g. red and blue (and at the same time nearly eliminate it for the other colours). The objective lenses in the spectrometer shown on page 361 are drawn as such lens pairs.

Locating images

Real images can be located by judging where they are best focused on a screen; but since there are the longitudinal aberrations mentioned above this is not a very precise process. If the eye can be used to look at an image and judge where it is relative to a similar object placed near the position at which the image is expected to appear, two advantages are immediately achieved:

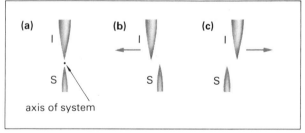

Figure 25.20 Parallax: in each of the diagrams it is assumed that the head is moved to the right. S is a search pin and I the image whose location is required. In (b) S is further from the eye than I, in (c) S is nearer to the eye than I and in (a), where there is no relative movement between S and I (no parallax) they coincide in space.

(i) The eye has a small aperture which reduces the effects of spherical aberration, and

(ii) the eye is relatively insensitive to wavelengths outside the range 5×10^{-7} m to 6×10^{-7} m in daylight (and outside slightly narrower limits at night), which reduces the effects of chromatic aberration. The eye, in other words, does rather better as a detector than one might at first expect.

This method of image location is known as the *no-parallax* method. The object is usually a brightly lit pin the point of which lies on the axis of the lens or mirror being studied. A similar search pin S is placed near the expected image position I (often best found roughly by some other method) and inverted with respect to the expected image. If the eye can now be placed so as to focus both I and S, the search pin can be moved until its tip coincides with the tip of the image by looking for no parallax (i.e. no separation) between I and S when the head is moved. In this way with care the position of an image can be located to within a millimetre. Figure 25.20 summarises the parallax effect.

26 Optical systems

26.1 Lenses

Lenses are important because they form images. Spherical lenses, both surfaces of which form parts of the surfaces of spheres, do not form perfect images. They are subject to aberrations (page 337) and there will be diffraction effects; we shall ignore both in what follows.

Figure 26.1 consists of two ray diagrams which we can use to show that all the light from a point object passing through a spherical lens passes through one image point. Consider in (a) a ray from an object at O which passes through the lens a distance h from its *optical centre* P (the point through which a ray is refracted without deviation). It is deviated by an angle D given by

$$D = (n - 1)A \qquad \text{(page 334)}$$

where A is the angle of the 'prism' formed by the sides of the lens at h. D is independent of the angle of incidence, so the ray in (b) is also deviated through an angle D. This ray is parallel to the *principal axis* (a line through P perpendicular to the lens surfaces) and is refracted so as to intercept the axis a distance f from P; f is the focal length of the lens – see figure 26.2.

In (a) $\qquad D = a + \beta \qquad$ (exactly)

and $\qquad \dfrac{h}{u} = \tan a \approx a \qquad$ (appendix B5)

Similarly $\quad \dfrac{h}{v} \approx \beta$, and from (b) $D \approx \dfrac{h}{f}$

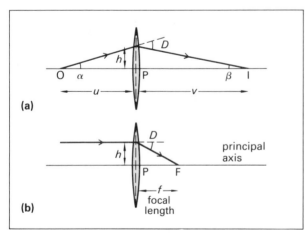

(a)

(b)

Figure 26.1 The object distance $OP = u$, image distance $IP = v$ and focal length $PF = f$ for a lens are related by the formula $1/u + 1/v = 1/f$.

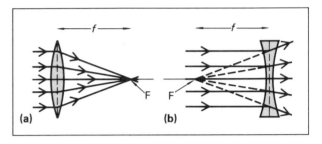

(a)   **(b)**

Figure 26.2 The principal focus F and focal length f of (a) a converging and (b) a diverging lens. Each lens has a second principal focus (not shown) on the other side of the lens.

Thus $\qquad \dfrac{h}{f} = \dfrac{h}{u} + \dfrac{h}{v} \qquad$ (approximately)

or $\qquad \dfrac{1}{f} = \dfrac{1}{u} + \dfrac{1}{v} \qquad\qquad$ (1)

The significant point about the result is that the relationship between u, v and f is independent of h. A similar proof of this *lens formula* can be produced for a diverging lens and also, for both types of lens, for all types and positions of object. In order to apply equation (1) in all cases we measure u, v and f in the following ways (from the optical centre P of the lens): distances to *real* objects and images are taken to be *positive*; to *virtual* objects and images they are taken to be *negative*. Because the image of an object at infinity produced by a diverging lens is virtual (figure 26.2b) the focal length f of such a lens must be represented in the lens formula by a negative number.

If u and v in the lens formula are interchanged $1/u + 1/v$ is still equal to $1/f$, i.e. object and image positions can be interchanged. This is typical of all optical systems and is the result of the *principle of reversibility of light*. Any ray of light will, if reversed, travel back along its original path. For instance in figure 26.2a a luminous object placed *at* F will produce a parallel beam emerging from the converging lens.

► A real object is placed on the axis of a converging lens of focal length 0.20 m at a distance of 0.40 m from the lens. On the far side of the lens and 0.20 m from it is placed another lens, this one being a diverging lens with a focal length of 0.30 m. Locate the image formed by this two-lens system.

It is always best to draw a sketch showing the formation of the image using two constructional rays: (i) an undeviated ray through the optical centre of the lenses and, (ii) a ray parallel to the principal axis which is refracted through F.

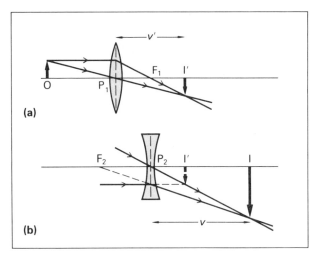

Figure 26.3 The two stages of the problem. The real image I' formed by the converging lens in (a) becomes a virtual object for the diverging lens in (b).

Such sketches help us to gain an understanding of the problem (figure 26.3). We usually draw an extended object rather than a point on the axis and assume that an object perpendicular to the axis at O becomes an image perpendicular to the axis at I. This also gives us an idea of the magnification. The calculation now runs:

For the first lens

$$\frac{1}{+0.20 \text{ m}} = \frac{1}{+0.40 \text{ m}} + \frac{1}{v'}$$

which gives $v' = +0.40$ m

i.e. the image I' is real and 0.40 m from the first lens (figure 26.3a). It is therefore $(0.40 - 0.20)$ m $= 0.20$ m *beyond* the second lens, the diverging lens. The rays never pass through I' as the second lens interrupts them and so I' now acts as a *virtual object* for the diverging lens and $u = -0.20$ m.

For the second lens

$$\frac{1}{-0.30 \text{ m}} = \frac{1}{-0.20 \text{ m}} + \frac{1}{v}$$

which gives $v = +0.60$ m

and the final image is a real image lying 0.60 m beyond the diverging lens. Figure 26.3b predicted this roughly.   ◀

The power of a lens

One important property of lenses which can be readily shown with the ray streak apparatus of figure 25.2 (page 328) is that parallel beams of light from very distant objects which enter a lens at small angles to the principal axis are all brought to a focus in one plane. This is called the *focal plane* of the lens and lies perpendicular to the principal axis. The *principal focus* F is the point at which the focal

plane and principal axis meet and the *focal length f* is defined as the distance between the principal focus and the optical centre P of the lens (figure 26.4).

A lens is powerful if *PF* is small. We define the *power of the lens* as F, where $F = 1/f$. The unit of F is m^{-1}, which in this situation is called the *dioptre* (D). A typical pair of spectacle lenses for a short sighted person might be: left eye $- 2.0$ D, right eye $- 1.5$ D. When lenses are placed in contact the effective power is the sum of the individual powers; adding a $+ 3.0$ D lens to a $- 1.0$ D lens produces a lens combination of power $+ 2.0$ D.

Magnification

For an extended object the magnification m measured perpendicular to the principal axis (the *linear magnification*) is given by

$$m = \frac{\text{height of image}}{\text{height of object}}$$

m is obviously equal to v/u as we can see by looking at any ray diagram for a single lens such as (a) or (b) of figure 26.3 Thus in the above example, the magnification produced by the converging lens $= v'/u = 0.40$ m$/0.40$ m $= 1.0$, i.e. the image is the same size as the object. The magnification produced by the diverging lens $= v/v' = 0.60$ m$/0.40$ m $= 1.5$, i.e. one-and-a-half times. The total transverse magnification is now $1.0 \times 1.5 = 1.5$ times. For a system of lenses it is generally true that the total magnification is the *product* of the separate magnifications.

To measure the magnification produced by a lens or lens system directly it is best to use a brightly-lit millimetre scale as the object and to focus the image on a screen covered with millimetre graph paper, but such experiments are not very precise and if a careful measurement is needed it is best to measure the ratio v/u and not the ratio of the heights.

Measuring focal lengths

Before attempting to measure the focal length of a lens carefully we should always try first to get a rough idea of its focal length. We can let light from a bright distant object fall on a converging lens and measure the image distance. For a diverging lens we find the weakest converging lens together with which it becomes (just perceptibly) converging. In each case a very quick experiment tells us f to $\pm 20\%$: quite good enough to design any subsequent experiment sensibly.

(i) *Converging lenses: the plane mirror method.* A point object placed at the principal focus will coincide with its image if a plane mirror is placed behind the lens and perpendicular to the axis. Using a brightly-lit object pin and the no parallax technique (page 338) to see when object and image coincide, the focal length of a lens for which $f \approx 200$ mm can be measured with an uncertainty of about

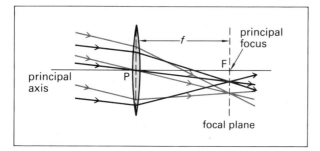

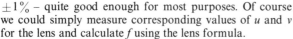

Figure 26.4 The focal length f and the focal plane of a converging lens.

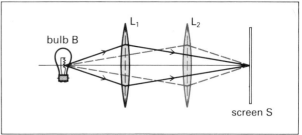

Figure 26.5 B and S are fixed. Lens positions L_1 and L_2 produce real images of the filament on the screen.

$\pm 1\%$ – quite good enough for most purposes. Of course we could simply measure corresponding values of u and v for the lens and calculate f using the lens formula.

(ii) *Diverging lenses: the virtual object method.* This involves setting up the situation described in the example on page 341. First we locate the position of the image I' formed by the converging lens (figure 26.3) and then, after inserting the diverging lens, locate I. We then have the values of the (virtual) object distance and the (real) image distance for the diverging lens, and can calculate f using the lens formula.

▶ The following measurements were taken in order to determine the length of the filament of a 12 V, 24 W bulb. The bulb B was set at a distance of about 1.20 m from a fixed screen S. A converging lens was moved between B and S in order to produce a real image of the bulb on the screen. At two positions, L_1 and L_2 of the lens separated by 0.69 m, clearly focused images were produced (figure 26.5). With the lens at L_1 the image was 22.5 mm long and with the lens at L_2 the image was 8.5 mm long. What is the length l of the filament?

The positions L_1 and L_2 must be symmetrically spaced between B and S; the values of u and v for the lens position L_1 become numerically those for v' and u' respectively for L_2.

For L_1 $\quad \dfrac{v}{u} = m_1 = \dfrac{22.5 \text{ mm}}{l}$

and for L_2 $\quad \dfrac{v'}{u'} = m_2 = \dfrac{8.5 \text{ mm}}{l}$

But $\quad \dfrac{v}{u} = \dfrac{u'}{v'}$

so that $\quad \dfrac{22.5 \text{ mm}}{l} = \dfrac{l}{8.5 \text{ mm}}$

and $\quad l^2 = (8.5 \text{ mm})(22.5 \text{ mm}) \Rightarrow l = 13.8 \text{ mm}$

Note that it was not necessary to know anything but the two image lengths in order to find l. It was not, for instance, necessary for us to know the exact position of the filament inside the bulb.

This experiment only works if BS is chosen to be greater than four times the focal length of the lens. We cannot get any images of the filament on the screen if BS is less than this value. When $BS = 4f$ a single image is formed; this is with the lens half way from bulb to screen, i.e. when $u = v = 2f$. ◀

26.2 Curved mirrors

When light is reflected from a curved mirror the law of reflection still applies but, as with lenses, we seldom get a perfect image because of spherical aberration (page 337). Object and image positions for *spherical* mirrors (the reflecting surface forms part of a sphere) are related by the same formula as for lenses. The proof for a concave mirror is based on figure 26.6. The ray reflected at R makes equal angles θ with the perpendicular at R which is the line from R to the centre of curvature C of the mirror surface. We then have

$$\theta = \gamma - a = \beta - \gamma,$$

or $\quad a + \beta = 2\gamma \qquad\qquad\qquad \text{(exactly)}$

This ray strikes the mirror a distance h from the axis.

$$\frac{h}{u} = \tan a \approx a \qquad\qquad \text{(appendix B5)}$$

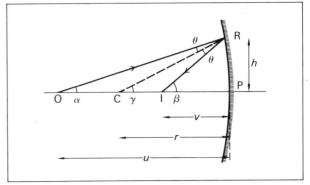

Figure 26.6 $1/u + 1/v = 2/r$ for a concave mirror.

Similarly $\dfrac{h}{v} \approx \beta$ and $\dfrac{h}{r} \approx \gamma$

where r is the *radius of curvature* of the mirror.

Thus $\quad \dfrac{h}{u} + \dfrac{h}{v} = \dfrac{2h}{r} \qquad$ (approximately)

or $\quad \dfrac{1}{u} + \dfrac{1}{v} = \dfrac{2}{r}$

If u becomes very large, $1/u = 0$ and $v = r/2$. We call this image distance f, the *focal length* of the mirror, and so we can write

$$\frac{1}{u} + \frac{1}{v} = \frac{2}{r} = \frac{1}{f} \tag{2}$$

The significant point about the result is that the relationship between u, v and f is independent of h. A similar proof of this mirror formula can be produced for a convex mirror and also, for both types of mirror, for all types and positions of object. Again in order to apply equation (2) in all cases we take u and v as positive to *real* objects and images (i.e. for objects and images *in front* of the mirror). Likewise for a concave mirror f and r are taken as positive, and for a convex mirror as negative.

To measure f or r it is again worth trying to get a rough idea of their size before attempting a careful experiment. E.g. with a concave mirror we can reflect a distant scene onto a piece of paper; we then have $f = r/2$. To measure r for a concave mirror more precisely we place the mirror on a low horizontal surface (e.g. the floor) and locate the centre of curvature, C, using a brightly-lit pin supported above it. When the object is at C the inverted image will be at C also so that we can use a no-parallax technique, moving the object pin until its position coincides with that of its image. The distance from C to the pole of the mirror P is then r; and $f = r/2$. We can measure r and f with an uncertainty of $\pm 1\%$ for mirrors of $f \approx 100$ mm. To find f for a convex mirror is not so easy. One method is to use a return image technique such as that outlined in the example below in which it is not necessary to know the focal length of the added converging lens.

▶ A converging lens forms a real image 260 mm from it. When a convex mirror is placed between the lens and this image it is found that the reflected image now coincides with the object. Draw diagrams to illustrate this situation and, if the mirror is 78 mm from the lens, deduce the focal length of the mirror.

Figure 26.7 shows what is happening. The rays are on this occasion drawn from a point on the axis as this is usually easier with return image situations. In (a) a search pin S′ is shown locating the position of the real image I′ formed by the converging lens. In (b) with the convex mirror in place the light returns along its path to form an image I which coincides with O, the original object pin. Because the light is reversed it must have been incident perpendicularly on the mirror, so that the mirror's centre of curvature C must be at the point occupied by the tip of the search pin S′ in

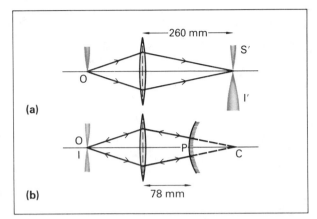

Figure 26.7 Finding the radius of curvature PC of a convex mirror.

(a). Clearly CP is equal to the radius of curvature r of the mirror, and so

$$f = \frac{r}{2} = \frac{(260 - 78)\,\text{mm}}{2} = 91 \text{ mm}$$

The lens-maker's formula

The surfaces of spherical lenses *reflect* some light, and form faint images. We can measure the radius of curvature of a concave lens surface optically just as for a concave mirror using a black background behind the lens. A non-optical method which quickly enables us to measure r for concave or convex surfaces uses a tripod device which rests on the lens surface. An adjustable leg is then pushed down to touch the surface and r is read directly. An interference method of testing a lens surface for sphericity is mentioned on page 335.

The radii of a lens's surfaces, together with the refractive index n of the material of which it is made, can be used to calculate its focal length. To reverse the argument it is possible to predict the radii of curvature needed to produce a lens of given strength from a given material, and a lens maker would require to know the relationship. It is readily shown that the angle A of the 'thin prism' formed by the lens surfaces at a distance h from the axis of the lens is given by

$$A = \frac{h}{r_1} + \frac{h}{r_2} = h\left(\frac{1}{r_1} + \frac{1}{r_2}\right)$$

and the deviation of a ray passing through the lens at height h is given by $D = h/f$. Thus using

$$D = (n - 1)A \qquad \text{(page 334)}$$

we have $\quad \dfrac{h}{f} = (n - 1)h\left(\dfrac{1}{r_1} + \dfrac{1}{r_2}\right)$

$$\Rightarrow \qquad \frac{1}{f} = (n-1)\left(\frac{1}{r_1} + \frac{1}{r_2}\right) \qquad (3)$$

This is sometimes called the lens maker's formula or the *full lens formula*. Thus to produce a lens with a focal length of 0.50 m using a plano-convex shape and glass of refractive index of 1.5, we have

$$\frac{1}{0.50\ \text{m}} = (1.5 - 1)\left(\frac{1}{r_1} + \frac{1}{r_2}\right)$$

As the plane side of the lens has an infinite radius of curvature, $r_1 = \infty$ and $1/r_1 = 0$; so we get $r_2 = 0.25$ m.

26.3 The eye

The Moon looks about the same size as the Sun. But so does a disc 20 mm in diameter (a penny) held just over 2 m away from the eye. All three form images on the retina which are the same size; each subtends the same angle at the eye – about 10^{-2} rad or 0.5°; figure 26.8a illustrates this idea.

The angles subtended by an object or by its image when viewed through an optical instrument, e.g. a magnifying glass, microscope or telescope, are called the visual angles of the object and the image respectively. The *visual angle* represents the apparent size of the object or image and hence determines the detail which can be seen. Clearly an object (or image) of given size has a larger visual angle if it is close to the eye. The penny mentioned above is seen in greater detail if it is placed 0.5 m rather than 2 m away and in even greater detail 0.1 m away *provided* the eye is then able to produce a focused image on the retina.

Figure 26.8b shows how the eye forms an image on the retina; (c) shows how we normally only draw rays from the top of the object and how the eye is usually represented symbolically without showing the formation of the retinal image. Finally (d) illustrates the principle for a *magnifying glass* held close to the eye. The rays we have chosen to draw go (i) straight through the optical centre of the converging lens and (ii) parallel to the axis so that they are refracted through the principal focus of the lens which is off the diagram to the right.

Diagrams of type (d) are usually drawn when studying optical instruments. The eye is taken to be very close to the optical centre of the lens through which it is looking, so that a_i (the visual angle of the image at the eye) is also the angle subtended by the image at the lens. In figure 26.8c and (d) only the top half of the ray diagram is drawn. This again is common practice and the visual angles a_o and a_i are, as here, taken to be half the angles subtended by the whole object or image at the eye.

Suppose in figure 26.8c the object O is placed as near to the eye as it can comfortably focus in order to study O. This position is called the *near point* and we take it to be 0.25 m from the eye. In practice this distance $D(= 0.25\ \text{m})$, called the *least distance of distinct vision*, varies from person to person and with age for any one person; 0.25 m is an agreed value for what we call a 'normal' eye. The visual angle of the object a_o is given, for small angles (appendix B5) by

$$a_o = \frac{h_o}{D} \qquad (4)$$

When the object O is viewed through a lens, a simple magnifying glass (figure 26.8d), O can be placed much closer to the eye. The virtual image I is usually arranged to be at the near point to get the clearest view. O and I subtend equal angles a_i at the eye, where

$$a_i = \frac{h_o}{u} \qquad (5)$$

and u is the distance of the object from the lens. Using this simple magnifying glass enables more detail to be studied than with an unaided eye because a_i is bigger than a_o.

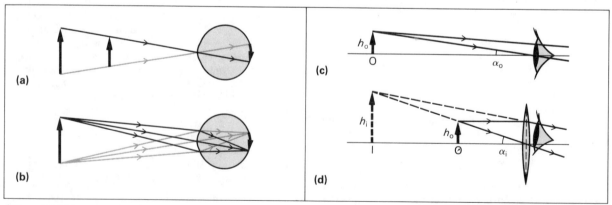

Figure 26.8 The eye and optical instruments. (a) Two objects of different size produce the same retinal image; (b) shows how the eye forms the image. (c) and (d) explain the action of a simple microscope or magnifying glass with the eye reduced to a representational symbol.

Accommodation

The power of a normal eye is $\approx +40$ D when it is relaxed (page 341). The shape of the eye's lens can be altered so that images of objects at different distances from the eye can all be formed on the retina. This adjustment of shape, which is achieved muscularly, is called *accommodation*. The power must increase to about $+44$ D to focus an object at 0.25 m – the near point. The range over which we can accommodate decreases with age falling from about 12 D in a baby to 4 D for the normal eye and to perhaps 1 D in old age.

A *short-sighted* person's eyes are too powerful. He will be able to focus objects at less than 0.25 m, perhaps only 0.10 m away ($+50$ D), but will be unable to focus distant objects clearly. His *far point* may be only 0.30 m from the eye and to see distant objects he wears diverging spectacle or contact lenses. On the other hand the eyes of a *long-sighted* person are not powerful enough. He can usually see distant objects clearly but his near point may be 1.0 m away ($+41$ D) or sometimes so far away that any object is difficult to focus. This defect is corrected by using converging lenses.

Angular magnification

The angular magnification or magnifying power of an optical system M is defined by the equation

$$M = \frac{a_i}{a_o}$$

The eye then judge the image to be M times larger than the object. In the discussion of a simple magnifying glass above $a_o = h_o/D$ (4) and $a_i = h_o/u$ (5) so that for a magnifying glass with the object a distance u from the lens and the (virtual) image at the near point

$$M = \frac{a_i}{a_o} = \frac{h_o/u}{h_o/D} = \frac{D}{u} \qquad \text{(image at near point) (6)}$$

The (virtual) image is arranged at the near point when the *maximum* value of M is required. For telescopes it is usual to arrange for the final image to be at infinity so that a relaxed eye may view it. We then make u equal to the focal length f of the lens through which the eye is looking. Equation (6) then becomes

$$M = \frac{D}{f} \qquad \text{(image at infinity)} \qquad (7)$$

▶ A single converging lens of focal length 50 mm is placed 50 mm from a pollen seed which measures 0.20 mm across (the person looking through the lens wants to examine the pollen for a long time). (i) What is the angular magnification achieved? (ii) What would it have been if the final image had been arranged to be at the near point?
(i) In figure 26.9a the object is placed 250 mm (0.25 m) from the eye, this being the least distance of distinct vision, and hence the visual angle of the object is

$$a_o = \frac{0.20 \text{ mm}}{250 \text{ mm}} = 8 \times 10^{-4} \text{ rad}$$

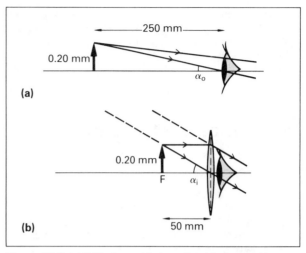

Figure 26.9 (a) With the unaided eye the object cannot be nearer to the eye than the least distance of distinct vision. (b) If the object is placed at the principal focus of a converging lens the final image is at infinity.

With the lens in position (b) and the object placed 50 mm from it (at its principal focus) the lens produces a parallel emergent beam with the image at infinity. The visual angle of the image is

$$a_i = \frac{0.20 \text{ mm}}{50 \text{ mm}} = 40 \times 10^{-4} \text{ rad}$$

so that the angular magnification is

$$M = \frac{a_i}{a_o} = \frac{40 \times 10^{-4} \text{ rad}}{8 \times 10^{-4} \text{ rad}} = 5$$

The retinal image is thus 5 times bigger using the lens. (ii) If the final object were to be at the near point then $v = -0.25$ m (a virtual image – see figure 26.8d); $f = +0.05$ m, and we can find the object distance u from

$$\frac{1}{u} + \frac{1}{-0.25 \text{ m}} = \frac{1}{+0.05 \text{ m}}$$

which gives $u = 0.042$ m or 42 mm. The visual angle of this image is now

$$a_i' = \frac{0.20 \text{ mm}}{42 \text{ mm}} = 48 \times 10^{-4} \text{ rad}$$

and the new angular magnification is

$$M' = \frac{48 \times 10^{-4} \text{ rad}}{8 \times 10^{-4} \text{ rad}} = 6$$

which is the largest angular magnification that can be achieved in viewing the pollen seed.
Note that:
(i) the values of M and M' do not depend on the size of the pollen seed (here 0.20 mm).
(ii) $M' = M + 1$; this is generally true for pairs of images at infinity and at the near point. (The difference between M' and M is hardly noticeable for large angular magnifications, say $M > 10$).

(iii) the eye can resolve two objects which are separated by about 3×10^{-4} rad (page 358) so that not much detail can be seen without the lens, since a_o is only 8×10^{-4} rad. Using the lens, when a_i is $> 40 \times 10^{-4}$ rad, quite a lot of detail can be distinguished. ◀

26.4 Optical instruments

The microscope

A single lens (a magnifying glass) is not very successful for angular magnifications of greater than about 5 because the image becomes distorted owing to various lens aberrations (page 337). A *compound microscope* consists essentially of (i) an objective lens which produces an enlarged real image of the object, and (ii) a magnifying glass (called the eyepiece lens), which is used to view the real image. Figure 26.10 shows the passage of two rays from the bottom of an object to the eye (the bottom is chosen so as to make the magnifying glass part of the diagram more obviously like figure 26.8d). We shall assume that the final image is at the near point; then the angular magnification is a maximum and this is the usual adjustment for a microscope. Referring to the microscope diagram we have

$$\text{visual angle of image} = \frac{\text{height of image}}{\text{eye-image distance}}$$

The distance from the eyepiece to the eye is exaggerated in the diagram and we shall assume that the eye-image distance $\approx D$.

Thus $\quad a_i = \dfrac{h_i}{D} \quad$ and $\quad a_o = \dfrac{h_o}{D}$

We obtain a_o from equation (4) on page 344 which explains how best to view the object with the unaided eye. The angular magnification M is given by

$$M = \frac{a_i}{a_o} = \frac{h_i/D}{h_o/D} = \frac{h_i}{h_o}$$

As $\quad \dfrac{h_i}{h_o} = \left(\dfrac{h_i}{h}\right)\left(\dfrac{h}{h_o}\right)$

so $M = \begin{pmatrix} \text{linear magnification} \\ \text{produced by eyepiece} \end{pmatrix} \times \begin{pmatrix} \text{linear magnification} \\ \text{produced by objective} \end{pmatrix}$

As this basic compound microscope consists of only two lenses each producing an image at a finite distance we can always approach microscope calculations from first principles and use the lens formula separately for each lens.

Spherical aberration limits the useful magnification, particularly for a very powerful objective lens which may have a focal length of less than 10 mm. A less powerful lens could produce the same objective lens magnification h/h_o, but would make the microscope longer and very unwieldy. A standard arrangement is to provide three separate objectives giving overall magnifications M of perhaps $\times 10$, $\times 40$ and $\times 100$. Only the first two of these could be achieved using the single objective lens of figure 26.10; an objective consisting of a number of lenses is needed to reach the highest M. If the magnification is increased diffraction eventually limits the ability of the microscope to *resolve* detail in an object. On page 359 the problem of diffraction-limited resolution is explained and the principle of the electron microscope outlined. An electron microscope may be able to give a useful magnification of 100 000 times and in so doing resolve down to about 10^{-9} m.

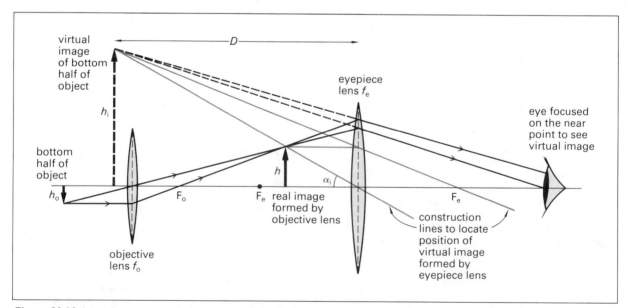

Figure 26.10 A two lens compound microscope with the final image at its usual position — the near point of the user's eye.

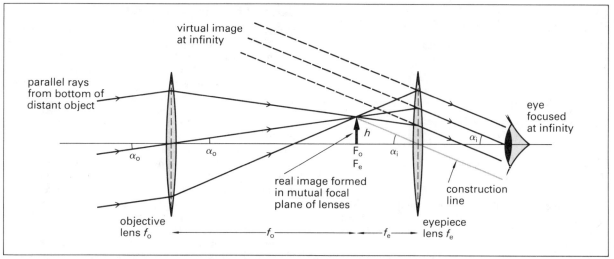

Figure 26.11 A two lens refracting astronomical telescope with the final image at its usual place – infinity.

The telescope

With microscopes the object can be illuminated strongly and placed wherever we wish but the purpose of an *astronomical telescope* is to study inaccessible objects. They are so far from the telescope that we can draw a beam of rays diverging from a point on the object (e.g. a lunar crater) as a parallel beam when it enters the telescope. As telescopes are generally used for long viewing periods the usual adjustment is with the final image at infinity.

In a refracting telescope a weak converging objective lens produces a real image of the object in its focal plane (figure 26.11); the focal length is typically a few metres in large instruments. The size of this image is $h = f_o a_o$ and so the weaker the lens the larger is the image. It is viewed by a magnifying glass (called the eyepiece); the final image is arranged at infinity. The figure has these two optical arrangements put together; it shows that the lens-to-lens distance, the length of the telescope, is $f_o + f_e$. The eye is placed as shown, since this is the place where it can collect the light passing through all parts of the objective lens.

If h is the size of the intermediate real image then the angular magnification M is given by

$$M = \frac{a_i}{a_o} = \frac{h/f_e}{h/f_o} = \frac{f_o}{f_e}$$

The most powerful refracting telescope in the world has $f_o \approx 20$ m and $f_e \approx 6.5$ mm, so that the greatest available value of M is about 3000. Its objective lens is about 1 m in diameter.

The ability of a telescope to produce sharp and easily resolved images of two close objects is affected by (i) the chromatic and spherical aberration in its lenses and (ii) diffraction (wave) effects which are determined by the diameter of the objective lens (page 359). To overcome chromatic aberration a combination of lenses made of materials with different optical properties can be used (page 338).

Reflecting telescopes

The largest optical telescopes all use a concave objective *mirror* rather than a converging objective lens. The reasons are that
(i) there is then no chromatic aberration,
(ii) spherical aberration can be eliminated for objects on the axis of the mirror by making the reflecting surface paraboloidal,
(iii) the mirror can be supported from beneath over all its area (a lens can be held only round its edge),
(iv) it is difficult to produce a large lens free from internal variations in refractive index whereas for a mirror only its surface needs to be 'perfect'.
The large apertures possible with reflecting telescopes both increase their ability to resolve and also enable fainter stars to be seen as they gather more light from a given source.

The camera

Figure 26.12 shows a simple camera. The converging lens forms a real image on the film. A better camera will have a system of lenses in place of the single meniscus lens shown (which is meniscus-shaped in order to reduce the distortion of images by spherical aberration). The diameter of the aperture stop is specified as a fraction of the focal length f of the lens, and can be set at $f/1.4$, $f/2$, $f/2.8$, $f/4$, $f/5.6$, $f/8$, $f/11$, $f/16$, $f/22$ or $f/32$. The numbers are chosen so that moving between adjacent *f-numbers* changes the area of the stop by a factor of 2; the larger the f-number the smaller the aperture. The size of the aperture stop not only controls the light entering the camera when the shutter is open but also the *depth of field*. For large f-numbers both the foreground and background will be well enough focused to be

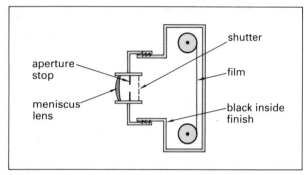

Figure 26.12 A simple camera.

seen on the photograph (so large *f*-numbers are used for landscapes) but for small *f*-numbers only a small range of distances will appear in focus (so small *f*-numbers are used for portraits).

To change the object distance at which a properly focused image is produced on the film the distance between the lens and the film is adjusted, usually with a screw as indicated in figure 26.12.

The shutter opens for a small time interval Δt when a picture is taken. The correct combination of exposure time Δt and *f*-number depends on the brightness of the object and the type of film used.

When using a telescope to photograph a distant object the film or plate is placed to record the real image formed by the objective lens. The telescope is then nothing more than a very long focal length camera. We can then examine the photograph with a magnifying glass or a low-powered microscope. Similarly a microscope objective can be used as a very short focal length camera with the film or plate placed to record the intermediate real image. The size of the photographed image depends only on the objective lens in both cases.

26.5 The spectrometer

Light can be dispersed using a prism or a diffraction grating. The glass of which a prism is made has a refractive index which varies with wavelength (page 338), so that different colours are deviated through different angles. Diffraction phenomena also depend on wavelength so that gratings produce colour maxima in different directions (page 360). To examine a *pure spectrum* of light from a source and to measure its properties requires an instrument (i) that produces a plane wavefront (a parallel beam of rays) incident on the prism or grating, (ii) that enables the direction of plane emerging wavefronts to be examined, and (iii) that provides an adjustable support for the prism or grating so that it can be correctly orientated relative to the incident light.

The *optical spectrometer* is designed to achieve all three of these requirements. Figure 26.13 shows how light of two colours from one edge of the slit forms two images in the plane of the crosswires after deviation through the prism. (Figure 27.17 (page 361) shows the general arrangement using a grating.) Notice that the spectrum seen is really a series of images of the slit. The narrower the slit the greater the possibility of separating two colours in the spectrum providing the optical system can then resolve the two images (page 359). A diagram very like figure 26.13 could be drawn with a grating in place of the prism; for a grating the image for the 'redder' light would be deviated most.

Measuring refractive index

If a sodium lamp is viewed through a prism a series of coloured images of the lamp is seen, the most prominent one being yellow. Hold a prism and, viewing with the naked eye, notice what happens as the prism is rotated. There is one position of the prism at which the deviation D of the beam is a minimum for the yellow light. We have shown

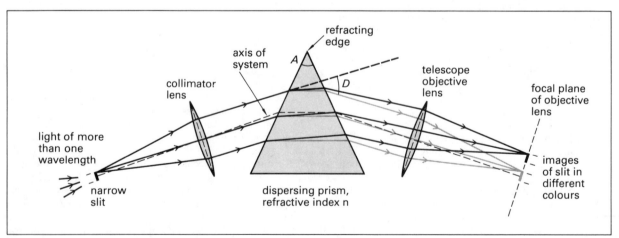

Figure 26.13 A ray diagram (not to scale) for a prism spectrometer showing the formation of two images of the slit. The image of the 'bluer' light is at the centre of the crosswires which lie in the focal plane of the eyepiece lens. The telescope eyepiece system is not shown. Note the axis of the system which is shown by the dashed line.

previously (page 333) that this minimum deviation $D_{\min}$ occurs when

$$n \sin \frac{A}{2} = \sin \left(\frac{A + D_{\min}}{2} \right)$$

where A is the refracting angle of the prism. This provides a basis for the precise measurement of the refractive index n of the material of the prism for any given wavelength. A spectrometer is used and the procedure is as follows (refer to figure 27.17, page 361).

(i) Adjust the *eyepiece* so that the crosswires C are clearly focused.

(ii) Look through the *telescope* at a distant object O and adjust the distance between the objective lens and the crosswires so that there is no parallax between C and the image of O. C now lies in the focal plane of the objective lens.

(iii) Illuminate the slit with the light to be used, e.g. from a sodium lamp. Swing the telescope (it can rotate about a vertical axis through the centre of the table) to accept light straight from the *collimator*. Adjust the distance between the collimator slit S and the collimator lens until an image of the slit is seen clearly in focus. S now lies in the focal plane of the collimator lens.

(iv) Place the *prism* on the table and, by swinging the telescope to receive light reflected from the side of the prism, adjust the table until the refracting edge of the prism is vertical and remains so as the table is rotated.

(v) Measure the refracting angle A of the prism. This is done by swinging the prism round so that the refracting edge roughly faces the collimator (figure 26.14). The angle between the telescope positions for light reflected from the sides of the prism is $2A$. It can be measured on the circular scale using the vernier which travels with the telescope.

(vi) Measure the minimum deviation $D_{\min}$ for the light of the required colour refracted through the prism. The right position for the prism and telescope are first roughly found and then the table rotated slowly until the refracted image just reaches the centre of the crosswires before retreating. $D_{\min}$ can be measured from this position to the 'straight through' position on the circular scale, again using the telescope vernier.

The value of n can now be found using the relation for minimum deviation given above, and may be measured with an uncertainty of about $\pm 0.1\%$ – a very precise measurement. The refractive index of a liquid can similarly be found by using a hollow prism filled with the liquid.

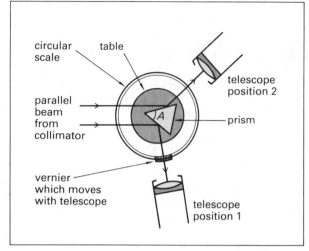

Figure 26.14 Using a spectrometer to measure the refracting angle of a prism. The telescope swings through an angle $2A$ between positions 1 and 2.

Measuring wavelength

To measure wavelengths we must either calibrate a prism spectrometer using a series of spectral lines of known wavelength e.g. those corresponding to certain absorption lines (Fraunhofer lines) in the Sun's spectrum, or use a diffraction grating with a known number of slits per unit length. When using a grating it must be set on the table with its fine slits or lines parallel to the collimator slit and its plane perpendicular to the beam of light from the collimator. Then, for the first order spectrum, $\lambda = s \sin \theta$ (page 360). If s is not known only one known wavelength is needed to find s and thus to calibrate the spectrometer.

In the above uses of a spectrometer all the spectra have been examined directly by the observer looking into the eyepiece. It is often more convenient, perhaps for light from a faint source (a star) or for wavelengths just beyond the visible, to take photographs by placing a photographic plate in the focal plane of the telescope objective lens in place of the crosswires, after the spectrometer has been properly adjusted.

In section 28.5 (page 378) various types of optical spectra are described and their significance in interpreting the structure of the atom or the constituents of stars and other extra-terrestrial matter is discussed.

27 Patterns of superposition

27.1 Coherence

All points on a wavefront from a single source, e.g. a loudspeaker, have the same phase (page 328). The output frequencies of two oscillators driving two loudspeakers will probably drift apart slowly and so the phase difference between them will be changing. If two wave sources are driven from the *same* supply, however, then there will be a constant phase difference between them and the two *sources* are said to be *coherent*. They will remain coherent sources if the single oscillator does drift or even if it undergoes occasional random fluctuations, for when the phase of one of the sources S_1 changes then so will the phase of the other source S_2. We can also have a constant phase difference between two sources which is not zero, e.g. two ripple tank dippers, where S_1 moves up as S_2 moves down, rather like a drummer's two sticks during a roll. In this case S_1 and S_2 are in antiphase (180° out of phase); so are two loudspeakers when the connections from a single oscillator to one of them are reversed.

For electromagnetic radiation in the visible spectrum it needs some care to devise a pair of coherent sources. Atoms emit light when they make transitions from one energy level to another. The time which an atom takes to undergo such a change is of the order of 10^{-8} s which is long compared with 10^{-15} s, the period of the emitted wave. An atom thus produces a pulse of light, or *wavetrain*, about $(10^{-8}$ s$)(3 \times 10^8$ m s$^{-1}) = 3$ m long. As atoms are usually excited by collisions during their random thermal motion in a discharge tube or a filament lamp, the light from a large number of atoms, each radiating over a time of about 10^{-8} s, is made up of a *random* collection of wavetrains. The atoms form a non-coherent set of sources. Two light sources S_1 and S_2 are made up of different sets of atoms, so that even if the radiation is all of the same wavelength (monochromatic) S_1 and S_2 will inevitably be non-coherent. (Of course, any *two* atoms emitting light of the same wavelength can be coherent over a time interval of less than 10^{-8} s). When we apply the principle of superposition (page 312) to light from two non-coherent sources the resulting interference pattern changes very rapidly (millions of times a second) and there is no observable pattern of maxima and minima.

Coherent light

In order to produce two or more effectively coherent sources with light it is necessary to take a wavefront from a single source and to use it twice. There are essentially two different ways of achieving this: they are shown in figure 27.1

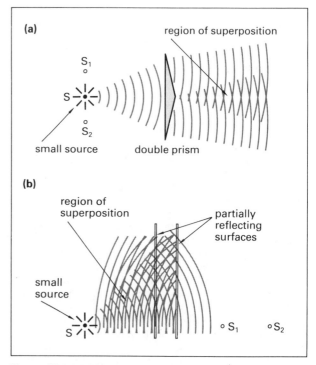

Figure 27.1 (a) Division of wavefront: the upper and lower parts of the incident wavefront refract to form (virtual) coherent sources at S_1 and S_2. (b) Division of amplitude: a fraction of the energy of the whole incident wavefront is reflected to the left at each surface to form (virtual) coherent sources at S_1 and S_2

In (a) a double prism, or biprism, refracts the upper and lower *parts* of the incident wavefronts so that they overlap to the right of the prism. Two different parts of the wavefront are used to produce a superposition pattern here; the technique is called *division of wavefront*. Another example is provided by two slits in a screen; these produce overlapping diffracted wavefronts. In (b) a fraction of the energy of the *whole* wavefront is reflected at each of two partially reflecting surfaces. The reflected wavefronts overlap to the left of the surfaces. This technique for producing a superposition pattern is called *division of amplitude*. As all points on the *same* wavefront must be in phase at all times the two derived (virtual image) sources S_1 and S_2 in cases (a) and (b) must be coherent.

The laser

A laser (the acronym for Light Amplification by the

Stimulated Emission of Radiation) produces a coherent beam of light. Unlike the atoms in a filament lamp which emit in a random way the light from one atom in a laser source is made to trigger or stimulate the emission of light from another atom and so on through the length of the laser tube. Each emission is in phase with the stimulating light and in this way the amplitude of a single wavefront is increased (hence light *amplification* in the name).

A typical output from a small helium-neon gas laser is about 5×10^{-4} W at a wavelength of 633 nm. The laser tube might be 1.0 mm in diameter and 200 mm long so that the coherent emerging beam is itself a millimetre across. This is the reason why laser light spreads out very little and can be beamed over long distances. The coherence of the beam also allows the possibility of using it as a high frequency ($\approx 10^{15}$ Hz) carrier wave for the transmission of information using frequency modulation (f.m.) techniques.

Two-source superposition patterns

(You should refer to sections 23.3 and 23.4 as an introduction to what follows.) When two coherent sources S_1 and S_2 of about the same amplitude are placed a few wavelengths apart a pattern of superposition is produced. Such a pattern is usually referred to as an interference pattern. The simplest two-dimensional demonstration is with ripples as shown in figure 27.2, but the sources could equally well be producing waves in three dimensions, e.g. using two small speakers driven from the same signal generator with a microphone (or the human ear) as the detector.

Consider the point P in figure 27.2b. If the sources are in phase then the phase difference between the two waves arriving at P from S_1 and S_2 is given by (page 314)

$$\text{phase difference} = \frac{2\pi}{\lambda}(S_2P - S_1P)$$

where λ is the wavelength and S_1P and S_2P are the distances of P from the two sources. $S_2P - S_1P$ is called the *path difference* for the two waves from S_2 and S_1 arriving at P, so that

$$\text{phase difference} = \frac{2\pi}{\lambda}(\text{path difference})$$

At places where the phase difference is 0 or 360° (2π rad) or a whole number multiple of this, i.e. $2\pi n$ radians, the waves will superpose to give a *maximum* displacement amplitude. When the phase difference is 180° (π rad) the superposition will be such as to produce a *minimum* which will be zero if the waves have equal amplitude at P. The waves at P are also in antiphase when the phase difference is 180° + 360° etc., i.e. $(2n + 1)\pi$ radians.

Therefore, for *maximum* $\frac{2\pi}{\lambda}(\text{path difference}) = 2n\pi$

i.e.   path difference $= n\lambda$

and similarly, for *minimum*   path difference $= (n + \frac{1}{2})\lambda$

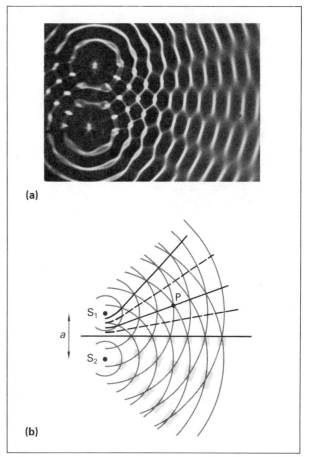

(a)

(b)

Figure 27.2 (a) A typical two-source pattern in a ripple tank and (b) two ways of sketching the antinodal areas (full lines or rows of shaded zones) where the amplitude of the displacement is a maximum, i.e. lines of maximum intensity.

For a given value of n, P will lie on a curve (a hyperbola) like those in figure 27.2a.

For ripple tank dippers at S_1 and S_2 we might find that $S_2P = 360$ mm and $S_1P = 210$ mm at the second minimum away from the central maximum. This gives a path difference of 150 mm and, as $(n + \frac{1}{2})$ is here equal to 1.5, a wavelength of 100 mm. Choosing other positions for P enables us to make several independent measurements of λ.

Superposition and energy

Figure 27.3 shows a two-source experiment using sound waves (5 kHz would be a suitable frequency). S_1 and S_2 are identical small loudspeakers and M is a sensitive microphone. Suppose M is placed at a point on the line bisecting S_1S_2; S_2 is switched off by disconnecting it from the signal generator, and we find that the height of the trace on the c.r.o. ($\propto$ wave amplitude) is h. If S_2 is now switched on the length of the trace increases to $2h$. This tells us that the

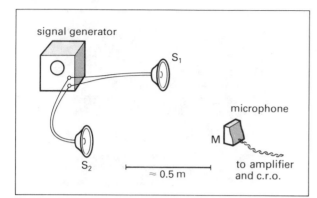

Figure 27.3 A two-source experiment using sound waves of about 5 kHz. S_1 and S_2 should be separately connected to the signal generator.

amplitude of the wave detected by M has doubled when the waves arriving at M are in phase. As the intensity of the wave $\propto$ (amplitude)^2, then the intensity with both S_1 and S_2 connected is four times that with S_1 alone (see also page 314). At places where the waves from S_1 and S_2 drive in antiphase very little sound energy is detected by the microphone.

The result of superposing the waves is thus to redistribute the energy in the interference pattern. There is four times the energy from one source at a maximum and zero energy at a minimum. The average energy taken over all positions for M is just twice the energy from one of the sources; as we would expect. When the microphone is moved about in front of S_1 and S_2 the trace oscillates in size. The extreme values of $2h$ and zero occur only near the centre line as at other points M is closer to one of the sources than the other and the amplitudes of the two incident waves are then not equal.

The same redistribution of energy occurs in the interference pattern produced by any pair of coherent sources. e.g. in the light experiment using a biprism (page 350). But when the sources are non-coherent, e.g. two violins playing the same note or two sodium lamps, there is simply twice the intensity produced by one of the sources at all points, and a uniform intensity pattern of sound or light with no maxima or minima, i.e. no interference pattern. For both coherent and non-coherent sources energy is, of course, conserved.

27.2 Young's experiment

The British scientist Thomas Young (1773–1829) was the first to perform and appreciate the significance of a two-source superposition experiment with light. To repeat the experiment in its essentials all we need is a piece of aluminium foil and some thin copper wire. We pull the wire until it breaks and use the point to punch two holes as close

together as possible in the foil. Then we hold up the foil to the eye and view a small lamp some distance away.

To perform the experiment in such a way that we can make measurements we use the arrangement shown in figure 27.4a. If the adjustable slit S is small then S_1 and S_2 form coherent sources. For a well-defined interference pattern to be observed it is also necessary that
(i) S_1 and S_2 are themselves narrow slits of roughly equal width,
(ii) the source produces light within a small wavelength range, e.g. a sodium lamp,

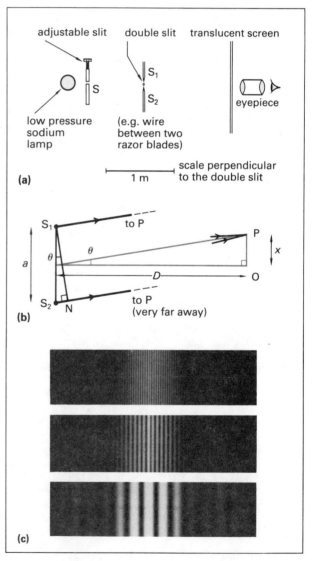

Figure 27.4 Young's experiment (a) in practice – note the horizontal scale; (b) the theory. (c) What is seen: in each case the width of the slits is 0.07 mm while a is varied from about 1.4 mm in the top to 0.2 mm in the bottom photograph.

(iii) S, S_1 and S_2 are parallel and

(iv) the screen is not too close to the double slit (the slits are not shown to scale in the figure; S_1 and S_2 are much closer together than they appear).

A typical value for the width d of each slit S_1 and S_2 is 0.10 mm, while their separation $S_1 S_2 (= a)$ might be 1 mm (i.e. $a \approx 10d$).

The interference pattern consists of light and dark lines parallel to the slits. They are called *Young's fringes* and, with the left-to-right scale suggested in figure 27.4a, the pattern may cover a centimetre or so of the screen.

Consider figure 27.4b. For a maximum at P we need

$$S_2 P - S_1 P = n\lambda$$

If P is a very long way from the slits (i.e. $D \gg a$), then $S_1 P$ and $S_2 P$ are almost parallel

and $\quad S_2 P - S_1 P = S_2 N = n\lambda$

where N is the base of the perpendicular from S_1 to $S_2 P$.

Now $\quad \dfrac{S_2 N}{a} = \sin \theta \approx \theta \qquad$ (appendix B5)

and $\quad \dfrac{x}{D} = \tan \theta \approx \theta \qquad$ (appendix B5)

so that $\quad S_2 N \approx \dfrac{ax}{D}$

Thus the condition for a maximum at P is that

$$\frac{ax}{D} = n\lambda \qquad (1)$$

If we now move the eyepiece (e.g. with a micrometer screw) from one maximum, say $n = 3$, to the next maximum, $n = 4$, it moves through a distance Δx called the *fringe separation*.

As $\quad \dfrac{ax}{D} = n\lambda$

and $\quad \dfrac{a(x + \Delta x)}{D} = (n + 1)\lambda$

then $\quad \dfrac{a\Delta x}{D} = \lambda$

i.e. $\quad \lambda = \dfrac{a}{D} \text{(fringe separation)} \qquad (2)$

► In a Young's slit experiment with sodium light, the eyepiece was moved 3.2 mm in counting from 0 to 4 fringes. If the distance from the double slit to the eyepiece was 0.96 m calculate the slit separation. Take the wavelength of the light to be 590 nm.

The fringe separation = 3.2 mm/4 = 0.80 mm. Using equation (2) above gives

$$590 \times 10^{-9} \text{ m} = \frac{a}{0.96 \text{ m}} (0.80 \times 10^{-3} \text{ m})$$

$$a = \frac{590 \times 10^{-9} \times 0.96}{0.80 \times 10^{-3}} \text{ m}$$

$= 0.71$ mm ◄

The above is a straightforward but not very precise method for finding the wavelength of visible electromagnetic radiation. We cannot measure a and Δx precisely and so we must expect an uncertainty of at least $\pm 5\%$ even if the experiment is performed with care. What can be quickly achieved with Young's fringes is a genuine superposition experiment with light and a quick check on the relative wavelengths of two colours. E.g. it is easy to show that $\lambda_{\text{red}} > \lambda_{\text{green}}$ by using a white light source and placing first a red and then a green filter in front of the eyepiece.

Equation (2) tells us that the fringe separation Δx is proportional to D, the distance from $S_1 S_2$ to the screen, and inversely proportional to a. The fringes can be seen anywhere to the right of $S_1 S_2$ where the diffracted beams overlap.

If the source is not monochromatic, e.g. a white light source, then there is a set of overlapping fringe systems beyond the slits. As the eye is very much more sensitive to wavelengths in the range 500 nm to 600 nm than in the rest of the visible spectrum, a set of Young's fringes in white light *is* observable for values of n up to about 3 or 4 before the overlapping of different colours obscures the pattern. The inner fringes have detectably coloured edges.

Young's experiment and diffraction

The experiment depends upon diffraction to achieve two coherent sources. The light waves from S_1 and S_2 spread into the space between the slits and the screen in the same way as the ripples on the water surface in figure 25.13. The intensity of the diffracted wave is plotted against θ in figure 27.12 (page 358). The central diffraction maxima from the two sources in the Young's slit experiment overlap and the wave amplitudes *within the overlapping region* then superpose to provide the interference pattern.

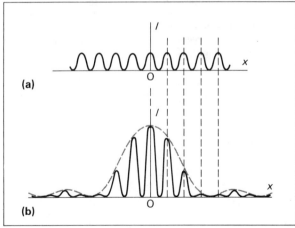

Figure 27.5 Intensity of Young's fringes across the screen (a) assuming uniform wave amplitudes from S_1 and S_2 and (b) with diffraction-limited wave amplitudes from S_1 and S_2.

If the wave amplitudes are independent of the angle of diffraction in the overlap region then the resulting intensity curve is sinusoidal as in figure 27.5a. Because of the modulating effect of diffraction, however, curve (b) is what we actually see, as in the photographs of figure 27.4c. A careful study of the photographs of figures 27.4 and 27.12 should help to make clear the effect of diffraction at a slit in the formation of Young's fringes.

A large-scale version of the Young's slit experiment can be performed with microwaves. The transmitter is placed about 0.5 m from a double slit consisting of two metal sheets with a narrow strip of metal (about 50 mm wide) between them. The receiver detects the characteristic pattern of superposition beyond the double slit and, by covering and uncovering one of the slits, both the diffraction effect and the redistribution of energy in the interference pattern can be studied. As D is *not* very much larger than $S_1 S_2$ in this demonstration we cannot make the approximation which led to equation (2).

Lloyd's mirror

It is possible to divide a wavefront to produce two side-by-side coherent sources in other ways than by diffraction at a slit or using a double prism. For example, figure 27.6 shows the principle of a method using microwaves; when used with light this is called the Lloyd's mirror experiment. Half the wavefront from S_1 is reflected thus producing a virtual source S_2. The receiver R is moved as the arrows indicate. We find the characteristic pattern of interference fringes because the geometry of the situation is exactly the same as in Young's experiment. We can find a rough value for the wavelength by measuring the fringe width which is very roughly equal to $\lambda D/a$. When R is placed right up to the reflecting surface (the metal sheet) it is found that the result of superposition is to produce a *minimum*, effectively a signal of zero amplitude. The path difference is, however, zero at that point. We can only conclude that the reflected wave has undergone a *phase change* of 180° at the metal sheet and so the waves reaching R are in antiphase. (A similar phase change on reflection for mechanical waves is discussed on page 313.) The experiment can be performed on a much smaller scale with light waves if we use a front-reflecting mirror of good quality. In this case we see a dark fringe at the mirror surface.

In using microwaves to demonstrate superposition effects we see that it is the (vector) addition of E-fields and B-fields to which we refer when discussing the superposition of electromagnetic waves; see section 28.2.

▶ A Young's slit arrangement with light of wavelength 590 nm is set up and viewed through a calibrated eyepiece. The light from one of the slits is arranged to pass through a tube 20 mm long from which the air can be removed with a vacuum pump. As the air is pumped out the fringe system is observed to shift sideways through a distance equal to eight times the fringe width. What can you deduce from this experiment?

As the fringes move sideways then the phase of the wave arriving at the eyepiece through the air cell has been altered. This movement is here equivalent to a phase shift of $8 \times 2\pi$ rad. Suppose that in 20 mm of air there are n wavelengths of light

i.e. $n\lambda_{air} = 20 \times 10^{-3}$ m

then in 20 mm of vacuum there are $(n - 8)$ wavelengths ($\lambda_{air} < \lambda_{vac}$, since the speed of light in air is less than the speed of light in a vacuum).

$\Rightarrow \qquad (n - 8)\lambda_{vac} = 20 \times 10^{-3}$ m

As $\lambda_{vac} = 590$ nm $= 590 \times 10^{-9}$ m

then $n - 8 = \dfrac{20 \times 10^{-3} \text{ m}}{590 \times 10^{-9} \text{ m}} = 33\,898$

and so $n = 33\,898 + 8 = 33\,906$

$\Rightarrow \qquad \dfrac{\lambda_{vac}}{\lambda_{air}} = \dfrac{n}{n - 8} = 1.00024$

But as the frequency f is constant this is the ratio of the speeds of light in the two media, which is the refractive index of air (page 331). Thus we can deduce from the experiment that under normal laboratory conditions the refractive index of air is $1.000\,24$. ◀

27.3 Thin film patterns

The most commonly observed optical interference patterns are the colours seen on puddles which have a film of oil on their surface or the colours which are seen in soap bubbles. These patterns differ from those described in the previous section in that (i) the sources are not point sources and (ii) we have to focus our eyes on the surface of the oil or soap film in order to see the interference patterns. We can produce similar effects in a simple laboratory experiment by placing two clean glass blocks on top of one another and looking at the reflection of a white fluorescent light in the blocks (a sodium lamp is even better). After a moment or two the eye accidentally focuses on the plane of the gap

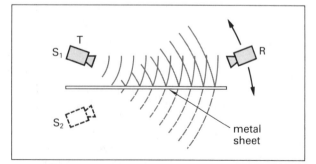

Figure 27.6 A 30 mm microwave 'Lloyd's mirror' experiment.

between the blocks and a yellowy-orange pattern rather like a contour map is seen which changes if the blocks are pressed together or twisted while in contact. Once found it is easy to study the pattern in detail.

Wedge fringes

Consider a uniform wedge of air trapped between two perfectly flat surfaces such as two glass blocks touching at one end and separated by a strip of thin paper, i.e. a very small distance, at the other. Suppose the air wedge is illuminated from above with light of wavelength λ and can be viewed from above as in figure 27.7a.

At a distance x from the apex of the wedge consider the beams which are reflected (i) from the lower surface of the upper block and (ii) from the upper surface of the lower block (the beams p and q in figure 27.7b). They are coherent. They enter the eye and are superposed at the retina. We shall ignore the bending of the light as it passes from glass to air or from air to glass because the angles of incidence are all very close to zero. The *extra* distance travelled by wave q which crosses the air wedge before being reflected is $2t$, where t is the thickness of the wedge at that point. This extra path is equivalent to a phase difference ϕ, where

$$\phi = \frac{2\pi}{\lambda}\,(\text{path difference}) = \frac{2\pi}{\lambda}\,(2t)$$

The superposition of the two wavefronts thus leads to bright and dark lines across the wedge for different values

Figure 27.8 Wedge fringes observed from the side. The blocks are in contact at the right edge and are separated by only about 1.5 μm at the left.

of t (figure 27.8). At the right edge of the photograph, where $t = 0$, there is a dark line. We know that there is a phase change for wave q (the Lloyd's mirror experiment tells us this), so wave p must be reflected *without* any phase change. So p and q are in antiphase close to 0 and produce a dark line there.

Considering the superposition of the two wave-fronts, we thus have

$$2t = n\lambda \qquad \text{for } minimum$$
$$\text{and} \quad 2t = (n + \tfrac{1}{2})\lambda \quad \text{for } maximum$$

E.g. for light of wavelength 660 nm there will be dark lines at $t = 0$, $t = 0.33\,\mu m$, $0.66\,\mu m$, $0.99\,\mu m$ etc. If the twentieth dark fringe is 10 mm from the line of contact, then the value of t there is 6.6 μm and the angle a of the wedge in figure 27.7b is given by

$$a \approx \tan a = \frac{6.6 \times 10^{-6}\text{ m}}{10 \times 10^{-3}\text{ m}} \qquad \text{(appendix B5)}$$

$$\Rightarrow \qquad a = 6.6 \times 10^{-4}\text{ rad or } 0.038°$$

If the lower block is known to have a perfectly flat upper surface we can study the flatness of other surfaces placed on it by observing the patterns of fringes. The lines of the interference pattern act as 'contour' lines, i.e., lines of equal t. The upper block has a lower surface which is flat to better than one wavelength (refered to as *optically flat*) if the wedge fringes are quite straight and are equally spaced.

A convex spherical surface, e.g. a weak converging lens, produces a circular pattern of fringes. These are not equally spaced but get closer and closer the further they are from the centre. Any deviation from circles indicates a lack of symmetry of the lens surface. Such circular interference patterns are sometimes called *Newton's rings*.

With monochromatic light (e.g. a sodium lamp) up to hundreds of fringes can be seen but with white light illumination separate fringes can only be seen up to about $n = 5$.

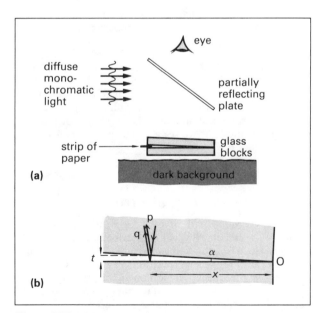

Figure 27.7 (a) The experimental set up for studying wedge fringes; (b) shows the essential geometry.

Whenever thin film fringes are observed in daylight, we know that we are dealing with surfaces which are less than a micrometre apart.

Bragg reflection

To simulate the thin film situation with 30 mm microwaves we need two reflecting surfaces, the first of which is only a partial reflector (figure 27.9). Suppose the two reflectors R_1 and R_2 are arranged to be *parallel* and that the microwave receiver is positioned to receive reflected waves from both. The simplest case is when $\theta = 90°$, i.e. when the incident beam strikes the reflectors at right angles. The extra distance travelled by the wave reaching R_2, the path difference, is then $2t$. This leads to the condition

$2t = n\lambda$ for maximum,

there being a phase change of 180° at *both* surfaces in this case.

If the separation t of the reflectors is changed, keeping $\theta = 90°$, R alternates between registering maximum and minimum signals. The experiment is easily performed and is the simplest way of finding the wavelength of the microwaves; t changes by $\lambda/2$ between successive minima.

Suppose, however, that t is fixed, i.e. the two reflectors are rigidly spaced a few centimetres apart. The extra distance travelled by the wave reflected from R_2 can now be altered only by altering θ (i.e. by making the incident beam strike

the reflectors obliquely) while arranging for the receiver to move so as to pick up the reflected waves. Figure 27.9b shows the geometry of the situation. FCG (*not* BCD) is the path difference between the two reflected waves as we can see by noting that the incident wavefront up to MF and the reflected wavefront from MG travel equal distances. If the angle between the incident direction and the reflector is θ (n.b. the glancing angle and not the usual 'angle of incidence' which would be $90° - \theta$),

then $FC = GC = t \sin \theta$
and $FC + CG = 2t \sin \theta$ = path difference

and so there will be a maximum at the receiver if

$$2t \sin \theta = n\lambda \qquad (3)$$

i.e. $\theta = \arcsin \dfrac{n\lambda}{2t}$

(arcsin x means the angle whose sine is x.)

For a fixed value of t there may therefore be a number of values of θ for which the receiver registers a maximum, i.e., for which the receiver registers strong 'reflected' waves. If, for instance $t = 1.4\ \lambda$, then the maxima will occur at

for $n = 1$: $\theta_1 = \arcsin (1/2.8) = 21°$
for $n = 2$: $\theta_2 = \arcsin (2/2.8) = 46°$

At other angles the received signal will vary between the maximum and zero. For this value of t there is no value of θ for $n = 3$, as then $\sin \theta$ is greater than 1. This type of interference is sometimes called *Bragg reflection* (after Sir William and Sir Lawrence Bragg, who developed the idea in 1913 and used it with X-rays to study crystal structure page 362).

When the incident radiation is not from a single source, i.e. the incident beam is not a coherent wave, the situation is much more complex. This is the case when we look at a thin parallel-sided film of oil on the surface of water, at an insect wing, or at mother of pearl, all of which exhibit *colour effects*. As we look at different parts of the film the angle θ changes. For a given value of θ light of one or more wavelengths superposes to give a maximum and for other wavelengths leaves a minimum. The result is that instead of reflected white light we see a colour which changes for different values of θ.

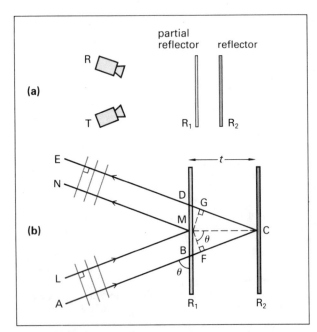

Figure 27.9 A double reflection experiment with microwaves. R_1 is a partial reflector (hardboard) and R_2 a full reflector (aluminium) of microwaves.

Stationary wave patterns

In section 23.5 the pattern of superposition known as stationary (or standing) waves is described for mechanical waves on strings and springs. Figure 23.13 (page 316) shows the result of the superposition of two waves of equal amplitude and frequency. In section 24.3 stationary waves in sound are discussed. The stationary wave pattern of *nodes and antinodes* using sound is produced by reflecting sound waves back along their path (page 324); we can similarly produce a stationary wave pattern using 30 mm

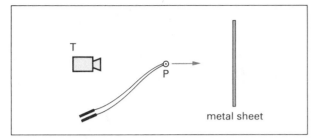

Figure 27.10 Stationary waves using 30 mm microwaves. P is a receiver probe which can detect waves from all directions.

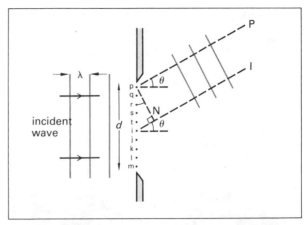

Figure 27.11 Diffraction at a slit: p, q, r, etc. act as in-phase secondary sources (Huygens's principle) in the plane of the slit.

electromagnetic waves. Figure 27.10 shows the microwave transmitter T facing a metal plate. Instead of the usual receiver with its directional horn we need a small probe P which receives equally well from all sides. This is connected to an audio amplifier and loudspeaker to detect the modulated output from T. As P is moved along the perpendicular from T to the reflecting surface maxima and minima can be detected at regular intervals. The distance between successive minima (nodes of variation of the electric field) is $\lambda/2$ and by measuring the distance across (say) 10 nodes the wavelength of the microwaves from the transmitter can be found. The uncertainty is only ± 1 mm in 300 mm or about $\pm 0.3\%$. At the nodes the received signal is not zero as the amplitude of the reflected wave is less than that of the wave direct from T. If the frequency f of the microwaves is known then measuring λ in this way enables the speed $c\ (= f\lambda)$ of the microwaves to be found.

27.4 Diffraction at a slit

Section 25.4 outlines the phenomenon of diffraction and figure 25.13 (page 336) shows the diffraction of water waves through a gap. Figure 27.11 shows a plane wave of wavelength λ incident at a slit of width d, the incident rays being perpendicular to the slit. In order to predict where the wave energy transmitted through the slit will go we need to imagine each point on the wavefront in the gap as a secondary source sending waves to the right of the gap (Huygens's principle). By superposing these secondary waves we can, in principle, find the resulting wave amplitude at any point. In practice this can be very difficult. The only case where it is possible to make any simple prediction is when we consider the contributions from each source in a given *direction* θ from the straight through direction. This special case is called *Fraunhofer diffraction*.

Suppose we consider ten sources p, q, r, s, t and i, j, k, l, m as shown. The waves from p and i, along pP and iI respectively, will have a phase difference which depends on their path difference iN. If $iN = \lambda$ these two secondary waves will be in phase. But if $iN = \lambda/2$ they will be in antiphase (180° out of phase) and the contributions from

p and i will superpose so as to produce zero displacement amplitude. Let us write $\theta = \theta_1$, when $iN = \lambda/2$, so that

$$iN = \frac{\lambda}{2} = \frac{d}{2}\sin\theta_1$$

i.e. there is a minimum where $\lambda = d\sin\theta_1$

What of the other secondary sources in the gap? If we pair them off as q and j, r and k, ect., then each pair will send waves in the direction θ_1 which are 180° out of phase (as for each pair the path difference is $\lambda/2$). There are five pairs of sources (it could be 10 or 100) in figure 27.11 and as *each pair* can be dealt with as p and i above, the total amplitude of the wave in the direction θ_1 is zero. For zero intensity therefore

$$\sin\theta_1 = \frac{\lambda}{d}$$

For visible light of wavelength $\lambda (\approx 5 \times 10^{-7}$ m) and a slit of width 4×10^{-6} m, we get $\sin\theta_1 = 5/40$, i.e. $\theta_1 = \arcsin 0.125 = 7°$ for a minimum.

The full theory gives minima at $\sin\theta = n\lambda/d$.

Figure 27.12a shows the result of varying the slit width d while keeping λ constant. The photographs were taken using a slit mounted on a spectrometer table with the film at the cross-wires of the telescope system (see figure 27.17 on page 361). Except for the top and bottom photographs, there is more than one minimum on either side of the central diffraction peak. The most important feature of these diffraction patterns is, however, the angular width $2\theta_1$ of the central peak. Figure 27.12b shows a graph of intensity against θ for the case where $d = 3\lambda$. The angle θ_2 of the second minima is about twice θ_1; the intensity of the side maxima are less than 5% of the central peak, and they are of angular width θ_1 *half* the width of the central maximum. To summarise,

(i) when $d \gg \lambda$ very little diffraction is noticeable, the edge

of the central diffraction peak more or less coincides with the edge of the geometrical shadow of the slit,
(ii) when $d \approx \lambda$ the central maximum spreads over most of the region in which we would expect a shadow, and
(iii) at intermediate values there is a central diffraction peak of width $2\theta_1$ where $\theta_1 = \arcsin \lambda/d$.

Diffraction patterns very similar to the photographs of figure 27.12a can easily be seen by mounting two razor blades so as to make a narrow adjustable slit. Hold this up to the eye and view a line filament bulb a metre or more away with the slit parallel to the filament. The effect of using different wavelengths can be seen by placing filters

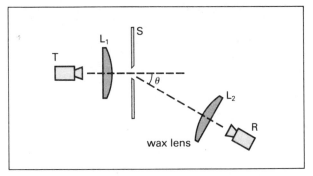

Figure 27.13 A single slit diffraction experiment using microwaves.

in front of the slit and the width of the slit can be varied by pushing the razor blades closer together.

► A microwave source of wavelength 30 mm is used to demonstrate diffraction at a slit (figure 27.13). The lens L_2 and the receiver can be swung together on an arm pivoted at the centre of the slit. If the width of the slit is 100 mm, at what angles will the received intensity be a minimum?

The first minimum occurs at θ_1,

$$\text{where} \quad \sin \theta_1 = \frac{\lambda}{d} = \frac{30 \text{ mm}}{100 \text{ mm}}$$

so that $\theta_1 = 17.5°$

Similarly $\sin \theta_2 = 0.6$ giving $\theta_2 = 37°$, (the full theory gives minima at $\sin \theta = n\lambda/d$), $\sin \theta_3 = 0.9$ giving $\theta_3 = 64°$. There are no further values as for $n > 3 \sin \theta$ is more than 1.0. Thus there will be three minima on each side of the central maximum. ◄

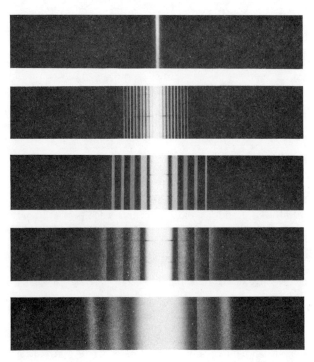

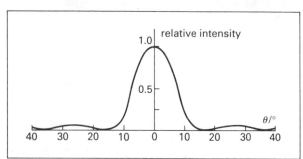

Figure 27.12 Single slit Fraunhofer diffraction patterns for light of wavelength λ using slits of different widths d. The top photograph uses no slit ($d = \infty$) and shows the width of the light source. The middle photograph shows $d = 3\lambda$ and the bottom one $d \approx \lambda$. Below is a graph of intensity against deviation θ for the middle photograph, $d \approx 3\lambda$.

Resolving objects

The eye

As we travel along a road the writing on a signpost only becomes decipherable when we are within a certain distance of the sign. In the laboratory two black lines drawn 2.0 mm apart on white paper cannot be separately distinguished by the eye from distances greater than about 6 m; if the lines are only 1.0 mm apart the distance is only 3 m and so on. It is the angle subtended at the eye

$$\frac{2 \times 10^{-3} \text{ m}}{6 \text{ m}} \approx 3 \times 10^{-4} \text{ rad or 1 minute of arc}$$

in this case which determines whether or not the two lines can be resolved. (In a road sign the two lines might be the sides of a letter H or a letter U.) If we think of the images of the two lines on the retina we realise that they are two diffraction patterns each like that shown in figure 27.12b. The smaller the aperture of the eye (the pupil) the wider each diffraction pattern and the more likely it is that the central peaks overlap. See also the example below.

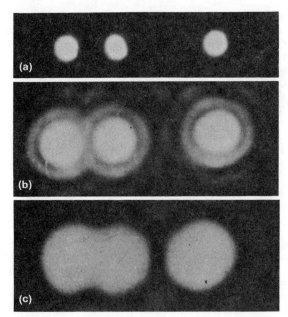

Figure 27.14 Three close point sources (e.g. stars) photographed using a telescope with three different apertures.

Telescopes

Figure 27.14 shows the image of three sources formed on a photographic film placed in the focal plane of a converging lens. The objects might be stars and the lens the objective of a telescope, or illuminated specks of dust and the lens the objective of a microscope. The three photographs are taken with different apertures (diameter d) in front of the lens. In (a) d is large and the images clearly resolved while in (c) d is small and the upper images can only just be detected as two separate sources. Magnifying (c) only makes the diffraction pattern bigger – it does *not* reduce the overlap of the images.

We know that the angular position of the first minima in a Fraunhofer diffraction pattern is arcsin (λ/d) for a slit

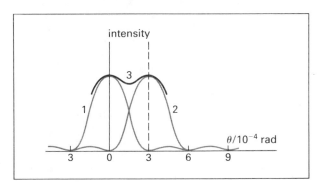

Figure 27.15 The overlapping diffraction patterns of two images which are just resolved. Curve 3, which the eye sees, is the sum of the intensities of curves 1 and 2.

of width d (page 357). For a circular aperture of diameter d the first minimum can be shown to occur at θ_1, where

$$\sin \theta_1 = \frac{1.22\lambda}{d}$$

In designing optical telescopes for use at the limits of resolution d must be made as large as possible. Thus in a 2 m or a 5 m astronomical telescope the number quoted is the *aperture d*; this is the telescope's most important and expensive feature. To manufacture, support and steer a concave optical mirror of diameter up to 5 m is a remarkable engineering achievement. In improving the ability of a telescope to resolve we are, of course, automatically collecting more energy and can thus detect fainter objects, for instance, more distant galaxies. For radio telescopes where the wavelength λ is of the order of centimeters, the reflecting dish must be very large (page 373). The same principle holds for microscope design but d cannot easily be increased without producing severe spherical aberration (page 337). The approach now is to reduce λ; and so we make use of the wave nature of electrons (page 244).

▶ The human eye can just resolve two objects with an angular separation of 3×10^{-4} rad. Is this consistent with the eye's resolving power being limited by diffraction?

The wavelength of visible light is of the order of 5×10^{-7} m and the diameter of the eye pupil is about 2 mm. Using these figures, the first minimum of the diffraction pattern of one object is at an angle θ_1 from the centre of the pattern, where

$$\sin \theta_1 = \frac{1.22(5 \times 10^{-7} \text{ m})}{2 \times 10^{-3} \text{ m}}$$

whence $\theta_1 \approx 3 \times 10^{-4}$ rad

Figure 27.15 shows that if the two objects subtend this angle they can just be resolved, so that this calculation suggests the eye is limited by diffraction rather than by the detailed structure of the retina. ◀

27.5 The diffraction grating

When a wave of wavelength λ is diffracted through a single slit the Fraunhofer intensity pattern is like that shown in figure 27.12, the angular spread of the central peak depending on the value of λ/d, where d is the width of the slit. If two such slits are used, the distance between their centres being s, then we have a Young's slit arrangement and the resulting intensity pattern is as shown in figure 27.5. Suppose the number of slits is increased keeping both d and s constant. The effect (we will not attempt to explain how) is to narrow the peaks of the Young's slit pattern and to make them more intense. The photographs of figure 27.16 show the result of increasing the number of slits – notice that the *spacing* of the maxima does not change. The peaks are so narrow in the bottom photograph that if a beam of light containing several different wavelengths

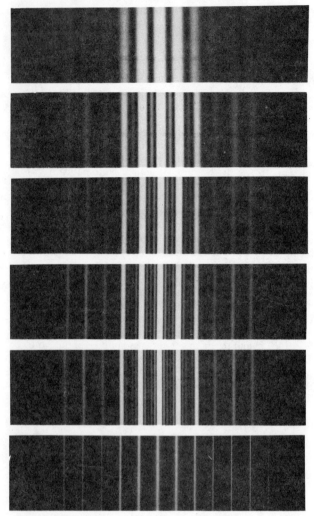

Figure 27.16 Multiple slit diffraction: from top to bottom: 2 slits, 3 slits, 4 slits, 5 slits, 6 slits and 48 slits. Note the faint maxima in the middle photographs and the gradual narrowing of the main maxima.

about 500 narrow parallel slits through which the light is diffracted. It might be 30 mm across and will thus contain a total of 15 000 diffracting slits spaced evenly with their centres 2×10^{-3} mm $(= 2 \times 10^{-6}$ m) apart. With this separation s the width d of each slit must be less than 2×10^{-6} m; in practice it is about 1×10^{-6} m, so light of wavelength $\approx 0.6 \times 10^{-6}$ m diffracts into a wide arc on passing through each narrow slit. The fact that there are as many as 15 000 slits makes the maxima or peaks in the resulting superposition pattern extremely narrow (the bottom photograph of figure 27.16 used a grating with only 48 slits). It is thus possible to distinguish images of a narrow slit source for two very close but separate wavelengths; the grating is said to *resolve* well.

The theory of the diffraction grating

The *spectrometer* (page 348) is ideally suited to studying the patterns of superposition from a grating. The light from the slit S is first collimated and then arranged to fall normally on the grating whose diffracting slits are adjusted to be parallel to S. The telescope is focused to receive parallel light and its eyepiece arranged so that the crosswires are in focus. Figure 27.17 illustrates the practical arrangement and shows the parallel beam of light incident on the grating together with one emerging beam.

A single wavelength of light λ gives rise to a set of sharply defined maxima at angles θ_1, θ_2 etc. to the straight through position. The beams diffracted at those angles *from adjacent slits* then arrive at the telescope (and hence at the eye) with a phase difference of 360° or 2π rad, 4π rad etc. (i.e. they are in phase). They will thus have a path difference of λ, 2λ etc. Consider the case where the path difference between adjacent diffracted beams is λ. The next slit will contribute a diffracted beam in the direction θ_1 with a path difference of 2λ, the next with a path difference of 3λ, the next 4λ etc., so that all these beams diffracted at θ_1 reach the telescope in phase (you should study the arcs from the slits to the first order line in figure 27.18a). (b) shows that the condition for this first maximum at θ_1 is

$$\lambda = s \sin \theta_1$$

The next maximum at θ_2 needs a path difference of 2λ between diffracted beams from adjacent slits, and so on. In general where there are maxima at angles θ_m on each side of the centre

$$m\lambda = s \sin \theta_m \tag{4}$$

For $s = 2.0 \times 10^{-6}$ m and $\lambda = 6 \times 10^{-7}$ m this gives

$m = 0$, $\sin \theta_0 = 0$, $\theta_0 = 0°$, zero order, brightest
$m = 1$, $\sin \theta_1 = 0.3$, $\theta_1 = 17°$, first order
$m = 2$, $\sin \theta_2 = 0.6$, $\theta_2 = 37°$, second order
$m = 3$, $\sin \theta_3 = 0.9$, $\theta_3 = 64°$, third order, faint
$m = 4$, $\sin \theta_4 = 1.2$, not possible, so no fourth order.

For $m \geqslant 4$, $m > s$ so that $\sin \theta_m > 1$ and no higher orders exist. The relative intensities of the maxima depend on the ratio λ/d, where d is the width of each individual slit. If d

is diffracted through the multiple slit arrangement (which is called a *diffraction grating*), each wavelength gives rise to a separate set of peaks to right and left of the central one. Each peak or maximum is an image of the light source (which in this case is a strongly illuminated narrow slit).

By simply holding a diffraction grating in the hand close to the eye and looking at a narrow gas-discharge tube, e.g. of neon, sharp images of the source in each of the colours (wavelengths) present in neon light can be seen. The grating is thus ideal for studying *spectra*; i.e. for analysing the wavelengths emitted by a source of light. The diffraction grating *disperses* the light into its component colours as does a refracting prism (page 348).

A typical grating will have, in each millimetre of its width,

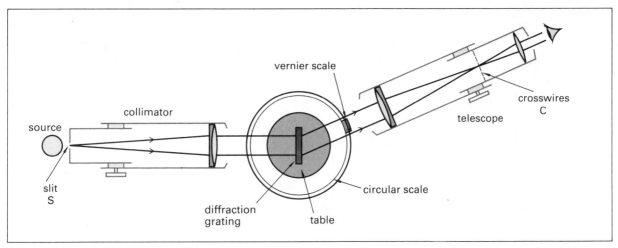

Figure 27.17 A spectrometer set for use with a diffraction grating. Only a single wavelength in the first order spectrum is shown.

were 0.8×10^{-6} m in this case, then there would be no light diffracted at

$$\sin \theta = \frac{\lambda}{d} = \frac{6 \times 10^{-7} \text{ m}}{8 \times 10^{-7} \text{ m}} \quad \Rightarrow \quad \theta = 49°$$

and the maxima would be as shown in figure 27.19 on page 362. For a slit width d which is only a fraction of λ the central diffraction peak will spread through 180° so that there will be no minima in the diffraction pattern.

When the incident light is not monochromatic the grating produces a series of spectra. In the first order each λ gives rise to a sharply defined maximum (on either side of the straight through line) at $\sin \theta_1 = \lambda/s$. Thus for white light a spectrum of colours from violet (smallest λ so smallest θ_1) to red is produced. The same is true for the other orders. Figure 27.20 shows that the orders beyond the first overlap and this point is reinforced in the example below. The study and significance of optical spectra (spectroscopy) is considered further in section 28.5.

▶ A helium discharge tube is found to emit yellow light of wavelength 590 nm which, in its second order, on being dispersed by a diffraction grating, coincides with the third order of a deep violet line. What is the wavelength of the violet light?

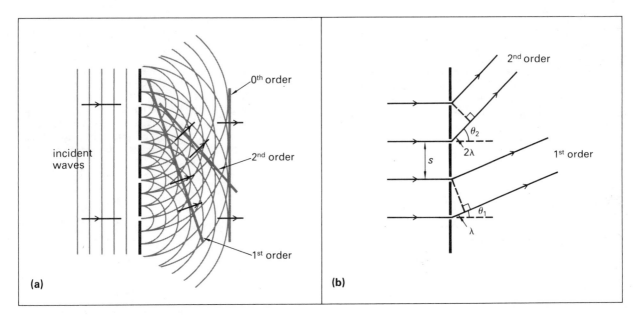

(a)

(b)

Figure 27.18 (a) Waves and (b) rays for (part of) a diffraction grating. In the lower part of (b) a path difference of λ is shown between rays deviated through an angle θ_1: clearly $\lambda = s \sin \theta_1$.

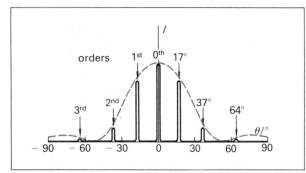

Figure 27.19 The sharp peaks of the pattern of superposition produced by a diffraction grating within the diffraction envelope produced by each slit (the dotted line).

The path difference in the second order for the yellow light is

$$2\lambda_y = 2 \times 590 \text{ nm} = 1180 \text{ nm}$$

and this must be the path difference also for the violet light.

so that $\quad 1180 \text{ nm} = 3\lambda_v$

$$\Rightarrow \qquad \lambda_v = \frac{1180 \text{ nm}}{3} = 390 \text{ nm}$$

to two significant figures; this is a violet light just within the visible part of the spectrum. Notice that in this calculation we did not need to know the grating spacing nor the angle at which these maxima occurred. This same overlap occurs with every grating. ◀

Reflection gratings

We can also make *reflection* diffraction gratings by scratching parallel grooves on a metal surface. The theory is just the same as for a transmission grating, each narrow strip of the reflector diffracting the light into a wide angle and the consequent superposition of the diffracted beams at a given angle giving rise to sharp maxima. A gramophone record acts as a simple reflection grating if it is held so that the light strikes the surface obliquely – this is probably the simplest demonstration of the wave nature of light.

27.6 Crystal diffraction

The spacing of the atoms in a crystal is of the order of 10^{-10} m. Soon after the discovery of X-rays the possibility of looking for superposition patterns formed by these very small-wavelength electromagnetic waves diffracted from a crystal lattice was suggested, first by Von Laue and then by Sir William Bragg (1862–1942) and his son Sir Lawrence (1890–1975). When a beam of X-rays is incident on a single layer of atoms in a crystal each atom scatters a minute proportion of the beam, i.e. acts as a secondary

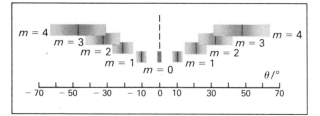

Figure 27.20 The spectra of white light produced by a diffraction grating. As actually viewed they would not be displaced. The central line in each order is for $\lambda = 550$ nm.

source of X-ray wavelets. In the direction for which the angle of incidence equals the angle of 'reflection' the wavelets superpose constructively, their path difference being then zero. Figure 27.21 illustrates this. In a crystal there are many such layers, parallel and regularly spaced (about 10^5 in a crystal 0.01 mm thick) and each layer produces a weak reflected beam. These beams from successive layers are usually out of phase, but when the beam strikes the crystal at certain angles the path difference between beams from adjacent layers can be λ, or 2λ, etc.

If t is the separation of the layers of atoms and θ is the glancing angle, then there is a strong reflection when

$$2t \sin \theta = n\lambda$$

as we saw on page 356. This result is known as *Bragg's law* for X-rays but is, of course, not confined to X-rays alone. Notice that θ is not, as is usual in diffraction phenomena, the angle measured from the perpendicular but the glancing angle (figure 27.22).

We can use this result to measure t for a crystal of, e.g. sodium chloride. An *X-ray spectrometer* is used, the principle of which is shown in figure 27.22. A very narrow beam of X-rays is defined by the slits (less than 0.1 mm wide) in the two lead screens S_1 and S_2. The crystal C is mounted on a table which can rotate about a vertical axis; this carries a graduated scale, so that the glancing angle θ of the beam on its faces can be measured. The reflected beam (deviated through an angle 2θ, as shown) passes through a slit in another screen S_3 and into an ionisation chamber with lead walls. For convenience a gearing system is used which automatically rotates the detecting system through twice

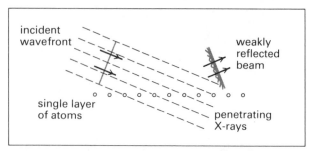

Figure 27.21 The 'reflection' of X-rays by a single layer of atoms.

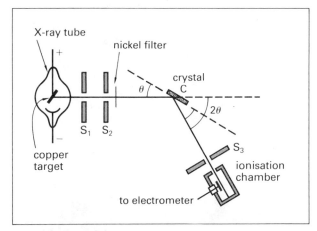

Figure 27.22 An X-ray spectrometer.

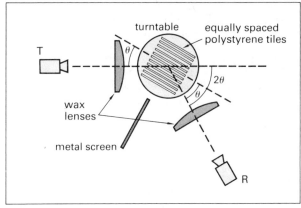

Figure 27.23 An X-ray analogue experiment using 30 mm microwaves.

the angle rotated by the crystal and about the same vertical axis. The ionisation chamber (or a special GM tube page 256) is connected to a suitable source of potential difference. An electrometer is used to measure the ionisation current, which is taken as a measure of the intensity of the reflected beam. The ionisation chamber and screen S_3 act just like the telescope of an optical spectrometer, the slits in the screens S_1 and S_2 take the place of the collimator, and the crystal and table replace the diffraction grating and table (compare figure 27.15).

The X-rays produced by the rapid deceleration of electrons at a copper target have a spectrum as shown in figure 28.26, page 380. The nickel filter absorbs a fraction of the long wavelength continuous spectrum, almost all the K_β line and most of the short wavelength end of the spectrum. The X-rays incident on C are thus essentially monochromatic of wavelength 0.154 nm together with a very weak continuous background. The ionisation current shows a continuous background, plus sharp maxima which occur at the angles defined by $n = 1, 2$ etc. in Bragg's law above.

Using this same X-ray source but different crystals allows us to compare their *lattice spacing t*. When these experiments were first performed there was no independent way of measuring X-ray wavelengths and λ was estimated by calculating t from the Avogadro constant L (page 38). Today the process is reversed; L is most precisely measured from X-ray diffraction experiments and the wavelength of the X-rays measured by using a ruled grating at very oblique incidence.

A microwave analogue experiment

It is not generally possible to perform X-ray experiments in a school laboratory, since the permitted sources of X-rays are so weak that their diffracted beams can be examined only by photography using very long exposure times. However, an analogue experiment can be carried out using electromagnetic waves of about 30 mm wavelength. Figure 27.23 shows how a model of an X-ray spectrometer may be

designed. The microwave transmitter T is mounted at the focus of a plane-convex wax lens so that a parallel beam of radiation is produced. The 'crystal' is placed on a large turntable centred in this beam. The receiver R is similarly mounted at the focus of a wax lens, and is carried (with its lens) on an arm that rotates about the same vertical axis as the turntable. A metal screen is usually needed to shield R from the edges of the direct beam from T. We can try several introductory experiments to demonstrate the principle of the full crystal diffraction technique. We can use:

(i) a single layer of hexagonally-packed polystyrene spheres to produce a partial reflection and to stress the behaviour shown in figure 27.21.

(ii) two parallel polystyrene tiles to demonstrate the Bragg condition explained on page 356. For $t = 40$ mm and $\lambda = 30$ mm there will be two faint and not very sharp maxima at $\theta_1 = \arcsin(3/8)$ and $\theta_2 = \arcsin(3/4)$.

(iii) many parallel tiles arranged as shown in figure 27.23. The maxima are now sharper than in (ii) and more intense. The most effective way of demonstrating this is to rotate the turntable at a steady speed (about one revolution per second) and use the modulated output of T. With R connected to an amplifier and loudspeaker an audible signal (a blip) is heard whenever the microwaves reach R. The receiver arm is now moved round very slowly while the turntable revolves. At most positions little or nothing is heard, but at one or two clearly defined values of 2θ strong reflections are received twice in every revolution of the turntable. The various values of θ so obtained are found obey Bragg's law.

The same apparatus may now be used with 'crystals' of polystyrene balls (like that shown in figure 4.15, page 44) mounted on the turntable. Crystal models may be constructed with these balls according to the different arrangements of atoms that are possible. Each structure gives its own characteristic pattern of reflections and a crystallographer can use the pattern to determine the structure of the crystal.

▶ A 'crystal' made from polystyrene spheres of diameter

50 mm is found to give strong reflections at 44° and 50° from the original direction when used in a microwave analogue of an X-ray spectrometer apparatus. If the wavelength used was 32 mm what is the ratio of the spacings between the layers responsible for these maxima?

The angle of 44° implies we have a glancing angle of 22°, and so using Bragg's law, equation (3) on page 356

$$32 \times 10^{-3} \text{ m} = 2d_1 \sin 22°$$

and for 50° $32 \times 10^{-3} \text{ m} = 2d_2 \sin 25°$

so that $$\frac{d_1}{d_2} = \frac{\sin 25°}{\sin 22°} = 1.13$$

Notice that only the angles are needed to calculate the answer; neither λ nor the diameters of the spheres need be given. This ratio is characteristic of a *face-centred cubic* structure, for which it can be shown that

$$d_1/d_2 = \sqrt{4/3} = 1.15$$ ◀

Powder photography

If a beam of monochromatic X-rays is incident on a thin wire of a metal element such as copper, then because of the very large numbers of tiny individual copper crystals involved (the copper is polycrystalline – page 44), there will be many planes of atoms which lie at any given angle to the beam. If there are certain Bragg angles θ_1, θ_2, etc. at which a strong reflection is to be expected then a series of cones of X-rays at $2\theta_1$, $2\theta_2$ etc. to the incident beam will emerge from the place where the X-ray beam strikes the wire.

Figure 27.24a shows the principle of the experiment in which the reflected cones are detected photographically in an X-ray powder camera. A strip of photographic film (wrapped in opaque paper) is placed around the inside of a cylindrical drum usually of diameter about 100 mm; there are gaps in the cylinder through which the X-ray

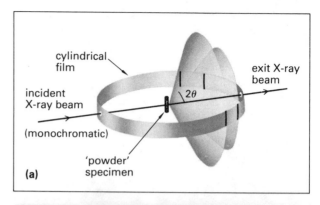

Figure 27.24 (a) A three-dimensional view of an X-ray powder camera. (b) shows a half circle of unrolled exposed film using a polycrystalline copper specimen. The X-rays used had a wavelength of 0.154 nm.

beam enters and through which the undiffracted beam leaves. The specimen in the form of a small tube of finely ground powder or, for a metal, a thin wire, is mounted at the centre of the drum; both are *polycrystalline*. After exposure the developed film shows a pattern of slightly curved lines which mark the intersection of the film with the cones into which the X-rays are diffracted. Figure 27.24b is a powder photograph for copper. The values of θ are calculated from the positions of the lines and the radius of the camera cylinder. A single exposure thus provides a complete record of all the Bragg angles of the material for the wavelength of X-rays used. The method is quick and simple, and is a routine experiment in many laboratories; but deducing the structure of the crystal from the Bragg angles can be very complex.

28 The electromagnetic spectrum

28.1 Electromagnetic waves

In chapter 23 the behaviour of mechanical waves is described. Electromagnetic waves from radio waves to γ-rays share many of the general wave properties of mechanical waves; for instance

- they transfer energy from place to place,
- they exhibit diffraction effects,
- they obey the principle of superposition,
- they can be described by drawing graphs,
- they sometimes undergo a phase change of $180°$ on reflection.

All *electromagnetic waves* also share the following properties:

- they are generated by accelerating charged particles,
- they travel at 3.00×10^8 m s^{-1} in free space,
- they obey the laws of reflection and refraction usually associated with light,
- they are transverse waves and exhibit polarisation effects,
- they can be emitted and absorbed by matter, the latter process raising the internal energy of the absorbing body,
- they show certain particle properties which demand a photon model instead of a wave model.

The above list can be followed through by consulting the index but it is as well to summarise some of the main properties of the various sorts of electromagnetic waves. The divisions, for instance into microwaves and infra-red waves, are characterised not by frequency ranges with clear-cut boundaries (the two types mentioned overlap) but in the way in which they are produced. Table 28.1 gives such information. It is not meant to be exhaustive but to give a broad survey of the whole range of electromagnetic waves from those of a 200 kHz long-wave broadcast to a high-energy γ-ray photon in cosmic radiation with a frequency of 10^{30} Hz. Electromagnetic waves of all wavelengths can be found beyond the Earth's atmosphere and are the result of cosmic, galactic or stellar interactions or explosions. The 'big bang' theory of the origin of our universe is supported by the existence of the 3 K background radiation, i.e. radiation which is characteristic of a body at a temperature of 3 K (page 376). In the second half of the twentieth century astronomy is gradually developing to include observation at all wavelengths. In the laboratory stray electromagnetic fields, especially those associated with the 50 Hz mains, are often picked up by a c.r.o. This *mains pick-up* is usually a nuisance. Though the fields link with the c.r.o. through inductive or capacitive effects (a hand holding an input wire is quite adequate), they can be thought of as very low frequency electromagnetic waves.

In reading the frequencies quoted at the left of the table on the next page we must remember that

(i) $\lambda = \dfrac{c}{f}$ and (ii) $E = hf$

so that: (i) wavelengths in vacuo can be found using $c = 3.00 \times 10^8$ m s^{-1}, and (ii) photon energies (page 242) can be found using $h = 6.63 \times 10^{-34}$ J s.

► At present the BBC broadcasts in one region using 247 m, 1214 kHz and 1500 m, 200 kHz. What do these numbers tell you about electromagnetic waves?

Using $c = f\lambda$ for each in turn yields firstly:

$$c = (1214 \times 10^3 \text{ Hz})(247\text{m}) = 3.00 \times 10^8 \text{ m s}^{-1}$$

and secondly:

$$c = (200 \times 10^3 \text{ Hz})(1500 \text{ m}) = 3.00 \times 10^8 \text{ m s}^{-1}$$

so that this information tells us that two electromagnetic waves different in their wavelengths by a factor of 6 travel at the same speed. ◄

The speed of light

The speed c at which light travels in a vacuum is $2.997\,925 \times 10^8$ m s^{-1}. It is one of the fundamental constants and can be measured with an uncertainty of ± 100 m s^{-1}, i.e. to better than 1 part in 10^6. All electromagnetic waves travel in a vacuum at the same speed. The simultaneous reception of waves from events which occur far away in space by both optical and radio telescopes provides us with very convincing evidence of this. We can thus measure c using whichever are the most convenient electromagnetic waves. A microwave of frequency 10 GHz has a wavelength λ of 0.030 m or 30 mm. By analogy with the example in which the speed of sound is measured (page 325), a stationary wave experiment using $f = 10$ GHz electromagnetic waves can lead to a measured value of their wavelength and hence to a value for $c \ (= f\lambda)$. To achieve the most precise measurement microwaves from caesium-133 atoms with an exactly defined frequency (appendix A2) are used and the dimensions of the evacuated resonating cavity established in terms of the wavelength of a line in the krypton-86 spectrum (appendix A2). In this way c is measured in terms of the two SI standards of time and length.

A knowledge of c enables large distances to be measured; e.g. a very short radar pulse transmitted from the Earth and reflected at the Moon's surface can be timed as taking 2.498 s for the round trip, thus determining the Earth-Moon surface-to-surface distance d at that moment as

$$2d = (2.998 \times 10^8 \text{ m s}^{-1})(2.498 \text{ s})$$
$$\Rightarrow \quad d = 3.745 \times 10^8 \text{ m or } 374\,500 \text{ km}$$

name and frequency range	source(s)	method(s) of generation	detected by	characteristic properties	selected applications
radio waves $10^4 - 10^9$ Hz	electrical or electronic equipment	oscillating electrons in conductors and dipole antennae	tuned oscillatory electric circuits	diffract around mountains, reflect from ionosphere,	radio communications, worldwide location systems (OMEGA), radioastronomy
microwaves $10^8 - 10^{12}$ Hz	klystrons, magnetrons, Gunn effect circuits, masers	oscillating electrons in resonant cavities	resonant or tuned cavities	conducted along waveguides, narrow beam dish-to-dish communication, dispersed by paraffin wax	radar, television communication, microwave cooking, analysis of molecular structure, vehicle speed detectors
infra-red $10^{10} - 10^{14}$ Hz	warm and hot bodies (including human beings)	random energy changes in molecules of liquids and gases, oscillating molecules in hot solids	special photographic plates, thermopile, photoconductive cell	cause sensation of warmth, not scattered much by atmospheric dust, dispersed by rocksalt	reconnaissance and tracking, domestic heating, beams for switching circuits, absorption spectroscopy for organic analysis
visible close to 10^{15} Hz	the Sun, fluorescent and hot filament lamps or lasers	energy transitions of outer orbital (valence) electrons in atoms and molecules	eyes, photographic plate, photocell	affect retina of eye, initiate chemical reactions, dispersed by glass	vision, photosynthesis, laser communication, all optical systems, identifying gaseous elements
ultra-violet $10^{15} - 10^{17}$ Hz	the Sun, very hot bodies, mercury vapour lamp, arcs and sparks	energy transitions of orbital electrons, randomly accelerating ionised particles	fluorescence, photoelectric cell or photographic plates	absorbed in upper atmosphere, damage living cells, produce chemical changes, sunburn, eject electrons from metals, dispersed by quartz	lamps used in medicine, sterilising food and utensils, detecting forgeries, fluorescence in washing powders, microscopy
X-rays $10^{15} - 10^{25}$ Hz	X-ray tubes, pulsars	rapid deceleration of high energy electrons, energy transitions of innermost orbital electrons	ionisaton chamber, photographic plates	very penetrating, ionise gases, cause photoelectric emission from metals, diffracted by crystal lattice	radiography, X-ray crystallography
γ-rays $10^{17} - 10^{32}$ Hz	nuclear reactions or cosmic radiation	changes of energy levels of nucleons, mass-energy conservation processes	GM tubes, phosphorescence	as for X-rays but more penetrating	non-destructive testing e.g. flaws in metals, potato sorting, induced mutations, research – interaction with materials

Table 28.1 The electromagnetic spectrum.

A variation of 200 ns in the time for the round trip represents a change of 60 m in the length of the path transversed, i.e. a variation of 30 m in d.

Foucault's rotating mirror

To measure c directly for light involves measuring the time t taken for light to travel a measured distance. In principle this is easy enough but c is so large that t is extremely small unless the light travels many thousands of kilometres. The apparatus for a laboratory determination of c is shown in figure 28.1a. In this arrangement $MR = l \approx 2$ m, so that the light travels only about 4 m and it is thus necessary to measure a time interval of only just over 10^{-8} s. The principle of the method is to rotate a small mirror R at a very high speed so that a pulse of light is reflected from R towards M as R passes through position 1 (figure 28.1b) and arrives back to find R in position 2. If the angle of rotation of the mirror θ can be measured and also its angular frequency n then the time taken t for the light to travel from R to M and back is given by

$$t = \frac{\theta}{2\pi n}$$

and hence $\quad c = \dfrac{2l}{t} = \dfrac{4\pi nl}{\theta}$

We can find θ for this n by setting up the source, crosswires C and mirrors (with R stationary) so that light passes

through the glass plate P, is reflected at R, M and then R again so as to form an image I_1 of the crosswires (after a partial reflection at P). This image is arranged to coincide with a scale in the focal plane of the eyepiece lens E. When R is rotated slowly the image I_1 is formed very briefly once every revolution but as R moves faster the eye sees the image all the time and when n becomes high a small displacement d of the image to I_2 can be measured from the scale.

Referring again to figure 28.1b.

$$2\theta = \frac{d}{s} \quad \text{i.e.} \quad \theta = \frac{d}{2s} \qquad \text{(appendix B5)}$$

so that $\quad c = \dfrac{8\pi nls}{d}$

As $d \approx 0.05$ mm in practice and is subject to an uncertainty of at least $\pm 20\%$, we can get only a rough value of c and need not try to measure the other quantities in this expression with any precision. To find n (≈ 500 s^{-1}) we can rely on comparing the pitch of its whine as it rotates with the note from a signal generator. Alternatively a photocell can be arranged so that the light beam from R gives a pulse of current once every revolution; the resulting voltage pulses are counted with a scaler.

On page 331 the refractive index n of a material is defined as c/c_m, where c_m is the speed of light in the medium. c_m can be measured directly using Foucault's rotating mirror as the path length is small enough to arrange for the light to pass through, for example, a tube of water. In the 1860's those who supported the wave theory of light saw this measurement (which showed that $c_m < c$) as conclusive evidence for a wave theory (rather than a particle theory) of light (page 331).

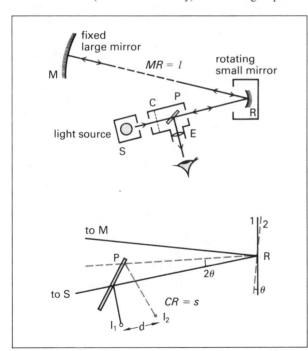

Figure 28.1 (a) A rotating mirror method for measuring c. (b) shows the image of the crosswires before and after being displaced.

The constant c

A knowledge of c is important in many calculations in physics; in $c = f\lambda$ for electromagnetic waves in a vacuum, of course, and also in the mass-energy relation $\Delta E = c^2 \Delta m$ (page 34). c turns out to be the *maximum speed* at which energy can be transmitted in our Universe. The speed at which electromagnetic waves propagate in a vacuum is independent of the speed of their source. This is extremely surprising; it means that if a star is moving away from the earth at a speed $v = 2 \times 10^8$ m s^{-1} (so that $v \approx 2c/3$) the speed of the light from the star measured on Earth is 3×10^8 m s^{-1} and not, as one might at first guess, $(3 - 2) \times 10^8$ m s^{-1}. This result has been rigorously tested in a number of ways. For example, we can measure the speed of γ-rays which come from sub-atomic particles (pions) moving through a laboratory at more than 99.9% of the speed of light. When such a pion emits a γ-ray in the forward direction the speed of the γ-ray is still c! This constancy of the speed of light is a cornerstone of the *theory of special relativity* from which is derived the mass-energy relationship mentioned above.

The Doppler shift

When a source of electromagnetic waves S and an observer O are in relative motion the frequency of the waves as measured by O is different from the frequency f at which they are emitted by S. The same is true for sound or water waves (page 323). If O and S are moving at a speed v either directly towards or away from one another then the change of frequency, Δf, is given by

$$\frac{\Delta f}{f} \approx \frac{v}{c}$$

provided $v \ll c$; Δf is called the electro magnetic Doppler shift. If v is a relative speed of *approach* then Δf is an *increase* in frequency, and there would be a consequent *decrease* in wavelength. The reverse is true if O and S are receding from one another.

The consequences of this Doppler shift include (i) the broadening of spectral lines as a result of the random motion of the emitting atoms in a hot gas, (ii) the galactic *red shift* which enables astronomers to measure the recessional speed of galaxies from our own (the Milky Way) and hence to expound a model of the Universe as an expanding one, (iii) the splitting of spectral lines from rotating sources of radiation such as our Sun, the rings of Saturn or double star systems, and (iv) the possibility of using electromagnetic waves for measuring speeds on Earth – e.g. in the radar speed trap.

To look quantitatively at this last example: suppose a car is approaching a source of radar waves of frequency 1.0×10^{10} Hz at 15 m s^{-1}. As $c = 3.0 \times 10^8$ m s^{-1}

we have $v/c = 5 \times 10^{-8}$ and so $\Delta f = 500$ Hz

An observer in the car would notice an increase of 500 Hz in the frequency of the received waves. The car acts as a reflector and some of the reflected waves are received by the waiting police. They too notice an increase of 500 Hz because the approach of the car in no way affects the speed of the reflected waves, only their frequency. If now the reflected waves are compared with the transmitted waves a *beat* frequency equal to the difference between the two will be detected, i.e. a beat frequency of 500 Hz (page 321). This can readily be turned into a dial reading giving the speed of the approaching car.

The Doppler relationship $\Delta f/f \approx v/c$ for light is very similar to those for sound (page 323) with $v_s \ll c$. The equation for light cannot, however, be derived in the way the sound one is.* This is because mechanical waves are swept along by a moving medium (air or water) but there is no analogous medium for light or other electromagnetic waves travelling in space. We cannot, therefore, talk of a moving observer *or* a moving source of light, only of the relative motion between them.

*For a derivation of $\Delta f/f \approx v/c$ see, for example, Akrill and Millar: *Mechanics, Vibrations and Waves*, John Murray 1974, pp 355–7.

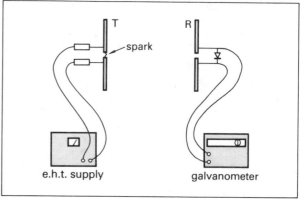

Figure 28.2 A spark transmitter similar to that first used by Hertz (not to scale).

28.2 The nature of electro-magnetic waves

In table 28.1 both radio waves and microwaves are said to be generated by oscillating electrons. The elementary transmitter T shown in figure 28.2, which consists of two brass rods (a dipole – page 371) connected to a 5000 V supply, can produce an oscillatory discharge between the polished ends of the rods. This sparking occurs when the electric field becomes big enough to cause ionisation in the air and for this to happen the field needs to be about 3 MV m^{-1} (page 190). A crackling sound is heard and the blue sparks can easily be seen if the room is darkened. With a similar receiving dipole R connected to a sensitive galvanometer a deflection is observed which changes with the distance between T and R, with their orientation and with the position of nearby conductors, e.g. the experimenter. Energy is being transmitted from T to R – the mechanism for this energy transfer we call an electromagnetic wave. (When the distance TR is small, i.e. a few wavelengths, the link between T and R is as much an electromagnetic induction effect as a wave. With TR more than 2 or 3 metres the

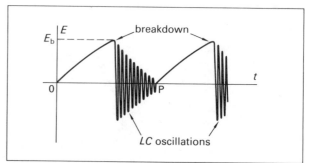

Figure 28.3 The electric field in the gap of a spark transmitter. After breakdown at $E \approx 3000$ V mm^{-1} decaying oscillations occur until the field no longer maintains ions in the gap at P; charging then starts again.

link is essentially wave-like.) This experiment demonstrates that it is the *motion* of the charged particles in the dipole rods and the spark gap which produces the wave effect for if the gap between the rods is increased and the potential difference raised to the previous value no sparking occurs and no signal is received by R.

A graph of the electric field between the brass rods of the transmitter looks like that shown in figure 28.3, the damped oscillatory nature of the field during the time when the air is ionised (and thus behaving as a conductor) being related to the inductive and capacitive properties of the transmitting rods and wires (page 282). The frequency of the oscillations is of the order of 10^9 Hz; this is not indicated on the graph which should have far more oscillations before each charging process restarts. The onset of the oscillations occurs irregularly as the exact moment of breakdown depends on the density of residual ions remaining in the gap. Spark transmitters are illegal because of the extensive interference with radio reception which they cause. That shown in figure 28.2 may be used for short periods within teaching laboratories.

Travelling fields

Suppose a positively-charged particle held at P (figure 28.4) is moved very quickly to Q. During this process it would of course speed up and slow down. At P there would be a radial *electric field* associated with the charge and zero *magnetic field* (it is at rest); similarly at Q. Of course the electric field lines extend in all directions from the point charge but the figure shows only a group of lines in two dimensions to the right. If we assume that during the motion of the charge from P to Q the field lines remain unbroken then they will bend in the way indicated unless the effect travels infinitely quickly. While the charge is moving it also produces a magnetic field which, to the right of PQ, will be into the page (you might work out what happens in other directions). The 'kink' in the electric field and its accompanying patch of magnetic field is an *electromagnetic wave pulse*. It is this kink that travels away from PQ at a speed which we can measure to be 3.00×10^8 m s^{-1}. As the magnetic field reaches a new region the change in B produces an electric field E, this being the familiar process of electromagnetic induction (page 222); in a similar sort of way the change in E produces a magnetic field. The latter is not easy to demonstrate directly but it was a suggested symmetry in the behaviour of electric and magnetic fields which led James Clerk Maxwell to predict in the 1860's the possibility of there being electromagnetic waves. The variations in the two fields are mutually self-supporting, the energy of the wave pulse being associated with them both just as the mechanical energy of a wave pulse on a rope is associated with its motion (kinetic energy) and its deformation (elastic potential energy).

This is a very elementary and unsupported picture of what we mean by an electromagnetic wave pulse but it

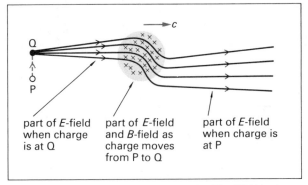

Figure 28.4 An electromagnetic wave pulse. The 'kink' in the E-field and its associated pulse of B-field move away from the source, an accelerating charged particle, at 3×10^8 m s^{-1}.

helps us to see how the spark transmitter of figure 28.2 (or the same dipole attached to a 1 GHz oscillator) produces changes in the surrounding electric and magnetic fields. Drawing the field variations near the dipole produces a much more complicated diagram than figure 28.4 but a long way from the source there is simply a plane wavefront. Figure 28.5a illustrates the situation then and (b) is a graph of how E and B vary if the charges in the source oscillate sinusoidally.

Maxwell's theory showed that the waves will travel in a vacuum at a speed c given by

$$c^2 = \frac{1}{\varepsilon_0 \mu_0}$$

for which $\varepsilon_0 = 8.85 \times 10^{-12}$ F m^{-1} and $\mu_0 = 4\pi \times 10^{-7}$ H m^{-1}. The value of μ_0 is fixed in the course of defining the ampere (page 213) but that for ε_0 comes from experiment (page 195).

Substituting into the above relationship gives $c = 3.00 \times 10^8$ m s^{-1}; you should check that the units are correct.

It was more than 20 years after Maxwell's predictions before Heinrich Hertz produced a spark transmitter not

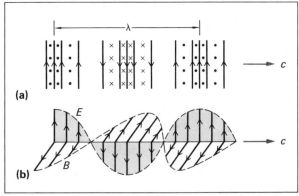

Figure 28.5 Representations of the E-field and the B-field in a sinusoidal electromagnetic plane wave: (a) the field lines, (b) the sinusoidally varying amplitudes.

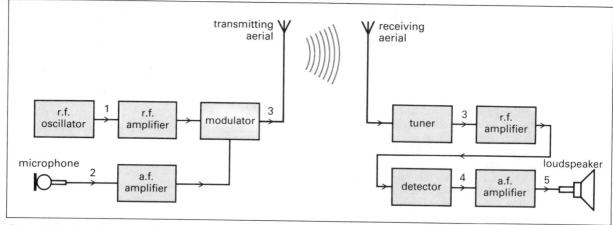

Figure 28.6 Block diagrams for (at left) an amplitude modulated radio transmitter at about 10^7 Hz and (at right) a receiver. The wave forms at 1 to 4 are shown in figure 28.7. See page 295 for the principle of a typical radio frequency (r.f.) oscillator.

unlike that of figure 28.2 and was able to produce electromagnetic waves demonstrably caused by accelerating charges. His receiver was a simple loop of wire with a tiny gap in it across which a fine train of sparks could be seen when the received magnetic field variations of the wave induced an e.m.f. in the loop. Hertz was able to demonstrate all the familiar properties of waves with his apparatus including superposition effects.

The wave pulse of figure 28.4 and the sinusoidal wave of figure 28.5 both contain electric fields in a vertical plane only; this is called the *plane of polarisation* of the wave (page 373). Notice that the two field vectors are at right angles so that a receiver which absorbs energy from the variations of E will have to be orientated differently from one which absorbs energy from the variations of B.

► In a radio wave of frequency 10^8 Hz and intensity 1.5×10^{-16} W m^{-2} the maximum values of the E- and B-fields are about 3×10^{-7} V m^{-1} and 10^{-15} T respectively. (The energies associated with these fields are each half the total energy flux). Calculate (i) the potential difference between the ends of an aerial of length 0.5 m and (ii) the induced e.m.f. in an air-cored coil of 2500 turns and cross-sectional area 1 cm^2, when each is in turn used to pick up the radio wave.

(i)   The maximum p.d. across the aerial (using $E = V/d$):

$$V_0 = (0.5 \text{ m})(3 \times 10^{-7} \text{ V m}^{-1})$$
$$= 1.5 \times 10^{-7} \text{ V or } 0.15 \,\mu\text{V}$$

(ii) The maximum magnetic flux through the coil (using $\Phi = NBA$):

$$\Phi_0 = (2500)(10^{-4} \text{ m}^2)(10^{-15} \text{ T})$$
$$= 2.5 \times 10^{-16} \text{ Wb}$$

so that the maximum induced e.m.f. will be

$$E_0 = \left(\frac{\mathrm{d}\Phi}{\mathrm{d}t}\right)_{max} = 2\pi f \Phi_0 \qquad \text{(page 224)}$$
$$= 2\pi(10^8 \text{ Hz})(2.5 \times 10^{-16} \text{ Wb})$$

$$= 1.6 \times 10^{-7} \text{ V or } 0.16 \,\mu\text{V}$$

The r.m.s. values of V_0 and E_0 will be about 0.11 µV in each case. ◄

Simple radio

To send and receive information it is necessary to vary or *modulate* the wave in some way. An unmodulated *carrier wave*, as it is called, conveys no information except that the transmitter is working. The simplest form of modulation is to switch the transmitter on and off in time with the dots and dashes of the Morse code. A tapping key may be arranged so that the waves are only sent out when it is pressed down. To transmit speech or music the modulation must take the form of varying some feature of the wave in time with the electrical signal from the microphone. Either the amplitude or the frequency may be varied. *Amplitude modulation* (a.m.) is used in medium-wave and long-wave broadcasting in Great Britain. Figure 28.6 shows in block diagram form an amplitude-modulated transmitting and receiving system. Figure 28.7 indicates the waveforms (V-t graphs) at different parts of the transmitting and receiving circuits. *Frequency modulation* (f.m.) gives higher quality reception, but needs the use of a large band of frequencies on either side of the main carrier frequency. Such bands are available only at very high frequencies (v.h.f.).

In the receiver the tuner is a resonant LC loop circuit (page 281) with a variable air capacitor such as that shown in figure 14.3 (page 178). The detector or demodulator is a diode suitable for high frequencies. Figure 28.8 shows a tuner and detector without any intermediate r.f. amplifier. C is a storage capacitor and the time constant RC is made to be about 5×10^{-5} s (page 184) which is very long compared with the period of oscillation of the carrier wave and hence of the aerial current. The storage capacitor therefore charges up to the peak potential difference across

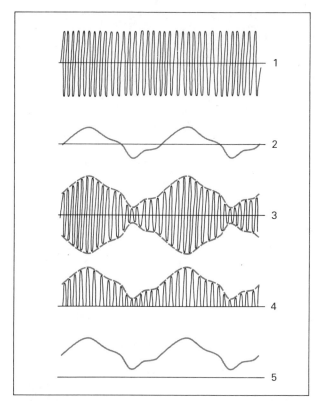

Figure 28.7 The waveforms at the positions numbered in figure 28.6

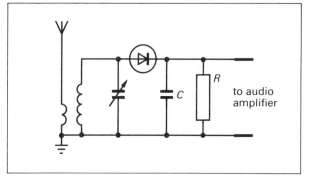

Figure 28.8 The basic components of an a.m. receiving and detector, circuit.

the tuned circuit. However, the time constant is short compared with the period of the audio modulation. The p.d. across C therefore follows faithfully the variations of amplitude of the aerial signal.

Aerials

A radio designed to pick up both f.m. ($f \approx 90$ MHz) and a.m. ($f \approx 1$ MHz) broadcasts will use (i) a conducting aerial for v.h.f. – the best position for the aerial is parallel to the E-field of the transmitted signal, and (ii) a coil wrapped

on a ferrite rod (to increase the B-field) for the medium wave – the best position for the rod is parallel to the B-field of the transmitted signal. A look inside a radio plus some simple experiments will enable you to establish the best orientations and hence the planes of the E-field and B-field of the broadcast signals.

We have only to look around to see the variety of aerials in use for radio and television reception. Two principles are employed. Firstly in order to match the aerial to the incoming wave a dipole made of two rods each $\lambda/4$ long is used (figure 28.9a and compare figure 20.16, page 283). For long wavelengths we usually dispense with the second quarter-wave element, and (b) join the receiver input between the aerial and earth. Secondly (c) a reflector is used behind the receiving dipole and $\lambda/4$ from it. The wave reaching this travels an extra distance $2 \times \lambda/4 = \lambda/2$, equivalent to a phase change of 180° and undergoes a phase change of a further 180° on reflection, so that it reaches the receiving dipole in phase with the main wave and superposes constructively to give a larger signal. The reflector also gives the aerial a directional property and this can be improved (d) by adding a set of conducting rods placed in front of the dipole. The dimensions of an aerial thus tell us the wavelength which it is designed to receive: a u.h.f. TV aerial has $\lambda/4 \approx 0.1$ m, so that $\lambda \approx 0.4$ m and $f \approx 750$ MHz. See also the example on the previous page.

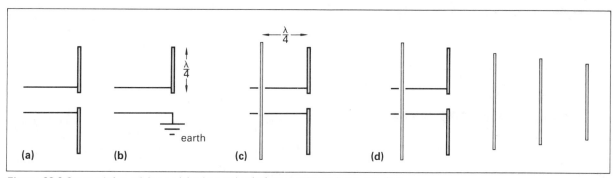

Figure 28.9 Some u.h.f. receiving aerials; the mechanical supporting rods are not shown.

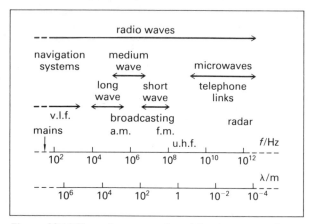

Figure 28.10 The radio wave section of the electromagnetic spectrum.

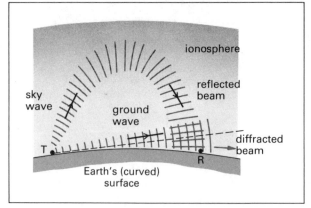

Figure 28.12 Two possible paths from a transmitter of radio waves at T to a receiver at R. Of course T transmits waves in all directions.

Radio waves and the ionosphere

Figure 28.10 shows the radio wave section of the electro-magnetic spectrum.

Ultra-violet radiation from the Sun produces strong ionisation of certain layers in the atmosphere, largely between 80 km and 500 km from the Earth's surface. These layers form what is called the *ionosphere* and behave like a partly conducting sheet which reflects long-wavelength radio waves. Figure 28.11 shows how the ionosphere and the atmosphere affect waves from space which strike the Earth. Astronomers on the Earth must use either visible light or radio waves of wavelength from just over 10 mm up to about 30 m if they are to receive information from space. On the figure these waves are shown reaching the Earth's surface in regions of the electromagnetic spectrum called the optical window and radio window respectively.

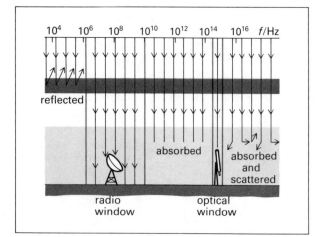

Figure 28.11 How the ionosphere (the upper layer) and the atmosphere affect incoming radiation from space.

Radio reception

Radio waves can reach a receiving aerial from a transmitter by two different paths (figure 28.12). One part of the wave travels directly over the surface of the ground; diffraction effects enable it to follow the curvature of the Earth and to be relatively unaffected by obstacles such as hills and buildings. Another part of the wave travels up into the air and is reflected from the ionosphere. Close to the transmitter the *ground wave* predominates but at distances of 100 km or more the *sky wave* may be of about the same amplitude; when the waves superpose the minima can then be very pronounced. Conditions in the ionosphere are changing continually. At one moment the receiving aerial may be at a maximum in the interference pattern and a strong resultant signal is received. At another the resultant signal may drop off almost to nothing. This is the fading in the medium wave band which is often noticeable at night (when the sky wave is more effectively reflected). The reflection is like the total reflection described for an optical mirage on page 331, i.e. the wave is refracted in a curve as shown and is thus totally reflected. For $\lambda < 30$ m, (i.e. v.h.f. radio at 90 MHz or u.h.f. TV waves) the sky wave is not reflected, except at very oblique angles of incidence, and the ground wave only is received. This prevents fading but limits the range of the signal as diffraction becomes less able to send the waves round mountains or even large buildings.

Radioastronomy

Radio signals from space were first identified by Karl Jansky in 1932. The problems of radioastronomy are similar to those in optical astronomy, namely the design of telescopes (i) which can collect large quantities of energy and (ii) which can resolve one radio source from another.
(i) The collection of enough energy to provide useful

information involves the design of large, sometimes steerable, dishes – reflecting telescopes such as that shown in figure 28.13 which has a bowl diameter of about 80 m. The example below illustrates the idea quantitatively.

(ii) The resolving ability (page 259) of radio telescopes is low because we cannot obtain a sufficiently low value of the ratio λ/d, where λ is the wavelength of the radio waves and d the diameter of the bowl; with the Jodrell Bank 1A telescope $\lambda/d \approx 10^{-2}$ for 0.8 m waves. Compare this with a value of λ/d of about 10^{-5} for an optical telescope of aperture 50 mm. To increase the diameter of a steerable dish beyond 100 m produces engineering problems of overwhelming complexity and instead radioastronomers use a number of separate dishes. The information from these can be synthesised to provide an effective diameter or aperture which is represented by their separation. The two parts of a telescope may be on different sides of the Atlantic and thus provide a diameter of about 8×10^6 m. Such techniques have enabled radioastronomers to resolve distant sources better than their optical colleagues have been able to do.

► An 80 m diameter dish collects energy from a radio-source known (from its electromagnetic Doppler or red shift) to be a hundred million light years away from the solar system. If the input power which the telescope measures is 10^{-18} W, estimate the power output of the source.

As there are about 3×10^7 s in one year, then 100×10^6 light-years is a distance r, where

$$r = (100 \times 10^6)(3 \times 10^8 \text{ m s}^{-1})(3 \times 10^7 \text{ s})$$
$$\approx 10^{24} \text{ m}$$

If the power of the source is P and we assume an inverse square law for electromagnetic wave energy spreading from it (page 329), the energy flux per unit area I at a distance r from it is given by $I = P/4\pi r^2$.

But the energy flux per unit area detected by the telescope is the input power divided by the area of the dish.

Thus
$$\frac{P}{4\pi(10^{24} \text{ m})^2} = \frac{10^{-18} \text{ W}}{\pi(40 \text{ m})^2}$$

which gives $P = 2.5 \times 10^{27}$ W

In calculating this value of P we assumed that all the energy setting off from the source towards the radiotelescope reached it and was detected. Of course the Earth's atmosphere and clouds of intergalactic dust will remove a very large proportion of this energy. It would thus be safe to say that the radio-source must be transferring energy to electromagnetic wave energy at the rate of at least 10^{30} J s^{-1}. ◄

28.3 Plane polarised waves

Radio waves are *plane polarised waves*. By this we mean that the variations in the electric field occur only in one plane as the wave propagates. For instance in figure 28.5 this plane, called the *plane of polarisation*, is vertical. Of course if the variations of E lie in one plane then the variations of B lie in a plane perpendicular to the variations of E, and so are perpendicular to the plane of polarisation. All radio waves and microwaves from dipole transmitting aerials are plane polarised and the variations of E occur in a plane formed by the dipole itself and the direction of propagation. For a 30 mm microwave transmitter (page 330) the plane of polarisation of the waves is vertical with the transmitter standing on the bench.

Both the 30 mm microwave transmitter and the spark transmitter used in the laboratory have a receiving dipole aerial which is vertical, so as to pick up the plane polarised waves. If, for instance, the microwave receiver is gradually rotated about a horizontal axis, i.e. turned on its side, the signal falls until at a 90° rotation no signal is received. On continuing the rotation until the receiver is upside down the signal rises again to its initial value. You can see by looking at TV receiving aerials that some programmes are broadcast using waves with a vertical plane of polarisation and some with a horizontal one.

A dipole receiving aerial absorbs the energy of the incident wave. Figure 28.14 describes the interaction of plane polarised microwaves with a grid of wires (only the electric field is shown). Absorption takes place when the incident E-field sets the free electrons in the wire oscillating *along* the length of the wires.

Polarised light

Infra-red waves, visible light, and all waves at the short wavelength end of the electromagnetic spectrum are generated (page 366) by random processes and consist of an incoherent collection of tiny waves each starting from a different source. Thus although the wavetrain from each excited molecule or atom is plane polarised the resulting

Figure 28.13 The Jodrell Bank mark IA radio-telescope. The dish is a parabolic reflecting surface formed on a metal framework.

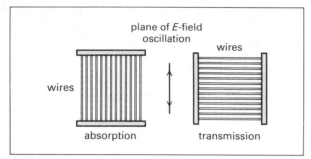

Figure 28.14 Selective absorption of plane polarised microwaves by a grid of parallel wires. The grid is seen head on with the waves travelling out of the page.

radiation consists of waves with E-fields oscillating in all possible planes. Such a wave is said to be *unpolarised*. There are certain sheets of crystals (e.g. polaroid, which consists of long molecules whose axes have been aligned during manufacture) which interact with light in the same way as the grid of wires (figure 28.14) with microwaves. After passing through such a sheet the light is plane polarised with only the resolved part of the E-field in one plane emerging. Rotating a sheet of polaroid through which light is passing can thus quickly establish its state of polarisation. If the transmitted light
(i) shows no change in intensity, the incident light was unpolarised,
(ii) varies slightly in intensity twice per revolution, the incident light was partly plane polarised,
(iii) is cut off completely twice per revolution, the incident light was plane polarised.

Polarised waves and reflection

If in figure 25.8 (page 332) the microwave receiver is positioned to receive reflected waves from the left side of the prism the usual law of reflection can be investigated. If, however, both the transmitter *and* the receiver are turned on their sides while correctly positioned for reflection a much weaker signal than expected is received. Figure 28.15 illustrates what is happening by showing the relative ampli-

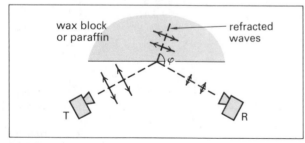

Figure 28.15 The 30 mm microwave transmitter and reflector are on their sides and the reflecting/refracting surface is vertical.

Figure 28.16 Glare, i.e. reflected light from the surface of the water, is cut out by the polaroid sun-spectacle on the right but not by the ordinary sun-spectacle on the left.

tudes of the E-field as little arrows in the refracted and reflected waves. There is one value of φ, the angle between these two beams, for which the amplitude of the reflected waves is zero. This value is 90° and the angle of incidence θ_a for which $\varphi = 90°$ will vary with the refractive index n of the material (wax or paraffin). It is called the *Brewster angle* and is given by $\tan \theta_a = n$ as can be shown by applying Snell's law in figure 28.15 for the case when $\varphi = 90°$.

The experiment can be repeated using plane polarised light but of more interest is the polarising effect of reflection on unpolarised light; for this is the key to understanding the use of polaroid to remove *glare*. If the glare is the result of reflection from a horizontal surface, e.g. a wet road or the sea, polaroid spectacles can cut off some (and at certain angles most) of the reflected light if the polaroid is orientated to absorb horizontal variations of E. This is readily tested by looking at the glare through one polaroid 'lens' while rotating the spectacles. Figure 28.16 illustrates the phenomenon. The refracted light is only partially plane polarised even when the reflected light is wholly plane polarised.

Scattered light

If, on a sunny day, we observe the sky through a sheet of polaroid held at various angles it is clear that the light from the sky is weakly plane polarised. So is that from cigarette smoke illuminated by a beam of light, e.g. in a cinema. Both are also blue or have a bluish tinge – the sky more obviously so. Both the atmosphere (the sky) and the smoke are seen because they scatter light (page 335). The light scattered at 90° to the incident beam is plane polarised and the strongest effect is observed when looking at right angles to the sunlight or across the illuminating beam. In the laboratory the effect can be demonstrated by passing light from a projector through a trough of water to which has been added a drop of milk.

Bees, ants and many other insects have eyes that are sensitive to the plane of polarisation of light. This helps them to navigate using the partly plane polarised sunlight scattered from the atmosphere.

Using polarised light

Photoelastic stress analysis

A major industrial and research application for the properties of polarised light is in the analysis of stresses in structures of all sorts. The structure, e.g. the beam gripped at two points shown in figure 28.17, is first reproduced on a small scale using perspex. The model is then viewed between crossed sheets of polaroid, i.e. light is first plane polarised by one sheet, passes through the perspex model and then is viewed or projected onto a screen after passing through a second sheet which is called the analyser. In unstressed perspex light of a particular wavelength travels at a fixed speed *v*. In perspex which is under stress, however, a plane polarised beam of light travelling through a region of stress can be treated as two plane polarised waves resolved parallel and perpendicular to the line of greatest stress. One travels at the unstressed speed *v* while the other travels at a slower speed *v'*; the phenomenon is called *double refraction* as the two speeds imply two different values for the refractive index. The effect is that the analysed light exhibits patterns of superposition (interference patterns) which reveal the *lines of equal stress* in the sample. Using monochromatic light and perspex of known thickness the number of the lines, starting from an obviously unstressed position, can be used to measure the stress at desired points in the structure. With white light a coloured pattern is seen which gives a dramatic qualitative idea of the stresses. Photoelastic stresses can be demonstrated in the laboratory by pulling a strip of polythene, perhaps with a nick cut in it, between two sheets of polaroid placed in the beam from a projector, see figure 5.16 page 56.

Many naturally-occurring crystals exhibit double refraction even when unstressed. An examination of a very thin transparent slice of a rock sample between two sheets of polaroid can quickly distinguish the many different crystals from which it is made.

Figure 28.17 Photoelastic stress patterns in a plastic beam gripped at the two points shown (it is held at right).

The polarimeter

Plane polarised light can also be used to measure the concentration of certain solutions, e.g. sugars. A beam of plane polarised light passing through the solution has its plane of polarisation *rotated*: the angle of rotation depending on the concentration of the solution and the length of the path of the light in the solution. The polarimeter consists of a tube full of liquid between a fixed sheet of polaroid and a second sheet which can be made to rotate through a measured angle.

28.4 Thermal radiation

The surfaces of *all* bodies, e.g. an electric fire or a human face, emit a continuous spectrum of electromagnetic radiation. The relative amount of energy in any small band of wavelengths depends on the temperature of the body and on the nature of its surface. The invisible radiation beyond the red end of the visible spectrum produces a sensation of warmth when absorbed by the skin; we call it *infra-red* radiation. However, thermal radiation includes the visible part of the electromagnetic spectrum and also goes well into the microwave region. For instance a human face emits 75% of its energy as thermal radiation in the (microwave) wavelength range 0.05 mm to 0.20 mm. It is readily noticeable if reflected back to the cheek by a large concave metal surface. Like all electromagnetic waves thermal radiation results from energy changes in the charged particles which make up matter; solids produce thermal radiation as a result of random molecular oscillations. (See table 28.1, page 366).

▶ Table 28.2 shows rough values for the emission of thermal radiation (expressed in W m^{-2}) by three bodies. Estimate (i) the rate of gain of energy by a human face when looking into an open oven door, and (ii) the rate of loss of energy from a face when looking into a freezer.

A human face has an area of about 20 cm × 30 cm = 0.06 m². Let us assume that the face absorbs half the radiation falling on it.

(i) power loss = (300 W m^{-2})(0.06 cm²)
= 18 W

With the face close to the open door,

power gain = $\frac{1}{2}$(2000 W m^{-2})(0.06 m²)
= 60 W

so that the *net gain* is (60 − 18) W ≈ 40 W which will feel pleasant rather than uncomfortable.
(ii) the power loss is the same as in (i) but the power gain (with the face close to the open freezer door) is now

= $\frac{1}{2}$(150 W m^{-2})(0.06 m²)
= 4.5 W

so that there is now a *net loss* of (18 − 4.5) W ≈ 13 W and the face will feel cold. The impression we get is that the

	temperature/K	power loss per unit area as thermal radiation/W m^{-2}
open deep freeze	240	150
human face	310	300
open warm oven	470	2000

Table 28.2 Typical radiative power losses per unit area from three surfaces at quoted temperatures.

freezer is radiating 'cold'. In fact our face is absorbing less thermal radiation than we are used to and so we become aware of a drop in its surface temperature as it continues to emit thermal radiation at the usual rate. When the net power gain is zero the face is absorbing energy at the same rate as it is emitting it.

The greenhouse effect

The selective absorption of ordinary glass explains how a greenhouse works. The radiant energy from the Sun which reaches the Earth's surface, i.e. which is not absorbed by the atmosphere, is mainly at visible wavelengths (the eye has evolved to make use of just this radiation). These waves pass through the glass of a greenhouse and are largely absorbed by the plants and soil resulting in an increase in their internal energy; they become warm. The plants and soil now radiate energy as waves in the far infra-red (microwave) region of the spectrum. Glass is opaque to these waves and thus the greenhouse acts as a sort of energy valve – a one-way filter for radiant energy.

Water vapour and carbon dioxide in the lower layers of the Earth's atmosphere have a similar effect. The ground emits thermal radiation, but much of this is absorbed by the cloud cover which re-radiates some of the energy back to the ground. A clear sky at sunset thus means that the temperature will drop rapidly as the thermal radiation is then emitted to space.

In both these examples we notice that the *surface* which emits thermal radiation drops in temperature, and not at first the air in contact with it. Thermal radiation passes through dry air without noticeably warming it.

Emitted power and wavelength

Figure 28.18 shows how the power of thermal radiation emitted varies with wavelength at two different temperatures. The vertical axis P needs some explanation (as does the phrase 'black body' used in the caption; see page 377). P is the power per unit area of surface emitted as electromagnetic waves between any wavelength λ and a wavelength higher by 1 μm, i.e. between λ and $\lambda + 1$ μm giving the unit W m^{-2} μm^{-1}. (In an experiment to measure P, the size of the wavelength interval is determined by the width of the dispersed radiation incident on the detecting instrument.) The 500°C graph is thus essentially a histogram

sampled, in figure 28.18, at 1 μm wavelength intervals. The power per unit area emitted between 7 μm and 8 μm is shown as 1400 W m^{-2} by the shaded rectangle.

▶ Refer to figure 28.18. What is the *total* power emitted per unit area by the surface at 500°C?

We must sum all the values of $P\Delta\lambda$. Thus, starting at the left P(0 to 1 μm) = 0, P(1 μm − 2 μm) = 300 W m^{-2}, P(2 μm − 3 μm) = 2300 W m^{-2}, etc.

The values are P/W m^{-2}: 0, 300, 2300, 3500, 3300, 2700, 1900, 1400, 1000, 800, 650, 500, 350, 300, 250, 200 (and then allowing for the part of the graph not shown beyond 15 μm), 100, 100, 50, 50, 25, 25. The total is

$$P_{\text{tot}} = 19\,800 \text{ W m}^{-2} \text{ at } 500°C$$

This is about 20 kW m^{-2} or 200 W from a surface measuring 10 cm by 10 cm, which at 500°C will just be beginning to glow a dull red.

Note that we could have counted the squares under the graph and interpreted each square as representing 100 W m^{-2}. ◀

Stefan's law

The total emitted power P_{tot} of a given body changes with temperature. The curves of figure 28.18 tell us that

$$\frac{P_{\text{tot}}(600°C)}{P_{\text{tot}}(500°C)} = \frac{32\,500}{19\,800} = 1.64$$

This is a very dramatic increase. Stefan's law states that

$$P_{\text{tot}} \propto AT^4$$

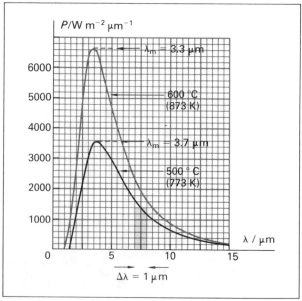

Figure 28.18 The continuous emission spectrum of a black body at 600°C and 500°C. The total power emitted per unit area at 500°C is represented by the area between the graph-line and the wavelength axis.

where A is the emitting area and T is the *kelvin* temperature i.e. 873 K and 773 K in figure 28.18. We can see that the law holds for these two curves as

$$(873/773)^4 = 1.63$$

Introducing a constant we have (for black bodies only – see below)

$$P_{tot} = \sigma AT^4$$

The constant $\sigma = 5.67 \times 10^{-8}$ W m^{-2} K^{-4} is called the *Stefan-Boltzmann constant*. You should recognise how rapidly P_{tot} rises with T; for instance the sum of $P\Delta\lambda$ for $T = 273$ K (0°C) is only 320 W m^{-2}. The corresponding graph would hardly show above the wavelength axis. Similarly a graph for $T = 2000$ K reaches such large values of P that it will not fit onto the page.

A second feature of the graphs is that λ_{max}, the wavelength at which there is a maximum power output, decreases with temperature. The exact relationship is a simple one and is known as *Wien's law*, namely

$$\lambda_{max} T = \text{constant}$$

where T is the kelvin temperature.

For figure 28.18 the values of λ_m are 3.3 µm at 873 K and 3.7 µm at 773 K, values which support the law, and show that the constant has the approximate value of 2.9×10^{-3} m K. This value can now be used to deduce the surface temperature of other (black) bodies. For instance the Moon's surface temperature can be shown to vary between 160 K and 220 K by locating the peak of its thermal spectrum. This temperature is sometimes called its 'radio temperature'. For an incandescent object the position of the peak of its thermal spectrum determines its *colour*. Thus a star which appears white has λ_m in the centre of the visible spectrum, i.e. $\lambda_m \approx 5 \times 10^{-7}$ m, which leads to a surface temperature of about 6000 K, like our Sun. Hotter stars are 'blue' and cooler stars 'red'.

▶ The Sun has a surface temperature of 6000 K. What is the power per unit area of the thermal radiation from the Sun at its entry to the Earth's atmosphere? Take the Sun's radius to be r_s and the radius of the Earth's orbit as $220r_s$. The Stefan Boltzmann constant = 5.7×10^{-8} W m^{-2} K^{-4}.

Stefan's law tells us that $P_{tot} = \sigma AT^4$

For the Sun's surface $A = 4\pi r_s^2$

and so $P_{tot} = 4\pi r_s^2 \sigma T^4$

This now spreads to an area of $4\pi(220r_s)^2$ at the Earth's orbit, so that the power per unit area there is given by

$$I = \frac{P_{tot}}{4\pi(220r_s)^2} = \frac{4\pi r_s^2 \sigma T^4}{4\pi(220r_s)^2}$$

$$= \frac{\sigma T^4}{220^2} = \frac{(5.7 \times 10^{-8} \text{ W m}^{-2} \text{ K}^{-4})(6000 \text{ K})^4}{220^2}$$

$$= 1500 \text{ W m}^{-2}$$

i.e. there is 1.5 kW m^{-2} of radiant energy from the Sun arriving at the Earth; more than a one-bar electric fire over

every square metre. But not all of this reaches the Earth's surface (see figure 28.11 page 372).

Black bodies

The discussion of Stefan's law, Wien's law and the numbers in figure 28.18 hold only for a so-called *black body*. Black bodies have two properties.
(i) A black body *absorbs all* the electromagnetic wave radiation which falls on it. It thus appears to be black when cold because no radiation such as light is reflected from it. A small hole in a windowless room is thus a black body and so approximately is the pupil of the eye – in both cases we are thinking of their properties at around 300 K.
(ii) At any temperature a black body *emits the maximum* power per unit area in any wavelength range which a body can emit at that temperature. The thermal spectrum of an ordinary body at 500°C will lie below the 500°C curve in figure 28.18. A black body appears to have no form, just as the radiation from deep within a coal fire does not enable us to see the shape of the individual pieces of hot material. Figure 28.19b shows the equivalent phenomenon using a hollow tungsten tube; we cannot see any features of the inside surface by looking at the radiation from the small hole. Similarly the Sun's radiation tells us nothing of the surface features of the Sun (a 'black' body); it could, for instance, be a flat disc. (When we see sunspots we are seeing an area of the Sun's surface which is at a lower temperature than the rest).

For a non-black or 'grey' body we can apply Stefan's law in the form $P_{tot} = \varepsilon\sigma AT^4$, where ε is called the total emissivity of the body and is a number always less than 1. For tungsten it is 0.26, and it is of a similar order for other metals.

A black body absorbs all the radiation falling on it so that when it is emitting at a rate σAT^4 it will at the same time be absorbing energy from its surroundings. Think what would happen if the black body was at the same temperature T_0 as its surroundings; it would then emit at

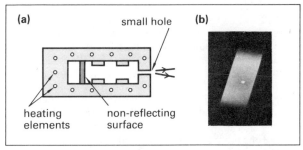

Figure 28.19 (a) A black body: the small hole at the right acts as a perfect absorber and energy emitted from the hole is black body radiation. (b) shows a red hot tungsten tube with a small hole drilled in its wall. The radiation from the hole is black body radiation and is characteristic of the temperature of the tube.

a rate $\sigma A T_0^4$ but its temperature would not change. Therefor it must be absorbing energy incident on it at the same rate, i.e. $\sigma A T_0^4$. Prévost was the first to propose this idea and also the idea that the *incident* energy would not change if the body's temperature became different from T_0. So the *net* rate of energy loss is, for $T > T_0$, given by

$$P_{\text{net loss}} = \sigma A (T^4 - T_0^4)$$

Figure 28.20 Part of the spectrum of atomic hydrogen from the visible (at right) to the near ultra-violet. The wavelengths of the lines in this series (the Balmer series) are given on the scale shown $\times 10^{-8}$ m.

28.5 Electromagnetic wave spectra

There are essentially two ways of investigating a beam of waves so as to discover the range of wavelengths (or frequencies) it contains. One, used at wavelengths near the optical part of the electromagnetic spectrum, is to disperse the waves using a prism or a diffraction grating. The other, used at radio and microwave wavelengths, is to use a tuned circuit. We shall want to investigate wavelengths right across the electromagnetic spectrum as most of the information we have for the structure of atoms and molecules and of the nature of the Universe beyond the Earth comes from the waves which atoms and stars emit.

Optical spectra

Figure 27.17 on page 361 shows a spectrometer and diffraction grating set to study an emission spectrum. The theory gives $m\lambda = s \sin \theta_m$. When s, the slit separation for the grating, is small compared with the wavelength of visible light then a widely-spaced (dispersed) first-order spectrum ($m = 1$) can be examined and the wavelengths measured. Alternatively a prism can be used on the spectrometer table though in this case we can only compare wavelengths with a set of known wavelengths. Figure 28.20 shows a photograph of part of the spectrum of atomic hydrogen.

Line spectra

The hydrogen spectrum is a series of lines, each being an image of the slit of the collimator which is placed close to the source being studied. When the source is an excited gas, the individual atoms are producing the radiation, as the distance between atoms is large compared to their diameter; the spectrum is then always a line spectrum *unique* to that element. Figure 28.21 shows diagrammatically the line spectrum of neon and indicates the relative intensities of the lines in two ways; as expected (think of a neon sign) many of them are in the red region of the spectrum. One of the commonest laboratory sources is a low-pressure sodium vapour lamp. In the sodium spectrum there are two prominent and very close lines in the yellow region with wavelengths of 589.0 nm and 589.6 nm – the source can be treated as a monochromatic (one-colour) source for many purposes.

The fact that each element produces an unique spectrum can be used to provide a chemical analysis of, for example, a sample of moon dust or a metal alloy. The sample must be vaporised with a high temperature arc or spark, e.g. an electric discharge across the gap between two carbon rods. Individual atoms are separated and are *excited*; they emit their characteristic light in returning to their ground state (page 246). Optical spectroscopy led, in the nineteenth century, to a detailed knowledge of the line spectra of most of the elements. The question 'how are line spectra produced?' was not at first asked, any more than 'how do the planets move as they do?' was asked in the sixteenth century, even though their positions could be accurately predicted. In each case a scientific revolution was needed before such matters became part of normal science; part of the stock-in-trade of a person educated in physics.

Band and continuous spectra

When excited atoms or molecules are not wholly independent of one another line spectra are not produced. Where two or more atoms form molecules the interactions of the atoms, e.g. in carbon dioxide (CO_2) or molecular hydrogen (H_2), produce band spectra with several bright bands each cutting off sharply at one side. Each band may, on further inspection, be seen to be made up of many close lines.

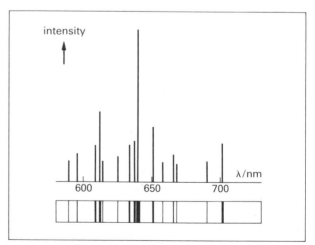

Figure 28.21 Some lines of the neon spectrum. Their relative intensities are shown in two ways.

In the case of incandescent solids such as tungsten filaments the spectrum becomes continuous; such thermal radiation was considered more fully in section 28.4.

Absorption spectra

When white light, e.g. the continuous spectrum from a filament lamp, passes through a filter and is then dispersed, a part of the spectrum is found to be absorbed. Figure 28.22 shows such an absorption spectrum. The transmitted light is a mixture of red, blue and violet, and so appears to be purple. This sort of absorption can be used to identify different blood groups, as a solution of blood of one group removes characteristic sections of a continuous spectrum. When white light is reflected from a red book, the redness is the result of absorption at the blue-green end of the spectrum leaving only red light to be reflected. It is the reflected light which enables us to see the book, or any object, in the first place and selective absorption at the surface which gives an object its colour in white light. Clearly the colour of a 'red' book illuminated by sodium light will not be its usual red.

Just as solids and liquids absorb a continuous part of the whole spectrum so band and line absorption occurs. The atmosphere of Jupiter is known (from a spectroscopic analysis of the sunlight reflected from it) to consist largely of methane and ammonia. The Sun's spectrum itself, which one would expect to be continuous, is found to be crossed by many narrow dark lines. These absorption lines correspond to elements in the Sun's chromosphere, the cooler outer gaseous atmosphere of the sun, and are called *Fraunhofer* lines.

Figure 28.23 shows how to demonstrate absorption in the laboratory. The sodium flame, a bunsen flame burning under a ring of asbestos which has been dipped in a strong salt (NaCl) solution, casts a shadow on the screen as energy at wavelengths characteristic of sodium atoms is absorbed from the incident light and then re-radiated uniformly in all directions. The shadow disappears almost completely if the ring is removed but the bunsen left in position.

To demonstrate the selective or resonant absorption of the sodium yellow light as white light passes through excited sodium atoms is more difficult. A sodium flame, like that in figure 28.23 is not rich enough in sodium atoms to produce any observable effect. The white light must pass through an intense yellow flame produced by the combustion of a piece of sodium or of a 'sodium pencil' in air and be dispersed by a prism or grating; we then see two very close dark lines crossing the continuous spectrum. These occur in the yellow at the wavelength characteristic of the two bright sodium emission lines. If the white light is removed, the emission

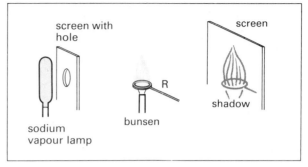

Figure 28.23 A shadow of the flame is cast when the asbestos ring R, soaked in salt solution, is placed as shown.

lines, especially the yellow, of sodium appear. The equivalent effect occurs in sunlight at a total eclipse when we can observe the emission spectra of elements in the Sun's chromosphere. It was in this way that helium was first discovered – spectroscopically in the Sun.

Spectroscopy beyond the optical region

Ultra-violet spectra

The general properties of ultra-violet radiation are outlined on page 366 where the absorptive property of the upper atmosphere is noted. Glass also absorbs all ultra-violet radiation of lower wavelength than 300 nm (the eye is sensitive down to the deep violet at about 400 nm). There is no fundamental difference between the optical spectra described above and ultra-violet spectra. Emission line spectra in the ultra-violet are important in chemical analysis and can be detected by photographic plates after being dispersed by a prism of quartz.

A simple way of demonstrating the ultra-violet spectrum of mercury is shown in figure 28.24. Where a reflection grating is used (page 362). The grating is stuck to a concave spherical surface and so will both disperse and focus the radiation it reflects. The position of the ultra-violet lines can be made visible by forming the spectrum on fluorescent paper. For mercury there is one prominent ultra-violet line of wavelength 254 nm, and there is a second-order spectrum of this line near the visible green first-order line of wavelength 546 nm. Placing a sheet of glass (opaque to ultra-violet light) in front of the grating enables us to distinguish the ultra-violet line and the overlap between orders enables comparisons of wavelengths to be made (see also the example on page 361).

The fluorescence of the paper in the above experiment, i.e. the emission of visible light when ultra violet radiation is absorbed, involves the same principle as that used in fluorescent lights (page 247). Mercury vapour is used in these lamps and the inner surface of the glass tube is coated with fluorescent material.

Figure 28.22 An absorption spectrum showing the light transmitted by a purple filter from red (at right) to violet.

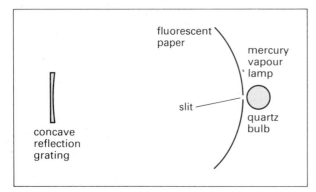

Figure 28.24 A plan view of a very simple reflection grating arrangement for ultra-violet light. The distance from slit to grating is equal to the radius of curvature of the reflecting surface.

Infra-red spectra

The thermal emission spectra of black bodies are discussed in section 28.4. The absorption spectra produced when thermal radiation from a source at about 800 K passes through solids, liquids or gases and is then dispersed are called *infra-red absorption spectra*. They can be used to identify particular gaseous molecules such as CO_2 or SO_2 but are more important than this in that they can give us information about the bonding in complex organic molecules. Infra-red spectra are examples of *resonant absorption*; the sample absorbs at the resonant frequencies of its bonds (acting a bit like two masses connected by a spring) and we study the radiation which is left over.

In principle an *infra-red absorption spectrometer* consists of a source which provides thermal radiation over a wide range in the infra-red. A prism or grating, with associated focusing system, sends a narrow range of wavelengths through the sample and is rotated to vary λ steadily. A detector then draws a graph such as that shown in figure 28.25. The different types of interatomic bonds, e.g. C-H, absorb at characteristic wavelengths and their absorption troughs show up even when the bond is a part of a more complex molecule.

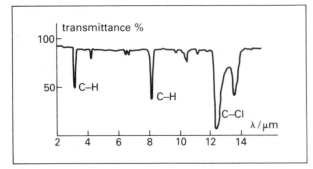

Figure 28.25 The infra-red absorption spectrum of trichloromethane (chloroform, $CHCl_3$). The main absorption lines are labelled with the bonds which produce them.

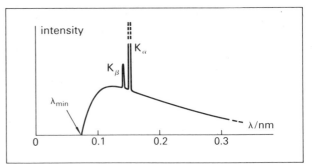

Figure 28.26 A typical X-ray spectrum for a solid target — here copper. The K_α line is an order of magnitude more intense than the K_β line. There is another series of lines, the L series, beyond 0.8 nm.

X-ray and γ-ray spectra

An X-ray spectrometer arrangement is shown on page 363. The X-ray source emits a continuous spectrum superimposed upon which there is a line spectrum consisting of two prominent lines as shown in figure 28.26 (for copper). Like the optical spectra of gases and vapours the line spectrum from an X-ray tube is characteristic of an element; in this case the element used as the target which slows down the electron beam in the tube. The general shape of the continuous spectrum is, however, the same for all elements. It shows a sharp cut-off at the low-wavelength end, the minimum wavelength λ_{min} being inversely proportional to the electron energy. This cut-off is independent of the element used as the target and is explained in quantum terms on page 251.

When the line spectra of a number of elements are compared they are found to be strikingly similar and simple, unlike optical line spectra which are complex and very different for different elements. The main feature is two prominent lines called K_α and K_β. Table 28.3 shows that these lines occur in slightly different positions for different elements in a way which is clearly related to the atomic number Z of the element in question. Before these experiments were first performed, by H.G.J.Moseley (the British physicist, 1887–1915) in 1913, the order in which the elements appeared in the periodic table was the order of their atomic mass, which was an unreliable guide; it was Moseley who first provided us with the concept of atomic number using the empirical rule.

$$(Z - 1)^2 \propto \text{frequency of the } K_\alpha \text{ line}$$

element	wavelength/nm		atomic number Z
	K_α	K_β	
iron	0.194	0.175	26
cobalt	0.179	0.162	27
nickel	0.166	0.150	28
copper	0.154	0.139	29

Table 28.3 Characteristic X-ray wavelengths for four elements.

The K_α line is in fact the first of a series, the K series, but further lines are difficult to resolve from K_β. At longer wavelengths, about 0.8 nm for copper, there is a second series of lines, the L series. The K and L series can be interpreted in terms of the inner electron shell structure of atoms (page 37).

All γ-rays originate in the *nucleus* of an atom. They are radioactive radiations but are identical with short wavelength X-rays (table 28.1 page 366). For a given γ-ray source only certain sharply defined wavelengths are observed, which are characteristic of the type of nucleus concerned, i.e. γ-ray spectra are line spectra. At these very short wavelengths it is usual to refer not to the wavelength, but to the γ-ray photon energy $W = hf = hc/\lambda$ (page 242), and to express the energies in electron-volts.

E.g. $^{212}_{83}\text{Bi}$ decays to $^{208}_{81}\text{Tl}$ by α-decay

but the resulting nucleus may be left in a number of *excited states*. It reverts to its ground state by the emission of γ-rays with energies of 0.04 MeV, 0.33 MeV, 0.47 MeV, 0.49 MeV, and 0.63 MeV, these forming a γ-ray line spectrum.

Radio and microwave spectra

Radiation from space is detected using aerials and tuned circuits. The spectrum of the radiation is thus plotted by scanning across the range of resonant frequencies of a tuned circuit. Clouds of atomic hydrogen emit a characteristic line at 0.21 m and complex molecules have been discovered in interstellar dust by their emission and absorption spectra. Since 1969 methanal (formaldehyde, HCHO), carbonyl sulphide (OCS) and isocyanic acid (HNCO) have been found by emission lines at wavelengths of 62 mm, 2.7 mm and 14 mm respectively. There are many more. Thus molecules containing H, C, O, N and S are all known to exist in forms from which more complex reproducing organisms, life itself, may well have evolved.

Appendix A

Measurements and units

A1 Physical quantities

When we define a new physical quantity, we do so by writing down an equation in which the new quantity occurs. Any other quantities appearing in the equation must of course be ones that are already defined. E.g. the defining equation for the pressure p acting on a surface is

$$p = \frac{F}{A} \qquad \text{(page 60)}$$

where F is the force acting on an area A of the surface. The equation enables us to calculate p if we know F and A. The unit of force is the *newton* (symbol N, page 14); and the unit of area is the metre squared (symbol m^2). The defining equation of pressure therefore fixes the unit of pressure as the newton per metre squared ($N\ m^{-2}$). We find it convenient to have a name for this particular combination of units, and it is called the *pascal* (Pa).

Thus $\quad 1\ Pa = 1\ N\ m^{-2}$

▶ Inside an artificial satellite there is a low air pressure of 5.0 Pa; the pressure in the space outside the satellite is effectively zero. Calculate the total force acting on a flat panel of area 0.40 m^2 forming one side-wall of the satellite.

Re-arranging the defining equation of pressure, given above, the force F is given by

$$F = pA$$

Substituting $p = 5.0\ Pa\,(= 5.0\ N\ m^{-2})$, and $A = 0.40\ m^2$,

we have $\quad F = (5.0\ N\ m^{-2})(0.40\ m^2) = 2.0\ N$ ◀

You should take note of an important point raised by this simple example. The *nought* after the decimal point in 5.0 Pa *really matters*. It shows what the uncertainty is in our knowledge of the pressure p; i.e. we know p to the nearest 0.1 Pa, or p is nearer to 5.0 Pa than to 4.9 Pa or 5.1 Pa. Similarly the area A is 0.40 m^2 to the nearest 0.01 m^2. In this case both quantities are given to 2 significant figures; and this carries through to the calculation of the force F, which must also be recorded to 2 significant figures (2.0 N); it would be wrong to write down 2 N or 2.00 N. You can soon convince yourself that this is so with the aid of a calculator. The maximum values of p and A allowed by the above data are 5.05 Pa and 0.405 m^2, giving $F = 2.045\,25\ N$. The minimum values are 4.95 Pa and 0.395 m^2, giving $F = 1.955\,25\ N$. Thus F lies between 1.95 N and 2.05 N, and it is correct to state it as 2.0 N (correct to the nearest 0.1 N).

physical quantity	symbol	SI unit	symbol
length	l	metre	m
mass	m	kilogram	kg
time	t	second	s
electric current	I	ampere	A
thermodynamic temperature	T	kelvin	K
amount of substance	n	mole	mol

Table A1 The SI base quantities and base units. There is one other recognised base quantity, *luminous intensity* (I_v) and corresponding base unit, the *candela* (cd); but we do not use these at all in this book.

A2 The SI system

To provide a comprehensive system of definitions and units we have to choose a small selection of quantities as fundamental; and all others have then to be defined ultimately in terms of these. Table A1 lists the *base quantities* and *base units* of the Système Internationale (SI). It is the only system now used in scientific literature. The system also allows certain standard prefixes to be placed in front of symbols for units to construct decimal multiples and sub-multiples of these; they are shown in Table A2. Multiples and sub-multiples should normally be converted to the basic unit for purposes of calculation (see the example in A3 below).

For reference purposes we quote here the definitions, agreed internationally, of the SI base units in Table A1. The arbitrary numbers in the definitions are of course chosen

multiple	prefix	symbol
10^{-18}	atto	a
10^{-15}	femto	f
10^{-12}	pico	p
10^{-9}	nano	n
10^{-6}	micro	μ
10^{-3}	milli	m
10^{-2}	centi	c
10^{-1}	deci	d
10	deca	da
10^2	hecto	h
10^3	kilo	k
10^6	mega	M
10^9	giga	G
10^{12}	tera	T

Table A2 The SI prefixes that may be placed before the symbol of any unit to represent decimal multiples and sub-multiples of the unit.

to make the units coincide with those in use before the adoption of SI. Nobody would expect you to memorise the details, though with the ampere, the kelvin and the mole the forms of the definitions are of some physical significance. These points are discussed in the relevant chapters of this book.

metre: *The metre is the length equal to 1 650 763.73 wavelengths in vacuum of the radiation corresponding to the transition between the levels $2p_{10}$ and $5d_5$ of the krypton-86 atom.*

This is equivalent to saying that the wavelength λ_{Kr} of the light produced by the particular transition of the krypton-86 atom is given by

$$\lambda_{Kr} = \frac{1}{1\,650\,763.73} \text{ m} \quad \text{(exactly)}$$

$$= 6.057\,802\,11 \times 10^{-7} \text{ m}$$

$$\approx 606 \text{ nm}$$

kilogram: *The kilogram is the unit of mass; it is equal to the mass of the international prototype of the kilogram.*

The international prototype of the kilogram is a certain cylinder of platinum-iridium kept by the Bureau International des Poids et Mesures at Sèvres near Paris. It may one day be possible to replace this definition by an atomic one; e.g. by specifying how many of a particular kind of atom form a mass of 1 kg. This would be equivalent to fixing the atomic mass of the particular atom. At present atomic masses cannot be measured with the precision necessary for this.

second: *The second is the duration of 9 192 631 770 periods of the radiation corresponding to the transition between the two hyperfine levels of the ground state of the caesium-133 atom.*

This is equivalent to saying that the frequency f_{Cs} of the electromagnetic radiation produced by the particular transition of the caesium-133 atom is given by

$$f_{Cs} = 9.192\,631\,770 \text{ GHz} \quad \text{(exactly)}$$

ampere: *The ampere is that constant current which, if maintained in two straight parallel conductors of infinite length, of negligible circular cross-section, and placed 1 metre apart in vacuum, would produce between these conductors a force equal to 2×10^{-7} newton per metre of length.*

This is equivalent to fixing the magnetic constant μ_0 (the permeability of a vacuum) so that (page 213)

$$\mu_0 = 4\pi \times 10^{-7} \text{ H m}^{-1} \quad \text{(exactly)}$$

kelvin: *The kelvin, the unit of thermodynamic temperature, is the fraction 1/273.16 of the thermodynamic temperature of the triple point of water.*

This is equivalent to saying that the temperature T_{tr} of the triple point of water (page 122) is given by

$$T_{tr} = 273.16 \text{ K} \quad \text{(exactly)}$$

mole: *The mole is the amount of substance of a system which contains as many elementary entities as there are atoms in 0.012 kilogram of carbon-12.*

This is equivalent to saying that the Avogadro constant L (page 38) is the number of atoms per 12 g of carbon-12. The best measurements at present give

$$L = 6.022\,169 \times 10^{23} \text{ mol}^{-1}$$

A3 Numbers in standard form

It is convenient to use powers of 10 to express very large and very small numbers. E.g. we prefer to write 3.4×10^6 rather than 3 400 000. Likewise we prefer to write 4.5×10^{-6} rather than 0.000 004 5. Numbers in *standard form* are written as above with the aid of powers of 10; they always have *one* significant figure before the decimal point.

When you handle numbers in standard form you need to remember the following basic rules of arithmetic:

multiplication: you *add* the powers of 10
division: you *subtract* the powers of 10

► On what area must a force of 7.5 mN act to produce a pressure of 2.5 kPa?

The prefix *milli* (m) has to be replaced by 10^{-3} and *kilo* (k) by 10^3 before we do the calculation. So in standard form the force F and pressure p are given by

$$F = 7.5 \times 10^{-3} \text{ N; and } p = 2.5 \times 10^3 \text{ Pa}$$

Re-arranging the defining equation of pressure we have

$$A = \frac{F}{p}$$

$$= \frac{7.5 \times 10^{-3} \text{ N}}{2.5 \times 10^3 \text{ Pa}} = 3.0 \times 10^{-6} \text{ m}^2$$

(This is about the area of a circle of diameter 2 mm.) In the final step of arithmetic above we divide 7.5 by 2.5 to give 3.0; and then we subtract the indices (3 from -3) to give the power of 10 in the result. ◄

If you use a 'scientific' type of electronic calculator, these processes are done automatically. Such calculators are equipped to handle numbers in standard form, the two parts of the number (multiplier and power of 10) appearing in different parts of the number register.

A4 Units of energy

The SI unit of energy is the *joule* (J); this is the name for the N m (page 22). This is the only unit which you should now find in scientific literature. But a number of other units of energy have been used and remain in use for particular purposes (in engineering, etc) for which they have some special convenience.

kilowatt-hour (kW h): The calculation of quantities of energy for commercial purposes is generally done in this unit in the electrical industry. (It is often referred to just as a 'unit' of electrical energy.) To calculate energy in this unit it is only necessary to multiply the power conversion in kilowatts by the time in hours for which it occurs.

Since   1 kW $= 10^3$ J s^{-1}
and   1 hour $= 3.6 \times 10^3$ s
therefore   1 kW h $= 3.6 \times 10^6$ J   (exactly)

British Thermal Unit (B.T.U.): This is used to some extent by British heating engineers.

   1 B.T.U. $= 1.054 \times 10^3$ J   (to 4 significant figures)

therm: This is used in the gas industry.

   1 therm $= 10^5$ B.T.U. $\approx 1.054 \times 10^8$ J

Q unit: This is sometimes used in discussions of global supplies of energy.

   1 Q $= 10^{18}$ B.T.U. $\approx 1.054 \times 10^{21}$ J

calorie (cal): This was a unit originally devised to measure quantities of energy transferred in heating processes. A variety of different definitions have been employed, giving slightly differing versions of the calorie; but to 2 significant figures we can write

   1 cal $= 4.2$ J

Calorie: This unit (spelt with a capital C) is the large Calorie used in calculations of energy received in food.

   1 Calorie $= 1$ kcal $= 10^3$ cal $\approx 4.2 \times 10^3$ J

erg: This is an older scientific unit of energy.

   1 erg $= 10^{-7}$ J   (exactly)

electron-volt (eV): This is a convenient unit for the calculation of particle energies in atomic and nuclear physics. It is the only additional unit of energy that we use regularly in this book, and then only for its own specialised purposes (page 191).

   1 eV $= 1.60 \times 10^{-19}$ J   (to 3 significant figures)

The *power* of some device is the rate at which it converts energy. Any of the above units of energy can be combined with a unit of time to give a unit of power. But there is one traditional unit of power that we give here for the sake of completeness:

horsepower (h.p.):

      1 h.p. $= 746$ W   (to 3 significant figures)

and so   1 h.p. $\approx \dfrac{3}{4}$ kW

A5 Graphs and tables

Graphs

It is important to learn the right way to state the units of the quantities plotted on a graph. Any statement of the value of a quantity must always include the unit in which it is measured. For instance, we might measure the length l of a room, and write down

   $l = 5.4$ m

This means that a metre (m) bar can be fitted into the length (l) 5.4 times. So we could write it

   $\dfrac{l}{m} = 5.4$

The figure 5.4 is a ratio of two lengths, and is therefore a pure number (with no units).

The figures that label the axes of a graph are normally pure numbers; and it is therefore essential that we attach a label to an axis that represents lengths measured in m in the above form:

   $\dfrac{l}{m}$ or l/m

The label on a graph axis must always be of the form

   (quantity plotted)/(unit used)

In figure B1 below, for instance, the numbers "50, 100, . . ." on the vertical axis are values of the ratio s/mm, the unit in this case being the millimetre. On the horizontal axis (giving the weight W hanging on the spring) the numbers "1, 2, 3, 4, . . ." are values of the ratio W/N, the unit of weight being the newton.

Tables

It is often convenient to arrange sets of values of related quantities in a table (e.g. the measured values of W and s from which figure B1 was plotted). We can, if we want, enter the unit after every figure in the table. But this is tedious and unnecessary. It is better to record only the *figures* in the columns of the table; but we must then remember that these numbers are values of the quantity tabulated divided by the unit used. The head of the column must therefore be labelled accordingly, in the same way as for graph axes. You will find plenty of examples of this throughout this book; and you should be just as particular in your own work if you want it to have the exactness of statement that science demands.

A6 Physical constants

In table A3 we give values of some of the constants that are discovered in the course of investigating the properties of the fundamental particles of matter. Some of these are properties of space itself (e.g. the speed of light c).

quantity	symbol	value
speed of electromagnetic waves (in vacuum)	c	2.998×10^8 m s^{-1} $\approx 3.00 \times 10^8$ m s^{-1}
gravitational constant	G	6.673×10^{-11} N m^2 kg^{-2}
magnetic constant (permeability of vacuum)	μ_0	$4\pi \times 10^{-7}$ H m^{-1} $\approx 1.257 \times 10^{-6}$ H m^{-1} (exactly)
electric constant (permittivity of vacuum)	$\epsilon_0 \ (= \dfrac{1}{\mu_0 c^2})$	8.854×10^{-12} F m^{-1} $\approx 1/(36\pi \times 10^9)$ F m^{-1}
electronic charge	e	1.602×10^{-19} C
unified atomic mass constant	m_u	1.661×10^{-27} kg
rest mass: of electron	m_e	9.110×10^{-31} kg $= 5.486 \times 10^{-4}\ m_u$
of proton	m_p	1.673×10^{-27} kg $= 1.0073\ m_u$
of neutron	m_n	1.675×10^{-27} kg $= 1.0087\ m_u$
specific charge: of electron	e/m_e	-1.759×10^{11} C kg^{-1}
of proton	e/m_p	$+9.579 \times 10^7$ C kg^{-1}
Planck constant	h	6.626×10^{-34} J s
triple point of water	T_{tr}	273.16 K (exactly)
Avogadro constant	L	6.022×10^{23} mol^{-1}
molar gas constant	R	8.314 J K^{-1} mol^{-1}
Boltzmann constant	$k\ (= \dfrac{R}{L})$	1.381×10^{-23} J K^{-1}
molar volume at s.t.p. (of the ideal gas)	V_m	2.241×10^{-2} m^3 mol^{-1}
Stefan-Boltzmann constant	σ	5.670×10^{-8} W m^{-2} K^{-4}

Table A3 Values of some fundamental physical constants. All the constants shown are in fact known to greater accuracy than this. But we give them mostly to 4 significant figure accuracy, since this is sufficient for the purposes of this book.

Appendix B

Useful mathematics

B1 Small changes

A useful way of describing small changes in a measured quantity makes use of the Greek letter *delta* (δ, or as a capital letter Δ). Placed in front of a symbol the letter delta means "a small increase" in the quantity. E.g. a spring is stretched so that its length *l* increases from 100 mm to 112 mm; we can then write

$$\Delta l = 12 \text{ mm}$$

which means "the increase in *l* is 12 mm".

In the same way Δv is the increase in velocity v of some body, Δp is an increase in pressure p, ΔI is an increase in electric current I. Of course, in a particular case ΔI might be negative (e.g. $\Delta I = -0.3$ A, which refers to a decrease in the current I of 0.3 A). Likewise the statement $\Delta I = 0$ implies that, whatever other changes are taking place, the current I remains constant.

B2 Straight line graphs

Proportionality

There are many physical processes in which two variables are *directly proportional* to one another. For instance, if we hang weights on the end of a spring, the extension s of the spring is directly proportional to the weight W; if we double W, we double s; treble W, and we treble s; and when $W = 0$, $s = 0$. We write this in symbols as

$$s \propto W$$

A graph of s against W is *a straight line passing through the origin* (figure B1).

In some cases the variation of two quantities works the opposite way. Consider, for instance, the pressure p and volume V of a given mass of gas (at constant temperature). In this case if we *double* the volume V, we *halve* the pressure p; and so on. Another way of describing this situation is to say that p and $1/V$ are directly proportional to one another. So we write this as

$$p \propto \frac{1}{V}$$

and we say that p is *inversely proportional* to V. A graph of p against V is a curve (called a hyperbola); but if we plot p against $1/V$, we again obtain a straight line through the origin (page 125).

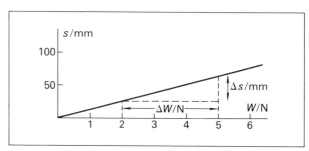

Figure B1 The extension s is directly proportional to the weight W. The gradient m is given by $m = \Delta s / \Delta W$.

Gradients

The slope or gradient of any straight line graph is constant, and may be calculated as follows (using figure B1 as an example). We consider any increase ΔW between two convenient points on the graph, and read off the corresponding increase Δs; then the gradient m of the line is given by

$$m = \frac{\Delta s}{\Delta W}$$

With a straight line graph it does not matter what value of ΔW we choose; all right-angled triangles that might be drawn under the graph line give the same value of $\Delta s / \Delta W$. E.g. when $W = 2.0$ N, $s = 28$ mm;

and when $W = 5.0$ N, $s = 70$ mm;

and so $\Delta s = 70 \text{ mm} - 28 \text{ mm} = 42 \text{ mm}$

when $\Delta W = 3.0 \text{ N}$,

giving $m = \dfrac{42 \text{ mm}}{3.0 \text{ N}} = 14 \text{ mm N}^{-1}$

In the case of a graph passing through the origin (like figure B1) it is probably easiest to take the origin itself as the starting point for our calculation of ΔW and Δs. In that case we are taking $\Delta W = W$, and $\Delta s = s$. You can readily check that this gives the same value of m, using either pair of values of W and s quoted above. Clearly for a straight line passing through the origin we can write down for any point on it

$$\frac{s}{W} = m \text{ or } s = mW$$

Both these equations are equivalent to the basic statement

$$s \propto W$$

A single constant m is sufficient to calculate values of s for given values of W.

To test whether two quantities are directly proportional to one another we therefore have two possible ways of proceeding:

(i) Plot a graph of one quantity against the other for a series of measured values, and see if it is a straight line *through the origin* within the limits of experimental uncertainty.

(ii) Tabulate measured values of one quantity *divided by* the other, and see if these are constant within the limits of experimental uncertainty.

When we suspect that two quantities are inversely proportional to one another, we may proceed similarly. Consider, for instance, as explained above, the pressure p and volume V of a gas; in this case

$$p \propto \frac{1}{V}$$

and we may test this by plotting p against $1/V$ which gives a straight line through the origin. Alternatively, we may express this as

$$p = \frac{k}{V}$$

where k is a constant (the gradient of the above graph). Therefore

$$pV = k$$

and so we may test the result by measuring values of the *product pV* and showing that this is constant within the limits of experimental uncertainty. Again a single constant k is sufficient to calculate values of p for given values of V.

Linear relations

In some cases a plot of two quantities one against the other gives a straight line graph that does *not* pass through the origin. The two quantities are *not* now proportional to one another. But we describe this by saying that there is a linear relation between them. For instance, using the same spring as in figure B1, instead of measuring the extension s we could measure the total length l of the spring for different values of the weight W hanging on it. A graph of l against W (figure B2) shows that l increases *linearly* with W. The gradient m is of course constant and may be calculated as previously. Thus, the triangle in the figure has been drawn for convenience so that $\Delta W = 4.0$ N (i.e. 5.0 N − 1.0 N). The corresponding values of l read off the vertical axis are 179 mm and 123 mm;

and so $\Delta l = 179 \text{ mm} - 123 \text{ mm} = 56 \text{ mm}$

Therefore $m = \dfrac{56 \text{ mm}}{4.0 \text{ N}} = 14 \text{ mm N}^{-1}$

This is the same gradient as the graph of s against W in figure B1; but every point is raised upwards on the graph by an amount l_0 compared with the corresponding value

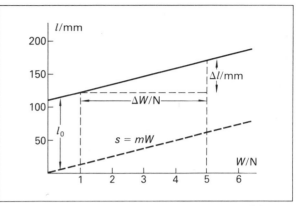

Figure B2 There is a linear relation between l and W of the form: $l = mW + l_0$. The gradient m is given by $m = \Delta l / \Delta W$.

of s. The theoretical relation between l and W is therefore

$$l = mW + l_0$$

In this case l is *not* directly proportional to W; *two* constants need to be known to calculate values of l, namely the gradient m and the intercept on the l-axis, which is the unstretched length l_0 of the spring (109 mm in figure B2). In general, if the graph of two measured quantities y and x proves to be a straight line, then the relation between them must be of the form

$$y = mx + c \qquad (1)$$

where m is the gradient, and c is the intercept on the y-axis (i.e. the value of y when $x = 0$). Notice that the constant term c in the equation does not affect the gradient. Only when $c = 0$ may we say that y is directly proportional to x.

B3 Rates of change

With a graph that is not a straight line the gradient varies from point to point along it. For instance figure B3 shows the variation of velocity v with time t for a car starting from rest and moving in a straight line. We can calculate the gradient at any point P by drawing a tangent to the curve at P and finding the gradient of that. Alternatively we may take two points *close* to P and build a small right-angled triangle on these as in the figure; the gradient m is then given, as usual, by

$$m = \frac{\Delta v}{\Delta t}$$

The increases Δv and Δt must obviously be kept small if this calculation is to give the gradient correctly – how small depends on how sharply curved the graph is at the point P. For this to be *always* a correct way of stating the gradient we must be sure that Δv and Δt are *infinitesimal*, i.e. so small that reducing them still further would make no

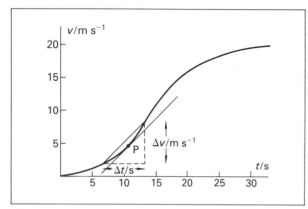

Figure B3 The gradient at P is the gradient of the tangent to the curve there. Alternatively, it is the gradient of a chord joining two points close to P.

detectable difference in the calculation of gradient. To express this idea, in place of $\Delta v/\Delta t$ we write

$$\frac{dv}{dt} \quad \text{(read as: "dv by dt")}$$

This is a *single* symbol which expresses the *rate of change* of v (i.e. the acceleration of the car). Similarly

$\frac{dp}{dt}$ is the rate of change of the pressure p in a sound wave;

$\frac{dV}{dt}$ is the rate of change of the p.d. V of the a.c. mains supply.

The branch of mathematics that deals with the calculation of such rates of change and the gradients of non-linear graphs is known as *differential calculus*.

B4 Arcs and angles

A part of the circumference of a circle is called an *arc*. The length s of an arc is directly proportional to the angle θ it subtends at the centre; doubling the angle doubles the arc also. For a given angle θ the length s of an arc is also directly proportional to the radius r; doubling the radius doubles the arc also (figure B4). If we measure the angle θ in radians (symbol rad), s is given by

$$s = r\theta \tag{2}$$

or, put another way, if we want to measure an angle θ in radians, we find the length s of an arc cut off by the angle from the circumference of a circle of radius r (figure B4).

Then $\theta = \dfrac{s}{r}$ rad

Since the total circumference of the circle $= 2\pi r$, the total angle ($= 360°$) round a point is

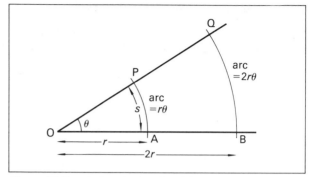

Figure B4 To calculate the lengths of the arc of a circle (PA or QB) we need to know the angle θ in radians.

$$\frac{2\pi r}{r} \text{ rad} = 2\pi \text{ rad}$$

Hence the connection between the radian and the degree (the more familiar unit of angle) is

$$\pi \text{ rad} = 180°$$

and the conversion factor $\pi/180$ enables us to convert angles from degrees to radians and vice versa.

▶ A tape is wrapped round the circumference of a wheel of radius 300 mm. The wheel is rotated through an angle of 45.0° by pulling the end of the tape. What length of tape is pulled off the wheel?

Changing the angle into radians,

$$45.0° = 45.0 \times \frac{\pi}{180} \text{ rad}$$

Hence   arc $= 300 \text{ mm} \times 45.0 \times \dfrac{\pi}{180} = 236 \text{ mm}$

and this is the length of tape that must be pulled off the wheel.   ◀

B5 Small angle calculations

Figure B5 shows four typical calculations that arise frequently in physics problems. In each of the first three we need to find the side of the triangle opposite the angle θ, given the length l from O to A. In the fourth we find the arc $l\theta$. In (a), (b) and (c) we have also shown the arc $l\theta$ as a dotted curve superimposed on the diagram. You will notice that, *if the angle θ is small*, in each of these cases the arc $l\theta$ is a fair approximation to the length required.

In (a) $AB = l \tan \theta \approx l\theta$
In (b) $AC = l \sin \theta \approx l\theta$
In (c) $AD = 2l \sin \tfrac{1}{2}\theta \approx l\theta$
So $\tan \theta \approx \sin \theta \approx 2 \sin \tfrac{1}{2}\theta \approx \theta$ (in radians)   (3)

It is often simpler to calculate θ (in radians) than to use

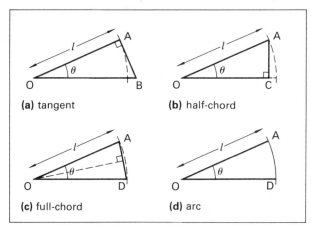

(a) tangent **(b)** half-chord

(c) full-chord **(d)** arc

Figure B5 If the angle θ is small enough, the arc $l\theta$ may be used as an approximation in the other three cases.

sines and tangents. The conversion factor $\pi/180$ enables us quickly to write down an angle in radians.

But how good are these approximations? Table B1 shows the values of $\tan \theta$, $\sin \theta$, $2 \sin \tfrac{1}{2}\theta$ and θ (in radians) for various angles. The shading of the columns shows the points at which the error in the approximations reaches the values shown. In the calculation of a full-chord (c) θ can be up to $30°$ before the error reaches 1%; and even up to $60°$ it is still less than 5%. An accuracy of 0.1% is obtained for angles up to about $5°$ in (a) and (b), and up to about $10°$ in (c).

► A beam of light from a projector shines approximately perpendicularly on the wall of the laboratory 5.40 m from the projector. When the beam is rotated through an angle θ the spot of light on the wall moves 1.90 m. Calculate the value of θ.

We assume that θ is small enough to justify the approximations above. So we write

$$\theta = \frac{\text{arc}}{\text{radius}} = \frac{1.90 \text{ m}}{5.40 \text{ m}} \text{ rad}$$

angle θ	**(a)** tan θ	**(b)** sin θ	**(c)** $2\sin\tfrac{1}{2}\theta$	**(d)** θ/rad	error
5°	0.0875	0.0872	0.0872	0.0873	0.1%
10°	0.1763	0.1736	0.1744	0.1745	
15°	0.2679	0.2588	0.2610	0.2618	
20°	0.364	0.342	0.347	0.349	1%
25°	0.466	0.423	0.433	0.436	
30°	0.578	0.500	0.518	0.524	
$\gtrless$	$\gtrless$	$\gtrless$	$\gtrless$	$\gtrless$	5%
60°	(1.732)	(0.866)	1.000	1.047	

Table B1 Values of trigonometric functions for small angles. The error involved in using the value of θ (in radians) instead of the other functions is often small. The shading in the table shows the approximate points at which the error reaches 0.1% (for 3 significant figure accuracy), 1% (for 2 significant figure accuracy), and 5%.

$$= 0.352 \text{ rad}$$

Hence $\quad \theta = 0.352 \times \dfrac{180}{\pi} \text{ degrees}$

$$= 20.2°$$

If the beam of light moves symmetrically on either side of the perpendicular from projector to wall (case (c) above) then table B1 shows that the error involved is about 0.5%, which justifies recording the value of θ to the nearest $0.1°$. ◄

B6 Sinusoidal oscillations

In physics we have often to deal with quantities that oscillate according to the relation

$$V = V_0 \sin 2\pi f t \qquad (4)$$

For instance, the p.d. V of the a.c. mains varies with time t in this manner (figure B6); V oscillates between the values $+V_0$ and $-V_0$ with frequency f (50 Hz in Europe). At the peaks and troughs of the curve, where V has these maximum and minimum values, the tangents to the curve are horizontal, and we have

$$\frac{dV}{dt} = 0$$

The maximum rates of change of V occur at the points where the graph crosses the time-axis, i.e. where $V = 0$; the origin is one such point.

We can calculate the gradient here as follows. Near the origin the angle $2\pi f t$ (which is in radians) is very small. We can therefore write

$$\sin 2\pi f t \approx 2\pi f t$$

and so we have $\quad V \approx V_0 2\pi f t = (2\pi f V_0)t$

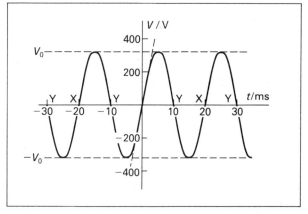

Figure B6 The maximum rate of change of a sinusoidally oscillating quantity V occurs at points where $V = 0$ (where the graph line crosses the time-axis). At these points the gradient is $2\pi f V_0$.

Very close to the origin this result is *exact*.

Thus, near the origin, V is directly proportional to t, and the curve approximates to a straight line whose gradient is $2\pi f V_0$. The gradient is of course the same at all similar points on the curve (marked X); and at the intermediate crossing points Y the gradient is $-2\pi f V_0$.

We thus have the important result for the maximum rate of change of a sinusoidal variation

$$\left(\frac{\mathrm{d}V}{\mathrm{d}t}\right)_{\max} = 2\pi f V_0 \tag{5}$$

This result we use frequently, not only for oscillating p.d.s but also for sinusoidally varying quantities of many other kinds.

▶ Suppose the peak value V_0 of the (50 Hz) a.c. mains is 325 V. Calculate (i) the maximum rate of change of V, (ii) the value of V at a moment 0.50 ms after the p.d. changes direction.
(i) Using equation (5),

$$\left(\frac{\mathrm{d}V}{\mathrm{d}t}\right)_{\max} = 2\pi\,(50\text{ Hz})\,(325\text{ V})$$

$$= 1.02 \times 10^5\text{ V s}^{-1}$$

(ii) Close to a point such as X (figure B6) we can write

$$\frac{\Delta V}{\Delta t} = 1.02 \times 10^5\text{ V s}^{-1}$$

If $\Delta t = 0.50 \times 10^{-3}$ s, we have

$$\Delta V = (1.02 \times 10^5\text{ V s}^{-1})(0.50 \times 10^{-3}\text{ s})$$
$$= 51\text{ V}$$

If you work out the angle $(2\pi f t)$ at this instant, you will find that it is less than $10°$; and table B1 shows that the approximation of assuming ΔV to be directly proportional to Δt is correct at this angle to within about 0.5%; this justifies our quoting the value of ΔV to the nearest 1 V. ◀

B7 Exponentials

There is an important class of physical processes in which we have some quantity z that varies with time t in such a way that the rate of change of z is directly proportional to z; in symbols

$$\frac{\mathrm{d}z}{\mathrm{d}t} = bz \tag{6}$$

where b is a constant. Figure B7a shows the way in which z varies with t; the gradient of the curve at any point is proportional to z. This is called an *exponential curve*. Figure B7b shows the result of plotting

$$\frac{\mathrm{d}z}{\mathrm{d}t}\text{ against }z$$

This is a straight line through the origin, whose gradient is b. The relation between z and t in figure B7a is of the form

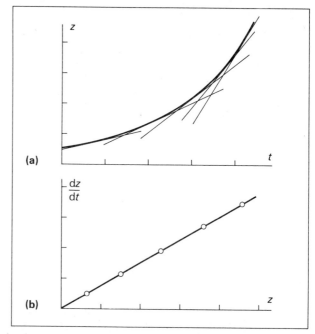

Figure B7 (a) An exponential curve. The quantity z is increasing in such a way that its rate of change is directly proportional to z, as shown in (b).

$$z = z_0 e^{bt} \tag{7}$$

In this equation e stands for a certain mathematical constant:

$$e = 2.72\quad\text{(to 3 significant figures)}$$

(It is actually a number like π with an infinity of non-recurring digits; its first few digits are $2.718\,281\ldots$) In equation (7) z_0 is the *initial value* of z; for when $t = 0$, $e^{bt} = 1$, and so $z = z_0$. Mathematical tables usually include tabulations of e^x for values of $x(= bt)$, and scientific-type calculators provide values at the touch of a button.

You should try plotting a typical exponential curve using equation (7). For instance, you could tabulate values of z and t, taking $b = 0.2$ s^{-1}, and $z_0 = 2$. You will be able to plot a useful part of the exponential curve by taking values of t from 0 up to 10 s. Draw a series of tangents to the curve, and test whether equation (6) applies by plotting also

$$\frac{\mathrm{d}z}{\mathrm{d}t}\text{ against }z\quad\text{(as in figure B7b).}$$

There is in fact a simpler way of testing whether a given experimental curve is exponential. We may re-arrange equation (7) by taking *natural logarithms* on both sides (i.e. logarithms to the base e, for which we use the symbol ln). Bearing in mind that $\ln e = 1.000$, we then have

$$\ln z = \ln z_0 + bt \tag{8}$$

This is of the form

$$y = mx + c \qquad \text{(equation (1) above)}$$

(We are seeing $\ln z$ as y, and t as x; $\ln z_0$ is a constant, taking the place of c.) It follows that a graph of $\ln z$ against t should be a straight line if z varies exponentially with t; the gradient of this line is b. You should also try plotting this graph with your tabulated values. Satisfy yourself that the gradient is indeed the value of b that you chose.

Equations (6), (7) and (8) above are exactly equivalent statements of the same result; they turn up frequently in all departments of physics. Theoretical considerations often show that a rate of change should be proportional to the quantity that is changing, and therefore yield an equation like (6) above. We can then at once write down the equation of the exponential curve (7); we can also make use of equation (8) if we need to test the result experimentally.

Exponential decay

Exponential processes most usually arise in physics with a *negative* value for the constant b in equations (6), (7) and (8) above. This means that the rate of change is negative, and the quantity z is *decaying* at a rate proportional to z. For instance, suppose we have a gas under pressure leaking *slowly* (so that the temperature stays constant) from a container; we might expect the excess pressure p of the gas to decrease at a rate proportional to p, so that

$$\frac{dp}{dt} = -kp$$

where k is a constant that depends on the size of the hole where the leak occurs. We can immediately write down

$$p = p_0 e^{-kt}$$

where p_0 is the initial value of p (when $t = 0$). This is represented in figure B8a. In (b) is the logarithmic graph by which we may test the result experimentally; this represents the equation

$$\ln p = \ln p_0 - kt$$

whose gradient is $-k$.

The power term $-kt$ in the exponential equation is always a pure number (without units). The constant k, known as the *decay constant*, is therefore measured in s^{-1}. The reciprocal of k is measured in seconds, and is known as the *time constant* τ (the Greek letter *tau*) of the process. Thus

$$\tau = \frac{1}{k} \qquad (9)$$

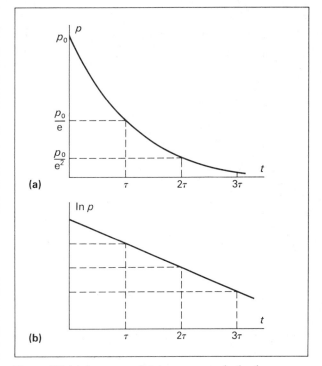

Figure B8 (a) An exponential decay process; in the time constant τ the pressure p falls in the ratio $e:1$. (b) The corresponding logarithmic graph, whose gradient is $-k$.

and $\quad p = p_0 e^{-\frac{t}{\tau}}$

Substituting $t = \tau$ in this equation, we find that in time τ the pressure p falls to the value

$$p_0 e^{-1} = \frac{p_0}{e}$$

In the next time interval τ it decreases again in the same ratio $(e:1)$ to the value

$$\frac{p_0}{e^2}$$

and after a time interval 3τ to the value

$$\frac{p_0}{e^3}$$

and so on. It falls towards zero without ever quite reaching it. However, after about 5τ the excess pressure p has fallen to less than 1 % of its initial value, and for practical purposes we should probably treat it as effectively zero after that point.

Index